Smart Innovation, Systems and Technologies

Volume 248

The Smart Innovation, Systems and Technologies book series encompasses the topics of knowledge, intelligence, innovation and sustainability. The aim of the series is to make available a platform for the publication of books on all aspects of single and multi-disciplinary research on these themes in order to make the latest results available in a readily-accessible form. Volumes on interdisciplinary research combining two or more of these areas is particularly sought.

The series covers systems and paradigms that employ knowledge and intelligence in a broad sense. Its scope is systems having embedded knowledge and intelligence, which may be applied to the solution of world problems in industry, the environment and the community. It also focusses on the knowledge-transfer methodologies and innovation strategies employed to make this happen effectively. The combination of intelligent systems tools and a broad range of applications introduces a need for a synergy of disciplines from science, technology, business and the humanities. The series will include conference proceedings, edited collections, monographs, handbooks, reference books, and other relevant types of book in areas of science and technology where smart systems and technologies can offer innovative solutions.

High quality content is an essential feature for all book proposals accepted for the series. It is expected that editors of all accepted volumes will ensure that contributions are subjected to an appropriate level of reviewing process and adhere to KES quality principles.

Indexed by SCOPUS, EI Compendex, INSPEC, WTI Frankfurt eG, zbMATH, Japanese Science and Technology Agency (JST), SCImago, DBLP.

All books published in the series are submitted for consideration in Web of Science.

More information about this series at https://link.springer.com/bookseries/8767

Tomonobu Senjyu · Parikshit N. Mahalle ·
Thinagaran Perumal · Amit Joshi
Editors

ICT with Intelligent Applications

Proceedings of ICTIS 2021, Volume 1

Springer

Editors
Tomonobu Senjyu
University of the Ryukyus
Okinawa, Japan

Thinagaran Perumal
Department of Computer Science
Faculty of CS and IT
Universiti Putra Malaysia
Seri Kembangan, Malaysia

Parikshit N. Mahalle
Sinhgad Technical Education Society
SKNCOE
Pune, India

Amit Joshi
Global Knowledge Research Foundation
Ahmedabad, India

ISSN 2190-3018 ISSN 2190-3026 (electronic)
Smart Innovation, Systems and Technologies
ISBN 978-981-16-4176-3 ISBN 978-981-16-4177-0 (eBook)
https://doi.org/10.1007/978-981-16-4177-0

This Springer imprint is published by the registered company Springer Nature Singapore Pte Ltd.
The registered company address is: 152 Beach Road, #21-01/04 Gateway East, Singapore 189721, Singapore

Preface

The Fifth International Conference on Information and Communication Technology for Intelligent Systems (ICTIS 2021) targets state of the art as well as emerging topics pertaining to information and communication technologies (ICTs) and effective strategies for its implementation for engineering and intelligent applications.

The conference is anticipated to attract a large number of high-quality submissions, stimulate the cutting-edge research discussions among many academic pioneering researchers, scientists, industrial engineers, students from all around the world and provide a forum to researcher; propose new technologies, share their experiences and discuss future solutions for design infrastructure for ICT; provide a common platform for academic pioneering researchers, scientists, engineers and students to share their views and achievements; enrich technocrats and academicians by presenting their innovative and constructive ideas; focus on innovative issues at international level by bringing together the experts from different countries.

The conference was held during 23–24 April 2021 digitally on Zoom and organized by the Global Knowledge Research Foundation in collaboration with KCCI and IFIP INTERYIT.

Research submissions in various advanced technology areas were received, and after a rigorous peer review process with the help of programme committee members and external reviewer, 160 papers were accepted with an acceptance rate of 16%. All 160 papers of the conference are accommodated in 2 volumes, and also, papers in the book comprises authors from 18 countries.

This events success was possible only with the help and support of our team and organizations. With immense pleasure and honour, we would like to express our sincere thanks to the authors for their remarkable contributions, all the technical programme committee members for their time and expertise in reviewing the papers within a very tight schedule and the publisher Springer for their professional help.

We are overwhelmed by our distinguished scholars and appreciate them for accepting our invitation to join us through the virtual platform and deliver keynote speeches and technical session chairs for analysing the research work presented by the researchers. Most importantly, we are also grateful to our local support team for their hard work for the conference. This series has already been made a continuous series which will be hosted at different locations every year.

Okinawa, Japan — Tomonobu Senjyu
Pune, India — Parikshit N. Mahalle
Seri Kembangan, Malaysia — Thinagaran Perumal
Ahmedabad, India — Amit Joshi

Contents

About the Editors

Dr. Tomonobu Senjyu received his B.S. and M.S. degrees in Electrical Engineering from the University of the Ryukyus in 1986 and 1988, respectively, and his Ph.D. degree in Electrical Engineering from Nagoya University in 1994. Since 1988, he has been with the Department of Electrical and Electronics Engineering, University of the Ryukyus, where he is currently Professor. His research interests include stability of AC machines, power system optimization and operation, advanced control of electrical machines and power electronics. He is Member of the Institute of Electrical Engineers of Japan and IEEE.

Dr. Parikshit N. Mahalle holds a B.E. degree in CSE and an M.E. degree in Computer Engineering. He completed his Ph.D. at Aalborg University, Denmark. Currently, he is working as Professor and Head of the Department of Computer Engineering at STES Smt. Kashibai Navale College of Engineering, Pune, India. He has over 18 years of teaching and research experience. Dr. Mahalle has published over 140 research articles and eight books and has edited three books. He received the "Best Faculty Award" from STES and Cognizant Technologies Solutions.

Dr. Thinagaran Perumal received his B.Eng., M.Sc. and Ph.D. in Smart Technologies and Robotics from Universiti Putra Malaysia in 2003, 2006 and 2011, respectively. Currently, he is Associate Professor at Universiti Putra Malaysia. He is also Chairman of the TC16 IoT and Application WG National Standard Committee and Chair of IEEE Consumer Electronics Society Malaysia Chapter. Dr. Thinagaran Perumal is Recipient of 2014 IEEE Early Career Award from IEEE Consumer Electronics Society. His recent research activities include proactive architecture for IoT systems; development of the cognitive IoT frameworks for smart homes and wearable devices for rehabilitation purposes.

Dr. Amit Joshi is currently Director of the Global Knowledge Research Foundation. He is Entrepreneur and Researcher and holds B.Tech., M.Tech. and Ph.D. degrees. His current research focuses on cloud computing and cryptography. He is Active Member of ACM, IEEE, CSI, AMIE, IACSIT, Singapore, IDES, ACEEE, NPA and several other professional societies. He is also International Chair of

InterYIT at the International Federation of Information Processing (IFIP, Austria). He has published more than 50 research papers, edited 40 books and organized over 40 national and international conferences and workshops through ACM, Springer and IEEE across various countries including India, Thailand, Egypt and Europe.

Performance Analysis of Secure Hash Algorithm-2 (SHA-) and Implementing on FPGA

Jyoti Patil Devaji, Nalini C. Iyer, and Rajeshwari Mattimani

Abstract In this modern and clumpsy world, security is more important than everything. Today, security of system is one of the fastest growing field in the world. With the goal of reduction in the accumulating area, throughput and power of the hardware performance of a crypto-analyzed system, secured hash functions of hardware working are extremely efficient and great importance. Now for most of the security services, hash function algorithm has become the default choice. SHA-2 (Secure Hashing Algorithm-2) hash family is a new security algorithm which is widely used in modern computer systems. Sometimes SHA-1 Hashing algorithm produce same hash for different files which is also known as a collision, as the number of hash functions increases, the smaller the number of chances that the standards will create the same hash function, and hence, SHA-2 algorithm is always being used in numerous cryptographic algorithms. Confidentiality, integrity, and authenticity are the major application areas of cryptography. The use of FPGA has vital advantages for implementation of cryptographic algorithms. The proposed work implements SHA-2 algorithm on cryptographic processor for efficient performance of cryptosystem using Verilog hardware description language (HDL).

1 Introduction

In the recent years of being everywhere wired and wireless communications, security is being the utmost priority. Confidentiality, integrity, and authenticity are the main factors to be relooked in security area. There are plenty of user applications regarding security, ranging from secure transactions to private communications, high-performance embedded applications, and protecting healthcare information [1]. One important aspect of safe and secure communications is that of cryptography, which includes code creation, analysis, and code breaking. Algorithms of cryptographic security algorithms may be slow but require very high computational sources and also implemented inadequately in general-purpose processors. This has encouraged

J. P. Devaji (✉) · N. C. Iyer · R. Mattimani
School of Electronics and Communication, KLE Technological University, Hubballi, India
e-mail: jyoti_patil@kletech.ac.in

T. Senjyu et al. (eds.), *ICT with Intelligent Applications*, Smart Innovation, Systems and Technologies 248, https://doi.org/10.1007/978-981-16-4177-0_1

the design and execution of dedicated computing structural designs that allow secure communication and to hasten the processing time and increase the speed expressed as megabits per second (Mbps).

A lot of secure hash functions have been developed till now, and MD5, SHA-1(SHA-160), SHA-256, SHA-384, and SHA-512 are the most popular among them [9, 11]. The oldest of them is the MD5 hash function [2]. These kinds of higher version hash algorithms are being used in a number of high-performance cryptosystems including RSA, DES, Rijndael (AES), and IDEA [3, 14, 15]. Sometimes SHA-1 Hashing algorithms produce same hash for different files which is also known as a collision, and security of cryptosystem can only be guaranteed as long as collisions do not occur [12]. SHA-2 uses higher bit lengths compared to lower hash algorithms like SHA-1 and MD5 [13].

This technical paper describes the design, performance analysis, architecture, and an efficient hardware implementation of the SHA-2 hash algorithm which is a new family of the Advanced Encryption Standard (AES) in cryptography. FPGA based SHA-2 design is implemented on a Xilinx Spartan6 XC6SLX16 device.

The cost of an FPGA is often higher than that of a comparable application-specific integrated circuits (ASIC), but field-programmable gate arrays are reprogrammable blocks of logic and their elasticity, hardware-timed speed, dependability, and parallelism makes it best suited for cryptographic applications [14]. Compared with ASIC, FPGAs offer shorter design time, greater flexibility, and lower non-recurring engineering costs. In order to get a very secure cryptographic portable device, the selected well-known algorithm must be trusted [4].

2 Literature Survey

Huffmire et al. [1] provides information about the problem that is associated with FPGA security such as design tool subversion that makes possible to load malicious design onto hardware making FPGA short circuit and also discusses bit stream protection techniques and compares ASIC and FPGA board in terms of IP theft problem. Mladenov et al. [5] show how hardware resources can be reconfigured to produce both SHA-256 and SHA-512 algorithms with hardware optimization; it reduces the memory usage and increases the performance. Wang et al. [6], the anticipated architecture was simulated using Verilog hardware descriptive language and implemented on Vivado Xilinx FPGA. The test vectors are used to verify the correctness of the algorithm. It is reported that area optimization is achieved with high throughput. Special hardware functionality is required to define that the device is not a fake. Arnold et al. [2] explain that the MD5 hash algorithm is mapped to NAPA architecture and MD5 uses one-way hash functions; these functions provide a "fingerprint" for a message or file, and NAPA architecture provides higher throughput for the same set of a clock cycle. Crenne et al. [7] discuss the implementation of Galois hash function is achieved on FPGA. It discusses the limits of GHASH function on the hardware and provides techniques to implement GHASH function in the

AES-GCM algorithm. Iyer et al. [4] describe the comparison between the SHA-1 hash function implementation with other related works and show that it achieves a higher throughput and also higher clock frequency. In this study, SHA-1 is designed on Verilog HDL. Separate modules are synthesized and analyzed using Xilinx ISE 14.2 tool. Test vectors has been used and observed that the design generates correct hash values.

Mironov et al. [8] narrate about cryptography hash functions. The algorithm defines computing from arbitrary length to a fixed length message digest cryptographic hash function which is used to map a message. As compared to application-specific integrated circuits (ASIC), FPGA offers way more flexibility in usage of cryptography including the following reasons:

1. FPGA can be reconfigured on site, so it is far more effortless to work with than ASIC.
2. Cryptographic applications can be implemented on FPGA with better optimization of size and time scale with the help of software like Xilinx.
3. FPGA based cryptographic processor have sequential-type hardware providing it way better speed performance than any software-based application [10].
4. FPGA based cryptographic processor have much lower cost and easy availability than ASIC.

3 Algorithm

The detailed algorithm is explained below with the following steps.

Step 1-First step is preprocessing the converting word to binary.

Step 2-initialization of hash values of 8. 1 to N range defines i. First 32-bit are defines by fractional parts of the square roots of first eight primes.

Step 3-creating the constants (k).

Step 4-the next step chunking the loop.

Step 5-creating the schedule of the message.

Step 6-compressing the data.

- Initialize variables **a, b, c, d, e, f, g, h** and set them equal to the current hash values, respectively, **h0, h1, h2, h3, h4, h5, h6, h7**.

Step 7-modifying final the hash values.

Step 8-concatenating the final hash to get a final digest values.

Figure 1 shows the block diagram of SHA-256 algorithm where Write data, Clock (Clk), Control Signal (CS), Reset_n, Write Enable (WE), and Address (7–0) are inputs to the model and Read_data and Error are the outputs. If the digest data is not equal to final computed hash value, then error flag is going to indicate Logic 1 (Figs. 2 and 3).

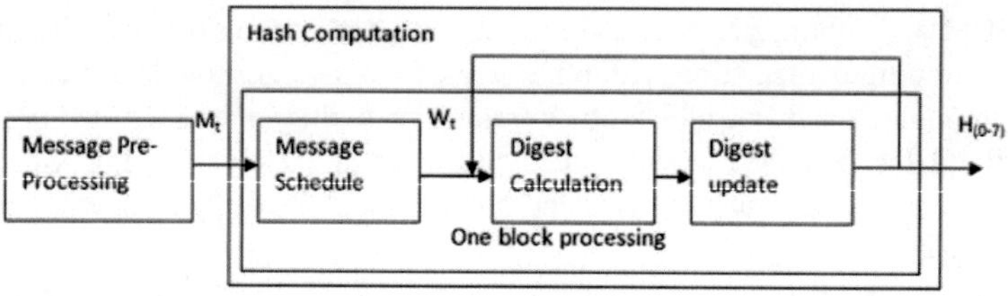

Fig. 1 SHA-2 algorithm and flow diagram

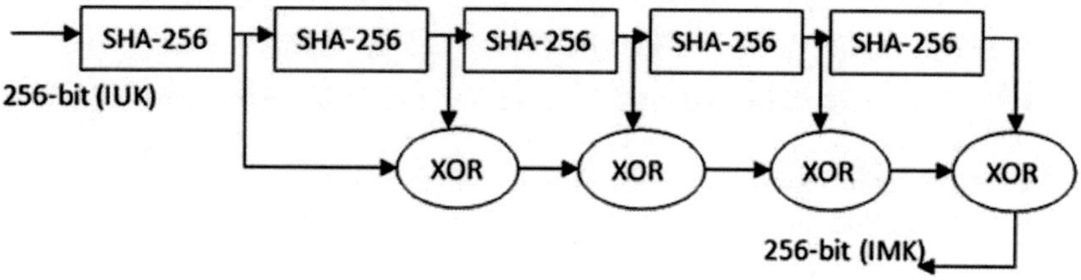

Fig. 2 Functions and constants

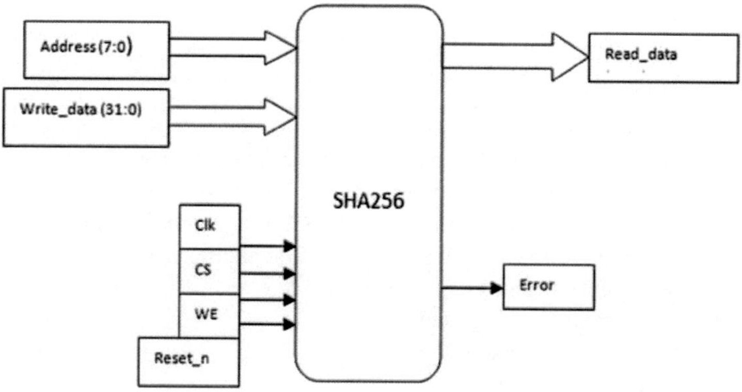

Fig. 3 SHA-256 block diagram

4 Details of Blocks

4.1 ROM Blocks

ROM blocks are stored with specific constant place carry $K_t(i)$ and used for every one hash task of the SHA-2 relations standard which is different. Chain of 80 constants equal to 64-bit uses two types of secure hash algorithm-2 and 3.

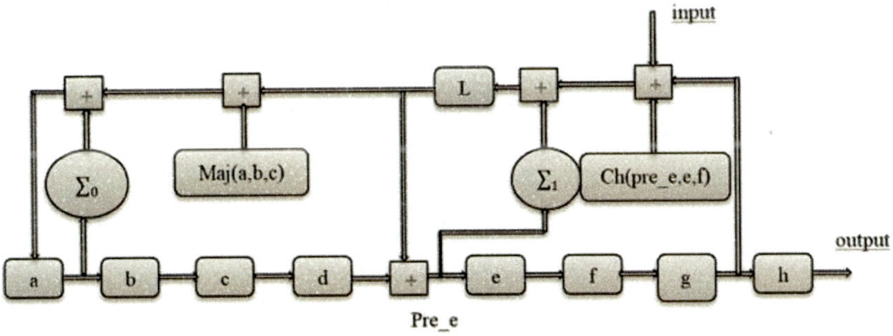

Fig. 4 Compression block of SHA-256

4.2 *Hash Computation Unit*

It gives the eight basic outputs by accepting eight basic data inputs ($A_{in}\ldots H_{in}$). During the initialization process, the constant input values (Acon...Hcon) are loaded. The other two inputs of unit are K_t (i) and W_t (i).

Figure 4 shows compression block of SHA-2 algorithm and shows how the hash value is calculated by making use of 32 bit registers and some intermediate values like Maj(a, b, c), Ch(pre_e,e,f).

4.3 *Padder and W_t Unit*

Multiple of 512-bit ensures the message length, with extra bits in the padder which are padded in the message. W_t unit is the input for output padder.

4.4 *Constant Units*

The two constants are used; they are one 32-bit and another one is 64-bit for both secure hash algorithm-2. The constants are equal to first 64-bits of the square roots of the first eight prime numbers.

Figure 5 shows the overall architecture of the SHA-2 algorithm and details of its subblocks.

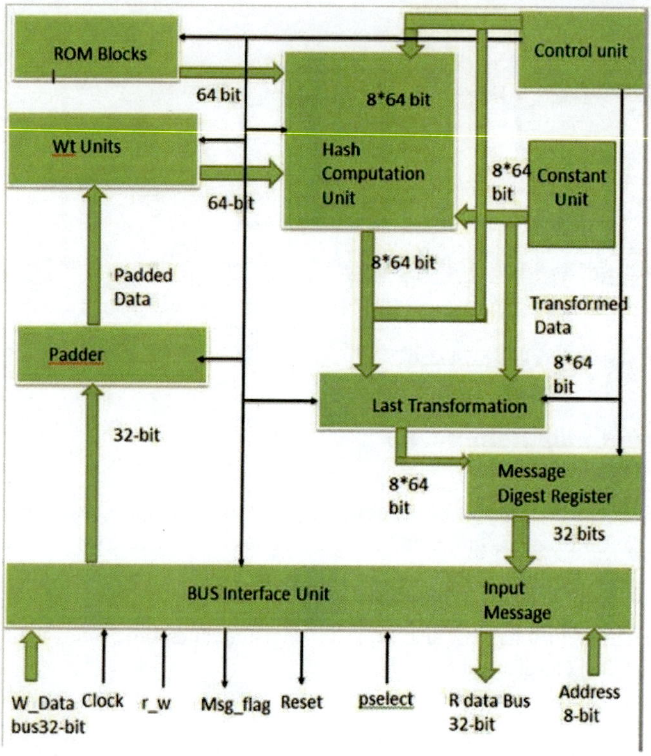

Fig. 5 Proposed design architecture of SHA-2

5 Results and Discussion

Table 1 describes the device Spartan 6 XC6SLX16 utilization summary, where this device is having 18,224 of slice registers, 9,112 of slice LUTs(Look Up Table), 2473 of LUT-FF (Look Up Table- Flip Flop) pairs, 232 of bonded IOBs (Input Output

Table 1 Device utilization summary after synthesis (Spartan6 XC6SLX16)

Utilization of logic (in numbers)	Already used	Presently available on board	% utilization
Slice Registers	1929	18,225	10%
Slice LUTs	2372	9113	25.5%
LUT-FF pairs	1791	2473	72%
bonded IOBs	77	232	33%
Occupied no of slices	724	2,278	31%

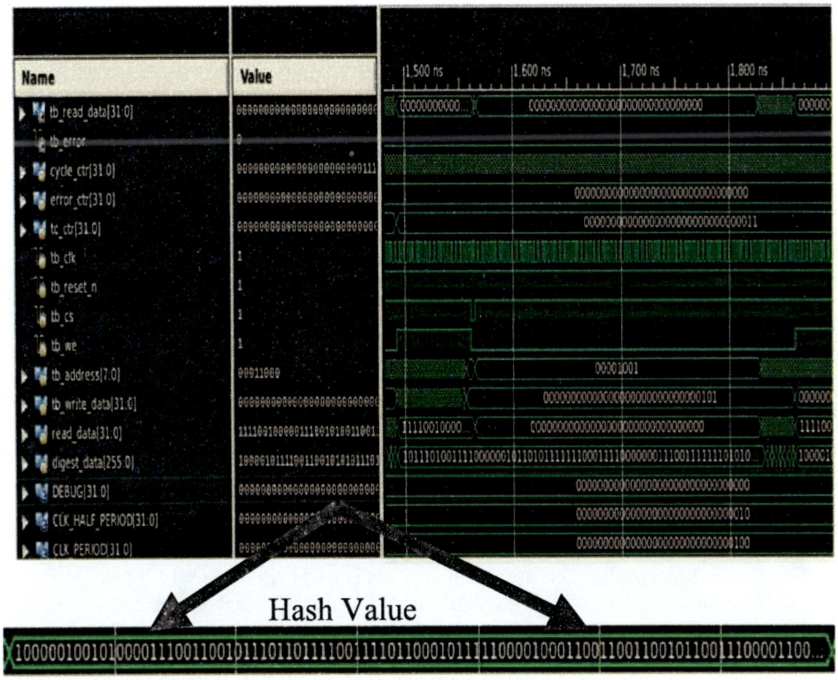

Fig. 6 Output waveform of SHA-256 Algorithm

Block), and 2,278 of occupied slices, but the SHA-2 algorithm uses 10% (1,929) of available slice registers, 26% (2373) of available slice LUTs, 72% (1791) of available LUT-FF pairs, 33% (77) of available bonded IOBs, and 31% (724) of occupied slices, respectively.

Figure 6 shows the output waveform generated by the SHA-256 algorithm in Xilinx 14.7 software. SHA-256 algorithm has been checked for different input combinations, and the result of one among them is given in this paper.

6 Conclusion and Future Scope

The objective of the paper to perform analysis and implementing on SHA-2 is done. Secure hash algorithm (SHA-2) is implemented successfully in this project using Xilinx with Verilog coding, designed, and simulated using ModelSim and verified using Spartan6 on FPGA board. For the reason of both efficiency and security, the much-secured algorithm is proposed for future scope, so the future work would be concentrating on the exploring and implementing of SHA-3 for IoT applications.

References

1. Huffmire, T., et al.: Managing Security in FPGA-Based Embedded Systems. In: IEEE Design & Test of Computers, vol. 25, no. 6, pp. 590–598, Nov.-Dec. 2008
2. Arnold, J.: Mapping the MD5 hash algorithm onto the NAPA architecture. In: Proceedings of the IEEE Symposium on Field-Programmable Custom Computing Machines (FCCM), pp. 267–268 (1998)
3. Deepakumara, J., Heys, H.M., Venkatesan, R.: FPGA implementation of MD5 hash algorithm. Proc. Canadian Conf. Electrical Comput. Eng. **2**, 919–924 (2001)
4. Iyer, N., Mandal, S.: Implementation of secure hash algorithm-1 using FPGA. Int. J. Information Comput. Technol. (2013)
5. Mladenov, T., Nooshabadi, S.: Implementation of reconfigurable SHA-2 hardware core. In: APCCAS 2008 - 2008 IEEE Asia Pacific Conference on Circuits and Systems, Macao, pp. 1802–1805 (2008)
6. Wang, G.: An efficient implementation of SHA-1 hash function. In: 2006 IEEE International Conference on Electro/Information Technology, East Lansing, MI, pp. 575–579 (2006)
7. Crenne, J., Cotret, P., Gogniat, G., Tessier, R., Diguet, J.: Efficient key-dependent message authentication in reconfigurable hardware. In: 2011 International Conference on Field-Programmable Technology, New Delhi, pp. 1–6 (2011)
8. Mironov, I.: Hash functions: theory, attack and applications. In: Microsoft Research, Silicon Valley Campus, November 14, 2005
9. Sklavos, N., Koufopavlou, O.: On the hardware implementations of the SHA-2 (256, 384, 512) hash functions. In: Proceedings of the 2003 International Symposium on Circuits and Systems, ISCAS '03, Bangkok, pp. V-V (2003)
10. Hua, Z., Qiao, L.: Hardware design for SHA-1 based on FPGA. In: 2011 International Conference on Electronics, Communications and Control (ICECC), Ningbo, pp. 2076–2078 (2011)
11. Xiao-hui, C., Jian-zhi, D.: Design of SHA-1 algorithm based on FPGA. In: 2010 Second International Conference on Networks Security, Wireless Communications and Trusted Computing, Wuhan, Hubei, pp. 532–534 (2010)
12. Hassan, M., Khalid, A., Chattopadhyay, A., Rechberger, C., Güneysu, T., Paar, C.: New ASIC/FPGA cost estimates for SHA-1 collisions. In: 2015 Euromicro Conference on Digital System Design, Funchal, pp. 669–676 (2015)
13. Ting, K.K., Yuen, S.C.L., Lee, K.H., Leong P.H.W.: An FPGA Based SHA-256 Processor. In: Glesner, M., Zipf, P., Renovell, M. (Eds.), Field-Programmable Logic and Applications: Reconfigurable Computing Is Going Mainstream. FPL 2002. Lecture Notes in Computer Science, vol 2438. Springer, Berlin (2002)
14. McLoone, M., McCanny, J.: High performance single-chip FPGA Rijndael algorithm implementations. In: Proceedings of the Cryptographic Hardware and Embedded Systems Workshop (CHES), pp. 65–76. LNCS 2162, Springer (2001)
15. García, R., Algredo-Badillo, I., Morales-Sandoval, M., Feregrino Uribe, C., Cumplido, R.: A compact FPGA-based processor for the secure hash algorithm SHA-256. Comput. Electrical Eng. **40**(1), 194–202, (2014)

Chat-It-Loud!: Real-Time Transcribing App for Communication for Hearing-Impaired

Pulin D. Shah, Prasanna J. Shete, Pravar U. Parekh, and Jaineel N. Shah

Abstract Text-it-Loud!, an Android application that renders the speaker's voice to text in real-time and displays it as a caption in the foreground of any application, it presents a viable solution to facilitate inclusive education of the hearing impaired. Although it improves the learning experience of hearing-impaired students in a classroom setting, it focusses only on one-way mode of communication, i.e., a single speaker to many listeners. It does not feature a way to respond and communicate with others verbally, to enable an inclusive group discussion. In this work, we propose "Chat-it-Loud!," a mobile application developed as an extension of Text-it-Loud! using Flutter to support Android as well as iOS devices that allow hearing-impaired people to concurrently communicate with other hearing-impaired people and everyone else in real-time. Chat-it-Loud! provides real-time speech to text and displays messages as in chat groups.

1 Introduction

As per Census 2011, persons with disability (PwD) constitute 2.21% (i.e., around 2.68 Cr people) of the Indian population. Among the total disabled in the country, 19% are with disability in hearing [1]. Out of 74,435 PwD students enrolled in higher education in our country, approximately 14,000 are hearing impaired [2].

In the paper "Text-it-Loud!: Real-Time Captioning and Transcribing App for Inclusive Teaching–Learning of Hearing Impaired," we displayed statistics and reasons for a new application to help hearing-impaired students. Text-it-Loud! provides continuous real-time speech recognition by working in the background and displaying as captions over any other application through an Android application [3]. We conducted testing at Rochiram Thadhani High School for Hearing Handicapped, Mumbai, where Text-it-Loud was presented to more than 30 teachers.

P. D. Shah (✉) · P. J. Shete · P. U. Parekh · J. N. Shah
K.J. Somaiya College of Engineering, Vidyavihar, Mumbai, India

P. J. Shete
e-mail: prasannashete@somaiya.edu

T. Senjyu et al. (eds.), *ICT with Intelligent Applications*, Smart Innovation, Systems and Technologies 248, https://doi.org/10.1007/978-981-16-4177-0_2

They suggested that two-way communication and support for multiple languages is crucial for entirely covering inclusive education of the hearing impaired.

To incorporate the suggestions mentioned above, we developed "Chat-it-Loud!" as an extension of Text-it-Loud! This allows hearing-impaired people to concurrently communicate with other hearing-impaired people and everyone else in real time. The speech recognition system continuously generates words in real time, and these words are continuously sent to all users, thus achieving real-time communication [4].

2 Chat-It-Loud!: Real-Time Transcribing and Verbal Chatting App

According to the feedback and the suggestions received for our previous application "Text-it-Loud!," we implemented the following features in order to enhance the experience of the users of "Chat-it-Loud!." Some features are inherited from Text-it-Loud!, while other features are novel:

1. Interactive two-way communication: Both students and teachers can listen as well as speak
2. Improved user interface: Chat like user interface to easily identify multiple speakers.
3. Not proximity restricted: Can be used from any corner of the world as its backend is on firebase.
4. Available for Android as well as iOS mobile devices.

3 Related Work

As technology is evolving, there are tools available for hearing-impaired people to help them communicate easily and thus making their lives easier.

One such application is Ava. Ava is an app designed to empower people who are deaf or hard-of-hearing by allowing them to follow conversations in real time. The app provides real-time captioning with up to 90% to 95% accuracy, on the smartphone. Ava allows the user to tap and correct words and teach Ava to transcribe them right. For group conversations where speakers are further away, Ava will show a real-time color-coded transcript of who says what. Ava provides the user with both free and paid options, but the free service does not provide two-way communication. The free option provides free usage of only 5 h per month with an additional 5 h of usage top-up for $4.99. The premium plan is priced at ~$14.99 per month [5, 6].

Live Transcribe is yet another application developed by research at Google to aid the hearing impaired. Since the use case of Live Transcribe is geared for real-time conversations where the speakers are at the same location, it only offers one-way communication. It does not distinguish between the speakers and can only work when the device is in proximity with the speakers.

4 Implementation

The implementation focuses on how to utilize the cell phone's speech-to-text technology to facilitate real-time communication between hearing-impaired users. Using Google's Flutter—an open-source software development kit, and "speech_to_text"—Flutter package developed by "csdcorp.com," we created a mobile application that satisfied both our goals—real-time communication between multiple users and a single codebase for Android and iOS applications [7, 8]. But this package still is not completely satisfying our need. Since it harnesses the device's speech-to-text technology, it is limited to one-time use. To achieve continuous speech recognition, we recursively use the single time speech transcription method. Since Text-it-Loud! already provides a quick, simple, and synchronized architecture to handle the flow of data, we modified it so that the mobile application would continue to facilitate the same performance but also allow multiple speakers to send and receive data, i.e., transcript generated from the speech, simultaneously.

"Chat-it-Loud!" uses Android's "SpeechRecognizer" and iOS's "SFSpeechRecognizer" and related classes for transcribing the speech input [9, 10]. To keep the UI simple, we use only the essential elements. The user interface includes "Home," "Join," "Create," and "Chat" screens. Description and interface of the four screens:

1. **Home:** The "Home" screen, that is loaded when the application is opened, includes buttons that redirect to "Create" and "Join" screens (Fig. 1).
2. **Create A Group**: To create a private group, "Create" screen is used which contains three input text fields and a "Group ID." The input text fields are username, password, and group name. Group ID is the first layer of security. It is unique and randomly generated every time. In contrast to "Text-it-Loud" which used a five alphabet "Group ID," "Chat-it-Loud" uses a six-letter alphanumeric combination. Username is required, so that others can recognize the user, password is to provide an extra layer of security. The password is required to join the group as a "Speaker" (Fig. 2).

Fig. 1 Home

Fig. 2 Create a group

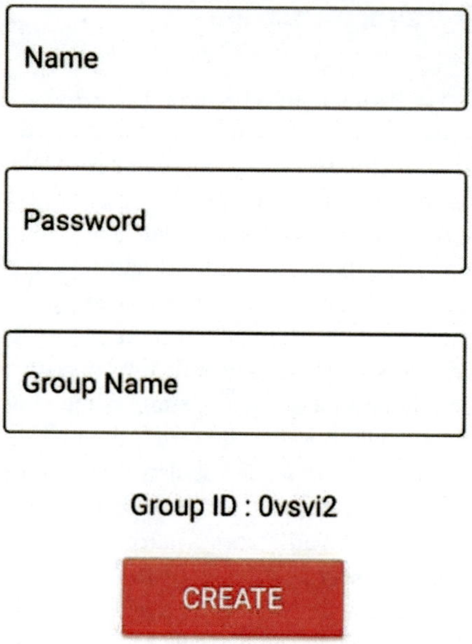

3. **Join A Group**: The user who has created a group will then share the password and the "Group ID" to other users. The users will be able to join the group using the "Join" screen. Similar to "Create" screen, it contains three text fields—name, password, and "Group ID." It also contains a dropdown which is used to join as a "Speaker" or a "Listener." A "Speaker" will be able to speak and communicate with others, while a "Listener" will just be able to see the "Chat" screen (Fig. 3).
4. **Chat**: The messages (transcribed by the speech-to-text technology) are visible on the "Chat" screen. Messages are labeled with the name of the sender. As per the design and requirement of the application, they appear in real time and

Fig. 3 Join a group

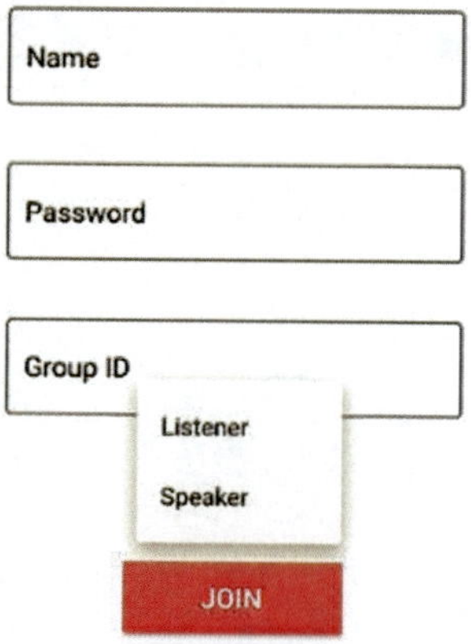

Fig. 4 Mobile application screenshots

in sequence. Sequence is determined using the timestamp. The "Chat" screen also contains a button that when pressed invokes the speech-to-text technology. It toggles to "listening" and back to "non-listening" state after a successful speech-to-text conversion or a device specific time limit. This time limit refers to the amount of time that the speech-to-text technology will listen with no audio input. After the time limit, it will automatically stop listening. This is done to reduce the device's battery consumption as the technology is rich and computationally intensive (Fig. 4).

The internal processing done by the speech-to-text technology is same, as described in "Text-it-Loud" [3]. Data synchronization and transfer between users is done using timestamps, a flag to indicate if the data is completely received or if it is still being received, and Firebase. All incoming transcribed messages are stored under one root node of the Firebase—real-time database tree structure. This root node is different for each group. All devices engaged in a group are listening to the changes done in the root node. Every new message sent by a user is added as a child in the root node, and it is then fetched by all the devices in a group. Based on the timestamp of the last message fetched by the device, in case of loss of internet or other reasons, all non-received messages are automatically synchronized when adequate internet in available (Fig. 5).

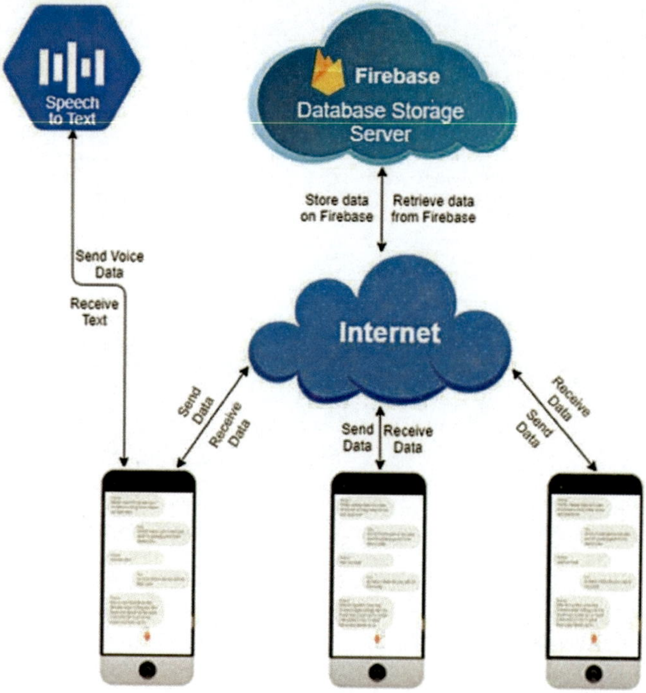

Fig. 5 System architecture

The speaker's device performs the speech recognition, displays text on the right side of its screen, and forwards the data in text format to all other devices via "Firebase" and displays text on the left side of the screen. While developing this particular application, care was taken to abide by the accessibility guidelines as suggested in.

5 Results

Since "Chat-it-Loud!" uses the same technology that is used by Text-it-Loud!, the accuracy of the speech recognition system is about 87–90% [1]. Even with a moderate internet connection, data is received by other users under 2 s after being spoken.

Figure 4 displays two screenshots of the app, wherein the last message is being received as the speaker is still speaking on the left screenshot, while it is completely received on the right screenshot.

6 Conclusion

This paper presented a Flutter application "Chat-it-Loud!," an extension to "Text-it-Loud!," that allows hearing-impaired users to communicate with each other and others verbally by converting their speech to text in real-time. "Chat-it-Loud!" can be used for group discussions, meetings, and other group communications with hearing-impaired people. The app can be used for quick and spontaneous interactions such as talking with colleagues at work or meeting people in public. It can also be used at events, classrooms, churches, doctor's appointments, business meetings, and much more.

7 Future Work

We intend to add support for multiple languages through which speech is transcribed in the respective language. Translation from one language to another is also planned to be implemented in the future.

References

1. Report on "Disabled Persons in India—A statistical profile 2016. Social statistics division, Ministry of Statistics and Programme Implementation, Government of India (2017)
2. Report on "All India Survey on Higher Education (2015–16), Ministry of Human Resource Development, Department of Higher Education, New Delhi (2016)
3. Shete, P., Shah, P., Parekh, P., Shah, J.: Text-it-Loud!: Real-time captioning and transcribing app for inclusive teaching-learning of hearing impaired. In: 2019 IEEE Tenth International Conference on Technology for Education (T4E), Goa, India, pp. 264–265 (2019). https://doi.org/10.1109/T4E.2019.00066
4. https://play.google.com/store/apps/details?id=com.google.audio.hearing.visualization.accessibility.scribe&hl=en_IN
5. https://play.google.com/store/apps/details?id=com.transcense.ava_beta&hl=en_IN
6. https://www.ava.me/about-us
7. Flutter—Beautiful Native Apps in Record Time. Flutter.dev. (Online). Available: https://flutter.dev/
8. speech_to_text|Flutter Package, Dart packages (2020) (Online). Available: https://pub.dev/packages/speech_to_text
9. Android Developers. (n.d.). Android.speech|Android Developers (online). Available at: https://developer.android.com/reference/android/speech/package-summary
10. Apple Developer Documentation, Developer.apple.com. (Online). Available: https://developer.apple.com/documentation/speech

ICT in Indian Agriculture: Opportunities and Constraints

G. S. Gill

Abstract Information and communications technology (ICT) has brought revolutionary changes in the lives of millions of Indian populace. Due to progress in fields of computer science and telecommunication, large amount of information can be acquired, processed, and delivered at any place and time to a person who needs it. Recent technological developments have made handling of enormous amount of information relatively easy. This information can be accessed by a huge number of users concurrently. These technological developments have provided vast prospects in improving utility and productivity of occupational technical areas like agriculture, extension, and educations. It has become a challenge for researchers to harness the ICT developments for improvising the quality of life of rural populace. In this paper, an attempt has been made to evaluate opportunities and constraints of associated with use of ICT to enhance productivity of agriculture and associated sectors in India.

1 Introduction

Economic transformation of a developing nation like India is significantly dependent upon the performance of agriculture sector. This sector plays a momentous role in rural living, employment, and food security. About 70% rural India is dependent on agriculture for their living [1]. In the recent past, thrust has been on exploring the potential of integrating ICT with agriculture to increase productivity and prosperity of rural households. ICT has made a considerable advancement in the last few decades and has touched almost all areas of life. Large advancements have taken place in the past few decades in hardware and software development. Also data handling and management have taken big strides in the recent past. Internet has connected people at global level [2]. It has become a challenge for researchers to harness the ICT developments for improvising the quality of life of rural populace. In this work, an attempt has been made to evaluate opportunities and constraints of ICT to enhance productivity of agriculture and associated sectors.

G. S. Gill (✉)
Kurukshetra University, Kurukshetra, Haryana, India
e-mail: gsgill@kuk.ac.in

T. Senjyu et al. (eds.), *ICT with Intelligent Applications*, Smart Innovation, Systems and Technologies 248, https://doi.org/10.1007/978-981-16-4177-0_3

2 ICT and Agriculture

To achieve sustainable development in agriculture, timely access to vital information along with efficient and reliable medium of communication is key elements. These attributes can be effectively used to increase information exchange and dissemination of knowledge and ideas in rural areas of the country. Till date, lack of timely information in the rural areas is still a big issue, and lack of its access leads to poorly timed decisions which further lead to drop in productivity and economic losses to the farmers. Flow of information through conventional media such as radio or television is predominantly supply driven, and the utility of such information is always very limited. Instead of this, the information should be requirement driven, and the focus should be on needs of the user. At this juncture, researchers have to focus on aptly utilizing the potential of ICT in improving the productivity of agriculture sector, in particular, and quality of life of rural societies, in general. ICT can bridge the divide between those having access to the information and those not having the same. In this information age, ICT can be a significant attribute in addressing many social and economic issues like poverty, caste disparity, and gender discrimination being faced by rural populace. The researchers and policy makers will have to accept the challenge to create requisite infrastructure at village level, so that the rural population is not excluded from the benefits of information age. Moreover, an agrarian economy like India cannot afford to keep its farmer information deficient when the world is taking giant strides in ICT. The following section presents an overview of ICT in reference to various aspects of agriculture.

Gargantuan development can be brought to rural India if the government bodies and agencies, the private players, education and research organizations, and the welfare bodies operate with collaborative approach and a single-point agenda of uplifting the rural life. An effective communication is the basic ingredient for these bodies to work cohesively. All the stakeholders can contribute their part by effectively sharing and exchanging their ideas and views on various aspects related to agriculture and allied sectors. In addition to conventional farming, farmers can be encouraged to venture out in allied areas such as bee keeping, dairy farming, pisciculture, and silviculture. They should be provided with information and other resources related to this and should be provided barrier-free environment and suitable platforms to discuss and present their experiences and success stories with their peers. This would also provide an opportunity to policy makes to have direct communication with the stakeholders.

Internet has provided an information highway to bridge the physical gap between the various users located at distant locations. Now a farmer can make a video call to an expert and present him the problem he is facing though the video call. The subject experts in any of the 64 state agriculture universities can be contacted through telephone or video call by scheduling a convenient time slot. Problems can be mailed, and customized solutions can be provided to the farmers in real time. The Internet being a source of vast and appropriate information can help in making the rural populace well informed about the global, national, and regional developments taking place in all spheres of life. With the suitable information at their disposal, they can make better decisions in their life and can enhance their standard of living. The exposure to relevant information would help them to achieve occupational growth, provide educational opportunities, access to better healthcare, and achieve the community development goals. Contribution of Internet in providing barrier-free communication across the globe is noteworthy for all the stake holders involved in agriculture sector [3–5].

3 ICT Projects in Agriculture

With global communication revolution impacting almost all spheres of life in India, there has been a radical change in agriculture sector as well. Like other people, farmers also want to keep themselves updated on the global issues that are likely to have an impact on them. Earlier such real-time information exchange was a distant dream, but now in the information age, such facilities are easily available to the individuals who need them. These revolutionary developments in field of ICT are aptly exploited to their fullest potential by the professionals engaged in the area of agriculture extension services to reach out to the farmers on their fields. A few such ventures like Village Knowledge Center, Kisan Call Centers, Bhoomi, and Ikisan are excellent examples of such initiatives [6].

3.1 Village Knowledge Centre (VKC)

One of the pioneer bodies to first venture out in this area was MS Swaminathan Research Foundation (MSSRF). In the year 1998, they came up with a novel concept named Village Knowledge Center (VKC). This project was started in Puducherry with the aim to provide relevant information to villagers thought ICT mode, so that they may be able to make a better living for themselves by optimal utilization of resources available at their disposal. In this setup, there is one nodal center called Value Addition Center and ten VKCs are attached to it. The centers have suitable hardware in the form of computers, radios, and telephones that is connected over a wireless communication link. The prime focus in setting up such network is aimed at achieving food security in rural belt of the catchment area of the centers. In order to achieve the desired outputs, the farmers are supplied with structural technical inputs to enhance the productivity. Farmers are provided complete solutions related to crop selection, crop production, crop protection, crop management, pre-harvest and post-harvest practices, and procedures. The information is created and supplied in local languages using native understanding, complemented by general facts as well as some inputs from domain specialists. Some incremental improvement is done by incorporating the use of audio-visual aids. VKCs are assigned the responsibility to pass on such information to the stakeholders in any form they deemed to be suitable, like use of charts, drawings, audio-visual aids, audio announcements, or small pamphlets.

3.2 *Kisan Call Centers (KCCs)*

Kisan call centers were started in 2004 by Ministry of Agriculture and farmers' welfare with the objective of providing answers to the queries of the farmers in their native language. These centers operate with a very nominal hardware and operating costs. The infrastructure includes a computer connected to Internet and a teleconferencing system. A farmer can dial a toll free number, 18001801551, and present his issue to the technical executive handling the query. The technical executive is a competent professional with a post-graduate or a doctorate degree in field of agriculture or allied disciplines and has a command over the local languages as well as Hindi and English. He listens to the query and tries to provide the best possible answer to it. However, if he is not able to satisfy the caller or is not able to respond to it, he can further transfer the query to next higher level of expert, a professional working with department of agriculture or in state or central agriculture university. As many as 25,000 calls are handled by these centers on daily basis. Currently, Kisan Call Centers are operating from 14 locations and cover the entire country.

3.3 *Bhoomi Project*

In the first initiative of its type, in year 1998, Karnataka government came up with a project involving computerization of land records of the entire state named "Bhoomi" project. Till date, Karnataka state revenue department has claimed to computerize over two crore records of land ownership of over 65 lakh farmers. A farmer can visit any of the nearest *taluka* (tehsil) office and after a nominal payment of Rs. 15 can obtain a computerized record of rights, tenancy, and crops called the RTC, a document that is required and accepted by many departments and banks for loan purposes. This project facilitated farmers and rural populace in obtaining the access to their land records. Also the various departments of the state government and the judicial system were provided access to this software, so that they can check the land records in real time as and when required by them to settle the issues and disputes pertaining to their respective departments.

3.4 *Ikisan*

Ikisan was incepted by Nagarjuna group in year 2000, and it aimed at rural prosperity and increase in agro-throughput. As a part of this program, a Web-portal was launched along with installation of information kiosks at villages. This facility was provided to users free of cost. It supplies the inputs related to crops, agricultural practices, crop marketability, and metrological data in eight regional languages. These kiosks are operated with a minimum level of infrastructure and manpower. They have an Internet-backed computer and a skilled person to help farmers access the information. The person available at kiosk helps the rural populace in accessing information relevant to their requirements. The portal offers complete solution to from crop selection to marketability. It also gives them ideas to process their produce in order to get enhanced profits according to the needs and demands of local market. It provides them a platform to interact and exchange their ideas with their peers in the same sector who may be distantly located.

3.5 National Agriculture Market (NAM)

A pan-nation online Web-based platform was inaugurated by agriculture ministry in year 2016 to connect agriculture produce market committees or *mandis* across the nation. This platform was named NAM, and its online model has been named as eNAM. A total of 962 mandis have been brought under eNAM scheme. The objective was to integrate the otherwise scattered agro-*mandi* network and to bring transparency and uniformity in its working.

NAM has been conceived with an objective to provide a single platform facilitating online trading of agro-produce free from the boundaries of state and regions. The synergy provided by such massive integration could benefit all the stakeholders from farmers to consumers. An organized market of agro-products would help in streamlining the otherwise haphazard existing unstructured marketing system in addition to bringing transparency and efficiency to it. NAM is more focused to protect the interests of producers by ensuring better marketability of their produce. It is a farmer-intensive model marketing rather than a market-intensive model of marketing. It provides a direct interface between seller and buyer, with the seller directly able to sell his produce to the one who quotes the highest bid for his commodity. From the buyer's perspective, this system allows him direct access to the quality material at the competitive price, directly from the producer, leaving aside the intermediaries.

At present, almost all the categories of crops are being sold and bought on this platform ranging from staples to cash crops. As per latest inputs from agriculture ministry, about 16 million farmers are using this platform for selling their produce directly to the potential buyers. Agro-produce worth one thousand billion was merchandized on this portal in year 2018–19. NAM/eNAM can be visualized as a giant milestone toward rural prosperity and well-being [7].

4 Constraints in ICT Implementation

In addition to the numerous advantages that are cited for IT implementation in agriculture, the constraints for the same cannot be completely relegated. The ICT adoption process has elicited a long list of successes, failures, and bottlenecks. Closer scrutiny brought to light the following issues and constraints:

Financial Implications. It is a big challenge to bring rural India on a communication highway keeping in mind the diversity that they have in terms of language, culture, geographical limitations, and poor knowledge base. Moreover, with limited exposure and expertise of rural populace, it is a momentous challenge to educate them in handling and usage of modern communication tools and services. It requires a concrete planning combined with efficient implementation, backed by sound financial support from the establishment to fulfill the vision of sustainable rural growth. This effort requires initial investment for setting up appropriate infrastructure along with expenditure to be incurred toward maintaining and updating the same. Government can seek proposals to set up such projects in public private alliance.

Inadequate access to telecommunication infrastructure. Marred by poor infrastructure and flawed implementation, rural populace is still years behind their urban counterparts when it comes to comparison of quality and access level of ICT services. With less number of users per unit area in rural areas, the bandwidth expenditure along with operating cost becomes exceedingly high, and these costs are eventually percolated down to the users. This further leads to increase in cost of such services and creates an environment where only an effluent few can afford them. This leads to another division in society by creating two sections with one being information rich and the other being information poor.

Access to ICT and Cost of access. The access to ICT is primarily dependent on the availability and access to hardware equipment like smartphones and computers along with the network connectivity. Further, if the equipment is available, the information access should also be available at equitable cost. This creates a digital divide between those having the access and those not having the access to the digital services due to the financial constraints. When the world is battling COVID-19 pandemic and almost every sector was pushed to work from home, it has been observed that the people having access to ICT resources have substantially increased their efficiency while those who do not have access to these resources have been sidelined, especially in rural areas. The poor access level to ICT resources in case of rural society was highlighted when students in rural areas could not continue their online classes and had to suspend their schooling as they could not afford the cost of ICT equipment and services.

Language and content limitations. Almost 82% people involved in agriculture are small and marginal farmers who are either illiterate or have inadequate education level. With such level of population, utility of ICT methods for agriculture is a big challenge in itself. The main hurdle in utilization of information by rural populace is the absence of content in their regional languages. ICT solutions can only be beneficial and persuasive to the rural folks if they are customized to meet their requirements. Audio-visual aids should be developed to deliver information in the form that can be easily assimilated by the masses in their own languages.

5 Summary

Without any doubt, it can be concluded from the success of various ventures that ICT can play a pivotal role is sustainable rural development. It can help farmers in providing timely information on the topics relevant to their needs on areas ranging from crop selection to post-harvest advisories. It can provide him suitable inputs on the issues such as agro-marketing and educate them about the demand and supply forces that determine the pricing of commodities produced by him. ICT has revolutionized extension services by facilitating farmers to seek customized solutions on the problems encountered by them from the subject experts located at distant places, which is otherwise an uphill task keeping in view the geographical distance that keeps them apart. Rural India is moving at a slow, but gradual, pace toward inclusive growth by harnessing technical advancements taking place in information age. Governments at state and national levels have implemented various ICT schemes to deliver the information to the remotest areas. Almost all the villages are having mobile Internet connectivity which is backbone for information exchange. A lot has been done in this area, but still there is lot more has to be done. Focus should be on need-based information delivery rather than supply-based content delivery in the local or regional languages. Instead of one way "information delivery," researchers in extension services should develop communication models that should rather focus on "information exchange." Involvement of progressive farmers in policy planning and implementation would provide keen insight on the constraints encountered on the field and the issued resolved in such manner would be more conclusive and effective. During COVID-19 when economic activity in almost every sphere has shown negative growth, agriculture has been the only sector that has exhibited positive growth. So, India being an agrarian economy nation must be committed towards inclusive growth of people engaged in this sector.

References

1. Agriculture and food management—Chapter 07, Economic Survey 2019–20, vol. 2, pp. 193–217 (2019)
2. Reddy, R.: Computing—the next 10 years. Talk presented at Georgia Tech 10th Anniversary Convocation, 6 April 2001
3. Reddy, P.K.: A Framework of information technology based agriculture information dissemination system to improve crop productivity. Curr. Sci. **88**(12), 1905–1913 (2005)
4. Mittal, S.C.: Role of Information Technology in Agriculture and its Scope in India (2001). Available at http://25.19.12.220/applications/Brihaspat.nsf/.../$FILE/it_fai.pdf

5. Shaik, N.M., Jhamtani, A., Rao, D.U.M.: Information and Communication Technology in Agricultural Development: A Comparative Analysis of Three Projects from India, Agricultural Research and Extension Network Paper No. 135 (2004)
6. Jorge, S.N.: The economics of ICT: challenges and practical strategies of ICT use for women's economic empowerment. In: Paper Presented in UN Meeting on ICTs and their Impact on and Use as an Instrument for the Advancement and Empowerment of Women, Seoul, South Korea (2002)
7. e-NAM Manual for District Level Functionaries (2017)

Estimation of Economic Impact of National Open Digital Ecosystems

Jaijit Bhattacharya, Rishav Bhattacharyya, Henam Singla, and P. Vigneswara Ilavarasan

Abstract Using a case study of Indian Railways Catering and Tourism Corporation (IRCTC), an attempt to capture the economic impact of open digital ecosystems is made. The study indicates that the economic value is around USD 147.59 b for 2019, approximately 5% of GDP. Need for strengthening open digital ecosystems is highlighted with few policy suggestions.

1 Introduction

Public infrastructures generate significant amount of data which can be unlocked to generate further value to the public and government. There are many examples of the government systems opening up to private sector, to provide value add CivicTech services to citizens and businesses. For instance, in India, Open Applications Programming Interface (API) policy mandated by the government [1]. However, many others such as passports or driving licenses are online but not compliant with the mandatory open API policy. This deprives the citizens from innovative CivicTech solutions and prevents unlocking of economic value from the real-time data of the government. The present paper makes an attempt to estimate the potential value of open digital ecosystems.

Open access to government's digital ecosystems can potentially lead to new innovations and creation of new systems that benefits businesses and consumers and increases the efficiency of the economy [2–4]. The more government digital systems are opened and accessible, and the more it will be used by the industry for creating innovative offerings that would lead to further economic growth. One of the critical success factors for making the open data initiative [5] is to engage with the broader

J. Bhattacharya (✉) · R. Bhattacharyya · H. Singla
Centre for Digital Economy Policy Research, New Delhi, India
e-mail: president@c-dep.org

P. V. Ilavarasan
Indian Institute of Technology Delhi, Delhi, India
e-mail: vignes@iitd.ac.in

T. Senjyu et al. (eds.), *ICT with Intelligent Applications*, Smart Innovation, Systems and Technologies 248, https://doi.org/10.1007/978-981-16-4177-0_4

set of stakeholders. The open datasets enable people to come together to explore data and learn from each other.

India started its 'Open Government Data' initiative by opening up government-owned shareable information in machine-readable formats for the use of the common public in October 2012 [6]. Apart from private enterprises, it can result in innovations in the GovtTech space, when certain principles for data sharing are followed [6]. According to the Digital India Initiative of Open Government Data (OGD) Platform [7], there are 173 departments that have contributed to open data (87 central and 86 State Departments) across 35 sectors. The Indian government provides access to various datasets such as agriculture, banking, economy, telecom, air pollution (industrial), etc. However, the datasets under many departments are not updated periodically [8]. This would inevitably create significant hurdles in the data explorations for new businesses. In contrast, the Indian government has kept certain datasets updated which includes Railway Data, Air Quality Data, India Weather Data, Aadhar Metadata, and Gateway to Indian Space Research Organization (ISRO). As a result, 68 organizations in India have been able to utilize this data from the platform to create services as per the Open Data Impact Map [9].

Using the case of Indian Railways Catering and Tourism sector of Indian Railways Catering and Tourism Corporation (IRCTC), this paper attempts to estimate the value created by opening up of government digital systems. By focusing on quantifying the economic value that can accrue from businesses leveraging data from this sector, we make a case for prioritizing the opening up of digital ecosystems from other sectors of government by providing a cost-benefit analysis of doing so.

2 Literature Review

Government of India's position on the issue is captured in MeitY's white paper on the national open digital ecosystem (NODES) [6]. However, it does not quantify the benefits of having a national open digital ecosystem. Other studies [10] indicate that India can create $500 billion in new economic value via open digital ecosystems by the year 2030. However, the report's projection is based on subjective assumptions on the impact of open digital ecosystems on the economy. From a global perspective, a report published in 2013 by McKinsey Global Institute [11] aimed to quantify the potential economic value of Open Digital Systems. The analysis focused on seven domains of the global economy, such as Education, Transportation, Consumer Products, Electricity, Oil and Gas, Health care, and Consumer finance. Each domain was subjectively assigned a monetary 'Potential Value' (lower and upper) that could be unlocked based on its ability to use relevant open datasets effectively. Based on their analysis, they estimated that open data has the potential to enable more than $3 trillion in global GDP value.

Based on the McKinsey study, in 2013, the Australian government calculated the total potential value of open digital ecosystems in Australia, for sectors within its economy [12]. It used the following formula for calculating the economic impact by extrapolating the global guesstimates from the McKinsey study.

Economic impact for each Industry Segment of Australia = ((Global Open Data Value of the Industry Segment)/(Global GDP Value in Year X)) * Australia's GDP Value in Year X.

The formula was used by Gruen [4] to get the lower and upper values for 'Australia's sector-wise' potential value of open digital systems. The present paper extends this approach.

3 Methodology

To estimate the economic impact of opening up of digital systems of the government, we take the Indian railways as a proxy sector and calculate the revenues generated by start-ups that have leveraged data from the Indian Railways Catering and Tourism Corporation (IRCTC) [13]. This is because this platform is run by the government and has already considerably opened up its systems and data, which is updated real-time, for private commercial usage.

Many start-ups such as Trainman and Confirmtkt have initiated offerings to bring in value add to the data coming out of IRCTC digital systems. This includes using advanced predictive analytics (used by Railyatri) in finding confirmed tickets and other associated service like tracking PNR numbers and even have meals delivered to their seat numbers. They have been able to track trends on train availability and delays and provide information to aid the travelers.[1]

The start-ups considered for the study are those whose revenues are available in the public domain and use either passenger related data or are involved with food for railway passengers or both. This is to maintain uniformity in the comparisons. There are two broad areas: Search and prediction (Railyatri, Confirmtkt, Trainman, Ridlr) and food on trains (TravelKhana, YatraChef, TravelZaika, Mera Food Choice, OMitra, Rail Rider, Rail Restro, and Comesum). We found total annual revenues, S, for the year FY 2019, from publicly available data for these companies.

$S = \Sigma(s_i)$, where '*s*' is the revenues from start-ups and '*i*' refers to all the start-ups listed. Based on the above-computed annual revenues, the ratio (r) of revenues from start-ups (S) that use IRCTC data to the total revenue (I) of IRCTC in Catering and Ticketing domains is computed ($r = S/I$).

Using the ratio r, we estimate the gross economic value, R, that will accrue from start-ups that can potentially be based on the open digital systems of the government ($R = r * G$).

G is the total government revenues in India, summed up across the central government, 28 state governments, and 3 Union territories [14].

[1] The Trainman app shares live information from the required destination or arrival points periodically.

That is $G = \Sigma(g_j)$, where 'g' comprises the revenues of each state government and central government in India and 'j' refers to the revenues of the central government, 28 state governments, and 3 Union territories [14].

The G was calculated as USD 65.4 million [15, 16].

To compute the value of value of I, we take the revenue of IRCTC from food (Catering and Rail Neer) and Internet ticketing for the year ended.

Therefore, I = USD 0.19 billion [17, 18].

By using these values, we get the value of r.

$r = 0.0654/0.19 = 0.344$ (approximately).

To calculate G, the revenues of States, Union, and Central government for FY 2019 have been totaled. Total revenue for States and Union Territories for FY 2019 come out to be USD 167.93 billion, while total central government revenue was USD 260.87 billion.

Therefore, G = USD 428.8 billion.

Using these values, the value of R was computed as below:

$R = r * G = 0.344 * 428.8$ = USD 147.59 Billion.[2]

4 Results

The total revenue of start-ups using IRCTC data on Ticketing and Catering domain is USD 0.0654 billion. The IRCTC revenue (I) is USD 0.19 billion. The ratio, $r = 0.344$ indicates that, start-ups have a potential to generate approximately 34% of the IRCTC revenue using IRCTC data. Extrapolating this to the government revenues (G), we find that the estimated value of start-ups leveraging open digital ecosystems is USD 147.59 billion.

GDP for India in 2019 was USD 2875.14 billion [19]. Therefore, from an empirical standpoint, the potential additional economic value that could have been obtained by opening up of government digital ecosystems is (147.59/2875.14) * 100 = 5.1336% or approximately, 5% of GDP in 2019.

As per the trading economics forecast, India's GDP is expected to be approximately USD 3250 billion in 2025, under normal circumstances (this excludes the government's published policy of propelling the economy to USD 5000 billion by 2025).

Therefore, by 2025, the potential additional economic value of opening up of government digital systems is 5% of USD 3250 billion = USD 162.5 billion.

The GDP is expected to grow at a faster rate due to government's focus on trying to achieve a USD 5 trillion economy by 2025. Therefore, the actual economic benefit from opening up of government's digital systems would be higher than USD 162.5 billion by 2025.

---

[2] The study is conducted for a particular year, i.e., 2019. However, if it is conducted over year-on-year, it will generate a positive trend indicating that each year this value will increase.

Table 1 Estimation of economic value add to Indian GDP (USD Billion)

Domain	Lower value	Upper value	India's lower value	India's upper value
Education	890	1180	21.38	28.34
Transportation	720	920	17.29	22.10
Consumer products	520	1470	12.49	35.31
Electricity	340	580	8.17	13.93
Oil and gas	240	510	5.76	12.25
Health care	300	450	7.21	10.81
Consumer finance	210	280	5.04	6.73
Total	**3220**	**5390**	**77.34**	**129.46**

Note Based on extrapolation of the McKinsey global guesstimates for 2013

5 Comparative Analysis with Similar Studies

We also estimated for different sectors following the McKinsey study. The formula used was:

Economic impact for each Industry Segment of India = ((Global Open Data Value of the Industry Segment)/(Global GDP Value in 2013)) * India's GDP Value in 2013.

Upper value of economic impact for each Industry Segment of India = ((Upper Value of Global Open Data Value of the Industry Segment reported by McKinsey Study)/ (Global GDP Value in 2013)) * India's GDP value in 2013.

Similar calculation was made for lower values as well.

Table 1 indicates that in 2013, the potential economic value add of open digital systems for sectors within the Indian economy was in the range of USD 77.34–129.46 billion in 2013. This was approximately, 4.17–6.97% of the country's GDP for that year. This is in line with the estimation made in this paper; i.e., opening of government digital systems has a potential economic value add of around 5% GDP.

Similarly, the study conducted by Omidyar Network and BCG guesstimates that the potential value add due to opening up of government digital ecosystems in India is about 5.5% of GDP. These guesstimates of the Omidyar Network and BCG paper are in line with our findings that are based on figures of actual economic value add emanating from opening up of current government digital systems.

6 Implications

According to the Global Open Data Index (GODI), India is ranked 32nd in the world. India has a mere 13% 'Open' against Taiwan and Australia, which top the index and are 80% and 53% open, respectively [20, 21]. Taking cues from other nations in this index, some potential areas which stand to benefit from having open data include common service centres (CSCs), weather forecasting, election results, government spending, and land ownership. The next steps for the implementation of open systems are:

- Identification of real-time datasets and update them periodically. Then release as open data.
- Enable accessibility of data through non-discriminatory licenses.
- Generate awareness about the open systems.
- Incentivize organizations that explore the open digital ecosystems for business creation and public good.
- As an example, government could consider training common service centre (CSC) operators to use such platforms. This will help to generate awareness among the public to use these systems.

7 Conclusion

The findings indicate that there is very significant potential economic value add if national digital ecosystems are opened up for the start-ups and the private sector, through open APIs [22]. By opening up other government sectors to private players, the Indian government has the potential to unleash transformative impact at scale, as the former bring in innovation and deep technology solutions, while the latter brings in enormous scale. Access to such data may have commercial models built-in, in order to make the infrastructure sustainable and enable the government to have additional non-tax revenue growth. Data mashing is encouraged, which will help to access multiple data at a single interface.

Open digital ecosystems need to have the digital infrastructure, active citizen participation, and robust governance to ensure fair outcomes for all. API-based connects to open government systems are required which are essential for the CivicTech innovations [23]. Government can also help build the tools to access and interpret government open digital systems, in order to reduce the cost of build-out of a national open digital ecosystem, wherein government and private sector data are freely available to commercial and social stakeholders.

Fortunately, the Indian government already has a policy framework for adopting Open data systems through the open API policy [23]. However, this policy needs to be aggressively implemented. In addition to providing open APIs, for the estimated benefits to actually accrue, other supporting frameworks also need to be put into place such as the government providing a minimum guarantee service-level agreement (SLA) for the open APIs. Some of the systems like passport, Aadhar, and flagship G2C schemes of the government have good SLAs and can follow the IRCTC, PAN, and ITR model of G2C services. Other government systems may be strengthened to bring them at a level where they can provide high levels of SLA.

References

1. Open API Policy of Ministry of Electronics and Information Technology, Government of India, May 19, 2015. https://meity.gov.in/writereaddata/files/Open_APIs_19May2015.pdf. Last accessed on 2020/11/30
2. Digital Future Society, Toward Better Data Governance for All: Data Ethics and Privacy in the Digital Era, Future Society (2019). Available at: https://digitalfuturesociety.com/app/uploads/2019/08/060819_Toward_better_data_governance_for_all_dfs_mwcapital_DIGITAL.pdf
3. Buteau, S., Larquemin, A., Prasad Mukhopadhyay, J.: Report on Open Data System from a Symposium of IFMR, 27th May 2015, Ottawa, Canada
4. Gruen, N., Houghton, J., Tooth, R.: Open for Business: How Open Data Can Help Achieve the G20 Growth Target, Lateral Economics for Omidyar Network, Australia (2014). Retrieved from: www.omidyar.com/sites/default/files/file_archive/insights/ON%20Report_061114_FNL.pdf1. Last accessed 2020/11/30
5. Prakash, P., Abraham, S., Shah, N.: Report on Open Government Data in India, Glover Wright, Centre for Internet and Society, Bangalore
6. Strategy for National Open Digital Ecosystems, MeitY, Consultation White Paper. Government of India. https://www.medianama.com/wp-content/uploads/mygov_1582193114515532211.pdf. Last accessed 2020/11/30
7. OGD Platform. https://www.nic.in/projects/open-government-data-ogd-platform-india/. Last accessed 2020/11/30
8. Sectors category on official India OGD platform. https://data.gov.in/sectors. Last accessed 2020/11/30
9. Open Data Impact Map. https://www.opendataimpactmap.org/index. Last accessed 2020/11/30
10. Building India's Digital Highways. The Potential of Open Digital Ecosystems, Omidyar Network and Boston Consulting Group (2020). Available at: https://opendigitalecosystems.net/pdf/ODE-Report.pdf
11. Manyika, J., Chui, M., Groves, P., Farrell, D., Van Kuiken, S., Almasi Doshi, E.: Open Data: Unlocking Innovation and Performance with Liquid Information. McKinsey Global Institute, McKinsey Center for Government, McKinsey Business Technology Office, October 2013
12. Open Government Data and Why it Matters, a Critical Review of Studies on the Economic Impact of Open Government Data, Bureau of Communications Research, Australian Government, February 2016
13. IRCTC Homepage. https://www.irctc.co.in/nget/train-search. Last accessed 2020/11/30
14. RBI Website for revenues of State and Center. https://m.rbi.org.in/Scripts/PublicationsView.aspx?id=18930. Last accessed 2020/11/30
15. Owler Homepage. https://corp.owler.com/. Last accessed 2020/11/30
16. Zoominfo Homepage. https://www.zoominfo.com/. Last accessed 2020/11/30
17. Financial Statements of IRCTC for the year ended 2020. https://www.irctc.com/assets/images/Audit%20Financial%20Results%20for%20the%20quarter%20and%20financial%20year%20ended%20March%2031,2020.pdf. Last accessed 2020/11/30
18. Budget of India by PRS (giving Union as well as States separately). https://www.prsindia.org/sites/default/files/budget_files/Union%20Budget%20Analysis%20-%202020-21.pdf. Last accessed 2020/11/30
19. Reported by World Bank and Trading Economics. Available at https://tradingeconomics.com/india/gdp
20. Durgaprasad, M., Mishra, A., Babbar, S., Gupta, V.: Open government data policy and indian ecosystems. In: ICEGOV '17: Proceedings of the 10th International Conference on Theory and Practice of Electronic Governance, pp. 218–227, March 2017. https://doi.org/10.1145/3047273.3047363
21. The Economic Impact of Open Data, Opportunities for value creation in Europe. Last update: 26.01.2020. https://europeandataportal.eu/. https://doi.org/10.2830/63132 OA-04–20–043-EN-N, European Union, 1998–2020

22. Implementation Guidelines for National Data Sharing and Accessibility Policy (NDSAP), ver 2.4, Open Government Data Division National Informatics Centre Department of Electronics and Information Technology Ministry of Communications and Information Technology Government of India, November 2015. Retrieved from https://data.gov.in/sites/default/files/NDSAP%20Implementation%20Guidelines%202.4.pdf
23. Implementation guidelines for Open API policy for e-Governance (National Data Highway), Government of India Ministry of Electronics & Information Technology, NeGD, STQC and MeitY, NeST-GDL-OAPI.01, V1.0, 04/03/2020

Analysis and Prediction of COVID-19 with Image Super-Resolution Using CNN and SRCNN-Based Approach

Purab Alok Jain and Pranali K. Kosamkar

Abstract The advancement of technology has created a huge scope and requirement for transforming the image into a high-visibility image. Here, transforming an image into a high-visibility image means to convert a low-visibility image into a high-visibility image. Super-resolution has many applications worldwide like in medical industries, surveillance, satellite photography, the study of the galaxy, etc. Also, COVID-19 is a monstrous threat to earth. Doctors are predicting whether the patient is having the coronavirus or not via X-Rays and CT-Scans. These scans sometimes miss little details because of the blurriness/low-visibility of the image. This problem can be overcome by using the Super-resolution convolutional neural network (SRCNN). The purpose of the study is the classification of whether the person has the coronavirus or not becomes very accurate by using the SRCNN model. For the transformation of a low-visibility image into a high-visibility image, SRCNN and for classifying whether the person is having coronavirus or not a convolutional neural network (CNN) is used. Our models are trained and tested on four datasets, which are Set5, Set14, COVID-chest-X-Ray dataset, and chest-X-Ray-pneumonia. Our results depict that after applying super-resolution on the X-Rays or the CT-Scans, the classification of COVID-19 attained an accuracy of 95.31% which is higher if compared to the classification of COVID-19 without image super-resolution that attained an accuracy of 92.19%. These were the results after running the model for 20 epochs. Hence, with the help of the SRCNN model, the classification of COVID-19 is much easier and accurate as compared to without the image super-resolution technique.

1 Introduction

Doctors are predicting if the patient has coronavirus or not by these X-Rays and CT-Scans. But the major problem is, these images miss some little details which might

P. A. Jain (✉) · P. K. Kosamkar
MIT World Peace University, Pune, Maharashtra, India

P. K. Kosamkar
e-mail: pranali.kosamkar@mitwpu.edu.in

T. Senjyu et al. (eds.), *ICT with Intelligent Applications*, Smart Innovation, Systems and Technologies 248, https://doi.org/10.1007/978-981-16-4177-0_5

lead to a wrong prediction of the disease. Here comes image super-resolution, which helps convert those images into a high-resolution image, and the system or doctors will be able to predict the disease more accurately. In this paper, we suggested a model which trains on Set5 and Set14 datasets for image super-resolution, and after that COVID-chest-X-Ray dataset and chest-X-Ray-pneumonia dataset are passed through the SRCNN model which generates and saves a high-visibility picture. These high-visibility pictures of X-Rays and CT-Scans are then passed through the CNN model to classify whether the X-Ray or CT-Scan shows COVID-19 traces or not. Transforming an image into a high-visibility image means to convert a low-visibility image into a high-visibility image. An image may have a low-visibility due to small size or blurriness. A low-visibility image is a degraded high-visibility image. In this paper, Part-1 describes about the image transformation and classification, Part-2 describes about related work, Part-3 will brief about system architecture, Part-4 talks about the outcome of the study, and, finally, Part-5 for conclusion.

2 Literature Survey

X. Niu has studied work related to the development in the field of transforming a low-visibility picture into a high-visibility picture and also provides a progress in this domain [1]. Thanh Thi Nguyen highlighted 13 group issues related to the COVID-19 pandemic and promised Artificial Intelligence techniques to address issues. The study aims to provide an overview of the current state of AI applications for AI researchers and the general public and to motivate researchers to use AI capabilities in the fight against COVID-19 [2]. They have built four-layer CNN-based image restoration method which surpasses the current methodologies [3]. For super-resolution first, CNN was used which is called SRCNN. This SRCNN model implements a complete image restoration. The major drawback in this model is it does not consider self-similarities [4]. The optimization is done to diminish errors between the obtained high-visibility image and the original image. The mean square error (MSE) and peak signal-to-noise-ratio (PSNR) are common methods used to test and compare the algorithms used for transforming image into high-visibility image [5]. Authors stated that hyperspectral image (HSI) method produces high-visibility pictures by amalgamation of low-visibility HIS and a high-visibility ordinary image [6]. Methods should be combined with actual requirements. The boom in the development of hardware devices will also bring new challenges to the application of super-resolution [7]. The authors showed that the ordinary scant coding-based super-resolution methods can be corrugated into a deep CNN. The SRCNN has aggrieved a superior performance than the modern methods [8]. The authors proposed a very shallow residual channel attention network (RCAN) that produces an accurate high-visibility picture with the help of residual-in-residual (RIR) model to reach deeper levels [9]. In image super-resolution, textures and clear edges are one of the major aspects. For maintaining the textures and clear edges, the authors suggested an iterative interpolation method in which the non-sub-sampled contourlet transform (NSCT) is used to produce better high-visibility pictures [10]. Authors concluded that for image super-resolution reconstruction, there are only some considerations that researchers take, that degraded model is linear, and noise is neglected, and systematic analyzing methods and filter designing methods have not been formed yet [11]. There exists a high affinity between a low-visibility picture and a high-visibility picture. For this high affinity between low-visibility pictures and high-visibility pictures, the skip connections are used [12]. The CNN models have perceived fortunes in single image super-resolution (SISR). These CNN models learn an effective nonlinear mapping from low-visibility picture input high-visibility target picture at a cost of requiring a huge number of parameters [13]. The authors proposed an optimized extreme learning machine (ELM) which perforce the CNN and distinct other classification techniques in classification problems as ELM has deconvolution layers [14]. The authors proposed a deep CNN within the Laplacian pyramid model for quick and precise high-visibility image transformation. This model gradually predicts high-frequency residuals. Using parameter sharing methods, they adapted very few parameters than the current methods [15]. Authors Shi J et al. results show that their proposed algorithm produces good result as compared to other CNN-based super-resolution. [16]. While there are uncertainties when identifying demographic data due to incomplete and inaccurate data, authors introduce the most popular model for epidemic assessment, namely susceptible, exposed, infectious, and recovered (SEIR) initially compartmental model for assessment [17]. Authors discuss different challenges which can be diagnosed as an infection and also provide a set of recommendations for those currently battling the global COVID-19 pandemic [18]. The authors used DeTraC deep CNN technique which depends on a class demolition for the classification of coronavirus pictures in the complete dataset of chest-X-Ray images. DeTraC displayed useful solutions and also coped up with the irregularity of the data and a limited number of images [19]. Authors proposed decision amalgamation-based approach which conflates the predictions generated by deep CNN networks to augment the performance. On using decision amalgamation, the false positive coronavirus cases were reduced. Authors also stated that this technique can control the widespread of coronavirus [20].

3 Methodology

The purpose of the study is to build a model that classifies whether a patient has the coronavirus or not, efficiently and as fast as possible. For this efficient behavior, we integrated the classification problem with image super-resolution. For that purpose, we use two phases. First is the training and 2nd is the testing phase. In the first phase, that is the training phase, image pre-processing, image degradation, image super-resolution, and CNN-based training occur. In the second phase, that is the testing phase, image super-resolution followed by CNN for classification takes place. For training super-resolution model or the SRCNN model, Set5 and Set14 datasets are used, and for training and testing ConvNet model, COVID-chest-X-Ray dataset and

Fig. 1 Sample images of Set5 and Set14 datasets (high-resolution images) [21], COVID-chest-X-Ray dataset [22] and chest-X-Ray-pneumonia dataset [23] of both COVID positive and normal cases

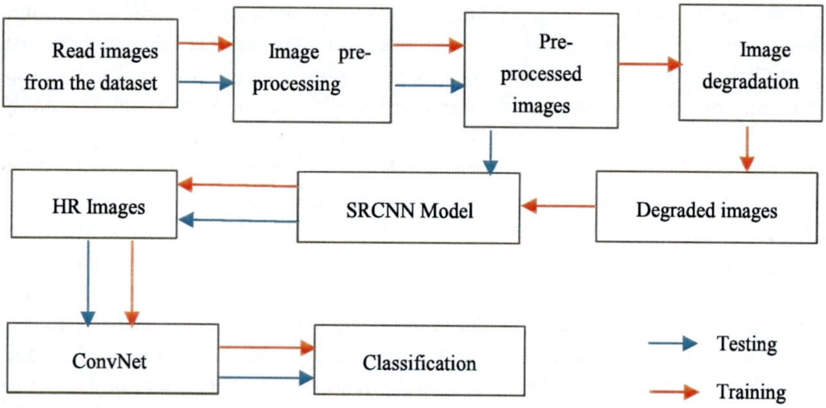

Fig. 2 System architecture for training and testing phases

chest-X-Ray-pneumonia dataset are used which contains the X-Rays and CT-Scans of lungs categorized into two classes, namely COVID and normal (Figs. 1 and 2).

3.1 *Working of the Model*

For both training and testing phases, distinct datasets are created, named as "training" and "validation" which is further divided into two "COVID" and "normal." Images are pre-processed, and the dimensions are converted into 224 × 224 × 3 as RGB images have three channels. Also, for the X-Rays and CT-Scans, we had to filter out the different viewing angles images. Only the front-view scans were considered for better results. After this, for the training of the SRCNN model, Set5 and Set14 datasets were used. These images are high resolution, and we had to degrade the images to train our SRCNN model. These degrades are passed to SRCNN model for training purposes, and the SRCNN model converts these LR images into HR images. These HR images are passed to ConvNet. This model classifies whether the patient has COVID-19 or not by analyzing the X-Rays and CT-Scans of the lungs. After getting the result, doctors can suggest the required medications to the patient for a cure.

4 Performance Analysis and Discussions

The performance for analyzing and prediction of COVID-19 using transformation of a low-visibility image into a high-visibility image is calculated on the basis of two different models, that are SRCNN model, which is used for the transformation of the image that means it is converting the X-Rays and CT-Scans to high-visibility pictures, and then by feeding these high-visibility pictures into the CNN network, we are predicting whether the person has COVID-19 or not more precisely. All this process required four datasets that are Set5, Set14, COVID-chest-X-Ray dataset, and chest-X-Ray-pneumonia. The result of the classification model (CNN model) has two classes that are "COVID" and "normal." For training purposes, 457 images are used, and for testing, 118 images were used. We hyper tuned the parameter with loss (binary cross-entropy) and optimizer (ADAM) per epoch for 20 epochs.

Table 1 depicts the performance of the model with respect to epoch.

From the above table that is Table 1, one can easily depict that after applying super-resolution to the X-Rays and CT-Scans and then classifying whether the patient has coronavirus or not is much better.

From the above graph or Fig. 3b, the training accuracy reaches 94.72% and testing accuracy to 92.19% after 20 epochs if the image super-resolution technique is not used.

From the above graph or Fig. 4b, the training accuracy reached 96.88%, and testing accuracy reached to 95.31% after 20 epochs if image super-resolution is used.

Therefore, from this performance analysis and discussions, we can say that the accuracy of training as well as testing is better if image super-resolution technique using SRCNN model is used prior to the classification of COVID-19 using the ConvNet.

5 Conclusion

In this paper, for the final classification of COVID-19, a CNN model is used, and for the transformation of high-visibility picture, a SRCNN model is used. We classified whether the person has COVID-19 or not using two different models: the first is without applying the transformation of low-visibility image into a high-visibility image or without using the SRCNN network prior to classification of COVID-19 and other which performs image super-resolution on the X-Rays and CT-Scans prior the classification of COVID-19. Our results show that if the classification of COVID-19 is done without the image super-resolution method, then the testing accuracy reaches to 92.19%; but if the image super-resolution method is applied before classification

Table 1 Performance of both models, that is, with SRCNN and without over 20 epochs

Epochs	Training accuracy (with SRCNN)	Training accuracy (without SRCNN)	Testing accuracy (with SRCNN)	Testing accuracy (without SRCNN)
1	0.6875	0.6098	0.4844	0.5156
2	0.7389	0.7114	0.7031	0.4844
3	0.8049	0.7480	0.8438	0.8281
4	0.9062	0.7500	0.8906	0.8438
5	0.9268	0.8496	0.9375	0.9219
6	0.9553	0.8821	0.9375	0.8750
7	0.9634	0.9062	0.9531	0.9531
8	0.9648	0.9065	0.9375	0.8750
9	0.9570	0.9336	0.9688	0.8906
10	0.9492	0.9431	0.9688	0.8594
11	0.9675	0.9350	0.9375	0.8750
12	0.9797	0.8780	0.9375	0.9219
13	0.9593	0.9390	0.9375	0.9219
14	0.9453	0.9431	0.9531	0.9062
15	0.9648	0.9187	0.9375	0.9062
16	0.9766	0.9570	0.9375	0.8750
17	0.9512	0.9453	0.9219	0.9531
18	0.9414	0.9570	0.9375	0.9219
19	0.9512	0.9593	0.9219	0.9688
20	0.9688	0.9472	0.9531	0.9219

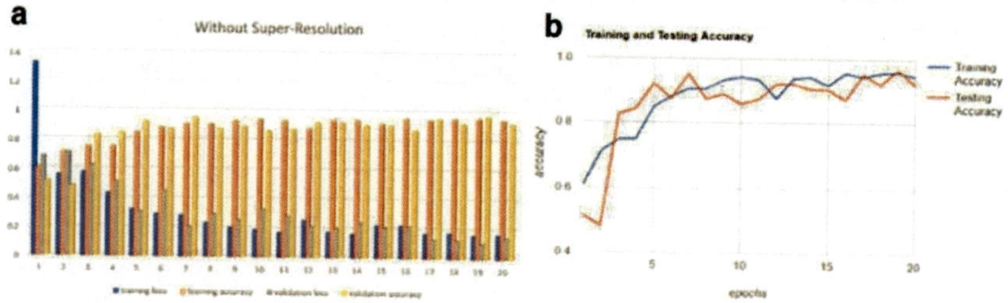

Fig. 3 **a** Comparative analysis of loss and accuracy without image super-resolution, **b** training and testing accuracy without image super-resolution

of COVID-19, then the testing accuracy reached to 95.31% after 20 epochs. Four different datasets were used, that are Set5, Set14, COVID-chest-X-Ray dataset, and chest-X-Ray-pneumonia. Image super-resolution holds a very important role in the classification of the disease here as the results are explaining, and if we see from a general perspective also, transformation of a low-visibility or ill-defined picture

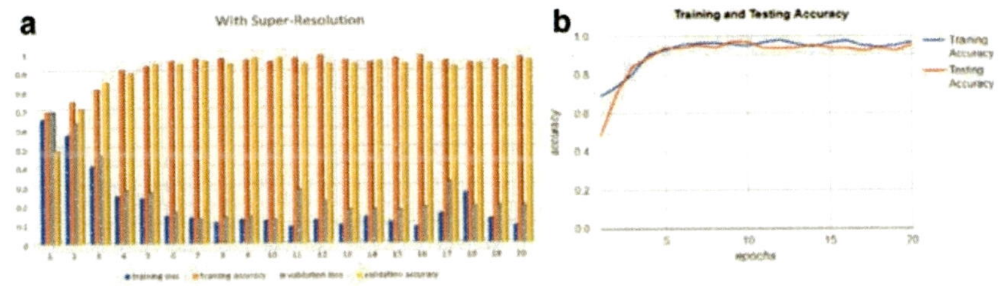

Fig. 4 **a** Comparative analysis of loss and accuracy with image super-resolution, **b** training and testing accuracy with image super-resolution

into a high-visibility or well-defined picture has an ample of scope in the field of image-processing and is capable to make a difference in future.

References

1. Niu, X.: An overview of image super-resolution reconstruction algorithm. In: 11th International Symposium on Computational Intelligence and Design (ISCID), Hangzhou, China, pp. 16–18 (2018). https://doi.org/10.1109/ISCID.2018.10105
2. Nguyen, T.: Artificial Intelligence in the Battle against Coronavirus (COVID-19): A Survey and Future Research Directions. arXiv:2008.07343 pp. 1–13 (2020). https://doi.org/10.13140/RG.2.2.36491.23846/1
3. Lin, G., Wu, Q., Huang, X., Lida, Q., Chen, X.: Deep Convolutional Networks-Based Image Super-Resolution, pp. 338–344 (2017). https://doi.org/10.1007/978-3-319-63309-1_31
4. Hou, J., Si, Y., Yu, X.: A novel and effective image super-resolution reconstruction technique via fast global and local residual learning model. Appl. Sci. **10**(5), 1856 (2020). https://doi.org/10.3390/app10051856
5. Ledig, C., et al.: Photo-realistic single image super-resolution using a generative adversarial network. In: IEEE Conference on Computer Vision and Pattern Recognition (CVPR), Honolulu, HI, pp. 105–114 (2017). https://doi.org/10.1109/CVPR.2017.1
6. Wei, W., Nie, J., Li, Y., Zhang, L., Zhang, Y.: Deep recursive network for hyperspectral image super-resolution. IEEE Trans. Comput. Imaging **6**, 1233–1244 (2020). https://doi.org/10.1109/TCI.2020.3014451
7. Yue, L., Shen, H., Li, J., Yuan, Q., Zhang, H., Zhang, L.: Image super-resolution: the techniques, applications, and future. Signal Process **128**, 389–408 (2016). https://doi.org/10.1016/j.sigpro.2016.05.002
8. Dong, C., Loy, C.C., He, K., Tang, X.: Image super-resolution using deep convolutional networks. IEEE Trans. Pattern Anal. Mach. Intelligence **38**(2), 295–307 (2016). https://doi.org/10.1109/TPAMI.2015.2439281
9. Zhang, Y., Li, K., Li, K., Wang, L., Zhong, B., Fu, Y.: Image super-resolution using very deep residual channel attention networks. In: Ferrari, V., Hebert, M., Sminchisescu, C., Weiss, Y. (eds) Computer Vision—ECCV 2018. ECCV 2018. Lecture Notes in Computer Science, vol. 11211. Springer, Cham (2018). https://doi.org/10.1007/978-3-030-01234-2_18
10. Zhang, Y., Wang, P., Bao, F., Yao, X., Zhang, C., Lin, H.: A single-image super-resolution method based on progressive-iterative approximation. IEEE Trans. Multimedia **22**(6), 1407–1422 (2020). https://doi.org/10.1109/TMM.2019.2943750

11. Liu, H.Y., Zhang, Y.S., Ji, S.: Study on the methods of su-per-resolution image reconstruction. In: International Archives of the Photogrammetry, Remote Sensing and Spatial Information Sciences, vol. XXXVII. Part B2. Beijing, pp. 461–466 (2008)
12. Hou, J., Si, Y., Li, L.: Image super-resolution reconstruction method based on global and local residual learning. In: IEEE 4th International Conference on Image, Vision and Computing (ICIVC), Xiamen, China, pp. 341–348 (2019). https://doi.org/10.1109/ICIVC47709.2019.8981305
13. Tai, Y., Yang, J., Liu, X.: Image super-resolution via deep recursive residual network. In: IEEE Conference on Computer Vision and Pattern Recognition (CVPR), Honolulu, HI, pp. 2790–2798 (2017). https://doi.org/10.1109/CVPR.2017.298
14. Li, J., Zhao, X., Li, Y., Du, Q., Xi, B., Hu, J.: Classification of hyperspectral imagery using a new fully convolutional neural network. In: IEEE Geoscience and Remote Sensing Letters, vol. 15, no. 2, pp. 292–296 (2018). https://doi.org/10.1109/LGRS.2017.2786272
15. Lai, W., Huang, J., Ahuja, N., Yang, M.: Fast and accurate image super-resolution with deep Laplacian pyramid networks. IEEE Trans. Pattern Anal. Mach. Intelligence **41**(11), 2599–2613 (2019). https://doi.org/10.1109/TPAMI.2018.2865304
16. Shi, J., Liu, Q., Wang, C., Zhang, Q., Ying, S., Xu, H.: Super-resolution reconstruction of MR image with a novel residual learning network algorithm. Phys. Med. Biol. **63**(8), 085011, 1–29 (2018). https://doi.org/10.1088/1361-6560/aab9e9
17. Wagh, C., Mahalle, P., Wagh, S.: Epidemic peak for COVID-19 in India, 2020. Int. J. Environ. Res. Public Health. **17**(15), 5330 (2020). https://doi.org/10.20944/preprints202005.0176v1
18. Shinde, G.R., Kalamkar, A.B., Mahalle, P.N., et al.: Forecasting models for coronavirus disease (COVID-19): a survey of the state-of-the-art. SN Comput. Sci. **1**, 197 (2020). https://doi.org/10.1007/s42979-020-00209-9
19. Abbas, A., Abdelsamea, M., Gaber, M.: Classification of COVID-19 in chest X-ray images using DeTraC deep convolutional neural network. In: Researchgate on March (2020). https://doi.org/10.1101/2020.03.30.20047456
20. Mishra, A., Das, S., Roy, P., Bandyopadhyay, S.: Identifying COVID19 from chest CT images: a deep convolutional neural networks based approach. J. Healthcare Eng. 1–7 (2020). https://doi.org/10.1155/2020/8843664
21. https://www.kaggle.com/ll01dm/set-5-14-super-resolution-dataset. Accessed on 19–12–2020 at 11:47PM
22. https://github.com/ieee8023/covid-chestxray-dataset. Accessed on 10–12–2020 at 7:03PM
23. https://www.kaggle.com/paultimothymooney/chest-xray-pneumonia. Accessed on 10–12–2020 at 8:21PM

Novel Approach to Improve Security Systems

A. Sneha Jain and Lakshmi Satya Vani Narayanam

Abstract Aadhaar with IoT-based security system in the AC compartments to improve passenger security by ensuring only reserved and authentic passengers is granted entry-exit permission to the compartments. In order to solve the problem of unauthorized and unchecked entry of people in and out of the compartments due to non-existence of any form of automatic security systems or automated supervision in trains. The novel approach presented in the paper lifts-up the travel experience with assured protection, safe and secure journey for passengers and their belongings. The solution proposed is split into multiple phases in accordance with the real-time scenarios of ticket booking with Aadhar and verified boarding at the time of travel, each of which ensure identification, authentication and validation of the person.

1 Introduction

The paper presents an Aadhaar-based IoT security system for improvement of passenger security in Indian trains as highly safe and feasible mechanism, linked with Aadhaar QR code, utilizing the power of world's largest biometric ID system.

Indian railways cover almost all parts of the nation and is one of the most sought after mode of transport due to its affordable rates, lively experience.

Reaching railway stations are much faster compared to airports for a vast majority, and one can reduce carbon footprint by travelling in trains. But one of the main consumer preference and market demand is passenger security during the journey which at the present condition has scope of upgradation and enhancement.

The security threats of travelling in trains cause negative effects on the mobility of individuals and groups of people. This leads to reversal in the positive economic impact of an efficient transportation system. So, it is necessary that we establish a

A. Sneha Jain (✉) · L. S. V. Narayanam
PES University Electronic City Campus, Bengaluru, Karnataka, India

L. S. V. Narayanam
e-mail: sathyavaninl@pes.edu

T. Senjyu et al. (eds.), *ICT with Intelligent Applications*, Smart Innovation, Systems and Technologies 248, https://doi.org/10.1007/978-981-16-4177-0_6

system such that when incorporated by government can provide the regular facilitates and in addition guarantee maximum security to the passengers.

The motivation of the paper is to minimize the problem of unauthorized and unchecked entry of people in and out of the compartments due to non-existence of any form of automatic security systems for automated supervision in trains that has left robbery, thefts, loot and killing to rise at an alarming rate, especially during night times.

The paper ideates to establish this system, with the prototype formulated specifically for the AC compartments utilizing the existing gated compartments to ensure only reserved and authentic passengers are granted entry-exit permission to the compartments.

The solution proposed is split into multiple phases in accordance with the real-time scenarios of booking and boarding, each of which ensure identification, authentication and validation of the person through the steps indicated in the section of proposed model.

The sensor-based authentication model in the entrance of every compartment gates act as automatic check-in and check-out system to allow only one verified individual at a time, exempting unauthorised multiple people entry under a single valid authentication.

Hence, the paper provides a novel, reliable and tamper-proof solution that ensures solo-travellers, single woman, elderly passengers and children a relatively safe, assured, concern-free and convenient mode of travel in the Indian Railways.

2 Literature Survey

Aadhaar card being taken as the most common justification for a national identity project was "better inclusion," while various identity documents (IDs) already existed. No single identity card was accepted across the board for public and private services, with different service providers demanding different sets of documents and verification processes [1]

Additionally, unique identification numbers issued by UIDAI carry many advantages. They do create a trustworthy social identity of a citizen, whom everybody can trust. This kind of identity is very helpful while making applications for bank loans, phone connections, passports, visas and admissions at educational institutions [2]

The concept of authentication is used to explain if someone or something is who or what it is declared to be. In other words, it is a mechanism to confirm if the item claimed by the claimant is actually theirs. It is done at the starting of any process and is used to check the identity of the user before any other action takes place [3]

In order to build android applications, Firebase is an ideal choice for backend because it is efficient as no third-party language to communicate with the database. It provides a secure channel to communicate with the database directly from the android application [4].

Implementation for several locations in a train needs to highly cost efficient so a Wi-Fi based system can eliminate the need to buy expensive auxiliaries for IoT. A small Nodemcu programmed with Firebase functions as the cloud hosted real-time database [5] helps in data synchronization and fulfils the necessities of the proposed solution.

3 Proposed Model

The model developed maintains zero intrudance in AC coaches while ensuring security checks at every point along with passenger convenience. The proposed system is implemented in three main stages to establish a tighter security and reduce the possibility of any form of robbery in AC coaches:

A. Stage I—Ticket Booking with original Aadhaar card QR code scanner
B. Stage II—Ticket and Aadhaar Verification
C. Stage III—Authenticated Entry-Exit System.

3.1 Stage I: Ticket Booking with Original Aadhaar Card QR Code Scanner in Detail

A new android application is built to aid ticket booking process along with scanning of Aadhaar card QR code of individual passengers. The details gathered at this stage are recorded in Firebase database as it provides real-time and free of cost storage facility. The activities in stage I as seen in Fig. 1, are:

i. Reading general travel details

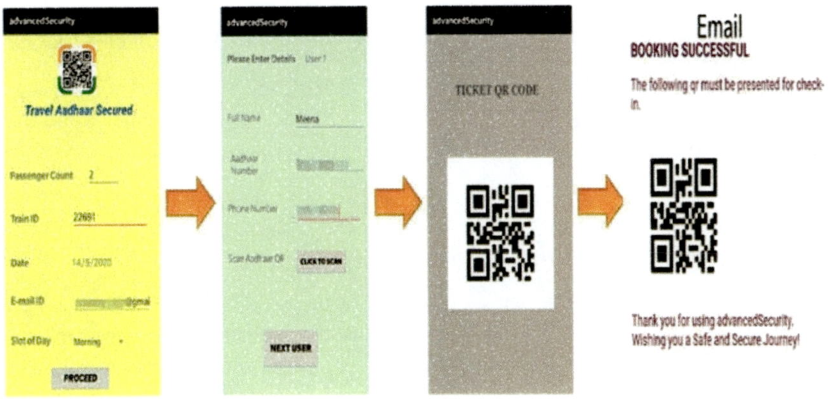

Fig. 1 Stage I outcome flow

- Number of passengers—to allow family booking under one ticket.
- Train number; date of travel; time slot of travel day—facilitate all possible travel combinations for a person.
- E-mail address—main e-Ticket that contains unique ticket QR code and travel details is provided in paperless manner through mail.

ii. Reading individual passenger details

- Personal details include full name, Aadhaar number and mobile number
- Scanning the original Aadhaar card QR code for authentication while boarding the AC coaches.

iii. Finally, when all details are filled and booking is confirmed

- The ticket QR code is displayed which acts as the main pass.
- The same is mailed to the e-mail provided in the beginning.
- Simultaneously, it is also downloaded to the mobile device as a backup.

3.2 *Stage II: Ticket and Aadhaar Verification in Detail*

Another android application is made to specifically scan the e-Ticket QR codes and passenger's Aadhaar QR code.

This is to verify QRs and supervise passenger movements in and out of the coaches. The train number can be set officially. The current date is automatically updated to allow validation of designated tickets and commuter identity.

The e-Ticket QR made available paperless, which can be printed if needed. The Aadhaar card being mandatory for every citizen is used so that no additional proof of identity is required to be carried by the passenger. The steps in this stage are:

- Step 1: Scan the e-Ticket QR code.
- Step 2: Enter the OTP sent at the registered mobile number for e-Ticket verification.
- Step 3: Either scan Aadhaar QR or enter Aadhaar number to obtain entry access for each passenger.

A success at every step authenticates the person and verifies him/her to be the reserved passenger for the specific travel.

3.3 *Stage III: Authenticated Entry-Exit System*

The Wi-Fi connected NODEMCU with IR sensor-based automatic check-in and check-out system that is updated with train number by the officials works as follows:

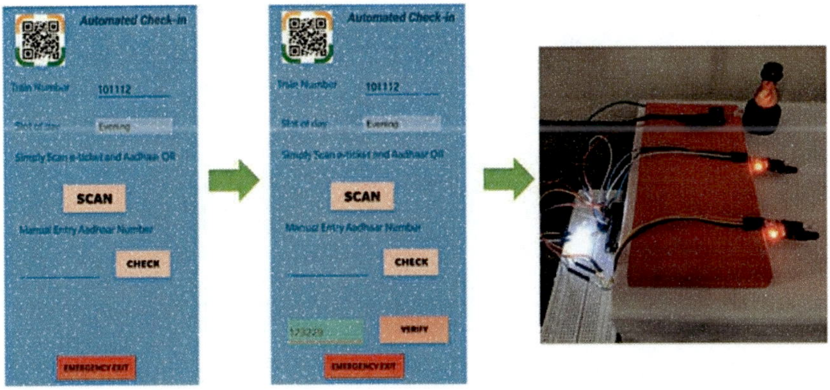

Fig. 2 Valid e-Ticket QR code scanned, OTP verified and white LED glows

- When the main ticket QR and OTP verification check, results as valid—white LED turns on as seen in Fig. 2, indication that passenger can now scan Aadhaar QR
- Scan of Aadhaar QR confirms passenger's authenticity—green LED glows and the individual can enter as seen in Fig. 3.
- When a passenger is standing at the entry point, the first sensor detects and glows blue LED to indicate the path is clear, and the person can move ahead.
- Once the person reaches the second sensor, blue LED glows again, indicating the next passenger can step forward.
- When any invalid QR is scanned and there is an unauthorized attempt to enter—red LED and buzzer turn on; the person is exempted from entering.
- This stage is IR sensor-based backed up by the developed authentication model, to be implemented in both entry and exit gates of the AC coaches to avoid congestion and human-traffic by allowing one person at a time.
  - At the time of entry—the alarm is beeped and red LED glows as seen in Fig. 4 when more than one person is attempting to enter in a single scan. Only once the second person steps back and his/her Aadhaar is validated to allow entry.

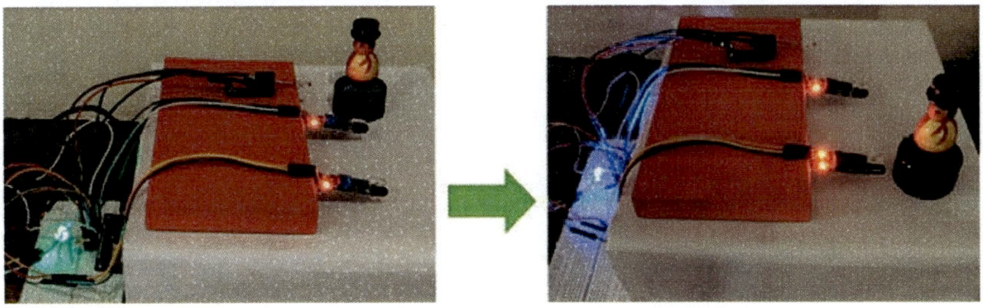

Fig. 3 Aadhaar validation and IR sensors response for person detection

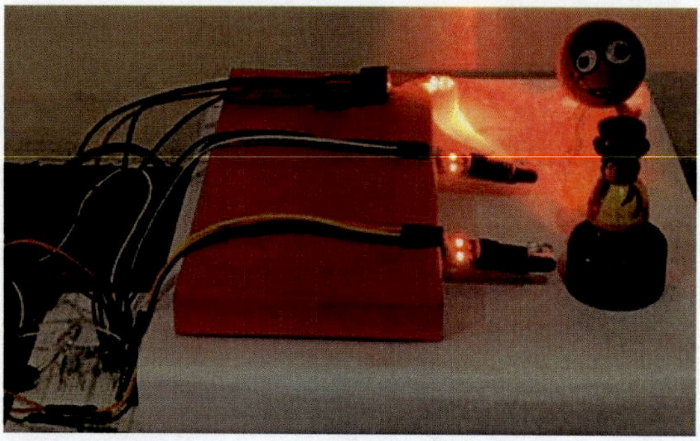

Fig. 4 Multiple entry detection at entry/exit points

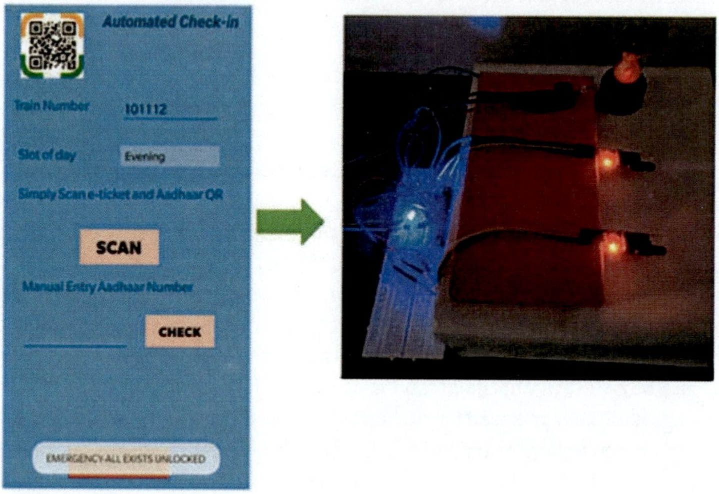

Fig. 5 Emergency unlock in case of accidents

- At the time of exit—the alarm and red LED turn on as seen in Fig. 4, so that no outsider can attempt to make a forceful entry when a passenger is about to exit.

- In case of emergencies like fire accident—a passenger inside can press the emergency exit button to unlock all gates and move out as seen in Fig. 5.

4 Results Obtained

See Figs. 2, 3, 4 and 5.

5 Conclusion

The solution approached in the paper establishes a tighter security and reduces the possibility of any form of robbery, theft and loot to the lowest.

- Ticket booking process aided with easy scan and offline available feature of Aadhaar QR code scan through mobile application for every passenger. The unique generated ticket is e-mailed to the user mail-ID, and QR code image is saved in the mobile device.
- At the time of travelling
- Scan the main e-Ticket QR and enter the OTP for verification and authenticity of passenger.
- Scan of Aadhaar QR to confirm passenger's identity.
- When a passenger is standing at the entry point, the first sensor detects to indicate the path is clear, and the person can move ahead.
- Once the person reaches the second sensor, the next passenger can step forward.
- When any invalid QR code is scanned and an unauthorized attempt is made to enter—alarm raises and the person is exempted from entering.
- The sensor-based check-in backed up by the developed authentication model, ensures that the alarm is beeped and red LED glows at the time of:
- Entry—When more than one person is attempting to enter in a single scan. Only once the second person steps back and his/her Aadhaar is validated, he/she will be allowed entry.
- Exit—No outsider can attempt to make a forceful entry when a reserved passenger is about to exit.

Hence, the system provides a tamper-proof security that will lift-up the travel experience with assured protection, safe and secure journey for passengers and their belongings in AC coaches of Indian Railways.

6 Future Work

In future, the existing fire sensors and smoke sensors in the AC coaches can be integrated to activate emergency condition and unlock gates automatically to save lives.

During the booking phase, government supplied STQC scanners could be used by registered booking agencies for passenger identification and complete shutdown of black market.

The railway officials can be provided with unique ID proofs. Ideally only the designated in-charge should be allowed to enter to minimisize internal tampering of security systems.

This system can be applied in all locations where booking and ticket-based entry is required like movie-theatres, airports, auditoriums and others. Additionally, places of high confidentiality that require guranteed security like bank locker rooms, special care centers, conference halls, research labs to name a few.

Thus, the proposed method provides a simple, fast and effective means of security through authentication, verification and validation process.

References

1. Sen, S.: A Decade of Aadhaar: Lessons in Implementing a Foundational ID System, ORF Issue Brief No. 292, May 2019, Observer Research Foundation
2. Int. J. Interdisc. Multidisc. Stud. (IJIMS) **2**(5), 88–95 (2015)
3. Securing IoT devices by generating QR codes. Int. J. Pure Appl. Math. **119**, No. 12 2018, ISSN: 13743-13749 (on-line version)
4. Khawas, C., Shah, P.: Application of firebase in android app development—a study. Int. J. Comput. Appl. **179**, 49–53 (2018). https://doi.org/10.5120/ijca2018917200
5. Shakthidhar, S., Srikrishnan, P., Santhosh, S., Sandhya, M.: Arduino and NodeMcu based ingenious household objects monitoring and control environment. In: 2019 Fifth International Conference on Science Technology Engineering and Mathematics (ICONSTEM), vol. 1, pp. 119–124 (2019)

Digital Forensic Tools for Cloud Computing Environment

Swapnil P. Bhagat and Bandu B. Meshram

Abstract In general, digital forensics process can be stated as a process of identifying, analyzing and presenting the digital evidence. It comprises activities like identification, preservation, examination, analysis. Complication and complexity in the execution of these activities increase with the size of data. Cloud computing technology exacerbate the situation. Various forensic tools have been developed to improve the performance and efficiency of forensic activities. This paper highlights such digital forensic tools and provides brief information about the forensic tools. This paper examines traditional digital forensic tools helpful for the cloud forensic as well as tools that are specially designed for the cloud environment. It includes commercial as well as open-source tools.

1 Introduction

Outsourcing the personal and business activities to cloud computing provides benefits at a various level including financial, management and maintenance. Cloud computing opens the doors for individuals and small-scale businesses to get the high-power resources which were unreachable earlier due to its cost and maintenance. Cloud computing environment provides different services to their customer based on the pay-as-you-go model. This advantage of cloud computing technology compels many organizations toward it.

Nevertheless, unplanned, unmanaged and insecure shifting to cloud computing may have catastrophic consequences. Besides, cloud architecture may have some vulnerabilities which may lead to security breaches. Cloud computing technology can also be utilized to perform malicious activities or storage for actionable data. So, industry and academy both are working to provide a secure cloud computing

S. P. Bhagat (✉) · B. B. Meshram
Veermata Jijabai Technological Institute (VJTI), Mumbai, India
e-mail: spbhagat_p17@ce.vjti.ac.in

B. B. Meshram
e-mail: bbmeshram@ce.vjti.ac.in

T. Senjyu et al. (eds.), *ICT with Intelligent Applications*, Smart Innovation, Systems and Technologies 248, https://doi.org/10.1007/978-981-16-4177-0_7

environment. Along with the security, digital forensics is also necessary to make it robust. Digital forensic assists to identify the answer of key question, i.e., what, why, how, who, when and where regarding the occurred malicious incident [1].

Investigators encounter many issues while applying digital forensics on the cloud computing environment. NIST has catalogued distinct 65 challenges faced during cloud forensics [2]. In addition, forensic investigation varies based on the service and deployment model [1]. Virtualize nature of data makes it challenging to recover data physically [3]. Consequently, tradition forensic tools which work efficiently with a standalone system, may not necessarily work efficiently in the cloud environment. First-generation (general purpose) forensic tools are failing to deal with the increasing complexity and evidential volumes [4]. In the case of cloud forensic, traditional forensic tools work correctly up to some extent [5]. Developers have endeavored to resolve these issues and make the forensic tools compatible with different computing environments. Researchers also developed new forensic tools that specifically deal with cloud forensics challenges.

This paper attempt to identify the traditional forensic tools which can be used during cloud forensics. It also scrutinizes some digital forensic tools which are specially designed for the cloud computing environment. This paper is arranged as follows. Section 2 presents detail of traditional digital forensic tools which can be helpful during cloud forensics. Section 3 reviews forensic tools specially developed for the cloud environment. Finally, the work is concluded in Sect. 4.

2 Traditional Forensic Tools

This section presents various forensic tools which are used for investigation of a standalone computer system. This section comprises only those forensic tools which are helpful for the cloud investigation. It includes proprietary as well as open-source tools.

2.1 EnCase

EnCase is one of the known digital forensic tools. It produces a forensically sound copy of data. It preserves the integrity of the data with hash and Cyclic Redundancy Check (CRC) values. It assists in recovering the hide, cloak or deleted file. It also analyzes data from various sources, including disk, memory, browser data, email, etc. EnCase provides facilities to collect data from various mobile devices, including smartphones, tablets, PDAs, and GPS navigation devices. It also supports data acquisition from cloud services, such as Facebook, Twitter, Gmail, and Google Drive. EnCase support various platforms. It includes Windows, Linux, AIX, OS X, Solaris [6].

2.2 FTK

Forensic Toolkit (FTK) is a digital forensics solution developed by AccessData. FTK suite amalgamates different forensic tools in one software like email analysis, file decryption, data carving engine, data visualization, web viewer, Cerberus malware analysis. It is used for different digital forensic activities, like data collection, hashing, email analysis, indexing, searching, reporting, etc. FTK tools are useful in multiple phases of digital investigation, i.e., acquisition, preservation, examination, analysis and reporting. For data acquisition, it uses FTK imager tool. It provides different features to acquire static evidence, live evidence and remote evidence. The acquired data is preserved by hash value for the assurance of the data integrity. There are three hashing functions available, i.e., MD5, SHA-1 and SHA-256. Hash of individual files can also be utilized to identify duplicate and known files. The Known File Filter (KFF) is an FTK forensic tool utility that automatically classifies ignorable files (known system and program files), files with alert status (malicious and suspicious) and disregarded files. FTK is equipped with dtSearch searching tool. FTK also aids to create the report comprising evidence and investigation results. It supports distributed processing. It permits to install the Distributed Processing Engine (DPE) for parallel processing. It supports up to three additional computers at a time for the processing of cases. It improves the processing time. FTK support windows platform including Windows Server 2016, Windows Server 2012 R2, Windows 7, 8.1, 10 [7].

2.3 SANS SIFT

SANS SIFT is an incident response and forensic tools suite. It is a free, open-source tool. It supports the Linux platform. It can be installed on Windows 10 [8]. It supports different file system like NTFS, raw, FAT12/16/32, EXT2/3/4, UFS1/2, VMDK. It incorporates various packages like log2timeline (Timeline Generation Tool), Rekall Framework (Memory Analysis), Volatility Framework (Memory Analysis), 3rd Party Volatility Plugins, bulk_extractor, autopsy, SleuthKit [9].

2.4 The Sleuth Kit (TSK)

The Sleuth Kit (TSK) is another famous forensic tool. It is a set of command-line tools and C library. It is utilized for disk image analysis and file recovery [10]. It facilitates the analysis of raw images and AFF file system. It provides supports for several file systems. It includes NTFS, FAT, ExFAT, UFS 1, UFS 2, EXT2FS, EXT3FS, Ext4, HFS, ISO 9660, and YAFFS2 file systems. It can also identify the files which are hidden by the rootkit. It also creates a timeline for file activities. It supports UNIX,

Mac OS and Windows platform. It is an open-source tool. Autopsy is a GUI forensic tool which runs The Sleuth Kit in background [11].

2.5 *Volatility*

Volatility is used to analyze the memory (RAM) of 32-bit and 64-bit Windows, Linux, Mac and Android system. It is an open-source tool. The Volatility Framework was released at Black Hat DC in 2007. It is utilized to extract information from running processes, network sockets, network connection, DLLs and registry hives [10]. It is written in Python language. It supports Windows, Linux and Mac platforms. It also provides extensible and scriptable API to extend its utilization [12].

2.6 *Xplico*

It is a network forensic analysis tool. It extracts and reconstructs the content from the packets captured by packet sniffer like Wireshark, tcpdump. Xplico is an open-source and free network forensics analysis tool. Xplico comprises four components, i.e., a Decoder Manager (DeMa), IP decoder (Xplico), a set of data manipulators (msite, mpaltalk, mfbc, mfile, mwmail, etc.), a visualization system (Xplico Interface) [13]. It supports various protocols, including HTTP, SIP, IMAP, POP, SMTP, TCP, UDP, IPv6, etc. It supports the Linux platform. Gebhardt and Reiser [3] designed a generic model for network forensic in a cloud environment. They used Xplico in the generic network forensic model for analysis.

2.7 *CAINE*

Computer Aided Investigative Environment (CAINE) provide an interoperable environment which assists investigator during acquisition, preservation, analysis and reporting. It is a Linux live distribution. It integrates different forensic tools like Nirsoft suite + launcher, WinAudit, MWSnap, Arsenal Image Mounter, FTK Imager. CAINE also provides a Windows IR/Live forensics tools [14].

2.8 *X-Ways*

X-Ways is a powerful disk analysis tool. It provides various features like capturing free space, slack space, inter-partition space, and text. It produces detailed drive contents table with all existing and deleted files and directories. It creates a sector-wise copy of the disk. It tracks all the activities of the investigator, which assist ensure the integrity of the evidence. It also protects any accidental changes. It displays the content based on the selected file type category. It supports various file systems including Native support for FAT12/16/32, exFAT, TFAT, NTFS, Ext2/3/4, etc. It utilizes various hash and checksum methods for the acquired images like SHA-1, SHA-256, MD5 PSCHF, CRC16, CRC32. It works with windows platform [15].

2.9 ProDiscover

ProDiscover is a disk forensic tool which provides disk acquisition and analysis features. It utilizes MD5, SHA1 and SHA256 hash techniques to ensure the integrity of the data. It supports various file system like FAT12/16/32, NTFS, HFS, HFS+, EXT2/3/4, UFS. Graphic viewer, registry viewer, Outlook email viewer, Internet History viewer, Event log viewer are integrated into the ProDiscover forensic tool. It also supports VMware to run a captured image [16].

3 Cloud Specific Forensic Tools

Forensic tools discussed in this section are specially designed for the cloud forensics. These tools work at different level like hardware, virtualization or inside the virtual machine. It includes proprietary as well as open-source tools.

3.1 OWADE

Offline Windows Analysis and Data Extraction (OWADE) is an open-source forensic tool. It analyzes the data to recover valuable data stored by Windows, browsers and instant messaging software. It extracts WiFi activity data like an access point, last access time, etc. It also assists in extracting credentials stored in popular browsers (like Internet Explorer, Firefox, Safari, and Chrome) and instant messaging software (like Skype, MSN live, Gtalk, etc.) [17]. User activities are reconstructed using the recovered data. It tries to identify the history of the computer network connection, accessed online services and online identities used. It identifies the list of Web sites that the user had visited and whether he has stored any data on the cloud [18]. OWADE is developed in Python language. It works on Linux platform.

3.2 FROST

Forensic OpenStack Tool (FROST) is designed to collect the data from the OpenStack cloud platform. It provides the facility to acquire virtual disk, API logs, firewall logs from Infrastructure as a Service model of OpenStack. FROST is integrated into the management module of the OpenStack. After retrieving data, the hash value is calculated, and data is arranged in a tree structure. Continuously calculating the hash value creates overhead on the system. It is an open-source forensic solution [19].

3.3 FAAS in IaaS

Forensic Acquisition and Analysis System (FAAS) for Infrastructure as a Service allows to acquire data from IaaS environment and perform its analysis. It collects data that are under the control of a cloud customer. So, it removes the dependency on the CSP. It suggests installing agents in a virtual machine which will collect non-volatile memory, volatile memory, network traffic and activity logs. This data is transferred to FAAS system by securing it using encryption. The data is stored in forensic storage and only accessed by authorized person. It integrates various forensic tools for analysis, including EnCase, NFAT, FTK [18].

3.4 UFED Cloud Analyzer

UFED Cloud Analyzer is a digital forensics tool that is used to acquire and analyze the data from various cloud services [20]. It allows the investigator to collect the data from over 40 social media and cloud-based sources including Facebook, WhatsApp, Twitter, Google (location history, activity, photo, calendar, contact, Drive, etc.), Gmail, OneDrive, Instagram, Yahoo mail, iCloud (App, location, Drive, photo, notes, etc.), Dropbox, etc. [21]. It extracts data based on various parameters like a mobile device, username. It also preserves the data using Hash technology. It also uses a unified format for visualizing the data [22].

3.5 ForenVisor

ForenVisor is a special purpose hypervisor used to perform live forensics directly on the hardware in a virtualized environment. It utilizes the hardware-assisted virtualization provided by the intel VT-x. It performs data collection and secure storage at the hypervisor level. It acquires process data, raw memory and I/O data by placing itself between the operating system and hardware. ForenVisor emphasizes on the reliability of data. It tries to achieve this with lightweight hypervisor architecture, acquiring data directly from hardware and protecting the data by Filesafe module. It has some limitations like it stores the data locally, which may destruct the original evidence. It is also compatible only with the FAT32 file system. It uses a Windows platform for its prototype [23].

3.6 *CloudTrail*

Amazon Web Services (AWS) began a service which provides the logging services. It makes various logs of cloud services available for the customers. CloudTrail is primarily designed for monitoring and securing cloud resources and services [24]. However, customers can use it to retrieve useful important information for digital forensics [18]. Since it is AWS cloud-based service, it is available only to the AWS cloud customers for AWS cloud resources and services.

Table 1 summarizes the tools discussed in previous sections. The table maps the forensic tools with the digital forensic framework phases. It also provides information regarding the supporting platforms of each forensic tool. This information can be useful while selecting forensic tools during cloud forensics.

4 Conclusion

Investigators confront various technical challenges during cloud forensics. Cloud architecture can also influence the reliability of the collected evidence. Investigators face most of the challenges during the acquisition phase. This paper overviews some essential digital forensic tools that can be useful during the investigation in the cloud environment. These digital forensic tools also endeavor to enhance the reliability of collected evidence. Cloud specific tools are also helpful up to some extent to eliminate cloud forensics challenges.

Tools discussed in this paper are used in a subset of the digital forensic phases. There are very few digital forensic tools that are used in all phases of digital forensics framework. This issue has also seemed in cloud forensic. So, there is a need for comprehensive cloud forensic tools which will perform most of the activities.

Table 1 Summary of digital forensic tools for Cloud Computing environment

Forensic tool	Used for	Platform	Proprietary/Open-source
EnCase	Acquisition, Preservation, Examination, Analysis	Windows, Linux, Unix	Proprietary (Paid)
FTK	Examination, analysis, Reporting	Windows	Proprietary (Paid)
FTK Imager	Acquisition	Windows	Proprietary (Free)
SANS SIFT	Analysis	Linux, Windows	Open-source
The Sleuth Kit	Analysis	Unix, Windows	Open-source
Volatility	Analysis (RAM)	Linux, Windows, Mac	Open-source
Xplico	Analysis (Network)	Linux	Open-source
CAINE	Acquisition, Preservation, Analysis, Reporting	Linux	Open-source
X-Ways	Acquisition, Analysis	Windows	Proprietary (Paid)
ProDiscover	Acquisition, Analysis	Linux, Windows, Mac	Proprietary (Paid)
OWADE	Analysis	Linux	Open-source
FROST	Acquisition, Preservation	OpenStack	Open-source
FAAS	Acquisition, Analysis	Any Cloud Platform	Proprietary
UFED Cloud Analyzer	Acquisition, Preservation	Windows	Proprietary (Paid)
ForenVisor	Acquisition, Preservation	Windows and Intel VT-x	Open-source
CloudTrail	Acquisition	AWS Cloud	Proprietary (Paid)

References

1. Rahman, N.H.A., Glisson, W.B., Yang, Y., Choo, K.K.R.: Forensic-by-design framework for cyber-physical cloud systems. IEEE Cloud Comput. J. **3**(1), 50–59 (2016)
2. NIST Cloud Computing Forensic Science Challenges (Draft NISTIR 8006). National Institute of Standards and Technology, U.S. Department of Commerce (2014)
3. Gebhardt, T., Reiser, H.P.: Network Forensics for Cloud Computing. In: Dowling, J., Taïani, F. (Eds.), Distributed Applications and Interoperable Systems. DAIS 2013. Lecture Notes in Computer Science, vol 7891. Springer, Berlin (2013)
4. Ayers, D.: A second generation computer forensic analysis system. Digit. Investig. **6**, 34-S42 (2009)
5. Povar, D., Geethakumari, G.: Digital forensic architecture for Cloud Computing systems: methods of evidence identification, segregation, collection and partial analysis. In: Satapathy, S., Mandal, J., Udgata, S., Bhateja, V. (Eds.), Information Systems Design and Intelligent Applications. Advances in Intelligent Systems and Computing, vol 433. Springer, New Delhi (2016)
6. Guidance Software.: EnCase Forensic v8.07 User Guide (2018). http://encase-docs.opentext.com/documentation/encase/forensic/8.07/Content/Resources/External%20Files/EnCase%

20Forensic%20v8.07%20User%20Guide.pdf
7. AccessData.: Forensic Toolkit (FTK) User Guide (2018). website: https://ad-pdf.s3.amazonaws.com/ftk/7.x/FTK_UG.pdf
8. Kovacs, F.: Windows 10 as a Forensic Platform. The SANS Institute (2018). https://digital-forensics.sans.org/community/papers/gcfa/windows-10-forensic-platform_13102
9. SIFT Workstation Documentation & Links. https://digital-forensics.sans.org/community/downloads., Last accessed 2019/12/15
10. Lokhande, P.S., Meshram, B.B.: Digital forensics analysis for data theft. Int. J. Forensic Comput. Sci. **10**(1), 29–51 (2016)
11. Open Source Digital Forensics. https://www.sleuthkit.org/. Last accessed 2019/12/15
12. Case, A.: Volatility. Available at: https://github.com/volatilityfoundation/volatility/wiki. Last accessed 2019/12/15
13. Xplico: https://www.xplico.org/. Last accessed 2019/12/15
14. CAINE Computer Forensics Linux Live Distro. https://www.caine-live.net/index.html. Last accessed 2019/12/15
15. X-Ways Software Technology AG.: X-Ways Forensics White Paper. https://www.x-ways.net/forensics/X-Ways_Forensics_White_Paper.pdf
16. ProDiscover Forensic. https://www.prodiscover.com/products-services. Last accessed 2019/12/15
17. Bursztein, E., Cassidian, V.F., Martin, M., Picod, J.M.: Doing Forensics in the Cloud Age OWADE: Beyond Files Recovery Forensic. Black Hat USA (2011) https://elie.net/static/files/beyond-files-recovery-owade-cloud-based-forensic/beyond-files-recovery-owade-cloud-based-forensic-paper.pdf
18. Alqahtany, S., Clarke, N., Furnell, S., Reich, C.: A forensic acquisition and analysis system for IaaS. Clust. Comput. **19**(1), 439–453 (2016). https://doi.org/10.1007/s10586-015-0509-x
19. Dykstra, J., Sherman, A.T.: Design and implementation of FROST: digital forensic tools for the OpenStack cloud computing platform. Digit. Investig. **10**, S87–S95 (2013)
20. Cellebrite.: Extracting Legally Defensible Evidence from The Cloud. Whitepaper (2015)
21. Cellebrite.: UFED Cloud Analyzer Supported Devices. https://cf-media.cellebrite.com/wp-content/uploads/2017/08/UFED_CloudAnalyzerSupportedDevices.pdf
22. Naaz, S., Siddiqui, F.A.: Comparative Study of Cloud Forensic Tools. Communication on Applied Electronics (CAE), vol. 5, issue 3 (2016)
23. Qi, Z., Xiang, C., Ma, R., Li, J., Guan, H., Wei, D.S.L.: ForenVisor: a tool for acquiring and preserving reliable data in cloud live forensics. IEEE Trans. Cloud Comput. **5**(3), 443–456 (2017)
24. What Is AWS CloudTrail? https://docs.aws.amazon.com/awscloudtrail/latest/userguide/cloudtrail-user-guide.html. Last accessed 2019/12/15

Privacy Preserving Scheme for Smart Transportation in 5G Integrated IoT

K. Shamshuddin and G. N. Jayalaxmi

Abstract The Internet of Things (IoT) is widely used in smart transportation and 5G technology is used as a mode of communication. Many devices are involved in monitoring, collecting and processing of information from various users and also sharing the information between devices as well as users. Security and privacy preservation is the biggest challenge in smart transportation because devices are more in number and also connected wirelessly. In this paper, we consider security and privacy issues in smart transportation and focus on improving privacy preservation. Also we analyzed the technique using the formal verification tool Scyther. The result shows that privacy is preserved.

1 Introduction

In most of the country, 60% of the population will be living in urban areas by 2030. Many countries are planning for smart transportation to ease the transportation and to provide good services to the public [1]. Smart transportation needs new technologies such as 5G and Internet of Things (IoT). The new technology integration with the transportation system opens the door of new security and privacy threats.

In IoT, device-to-device communication is involved and many sensors are used to gather information from surroundings, also involve the processing of information to make a useful decision. The various sensors which monitor and gather information about machines, as well as human interaction, are mounted on physical devices. Using IoT devices, complicated tasks can be performed with more intelligence when they are connected to a network [2]. One such application of IoT is Transportation intelligence, which is the important entity in the smart city. Smart transportation involves physical devices and access to Wi-Fi networks, therefore the possibility of

K. Shamshuddin (✉) · G. N. Jayalaxmi
SoECE, KLE Technological University, Hubballi, Karnataka, India
e-mail: baushamshuddin@kletech.ac.in

G. N. Jayalaxmi
e-mail: jaya_gn@kletech.ac.in

T. Senjyu et al. (eds.), *ICT with Intelligent Applications*, Smart Innovation, Systems and Technologies 248, https://doi.org/10.1007/978-981-16-4177-0_8

many attacks and the risk is very high. This paper further addresses the need for security and privacy issues in smart transportation.

The 5G wireless network is used in smart transportation as a mode of communication to provide very high data rates, higher coverage with significantly improved Quality of Service (QoS), and extremely low latency [3]. Due to the dense deployment of base stations in 5G, it will be possible to provide ultra-reliable, affordable broadband to machine-to-machine (M2M) and IoT devices [4]. This opens the door for many security vulnerabilities and needs a more robust security mechanism.

The user information such as identity and location leakage may result in a breach of privacy and confidentiality. Many researchers have proposed techniques to preserve privacy and protect user information. In this paper, we describe one such technique to improve privacy preservation in smart transportation.

The organization of the paper is as follows: Sect. 2 describes related work, Sect. 3 Illustrates the proposed method, Sect. 4 explains Scyther tool specification language and also implementation, Sect. 5 Shows simulation results and Sect. 6 discusses the conclusion and future scope.

2 Related Work

Security and privacy are always an important topic of discussion and also the main concern for smart transportation. The number of smart cities around the world is increasing thus researchers are paying more attention to security and privacy in smart transportation.

We reviewed several types of research that discuss security and privacy challenges related to smart transportation. Smart transportation can have similar attack types as that of wireless network devices [5]. Security issues such as availability, authentication, authorization, confidentiality, and data integrity are the key requirement to protect any network from security attacks. Also, the security issues in IoT can be classified as Legal or accountability, privacy, security, and general [6]. Main security challenges in smart transportation can be classified as verification, authorization, privacy, access control, management, information storage, and system configuration.

Privacy means that the user data should be kept secret to protect them from others. A huge amount of data is transferred between IoT devices using a communication system such as 5G and needs data protection to avoid the attack on the user. The heavy traffic in the 5G network can be reduced by using device-to-device (D2D) communication [7]. However, there are several security issues due to the open communication nature of D2D communication. Therefore D2D communication has various threats such as privacy violation, data modification, jamming and free-riding.

Providing security and privacy is very challenging in IoT as it involves many devices. In smart transportation, for privacy preservation the data such as location trajectory should be protected, from the location data it is easy obtained the relationship and interest kind of personal information. Similarly any data exchanged between devices and users should not disclose any user information, therefore to protect user data cryptographic techniques are used in network [8], one such technique is presented in this paper using asymmetric key [9].

3 Proposed Method

In smart transportation, all the IoT services are going to use 5G as the mode of communication or it will be using wireless technologies for D2D communication. As shown in Fig. 1, the different service types, such as D2D, V2V and Mobile Communication. The user equipment (DEVICE) means a vehicle or Mobile can connect to each other using wireless network, to avail different services from a smart transport server. Usually, in a smart transport system, a user equipment (mobile) will be communicating with the server. The server will be responsible for service allocation after the verification of some authentication data. As communication in a

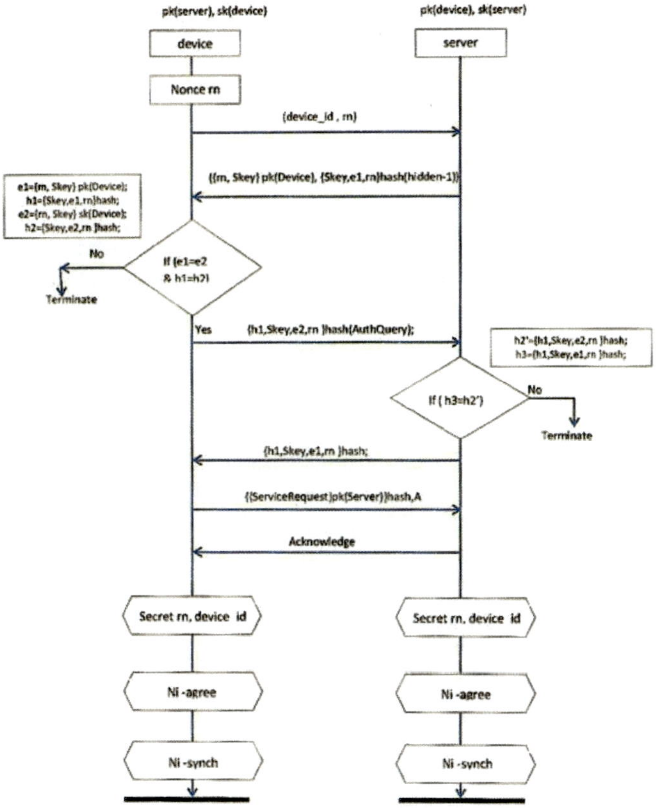

Fig. 1 Steps of proposed security protocol

smart transport system takes place in an open environment, an attacker can passively capture the information such as type of service and user credentials. There is the possibility of breach of user privacy and also the attacker may use the captured information for a replay attack. Therefore security protocol is needed to protect privacy and to provide security to the users, apart from this confidentiality and integrity of the data also needs to be preserved. The security protocol proposed here is an improvisation of the protocol proposed in the paper [10], the author has used a symmetric key for encryption for information exchange between nodes. Symmetric key encryption can be easily hacked or a replay attack can be easily launched by using passive listening, therefore we proposed using a more secured approach that is asymmetric key cryptography. Asymmetric key encryption is difficult to hack and provide stronger protection than symmetric key encryption. The disadvantages of using symmetric keys are, maintaining the secrecy of the keys used for the communication by the two-party on both ends, also in a large network many key pairs need to be managed properly, which needs an unconditionally trusted a trusted third party (TTP), and key needs to be changed for every session.

In the security protocol, the User or vehicle (DEVICE) sends a message to initiate the communication, which involves 2 fields to the smart transport server (SERVER), the steps of security protocol is explained in Fig. 1.

4 Scyther Tool Specification Language

Scyther tool uses Scyther Security Protocol Description Language (.spdl) to model protocols specification such as roles, messages (sending and receiving), and parameters types [11]. Syntax used in Scyther is shown in Fig. 1.

Example given in Fig. 2 describes a protocol with name pn and two node n1 and n2 communication to each other by exchanging messages. Activity of a node is defined inside the role. User type is used to declare user-defined type such as SharedSecret, using this it is possible to define new types. The predefined variable type fresh is used to assign a value which is generated in the same role. Another predefined term nonce is used to define a constant. The term var is used to store a value which is received from a different role. Global definition for hash is written as hashfunction to provide access to all role [12].

Symmetric key is defined as k(n1;n2) and asymmetric key as sk(n1) and pk(n1) for private key and public key of role n1. Send() is used to define send message and recv() to receive a message. To compare two values match function is used and claim is used to check security property as shown in Fig. 1 [12].

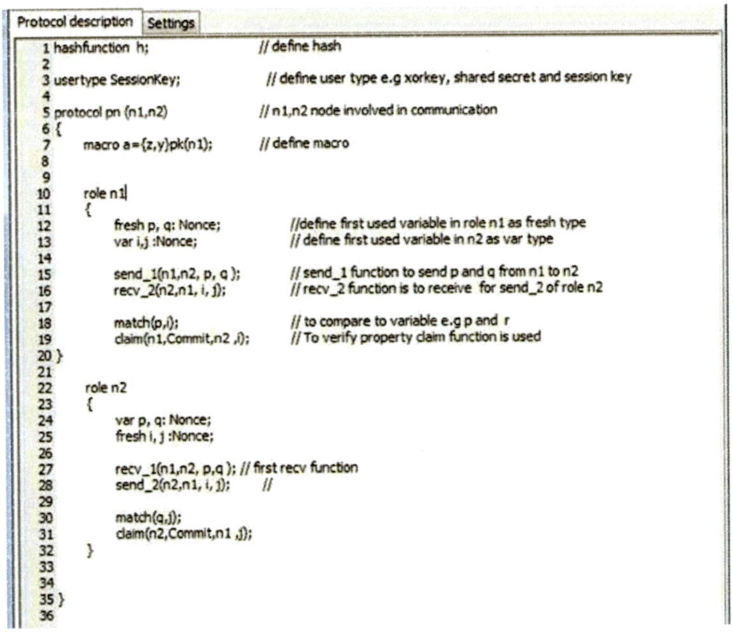

Protocol description | Settings

```
hashfunction h;                      // define hash

usertype SessionKey;                 // define user type e.g xorkey, shared secret and session key

protocol pn (n1,n2)                  // n1,n2 node involved in communication
{
    macro a={z,y}pk(n1);             // define macro


    role n1
    {
        fresh p, q: Nonce;           //define first used variable in role n1 as fresh type
        var i,j :Nonce;              // define first used variable in n2 as var type

        send_1(n1,n2, p, q );        // send_1 function to send p and q from n1 to n2
        recv_2(n2,n1, i, j);         // recv_2 function is to receive  for send_2 of role n2

        match(p,i);                  // to compare to variable e.g p and  r
        claim(n1,Commit,n2 ,i);      // To verify property claim function is used
}

    role n2
    {
        var p, q: Nonce;
        fresh i, j :Nonce;

        recv_1(n1,n2, p,q ); // first recv function
        send_2(n2,n1, i, j);      //

        match(q,j);
        claim(n2,Commit,n1 ,j);
    }


}
```

Fig. 2 Example of Scyther Security Protocol Description Language (.spdl)

4.1 Syntax

```
. . /*      * comment type -1         */              //
comment type-2
// hash function
// user type
protocol protocol_name (n1,n2)   // node involved in com-
munication
{     // macro
  role role_name_1
  {     // variables
  // send function
  // receive function
  // compare
  // test parameter
}
role role_name_ni
{          }
}
```

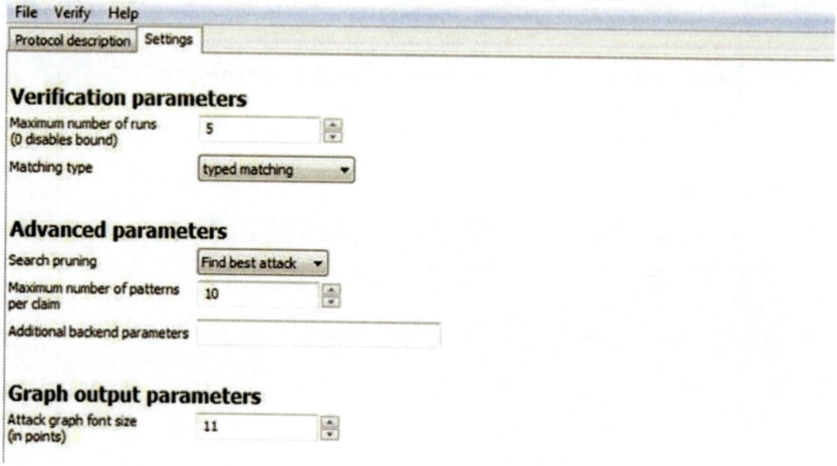

Fig. 3 Settings used in verification of the security protocol

4.2 *Example*

See Figs. 2, 3 and 4.

5 Simulation Results

Using the formal verification tool—Scyther, the Security protocol is simulated and Security Protocol Description Language (.spdl) is used for modeling the protocol [13]. Figure 4 displays the result of various claims achieved from the simulation tool Scyther, with the predefined setting as shown in Fig. 3, also Fig. 5 displays results of autoverified list of parameters. The security protocol is subjected to a test for Aliveness, Non-Injective Synchronization, Secrecy and Non-Injective Agreement. The security terms mentioned above are given in detail. It can be clearly seen from the result that the protocol is resistant to various attacks, such as Confidentiality, Integrity and Availability. The information is secretly communicated between the nodes therefore preserving privacy.

6 Conclusion

Smart transportation uses IoT, requiring more bandwidth to accommodate various services of IoT. For connectivity, mobility and fast- communication the 5G network provides the support needed for IoT services. In this paper, we surveyed the types of

Device_Server	Device	Device_Server,Device2	Secret _Hidden_2	Ok	Verified	No attacks.
		Device_Server,Device3	Secret _Hidden_1	Ok	Verified	No attacks.
		Device_Server,Device4	Secret A	Ok	Verified	No attacks.
		Device_Server,Device5	Secret Deviceid	Ok	Verified	No attacks.
		Device_Server,Device6	Secret m	Ok	Verified	No attacks.
		Device_Server,Device7	Secret SR	Ok	Verified	No attacks.
		Device_Server,Device8	Secret AuthQuery	Ok	Verified	No attacks.
		Device_Server,Device9	Secret Ack	Ok	Verified	No attacks.
		Device_Server,Device10	Secret Skey	Ok	Verified	No attacks.
		Device_Server,Device11	Secret hidden_1	Ok	Verified	No attacks.
		Device_Server,Device12	Alive	Ok	Verified	No attacks.
		Device_Server,Device13	Weakagree	Ok	Verified	No attacks.
		Device_Server,Device14	Niagree	Ok	Verified	No attacks.
		Device_Server,Device15	Nisynch	Ok	Verified	No attacks.
	Server	Device_Server,Server2	Secret _Hidden_3	Ok	Verified	No attacks.
		Device_Server,Server3	Secret Ack	Ok	Verified	No attacks.
		Device_Server,Server4	Secret Skey	Ok	Verified	No attacks.
		Device_Server,Server5	Secret hidden_1	Ok	Verified	No attacks.
Done.		Device_Server,Server6	Secret Deviceid	Ok	Verified	No attacks.

Fig. 4 Autoverified results of the protocol from Scyther tool

security and privacy attacks in smart transportation. A security protocol for communication between user equipment (Device) and smart transport server (Server) is tested for various security parameters, using formal verification tool—Scyther. The protocol is verified for various security requirements and it is found that protocol preserved Secrecy, Non-Inactive Synchronization (Nisynch), Aliveness, and Non-Injective Agreement (Niagree) the details of these terms are given in [13]. It is assumed that the devices involved in communication are trusted and free from malware, in future we will be working on to improve device trust and to identify malicious devices.

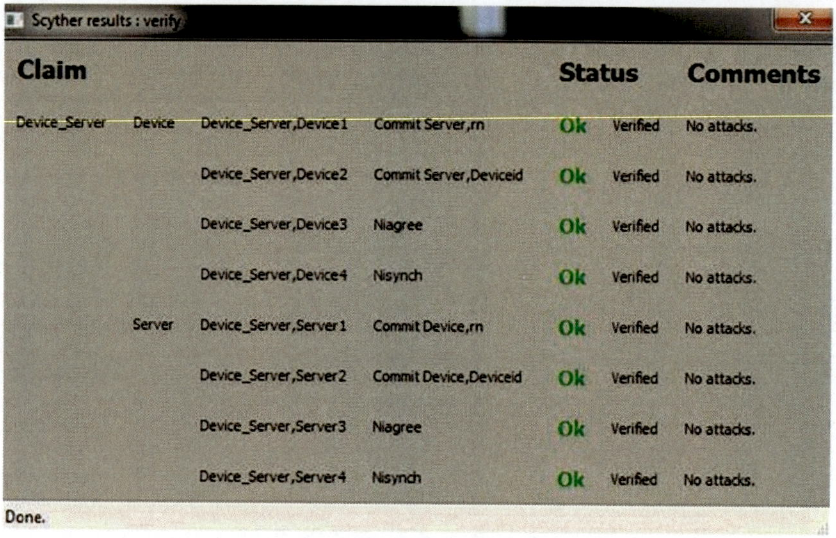

Fig. 5 Security claim verification results of the protocol from Scyther tool

References

1. Sookhak, M., Tang, H., Yu, F.R.: Security and privacy of smart cities: issues and challenge. In: Proceedings 20th International Conference High Perform. Comput. Commun. 16th Int. Conf. Smart City 4th Int. Conf. Data Sci. Syst. HPCC/SmartCity/DSS 2018, vol. 21, no. 2, pp. 1350–1357 (2019). https://doi.org/10.1109/HPCC/SmartCity/DSS.2018.00224
2. Hamid, B., Jhanjhi, N., Humayun, M., Khan, A., Alsayat, A.: Cyber security issues and challenges for smart cities: a survey. In: MACS 2019 - 13th Int. Conf. Math. Actuar. Sci. Comput. Sci. Stat. Proc. (2019). https://doi.org/10.1109/MACS48846.2019.9024768
3. Pan, F., Wen, H., Song, H., Jie, T., Wang, L.: 5G security architecture and light weight security authentication. 2015 IEEE/CIC Int. Conf. Commun. China - Work. CIC/ICCC **2015**, 94–98 (2017). https://doi.org/10.1109/ICCChinaW.2015.7961587
4. Fang, D., Qian, Y.: 5G wireless security and privacy: architecture and flexible mechanisms. IEEE Veh. Technol. Mag. **15**(2), 58–64 (2020). https://doi.org/10.1109/MVT.2020.2979261
5. Pattanashetty, V.B., Iyer, N.C., Dinkar, A., Gudi, S.: Self-coordinating bus route system to avoid bus bunching. In: Satapathy, S., Bhateja, V., Joshi, A. (Eds.), Proceedings of the International Conference on Data Engineering and Communication Technology. Advances in Intelligent Systems and Computing, vol. 469. Springer, Singapore (2017). https://doi.org/10.1007/978-981-10-1678-3_50
6. Alsaffar, N., Ali, H., Elmedany, W.: Smart transportation system: a review of security and privacy issues. In: 2018 Int. Conf. Innov. Intell. Informatics, Comput. Technol. 3ICT 2018, pp. 2018–2021 (2018). https://doi.org/10.1109/3ICT.2018.8855737
7. Li, Y., Cai, L.: Security aware privacy preserving in D2D communication in 5G. IEEE Netw. **31**, 56–61 (2017)
8. Anajemba, J.H., Iwendi, C., Mittal, M., Yue, T.: Improved Advance Encryption Standard with a Privacy Database Structure for IoT Nodes, pp. 201–206 (2020). https://doi.org/10.1109/csnt48778.2020.9115741

9. Al-Riyami, S.S., Paterson, K.G.: Certificateless public key cryptography. Lect. Notes Comput. Sci. (including Subser. Lect. Notes Artif. Intell. Lect. Notes Bioinformatics) **2894**, 452–473 (2003). https://doi.org/10.1007/978-3-540-40061-5_29
10. Sharma, S., et al.: Secure authentication protocol for 5G enabled IoT network. In: 2018 Fifth International Conference on Parallel, Distributed and Grid Computing (PDGC), Dec. 2018, pp. 621–626. https://doi.org/10.1109/PDGC.2018.8745799
11. Pradeep, R., Sunitha, N.R., Ravi, V., Verma, S.: Formal verification of authentication and confidentiality for TACACS+ security protocol using scyther. 2019 10th Int Conf. Comput. Commun. Netw. Technol. ICCCNT **2019**, 6–11 (2019). https://doi.org/10.1109/ICCCNT45670.2019.8944623
12. Yang, H., Oleshchuk, V., Prinz, A.: Verifying Group Authentication Protocols by Scyther, vol. 2, no. June, pp. 3–19 (2016)
13. Cremers C (2016) Scyther: semantics and verification of security protocols. In Tech-nische Universiteit Eindhoven. https://doi.org/10.6100/IR614943

Classification of Medical Images Using Deep Learning to Aid in Adaptive Big Data Crowdsourcing Platforms

C. Jaya Sudha and Y. S. Sneha

Abstract Deep Learning has radically transformed the field of Medical Image Analysis by surpassing human performance in several challenging classification tasks and thus has widespread applications in industry and academia. In this work, we employ a Deep Convolutional Neural Network (DCNN) that uses Transfer Learning and is trained using the Open-access Medical Image Repositories and OpenI repository. We use DCNN to assist in a crowdsourced Big Data platform by conducting two crucial classification tasks: (1) Interclass classification of the type of the medical image—MRI, Ultrasound or X-Ray; (2) Intraclass classification of the X-Ray images into chest or abdomen X-Rays. Our approach automates the classification process of medical images, to provide certified medical experts with medical images of individuals on a crowdsourced platform, along with similar medical images of the suspected medical condition for validation, in order to enhance diagnostic effectiveness. In this work, we focus primarily on the classification tasks using DCNN; the elucidation of the privacy-preserving architecture of our cloud-based crowdsourcing platform and classification using conventional clustering approaches in the absence of large training images is deferred to future work.

1 Introduction and Related Work

Deep Learning (DL) [1, 2] based methods such as Artificial Neural Networks [3] have received a lot of attention from various medical fields in the past decade, due to their ability to discover complex patterns that are generally not found with conventional methods. The accuracy and capabilities of DL methods [4, 5] in dealing with varied data depends on the presence of substantial amount of training images. Since the advent of Big Data [6], DL methods have access to an exorbitant amount of images, that are easily available in an organized manner and with drastic increase in the availability of processing power and GPUs, the effectiveness of Deep Neural

C. Jaya Sudha (✉) · Y. S. Sneha
JSS Academy of Technical Education, Bengaluru, India

T. Senjyu et al. (eds.), *ICT with Intelligent Applications*, Smart Innovation, Systems and Technologies 248, https://doi.org/10.1007/978-981-16-4177-0_9

Networks is constantly getting better [5, 7]. Deep Neural Networks are widely used in Computer Vision, Natural Language Processing and Robotics [4, 5].

Deep Learning received substantial attention from Computer Vision and other interrelated fields [2, 4, 5], when neural networks surpassed the performance of conventional methods on many high-profile image analysis benchmarks [5]. One such famous benchmark that neural networks performed exceptionally on was the ImageNet Large-Scale Visual Recognition Challenge (ILSVRC) [8]; in 2012, Convolutional Neural Networks (CNN) [9] achieved half the error rate of the second-best method in performing classification [8] and they even surpassed human performance [5].

The amount of data that is generated daily by healthcare providers is so massive that conventional methods cannot deal with them competently [6], while DL based methods can effectively integrate, analyze and predict from these Big Data heterogeneous data sets and aid in Health Informatics [5, 10]. The applications of Deep Learning in Healthcare range from Medical Diagnosis [11], Drug Discovery [12], Medical Image Diagnosis [4], Supporting Clinical Decision [13], Outbreak Prediction [14], Health Record Analysis [15] among others.

Crowdsourcing in medical field [16] has grown exponentially over the last few years, as it promotes researchers and doctors to have access to data that is voluntarily uploaded by users across the globe. This process usually involves a large group of experts that get together to attempt to solve a crucial problem and then share potential solutions with the broader public. Apple Research Kit [17] permits users to access interactive apps to employ deep learning algorithms to perform face recognition and gain insights into Parkinson and Asperger's diseases; IBM has in partnership with Medtronic consolidated and made available large amounts of data on diabetes and insulin [18]. Hence with the prevalence of high-performance computing, networking infrastructure, novel ways for users to participate and contribute online and the proliferation of electronic devices capable of easily transmitting medical data, crowdsourcing has a lot of potential in helping solve many of the critical problems in medical science [4, 5, 19].

In our proposed approach, we gather and store in a private cloud-based container, large amounts of medical image data from open-access medical repositories [20], in order to perform classification on them using a Deep Convolutional Neural Network (DCNN) [21]. Since our database would consist of enormous amounts of assorted medical images that are generally not labelled, it is a crucial task, that is quite cumbersome and may not be executable accurately when done manually. Hence our DCNN algorithm is trained using open-access medical datasets [20] to be able to differentiate between different types of medical images: MRI, Ultrasound and X-Ray, i.e. Interclass Classification. We also perform Intraclass classification of X-Ray images in this work, where we classify X-Ray images into Chest X-Ray or Abdomen X-Ray, for easy diagnoses on the crowdsourcing platform.

Furthermore, the data is anonymized by removing the name labels from the images and substituting it with a unique ID. Hence the user's name would not be publicly shown in the crowdsourced platform and after we get a diagnosis from the platform, the diagnosis can be relayed to the individual by cross-referencing the unique ID with the name.

We will not elaborate on the architecture of the cloud-based container (scheduling, buffer management, etc.) or the technical details of crowdsourcing platform in this work and defer it to future work. We will primarily focus on two classification tasks with DCNN: (1) Interclass classification based on medical image type—MRI, Ultrasound or X-Ray; (2) Intraclass classification to classify the X-Ray images into either Chest X-Ray or Abdomen X-Ray.

The rest of the sections are divided into the following: Sect. 2 provides an outline of our proposed methodology, Sect. 3 provides an overview of the working of DCNN and Sect. 4 provides the results and database design of our approach and finally Sect. 5 provides a conclusion of the work and outlines future directions.

2 Methodology

The framework of our proposed model is presented in Fig. 1. Our approach involves (1) gathering medical images from the Open-access Medical Image Repositories [20] and OpenI repository [22], (2) storing them in a private cloud-based container and then (3) employing a neural network that is pre-trained with Google Inception V1, V3, V4 (Transfer Learning) and then training the last layer of the DCNN using open-access medical repository images and images from other medical sources (4) Using DCNN thus obtained to classify assorted medical images into different interclass types and further into intraclass types (chest or abdomen X-Rays) (5) finally these are uploaded into different discussion threads of the crowdsourcing platform in an anonymized manner (for instance chest X-Ray would be uploaded on a section of the platform with a unique ID and experts in respiratory disorders or other general MDs can provide diagnoses).

The privacy-preserving architecture of our cloud-based platform is based on the works of Joongheon et al. [19] and it involves employing Queue Backlog and Rate-aware Scheduling algorithms to perform effective buffer management and securely

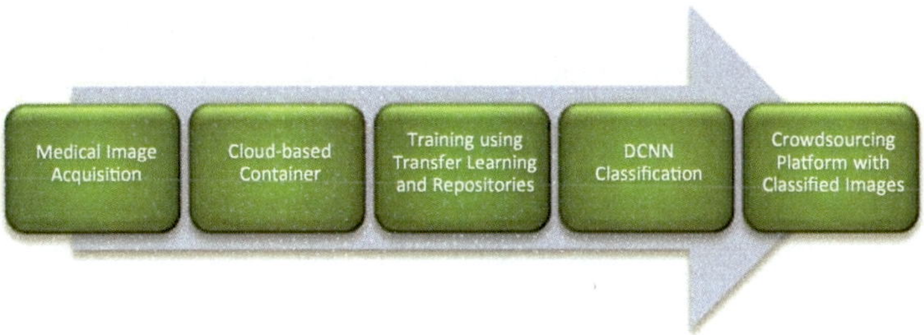

Fig. 1 Framework of our method

handle the private data. Our classification pipeline is inspired by the works of Khosravi et al. [23] as their CNN was quite effective in differentiating pathology images. Here we use Transfer Learning [23] for training, which involves pre-training a network architecture using a large dataset and the trained model thus obtained can be used to classify new datasets that are of a smaller size. Hence here we can use all the knowledge that is already available in models such as Google's Inception V3 [24] and Inception V4 [24], which is already trained and then train the last layer of the neural network with our medical repository to render a robust DCNN model capable of classifying most of the medical images.

3 DCNN Background

Deep Convolutional Neural Network (DCNN) [21] consists of a stack of Convolution layers, Pooling Layers and Fully-connected layers; in each of the convolution layers, all of the common patterns in the local regions of the inputs are extracted by performing convolution of a filter over these inputs [7]. Subsequently, the result of the convolution thus performed is stored in a feature map to gauge as to how well the filter is able to match each part of the inputs; then sub-sampling is carried out by the pooling layer in order to aggregate statistics of these features, perform dimensionality reduction of the data and deal with any over-fitting problems that might occur [7, 21]. Then a non-linear activation function is employed to obtain the output of the layer. After the stacks of convolution and pooling layers, we have the fully-connected layers which perform additional aggregation of the local information that is learned in the previous two layers, in order to effectively differentiate between different classes, i.e. classification.

DCNNs consist of a neuron as the basic building block, which carries out convolution, pooling and activation operations [7]. Convolution operation involves determining the product of the input $x_i's$ and its corresponding weights $w_i's$ and then the pooling step takes these inner product results and sub-samples them to obtain a single result [25]; pooling can be done in two ways, we can either select the largest element in each of the pooling regions—known as max pooling, or we can determine the arithmetic mean—known as average pooling [7]. Subsequently, an activation function is applied before the output; we can choose Hyperbolic Tangent (Eq. 1), Logistic Function (Eq. 2) or Rectified Linear Unit function—ReLU (Eq. 3) [7, 25].

$$\tan h\ f(x) = \tan h(x) \tag{1}$$

$$f(x) = \left(1 + e^{-x}\right)^{-1} \tag{2}$$

$$f(x) = \max(0, x) \tag{3}$$

If the DCNN is not appropriately designed and optimized, it can result in accuracy decline, hence it is necessary to optimize the activation function properly in accordance with the classification task, to ensure adequate accuracy [7, 25]. An exhaustive overview of the working of DCNN is provided in [7].

4 Results

4.1 Database Design

The presence of rotation, translation, horizontal flip and shear, in a medical image may affect the recognition rate and hence classification; so during training, we create different transforms from a given image and train our DCNN model to be effective in recognizing all the variations that may occur in real-world medical images [26]. This is known as Augmentation and this helps train the model for different variations, by creating multiple transformations from a single image, without the added expense of procuring more training datasets. Augmentation helps make the model more accurate and enhances the generalization of the CNN [26, 27]. Keras generator [27] can be used for Augmentation as illustrated by Lakhani et al. in [26] and a few transformations created from augmentation of a chest X-Ray image is shown in Fig. 2 [26].

Our classification approach includes a basic DCNN architecture and Google's Inceptions with 3 training steps (V1, V3, V4) and a combination of two Inception and ResNet algorithms. During training, we train the last layer of Google's Inceptions V3 and V4 with the Open-access Medical Image Repositories [20] and OpenI repository [22] and adjust the data by employing two pre-trained versions—Inception-V1 and Inception-V3. This classification pipeline is inspired by the works of Khosravi et al. [23]. The Interclass datasets are procured from Open-access Medical Image Repositories [20] and Intraclass datasets from OpenI [22], a repository of medical images from published PubMed Central articles, hosted by the National Institutes of Health. Additionally, our intraclass testing model consisted of 200 images, 100 for chest X-Rays and 100 for abdomen X-Rays.

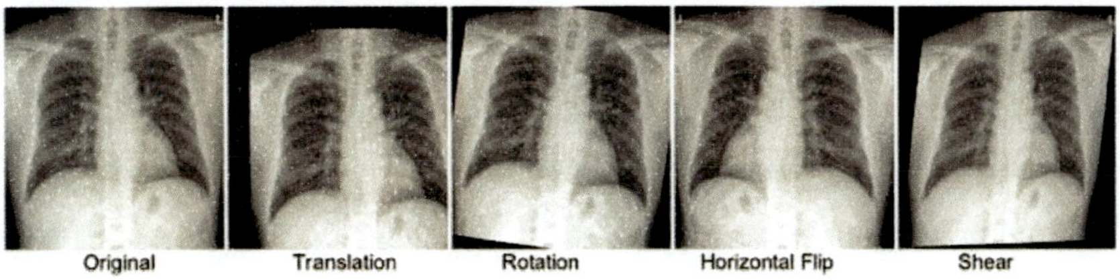

Fig. 2 Augmentation to include common real-time variations [26]

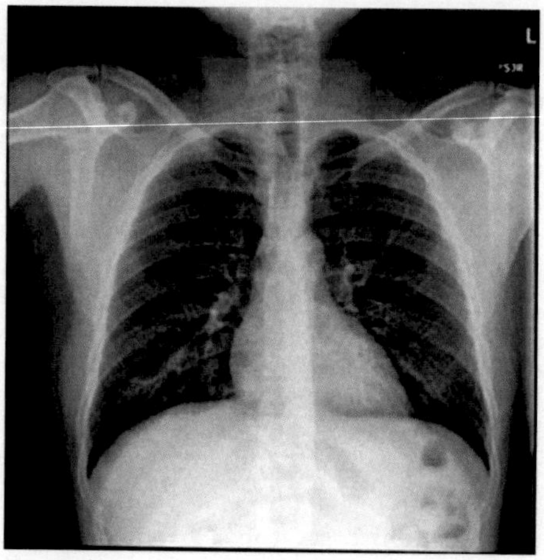

Fig. 3 Image 101—Chest X-Ray

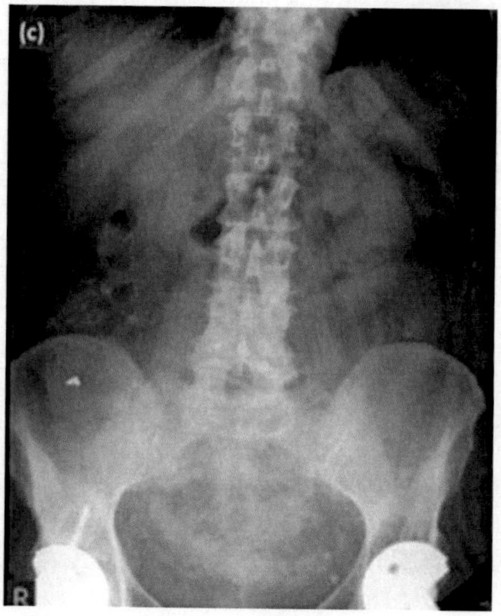

Fig. 4 Image 201—Abdomen X-Ray

4.2 Experimental Results

The classification results of two X-Ray images 101 and 201 are illustrated in Figs. 3 and 4 and their actual and predicted values are shown in Table 1.

In Table 1 we provide classification information about two images—Image 101 (Fig. 3) and Image 201 (Fig. 4). We can observe that with Image 101, it was correctly classified first as belonging to X-Ray class (Interclass) and furthermore it was predicted to be a chest X-Ray (Intraclass). Similarly, with Image 201, it was correctly classified as belonging to X-Ray class and then as an abdomen X-Ray.

Table 1 Image classification results

Image type	Image number	Detected class	Actual	Predicted
X-Ray	Image 101 (Group 1)	X-Ray	Chest X-Ray	Chest X-Ray
X-Ray	Image 201 (Group 2)	X-Ray	Abdomen X-Ray	Abdomen X-Ray

In our experimentations, we have 100 testing images for each group, i.e. 1–100 for chest X-Rays (Group 1) and 10–200 for abdomen X-Rays (Group 2). Our results showed that 96 images were correctly classified as chest X-Rays in Group 1 and 94 abdomen X-Rays were correctly classified as abdomen X-Rays in Group 2.

5 Conclusion and Future Work

In this work, we have provided a Deep Learning based Deep Convolutional Neural Networks (DCNN) that was trained using Open-access Medical Image Repositories and OpenI to perform classification, in order to aid in Big Data Crowdsourcing platforms. Our DCNN was employed to perform two classification tasks, Interclass and Intraclass; in Interclass, it was capable of effectively classifying MRI, Ultrasound and X-Ray images and in Intraclass, it was able to effectively classify two types of X-Rays—chest X-Rays and abdomen X-Rays.

For future work, we aim to present an effective privacy-preserving cloud-based crowdsourcing platform, which includes appropriate rate-aware scheduling, queue backlog and buffer management; subsequently, we hope to develop an agglomerative clustering method to perform classification in scenarios, where there are limited amount of medical images for training. Currently work is also underway to effectively classify knee MRI images for abnormalities (tears) using our proposed DCNN system on the Stanford MRI Dataset [28], that consists of 1370 knee MRI exams obtained from Stanford University Medical Centre, wherein there are 1104 abnormal exams with 319 ACL tear images and 508 meniscal tear images.

References

1. Razzak, M.I., Naz, S., Zaib, A.: Deep learning for medical image processing: overview, challenges and the future. In: Classification in BioApps, pp. 323–350. Springer, Cham (2018)
2. Gibson, E., Wenqi, L., Carole, S., Lucas, F., Dzhoshkun, I.S, Guotai, W., Zach, E.-R., et al.: NiftyNet: a deep-learning platform for medical imaging. In: Computer Methods and Programs in Biomedicine, vol. 158, pp. 113–122 (2018)
3. Jiang, J., Trundle, P., Ren, J.: Medical image analysis with artificial neural networks. Comput. Med. Imaging Graph. **34**(8), 617–631 (2010)
4. Maier, A., Syben, C., Lasser, T., Riess, C.: A gentle introduction to deep learning in medical image processing. Zeitschrift für Medizinische Physik **29**(2), 86–101 (2019)

5. Lundervold, A.S., Lundervold, A.: An overview of deep learning in medical imaging focusing on MRI. Zeitschrift für Medizinische Physik **29**(2), 102–127 (2019)
6. Belle, A., Thiagarajan, R., Soroushmehr, S.M., Navidi, F., Daniel, A.B., Najarian, K.: Big data analytics in healthcare. BioMed. Res. Int. 2015 (2015)
7. Li, J., Yuan, Z., Li, Z., Ding, C., Ren, A., Qiu, Q., Draper, J., Wang, Y.: Hardware-driven nonlinear activation for stochastic computing based deep convolutional neural networks. In: 2017 International Joint Conference on Neural Networks (IJCNN), pp. 1230–1236. IEEE (2017)
8. Krizhevsky, A., Sutskever, I., Hinton, G.E.: ImageNet classification with deep convolutional neural networks. In: Pereira, F., Burges, C.J.C., Bottou, L., Weinberger, K.Q. (eds.) Advances in neural information processing systems 25, pp. 1097–1105. Inc., Curran Associates (2012)
9. Kayalibay, B., Jensen, G., van der Smagt, P.: CNN-based segmentation of medical imaging data. arXiv preprint arXiv:1701.03056 (2017)
10. Ravi, D., Wong, C., Deligianni, F., Berthelot, M., Andreu-Perez, J., Lo, B., et al.: Deep learning for health informatics. IEEE J. Biomed. Health Inform. **21**, 4–21 (2017)
11. Kermany, D.S., Goldbaum, M., Cai, W., Valentim C.C.S., Liang, H., Baxter S.L., et al.: Identifying medical diagnoses and treatable diseases by image-based deep learning. Cell **172**, 1122–1131 (2018). e9
12. Jimènez J., škalič, M., Martínez-Rosell, G., De Fabritiis, G.: KDEEP: protein-ligand absolute binding affinity prediction via 3D-Convolutional Neural Networks. J. Chem. Inf. Model. **58**, 287–296 (2018)
13. Katzman, J.L., Shaham, U., Cloninger, A., Bates, J., Jiang, T., Kluger, Y.: DeepSurv: personalized treatment recommender system using a Cox proportional hazards deep neural network. BMC Med. Res. Methodol. **18**, 24 (2018)
14. Chen, M., Hao, Y., Hwang, K., Wang, L., Wang, L.: Disease prediction by machine learning over big data from healthcare communities. IEEE Access **5**, 8869–8879 (2017)
15. Shickel, B., Tighe, P.J., Bihorac, A., Rashidi, P.: Deep EHR: a survey of recent advances in deep learning techniques for electronic health record (EHR) analysis. IEEE J. Biomed. Health Inform. (2017)
16. Riedl, J., Riedl, E.: Crowdsourcing medical research. IEEE Comput. **46**(1), 89–92 (2013)
17. Jarrett, J., Blake, M.B., Saleh, I.: Crowdsourcing, mixed elastic systems and human-enhanced computing—a survey. IEEE Trans. Serv. Comp. **11**(1), 202–214 (2017)
18. Hughes, A.S.: Seeking Health Information in the Diabetes Online Community (DOC): The Role of Health Literacy, Social Support and Community Engagement on Treatment Adherence. PhD dissertation, The University of Texas at El Paso (2018)
19. Kim, J., Lee, W.: Stochastic decision making for adaptive crowdsourcing in medical big-data platforms. IEEE Trans. Syst. Man Cybern.: Syst. **45**(11), 1471–1476 (2015)
20. Open-Access Medical Image Repositories (accessed November 25, 2020). http://www.aylward.org/notes/open-access-medical-image-repositories
21. Shin, H.-C., Roth, H.R., Gao, M., Le, Lu., Ziyue, Xu., Nogues, I., Yao, J., Mollura, D., Summers, R.M.: Deep convolutional neural networks for computer-aided detection: CNN architectures, dataset characteristics and transfer learning. IEEE Trans. Med. Imaging **35**(5), 1285–1298 (2016)
22. OpenI repository from National Institute of Health (accessed November 25, 2020). https://openi.nlm.nih.gov/faq#collection
23. Khosravi, P., Kazemi, E., Imielinski, M., Elemento, O., Hajirasouliha, I.: Deep convolutional neural networks enable discrimination of heterogeneous digital pathology images. EBioMedicine **27**, 317–328 (2018)
24. Szegedy, C., Ioffe, S., Vanhoucke, V., Alexander, A.A.: Inception-v4, inception-resnet and the impact of residual connections on learning. In: Thirty-First AAAI Conference on Artificial Intelligence (2017)
25. Qayyum, A., Anwar, S.M., Awais, M., Majid, M.: Medical image retrieval using deep convolutional neural network. Neurocomputing **266**, 8–20 (2017)

26. Lakhani, P., Gray, D.L., Pett, C.R., Nagy, P., Shih, G.: Hello world deep learning in medical imaging. J. Digit. Imaging **31**(3), 283–289 (2018)
27. Chollet, F.: Building powerful image classification models using very little data. Keras Blog (2016)
28. Stanford MRNet Knee Dataset (accessed December 2, 2020). https://stanfordmlgroup.github.io/competitions/mrnet

EZAdopt

Chirayu Lalwani, Prerna Gurbaxani, Snehal Jagatap, and Maaz Shaikh

Abstract If anyone wants to adopt a child or a pet, they have to go looking from organization to organization. Also, most adopters are unaware of the organizations and the processes to be carried out to be a legal adopter. Most of them are not aware of the eligibility norms. But even after considering all the valid conditions, the process goes on for a while that makes it tidier to manage. Therefore, we propose a system that effectively manages the data of organizations that facilitate adoption and helps the users search for one. We aim at providing a user-friendly platform for searching organizations for the adoption of human babies, cats, or dogs depending on the respective eligibility criteria provided by the government. The system collects data for the adoptee and government documents for verification from users and searches for the nearby foster homes depending upon the location preferred by the user via a Web-based application. If the users satisfy the eligibility criteria, and if their desired adoptee is available, as a result, the organization details are displayed to the user. A GPS provided in the application will help the user get details of foster homes nearby. The users that are not able to take care of their pets can also get help from the nearest organization. Additionally, a security feature will be provided to keep the user's and the organization's data safe. The entire client's data will be uploaded to a cloud platform in encrypted form for safe, easy, and dynamic access.

1 Introduction

In the growing world, adoption has become an important part of maintaining a healthy and loving relationship between humans and humans and in between humans and animals. An adoption is an act that not only impacts the child who is being adopted, but also their family, extended family, and community. Adoption is important for the child/pet who is adopted. Children and pets in foster homes do not receive the necessary love and care which can be provided by the adopter family. Adoption is equally important for the adoptive parents as it leads to improvement in the family. The

C. Lalwani (✉) · P. Gurbaxani · S. Jagatap · M. Shaikh
PES's Modern College of Engineering, Pune 411005, India

T. Senjyu et al. (eds.), *ICT with Intelligent Applications*, Smart Innovation, Systems and Technologies 248, https://doi.org/10.1007/978-981-16-4177-0_11

adopted could get a proper education. While proceeding for adoption, the adopters need to find suitable and easily accessible foster homes depending upon their choices of location. The adopter needs to get the proper and true information about the adoptee and the organization after checking if he/she is eligible legally to adopt. The first point while moving towards adoption has to be the selection of the adoptee type and then depending on the type, the selection of their age, their gender, and breed in the case of pets. After the above-mentioned choices have been selected, it is important to find a foster home which sited near to the adopter. This system helps the user to find the nearest foster homes in their locality where they could visit and find the child/pet they wanted to adopt. The adoption criteria specified by the Government of India are verified by the user details, and if satisfied, the user gets the best match. Organizations are provided with the privilege of modifying their data. The system provides an efficient way to handle a large amount of data of all the organizations and provides required information without much effort. Also, it manages the user details. Therefore, we believe that the system proposed will help in digitizing some steps of adoption with providing high-level security to the accounts and user's data by storing it on the cloud in an encrypted format along with providing easy access to organizations' location with the help of GPS provided via a Web-based application. Along with all this, a GPS will be attached to the pets which will be accessible to the foster homes for further parental verification, and follow-up of the child's upbringing will be taken by the organization provided with periodic reminders, till a year. The adoption process might involve long waits and has many procedures and verifications at every step. The project is intended to provide help at the initial stage of the process. The study was conducted, and the information related to the process was collected by visiting the organizations.

2 Literature Survey

The regulatory body for the process of adopting a child in India is monitored by the Central Adoption Resource Authority (CARA). It works at intra-city to inter-state all over India and is responsible for all activities under the Ministry of Women and Child Care. The one who wants to adopt needs to first get registered with an authorized institute, after which the person from the organization visits home for a family check as well as counseling post-adoption. Further, a petition is presented to the court, and an order is generated which decides whether the child can be adopted or not [1].

Applications such as Patronization for Orphans and Adoption using Android Application by B. Kamala B. Rakshana B. Reshma B. Deivasree, are used where people can donate food, clothes, and money. Besides this, there is also a process of helping in adopting a child. The orphanage or organization will enter details of their children from where the prospective adoptive parents can search and choose the kid they want to adopt.

However, certain criteria must be satisfied when someone decides to adopt a child (e.g., whether he/she is married and financial status). Many people are unaware of such laws. There is no such system or feature inside any application available that makes the user aware of such laws [2] for pets (dogs and cats); however, there is no such complicated process. One process of adopting a pet is that the interested adoptees visit a nearby or any animal shelter; from there, they decide which pet they want to adopt. The shelter owner briefs them about the pet and its history. An agreement is made, and the pet can be adopted. Another approach is that a person finds a stray pet and makes a decision to adopt it or decides to put it for adoption. He/she uploads the pet's photographs on social media and other adoption sites from where an interested person can approach them. JSP-based Pet Adoption System by Haoran Liu and Xiue Meng makes use of MVC architecture and is a module that is solely based on adopted/to-be-adopted pets.

There are no such existing systems which use combined search for both pets and child. Also, there is a need for enhanced search features like GPS because sometimes, for example, people are unaware that there is a certain pet for adoption nearby or in a desirable location. There is no feature of transparency between the user and the organization.

3 Proposed System

The system we propose is designed to bring technological advancement to make a society a better place for the orphans. The system will foremost have two main modules, namely user module and organization module. The user module will allow the user or the society to see if they are fulfilling the norms to adopt a kid or pet. If the user is fulfilling all the criteria according to the norms suggested by the national governing organization, then they are accessible to check if anyone is available for adoption in their nearby locality that fits their match. To get verified and fulfill all the criteria, the user must upload all their documents by a valid authority. Further, the user just needs to enter their criteria for adoptable, and all the nearby adoption centers/foster homes would be visible to the user with the help of a live Global Positioning System (GPS). A GPS facility can be provided in the application to help the users find the required organization. When a particular organization is suggested to users, the user will be provided with a route to the organization with the help of GPS. After the adoption of a pet, a wearable GPS device will be attached to the pet for a year to track its activity with the adopter. The device's location will be accessible to the authorized organization, and with its help, the authority can keep a check on the adopter if they are taking proper care of the adoptee or not. Through this device, the organization can check if the pet is at a safe location or not and can help the adopter to find the pet in case if he/she is lost. Also, the user has to be registered on the application as they will be notified of the scheduled visit dates that would be conducted to check whether the adopted is being treated well and is comfortable and well to do. These checks are conducted until the adulthood of the adopted, whereas the organization has the task to update its portal with the number of adoptable they have and also their detail and pictures (in case of pets). Also once a verified user visits the organization, the organization needs to update its portal with all the details in case of further references.

The database that will be used to store the data will be a cloud server that will enable faster and easier access to the data. It will also help to scale the system depending upon the number of users.

Considering the data to be highly sensitive for the user as well as the organization, these documents hold delicate information that needs to be provided high-level security while in the database. Appropriate care must be taken to maintain the integrity, confidentiality, and availability of the data. To address the data security challenges, we suggest data encryption to protect sensitive data from data breaches; this data is passed through a cryptographic algorithm that will encrypt the data before it is stored on the cloud server, and the data is passed through a decryption algorithm before retrieving it from the cloud server. Also an additional feature is present for users who have a pet but is unable to take care of it, and the user can send it to the nearest pet home that will take care of it. This will in turn help in saving the kids and provide them a suitable family and make the world a better place to live for them. So the technologies used in the system are as follows:

This system uses a Web application to build the application and a cloud database as its server back end. Web application reduces the use of system memory and the maintenance cost. It increases the accessibility across devices with an increase in data security (Fig. 1).

A. Components.

- *Document Verifier:*

The use of physical documents is a difficult task and easier to forge, and the world is moving toward a digital life. Digital documents are convenient to use and easier to store once it is verified. Since there is no methodology to validate and verify digitized

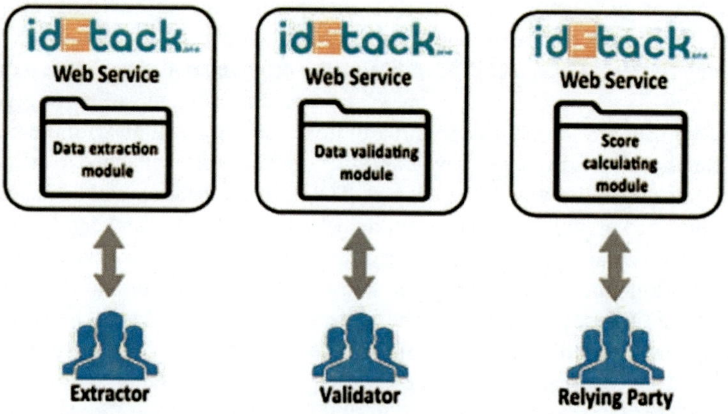

Fig. 1 ID stack architecture

documents, the process of analyzing the documents in private organizations is difficult. Using a score to determine how much a document is validated and how much a set of documents related to a user can be useful. It would provide alternative support for the user to determine whether to trust the document or not. This paper discusses a proposed solution for the above-mentioned problems. ID stack is a technology that uses digital signatures to verify the documents uploaded by the users, and all the documents are provided with some rating-based mechanism through which the authenticity of the document can be measured [3].

- *GPS:*

The GPS consists of both front end and back end which uses location to provide new content to the user. The back end can be written in any suitable scripting language to hold a list of articles and their related location and their coordinates. The front-end Web page uses the location services supported by the chosen browsers to get the GPS coordinates of the user. It then sends those coordinates to the back end in a request. The back-end system responds with a listing of articles, which the front end then renders dynamically.

In this process, the location is used in two different ways as firstly when the articles are being added into the database and then they are converted into the respected coordinates, and secondly while accepting a request from the user and then responding to it using the database [4] (Figs. 2 and 3).

- *Security:*

The data is passed through an algorithm that will encrypt the data before it is stored on the AWS, and the data is passed through a decryption algorithm before retrieving it from the AWS. This uses the block-level data encryption using a 256 cryptography algorithm in rotation. Besides, data users can reconstruct the requested data from the cloud server using a shared secret key. We analyze the privacy protection of outsourced data using an experiment that is carried out on the repository of text files with variable size. The security and performance analysis shows that the proposed

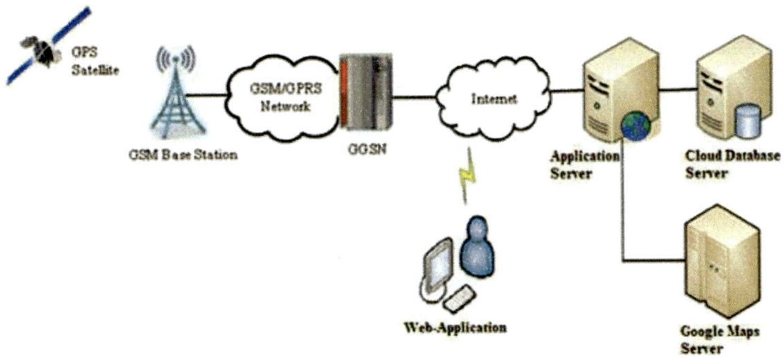

Fig. 2 Location-tracking architecture

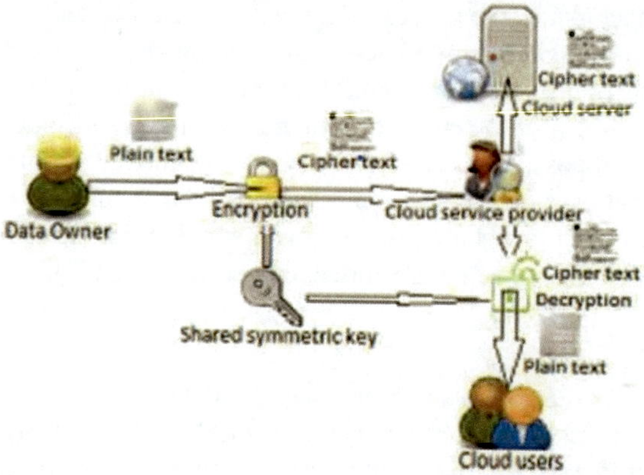

Fig. 3 Block diagram or data encryption and decryption in cloud system

method is highly efficient than the performance of the existing methods. The encryption algorithm has several steps and is composed of a key Chooser, Circular Array Inverter, and Circular Array Shifter. The encryption algorithm is designed such that the information at the highest factor is selected by applying a series of rotations on every block character, and the key is rotated for every character. From this, it is ensured that the same key is not used for encrypting every character, and hence, this algorithm is called a key motor encryption algorithm. The file is divided into blocks, and confidentiality is emphasized on every character level of a block [5].

- *Server:*

AWS Lambda is an event-based platform that does not use a server as so and is provided as a part of Amazon Web Services. It is a technology that gives a response to a different event, and based on those events, it computes the outcome and immediately gives out the response. Lambda is used as it responds to events in milliseconds. The purpose of Lambda, as compared to AWS EC2, is to simplify building on-demand applications that respond to events and new information (Figs. 4 and 5).

B. Architecture

- The user is the one who wishes to adopt, and he/she must get themselves verified from the Web application. Verified users will be able to search according to the

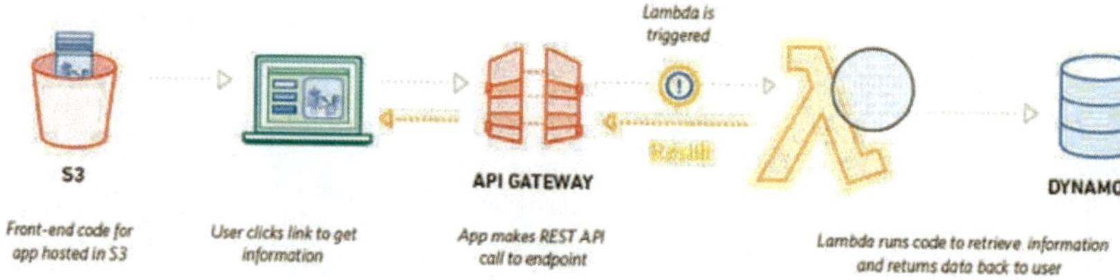

Fig. 4 AWS Lambda architecture

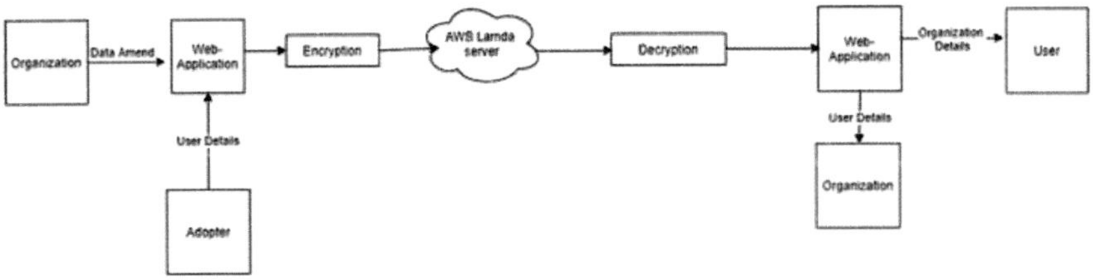

Fig. 5 Architectural diagram for EZAdopt

criteria entered at the time of verification, based on which the desired result may be provided.

- The organization is responsible to maintain the data that will be used while filtering the needs of the user. The organization has to amend the data of the adoptee and the ones which are already adopted. Also once a user physically visits the organization is required to maintain the physical data of the user.
- Web application is the medium through which the user and organization share their data. It is a portal where the user gives out their requirements, and if the organization meets them, the data is shared with the user.
- Encryption/Decryption algorithm is used to provide security to the highly sensitive data of the user as well as the organization which is being sent on the server.
- The server is responsible to perform filtration over the data and send it back to the Web application. It also includes a database to store data from the organization as well as from the user.

c. *Module.*

The system is sub-partitioned into a couple of modules listed below. The modules listed below only demonstrate an overview of it.

- For user: This module for users consists of user verification components, and after successful verification, the user can filter the type of pet/kid they want to adopt. The user gets different search results based upon the user details and family conditions. It also consists of a component through which the user can find a pet home if they cannot take care of the pets.
- For organization: This module consists of a component to alter the data of the organization as well as add the data of the pets/kid available to adopt. Also if pet/kids get adopted, organization can alter the data accordingly.

3.1 Result

Table 1 represents the result for various categories of users those wish to adopt a child, and their validations of their search criteria are displayed with the help of Web application.

Table 1 Pet criteria

Gender	Marital status	No. of years of marriage	Composite age or age of single parent (years)	Family income (lakhs per annum)	No of children	Age of child (years)	Can adopt	
							Boy child	Girl child
Male	Single	–	<30	<3	0	All age-groups	No	No
Male	Single	–	up to 45	>3	0	<4	Yes	No
Male	Single	–	up to 50	>3	0	4–8	Yes	No
Male	Single	–	up to 55	>3	0	8–18	Yes	No
Female	Single	–	<30	<3	0	All age-groups	No	No
Female	Single	–	up to 45	>3	0	<4	Yes	Yes
Female	Single	–	up to 50	>3	0	4–8	Yes	Yes
Female	Single	–	up to 55	>3	0	8–8	Yes	Yes
Male/Female	Married	<2	up to 90	<3	0–3	All age-groups	No	No
Male/Female	Married	>2	up to 90	<3	0–3	All age-groups	No	No
Male/Female	Married	>2	up to 90	>3	0–3	< 4	Yes	Yes
Male/Female	Married	>2	up to 100	>3	0–3	4–8	Yes	Yes
Male/Female	Married	>2	up to 110	>3	0–3	8–18	Yes	Yes

Table 2 represents the result for various categories of users those wish to adopt a pet, and their validations of their search criteria are displayed with the help of a Web application.

The above-mentioned literature survey and tables explain the need for automating the process of adoption and the criteria for adoption depending upon adopter and

Table 2 Pet search table

Gender	Age (in years)	Family income (in lakhs)	Total children in family	Total no. of already owned pets	Can adopt (cat/dog)
Male/Female	<18	N/A	N/A	N/A	None
Male/Female	>18	<3	≤3	0	None
Male/Female	>18	>3	≥3	≥1	None
Male/Female	>18	>3	≤3	≥1	Both
Male/Female	>18	>3	≤3	0	Both
Male/Female	>18	<3	≤3	0	None
Male/Female	>18	<3	≥3	≥1	None

adoptee's ages. The data contributes a clearer understanding of how a Web application related to adoption will help the users to decide their level of qualification when deciding to adopt online and approach the respective organization.

4 Limitations

The application cannot completely digitalize the adoption process, as the adopters have to file the petition for adoption in court and also be physically present for the court hearing. Also, the organization has to conduct counseling sessions of parents as per the rules and conduct follow-up for the well-being of the child or the pet. The AWS server used also has incurred costs. The organization needs to update data for each kid/pet they have in the foster home. The organization needs to put a one-time effort to make all the adoptee's detail available digitally.

- Dependent on the server.
- The effort to make data available digitally.
- Physical availability in the courtroom is required.
- Incurring cost for scaling up.

5 Conclusion

The main aim of the application is to provide the foster homes a better data management facility and to simplify the adopters' search. The proposed application aims to digitalize the search of the adopters for adaptability and also to scrape off the paperwork of organizations for user verification. The user details and documents are verified according to government organization provided eligibility criteria's. The application provides personalized results to verified users. The users that are not able to take care of their pets can also get help from the nearest organization. The user experience is improved by GPS facility to provide the nearest suitable organization and to facilitate the user to the organization if he/she wishes to. The organization data and user data are secured by providing additional encryption to data while sending the data on the server. The application provides reminders to the parents/adopters as well as the organization about the visits for post-adoption checks until the adulthood of the child/pet. The application thus simplifies most of the steps for adoption and performs most of the activities digitally.

References

1. Liu, H., Meng, X.: JSP-based pet adoption system (2019). https://doi.org/10.1109/ICVRIS.2019.00064

2. Kamala, B., Rakshana, B., Reshma, B., Deivasree, B.: Patronization for orphans and adoption using android application. https://doi.org/10.1109/ICCCT2.2019.8824946.3
3. Lakmal, C., Dangalla, S., Herath, C., Wickramarathna, C., Dias, G., Fernando, S. IDStack—the common protocol for document verification built on digital signatures. https://doi.org/10.1109/NITC.2017.8285654
4. Mohammad, O.A., Hameed, R.T., Tapus, N. Design and implementation of real time tracking system based on Arduino Intel Galileo. https://doi.org/10.1109/ECAI.2016.7861114
5. Prakash, G.L., Prateek, M., Singh, I.: Data encryption and decryption algorithms using key rotations for data security in cloud system. https://doi.org/10.1109/ICSPCT.2014.6884895

5G Security and Privacy Issues: A Perspective View

Tahamina Yesmin, Sarmistha Agasti, and Kisalaya Chakrabarti

Abstract Subsequent to tremendous prominence and development of cell innovation, 5G network is being considered as a focal area for future correspondence design. 5G will be supporting a great many existing gadgets without satisfactory inherent security, just as new gadgets whose outrageous processing force will make them appealing focuses for programmers. 5G will relate normal structures that will require greater security to undertake well-being of the basic foundation as well as security of the general public all in all. Here, a portion of the identified security problems of 5G networks is being discussed. The majority of the data depends on writing survey and reports distributed by unmistakable media transmission giants and different books part. Subsequently, this article edifies its per users on what is anticipated from 5G network and what are the fundamental security and protection issues in accomplishing those objectives.

1 Introduction

The first generation (1G) flexible networks had been willing to the complexity of illegitimate interferences, cloning, and disguising [1]. Message spamming for inevitable attacks, concoction of bogus records, or broadcasting unwelcome statistics were given basic in the second generation (2G) of versatile networks. The most important intermission was indoctrinated within the third generation (3G) of flexible networks in which IP-primarily based correspondence legal the movement of Internet safety weaknesses and difficulties into transportable networks. The security hazard scene got moreover broadened and convoluted in the fourth-generation (4G) transportable networks with the increased utilization of IP-based totally correspondence essential for new devices and new administrations [2]. The union of substantial variety of IoT devices and the arrangement of new administrations, for example, for devoted houses, medical clinics, delivery, and electric matrix frameworks in 5G will furthermore stimulate the security demanding situations.

T. Yesmin (✉) · S. Agasti · K. Chakrabarti
Haldia Institute of Management, Haldia 721657, India

T. Senjyu et al. (eds.), *ICT with Intelligent Applications*, Smart Innovation, Systems and Technologies 248, https://doi.org/10.1007/978-981-16-4177-0_12

The security arrangements and designs applied in past ages (for instance 3G and 4G), obviously, might not do the trick for 5G. The customary explanation behind new security preparations and engineering are the elements of recent administrations and innovations in 5G [3]. For delineation, virtualization and multioccupancy where in abrupt and conceivably conflicting, organizations extent an equal adaptable organization premise had been only one out of every odd day some time ago. The inertness necessities, as an example, corroboration inactivity in vehicular correspondence or "Unmanned Aerial Vehicles" (UAVs), were not a lot imperative. Safely, the security systems of the beyond a while arise quickly on the refinement expected to make sure about 5G networks. The Next Generation Mobile Networks (NGMN) support that 5G necessities to supply more than take off via sonar and radio help security, which turned into a normal in 4G and prior some time of cell organizations.

5G will elucidate the obscurities instigated from the genuinely precise presentation provisions in an intensive scope of broadened situations, in which various "Key performance Indicator" (KPI) customer achieved transmission price, freight volume density, idleness, vitality productivity, and affiliation thickness may be testing cases. 4 time-honored specialized conditions of fifth-generation networks are obtained from the essential usage situations, management requirements, and key difficulties of transportable Internet and net of factors: high-restrict hassle region scenario, low-strength big association's situation, consistent wide-zone inclusion situation, and low-inactivity high-dependability situation [5, 6].

The transition to 5G is to a superior extent of a structural move than an advancement of current innovation. 5G guarantees an assortment of enhancements over the 3G and 4G cell administrations with fundamentally higher data transmissions and lower inactivity (delay). These capacities will reinforce extended applications going from HD video spilling to telemedicine to self-driving vehicles and brilliant urban communities under the "Web of Things" rubric. Data transfer speeds of up to 20 Gbit/s are plausible, utilizing millimeter wave recurrence groups (at 28 GHz and up to 60 GHz) and multiple input multiple output (MIMO) pillar shaping radio wires strategies in the center groups (3.5–4.2 GHz) and dependable spread in the 600 MHz band. Various transporters are arranging the sending of 5G innovations for both fixed and versatile applications, mostly superseding nearby fiber and Wi-Fi. Prearranging the reorganization of both remote and wired foundation, many perceive 5G as the new Internet.

5G will prior to the extended advancement of express multiplication of IoT, correlating billions of gadgets to billions of individuals. 5G's strong and solid information transport limit is on different occasions speedier than 4G and will pass on monstrous proportions of information where region, character, and individual information spillage transforms into the new security challenge. In this manner, it is very significant to portray every security threat that are negotiated not only just because of the remote idea of versatile networks but where prerequisites meet in every possible progression which are outstandingly significant for 5G networks.

Fifth generation of networks will make use of a large number of portions than previous generations of distant networks, and the extension of 5G framework may give malicious entertainers more offensive direction. The viability of 5G's security improvements will lead to a limited stretch that depends on legitimate escalation and disposition. In any case of security improvement over past decades, it is very uncertain what new weaknesses might be found in 5G networks. Additionally, 5G expands upon past decades of remote networks and will at first be integrated into 4G Long Term Evolution (LTE) networks that contain some delicateness endowment.

The 5G remote networks will give exceptionally more channel capacity and excessive inclusion with all in all improved QoS and incredibly low inertness [7]. With extremely intense uses of base stations, 5G will endow with especially steadfast and inexpensive broadband access everywhere not only to cellular hand-held devices, but also to a huge number of newly launched devices to machine-to-machine communication (M2M), Internet of things (IoT), and cyber-physical systems [8]. Similar improvement infers that 5G is certainly not an insignificant gradual progression of 4G as one would naturally think, however a reconciliation of new complicated modification to fulfill the regular constraint of traffic subsists with upcoming Internet of things gadgets [9]. Along every one of these capacities, 5G will interface almost all parts of the human life to compatibility networks, and this underscores the requirement for strong security systems over all network particles of the 5G. Distant networks have especially been a remarkable objective for most security deficiency from the very commencement.

This review article contemplates the cutting edge various security challenges.

2 A Short Review of Literature

Afrin et al. [10] investigated the test of planning diversified emphasis from remote radio heads to baseband processing units while reasoning about the two standpoints, i.e., end-client quality of experience (QoE) and specialist organizations' benefit. In the suggested jobs, a multifaceted devious programming ordering is created, along center around the boost of QoE and advantageous sentient asset designation. Two estimating realistic purpose calculations, named first fit fulfillment and first fit benefit calculations, are produced for the ideal compromise in the middle of QoE and benefit. The reproduction outcomes did on Cloud Sim SDN, reveal the demonstration upgrades as far as client QoE, QoS fulfillment, normal holding up time, and specialist organization benefit.

Sodhro et al. [11] basically analyze the two most important technology-based features that are Fog processing and 5G-based IoT technology. Here author featured the weakness of Fog registering as far as less accumulated restriction and more power medium. In contrast to additional interconnected writing, the researches nature of administration space of correspondences among Internet of things and Fog registering. The author suggested the Transmission Power Control (TPC)-based QoS advancement calculation (QoS-TPC). The suggested Fog-IoT-TPC-QoS design thinks about the unvarying and vigorous conduct of the medium. Significantly, a correlation with the traditional Automatic Transmit Power Control (ATPC), Team Game (TG), and steady TPC strategies is appeared in the suggested work. Be that as it may, the suggested work conveys unquestionable constraints as well as necessity of most getting indication in power measure qualities, high postponement, and jitter esteems [12, 13], and Davies et al. [14] convey the reduction of security risk is a continuous concern. Countless familiar procedure subsists, i.e., utilization of encryption. Here, author represents a diagram hypothesis way to deal with anticipates attend in sending data at the same time with various methods. This is a decent system in packed city conditions, in light of the fact that different ways are probably going to exist, numerous gadgets are probably going to be having low power handling capacities, and possible spies are various. It is tested and thought about three diverse multiway transfer procedures for their protection from spies. This is practiced by running a reenactment utilizing Davies' "City Section Mobility Model" a reflection of how individuals and gadgets move in the city.

One US transporter conveyed a fixed 5G network to four US urban areas in October 2018, which will have the advantages of 5G flags yet will not give portable access to a 5G network [15]. The home Internet help is a preguidelines organization and does not touch the universally perceived 5G quality, in spite of the fact that the organization says it will receive the business quality in 2019 although it turns out 5G versatile administrations for telephones. A second US transporter sent a versatile 5G network in twelve US urban communities in December 2018, even though restricted accessibility 5G portable hotspots were the main gadgets equipped for utilizing the 5G network; 5G proficient smartphones were relied upon to be accessible in 2019. Two extra US transporters were focusing on 2020 for full across the country 5G inclusion with constrained spread-out starting in 2019 [16, 17]. Many of these gadgets were empowered by a 5G modem created by a main US chip producer, which eighteen worldwide Original Equipment Manufacturer (OEM) organizations and nineteen remote network administrators had focused on utilizing in preliminaries [18].

3 How Fast 5G is (Diagrammatic Approach)

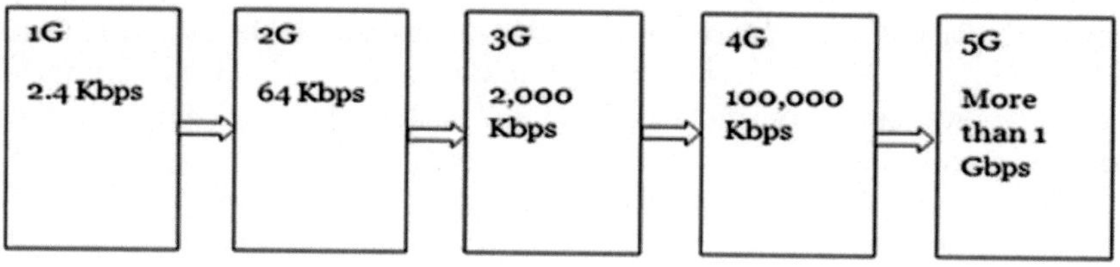

4 Key Security Challenges in 5G

4.1 5G There at Mitigation Controls: Distributed Denial of Services (DDoS) Attacks

Network cutting is the ability of the community to manifestly layout and run several legitimate networks as essentially unfastened commercial agency approach on an average physical foundation. Lowering, albeit scantily applied nowadays for enterprise use instances, is relied upon to be a maximum vital engineering a part of the 5G network, pleasant maximum of the 5G use times. In keeping with the 0.33 generation partnership venture particulars, the network reduce is a completed realistic community (giving telecommunication offerings and network skills) which includes get proper get right of entry to access network (AN) and core network (CN). Network diminishing engages the administration of various ordinary organizations concerning all expectations and capacities free business association approach on a middle substantial structure through and with the guide of, this looks at to the chance that the movable organization might be distributed pretty a couple of houses that might be virtual. Absolutely everyone is known as a "cut" that can be assigned for several purposes, for example, a cut may be allocated to a mobile versatile network operator (MVNO), an enterprise customer, an IoT vicinity, or a few other useful arrangements of administrations. A community cut expands the passage call get right of entry to component call idea utilized inside the versatile community these days.

4.2 IoT and DDoS Threat Mitigation

Network cutting is the restriction of the community to commonly plan and run numerous actual networks as essentially loose commercial enterprise gadget on a median real established order. Slicing, but scantily used today for corporation use instances, is depended upon to be a prime designing piece of the 5G network, pleasing the general public of the 5G use instances. As in line with the third era partnership undertaking factors of hobby, the community reduce is a finished affordable community (giving telecommunication services and network competencies) which includes get right of access network (AN) and core network (CN). Network cutting permits the company of diverse dependable networks with appreciate to all expectancies and purposes free enterprise system on a not unusual real shape. Before lengthy, this analyzes to the probability that the flexible community will be divided a ton of resources that is probably digital. All and sundry are known as a "cut" that can be allocated for different functions. As an example, a slice can be disbursed to a mobile versatile virtual network operator (MVNO), a venture consumer, an IoT area, or a few other supportive associations of companies. A community reduce expands the section call Access Point Name (APN) idea used within the versatile network today [19].

4.3 Insufficient Technology Readiness Level (TRL)

Sensitive records in non-secure physical devices regions ought to be scrambled and its respectability ensured. Devices should cryptographically take a look at software bundles at machine boot or replace, simply as keep up the capacity to get a long way off device refreshes even in the event of malware infection. Adequate capability needs to be accommodated programmed rollback in case of update sadness. Anyhow, pernicious rollback to extra set up programming forms that reintroduce vintage weaknesses should be forestalled. The need for security isolation between tool-resident packages is critical. One opportunity is to provide system primarily based confinement between programs. Despite the truth that this usefulness has frequently been given by way of dedicated equipment, it may likewise be mentioned with a Trusted Execution Environments (TEE). The TEE is secluded from the purchaser side execution situation and evaded to as Rich Execution Environment (REE) in like way processors.

4.4 Security Threats in 5G-Enabled IoT

1. IoT with 5G enabled carries plentiful advantages to the end clients. It additionally conveys some security and protection challenges. In spite of the basic customary security and protection issues, there additionally have some extraordinary security and security challenges in 5G-empowered IoT. A recent report to research the effect of Internet of things security on information technology and line-of-business (LoB) pioneers uncovered IT and LoB pioneers' worries concerning Internet of things security since assaults can altogether influence basic business activities. One alarming truth including Internet of things uncovered most of associations cannot give a total record of all their network-associated gadgets. Each new gadget that comes online speaks to another extension (another assault vector) of the general danger surface. In any event, for distinguished IoT elements, the possession, from a security perspective, much of the time stays dim, further aggravating the issue [20].
2. Safeguarding Internet of things should not to be an untimely idea. Internet of things security should be tended to at the structure stage, not included post-organization.
3. Various layers of securities include IoT like hardware, software, in-transit data, stockpiling, network, and applications. The significance and interaction between these layers are profoundly logical. Large IoT security configuration must consider this reality.

4. Security must be very strong at the vulnerable point in IoT. Noteworthy consideration is regularly paid toward making sure about a cell phone without recognizing what occurs inside the sprinkler control or vehicle key applications that live on it.
5. Complex IoT gadgets (for instance, modern hardware and associated vehicles) are the most troublesome IoT situations to make sure about. For instance, the outcomes of a hacked associated vehicle can be considerably more adverse contrasted with that of an associated electric meter or fridge.

4.5 Cloud Computing and Edge Paradigms

Cloud computing was the very basic piece of IoT in past decade of time. It utilizes some standard systems access to different configurable assets, for example, application, storage, networks, workers, and administrations. The first run through distributed computing referenced to the world is in a Compaq inward archive as ahead of schedule as 1996.

In 2006, Amazon presented the made Elastic Compute Cloud (EC2) which can boost Amazon Web Services. After two years, Google conveyed the Google App Engine. During 2010, there were different smarter computing establishment showed up after IBM discharged IBM Smart Cloud. In these more brilliant registering structures, distributed computing was the basic part. There were numerous administrations models for distributed computing; three models had been opened. Cloud computing has numerous focal points, for example, it empowers to lessen the cost, it very well may be negligible exertion of the deal with the framework, and the clients possibly need to pay when they need procedure and store data in a private or open cloud [21].

5 Privacy Challenges in 5G

From the client's point of view, the significant protection concerns could emerge from data, area and character [22]. Most personal digital assistant (PDA) applications require subtleties of supporter's very own data before the establishment. The application engineers or organizations once in a while notice that how the data is put away and for what purposes it will be utilized. Dangers, for example, semantic data assaults, timing assaults, and limit assaults for the most part focus on the area security of supporters [23].

On the physical layer degree, location safety can be spilled by way of using passage manner choice calculations in 5G bendy networks [24]. International mobile subscriber identity (IMSI) getting attacks can be applied to uncover the character of a supporter via getting the IMSI of the endorser's user equipment (UE). Such attacks can reason be conveyed about through placing in a fake base station it truly is considered as wanted base station through method of the UE and thusly allies will respond with their IMSI. Besides, 5G networks have different factors, for example, virtual mobile network operators (VMNOs), communication service providers (CSPs), and network framework suppliers. These on-screen characters have diverse wishes for safety and protection. The harmonization of befuddling safety procedures among those on-screen characters can be a take a look at in 5G networks [25]. Within the past ages, portable administrators had direct access and manipulate of all the framework parts. Anyhow, 5G versatile directors are dropping the entire control of the frameworks as they will depend upon new on-screen characters such communications provider carriers. On this way, 5G directors will lose the full management of safety and safety [26]. Customer and information protection are clearly tested in shared conditions wherein a similar foundation is shared among distinct on-screen characters, for instance virtual mobile networks operators (VMNOs) and one-of-a-kind contenders. Additionally, there are not any bodily limits of 5G community as they use cloud-based totally records stockpiling and network function virtualization (NFV) highlights. Therefore, the 5G administrators have no instant manipulate of the facts putting away spot in cloud conditions. As diverse international locations have various diplomas of records security contraptions counting on their desired putting, the safety is examined if the client statistics is positioned away in a cloud in an exchange state [27]. The incredibly rapid of 5G remote let an ever-increasing number of physical articles be the IoT gadgets, for example, microwave, robot vacuums, a material, even a little ring. This keen IoT gear fundamentally improves human's life quality. Interim, the across-the-board affectability individual data additionally will be gathered by these gadgets. Here and there, the associated gadgets ask end clients to include individual affectability data, for example, name, sex, age, postal division, e-mail address, and so on. Despite the fact that occasionally, the data simply communicated by a given endpoint, there still have security hazard. On the off chance that the aggressor integrates and gathers and uses propelled data mining calculations to investigate the divided data from different endpoint; the delicate individual data despite everything can be accomplished [28]. Here and there, the associated gadgets ask end clients to include individual affectability data, as for example, name, sex, age, postal division, e-mail address, and so on. Despite the fact that occasionally, the data simply sent by a given endpoint, there still have protection chance. It is summarized in Table 1.

6 Conclusions

From this paper, it has been featured the fundamental safety and confidentiality encounters that can turn out to be all the more undermining in 5G, except if appropriately tended to. We have likewise introduced the security systems. However, it may, because of the constrained independent and incorporated sending of these advances in 5G, the safety danger vectors cannot be completely acknowledged as of now. We have concentrated to the privacy and security aspects mainly. With the advancement of 5G remote and IoT gadget, disseminated-based registering is turning into a

Table 1 Security challenges for privacy of data in 5G technologies

Security threats	Target points/network elements	Privacy
Malware	User clicks on a malicious link or attachment	√
Boundary attacks	Subscriber's locations	√
Denial of service	Originates from a computer network	√
User identity thefts	User information data basses	√
Phishing	Use fake communication	√
SQL injection	Inserting malicious code into a server that uses SQL	√
International mobile subscriber Identity catching attacks	Subscriber identity	√

proficiency and conceivable innovation answer for handle the billions of the 5G-IoT gadgets. The new methods bring us different accommodation and high life quality. Be that as it may, they likewise produce new security dangers and various clients' private data in danger. We need to investigate different dangers and protection for security reason to show signs of improvement secure network association and innovation. Security is an ongoing theme among the utilization cases just as the vertical administrations and Internet of things to be utilized by the 5G cell principles.

References

1. Wey, .J.K., Chang, H.T., Sun, L.F., Yang, W.P.: Clone terminator: an authentication service for advanced mobile phone system. In: 1995 IEEE 45th Vehicular Technology Conference. Countdown to the Wireless Twenty-First Century, vol. 1, pp. 175–179 (1995)
2. Ahmad, I., Kumar, T., Liyanage, M., Okwuibe, J., Ylianttila, M., Gurtov, A.: 5G security: Analysis of threats and solutions. In: 2017 IEEE Conference on Standards for Communications and Networking, pp. 193–199
3. Arfaoui, P., Bisson, P., Blom, R., Borgaonkar, R., Englund, H., Flix, E., Klaedtke, F., Nakarmi, P.K., Nslund, M., et al.: A security architecture for 5G networks. IEEE Access **6** 22466–22479
4. Alliance, N.: NGMN 5G White Paper, Next Generation Mobile Networks, White paper (2015)
5. IMT-2020, P.G. 5G Concept. IMT-2020 (5G) Promotional Group, China. Volume 2, 2015. Available online: http://www.imt-2020.org.cn/en/documents/download/3. Accessed on 11 Oct 2018
6. Bajracharya, R., Shrestha, R., Zikria, Y.B., Kim, S.W.: LTE in the Unlicensed spectrum: a survey. IETE Tech. Rev. **35**, 78–90 (2018)
7. Agiwal, M., Roy, A., Saxena, N.: Next generation 5G wireless networks: a comprehensive survey. IEEE Commun. Surveys Tutor. **18**(3), 1617–1655 (thirdquarter 2016)
8. Kutscher, D.: It's the network: towards better security and transport performance in 5G. In: 2016 IEEE Conference on Computer Communications Workshops (INFOCOM WKSHPS), April 2016, pp. 656–661
9. Andrews, J.G., Buzzi, S., et al.: What will 5G be? IEEE J. Sel. Areas Commun. **32**(6), 1065–1082 (2014)

10. Afrin, M., Razzaque, M.A., Anjum, I., Hassan, M.M., Alamri, A.: Tradeoff between user quality-of-experience and service provider profit in 5G cloud radio access network. sustainability **9**, 2127 (2017)
11. Sodhro, A.H., Pirbhulal, S., Sangaiah, A.K., Lohano, S., Sodhro, G.H., Luo, Z.: 5G-based transmission power control mechanism in fog computing for internet of things devices. Sustainability **10**, 1258 (2018)
12. Li, S., Xu, L.D., Zhao, S.:5G internet of things: a survey. J. **10**, 1–9 (2018)
13. Mouradian, C.,Naboulsi, D.,Yangui, S., Glitho, Polakos, M.J.: A comprehensive survey on fog computing: state-of-the-art and research challenges. IEEE, **20**, 416–464 (2018)
14. Davies, V.A., et al.: Evaluating mobility models within an ad hoc network. Master's thesis, Citeseer (2000)
15. McGarry, C.: The truth about 5G: what's coming (and What's Not) in 2018, Tom's Guide. https://www.tomsguide.com/us/5g-release-date,review-5063.html. Accessed on June 5, 2019, Ibid.
16. AT &T.: AT&T first to make mobile 5G service live in the U.S. on Dec. 21. https://about.att.com/story/2018/att_brings_5g_service_to_us.html. Accessed 5 June 2019.
17. Cheng, R.: Sprint targets first mobile 5G nationwide network by early 2019. https://www.cnet.com/news/sprint-targets-first-mobile-5g-nationwide-network-by-early-2019/. Accessed on 5 June, 2019
18. Michaels, P.: Qualcomm's blazing 5G modem coming to phones in 2019. https://www.tomsguide.com/us/qualcomm-x50-5g-modem,news-26584.html. Accessed on 5 June, 2019
19. The Evolution of Security in 5G- 5G Americas White Paper
20. IoT and OT Security Research Exposes Hidden Business Challenges, Forrester Consulting report commissioned by Forescout Technologies, Inc. 2017. https://www.forescout.com/iot_forrester_study/
21. Peter, M., Tim, M., et al.:Thenist definition of cloud computing (2011)
22. Liyanage, K.M., Braeken, A., Ahmad, I., Ylianttila, M.: From gadget to gadget-free hyper connected world: conceptual analysis of user privacy challenges. In: 2017 European Conference on Networks and Communications (EuCNC), pp. 1–6 (June 2017)
23. Wang, P., Move, O.A., Liu, Y., et al.: A location cloaking algorithm based on combinatorial optimization for location- based services in 5G networks. IEEE Access **4**, 6515–6527 (2016)
24. Farhang, S., Hayel, Y., Zhu, Q.: PHY-layer location privacy preserving access point selection mechanism in next-generation wireless networks. In: 2015 IEEE Conference on Communications and Network Security (CNS) (Sept 2015), pp. 263–271.
25. Yan, Z., Zhang, P., Vasilakos, A.V.: A security and trust framework for virtualized networks and software-defined networking. Secur. Commun. Netw. **9**(16), 3059–3069 (2016). sCN-14-0760.R1. [Online]. Available: https://doi.org/10.1002/sec.1243
26. Sorensen, L.T., Khajuria, S., Skouby, K.E.: 5G visions of user privacy. In: 2015 IEEE 81st vehicular technology conference (VTC Spring), May 2015, pp. 1–4
27. Kemmer, F., Reich, C., Knahl, M., Clarke, N.: Software defined privacy. In: 2016 IEEE International Conference on Cloud Engineering Workshop (IC2EW), April 2016, pp. 25–29
28. Bree Fowler. Gifts That Snoop? The Internet of Things Is Wrapped in Privacy Concerns, gifts-that-snoop-internet-of-things-privacy-concerns/ (2017)

Design and Development of an Algorithm to Secure Big Data on Cloud Computing

Ankur N. Shah and Jay A. Dave

Abstract In any organization, data is coming from various online sources and various offline sources. It is very difficult to equipment huge amount of data. Big data is one of the technologies who tackle huge amount of data easily. Cloud computing is a platform for big data. Although there are vast number of advantages for big data and cloud computing, they failed to make their place in people heart as people are anxious about security of their data. They are huge number of advantages we gain if we move big data on cloud computing like on-demand service availability, availability of data and information on Internet, resources grouping, easy to manage, easy to analyze large volume of big data, and most important is cost effective. Although there are numerous advantages to moving big data on cloud computing, we cannot avoid challenges to big data and cloud computing. The various challenges to big data are cost, data quality, rapid change, skilled man power requirement, infrastructure need, moving data onto big data platform, need for synchronization across data sources, and most important is security of data. These all challenges are not going to be easily solved; various researches are going on to solve all these problems. In this paper, we provide review about various research done in the area of security of big data on cloud computing, and finally, we provide research plan for our project regarding security of big data on cloud computing.

1 Big Data

1.1 Introduction

As the name, big data, it is having huge amounts of data which in TeraBytes (TB) or PetaBytes (PB) or ExaBytes (EB) or more. Data also may be structured form or semi-structured form or it is in unstructured form [1–3].

A. N. Shah (✉)
Indus University, Ahmedabad, India

J. A. Dave
Sankalchand Patel University, Visnagar, India

T. Senjyu et al. (eds.), *ICT with Intelligent Applications*, Smart Innovation, Systems and Technologies 248, https://doi.org/10.1007/978-981-16-4177-0_13

1.2 Vs of Big Data

Various Vs are: volume, variability, velocity, veracity, variety, value, validity, etc. [1, 4].

1.3 Advantages of Big Data

Various advantages of big data are cost, time, new product development, understand the market, fraud detection, better decision, faster decision, etc.

1.4 Challenges to Big Data

Various challenges to big data are cost, data quality, skill men power requirement, hasty change, need of infrastructure, moving data to big data platform, need to synchronization across various data sources, and most important is data security.

We organize this paper as follows: Sect. 2 presents basics about cloud computing. Section 3 presents literature survey about various papers. Section 4 is about proposed method. Section 5 presents algorithm for proposed method. Section 6 presents result and discussion. Section 7 presents the final conclusion.

2 Cloud Computing

2.1 Introduction

Cloud computing is the technology which offers various services like storage, database, networking, software, etc. The provider will take charges as per usage such as we are paying gas pipeline bill at our home whatever we use we are paying for that [1, 5].

2.2 Characteristics

Cloud computing has various important characteristics like resource pooling, on-demand capabilities, rapid elasticity broad network access and measured service, etc. All these make it different from other computing standards [1, 6].

2.3 *Advantages of Cloud Computing*

Various advantages of cloud computing are like easy to access, integrity, scalability, improved storage capacity, and most important is cost.

2.4 *Challenges in Cloud Computing*

Various challenges to cloud computing are like service provider trustworthiness, portability, computing performance, service quality, availability, and most important is data security [7].

3 Literature Survey

From literature survey (Table 1), we come to know that there are various methods already developed to secure big data but still this is not an end and need more techniques to secure big data efficiently. For our proposed method, we used AES algorithm because it is not easy to attacked it due to length of key used it 128, 192, or 256 bits even we are going to use 512 bits [23]. AES algorithm is faster than RSA [24]. So we decide to use AES algorithm in our proposed method as discussed in next section.

4 Proposed Method

As we know health records are very huge in size and it also contains personal information about persons and their health issues. This information is sensitive and embarrassing sometimes. So it is necessary to protect this kind of data. In our proposed system, we are planning to work on health records for these we have gathered real-time health records from TB hospital, Surendranagar which is of size about 2 GB. We will plan to secure it using our proposed method and I am very much thankful to the TB hospital for all kind their support.

Following Figs. 1 and 2 shows various steps of our proposed method to provide security to big data. As shown in Fig. 1, first we take big data as input where data can come from online shopping, sensors, smart phones, WhatsApp, YouTube, customer feedback, Facebook, Twitter, etc. Our next step is to convert it into semi-structure format. The data which have fix format for input and output is known as structured data. The data which have no format for input and output is known as unstructured data. The data which lives between structured data and unstructured data is known as semi-structured data.

Table 1 Literature survey

Reference no.	Work done	Further work
[8]	A SSGK is proposed by authors in this paper. SSGK means secret sharing group key management protocol. The shared data and communication is protected from unauthorized access by SSGK	Dynamic mechanism of group members that can be efficiently deployed can be implemented in future
[9]	PFP algorithm is proposed by this paper. It is time efficient, good privacy protection, and better utility	If we talk about large database then it is not feasible to grasp FP-tree in main memory. This can be solved by partitioning FP-tree into smaller database and construct FP-tree for all these
[10]	This paper gives idea about parallel and distributed mining to overcome difficulty that arises to extract pattern from large-scale data	Further work can be done on discussed challenges of parallel and distributed mining
[11]	The issues of quality of data at same level and the issues of security of data at the same level are focused inside this paper. The context of big data imposed the challenges and cause problems for quality and security	In future to assess and improve big data quality we can implement it via secure process
[12]	Firstly authors introduce big data security technique. Then illustrate founded on big data classification technique. Here using classification technique they recognize file and confidential make it secure previous to any data transmission happens over cloud nodes	An adaptive method for incorporating security constraints need to be considered during classification of big data in developing and designing cloud computing systems. For classification authors will go to consider image, video and audio files. It required a special treatment because of its nature is different from txt, csv, log, xls, and sql files. Therefore, new techniques are needed so that it will be able to deal with file attributes for such big data types

(continued)

Table 1 (continued)

Reference no.	Work done	Further work
[13]	The complexity of different network relevance on original data like irregularity detection can be reduced using proposed network traffic summarization method. Basic sampling is mainly used in this method. To find best summary from data summarization is always a problem. The quality of summary is affected by the size of summary. It is necessary to create summary for underlying data patterns which are reflected. The anomaly recognition with less time required when we use summary of data as a replacement for using original dataset that is also shown here	To detect collective anomaly is difficult future work can be done in that direction
[14]	This review paper gives idea about cloud services and deployment models, cloud security classification, security necessities and cloud architectural components with vulnerabilities	As per author point of view, flexible security method is required to develop in future
[15]	In this paper, author uses the methodology that takes the advantage of log files and tries to improve to increase the level of security and anomaly detection. In this approach, author used parallel and distributing processing using MapReduce for detection of SQL Injection and DDOS attacks	In future, suitable learning techniques can be applied to the machine logs by spotlighting machine learning part
[16]	Privacy preserving association rule mining algorithms are proposed in this paper. It does not allocate all privacy values by using set of rules and mainly premeditated for distributed servers. This method mainly secures privacy and it also looks for time do not increases for doing the same	Further work can be done by comparing with other tool and algorithms

(continued)

Table 1 (continued)

Reference no.	Work done	Further work
[17]	In this paper, author first discusses recent issues of big data on cloud and challenges for big data on cloud and then author proposed a model. Two algorithms were presented for cloud environment for detection of the external and internal hackers	Future work can improve to guard the insightful data and concept of honeypots can also be used
[18]	This paper describes various security issues in communication, computational and service level agreements are explored	As per author suggestion, more work need to be done for security at service level agreement
[19]	In this paper, author proposed a security method to find insider attacks fast with low overhead for big data. The method works in two different steps like attack detection and safe and sound communication protocol	For future work, hardware architecture of security chips can be prepared that can separately maintain this system
[20]	Here authors used big data-based security analytics (BDSA) approach to defending virtualized infrastructures in cloud computing beside superior attacks	In future work, this can be compared with other approaches as well
[21]	In this paper, authors first propose an improved NTRU cryptosystem. Also, present a secure and verifiable access control scheme to protect big data in cloud based on the improved NTRU	For further work, we can combine the (t; n) threshold in present scheme
[22]	For intrusion detection problems, the paper discusses and applies the association rules and the Apriori algorithm. To improve security with rapid development of Internet using data mining algorithm along with network security issues can be reduced	In future, this improved method can be compared with other one

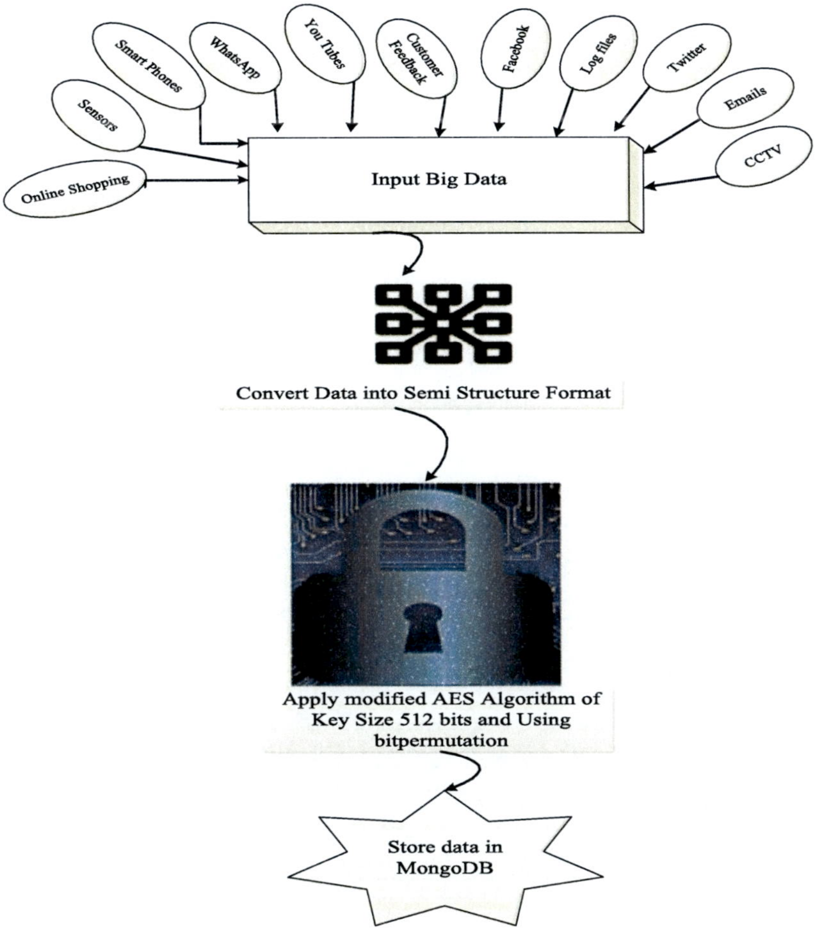

Fig. 1 Steps of proposed system

After that we apply modified AES algorithm as shown Fig. 2 on them for security purpose. In modified AES algorithm, we changed Mix Column Round with Bit Permutation so that it can reduce time execution specially when dealing with big data. Modified AES algorithm also used 512-bit keys [25] for more security purpose instead of 128/192/256-bit keys. Modified AES algorithm also used bit permutation instead of mix column to reduce complexity [26]. The various rounds of modified AES algorithm for encryption and decryption process are shown in Fig. 2.

In the final step, we store result in MongoDB. As shown in proposed method, we are going to apply security to data at application layer, i.e., SaaS layer of cloud computing.

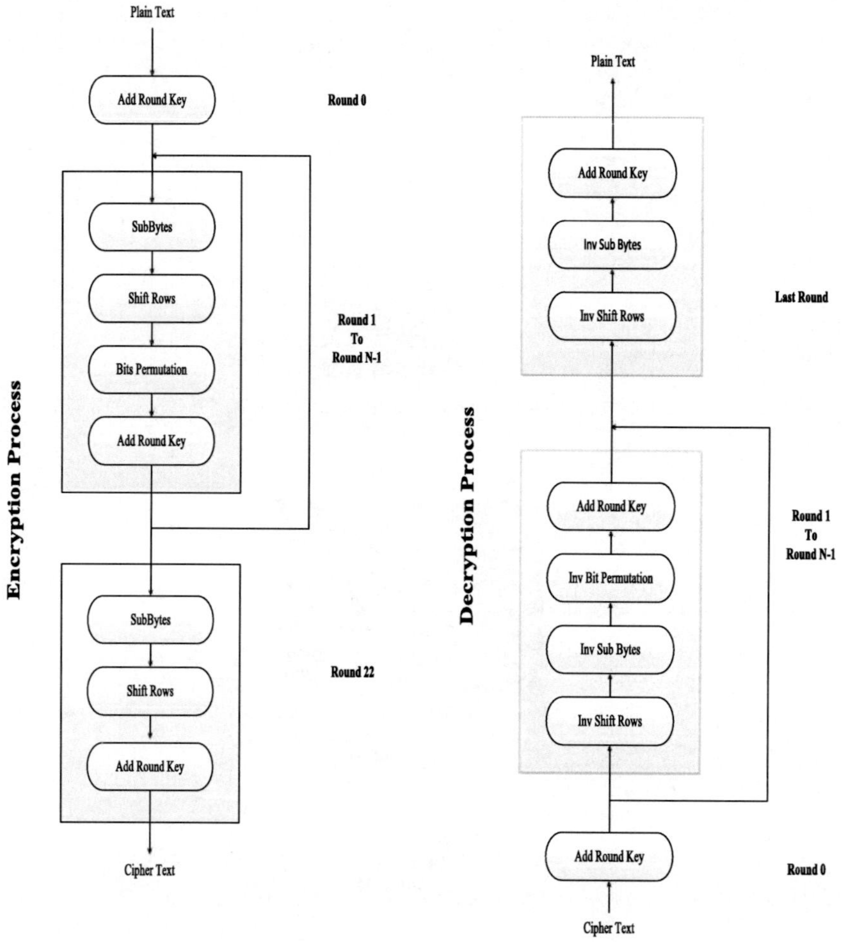

Fig. 2 Steps of encryption and decryption process for modified AES algorithm

5 Algorithm for Proposed System

Following are the various steps of proposed system for encryption process.

Step 1: Input big data.

Step 2: Convert data into semi-structured format.

Step 3: Apply modified AES algorithm: Encryption: for round 0, add round key to plain text.

Step 4: for 1 to N-1 perform below step 5 to step 8.

Step 5: Perform subbytes.

Step 6: Perform shift rows.

Step 7: Perform bit permutation.

Step 8: Perform add round key.
Step 9: For round 22, perform subbytes.
Step 10: Perform shift rows.
Step 11: Perform add round key and produce ciphertext.
Step 12: Store data in MongoDB.

As shown in the algorithm first we take big data as input then we process on them and convert it into semi-structured format. After that we apply the modified AES algorithm and perform encryption process where we used 22 rounds for 512 bits key as well as we used bit permutation instead of mix column operation in original AES algorithm because of this we can achieve time efficiency specially for big data. Final step is to store them on MongoDB.

Following are various steps of proposed system for decryption process.
Step-1: Take require data from MongoDB.
Step-2: Apply modified AES algorithm: Decryption: for round 0, add round key.
Step-3: for 1 to N-1 perform step-4 to step-7.
Step-4: Perform inverse shift rows.
Step-5: Perform inverse subbytes.
Step-6: Perform inverse bit permutation.
Step-7: Perform add round key.
Step-8: Perform last round, perform inverse shift rows.
Step-9: Perform inverse subbytes.
Step-10: Perform add round key and get required data.

As shown in algorithm first, we take required data from MongoDB. Then we apply modified AES algorithm for decryption process. For round 0, we add the round key. For round 1 to N-1, we perform inverse shift rows, inverse subbytes, and inverse bit permutation and add round key. For last round again we inverse shift rows, inverse subbytes, and add round key. Finally we get required data.

6 Results and Discussion

In future, we are planning to implement the proposed method, and we will compare with original method to check that result of proposed method is more secure to store big data on cloud computing.

7 Conclusion

In today's era, big data and cloud computing both are not just in demand they are essential as well. Various research has been already done to secure big data on cloud computing and various research still going on. But still people and organizations are in waiting for technique which can provide security to big data on cloud computing, and they can feel relax about their data. In this paper, we provide review to various works done in the direction to provide security to big data as well as we provide steps of our proposed approach. In future, we are planning to implement the proposed method, and we will compare with original method to check that result of proposed method is more secure to store big data on cloud computing.

References

1. Teng, L., Li, H., Yin, S., Sun, Y.: A Modified Advance Encryption Standard for Data Security. Int. J. Netw. Secur. 1–6 (2019)
2. https://www.geeksforgeeks.org/world-big-data/. Last Accessed on 11 Oct 2020
3. https://www.sqlshack.com/azure-data-explorer-for-beginners/. Last Accessed on 11 Oct 2020
4. Shah, A., Jay D.: Classification and security enforcement of the semi-structured health records on MongoDB leveraging hybridization of Modified AES algorithm with Apriory algorithm. Juni Khyat **10**(10), 91–95 (2020)
5. Jay, D.: An optimised robust model for big data security on the cloud environment: the numerous user-level data compaction, pp. 209–216. Springer (2020)
6. Golani, S., Sahal, S. Sonika, Vaishali: Technical survey on cloud computing and its security issues. IJERnD **2**(3), 11-14 (2019)
7. Dave, J., Vyas, D.: Multi-level data security model for big data on public cloud: a new model. IJANA **09**(06), 3654–3657 (2018)
8. Han, S., Han, K., Zhang, S.: A data sharing protocol to minimize security and privacy risks of cloud storage in big data era, pp. 60290–60298. IEEE (2019).
9. Kavitha, M., Kowsalya, K.: Development of big data security in frequent itemset using FP-Growth algorithm, pp. 120–123. IJERT (2019)
10. Kumar, S., Mohbey, K.K.: A review on big data based parallel and distributed approaches of pattern mining. Elsevier (2019)
11. Talha, M., Abou El Kalam, A., Elmarzoqui, N.: Big data: trade-off between data quality and data security, pp. 916–922. Elsevier (2019)
12. Hababeh, I., Gharaibeh, A., Nofal, S., Khalil, I.: An integrated methodology for big data classification and security for improving cloud systems data mobility, pp. 9153–9163. IEEE (2019)
13. Ahmed, M.: Intelligent big data summarization for rare anomaly detection, pp. 68669–68677. IEEE (2019)
14. Kumar, R., Goyal, R.: On cloud security requirements, threats, vulnerabilities and countermeasures: a survey, pp. 1–48. Elsevier (2019)
15. Azizi, Y., Azizi, M., Elboukhari, M.: Log files analysis using map reduce to improve security, pp. 37–44. Elsevier (2019)
16. Huang, K., Zhou, D., Xu, W., Cong, Y., Wang, W., Que, Z.: Distributed apriori on cloud plateform without exposing individual transaction data, pp. 397–401. IEEE (2018)
17. Reddy, Y.: Big data security in cloud environment, pp. 100–106. IEEE (2018)
18. Subramanian, N., Jeyaraj, A.: Recent security challenges in cloud computing, pp. 28–42. Elsevier (2018)
19. Aditham, S., Ranganathan, N.: A system architecture for the detection of insider attacks in big data systems, pp. 1–14. IEEE (2017)
20. Win, T.Y., Tianfield, H., Mair, Q.: Big data based security analytics for protecting virtualized infrastructures in cloud computing, pp. 1–15. IEEE (2016)
21. Hu, C., Li, W., Cheng, X., Yu, J., Wang, S., Bie, R.: A secure and verifiable access control scheme for big data storage in clouds, pp. 1–14. IEEE (2016)
22. Hanguang, L., Yu, N.: Intrusion detection technology research based on apriori algorithm, pp. 1615–1620. Elsevier (2012)

23. https://www.comparitech.com/blog/information-security/what-is-aes-encryption/. Last Accessed on 11 Oct 2020
24. Shehadeh, H., Obeidat, Q.Y., Darwish, D.: Comprehensive study on data security in cloud data store. ICSRS **07**(04), 53–66 (2014)
25. Anand, K., Sekar, A.C., Nagappan, G.: Enhanced AES algorithm using 512 bit key implementation. AIJSH **7**(2), 353–366 (2017)
26. Gamido, H.V., Sison, A.M., Medina, R.P.: Implementation of modified AES as image encryption scheme. IJEEI **6**(3), 301–308 (2018)

Curiosity-Driven Dynamic Reconfiguration of Swarm Robots by Using Simple Behaviours

K. R. Shylaja and Darryl N. Davis

Abstract This research paper focuses on designing evolving swarm robot groups that actively learn to optimize the goal achievement. The swarm groups mimic social insect swarm and can adopt optimal behaviours over a period of time by learning the effect of behaviours on their internal and external goal states. These swarms are made to dynamically reconfigure themselves into different topologies based on their capabilities and their desired goals. The collective behaviour of swarms can be driven by special cognitive states that are attained by individual robots at microscopic levels, such as curiosity or boredom. This paper discusses an algorithm for macroscopic swarm design that enables a robot at microscopic level to adopt a new behaviour. The curious cognitive state is exploited to explore and learn new approaches to improve the speed of convergence. This study can lead the future artificial swarms to evolve from naïve behaviours to self-learn complex behaviours with continuous evolution by adopting better learning algorithms and eliminate slowly the behaviours that are suboptimal.

K. R. Shylaja (✉)
Ambedkar Institute of Technology, Bangalore 560056, India
e-mail: Shylaja.kr@dr-ait.org

D. N. Davis
University of Hull, Hull, UK

ICE (Institute for Cognitive Engineering), Academie Musicale, Milan, Italy

T. Senjyu et al. (eds.), *ICT with Intelligent Applications*, Smart Innovation, Systems and Technologies 248, https://doi.org/10.1007/978-981-16-4177-0_14

1 Introduction

1.1 Background

The evolution of life on earth has always been an inspiration for researchers to come up with new inventions that have drawn global attention in the recent era. Artificial life or Alife has been a thrust research area in AI that studies either functioning of biological systems using computational methods or new computational techniques evolved to reflect some biological system(s) behaviours [1–3]. Swarm Intelligence is one of the area of focus in AI to study the social behaviour of a colony of insects and in some small animals and insects such as ants, termites, wasps, birds and fish. The SI concept was first coined by Beni, Hackwood and Wang [4] in the context of cellular robotics systems.

A tiny social insect that has simple internal structure, with a set of simple activities and limited local communication, can collaborate with hundreds of its peers to demonstrate seamlessly amazing, emerging collective behaviours. For example, ant colony optimization [20], artificial fish swarm algorithm [12], artificial bee colony [13], glow worm swarm optimization [17] and bat algorithm [28] were inspired by the swarm intelligence of social animals. In these insect colonies, most of the tasks are executed by worker insects that are specialized in a set of tasks, according to their morphology, age and chance [20]. For example, in the Pheidole species of ants, workers are divided into minor class and major class based on morphological developments. The minors feed the brood and clean the nest, whereas the majors defend the nest and cut big preys and transport them to the nest. If anytime the minors are removed from work, the majors get stimulated to perform and complete the task. However, most of the aspects of collective behaviours are self-organized, and the actions and interactions at a microscopic level among insects will self-emerge some patterns at the macroscopic level in the colonies [7].

Particle swarm optimization (PSO) is a biologically inspired computational method to solve search and exploration optimization problems [14]. PSO method has greater influence from a study of behaviour of insects in colonies that converge to a better source of food by randomly generating candidate solutions. PSO has been successfully adopted for wide range of applications to solve problems of searching and optimizations [23]. Kennedy and Eberhart [14, 16] introduced the concept of using PSO in optimizing nonlinear equations [15, 25].

1.2 Computational Traits of Social Insects for Artificial Swarm Groups

Self-organization, the main trait of all social insect colonies, is a set of dynamic processes that harnesses emerging behaviours at a global level of swarm colonies through direct or indirect communication among peers. These insects with limited local sensing and communication are capable of dynamically switching from behaviours that are not immediately helpful to those that benefit the attainment of desired states. In communicative swarms, the insects directly communicate through antennas to physically exchange the information, whereas in the indirect method of communication (or "stigmergy"), the insects make changes to the external environment, and the other insects respond to these changes. The stigmergy acts like a feedback system for swarms to emerge with self-organized behaviour [7]. Swarm intelligence is one such emergent outcome of collective, decentralized and self-organized behaviours in insect colonies in a natural way [4].

In every biological system, the behaviours are correlated to the sensing and acting abilities of the individual, and the actions are driven by the special cognitive states of animals that are contextual [18]. This class of behaviour fit to the left of the niche spaces shown in Fig. 1.

Swarm robotics is a novel approach of designing an artificial swarm of robots or a colony of robots. According to Sahin [24], "Swarm robotics is the study of how large number of relatively simple physically embodied agents can be designed such that a desired collective behaviour emerges from the local interactions among agents and between the agents and the environment."

The modelling of such artificial self-organized swarm systems requires the study of a mechanism that generates collective behaviours that are flexible, adaptive and robust with a decentralized control [7].

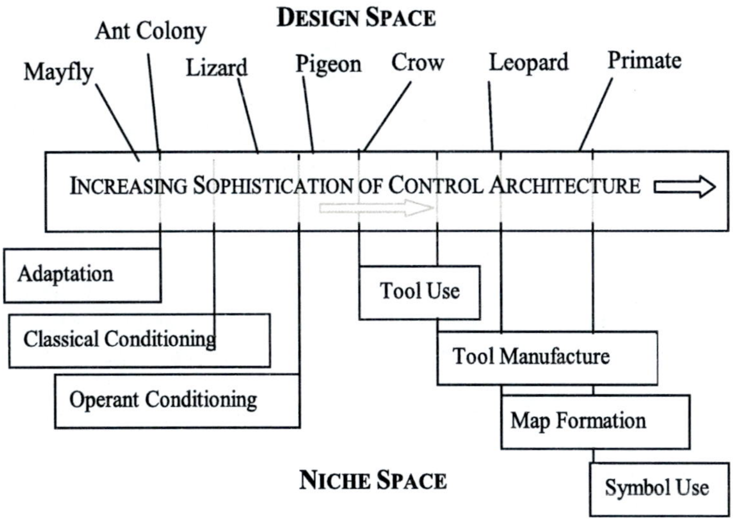

Fig. 1 Niche spaces of increasing sophistication with associated design space are examples of biological control architectures

1.3 Characteristics of Artificial Swarms

The robots in artificial swarm groups that are mounted with small number of sensors and motors are expected to exhibit the following characteristics:

- The individual robot should perform an independent activity autonomously.
- The robots must be situated in an environment by interacting with the environment and perceiving their internal and external needs.
- The robots must have a mechanism built on them for local sensing and local communication.
- The robots in swarm have a decentralized control, where they are not controlled or guided by any other entity or robot in the group and must be able to work in different group size.
- The robots have no global knowledge of evolving macroscopic behaviour or pattern of the group.
- Cooperation among robots to achieve the desired goals at swarm level must be exhibited, and they must be able to adapt to different work roles in loss of an existing member.

1.4 Cognitive Drives for Designing Artificial Swarms

Curiosity is a psychological construct that lends itself to exploratory behaviour in social animals [5, 6]. The learning and exploring behaviours in biological systems is generally driven by curiosity, a motivated desired cognitive state. Curiosity arouses when there is diminished coherence between the ongoing thought processes and one's understanding about those thoughts. It drives animals to explore and find new knowledge that can clear the gap if there is any such no coherence [21]. Artificial curiosity can be a rewarding drive that enables an artificial agent to improve the learning progress and quickly acquire new knowledge during exploratory behaviour [26].

The internal cognitive states such as emotions, motivational drives, memory and attention play a wider role in arousal of curiosity in social animals. Motivational behaviours are always goal-driven and are means-to-end approach to attain basic desired goal states [11]. Motivations and rewards are closely associated with curiosity as the intrinsic motivation to learn new knowledge is always to gain rewards. In other words, any biological system is motivated to attain pleasurable or rewarding states through their behaviours and learning skills.

Curiosity arouses to its peak if an unknown or novel stimulus has been perceived in the environment. Learning is the outcome of new knowledge explored that either enables an animal to avoid the undesirable state or to maintain the desired state. Selective attention in infants is a simple example of exploratory behaviour like searching for toys that make new sounds, previously unseen colourful toy, etc. Generally the stimuli that involves qualities such as surprise, uncertainity, conflicting with known information, abnormal changes that involves complexity leads to a state of attention. The more attention a stimulus draws it is more likely that a stimulus is in focus or driving curiosity [21].

In this research project, the swarm is designed by using curiosity as a cognitive state that is driven by emotions attached to their energy levels. The robots once attain a curiosity state, they tend to explore and learn new behaviours based on their previous experience and their peers' experiences.

2 Braitenburg Reactive Behaviours

Valentino Braitenburg, an Italian-Austrian neuroscientist and cyberneticist and Director at the Max Planck Institute for Biological Cybernetics, investigated creatures with simple internal structure that can exhibit seemingly complex and intelligent cognitive behaviours named as "Braitenburg Vehicles." These vehicles are mounted with few sensors and motors that are connected in various fashion to create many different complex behaviours.

The behavioural response of Braitenburg vehicles correlates to the speed of the motor that is controlled by the strength of the signal being sensed on sensor. The wheel rotations can be proportional or inverse to the signal strength. In proportional connection, the more the sensor excited, the faster the motor rotates, and in inverse connection, the more the sensor excited, the slower the motor rotates [8, 9].

These vehicles with left and right sensors and left and right motors can implement the following different behaviours:

- If the left sensor excites more, give more rotations on the left motor (parallel proportional).
- If the right sensor excites more, give more rotations on the right motor (parallel proportional).
- If the right sensor excites more, give more rotations on the left motor (cross proportional).
- If the left sensor excites more, give more rotations on the right motor (cross proportional).
- If the left sensor excites more, give less (−ve) rotations on the left motor (parallel inverse).
- If the right sensor excites more, give less (−ve) rotations on the right motor (parallel inverse).
- If the left sensor excites more, given less (−ve) rotations on the right motor (cross inverse).
- If the right sensor excites more, given less (−ve) rotations on the left motor (cross inverse).

The above connections of sensors and motors of these vehicles demonstrate a range of behavioural activities on the vehicles. The following four different behaviours are created in this experiment: Move Away and Oscillate (MAAO), Move Away and Slowdown (MAAS), Move Towards and Collide (MTAC) and Move Towards and Stop (MTAS).

The robots used in this project are running with layered reactive control architecture [10, 27] with the highest layer being Braitenburg reactive behaviours (MAAS, MTAC, MTAS and MAAO).

- MAAO and MAAS generally avoid the threat and try to escape. Hence, we can consider the avoidance as a collective behaviour at macroscopic level.
- MTAC demonstrates the behaviour of turning towards the object and colliding with high speed, and hence, it shows aggressiveness.
- MTAS makes the robot move fast towards the object but does not collide. This shows attraction towards the object as a collective behaviour at macroscopic level.

3 Micro-Agent Level State Transition Design for Individual Robot in Swarms

The individual robot in swarm is independent in making decisions and adopts to random behaviour without interacting with its peers. Each robot can make decisions based on their internal states that are affected by external events. The internal microscopic design of robots is represented as state transition diagram, diagrammatically shown in Figs. 2 and 3. The robots can experience attacks and collisions in the environment that can cause internal perturbations. These perturbations are sustained by robots as long as they are not above threshold levels. Once these negative feedbacks are perceived, the robots switch to other states.

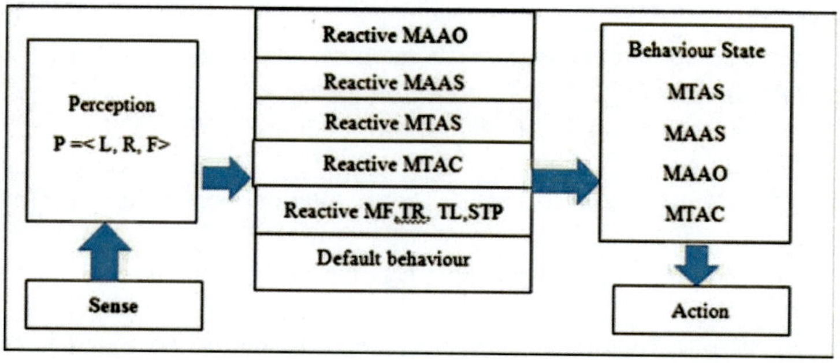

Fig. 2 Reactive micro-agent architecture

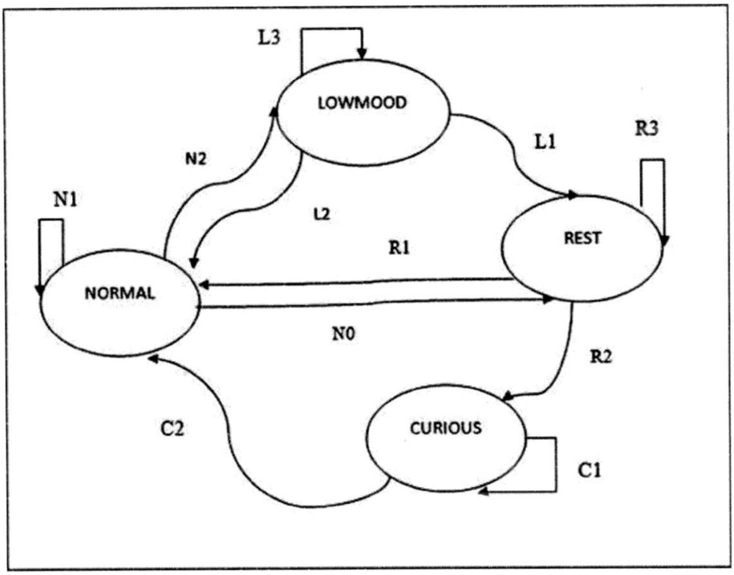

Fig. 3 Finite state machine for micro-agents

Algorithm for instantiating robots into swarm groups:

The algorithm uses a set of robot RS with a predefined action set a_R. The combinations of actions are used to create a set of behaviours b_R on each robot. Each behaviour b_{iA} is further defined as a combination of some actions. Each robot has internal parameter set IP_{iR} that are considered for performance evaluations. The robots respond to an event set EP_{iR}.

b_R is behaviour set given as {b_1, b_2,b_3,......,b_n}
R_S is a set of swarm robots {R1, R2, R3,.........,Rn}
a_R is an action set of robot {a1, a2, a3,.......,an}
b_{iA} is a behaviour-action set {bi1, bi2 , bi3, ...bin}
IP_{iR} is internal parameter set {IP1, IP2, IP3,.....,IPn}

Step 1: The robots are inducted into swarm group with the following initial set:
b_R: { MTAC,MTAS, MAAS, MAAO}
a_R: { Move Forward(MF), Move Forward Inverse(MFI), Move Forward Proportional (MFP), Turn Right Inverse(TRI), Turn Left Inverse(TLI), Turn Right Proportional(TRP), Turn Left Proportional(TLP), Turn Right Negative Proportional(TRNP), Turn Left Negative Proportional(TLNP), Stop (STP), No Stop(NSTP)}

IP_{iR}: {Internal-state, Initial-Energy, Threshold-Energy, Rest-cycles, Rest-Threshold}

Where Internal-state can take one of the states in the given set:
{NORMAL, LOWMOOD, REST, CURIOUS}
MTAC: {MF, TLI, TRI, STP, NSTP}
MTAS: {MF, TRNP, TLNP, STP}
MAAS: {MF, MFP, TRP, TLP, STP}
MAAO: {MF, MFI, TRI, TLI, STP, NSTP}

Step 2: Initialize parameter values for each robot
Internal-state: NORMAL (default state)
Initial Energy: E_{Ri} is the maximum energy that a robot virtually maintains in initial state, which is taken as 100 units.
Energy threshold: E_{Th} is always between 0 and E_{Ri}, i.e. $0 < E_{Th} < E_{Ri}$
Threshold energy: Threshold Energy (E_{Th}>= 50% of initial energy
Threshold rest cycles: (taken as maximum 3 rest cycles)
Threshold on attacks: Maximum attacks ($Attk_{Th}$= not more than 3 attacks in 10 cycles)

Step 3: Activate and run the swarm robots with the randomly chosen behaviour to respond to the events in the environment setup.
Step 4: Update the internal and external parameters that are affected by external events.
Step 5: Based on the previous 10 cycles, evaluate the fitness of the behaviour and update the behaviour fitness value by using genetic algorithm.
Step 6: Evaluate the average fitness value of each swarm group and by using sigmoid function either increase or decrease the probability value of each behaviour.
Step 7: Self-reconfigure swarm groups as per macroscopic design.
Step 8: Repeat steps 3 to 7 till the robots complete the defined number of cycles.
Step 9: Stop.

3.1 Detailed Microscopic Design of Swarm Robot with FSM

The robots initially executes with its internal state as NORMAL. In general, if the attacks are more in some cycles, the robots switch to LOWMOOD state from NORMAL state. The continuing attacks can result into drastic decrease in energy level of robots which make them to switch from LOWMOOD state to REST state. The robots regain energy level in REST state and switch back to NORMAL state. The robots perceive the number of REST cycles used, and when the count of rest cycles is more than rest threshold, they enter the CURIOUS state. In the CURIOUS state, the robot switches to explore other behaviours that are at their disposal and changes their response to external events to improve the chances of attaining their desired internal states. In the curious state, the robots switch to other behaviours based on the current highest probability value of behaviours.

In the NORMAL state, the robot runs the following cycle:

i. Each robot can join the group with the state being NORMAL.
ii. Check the desired behaviour of swarm group, and perform the group behaviour.
iii. If the number of attacks is less than the attack threshold and current energy level is greater than threshold energy, stay in the NORMAL state, and go with default group behaviour.
iv. If the number of attacks is greater than the attack threshold and current energy level is greater than threshold energy, switch to the LOWMOOD state.
v. If the number of attacks is less than the attack threshold and current energy level is less than threshold energy, switch to REST state.

Being in the LOWMOOD state, the robots run the following cycle:

i. Check the energy levels, and if it is below the energy threshold, it moves into the REST state; otherwise, it continues to execute the default behaviour with the LOWMOOD state.
ii. If the number of attacks is less than the attack threshold and the energy level is greater than the energy threshold, switch back to the NORMAL state.

Being in the REST state, the robots run the following cycle:

The energy is reinitialized to its maximum energy but keeps the number of attacks same and number of rest cycles same.

i. If the number of rest cycles are lesser than the rest threshold and the energy level is greater than the energy threshold, switch to the NORMAL state.
ii. If the number of rest cycles is greater than the rest threshold and the energy level is greater than the energy threshold, switch to the CURIOUS state.

The robots in this state have more chances of entering back to rest cycles as the attack count remains the same, and if the attack count has reached a threshold in which case, it will further increase the chances of switching to REST cycle more often.

Being in the CURIOUS state, the robot runs the following cycle:

i. If the number of REST cycles is greater than the rest threshold, change the state to CURIOUS.
ii. Choose the next behaviour based on highest probability value, and join the group with new behaviour with the NORMAL state with following modifications:

   a. Reinitialize its energy level.
   b. Reinitialize attack count to zero.
   c. Reinitialize number of rest cycles to zero.
   d. Change the behaviour code.

iii. Perform the tasks of new behaviour, and repeat the cycles.

Finite State Machines for microscopic design:

Lexicon: Input, State, Output (Perform Behaviour)

Input: N_a: Number of attacks in current cycle
$Attk_{Th}$: Attack Threshold (maximum attacks)
E_R: Current energy level
E_{Th}: Energy Threshold
N_R: Number of rest cycles till current cycle
R_{Th}: Rest cycle threshold (maximum number of rest cycles)

State: NORMAL, LOWMOOD, REST, CURIOUS

Output: Perform Actions of Behaviour

Functions: Adopt-Behaviour Function, Change-Behaviour

Behaviours:

• B-MTAC	MoveToAndCollide	Cross, Proportion
• B-MTAS	MoveTowardsAndStop	Parallel, Inverse
• B-MAAS	MoveAwayAndSlow	Parallel, Proportion
• B-MAAO	MoveAwayAndOscillate	Cross, Inverse

Similarly as in Table 1, Exhaustive State Transition Table For Lowmood State, Exhaustive State Transition Table For Lowmood State and Exhaustive State Transition Table For Curious State are prepared for robots to change their states.

4 Macroscopic Design of the Swarm

As explained in microscopic design, robots in CURIOUS state try to explore other behaviours and choose the behaviour based on the highest probability value generated

Table 1 Exhaustive state transition table for normal state

Transition number	Current state	Number of attacks	Energy-level	Number of rest cycle	New state	Behaviour
NO	NORMAL	$Na < Attk_{Th}$	$E_R < E_{Th}$	$N_R < R_{Th}$	REST	Default grp
NO	NORMAL	$Na < Attk_{Th}$	$E_R < E_{Th}$	$N_R >= R_{Th}$	REST	Default grp
N1	NORMAL	$Na < Attk_{Th}$	$E_R >= E_{Th}$	$N_R < R_{Th}$	NORMAL	Default grp
N1	NORMAL	$Na < Attk_{Th}$	$E_R >= E_{Th}$	$N_R >= R_{Th}$	NORMAL	Default grp
N2	NORMAL	$Na >= Attk_{Th}$	$E_R < E_{Th}$	$N_R < R_{Th}$	LOWMOOD	Default grp
N2	NORMAL	$Na >= Attk_{Th}$	$E_R < E_{Th}$	$N_R >= R_{Th}$	LOWMOOD	Default grp
N2	NORMAL	$Na >= Attk_{Th}$	$E_R >= E_{Th}$	$N_R < R_{Th}$	LOWMOOD	Default grp
N2	NORMAL	$Na >= Attk_{Th}$	$E_R >= E_{Th}$	$N_R >= R_{Th}$	LOWMOOD	Default grp

by the swarm groups to adopt new behaviour. Over a period of time, the robots converge into same behaviour to fulfil the desire of reducing or optimizing the number of attacks.

Particle swarm optimization (PSO) algorithm has been adopted to make swarm robots to converge based on the fitness value of the behaviour computed at the swarm level. This is done in two steps after every ten cycles:

STEP 1: Update Robot Performance

i. Robots update their individual fitness value for the behaviour that has been adopted in the last ten cycles, which will be considered by the swarm configuration process.
ii. In each cycle, the fitness value is calculated by each particle considering the number of attacks in that cycle.
iii. Takes an average of fitness value of the behaviour in last ten cycles and updates the behaviour fitness.

STEP 2: Reconfigure Swarm

i. Check the number of robots in each behaviour group and their fitness value of the behaviour.
ii. Calculate an average fitness value of the behaviour by taking the robot's average fitness value in the group.
iii. If the average fitness value of the behaviour is > F_{TH} (fitness threshold that is evaluated using sigmoid function on average fitness value), then decrease the probability value of the current behaviour by 0.1, and increase the probability

of the other two behaviours by 0.1. (The higher the fitness value, the lesser the chances of the robot choosing that behaviour in next cycles.)

The formula considered to calculate the fitness value:

$$\mathrm{Bh_{Fjk}} = \mathrm{Bh_{Fik}} + c1 \times R1 \times (P_{\mathrm{best}} - P_{\mathrm{present}}) + c2 \times R2 \times (G_{\mathrm{best}} - P_{\mathrm{present}})$$

Where

- $\mathrm{Bh_{Fjk}}$ is the behaviour fitness value of *K*th robot in the *i*th cycle.
- $\mathrm{Bh_{Fjk}}$ is the behaviour fitness value of *K*th robot in the *j*th cycle; i.e. j is $i + 1$.
- c_1 and c_2 are constants chosen to give weightage to the local and global fitness values of the behaviour. R_1 and R_2 are two random values between 0 and 1.
- P_{best} is the optimal collisions it has achieved so far in the previous number of configuration cycles.
- P_{present} is the current number of collisions it has encountered in the current cycle.
- G_{best} is the optimal number of collisions encountered by any of the robot in that group till the current configuration cycle.

4.1 *Swarm Control and Learning*

In Fig. 4, a swarm of five robots (R_1, R_2, R_3, R_4, R_5) is shown. Each of these robots is in one of the four possible states (NORMAL, LOWMOOD, CURIOSITY and REST) described above. The swarm organization changes over time as the robots and the macro-swarm agency modify their states. In the figure, just two of the possible swarm topology organizations are shown (VRi and VRj). The third level shows one meta-level macro-swarm agency (M1VRk); this one is the learning agency for the swarm. Other meta-level agencies are possible (see [10, 27] for further reading). Message passing between any two levels is time stamped; the accumulation of these messages becomes part of the swarm's history (or episodic memory).

Messages between the robots and the swarm organization ($\mathrm{VR_i}$) are defined as $M_{i,n,t} : <B_{i,r,t}, S_{a,r,t}>$

Where $M_{i,n,t}$ is message M_i from robot r at time stamp t defined as behaviour valence B_i (with B_i defined within the interval $\Re[0.0, 1.0]$) of robot r at time stamp t with the robot in one of four states as defined by state S_a of robot r at time stamp t.

The number of swarm robots allows the swarm topology organization (V_{Ri}) to determine a mode state for the swarm ($S_{a,t}$) and an average behaviour valency ($B_{i,t}$). These together with the reported environmental state $\mathrm{Env_t}$ are then compared to a "goal-swarm history" profile to determine corrections to the robots; i.e. the message from the swarm organization to each robot as $M_{i,n,t+1}$: $<B_{i,r,t+1}, S_{a,r,t+1}>$. The meta-level learning agency uses the "goal-swarm history" to find optimal behavioural

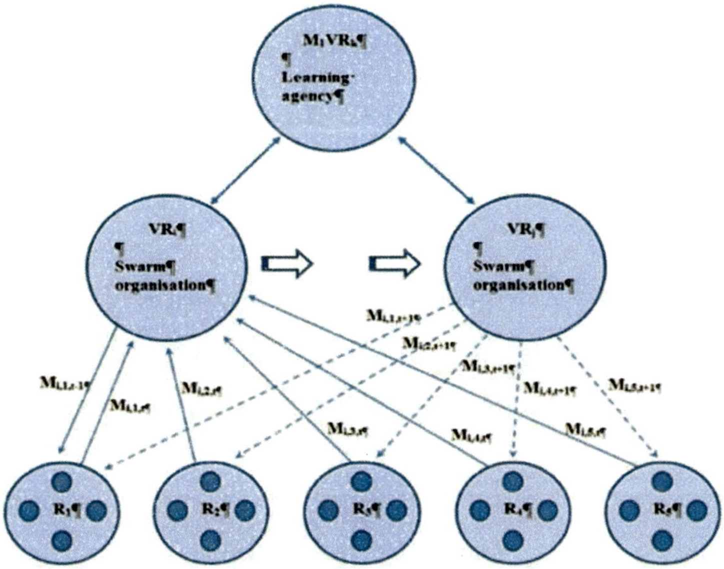

Fig. 4 Communication across three levels of a swarm topology, with a swarm consisting of five robots (R_1, R_2, R_3, R_4, R_5; each in one of four possible states). The second level (the VRi and VRj agency) is the swarm organization; the third level (the M1VRk agency) is the learning framework for the swarm

changes for the swarm as a whole and each individual within the swarm in order to further the goals of the swarm.

4.2 Curiosity-Driven Meta-level Learning for Dynamic Reconfiguration

The above message passing techniques adopted in swarm group enable swarms to self-learn to choose optimal behaviours and dynamically reconfigure into different groups. Each robot in the swarm group over a period of time learns to converge to a behaviour that drastically reduces the attacks and improves the performance. As shown in Fig. 4, the robots at very low levels change their states based on the number of attacks encountered, its energy level and rest cycles as per microscopic design. The robots in the CURIOUS state try to switch from their current behaviour to a new behaviour to form virtual swarm groups. These virtual swarm groups are qualified with states such as ATTACKING, ATTRACTING and AVOIDING along with internal microscopic robot state. The robots in each virtual swarm group will execute the same cycles as defined in the microscopic design.

Once the robots reach the CURIOUS state which indicates the lower performance of current behaviour, they tend to change their behaviour to explore other behaviours. The swarm robots just choose the behaviour that has a higher probability value. The meta-learner above these virtual swarm groups evaluates the performance of each virtual swarm group and sends a feedback to the swarm whether to exploit the current behaviour or explore the other behaviours by increasing or decreasing the probability value for each of the behaviour policy.

Algorithm for Meta-Learner:

Step 1: *Initialize* $Q(s,a) = 0$, $Best_{Avg} = P$ *(where P is acceptable best avg performance value).*

$\alpha = 0.1$

$\Upsilon = 0.9$

Step 2: *Evaluate Behaviour valency* $B_{(i,t)}$ *for each virtual swarm group.*

Step 3: *Calculate* $Q(s,a) += \alpha * (B_{(i,t)} + \Upsilon * (Best_{Avg} - Q(s,a)).$

Step 4: *if* $Q(s,a) > Best_{Avg}$

Decrease the probability value of the current behaviour which enables robots to explore other behaviour by sending a message $M_{i,n,t+1}$:- $<B_{i,r,t+1}, S_{a,r,t+1}>$ *to each individual robot*

Else if $Q(s,a) < Best_{Avg}$

Increase the probability value of the behaviour which increases the scope for robots to exploit the current behaviour by sending message $M_{i,n,t+1}$:- $<B_{i,r,t+1}, S_{a,r,t+1}>$ *to each individual robot*

$Best_{Avg} = Q(s,a)$

Else

Maintain same probability values for each of the behaviour so that the robots choose the same behaviour for next cycle

Step 5: $s = s'$ *and* $Q(s,a) = Q(s',a')$

Step 6: *Repeat the steps from 2 to 4 for each virtual swarm group*

The meta-learner generates q values for all possible transitions from one virtual swarm group to another, where each swarm group is considered to live in the virtual state. The transition is based on the maximum probability value of the behaviour. Initially, the robots choose greedy approach to select the behaviour based on which they form virtual groups. The robots in each group contribute to the average fitness value of the behaviour which is considered as a reward or punishment. If the average fitness value is greater than the best average guessed (the robots use a threshold value as 2 and above of average fitness value), the probability value of the behaviour is decreased so that the robots will not choose the behaviour.

5 Results and Discussions

The experiment was conducted by using seven parallax activity bots in a controlled environment within an arena of size 6' X 6'. The robots are initially placed in equal

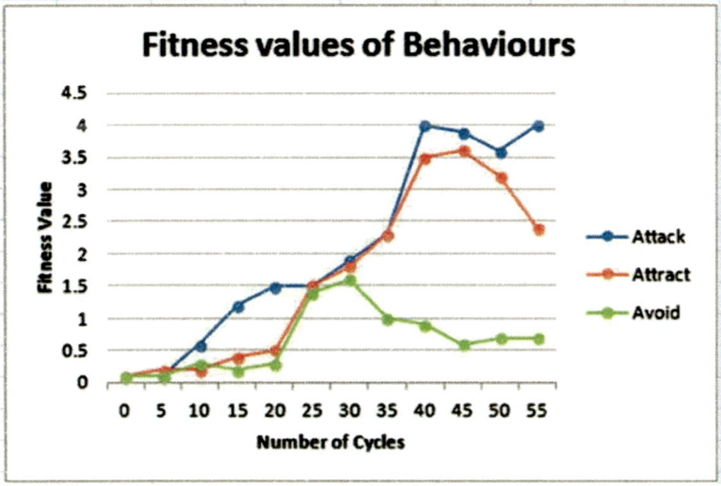

Fig. 5 Fitness value of swarm groups

distance facing random and in a safe range that no robot senses any obstacle. The robots randomly adopt any one of the Braitenberg behaviours coded in architecture. The preferred direction of move basically depends on the behaviour chosen; the attacking and attracting behaviours generally move towards the obstacle or other robot, whereas avoiding behaviour makes a robot move away from the robots or obstacles (Fig. 5).

Experiment conducted for 100 cycles demonstrated that at micro-level each robot in attacking swarm will experience the highest number of collisions and in turn will generate the highest average fitness value. Whereas the avoidance swarm will experience least collisions and hence the swarm generates smaller average fitness value and the attract swarm shows a behaviour of moving towards but not colliding, hence they will have moderate collisions. The fitness value of the behaviour of swarm is calculated by taking the collisions experienced by each robot in the group. The experiment was conducted by creating a swarm by making robots to chose random behaviour while instantiation of swarm. The graph given above shows the comparison of fitness values of all three behaviours up to 60 cycles. The trend remains same after 60 cycles.

In second case, the robots take different behaviour, and hence, the number of robots in each swarm group will be different. The attacking and attracting swarms experience the same number of collisions, and hence, their average fitness value will be greater. As the robots use sigmoid function on average fitness value of the swarm group, the fitness value above 2 will decrease the probability value of the behaviour by 1, and this ensures that robots in swarm group adopt the behaviour with highest probability value. The robots start converging to avoidance between 40th and 50th cycles and remain in avoidance till the 100th cycle.

Conclusion

The research experiment conducted and discussed in this paper is to study the natural foraging of insects and emulate the same on artificial swarm using parallax activity bots. The swarm robots at micro-level are designed to adopt random behaviours and self-learn to converge to an optimal behaviour over a time period. The genetic algorithm is designed to enable the robots to learn the fitness value of the behaviour based on their internal state attained with each behaviour. As discussed in Sect. 3, the robots internal states are affected by external events (Fig. 3) like number of attacks that makes the robot to switch to different behaviour. This indirectly affects the fitness value that is evaluated using genetic algorithm. The results obtained through this work demonstrate that the behaviour with higher fitness value is adopted by all the swarm robots and dynamically configure themselves into different swarm groups. The survival of strong behaviour that satisfies the internal goals and needs of the species and the curiosity that drives an insect to explore the new behaviour is well adopted in the experiment. The microscopic and macroscopic design of the swarm groups adopted in this experiment enabled the indirect communication among the robots which are otherwise independent in executing the task and can make their own decisions. The communication from microscopic level to macroscopic level will allow the robots to dynamically change into different topologies. This paper covers a detailed discussion of algorithms used for macroscopic and microscopic swarm design.

The study shows that the non-communicative robots in swarm groups with simple behaviours evolve themselves to converge adopting to optimal behaviour. By adopting different learning techniques like reinforcement learning, the robots can learn complex behaviours. Time complexity analysis for this experiment is not a performance index as its execution is purely based on dynamics of the robots and real-time environment.

References

1. Aguilar, W., Santamaría-Bonfil, G., Froese, T., Gershenson, C.: The past, present, and future of artificial life. Frontiers in Robotics and AI **1**(8) (2014). https://doi.org/10.3389/frobt.2014.00008
2. Bedau, M.A.: Artificial life. In: Matthen, M., Stephens, C. (eds.) Handbook of the Philosophy of Biology, pp. 585–603. Elsevier, Amsterdam (2007)
3. Bedau, M.A.: Artificial life: organization, adaptation and complexity from the bottom up. Trends Cogn. Sci. **7**(11), 505–512 (2003)
4. Beni, G.: The concept of cellular robotic systems. In: Proceedings 1988 IEEE International Symposium on Intelligent control, pp. 57–62. IEEE Computer Society Press, Los Alamitos, CA (1988)
5. Berlyne, D.E.: A theory of human curiousity. British J. Psychol. **45**(3), 180–191 (1954)
6. Berlyne, D.E.: Conflict, Arousal, and Curiosity. McGraw Hill, New York (1960)
7. Bonabeau, M. Dorigo and G. Theraulaz.: Swarm Intelligence from natural to artificial systems. Oxford University Press, ISBN 0-19-513159-2 (1999)
8. Braitenberg Vehicle, Wikipedia, accessed from https://en.wikipedia.org/wiki/Braitenberg_vehicle

9. Braitenberg, V.: Vehicles: experiments in synthetic psychology. Cambridge, MA: MIT Press. "Archived copy". Archived from the original on 2010–01–29. Retrieved 2012–06–18 (1984)
10. Davis, D.N. (2010) Cognitive Architectures for Affect and Motivation, Cognitive Computation Volume 2(3) (Sept 2010)
11. Franken, R.E.: Human Motivation, 3rd edn. Brooks/Cole, California (1994)
12. Gao, X.Z., Wu, Y., Zenger, K., Huang, X.L.: A knowledge-based artificial fish-swarm algorithm. In: 13th IEEE international conference on computational science and engineering, pp. 327–332. IEEE Computer Society, Hong Kong (2010)
13. Karaboga, D., Basturk, B.: A powerful and efficient algorithm for numerical function optimization: Artificial bee colony (ABC) algorithm. J. Global Optim. **39**(459), 471 (2007). https://doi.org/10.1007/s10898-007-9149-x
14. Kennedy, J., Eberhart, R: Particle Swarm Optimization. In: Proceedings of IEEE International Conference on Neural Networks. IV. pp. 1942–1948 (1995). https://doi.org/10.1109/ICNN.1995.488968
15. Kennedy, J.: The particle swarm: social adaptation of knowledge. In: Proceedings of IEEE International Conference on Evolutionary Computation. pp. 303–308 (1997)
16. Kennedy, J., Eberhart, R.C.: Swarm Intelligence. Morgan Kaufmann (2001). ISBN 1-55860-595-9
17. Krishnanand, K.N., Ghose, D.: Glowworm swarm optimization for simultaneous capture of multiple local optima of multimodal functions. Swarm Intell. **3**, 87–124 (2009). https://doi.org/10.1007/s11721-008-0021-5
18. Lallee, S., Verschure, P.: How? What? Who? When? Where? Why? Grounding ontology in the behavior of a situated agent, Robotics **4**(2), 169–193 (2015)
19. Dorigo, M., Stutzle, T.: Ant colony optimization. MIT Press, Cambridge, MA (2004)
20. Dorigo, M., Birattari, M.: Ant Colony Optimization, pp. 36–39. Springer publisher, Encyclopedia of machine learning (2011)
21. Macedo, L., Cardoso, A.: The exploration of unknown environments populated with entities by a surprise-Curiousity-based agent. Cogn. Syst. Res. **19–20**, 62–87 (2012)
22. Manguel, A.: Curiousity. New Haven, CT: Yale University Press (2015). ISBN 978-0-300-18478-5
23. Poli, R.: An analysis of publications on particle swarm optimisation applications (PDF). Technical Report CSM-469. Department of Computer Science, University of Essex, UK (2007)
24. Sahin, E.: Swarm robotics: from sources of inspiration to domains of application. In: Swarm Robotics, volume 3342 of Lecture Notes in Computer Science, pp. 10–20. Springer, Berlin, Heidelberg (2005)
25. Shi, Y., Eberhart, R.C.: A modified particle swarm optimizer. In: Proceedings of IEEE International Conference on Evolutionary Computation. pp. 69–73 (1998)
26. Schmidhuber, J.: Developmental robotics, optimal artificial curiosity, creativity, music, and the fine arts. Connection Sci. **18**(2), 173–187 (June 2006)
27. Vijayakumar M.V, Davis, D.N., Shylaja, K.R.: Design of Metacontrol and Metacognition mechanisms in SMCA Using Norms and Affect rules, International Conference on Artificial Intelligence and Pattern Recognition (AIPR-09), pp. 13–16. Orlando, FL, USA (July 2009)
28. Yang, X.S.: A new metaheuristic bat-inspired algorithm. Nature Inspired Cooperative Strategies for Optimization (NICSO 2010). Studies in Computational Intelligence **284**, 65–74 (2010)

Pre-stampede Monitoring and Alarm System

Rajeshwari Mattimani, Nalini C. Iyer, and Jyoti Patil Devaji

Abstract Bridge collapse and stampede are the major problems today. Due to this, there have been many casualties in the past. The Elphinstone Bridge collapse was the recent one; this happened in Mumbai. According to media reports, the main casualties were because of stampede at the middle of the bridge. This project focuses on to stop any possible stampede occurring and to alert the concerned department about the discovered stampede. This paper also concerns the regulation of the direction of track. This paper has also reviewed different studies about the human stampede over past many years and also its different applications of building a model beneficial to society. This paper concerns using Raspberry Pi as to solve human stampede. The system takes the crowd as measure, alerts, regulates track and prevents occurring stampede. The paper has also the information on using PIC microcontroller to solve stampede. This review provides information and details of system architecture and different types of optimization.

1 Introduction

There has been many such stampede events in the past, the major cause for this is infusing large public into narrow path. The recent events include Elphinstone Bridge stampede; according to the media reports, people continuously entered the bridge unaware of the stampede. So for this reason, movement of the crowd safely from one place to another should be the main concern of the authorities. Sometimes when the density of the crowd is very high, it becomes difficult for the authorities to manage public [1]. It may be also when the large track is regulated to narrow path, we need to design a system that take cares of the free movement of the track [2]. The crowd needs to be kept in check; if the crowd increases, they need to be directed to their way safely, and on event of stampede, the concerned authorities should be alerted [3]. There have been many such stampede events when the crowd is left unchecked, so the crowd needs to be kept in check and regulated to safe places [4].

R. Mattimani (✉) · N. C. Iyer · J. P. Devaji
School of Electronics and Communication Engg, KLE Technological University, Hubballi, India
e-mail: rajeshwari_m@kletech.ac.in

T. Senjyu et al. (eds.), *ICT with Intelligent Applications*, Smart Innovation, Systems and Technologies 248, https://doi.org/10.1007/978-981-16-4177-0_15

2 Objectives

The main objective of our project is as follows:

- To prevent any stampede;
- Regulate the direction of track;
- To alert concerned department on event of stampede.

3 Literature Survey

Human Stampedes: An Updated Review of Current Literature. Human stampedes are a major cause of mortality in mass gatherings, but they have received limited scientific attention. While the number of publications has increased, there is no recent review of new study results. This study compiles and reviews available literature on stampedes, their prevention, preparedness and response [2, 5, 6].

4 Stampedes

These pre stampedes are used to call an emergency by alarm so that they stop occurring accident. In the present situation stopping the accidents using stampede method is very important.

5 Functional Block Diagram

Raspberry Pi can be used to achieve the problem statement. Processor speed ranges from 700 MHz to 1.4 GHz for the Pi 3 Model B +, and onboard memory ranges from 256 MB to 1 GB random access memory (RAM) [7]. "Programmable Intelligent Computer" microcontrollers are used in many applications. This microcontroller is same as Pi, but the Raspberry Pi provides a much user-friendly interface. On the hardware side we have used GSM module, also called GSM modem, that can be interfaced with the Raspberry Pi. LCD is used to display the message and buzzer are used to show if there is any occurrence of stampede. Raspberry Pi 3 is one of the most preferred microcontrollers. It has 24 GPIO pins. LCD (16X2) is an electronic display module which uses liquid crystal to produce a visible image. The LCD will display the characters. The design of the circuit for Interfacing 16×2 LCD with Raspberry Pi is very simple [5] Fig. 1.

In this LCD, each character is displayed in a 57 pixel matrix. GSM-SIM800A is used which works on the frequencies of 850/900/1800/1900 MHz [8, 9]. The baud rate is configured to 9600; this module is used to send and receive message. This module

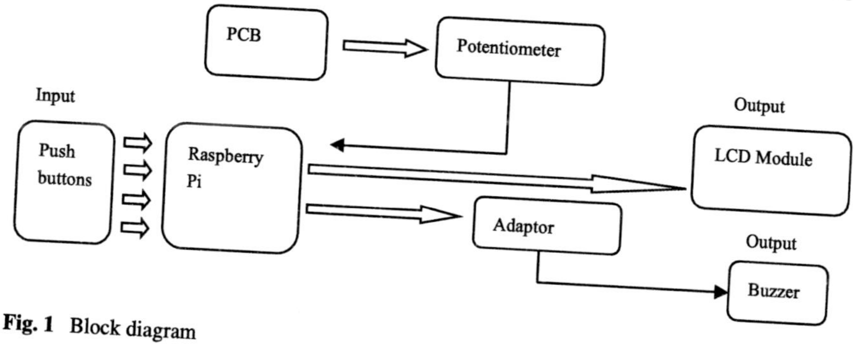

Fig. 1 Block diagram

is used for serial interfacing on Pi. The push button switches are connected to logic gate 7411; this gate has three inputs and gate. The output of these gates is connected to the pull-up resistor, to increase the op voltage [10]. The Pi is interfaced with the buzzer and the LCD, which displays the possible way to prevent the stampede. When all the logics are high, then the message is sent to concerned authorities [5].

6 Flowchart

See Fig. 2.

Fig. 2 Block diagram

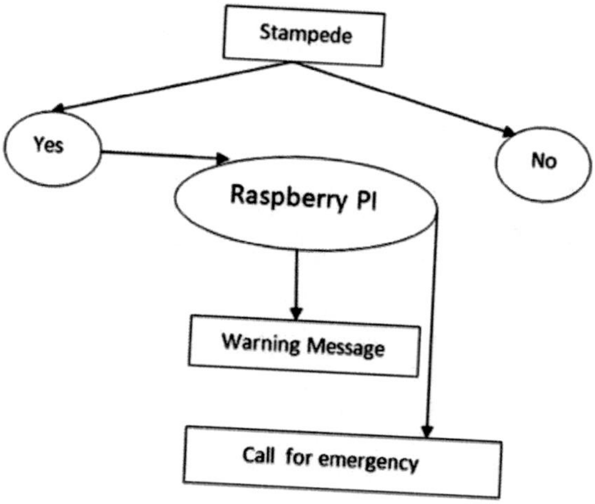

7 Optimization

7.1 *Introduction to Optimization*

Optimization is a method of finding the way to reduce the time of execution and to minimize the errors. The optimization has higher-quality results and is easy to debug.

7.2 *Types of Optimization*

Try and except is best way to replace the if/else statement, because the interpreter executes the sub() reset and throws an exception in the exceptional case, but in the reset case, the interpreter has to search each time the 'if statement' is executed, because the name could refer to something different since the last time the statement was executed. Under code optimization we have various techniques such as code motion, dead code elimination, strength reduction, loop unrolling, memory alignment etc. The system should also consume less power to give the required output. Last but not the least the system should also be cost effective.

8 Results

The LCD and buzzer are used for output result. Finally, the result is obtained by using Raspberry Pi. Stampede is defined by LCD and buzzer, and a call will be given or alert for emergency.

9 Conclusion

The model comprises GSM, LCD and buzzer; these modules decrease the chance of stampede. These can be further attached while construction for efficient use. The keys can be replaced with sensors that can measure the crowd density. The model is presented by Raspberry Pi.

10 Future Scope and Applications

Pre-stampede monitoring system has a large scope in the near future, as the population around the world is expected to grow and is growing at a rapid pace. Also most of the rural population is expected to migrate to cities in search of jobs. This means the human population is going to concentrate in cities in large numbers. Now all the public places, malls, halls, subways and bridges are going to be overcrowded because of the population explosion. This is where the pre-stampede monitoring is of utmost importance to stop any kind of misfortunes in public places. Also the present pre-stampede monitoring systems will be modified to suite the growing capacities and requirements. This system data can be attached to any database. And by using the Apache server, we can plot the real-time data on the Web server using the RRD tool. The RRD tool is a tool to graph the real-time data; using this, the public can choose which path to take in case of a stampede.

This system can be used in various crowded places like big shopping malls where the crowd footfall is high especially during festive seasons. It can also be used in public rallies where the crowd is huge. Other places where the pre-stampede monitoring can be used are the numerous public places like bus stands, railway stations, airports, public processions, political rallies, government, concerts, music and festivals [11, 12].

References

1. A Fast and Accurate Unconstrained Member, IEEE, Anil K. Jain, Fellow, IEEE, and Stan Z. Li, Fellow, IEEE
2. https://en.wikipedia.org/wiki/Category:Human stampedes in India
3. https://www.engpaper.com/stampede-detection.html
4. Face Detector Shengcai Liao, detection using a boosted cascade of simple features. In: Proc. IEEE Comput. Soc. Conf. Comput. Vis. Pattern Recog. I-511–I-518 (2001)
5. http://www.sciencentechnology.com/eshop/project-kits/149-pre-stampede-monitoring-and-alarm-system-using-pic-microcontroller
6. https://en.wikipedia.org/wiki/2017 Stampede, Mumbai
7. Raspberry Pi based interactive home automation system through E-mail Published in: Optimization, Reliability, and Information Technology (ICROIT). In: 2014 International Conference
8. GSM AT Command Set by Chris Cockings Date 09–04–2001 in Document Number BCO-00–0621-AN
9. Potnis, M., Chimnani, A., Chawla, V., Hatekar, A.: Home Security System Using Gsm Modem. Int. J. Eng. Res. Appl. **5**(4), (Part -6), pp.143–147 (Apr 2015)
10. https://acadpubl.eu/hub/2018-120-5/2/137.pdf
11. Why places of worship in India are vulnerable to stampedes by Nikhitha Sattiraju in SOCIALSTORY posted on 20th April, 2017
12. Why religious pilgrimages in India are conducive to tragdies. by The NEWS Minute Friday, Sept 25, 2015 - 17:20

Review of NOMA with Spectrum Sharing Technique

Mohamed Hassan, Manwinder Singh, and Khalid Hamid

Abstract Two new multiple access technologies in future wireless networking requirements for the effective use of bandwidth, non-orthogonal multiple access (NOMA) and Cognitive Radio (CR). The NOMA is one of the most promising performance improvement methods in new generation cells and CR will make the most of dynamic spectrum access and diversity throughout the broad spectrum possible, to alleviate the spectrum scarcity problem and meet the huge demands for wireless connectivity. This article discusses the basic principle of NOMA, cognitive radio fundamentals, spectrum sharing way, and the advantages and challenges of CR and NOMA spectrum sharing.

1 Introduction

Wireless networks have undergone a revolution in the past few decades through various approaches. Multiple users are assigned to wireless backbone networks for orthogonal radio services by time, frequency, and code field, or they are combined by conventional orthogonal multiple access (OMA) systems, such as Frequency Division Multiple Access (FDMA) and Time Division Multiple Access (TDMA). Code Division Multimeter (CDMA). Orthogonal frequency division multiple access (OFDMA) for 1G, 2G, 3G, and 4G respectively.

The FDMA frequency resource enables each user to send a single user signal to easily detect any user data within their Frequency Bands. In CDMA, several users can share the same frequency resources and orthogonal sequences can be correlated with symbols to various users. Intelligent FDMA and TDMA integration can be

M. Hassan (✉) · M. Singh
Department of Wireless Communication, Lovely Professional University, Phagwara, India

M. Singh
e-mail: manwinder.25231@lpu.co.in

K. Hamid
Department of Electronics and Communication, University of Science and Technology, Omdurman, Sudan

T. Senjyu et al. (eds.), *ICT with Intelligent Applications*, Smart Innovation, Systems and Technologies 248, https://doi.org/10.1007/978-981-16-4177-0_16

called the OFDMA, whereby the radio services are orthogonally distributed through the time flow network.

The number of supporting users in the conventional OMA diagrams is, however, strictly reduced, to the number of orthogonal resources that are available, which becomes difficult if a large 5G link is required. Furthermore, OMA has technically been shown, that for multi-use wireless systems it cannot always reach the maximum rate.

5G is mainly concerned with high data rates, low latency, low power usage, increased system capability, and most notably mass networking of devices. These needs can be satisfied by the inclusion of a NOMA technique in the 5G system, as NOMA can achieve multi-user capacity through time-sharing or rate-splitting if necessary. The key characteristic of NOMA is the use of more citizens than orthogonal capital. There are two groups of the NOMA family: the NOMA power domain and the NOMA code domain. Due to their channel quality allocated to the power-domain NOMA, various users have different power levels, while multiple users share the same time-frequency code resources. The power-domain NOMA uses the power difference of the users to distinguish different users based on successive disruptions. The NOMA code domain is a low density or orthogonal sequence that have no interrelation with CRDMA and multicarrier CDMA (MC-CDMA) [1].

Due to excellent development, improvement in communication systems, and technological progress, there is a problem with the available radio spectrum. All devices struggle hard for bandwidth, therefore. Considering all this, almost 50 billion wireless devices are expected to compete for wireless connectivity in the coming years. By 2030, the wireless networking market is expected to be 250 times higher [2].

The wireless range, sadly, is a finite resource and the wireless width available does not cover the required wireless bandwidth. Thus, it will be essential for wireless networks of the next decade to build new technologies to fulfill the increased demand for traffic and services to address the inevitable spectrum breakdown [3].

Sharing the spectrum where multiple consumers, whether in time or space, participate in spectrum bands represents a possible and very viable solution to the efficient usage and anticipated rise in transport demand for spectrum (i.e., counter bandwidth scarcity). The spectrum share can therefore be divided into two types: (i) unlicensed band sharing and (ii) allowed band share. Regulated band signals and traffic with special quality of services (QoS) specifications when less essential band traffic is downloaded.

To avoid interference while optimizing spectrum use, greater knowledge and coordination will be required for the next-generation radio networks. Such radio stations will need to collaborate smartly with their colleagues to use the organization's characteristics and maximize their valuable resources without any prior information. Besides, integrated, intelligent radios must be able to connect with other unknown radios to optimize the hybrid broadcasting spectrum. These radios must not only be intended to communicate effectively in congested and contested environments but also to communicate the spectrum band accessible to a wide range of heterogeneous radios without central coordination or prepared spectrum pre-allotment [2].

2 Noma Fundamental Principle

The 5G system of NOMA involves many users using the same subcarrier served by a base station (BS). Users on the same frequency will be doubled using different levels of transmission power, as shown in Fig. 1. This setting can also apply to a code or time domain [4].

A superimposed blend of messages from the BS is sent to the same subcarrier with different transmission power levels to the two multiplexed users. To ensure that every consumer achieves success, it needs the processing of a digital signal to distinguish the mixed signals.

Both multiplexes are sent to the same subscriber with a different amount of transmission power by BS to a superimposed mix of messages. To ensure that every consumer achieves success, it needs the processing of a digital signal to distinguish the mixed signals. The NOMA platform allows U users to count the entire W_T bandwidth and N_{SC} subcontractor. Besides, the NOMA SIC is used to ensure interference is avoided between two users of the same subcarrier.

The NOMA framework provides U users with the full W_T bandwidth and N_{SC} number of subcarriers. Moreover, the NOMA system uses the SIC to ensure the cancellation of interference between the two users of the same subcarrier. On the nth subcarrier ($n \in \{1, 2 \ldots, N_{SC}\}$) the BS transmits the $X_{u,n}$ signal to the consumer uth (u $\in\{1, 2 \ldots U\}$) with a transmitting power $P_{u,n}$. The received signal Y_u got from the uth user on n_q the subcarrier.

$$Y_u = \sum_{i=1}^{U} \sqrt{P_{i,n}} G_{i,n} X_{i,n} + N_i \tag{1}$$

where $G_{i,\,n}$ is the channel gain between BS and ith doubled users on n subcarriers, while N_i is the added Gaussian white noise.

In the context of a NOMA-based system, both users receive a common message. SIC is applied to multiplexed users with a stronger NOMA-based signal. This signal is a linear two-message sequence in which the SIC is applied on the receiver side.

As seen in Fig. 2, the signals sent to the two users with the intended message are multiplexed at the same frequency. In this case, it is believed that the channel conditions for user 1 are better than for user 2. User 1 would then enforce a SIC to interpret the message for User 2 first. User 1 will then cancel and decode the received

Fig. 1 NOMA diagram's transmission power levels

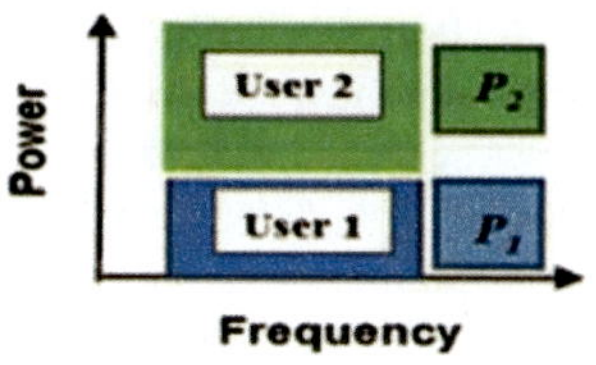

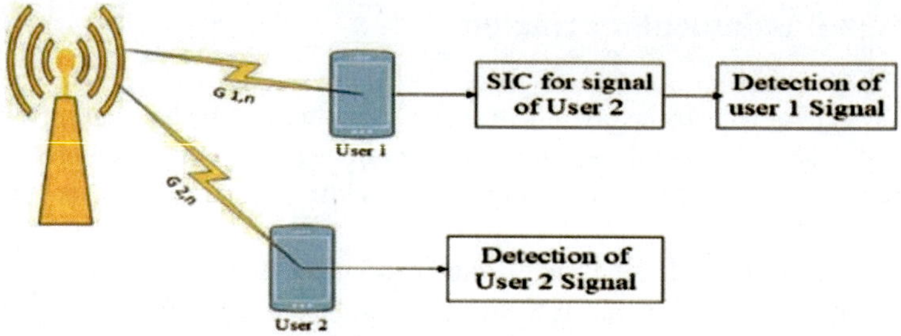

Fig. 2 Basic concept of NOMA scheme

signal as the intended message. User 2 would, on the other hand, consider a User 1 signal to be noise and decipher the message intended from the signal received. As shown below, you can express the results obtained by a NOMA method.

$$R_{\text{User1}} = B_{\text{sc}} \sum_{n=1}^{N_{\text{sc}}} \log_2\left(1 + \gamma_{1,n}\right) \tag{2}$$

$$R_{\text{User2}} = B_{\text{sc}} \sum_{n=1}^{N_{\text{sc}}} \log_2\left(1 + \gamma_{2,n}\right) \tag{3}$$

where the subcarrier bandwidth is $B_{sc} = W_T/N_{SC}$ and is $\gamma_{1,n}, \gamma_{2,n}$ the signal to interference plus noise ratio (SINR) for user 1 and user 2 respectively. According to (2) and (3), the transmit powers $P_{1,n}$ and $P_{2,n}$ and play a vital role in the throughput [5].

3 Cognitive Radio Fundamentals

The first cognitive radio definition was given by J. Mitola [6]. CR describes the stage of sufficiently organized radio and related IT assistants with their networks to ensure that: (a) the discovery of communication requirements by the user in the sense of their use, and (b) the provision of wireless radio tools and services. The following are identified. [6].

The CR senses and modifies its functional parameters so that bandwidth is efficiently utilized. The key objective is to maximize the spectrum usage as the available frequency channels are suitable for 90% of the time. The CR recognizes and transmits or defines a range of frequencies at a certain time and positions unused spectrum. Using this blank tape effectively for secondary use without intervening first with the approved user. The CR framework must also differentiate between the main user and the secondary user. A higher priority is given to approved users and principal users. Provides a broad bandwidth for enhancing data rate and content services. The sensing of the spectrum is a significant phase in CR, spectrum analysis, spectrum decisions, and spectrum versatility.

3.1 System for Spectrum Detection

During intra- and interframe data transmission silent times, spectrum sensing is performed. The search/communication protocol for frequencies is used to detect previously used channels with the nearest transmission of the channel [7].

The following methods are widely used for spectrum sensing.

(1) **Energy Sensing**: The fundamental technique for its technical simplicity and quick implementation. By comparing measured energy with the required threshold, a spectral hole is determined.
(2) **Coherent sensing**: The detector knows the primary user signals such as modulation form, set-up, pulse size, data frequency, and bandwidth accurately in this technology. The received signal is linked to a creative copy of the message, compared with a threshold value.
(3) **Cyclo-stationary Sensing**: The function (x), has other features that are not completely known, then this data is used to produce test data that fits the signal closely. It contains a periodic function, which is constant in the case of noise or broad sentences, of the subjective relationship of the received data signal.
(4) **The autocorrelation feeling**: the main data signal is static and is used to differentiate between white noises and the principal data signals with its autocorrelation.

3.2 Sharing of Spectrum

This is an important method to prevent interference between cognitive and simple radio licenses, which allows higher functionality of the radio frequency. There are two ways this can be achieved:

(1) **Spectrum Sharing Cooperative**
A data collector center and a control channel between them or through distributed networks are part of the frequency band. The first approach is effective, but the time is much longer, it is very difficult to implement, and energy usage is greater than the second [8].
(2) **Spectrum sharing non-cooperative**
This is not a way of sharing information between users, but it is a useful communication tool for certain perceptual networks of users. [9].

3.3 Access Types of Spectrum

The CR activity can be graded as follows according to the spectrum use scenario:

(1) **Interweave**: The SU can transmit only when the licensed spectrum does not take up the PU.
(2) **Underlay**: Concurrent primary and secondary transmissions are enabled provided the primary network interference is below the controllable level.
(3) **Overlay**: The SU unit supplies the essential transport network equipment and simultaneously transmits its signals.

4 Related Work

The examining the performance of the NOMA-based Collaborative Relay System (CRS) In a basic spectrum sharing scenario, the authors focus on looking at the PIC cover, in which secondary (unauthorized) network interference peaks over the primary user of the (authorized) recipient (PU-Rx). Less than the preset amount [10]. A new safe NOMA transmission strategy is established to handle interference among users and guarantees the quality of primary user services where primary and secondary users are balanced following their channel gains and NOMA power-domain transmissions are used [11]. A complete analysis and evaluation of existing NOMA technologies by several major companies for the fifth-generation wireless standard (5G) will be carried out by the authors in [12] and the overall rate efficiency of NOMA systems and their possible performance gains such as a comparison with OMA will be evaluated. Although the author in [13] focuses his research on the total rate of NOMA, the combined detrimental effect of practical obstacles is a combination of a multiple-input relationship and non-orthogonal distribution of the basic spectrum. The authors focus primarily on the NOMA power field, which uses transmitter overlays and cascade interference cancellation in the receiver and explains how NOMA works with several well-known wireless communication technologies, including collaborative communication, multiple-input outputs, beamforming, spacetime coding, and network coding [14].

5 NOMA with CR Spectrum Sharing

The principal concept of NOMA is the use of energy multiplexing, which is different from traditional OMAs, to promote the allocation of the spectrum between multiple users within a single resource network [15].

The combination of NOMA and CR can meet 5G standards for high performance, great connectivity, and low latency. Also, given these potential advantages, the successful cognitive construction of NOMA is a very difficult problem in practice, as both NOMA and CR Subject to interference, the outcome in NOMA multiplexing in the control domain, which inevitably leads to significant interference between networks Primary, secondary, and intra-network interference (also called co-channel interference) [16].

Therefore, to reduce interference and allow efficient use of the core tools of the system, NOMA and CR must be properly integrated. Remember that there is little interaction with both NOMA and CR, which may impair reception reliability. A reliability-enhancing collaborative migration strategy was designed to address this issue.

By combining CR and NOMA spectrum sharing, cognitive NOMA aims to better share the spectrum in a proactive way to increase spectrum utilization. The advantages of an insightful cognitive sharing of the NOMA spectrum

- **Enhanced spectrum use:** NOMA cognitive networks will enable both Pus and Sus at an appropriate level of reception.
- **Massive Connectivity:** Many smart devices are planned for 5G wireless networks. NOMA cognitive networks, where multiple PU and/or modules are supplied in a single source block at different power levels simultaneously, will satisfy this requirement [17].
- **Low latency:** In cognitive NOMA networks, transmission delays in Sus can be high, resulting in low latency. For example, several SU units can be connected simultaneously by making use of NOMA to support CR networks [18].
- **Better Justice:** NOMA Perceptual Networks will ensure better equality among consumers. This results in a healthy compromise between equity and secondary network efficiency [19].

NOMA helped the SRs to be exposed not only to secondary network interference but also from the interaction induced by the primary transmitters (PTs). Each NOMA SR powered is safeguarded by a d0 radius (called the protected area), and there is no PT in the circuit. Fitting error output is shown for secondary transmissions and full range commands are issued for each SR with the help of NOMA. Second, propagation of primary network emissions should not be interfered with when secondary access is permitted as a distinct facet of the base CR model. Meaning that the interworking that interferes with PR must be below a predetermined threshold, it means that the energy management of the ST is important to the NOMA backbone. Moreover, the power allocation of the STs must also be ensured, so that the various Quality of Service (QoS) demands of the SRs are completely fulfilled, and the intranet interference between the NOMA-supported SRs is carefully controlled [19]. The basis for conventional CR is restricted regularly to short-distance communication due to the limited transmission capacity of STs.

6 Conclusions

In reviewing the key rules of Non-Orthogonal Multiple Access and the concept of cognitive radio, we found that these are effective methods to solve the problems of a specific frequency spectrum and additional spectral efficiency benefits would be provided. Combining NOMA and CR can meet 5G standards for high performance, great connectivity, and low latency, but many challenges are facing the integration of NOMA with CR which to need to study and analyze the studies recommend that the wireless spectrum could be used more effectively by combining NOMA and CR technologies.

References

1. Sharma, S., Tripathi, C.C., Rishi, R.: Cognitive radio: an efficient and effective solution for future spectrum implications. In: 2017 International Conference on Computer, Communications and Electronics (Comptelix) IEEE, pp. 112–117. Jaipur (2017)
2. Yucek, T., Arslan, H.: A Survey of Spectrum Sensing Algorithms for Cognitive Radio Applications. IEEE Commun. Surv. Tutorials **11**(1), 116–129 (2009)
3. Zheng, G., Song, S., Wong, K., Ottersten, B.: Cooperative cognitive networks: optimal, distributed and low-complexity algorithms. IEEE Trans. Signal Process. **61**(11), 2778–2790 (1 June 2013)
4. Islam, S.R., Avazov, N., Dobre, O.A., Kwak, K.S.: Power-domain non-orthogonal multiple access (NOMA) in 5G systems: potentials and challenges. IEEE Commun. Surv. Tutorials 721–742 (2016)
5. Al-Abbasi, Z.Q., So, D.K.: Power allocation for sum rate maximization in non-orthogonal multiple access system. In: IEEE 26th Annual International Symposium on Personal, Indoor, and Mobile Radio Communications (PIMRC), pp. 1649–1653 (Aug 2015)
6. Mitola, J., Maguire, G.Q.:Cognitive radio: making software radios more personal. IEEE Personal Commun. **6**(4), 13–18 (Aug 1999)
7. Greenstein, L.J.: Principles of cognitive radio. Cambridge University Press (Jan 2012)
8. Khan, A., Rehmani, M., Reisslein, M.: Cognitive radio for smart grids: survey of architectures, spectrum sensing mechanisms, and networking protocols. IEEE Commun. Surv. Tutorials **18** (2015)
9. Tzeng, S., Lin, Y., Wei, L.: Spectrum hand-off for RF energy harvesting in cognitive radio based machine to machine communication networks. In: 2018 International Conference on System Science and Engineering (ICSSE), pp. 1–4. New Taipei (2018)
10. Kumar, V., Cardiff, B., Flanagan, M.F.:Fundamental limits of spectrum sharing for NOMA-based cooperative relaying under a peak interference constraint. IEEE Trans. Commun. **67**(12), 8233–8246 (Dec 2019)
11. Xiang, Z., Yang, W., Pan, G., Cai, Y., Song, Y.:Physical layer security in cognitive radio inspired NOMA network. IEEE J. Sel. Top. Signal Process. **13**(3), 700–714 (June 2019)
12. Wu, Z., Lu, K., Jiang, C., Shao, X.: Comprehensive study and comparison on 5G NOMA schemes. IEEE Access **6**, 18511–18519 (2018)
13. Li, Y., Aruma Baduge, G.A.: Underlay spectrum-sharing massive MIMO NOMA. IEEE Commun. Letters **23**(1), 116–119 (Jan 2019)
14. Islam, S.M.R., Avazov, N., Dobre, O.A., Kwak, K.: Power-domain non-orthogonal multiple access (NOMA) in 5G systems: potentials and challenges. IEEE Commun. Surv. Tutorials **19**(2), 721–742, Secondquarter (2017)

15. Liu, L., Zhou, Y., Tian, L., Shi, J.: CPC-based backward-compatible network access for LTE cognitive radio cellular networks. IEEE Commun. Mag. **53**(7), 93–99 (July 2015)
16. Naeem, M., Anpalagan, A., Jaseemuddin, M., Lee, D.C.: Resource allocation techniques in cooperative cognitive radio networks. IEEE Commun. Surv. Tutorials **16**(2), 729–744, Second Quarter (2014)
17. Zhong, B., Zhang, Z., Chai, X., Pan, Z., Long, K., Cao, H.: Performance Analysis for opportunistic full-duplex relay selection in underlay cognitive networks. IEEE Trans. Veh. Technol. **64**(10), 4905–4910 (Oct 2015)
18. Liau, Q.Y., Leow, C.Y.: Amplify-and-forward relay selection in cooperative non-orthogonal multiple access. In: 2017 IEEE 13th Malaysia International Conference on Communications (MICC), pp. 79–83. Johor Bahru (2017)
19. Zhang, X., et al.: Performance analysis of NOMA-based cooperative spectrum sharing in hybrid satellite-terrestrial networks. IEEE Access **7**, 172321–172329 (2019)

Review on Secure Ad-Hoc Networks for Wireless Sensor Network

Kanika Chauhan, Kalpana Yadav, and Ankita Singh

Abstract In a current scenario, wireless sensor networks (WSN) are the most common and trendy network technology which is used in wide areas of the real world. Recent results show that WSN has been used in many solicitation fields in the current world like many health organizations, supervision in the military, and many more. However, due to some cumbersome ideas of WSNs, protocols like Ad-hoc on-demand distance vector routing protocol (AODV) tend to cease malicious nodes and attacks. Therefore, the attack is known as a playback attack. An isolated sensor node has a restricted number of transposition and estimation. As we know, sensor nodes have a restricted amount of vitality, so through this attack, we have a major influence on the network performance. While using memory-efficient data, structures called as bloom filters in this work make the functionality of the network better.

1 Introduction

As we all know wireless sensor networks (WSNs) are considered as the most significant and ingenious technology nowadays, WSN has been used in many solicitation fields in the current world like many health organization, supervision in the military, and also to monitor the temperature of different countries. Sensor hub is a vast area of the network where n number of nodes are directly connected where the sensor nodes can easily communicate with the help of sensors as these nodes are connected in a self-structured topology which means nodes are free to move any time anywhere in a wireless channel. Therefore, there are various approaches in routing and protocols are there which are used to protect networks, and AODV is one of them which collects and monitors the data long with a central location in wireless sensors.

K. Chauhan (✉) · K. Yadav · A. Singh
Indira Gandhi Delhi Technical University for Women, New Delhi, India

K. Yadav
e-mail: kalpanayadav@igdtuw.ac.in

A. Singh
e-mail: ankita@igdtuw.ac.in

T. Senjyu et al. (eds.), *ICT with Intelligent Applications*, Smart Innovation, Systems and Technologies 248, https://doi.org/10.1007/978-981-16-4177-0_17

An ad-hoc routing protocol is one the common protocol mostly used in the area of WSNs. The network of AODV is based upon the request-reply mechanism which simply means that it helps the nodes to determine the best optimal path in the ad-hoc network. Protocols in the wireless network are specific to node tht how to communicate and deliberately delivered packets so that there should not be any loss of vitality, AODV improves the network's functionality in two manners, i.e., help the nodes to decide the optimal routing path and to transfer the packet through the network in a secure manner, so that there should be any kind of loss. Using an AODV routing protocol in sensor hub makes the network fully preserved and vulnerable to assaults [1]. One of the attack is a replay assault also called a playback attack. This type of attack plays a very common role in the ad-hoc sensor networks in a very subsequent manner like retransmission of multiple copies of the packet which is retransmitted repeatedly multiple times, so, due to this, attack assaults a network very effectively, and there is a loss in the vitality of nodes very badly. Therefore, to overcome this obstacle, bloom channels are used. In this paper, bloom filters are worn to recognize the validity of a data bundle, and bloom filters are a type of data structures which is known as memory-efficient structures that test all the entities with it and filter them. Till now, bloom filters are only used in the era of only spell checkers and to prevent from insecure passwords, but today bloom filters have again a very high tribute to many networking areas especially in wireless sensor networks. Here, bloom filter plays a very vital role to enhance the functionality of networks concerning different security parameters like delay, throughput, energy, etc. Here, a data is transmitted between the nodes and each outgoing data packet is matched with the stored data in structures and through this strategy, one can prevent playback attack and improves the network representation. The main ambition of this work is to study the different protocols which are used in sensor hub and also about the attacks that affect the wireless sensor network.

Further, the paper is as follows, Sect. 2 describes the literature survey, Sect. 3 states the review of the protocols in an ad-hoc network of different authors in the tabular format, Sect. 4 elaborates the protocols used in wireless sensor networks, Sect. 5 has a piece of detailed information about assaults in ad-hoc network, Sect. 6 have techniques used in ad-hoc network, and at the end, Sect. 7 is the conclusion.

2 Literature Review

Author states that AODV protocol is used to make the network prone to replay attack, and the data structure technique is further discussed to intensify the network functionality that is related to packet delay, vitality, and also improves the legitimacy of packets [2]. The main challenging problem in ad-hoc routing protocol is data transfer with the minimum loss; this paper resolved the issue with new a scalable AODV-ES. The aim of the paper is to reduce the delay of packets and boost the packet delivery of the packets through AODV-ES. The author also come up with the result that exhaustive simulation better results as compared to AODV in terms of delivery ratio, delay, and also preferably states AODV-ES is capable of wireless ad-hoc sensor networks where scalability concerning period is the dominant issue.

Intention of this paper is to decrease an end-to-end delay for delivery of the packet [3]. This paper describes the different WSN protocols in routing like (DSR) dynamic source routing, AODV, and destination-sequenced distance vector (DSDR). A comparative study is made based on performance metrics like packets delivery ratio and packets dropped [4]. The author has done a comparative study with simulation using NS2 between DSR and AODV routing protocol to overcome a load of nodes, and simulation shows the comparative performance of the routing protocols, and SONET is proposed for efficient routing scheme [5]. The triple-factor approach has a very similar security feature w.r.t. DSR, and AODV strategy is basically designed to transmit the data packet in a secured way. The author has described MANET as a most feasible network paper mainly focused on different security attacks like the black hole attack which assaults serious security threats based on routing services [6]. Routing is also one of the main challenges in a mobile ad-hoc network to overcome this challenge that hybrid routing mechanism has been proposed based on different topology (like star, mesh, bus) and a quality of service based on networks [7]. The interpretation of the mobile ad-hoc network (MANET) is mainly affected by the malicious nodes, and attack assaults to this network are DoS attack; further author has proposed an intrusion detection based upon support vector machine algorithm to detect and removes the malicious nodes from the network which does not affect the network scalability [8]. Like the MANET WSN routing protocol also used in vehicular ad-hoc network, the paper provides a precis about VANETs characteristics and security attacks on services like confidentiality, integrity, availability, and develop countermeasures to make the network more secure [9]. The main goal is to show how these attacks are badly affecting the network functionality in terms of those metrics. The author also proposed the countermeasures against node misbehavior attacks.

Further work will proceed in this domain that focusing another senor node attacks like flooding attack, wormhole attack, and many more in the sensor hub [10]. Various routing protocols are evolved to maximize the existence of the WSN where the main goal is to choose the routing scheme which is most probabilistic through this PEER probabilistic energy-efficient routing approach which is developed to stable the network vitality utilization and enhance the network life [11].

3 Review of Various Protocols in Ad-Hoc Network

This section provides information about different protocols and attacks that are used in various aspects of mobile networks. The description have been presented in Table 1.

Table 1 A comparative review of the literature work onto various assaults and protocols in an ad-hoc network

S. no	Author	Statement	Protocol used	Attacks	Improvements	Future work
1.	Shaikha Alkhuder [2]	Securing AODV routing Protocol in WSN	Ad-hoc on- demand distance vector routing protocol	Replay attack	Using bloom filters improve the legitimacy of a packet in terms of vitality	On applying any other technique we can improve any other security metrics like throughput, delay
2.	Rakesh Ranjan, Nirnemesh Kumar Singh, Mr. Ajay Singh	Security Threats in Blackhole attack	Adhoc on- demand distance vector	Blackhole attack	Work has been proposed to safeguard the network from black hole attack	Further research can also be proposed for gray and wormhole attack to safeguard the network
3.	Srinivas Sethi, Ashima Rout, Debajyoti Mishra	Routing Protocol is to be designed that can solve packet transmitting channel	AODV-ES(Effective and Scalable)	Not addressed	Improve the potency of AODV-ES in terms of delay & PTR	The spare protocol can also be designed to test the safety of the network
4.	Zeyad M. Alfawaer Belgaum Mohammad Riyaz	An Enhanced Multipath Strategy Protocol	DSR & AODV routing protocol	Not addressed	SONNET has been proposed to choose an optimum path for data transfer in a secured way	Research can be done based on mathematical analysis wrt to SONNET to improve security metics
5.	Adnam Ahmed, Umar Ali Khan, Adnam Rajpar	Analysis of AODV protocol in opposition to node misdeed attacks in WSN	AODV routing protocol	Blackhole, gray hole attack using AODV	Enhancement in-network functionality is improved using several countermeasures against missattendeed attack	Other nodes misdeed attack can be implemented like a flood, sbil attack & many more

(continued)

Table 1 (continued)

S. no	Author	Statement	Protocol used	Attacks	Improvements	Future work
6.	Mansi Rajgor, Prasanna Shete, R.N. Awale	Probabilistic energy-efficient routing protocol for WSN	PEER protocol	Not addressed	PEER is proposed to balance the network energy consumption & upgrade the network life wrt battery capacity over AODV	PEER can be combine with any other protocols to improve other security aspects
7.	Mohammad Riyaz Belgaum, Shahrulniza Musa, Mazliham Mohd Su'ud, Muhammad Alam, Safeeullah Somro, Zainab Alansari	Secured approach tends to reactive routing protocol using triple factor	DSR & AODV routing protocol	Not addressed	Improves the throughput in the network for the existing approach	Future research can be done for concerning protocols that simulate the architecture of the network based on the routing process
8.	S.Shruthi	Proactive routing protocol for MANET	Wireless routing, DSDV, global state routing protocol	Not addressed	To improve the scalability, efficiency and power management in ad- hoc network	Improve the performance of network wrt cross- layer mechanism

4 Protocols Used In Wireless Sensor Networks

A. AODV

In a current scenario, different technologies and routing protocols used in mobile networks are flooding, clustering hierarchy for low energy, neighboring distance, gossiping, directed diffusion, and AODV. AODV is one of the main protocol that can be used for the simulation purpose in the ad-hoc network because testing in this type of network can be done via various scenario which is based on real entities. Also, one of the commonly used protocols in the sensor hub for reactive routing (i.e., nodes in the network is only created when they are needed) is AODV. This type of protocol relies on the request-reply mechanism (RREQ RREP) to find the routes that significant to other nodes in the network. Packet in the network can be transmitted between the nodes using RREQ and RREP. Therefore, an initial node in the network broadcasts the packet with RREQ and waits for a specific frame of time to reply to the destination knot once the destination hop has acquired, then the RREP parcel starts communicating the information according to received input through routes [2]. Originally, for the transmission of packets between the AODV nodes, it creates a socket with user datagram protocol, i.e., UDP, with the kernel, so that transmission can be done in a more convenient way using port number 654. Other port numbers in the network can also be used to update the routes in the routing table, for multicast, and also for some specific event say node death. Sequence numbers are also used by AODV to get recent and new information (information like destination address, next hop, number of hops used to reach the packet from source to destination) on the routes [2].

B. DSR

Another illustration of responsive directing conventions is dynamic source routing (DSR) which has a similar working like AODV. The convention depends on the connection state calculation that implies one, and all node is capable to keep up the superior route to the target knot. The source hub characterizes the total way of the target knot from one jump to further, additionally utilized the source steering strategy where the ways are put away in the way reserve. Assuming interchange gather in the topology of the network, the organization will get data by flooding. In the event that there is any connection disappointment in the organization, at that point, the hub engenders a RRER message [4].

C. DSDV

An illustration of proactive routing convention is destination-sequenced distance vector (DSDV) convention. The steering table keeps up by every hub, which contains a list of all potential objections corresponding the number of hops in network. Routing table information is trade between nearby hubs, and directing information is updated with another information by every node. In the event that the package cannot discover its objective, at that point, the packages were stored temporarily. At that point, the information packages cannot get until the delivery report shows up from the destination [4].

D. PEER

In WSN, hub nodes have a limited amount of energy which is unfeasible to recharge or restore during a process. Therefore, the vitality constraint and transmission between the nodes in the network become condemning. This implicit in the deduction in the existence of power in ad-hoc networks. The battery is a very crucial factor in sensor networks. So, nodes have to use it very deliberately. To avoid the separation in network and conserve the connectivity of sensor hub, vitality innovation in sensor networks has a broad area of research in ad-hoc networks. The solution to the problem is to innovate a new and intensive protocol known as PEER, which saves the vitality of the battery in between the nodes and enhances the lives of the network [12].

5 Attacks in Ad-Hoc Network

Protocols in the sensor hub are undefended to various assaults in the network. Assaults in the network are mainly categorized into two types, i.e., active and passive. In passive, the aggressor only gets the data by attending the network traffic only, whereas the inactive striker changes and alters the information too. Few assaults interrupt the nature of the network, i.e., playback assault, black hole, wormhole, and gray hole attack [6] Replay attack is a kind of security assault to the information sent over an organization. In this assault, the hacker or any individual with unapproved access catches the traffic and sends correspondence to its original target, going about as the first sender. The recipient feels that it is a confirmed message yet it is the message sent by the aggressor. The primary component of the replay attack is that the customer would get the message twice, and consequently, the name, replay attack, also known as playback attack or copycat attack, in sensor hub. Playback attack is one of the greatest frequent attacks used by AODV reactive protocol. The main aim of this attack is to sending and receiving a duplicate packet to the next sensor hub and convincing the target knot to take in the copied packet as a legalized one. This should be possible inside a comparative show or in another distinctive show. A replay assault can be cultivated by controlling bundle substance such as headers to conceal the certified operator and emulate the underlying origin, or without developing some of the bundle substance and recommit a duplicate of the package in a substitute period, where case the group is sent with no guarantees. An occasion of controlling bundle substance to play out a playback attack is to supplant the center location area in the header of the package with a ridiculed MAC address of a credible sensor hub center point that is not basic for the main center point association course. By then, it dispatches the duplicate package from the sensor center with the copied address. Having bloom directs in the model of sensor associations can be used to perceive copied hubs that are replayed without adjustment. Regardless, this line of the shield is kept away from by controlling the substance of the packets [2]. Black assault is also of two types. One is a single black hole, and the second is a cooperative black hole assault. Therefore, black hole is a type of network layer assault where misdeed hop waits for a request message, and on receiving it one will send a reply message to the initial to the target node. Therefore, if the way of the network is chosen by the initial node and packet is sent through the chosen path, then the misdeed node silently engages all the data packet that was actually for the target node, whereas in a cooperative black hole attack here two misdeed hops work in cooperation and try to recognize the malicious hop even more laborious and these type of attacks also include more than two misdeed nodes. In Wormhole assault, the misdeed node hijacks the data packet from one position of hop and resends it to another malicious node in various space. This procedure has a serious security threat impact on the network. More than one misdeed node is needed in wormhole assault. These attacks are very minacious and extremely injurious to defend against the routing information. Lastly, the gray hole attack is one of the most special cases of the black hole which starts releasing the data packets very carefully [6].

6 Techniques in Ad-Hoc Network

Bloom filter is an array type memory-efficient data structure and becoming a more popular algorithm in the field of MANET networks. Bloom filters are used in the different networking problems where the listed structure is required and space is a crucial issue in the organization. Bloom filters are used in numerous applications like spell checker dictionary, Unix. Mathematically, bloom filters are represented in a set of arrays where utter elements are first set to 0, and bloom filters also use a hash function $h_1(x)$, $h_2(x)$,…., $h_m(x)$ which gives the resultant value from 0 to n. Also, if the resulting hash function specifies the array bit which convey the value of 1, then x is the element of the array set, and also if the array set indicates the value 0, then x is not a part of the set bloom filter. The bloom filter automatically exchanges to 1 to make a new documentation to the segment [2]. Nodes in the sensor hub can be freely moved, and the aim is to transfer the data from one initial node to another target node safely. The author's main focal point is on a different path for routing the data through routing convention in the ad-hoc networks, and here author investigates the path between the two protocols, i.e., DSR and AODV, which is based on network simulation. Using these two protocols in a sensor hub makes no nodes encumber in the network. Using the router discovery process nodes can be easily able to choose a feasible path to deliver the data in such a way that no node in the overburden. The author has designed the new table known as a search of the next node inquiry table which gives the best nearest node to choose an optimal path and also to distribute a load. Through this approach, the packet is being transmitted and solved the end-to-end delay problem. In future work using the SONNET approach, many security metrics can also be computed to enhance the scalability of the network [13].

7 Conclusion

In this research paper, we have carried out a review on assured ad-hoc networks for wireless sensor networks that concerning different types of protocol attacks in a sensor network. We have also performed a comparative analysis of the different author's work executing in different methods to enhance the lifetime of the sensors network, while using different improvement areas by the use of various protocols in the field of wireless network possibly enhance the quality of coverage and quality of service of the network. Based on our research, we have concluded that protocols and attack assaults provide real-time reporting that helps the network manager to manage the whole network functionality, and we can ensure that using the various approaches in ad-hoc networks one can easily enhance the functionality of the network.

References

1. Pravin, A., Sarika, S., Selvamini, K., Vijayakumar, A.: Security issues in MANET. In: Publications in Proceedings of 2nd International Conference on Intelligent Computing, Communication and Convergence (2016)
2. Alkhuder, S.: Securing Ad-hoc on demand distance vector routing protocol in wireless sensor network. In: Publications in Proceedings of International Conference on Computing Science and Engineering (2018)
3. Sethi, S.N., Rout, A., Mishra, D.: An effective and scalable AODV for wireless ad-hoc sensor networks. Int. J. Comput. Appl. **5**(4), (2010)
4. Hussien, A.A., Matloob, S.I.: The comparative study some of reactive and proactive routing protocols in the wireless sensor networks. J. Univer. Babylon **26**(4) (2018)
5. Alfawaer, Z.M., Riyaz, B.M.: An enhanced multipath strategy in mobile ad hoc routing protocols. In: Publications in Proceedings of 9th IEEE-GCC Conference and Exhibition (2017)
6. Ranjan, R., Singh, N.K., Singh, A.: Security issues of black hole attacks in MANETS. In: Publications in Proceedings of International Conference on Computing, Communication and Automation (2015)
7. Biradar, R.C., Walikar, G.A.: Review on hybrid routing mechanism in MANET. J. Netw. Comput. Appl. **77** (2016)
8. Shruthi, S.: Proactive routing protocols for MANET:a review. In: Publications in Proceedings of International Conference on IoT in Social, Mobile, Analytics and Cloud (2017)
9. Azees, M., Vijaya Kumar, P., Jegatha Deborah, L.: A comprehensive survey on security services in vehicular Ad-Hoc networks (VANETs). IET Intell. Transp. Syst. **10**(6),8 (2016)
10. Ahmed, A., Khan, U.A., Rajpa, A.: Analysis of ad-hoc on demand distance vector routing protocol against node misbehavior attacks in wireless sensor network. Quaid-E-Awam Univer. Res. J. Eng. Sci. Technol. **13**(2) (2014)
11. Mex-Perera, C., Del-Valle-Soto, R., J.A. Ramiro, Nolazco-Flores, S., Velazquez, A.C.: WSN energy model and its use in the optimization of routing protocols. Energies (2020)
12. Shete, P., Rajgor, M., Awale, R.N.: Probabilistic energy efficient routing protocol for WSN. In: Publications in Proceedigs International Conference on Communication, Information and Computing Technology (2018)
13. Belgaum, M.R., Musa, S., Mohd Su'ud, M., Alam, M., Soomro, S., Alansar, Z.: Secured approach towards reactive routing protocols using triple factor in mobile ad-hoc networks. Annals of Eng. Technol. Comput. **3**(2), (2019)

A Comparative Study on Securely Sharing and Searching Data Using Edge of Cloud-Supported Internet of Things

Kanika Gupta, Kalpana Yadav, and Ankita Singh

Abstract With many advances in the technology going on day by day, new needs have arrived. Internet is one of these. To make it more advance, we have come up with Internet of Things (IOT) which is used to link the digital gadgets to the Internet. To add more advantage to it, we use cloud with it where all the data produced by them is stored there which reduces the burden of smart gadgets. Additionally, edge servers are used by placing them close to the smart gadgets and can be used for communication and provide security operations. In this research paper, we have presented comparative literature review on the securely data searching and sharing using edge of cloud-supported Internet of Things. We have reviewed in terms of area of application, findings, and other relevant factors.

1 Introduction

Internet of Things also known as successor of Internet. It came into existence in the year 1999 by Ashton in MIT. With the help of Internet of Things, all smart physical gadgets can be linked to Internet. It has caught the eyes of many industries because of its many advantages. It can be very useful in the area of digital cities, health and fitness, digital cars, digital retails, and much more. It is said by Cisco that by the year 2020, 50 billion gadgets will be linked to the world of Internet. When smart physical devices are used in IOT environment, they tend to produce large amount of data which can be used for analysis and processing but for this they need storage. It is lacked by the smart physical devices as they do not have much capacity to store large volume of data. This problem is resolved with the help of cloud. Cloud provides abundance of storage facility with processing feature and approachability to the

K. Gupta (✉) · K. Yadav · A. Singh
Indira Gandhi Delhi Technical University for Women, New Delhi, India

K. Yadav
e-mail: kalpanayadav@igdtuw.ac.in

A. Singh
e-mail: ankita@igdtuw.ac.in

T. Senjyu et al. (eds.), *ICT with Intelligent Applications*, Smart Innovation, Systems and Technologies 248, https://doi.org/10.1007/978-981-16-4177-0_18

stored data from anywhere at any point of time. It is very helpful for devices in IOT applications as they have shortage of space to store data with them. But when these smart devices are running in IOT applications, they have few requirements such as high data rate, low latency, real-time data processing, fast data access, and mobility support which cannot be well taken cared by the cloud. To provide compatibility with above drawbacks, edge servers can be used. Edge servers are semi-trusted, freestanding network devices which are kept closed to the smart physical devices. They provide long distance connection to the smart devices which is required by them when they are running as part of the IOT applications. Apart from this, edge servers work in good coordination with cloud. They also provide security operations namely encryption, decryption, and key management. When the physical smart devices are a part of IOT applications, then it is required that they share data without which there existence is meaningless. This sharing of data raises the concern for security because of CIA triad, i.e., confidentiality, integrity, and availability. Confidentiality here refers that data is accessed by the ones who are authorized to use it. Integrity means that the data is not compromised and remains in its original form. Availability means that data is available to its users and are not denied from it. If any of it missing, then there might be chances of attack from the attacker which can lead to compromise, delete, stealing, or altering of data which can be dangerous. Suppose a device A wants to share data with other devices which are a part of the application. The data shared by the device will be in plaintext and due to which any malicious attacker can steal the sensitive data and data gets compromised. When the device B wants to search the desired data kept in storage in its original form, then anybody can steal the information from the storage by getting access to it and so it is required that the data is shared between the smart devices in an encrypted form and decrypted by the authorized device only as it will have the key to decrypt it. Due to the above-said reasons, a data-searching technique is also required, so that data can securely searched and retrieved only by the authorized devices and no one else. Sharing and searching go hand in hand, and so security concerns of both need to be addressed, so that security attacks can be minimized, and the confidential data remains safe. In the current scenario, only few options are available for security during data sharing and searching. In this, we have done an intensive literature survey for securely sharing data on cloud or Internet of Things-based servers. We have analyzed the security aspects with key arrangement technique in the form of survey and tried to improve the performance of the existing system. We have planned a cryptographic scheme, so that smart devices can share data among others at the edge of cloud-supported IOT, and edge server handles all the security operations. Also planned searching scheme for data where the authorized users can find their desired data from the stored data placed in an encrypted form in the cloud storage. These security parts are not taken cared by IOT smart devices as it would only increase the computational burden on them.

2 Internet of Things

Internet of Things also known as IOT is considered to be evolution of Internet where all the smart gadgets would be connected to it and human beings and control it from wherever they want. For example, person A can control the fan of his house from his mobile while he is sitting in his office. The concept was introduced as early as 1982 when CoCo CoLa machine was linked to the Internet and was able to provide data related to the stocks and if the new drinks added were cold or not at the Carnegie Mellon University. With the help of IOT, we can improve the security of the system, and it also helps in large amount of data collection which can be used for analysis. It saves time and minimizes human involvement. Although it is quite complex and can hamper privacy, the important technologies of Internet of Things are machine-to-machine interfaces (M2M), wireless sensor networks (WSN), embedded sensors, transfer of data, RFID, etc. [1]. The components involved are connectivity, sensors, process, and people. IOT can help the organizations to improve their performance with the help of IOT security and IOT analytics. In IOT analytics, large volume of data can be generated which helps in understanding the end users, and IOT security helps in protecting the sensitive information of the end users to gain there trust and maintain reputation in the market. Additionally, it also helps the organization to improve productivity and increase asset utilization and create better opportunities of business for the organization as well as people. It is used in various applications. It can be useful in consumer applications like health applications, connected vehicles, and home automation. It can be very useful in providing assistance to those with disabilities and the elderly which includes alert systems and voice control.

The user registers itself and then login it by entering the username and password in the application. After login, the user can upload, share, search, and download the data as per his need. The data is uploaded on the edge server and stored in an encrypted form in the storage, so that it can be shared among the authorized users. It is kept in an encrypted form, so that the malicious attackers cannot get access to the data in its original form. The user can search for the desired data with the help of keywords. For encryption of the data AES encryption is used which is based on substitution and permutation network. The operations are connected to one another as inputs are replaced by the outputs. SHA is a hashing algorithm which is used here for verifying the integrity of the data. With the help of this hashing algorithm, the user can check that the data received by him is in its original form or not. It is done with the help of edge servers as they provide security operations and help in maintaining communication with smart devices and cloud servers. The analysis shows that it provides better efficiency in terms of processing time as compared to other systems and can be deployed for practical use [2].

3 Security in IOT

3.1 Key Encryption Schemes

In secret key encryption, the device of the user generates the key with which the data to be sent is encrypted. The data is sent to the receiver. The receiver uses the same key to decrypt the data and recover the data. Only single key is used in encryption which is shared among the sender and the receiver through a communication channel which is secure. If the channel is not secure then the attacker can attack while the sender and receiver are communicating and can cause any malicious activity [2].

In public key encryption, two keys are involved: public key and secret key. For each user, both the keys are generated. The data is encrypted with the public key of the receiver, and when received by the receiver, it is decrypted with the secret key of the receiver. The public key of the user is publically available to all, so that one who wants to contact can use the public key and can reach out to that particular person. The secret key is also called as the private key and is only available to its user, and no other person can have access to it [2].

Searchable key encryption is a kind of secret key encryption in which a trapdoor is used to search a specific data on encrypted. The user has to share the secret key with every authorized user to produce the trapdoor. With the help of the user's secret key, the trapdoor is created and is loaded to the storage space with request to search the data [2].

3.2 One Way Hash Algorithm and Digital Signature

It is basically an integrity checking in which it is checked if the data is changed or not, as it is necessary that the data should be received in its original form by the receiver in which the sender has send. This type of verification is done with the help of hash function. The process is such that a known hash function is put in to the data with a fixed length and the output value is called is hash value. It is only a one-way process as the data cannot be generated back from the hashed output. The sender sends the hashed output along with the data. The receiver generates the hashed output in a similar way and matches with the one received by the sender. If both the hashed output matches, then the data is not modified and is in its original form. In other words, it can also be said the data received is authentic and not corrupted [2].

A certificate is issued by a certifying authority. Along with certificate, key pair is also given which includes the public key and the secret key. Anyone who has the certificate can be considered authentic user and can also be used to identify him. When the sender sends the data to the receiver, he sends the hash value signed with the help of the secret key and the data. When the receiver receives it, he checks the hash value and the certificate for verification [2].

3.3 Homomorphic Encryption

In this kind of encryption, we execute operations on the encrypted data itself which leads to encrypted results. Mathematical operations are executed on the encrypted data. The result obtained is equivalent to the one when the operation is executed on the plaintext when there is no encrypted text. There are many homomorphic encryption which support a particular type of mathematical operation only. Only fully homomorphic encryption is successful in using many mathematical operations without the loss of any accuracy in the result which is decrypted. The con of using fully homomorphic encryption is that the computational complexity is high [1].

3.4 Secure Data Sharing in Cloud (SeDaSc)

It is a methodology which provides access control, data confidentiality, insider threat security, forward and backward access control, and integrity. For each user, two keys are generated, and only one is provided to the user. In this methodology, the data is encrypted with one key only. The second is kept in the possession of the cryptographic server which is trusted third party. It is capable of access control, encryption, decryption, and access control. The second is kept with it, so that the insider threat can be avoided. This methodology can be used mobile computing as well as the conventional computing [3].

3.5 Fog Computing Architecture

Fog computing is an extended version of cloud computing in which there is a provision of resources at the edge of the network. It is extended version as it provides high bandwidth, low latency, and location awareness. Fog computing is compatible with cloud computing and has strong connection with it. It can be kept in the close proximity of the end users. It can support mobility as it is aware of it location. It helps in saving the bandwidth as it reduces the volume of data at the early stages. It can be used in various applications such as smart grids, smart homes, smart vehicle, healthcare management, and many more [4].

4 Literature Review

There are various researchers working on the issue of security while searching and sharing data on the cloud as it is the need of the hour. At the current scenario, only a few solutions are available for the same. Mollah et al. have introduced a lightweight cryptographic scheme for secure data sharing by smart devices at the edge of cloud-based Internet of Things and furthermore introduced a scheme for secure searching of data at the storage by authorized users [2]. Similarly, Thirumoorthy et al. have also proposed a structure to secure data while sharing data with others and searching data at Datain cloud-based services [5]. Yi et al. have done a survey regarding the security and privacy concerns connected to fog computing and also mentioned the solutions for the same [6]. Sridevi has used least processing cost first technique in Internet of Things to address the issue of security while searching and sharing data [7]. Rajarao et al. have introduced an improved system for securely sharing data at the edge of cloud-assisted Internet of Things [8]. Rameshwaraiah et al. have introduced a distribution plan for securely sharing and searching data when smart devices are used as a part of IOT applications [9]. Tao et al. have proposed a scheme for securely sharing and searching data in cloud edge collaborative storage (CECS) [10]. Pravallika et al. say that Internet of Things is considered to evolve Internet where all the digital shrewd gadgets can be connected to the Internet and provide data which can be useful. But these gadgets cannot store this data, so they take the help of cloud. The cloud provide them abundance of storage space and other useful resources. The paper proposes a plan for sharing and searching of data in a protected manner from the storage area with the help of edge servers placed closed to the IOT devices [11]. Singh et al. have discussed the issues of security w.r.t IOT while taking into consideration cloud-tenants, end users, and cloud providers [12]. Yi et al. have proposed the definition of fog computing while discussing its current definitions. In this paper, the issues and goals of fog computing are discussed with implementation of it [4]. Ali et al. have proposed a working prototype of SeDaSc methodology and evaluated on consumption of time considering many operations [3]. Kumarage et al. have discussed the way in which homomorphic encryption can be used to provide data security while data searching and sharing in cloud-based data analytics systems in Internet of Things [1]. Palattella et al. have discussed about the 5G technologies which can be used for IOT while taking into account the technology and standardization [13]. Bernabe et al. have presented a plan which is known as TACIoT which stands for trust-aware access control system for IOT. The plan is presented because in IOT applications multiple digital devices are connected to it which share information with each other through a weak network. So to provide access control and reliable communication, TACIoT is used. TACIoT extends access control system and is implemented with constraint and non-constraint digital devices [14]. Bhagat and Rathee have discussed various techniques which can be used for sharing data in a secure manner in cloud [15].

5 Review on Various Measures of Securing Data Using Edge of Cloud-Supported Internet of Things

This section provides information about various security measures taken while securing data using edge of cloud-supported Internet of Things. The description is presented in in Table 1.

6 Conclusion

In this paper, we have carried a comparative study on ways of securing data searching and sharing using edge of cloud-supported Internet of Things. While using different types of security measures in this field, it increases confidentiality, integrity, and availability of the data, and so it makes it feasible to apply IOT in different types of environments. Based on our research, we have concluded that security measures in Internet of Things can surely enhance the security and protect the privacy of various applications, so that the attack by the malicious attackers and loss of sensitive data can be prevented and minimized, though it still requires future research for improving the search time so as to improve the security in this field.

Table 1 Comparative review of literature work

S. no	Author	Problem	Area of application	Findings	Future research
1	Mollah et al. [2]	Security issue while searching and sharing of data by the smart devices	Secret key encryption and public key encryption	Provides better efficiency in case of processing time	Work on access control and authenticating issues
2	Tao et al. [10]	Leakage of data in CECS due to safety issue in cloud and edge servers	Public and private key pair and SE encryption	Ensures confidentiality and reduces computing overhead	Not addressed
3	Pravallika et al. [11]	Security of data at the edge of cloud-supported IOT	Implemented CTAC model	Assured security achieved with the model	Comparative analysis of presented CTAC model
4	Thirumoorthy et al. [5]	Framework required to provide security in Datain cloud-based services using IOT	Encryption and decryption algorithms	Better performance achieved and gadgets attached to the servers slowly get bottleneck	Not addressed
5	Bernabe et al. [14]	Need for reliable communication and adaptable mechanism between multiple devices	Architectural reference model	Considerable performance achieved in trust values	Proceed further to measure the accuracy
6	Li et al. [4]	Facing challenges in using cloud	Fog computing architecture	Low latency and high bandwidth	Plan for imposing full fog platform
7	Kumarage et al. [1]	Requirement for security cloud-supported data analytics system for IOT	Homomorphic encryption	Capable of performing analysis on ciphertext	Nature of data processing models and analysis tasks
8	Ali et al. [3]	Threats to the cloud storage which leads to leakage of data	SeDaSC	Access controls for malicious insiders, confidentiality	Extended by limiting the trust level

References

1. Kumarage, H., Khalil, I., Alabdulatif, A., Tari, Z., Yi, X.: Secure data analytics for cloudintegrated internet of things applications. IEEE Cloud Comput. **3**(2), 46–56 (2016)

2. Mollah, M.B., Azad, Md.A.K., Vasilakos, A.: Secure data sharing and searching at the edge of cloud-assisted internet of things. IEEE Cloud Comput. **4**(1), 34–42 (2017)
3. Ali, M., Dhamotharan, R., Khan, E., Khan, S.U., Vasilakos, A.V., Li, K., et al.: SeDaSC: secure data sharing in clouds. IEEE Syst. J. **99**, 1–10 (2015)
4. Yi, S., Hao, Z., Qin, Z., Li, Q.: Fog computing: platform and application. In: 2015 Third IEEE Workshop on Hot Topics inWeb Systems and Technologies (HotWeb), pp. 73–78. IEEE (2015)
5. Thirumoorthy, D., Kedir, T., Geleta, R., Kadir, A.: A secured frame work for searching and sharing of datain cloud based services using IOT. Int. J. Adv. Eng. Res. Sci. **6**(6), 503–507 (2019)
6. Yi, S., Qin, Z., Li, Q.: Security and privacy issues of fog computing: a survey. In: Publication in Proceedings of International Conference on Wireless Algorithms, Systems and Applications (2015)
7. Sridevi, B., Phaneendra, S., Manasvi, S.: Secure data sharing and searching at the edge of cloud—assisted using least processing cost first technique in internet of things. J. Emer. Technol. Innov. Res. **6**(3), 194–198 (2019)
8. Rajarao, P.B.V., Srinivasu, N.: Design of improved secure data sharing and searching at the edge of cloud-assisted internet of things. Int. J. Innov. Res. Technol. **4**(6), 180–184 (2017)
9. Rameshwaraiah, K., Babu, K.S., Reddy, S.S.: Data-sharing and - searching scheme to share and search data securely by IOT smart devices. Int. J. Recent Technol. Eng. **8**(4), 7558–7561 (2019)
10. Tao, Y., Xu, P., Jin, H.: Secure data sharing and search for cloudedge-collaborative storage. IEEE Access **8**, 15963–15972 (2019)
11. Pravallika, C., Prakash, I.B., Sukanya, P.: Secure data sharing and searching at the edge of cloud-assisted internet of things. Int. J. Adv. Res. Comput. Sci. Softw. Eng. **8**(4), 328–332 (2018)
12. Singh, J., Pasquier, T., Bacon, J., Ko, H., Eyers, D.: Twenty security considerations for cloudsupported internet of things. IEEE Internet of Things J. **3**(3), 269–284 (2016)
13. Palattella, M.R., Dohler, M., Grieco, A., Rizzo, G., Torsner, J., Engel, T., et al.: Internet of things in the 5g era: enablers, architecture, and business models. IEEE J. Selected Areas in Commun. **34**(3), 510–527 (2016)
14. Bernabe, J.B., Ramos, J.L.H., Gomez, A.S.: Taciot: multidimensional trust-aware access controlsystem for the internet of things. Soft. Comput. **20**(5), 1763–1779 (2016)
15. Bhagat, A., Rathee, N.: Addressing techniques for secure data sharing in cloud. In: 2018 Second International Conference on Inventive Communication and Computational Technologies (ICICCT), pp. 499–503. (2018)

Comprehensive and Empirical Evaluation of Classical Annealing and Simulated Quantum Annealing in Approximation of Global Optima for Discrete Optimization Problems

Madhan Kumar Srinivasan and Kishore Kumar Gajula

Abstract The classical annealing and simulated quantum annealing are heuristic algorithms, which simulate the natural dynamics of physical systems to solve hard optimization problems. The former makes use of thermal fluctuations to arrive at the best solution, while the latter either partially or completely ignores classical dynamics and uses quantum fluctuations instead. The literature says quantum algorithms are superior to their classical counterparts in terms of convergence speed and optimal solutions. The classical effects are relatively easy to simulate on classical computing machines, but simulating quantum effects on classical machines proves hard because of their intrinsic parallel nature. To simulate quantum effects, one should make use of quantum Monte Carlo techniques borrowed from quantum physics. Most of the current literature available is focused on finding better algorithms to simulate quantum effects on classical machines, and not much research has been conducted to evaluate the performance of the existing algorithms on optimization problems. In this paper, we try to address the effectiveness of simulated quantum annealing algorithm in finding the global optima for a distinctive set of combinatorial optimization problems and also compare the solutions obtained from simulated quantum annealing with the solutions obtained from its classical counterpart, simulated annealing algorithm.

1 Introduction

The world we live in is full of choices. These choices are not just limited to activities of daily living, instead they influence all aspects of human endeavor, every engineering design we make, and every program we write in the world of information technology. Computational techniques such as discrete optimization help us in choosing the right set of components in the right proportions while building a product or business

M. K. Srinivasan (✉) · K. K. Gajula
Cloud Analytics, ICI Innovation Center, Accenture Technology, Bangalore, India
e-mail: madhan.srinivasan@accenture.com

K. K. Gajula
e-mail: kishore.kumar.gajula@accenture.com

T. Senjyu et al. (eds.), *ICT with Intelligent Applications*, Smart Innovation, Systems and Technologies 248, https://doi.org/10.1007/978-981-16-4177-0_19

strategy. This has a dramatic impact on cost savings, the performance of products, and quality of services.

As the time complexity of these discrete optimization problems grow, they could not be solved by classical computers and algorithms in polynomial time. The necessity drove researchers to invent new field of computation called quantum computing. It is primarily focused on the development of information processing systems and algorithms based on quantum physics. The qubits are the basic building blocks of these quantum computers. A lot of work went into perfecting and scaling these qubits, yet these quantum computers are very noisy and notorious for being unstable. The actual quantum hardware progress is nowhere near expected despite all efforts and advancements in the field. Researchers are parallelly trying to simulate the behavior of these systems on the classical computer using algorithms and simulation software. Quantum annealing (QA) is one of the physical realizations of quantum computer, which is also a restricted form of adiabatic quantum computation focused on solving combinatorial optimization problems. Simulated quantum annealing, as the name indicates, is a simulated version of QA targeted for digital computers to study the nature of quantum effects without using any actual quantum hardware.

In this paper, we present two different discrete optimization algorithms and their implementations, both of these algorithms were inspired by physical systems, one is simulated annealing (SA) algorithm, which is a classical counterpart of simulated quantum annealing (SQA), and it primarily relies on the thermal agitations as the basis of computation. The second one is quantum simulated annealing, which makes use of quantum dynamics as basis for computation. Since both algorithms are simulated versions of actual physical systems, there are many unanswered questions such as can we really simulate a quantum transition and take advantage of it in obtaining better optima with speedups promised by quantum computation theory on a classical or digital hardware? Is every possible implementation of simulated quantum annealing algorithm superior to classical annealing algorithm?

Crosson et al. [1] argue that SQA is more effective than SA and some other results [2] shows that SQA has not much considerable advantage over SA. In all these cases, they have either performed theoretical analysis or performed very limited experimentation just considering only two or three optimization problems. The limitation of this approach is that not all heuristics obtain best solution for given optimization problem, so we have considered an empirical approach in evaluating the effectiveness of the algorithms. The contents are organized as follows: In Sect. 2, we introduce the discrete optimization problem, its nature, and applications. Section 3 introduces the theory of simulated annealing, its realization, and inner workings. Section 4 introduces the simulated quantum annealing algorithm and its inner workings. Section 5 provides the related research findings on various implementations of SA and SQA algorithms alongside their applicability and effectiveness. Section 6 introduces a standard set of combinatorial optimization problems used for benchmarking, and it also captures the solutions obtained by SA and SQA. The conclusions drawn from our work are presented in Sect. 7.

2 Discrete Optimization

Optimization is finding the maxima or minima of a given mathematical function that is defined on a certain domain. Classical optimization techniques that make use of calculus and numerical analysis deal with infinite domains. Combinatorial optimization is the discrete counterpart of the optimization that runs over real values. It is primarily concerned with the task of choosing an optimal solution from a given set of finite or countable infinite solutions.

Optimality is always defined with respect to some criterion function, which is to be maximized or minimized. Solving problems on finite domains seems trivial, but finding the optimal travel route for the traveling salesman or finding minimum vertex cover or finding the Hamiltonian cycles in a graph proves to be practically hopeless even for instances of very small or moderate sizes [3]. As per complexity theory, a problem is tractable only if one can find an optimal solution in polynomial time.

For many optimization problems with wiggly objective functions and complicated constraints, if the problem instance grows too large, it is almost impossible to find an optimum solution in a reasonable amount of time. Most of the real-world problems fall under this category. This category of problems is called as NP-Complete [4]. For these problems, where an exhaustive search is almost impossible and at times even optimal solutions are too hard to find, we have a new category of powerful algorithms called heuristic algorithms. These are designed to give away solution to a problem in a faster and more efficient way, just by sacrificing accuracy and precision for speed. We have considered a couple of NP-hard problems such as Max-Cut, vertex cover, integer programming, and graph coloring along with traveling salesman for our study.

Where and when optimum solutions are too hard to find, heuristics help to produce approximately optimal or near-optimal solutions. It is worth noting that these sub-optimal solutions have a guaranteed solution quality. For example, for a given problem, the value of the heuristic solution is at least 85% of the optimal value for every input. These limits on lower and upper bounds of optima help to analyze and compare the performance of the algorithm at hand with other available algorithms.

The most interesting fact about these metaheuristics is that majority of these algorithms were inspired by either physical or biological phenomena. The popular list of metaheuristic algorithms includes genetic algorithm, group search optimization, swarm intelligence, etc. In this paper, we try to evaluate the effectiveness of two of these algorithms, namely classical annealing and simulated quantum annealing in the context of discrete optimization.

3 Classical Annealing

Classical annealing (CA) or simulated annealing (SA) is yet another metaheuristic algorithm that approximates global optima for optimization problems with a large search space. It is inspired by annealing mechanism in materials science. This technique involves heating and controlled cooling of materials to reduce all possible material defects, which in turn increase the crystallinity. There is a strong correlation between the thermodynamic free energy and the quality of materials. The right heating and cooling cycle affect the thermodynamic free energy of a material. CA simulates annealing procedure to find an approximate global minimum.

In classical annealing, we establish a connection between problem instance and the physical system that assume a finite set of temperature states. This leads to a one-to-one correspondence between all possible solutions of a given problem instance and the energy states of the physical system, and this exercise transforms the objective of finding the best solution into finding the ground state of the system. With high initial temperatures, it is very easy to obtain the equilibrium state of the system. For successful annealing, we ensure that the initial temperatures are sufficiently high.

Consider a system at equilibrium, start lowering its temperature, drive it toward zero by following a right thermal scheme, provided very small ramp down rates, and the system maintains equilibrium and reaches the ground state when the temperature becomes zero. We thus obtain the solution. Figure 1 provides a clear insight into thermal dynamics of the system. The basic idea is to use thermal fluctuations to facilitate the system to arrive at global minima and not to get stuck at local minima of the cost function.

In the context of exploring the configuration space using classical annealing, the notion of controlled cooling is interpreted as a slow decrease in the probability of accepting worse solutions. At the cost of accepting a few worse solutions initially, we can search large function spaces, and this is the fundamental aspect of metaheuristics. There are very many implementations of classical annealing. States of thermodynamic system can be simulated either by using stochastic sampling techniques or by obtaining solution to a set of kinetic equations for probability densities. Since former is simpler in terms of computational complexity, we have considered latter for this study.

The Markov Chain Monte Carlo (MCMC) algorithms form a predominant subclass of MC sampling techniques. These algorithms can generate a Markov chain with the required invariant distribution. The classical annealing algorithm discussed

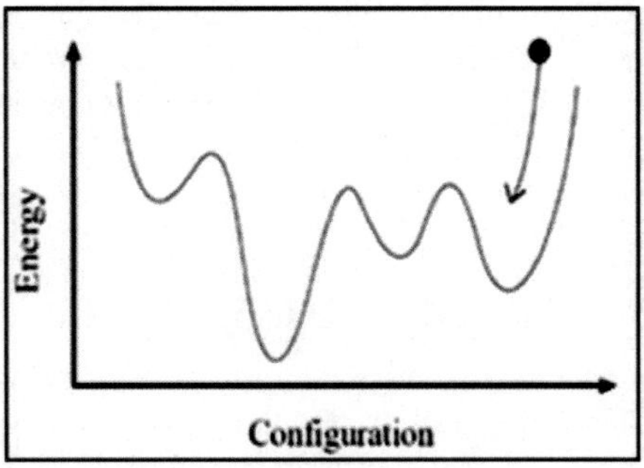

Fig. 1 Finding the global optima by classical means

in this paper makes use of Metropolis–Hastings (MH) sampler, which is one of the popular MCMC [5, 6] techniques.

Given a desired equilibrium distribution π, finding an appropriate (arbitrary) transition probability from state i to state $j : q_{ij}$.

Metropolis–Hastings is a three-step method.

Step 1: If the chain is currently in state i, then in the proposal stage, a new state j is proposed according to q_{ij}.

Step 2: A random number is drawn uniformly from [0, 1] to determine whether the proposed step is taken.

Step 3: If the number is less than the Metropolis acceptance probability, then the

$$p_{ij} = \min\left\{\left(\frac{\pi_j q_{ji}}{\pi_i q_{ij}}\right), 1\right\} \tag{1}$$

proposed step is accepted; otherwise, the proposed step is rejected, and the chain remains intact. This generates a Markov chain with equilibrium distribution π.

Classical annealing's objective is to find the most probable state of a Markov chain: $k, \pi_k > \pi_i$ for all $i \neq k$. Metropolis–Hastings algorithm has been used to evaluate the acceptance probability for classical annealing.

$$p_{ij} = \min\left\{\left(\frac{\pi_j}{\pi_i}\right)^{\frac{1}{\tau}}, 1\right\} \tag{2}$$

τ is call temperature.

If τ goes from ∞ to 0, it becomes harder to accept/make a move. Tuning τ helps to identify the optimal breakeven between exploration and exploitation. More details on the theory and implementation of Metropolis–Hastings sampling and classical annealing are available for interested readers. View references [1, 7–9].

4 Simulated Quantum Annealing

Simulated quantum annealing (SQA) is a quantum Monte Carlo (QMC) algorithm [6, 10, 11], which is yet another metaheuristic algorithm to simulate quantum annealing on a digital computer. The quantum annealing has its roots in Ising model, which explains ferromagnetism and phase transitions in materials. To perform annealing, the optimization problem under study is to be encoded as an interacting Ising spin model. Adding the transverse field to the Ising model is a trivial way to introduce quantum fluctuations. Since Hamiltonian of the transverse field and that of Ising spins do not commute, we assume strong initial transverse field and a ground state as initial state for initial Hamiltonian.

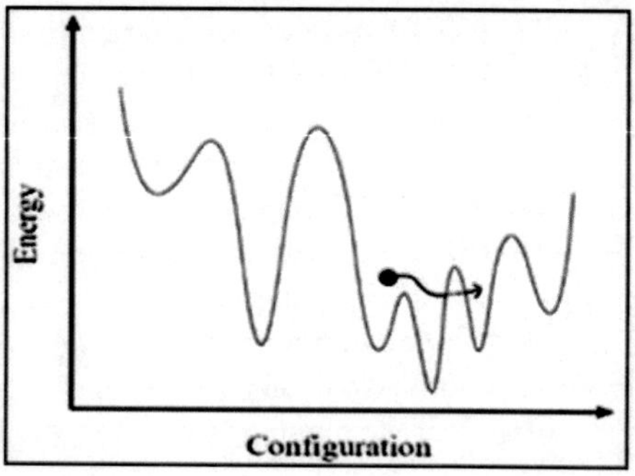

Fig. 2 Finding the global optima by quantum means

Initially, a one-to-one mapping is established between the energy states of the system and the candidate solutions of the optimization problem. A quantum mechanical superposition of those energy states is also assumed before we proceed further with annealing. All states are provided with equal weights before annealing. The evolution of system during annealing process follows the time-dependent Schrödinger equation. The amplitudes of all candidate states keep on changing in accordance with strength of time-dependent magnetic field.

The nature of quantum fluctuations is captured in Fig. 2. For a system to evolve adiabatically and to end in a ground state of the Hamiltonian as transverse magnetic field completely vanishes, the time rate of change of the magnetic field should be as slow as possible. By the end of annealing, the system is expected to have arrived at the ground state that corresponds to the solution of the optimization problem under study. Quantum tunneling between states forms the basis for selecting best candidate solution.

Quantum Monte Carlo (QMC) comprises a large family of algorithms that use classical Markov Chain Monte Carlo methods to estimate low-energy states of Hamiltonians [12]. There are many QMC methods, each of which uses Monte Carlo techniques in different ways to solve the quantum many-body problem. Zero-temperature QMC techniques include Diffusion Monte Carlo, Gaussian QMC, and likewise Finite-temperature Monte Carlo techniques [12] include Continuous-time QMC and Path Integral Monte Carlo. There is no general theory which quantifies the computational complexity and effectiveness of many of these algorithms. Most of the physicists tend to choose a technique for a quantum many-body simulation [10], which involves a family of quantum particles, for example, bosons, electrons, etc.

For this study, the effect of quantum superposition was approximated by parallel execution of different world-line, which can effectively simulate quantum systems. Path integral Monte Carlo (PIMC) based on Suzuki–Trotter matrix decomposition has been utilized.

4.1 Annealing Mechanics

Quantum annealing Hamiltonian with n qubits can be represented in one of its standard forms as

$$H(\Gamma) = H_P + \Gamma H_B \tag{3}$$

where the H_B is a uniform transverse field, which corresponds to beginning Hamiltonian.

$$H_B = -\sum_{i=1}^{n} \sigma_i^x \tag{4}$$

$f(z) : \{0, 1\}^n \rightarrow \mathbb{R}$ is the function to be minimized, and it is been encoded by the final Hamiltonian H_P, which is the diagonal in the computational basis.

$$H_P = \sum_{z \in \{0,1\}^n} f(z)|z\rangle\langle z| \tag{5}$$

Initially, the system is set in a uniform superposition of computational basis states. This corresponds to ground state of H_B. The field parameter (Γ) is reduced sufficiently slowly from a value $\Gamma \gg |H_P|$ to zero. This brings the system to the ground state of H_P.

The rate at which the field parameter is decreased determines the actual run time of the algorithm. The adiabatic theorem implies that a total run time, which is polynomial in $g_{\min}^{-2}$, must suffice.

$$g_{\min} = \min_{\Gamma} E_1(\Gamma) - E_0(\Gamma) \tag{6}$$

The term $g_{\min}$ represents the minimum energy gap between the ground/lowest state energy and first excited energy level during the system evolution.

4.2 PIMC Implementation

The path integral Monte Carlo method is based on the Suzuki–Trotter approximation.

$$\mathrm{e}^{A+B} \approx \left(\mathrm{e}^{\frac{A}{L}} \mathrm{e}^{\frac{B}{L}}\right)^L \tag{7}$$

Trotter approximation is used along with a quantum to classical mapping by which the QA thermal density matrix ρ

$$\rho = \frac{e^{-\beta H}}{tr\left(e^{-\beta H}\right)} \tag{8}$$

at inverse temperature β is approximated by a classical Boltzmann distribution. Rigorous derivation of the PIMC method, additional theoretical foundations, and implementation details are available at [13–15].

5 Related Work—CA & SQA

Whether we can perfectly simulate any arbitrary quantum system, which is inherently parallel on a simple sequential machine-like digital computer in polynomial time, is still an open question. A lot of research has been conducted on these lines [16, 17] and yet more expected to happen. Our current study primarily focuses on contrasting the classical annealing and quantum annealing-based algorithms and in determining, if quantum is always superior to classical in terms of performance.

We can draw analogies between classical and quantum annealing [1]. In classical annealing, temperature plays a similar role to quantum annealing's tunneling field strength. In classical annealing, the temperature determines the probability of moving to a higher energy state from the current state. In quantum annealing, the strength of the transverse field determines the probability of amplitude changes for all states in parallel. Some studies have compared CA and SQA results numerically [9, 18]. Most of them have reported in favor of SQA. However, we have no strong analytic evidence that shows the superiority of SQA over CA. In many cases, it remains an open challenge to determine the performance of both QA and SQA.

One evidence for the strength of SQA over CA comes from an example by Farhi et al. [19]. They have carefully selected a cost function with many thin and tall energy barriers to show an exponential separation between quantum and classical algorithms. The thermal fluctuations consume exponential time to climb the barrier, but the quantum fluctuations provide additional degrees of freedom such as tunneling through the barrier to circumvent the situation, and this maximizes the probability of arriving at global minima in polynomial time. Not all real-world optimization problems have thin and tall energy barriers in their cost function landscapes. So, quantum effects such as tunneling would not help much for those problems.

6 Experimentation and Results

Are all quantum annealing implementations equally powerful and superior to classical annealing? We did try to answer this question in this study. We have followed a statistical approach to evaluate the effectiveness of the algorithms by using experimental data obtained from applying two algorithms on various combinatorial optimization problems. The list includes traveling salesman, vertex cover, graph coloring, Max-Cut, job sequencing, binary integer linear programming, number partitioning, and prime factorization. We have considered small to medium problem instances for each problem class and chose the best performance metric values.

Since we analyze a specific quantum annealing implementation for this study, our evidence is not obviously decisive. However, the strategy applied here may be valuable for future studies on quantum annealing implementations to know the appropriate class of cost functions where quantum annealing has an edge over thermal annealing. This approach can be adopted and further expanded to heuristics in general. All experiments have been performed using given below parameter settings for algorithms.

For CA:

Initial temperature [Ts]: 10.
Final temperature [Tf]: 0.02.
Iterations [ite]: 1000.
Decreasing rate of temperature [R]: 0.95.

For SQA:

Initial strength of transverse magnetic field [Gs]: 10 and 100.
Final strength of transverse magnetic field [Gs]: 0.02.
Trotter slices: 16.

Performance comparisons were made based on the run time and lowest energy state possible. The solutions with the lowest energy levels are always superior or at least equivalent to solutions with higher energy values. Shorter run times are always preferred over longer run times, if energy levels of both solutions fall in the same range. In the case of SQA, we experimented with different transverse field strengths to find the right configuration value. The better solution has lower LEL. For all problem instances, run times are presented for Lowest Energy Levels (LELs) only.

6.1 Traveling Salesman

For a given list of cities and the distances between them, the goal is to find the shortest possible tour, which starts from one city and visits each city and returns to the origin city. The running time for brute-force approach lies within a polynomial factor of O (n!), even for few tens of cities, the factorial function grows a lot, and solution becomes impractical. It is one of the most intensively studied NP-hard problems, and ever since it was formulated in 1930, it is also considered as a standard benchmark problem to evaluate performance of many optimization algorithms. Table 1 shows that simulated annealing has little edge over SQA in this case.

Table 1 Lowest energy solution for traveling salesman

Algorithm	Lowest energy level (LEL)	Run time (min)
CA	−8.0	0.104
SQA	−7.42	1.270

Table 2 Lowest energy solution for vertex cover

Algorithm	Lowest energy level (LEL)	Run time (min)
CA	−9.5	0.086
SQA	−8.24	0.136

Table 3 Lowest energy solution for graph coloring

Algorithm	Lowest energy level (LEL)	Run time (min)
CA	−6.0	0.246
SQA	−2.94	0.484

6.2 Vertex Cover

Vertex cover (VC) is also known as maximum vertex cover. VC of a graph is a set of vertices such that each edge of the graph is incident to at least one vertex of the set. The objective is to find a smallest vertex cover in a given graph. The VC is yet another NP-hard problem, which has no known polynomial scheme. As shown in Table 2, CA outperforms the SQA for the given instance of vertex cover problem.

6.3 Graph Coloring

Graph coloring is a very common graph problem, and given n colors, it is an assignment of colors to the components of the graph, subject to certain constraints. It is a way of coloring the vertices of a graph 'G' such that no two adjacent vertices are of the same color. This is also called as vertex coloring. The graph coloring problem belongs to NP-complete class with run time complexity of O $(2^n n)$. This is one of the problem instances, where CA greatly outperforms SQA as shown in Table 3.

6.4 Max-Cut

For a graph, a Max-Cut or maximum cut is a way of maximizing the cut, whose size is at least the size of any other cut. The problem can be stated simply as follows. Partitioning the nodes of a graph G into two subsets such that the sum of the weights of the edges going from one subset to its complementary subset is maximized. There

Table 4 Lowest energy solution for Max-Cut

Algorithm	Lowest energy level (LEL)	Run time (min)
CA	−2.0	0.095
SQA	−1.50	0.091

Table 5 Lowest energy solution for job sequencing

Algorithm	Lowest energy level (LEL)	Run time (min)
CA	−9.34	0.141
SQA	−4.93	0.352

are no known polynomial time algorithms for Max-Cut in general graphs as it belongs to NP-hard complexity class. Table 4 shows CA slightly outperforms SQA.

6.5 *Job Sequencing*

Job sequencing is a way of distributing workload equally by assigning jobs to multiple computer processes based on their integer lengths. The time complexity of direct solution to this NP-hard problem is exponential. Table 5 shows that CA greatly outperforms SQA.

6.6 *Binary Integer Linear Programming*

The integer programming problems, which involve 'yes' or 'no' decisions and restricted to only binary variables are called binary integer linear programming (BILP) problems. A combination of binary variables (1 or 0) can be used to realize any bounded integer variable. The BILP problem belongs to NP-hard complexity class, which makes many instances of the problems intractable, and heuristic methods are the right choice to tackle problems of this sort. In this case, as shown in Table 6, interestingly SQA outperforms CA.

Table 6 Lowest energy solution for binary integer linear programming

Algorithm	Lowest energy level (LEL)	Run time (min)
CA	−35.0	0.152
SQA	−36.25	0.840

Table 7 Lowest energy solution for number partitioning

Algorithm	Lowest energy level (LEL)	Run time (min)
CA	−5041.0	0.099
SQA	−5040.1	0.248

6.7 Number Partitioning

In computer science, partition problem or number partitioning is the task of deciding whether a given set S can be partitioned into two subsets S1 & S2, such that sums of elements in both subsets are equal. Although the partition problem is NP-hard, there is a pseudo-polynomial time solution, and also, there are many heuristics, which solve the problem optimally in most instances, that make number partitioning the easiest hard problem. Table 7 shows that CA slightly outperforms SQA.

6.8 Prime Factorization

In number theory, prime factorization is a method of decomposing a composite number into a product of prime numbers. No algorithm has been known, which can factor all integers in polynomial time. The problem perfectly belongs to NP complexity class but has not been proved if it belongs to the NP-complete class. Researchers generally believe that prime factorization is not one of those NP-complete problems. There are couple of published algorithms, which can factor b-bit numbers faster than $O((1 + \delta)\ b)$ for all positive δ. Table 8 shows that CA outperforms SQA.

The eight problems we have discussed so far were introduced in decreasing order of complexity. Most of these problems belong to NP-hard class. We have carefully chosen these problems that are originating from different fields, which include computer science, graph theory, number theory, linear programming, etc., and this gives a lot of scope to empirically evaluate the effectiveness of any optimization algorithm. Computational complexity theory provides a general framework, which helps in understanding the nature of these problems. Even though many of these problems belong to class NP-hard, they have different run time complexities for different heuristics, because of unique iterations between applied heuristics and the landscapes of problems objective functions. Some have better approximate solutions compared to others. With this general framework in mind, we try to draw conclusions based on results obtained from the experiments performed.

Table 8 Lowest energy solution for prime factorization

Algorithm	Lowest energy level (LEL)	Run time (min)
CA	−196.0	0.108
SQA	−195.34	0.59

7 Conclusion

For this study, we have considered small- and medium-scale instances of combinatorial problems discussed in Sect. 6. Both algorithms converged to almost the same energy levels. Time evolution curves of SQA as depicted in Figs. 3, 4, 5, 6, 7, 8, 9, and 10 clearly show the effects of quantum transitions. Except for one or two spikes in each curve, most of the part seems smooth, which is not the case with CA. For five out of eight problem cases, the SQA run times are longer than CA, and this could be attributed to a higher transverse magnetic field. In seven out of eight cases, CA has lower energy values than SQA. The results of job sequencing and graph coloring clearly show the superiority of CA over SQA.

The previous studies marked the possibility of SQA outperforming CA. But our results show that, for this SQA implementation, the CA has a slight edge over SQA. This explains that not all SQA implementations are equally powerful in simulating quantum fluctuations on classical hardware. The results also indicate the need for detailed analytical studies on energy landscapes of optimization problems. The beforehand knowledge on the nature of energy landscape helps us to choose the right algorithmic tools for problem at hand.

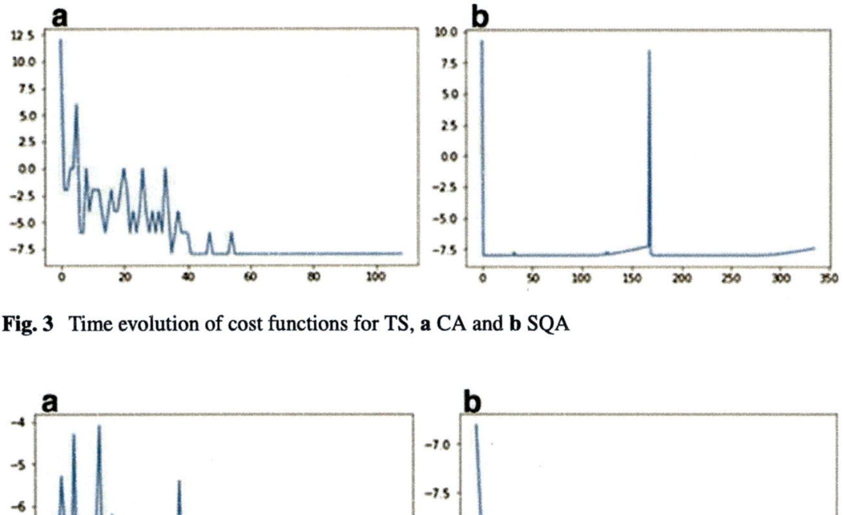

Fig. 3 Time evolution of cost functions for TS, **a** CA and **b** SQA

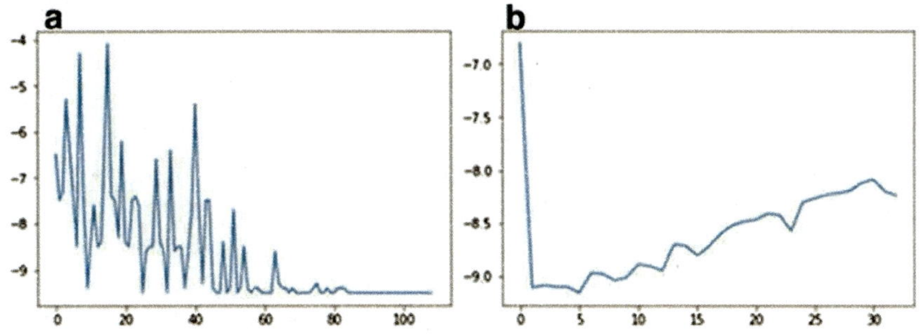

Fig. 4 Time evolution of cost functions for vertex cover, **a** CA and **b** SQA

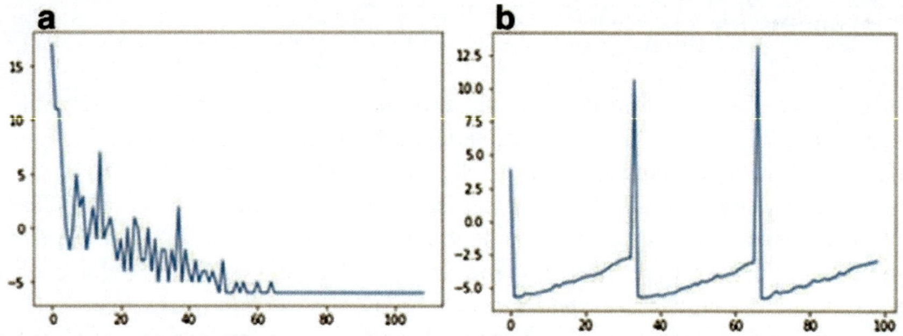

Fig. 5 Time evolution of cost functions for graph coloring, **a** CA and **b** SQA

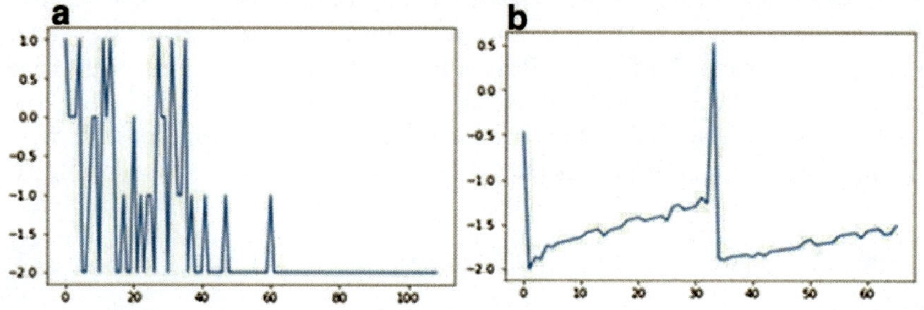

Fig. 6 Time evolution of cost functions for Max-Cut, **a** CA and **b** SQA

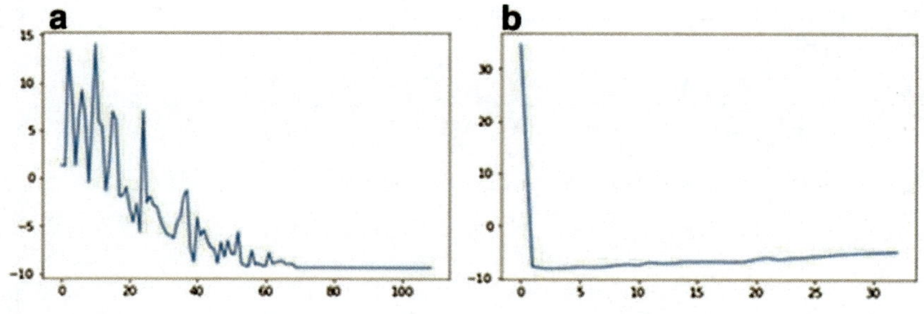

Fig. 7 Time evolution of cost functions for job sequencing, **a** CA and **b** SQA

Santoro et al. [2] provide a detailed and plausible explanation for SQA's poor performance compared to CA's. Reference [2] considers a hard 3-SAT problem to benchmark the performances of SQA and CA. According to reference [2], the worse performance of the SQA(PIMC-QA) over CA could be because of the intrinsic nature of the 3-SAT landscape. While exploring the energy landscape the flat, and open sectors selected by the quantum fluctuations pushes 3-SAT case towards dead ends.

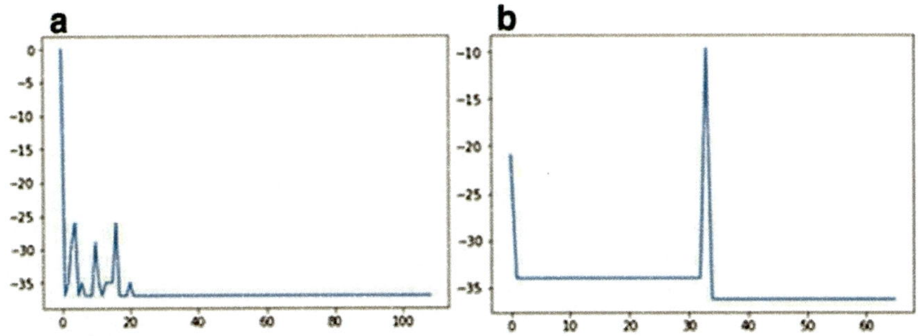

Fig. 8 Time evolution of cost functions for BILP, **a** CA and **b** SQA

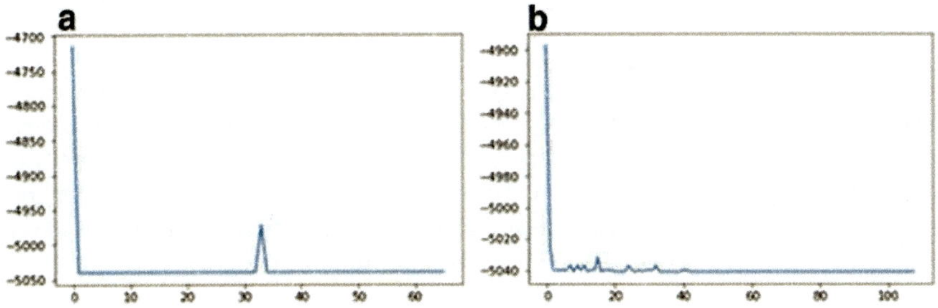

Fig. 9 Time evolution of cost functions for number partitioning, **a** CA and **b** SQA

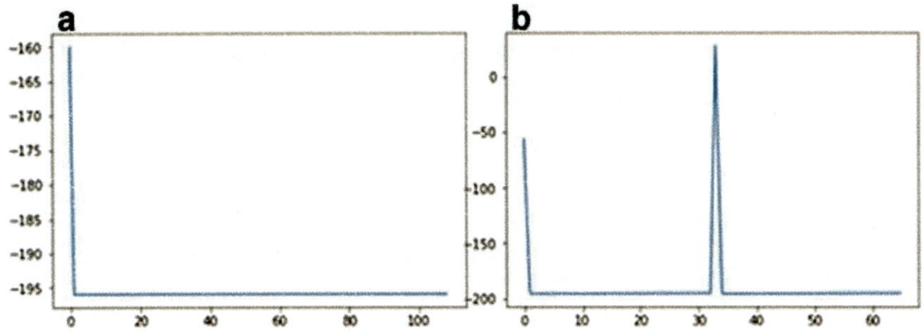

Fig. 10 Time evolution of cost functions for prime factorization, **a** CA and **b** SQA

Reference [2] also highlights the important differences between the CA and SQA dynamics by analyzing the energy landscape geometry, and the phase-space explored by the two algorithms is very different from each other, presumably because of the very fact that they resort to different physical dynamics for phase transitions. Reference [2] also shows that for the same value of energy, the quantum systems visit configurations with significantly larger number of flat directions and downhills.

Reference [2] also hypothesizes that classical annealing follows narrow potential wells, while the SQA prefers to explore the edges of mid-altitude plateaus.

The lowest energy levels attained by CA and SQA for most of the problem cases are very close in terms of their energy. We believe that this is because both CA and SQA are attaining solutions which correspond to very low-energy regions where clustering of states takes place. The comparison between the SQA and CA shows that the two algorithms explore different sectors of the phase-space characterized by different geometrical properties. The problem cases such as graph coloring and job sequencing record substantial difference in the solution quality obtained by both algorithms. We believe that broad basins of attraction present in energy landscape with considerably larger fraction of flat directions account for inefficient optimization in the case of SQA.

In conclusion, quantum optimization algorithms are not always better. Superiority of some quantum algorithms cited by other references [9, 18, 19] could be because of those algorithmic implementations. Different quantum implementations explore different sections of energy landscape. There are many questions yet to be answered. We need to understand the necessity of the reliable analytical theory, which explains the nature of the complex energy landscapes and right optimization algorithm for each landscape. We also believe that machine learning has a role to engineer the features of the energy landscape, cluster/classify optimization problems based on their energy landscapes, and predict the right algorithm for the given problem.

References

1. Crosson, E., Harrow, A.W.: Simulated quantum annealing can be exponentially faster than classical simulated annealing. In: 57th Annual Symposium on Foundations of Computer Science (FOCS), pp 714–723. IEEE (2016)
2. Crosson E, Deng M (2021) Cornell university arxiv homepage. https://arxiv.org/abs/1410.8484. Last Accessed 11 Jan 2021
3. Papadimitriou, C.H., Kenneth, S.: Combinatorial optimization: algorithms and complexity. Unabridged edn, Dover (1998)
4. Vazirani, V.V.: Approximation Algorithms, 2nd edn. Springer, Berlin (2003)
5. Kirkpatrick, S., Gelatt, C.D., Vecchi, M.P.: Optimization by simulated annealing. Science **220**, 671–680 (1983)
6. Morino, S.: https://github.com/shinmorino/wildqat. Last Accessed 11 Jan 2021
7. Otten, R.H.J.M., van Ginneken, L.P.P.P.: The Annealing Algorithm. Kluwer, Boston (1989)
8. Tadashi, K.: Cornell university arxiv homepage. https://arxiv.org/abs/quant-ph/0205020v1. Last Accessed 11 Jan 2021
9. Heim, B., Rønnow, T.F., Isakov, S.V., Troyer, M.: Quantum versus classical annealing of ising spin glasses. Science **348**(6231), 215–217 (2015)
10. Mazzola, G., Smelyanskiy, V.N., Troyer, M.: Quantum monte carlo tunneling from quantum chemistry to quantum annealing. Phys. Rev. B **96**(13), 134305 (2017)
11. Kadowaki, T., Nishimori, H.: Quantum annealing in the transverse Ising model. Phys. Rev. E **58**, 5355 (1998)
12. Mitchell, D.R., Adami, C., Lue, W., Williams, C.P.: A random matrix model of adiabatic quantum computing. Phys. Rev. A **71**, 052324 (2005)

13. Hadfield, S.: Cornell university arxiv homepage. https://arxiv.org/abs/1805.03265. Last Accessed 11 Jan 2021
14. Bapst, V., Semerjian, G.: Thermal, quantum and simulated quantum annealing: analytical comparisons for simple models. J. Phys.: Conf. Ser. **473**, 012011 (2013)
15. David, M.: Ceperley: an overview of quantum monte carlo methods. Rev. Mineral. Geochem. **71**(1), 129–135 (2010)
16. Suzuki, S: A comparison of classical and quantum annealing dynamics. J. Phys. Conf. Ser. **143,** 012002 (2009)
17. Kieu, T.D.: The travelling salesman problem and adiabatic quantum computation: an algorithm. Quantum Info. Process. **18**, 90 (2019) Springer
18. Farhi, E., Goldstone, J., Gutmann, S.: Cornell university arxiv homepage. https://arxiv.org/abs/quant-ph/020103. Last Accessed 11 Jan 2021
19. Battaglia, D.A., Santoro, G.E., Tosatti, E.: Optimization by quantum annealing: lessons from hard satisfiability problems, Phys. Rev. E **71**, 066707 (2015)

Enhancing Signal in Noisy Environment: A Review

Ganta Naga Sandhya Devi and Vinay Kumar Mittal

Abstract Noise is present in all environments. In the signal processing domain also, there is a presence of noise. In any application, after processing, we need a signal which is free from noise. So to obtain a high-quality signal, elimination and reduction of noise are important. The different noises include Salt-Pepper, Gaussian, Speckle, Poisson, etc. So in order to eliminate this noise, some of the noise reduction techniques or filters can be employed in between. This survey paper presents the elimination of noise in images during duct, rain, snow, etc., elimination of noise in speech signals or mobile audio in high noisy environment such as conferences, meetings, heavy traffic conditions, and elimination of noise in biomedical signals in the presence of other noisy signals. For this purpose, many filtering techniques can be used, and depending on the filtered image obtained, we can find out which is best suited for the specific purpose.

1 Introduction

In processing a signal, noise is one of the main issue. However, we tend to face many problems while processing a signal. There exists noise in between. There are many sources of noise which occur before processing, while processing and after processing. Various noises like Gaussian, Poisson, Blurred, Speckle, and Salt-Pepper noise occur. So in signal processing applications, removing noise is necessary for obtaining the high-quality signal, and for this purpose, the filters like Median filter, Gaussian filter, and Weiner filter can be used [1]. The standard filters or the best filters can be used to remove the noise in order to smoothen the image. The important aspect to be considered while removing the noise is that the information in the edges of the image also has to be retained. By taking these points into consideration, one can know which method is good and which is the best filter for the concerned

G. N. S. Devi (✉)
Gudlavalleru Engineering College, Gudlavalleru, India

V. K. Mittal
KL University, Vijayawada, India

T. Senjyu et al. (eds.), *ICT with Intelligent Applications*, Smart Innovation, Systems and Technologies 248, https://doi.org/10.1007/978-981-16-4177-0_20

application. The important aspects to be considered are eliminating the noise from the signal and preserving the originality of the signal. In this review paper, we focus on noise reduction applications in the domains of signal processing, i.e., image signal processing, speech signal processing, and biomedical image processing.

In image processing, after the processing is done on the signal, noise must be removed and the high-quality signal must be obtained. In speech processing, speech recognition and speech enhancement are done on the speech signal [2]. The goal of speech recognition is to recognize the speech of a user, whereas speech enhancement is focused on removing noise and enhancing the required speech signal. Recently, more research has been in multimodal speech processing. Human speech is multimodal, and there exists coherence between audio and visual signals [3]. This paper is organized as follows: Section 2 presents a description of multimodal signals. Section 3 presents how the speech signal is recognized in a highly noisy environment. Section 4 presents how the mobile audio signal is recognized in a highly noisy environment. Section 5 focuses on how an image signal is recognized in snow or duct. Section 6 focuses on how a biomedical image is recognized from other noisy images. Section 7 presents applications. Section 8 presents a summary and conclusions.

2 Multimodal Signals

Multimodality refers to using more than one imaging technique for production of signals. Multimodality imaging involves the combination of two or more imaging modalities. Multimodal signal processing field is a field that processes signals and combines information from different domains which significantly increase the performance of interaction of human–computer or increases the performance of human–human communication.

2.1 Multimodal Signals in Image Processing

Multimodality in image processing pertains to the multiple characteristics in the image. Whenever we convert image from one form to another or do some processing on the image, deterioration occurs at the output [4]. This noise is "salt and pepper." Salt and pepper lowers the image quality [5]. So denoising methods must be applied with the help of filters [6].

2.2 Multimodal Signals in Speech Processing

The nature of how we hear, interpret, and understand speech is multimodal in nature. Any observer will use the visual speech information to readily combine with auditory speech for better understanding. Speech perception is, however, inherently multimodal [7]. If we tend to speak something, the expression on the face also matters to get the exact meaning of the message to be conveyed [8].

2.3 Multimodal Signals in Biomedical Processing

In biomedical signal processing, computed tomography (CT) techniques are multimodal in nature. The best example of multimodality imaging is the PET modality which is mostly used as the "multimodal" imaging method [9]. Multimodal clinical PET and MRI techniques are now the latest fields. Various techniques (CT, MRI, PET, etc.) encompass abundant information. Therefore, by combining the multiple modality images, an accurate estimation can be done. To improve the accuracy of locating a tumor or any other problem, multimodal medical imaging is greatly helpful [10]. Deep learning has achieved great success in image recognition and also has a great future in multimodal medical imaging analysis [11]. Hence, these techniques are being used now. Near-infrared spectroscopy (NIRS) is one of the new methodology which uses multimodal utility [12]. This is a neuroimaging methodology. The other new field is two-dimensional electroencephalography (EEG) and magnetoencephalography (MEG) techniques.

3 Speech Signal in High Noise Environment

In this section, the focus is on obtaining the speech signal clearly when present in high noisy environment. This comprises of voice activity detection, noise estimation, removal techniques, and an optimal filter [13, 14].

Voice activity detection (VAD) is the process of detecting whether any speech signal is present or absent in the given signal. It is basically used in speech recognition process. In order to evaluate VAD, the signal-to-noise ratio (SNR) is calculated in terms of speech hit rate (HR1) and non-speech hit rate (HR0). The comparison of different signal processing methods used for noise reduction in speech signal are listed in Table 1.

3.1 Noise Estimation

Noise estimation is the first step to be done while processing a speech signal. There are many types of noise estimation algorithms [15]. This is based on different Weiner algorithms, the calculation of SNR, and the noise covariance matrix. To enhance the quality of the signal, the noise estimate is important, i.e., if the value of noise estimate

Table.1 Comparison of the different signal processing methods used in speech processing

S.no	Signal processing methods used	Overall quality	Reference papers
1	Individual algorithms such as MMSESPU, logMMSE, logMMSE-ne	Good	[24]
2	Fusion algorithms	Best	[21]
3	Subspace algorithms	Poor	[16]
4	Mel frequency cepstral coefficients	Better. Also used in NLP for noise reduction	[44, 58, 59]
5	Gammatone frequency cepstral coefficients (GFCC)	Better compared to MFCC	[45]
6	Adaptive LMS filter and ICA method, KNX technology, and LabView software tool	A success rate of 95% has been achieved	[46]
7	Spectral enhancement and temporal enhancement of speech signal	55.7%	[47]
8	Eigenspace normalization method	Relative error rate reduction greater than 16%	[48]
9	Depending on fundamental frequency	Accuracy up to 97% has been achieved	[55–57]
10	Classifiers such as SVM, DNN	Performance of SVM is 87.46% and more than DNN	[53, 54]

is very less, a residual noise will be heard in the background, and if the noise estimate is very high, the entire speech signal will be lost.

3.2 *Noise Reduction*

The process of decreasing the noise level is noise reduction. The algorithms for noise reduction are: spectral subtractive, subspace, statistical-model-based, and Wiener-type [16, 17]. The log minimum mean square error logMMSE and the spectral subtraction are the mostly used noise reduction algorithms, and many are in existence [18]. If any of the additional noise has been added, then by means of adaptive noise cancellation (ANC), it can be reduced [19]. The subspace algorithms are poor in enhancing the quality of the signal when compared to remaining algorithms [20].

3.3 *Fusion Techniques*

The combination of dynamic time warping (DTW) and hidden Markov model (HMM) is called recursive least squares (RLS) algorithm which improves the speech by eliminating the background noise [21]. The other fusion technique is the combination of independent component analysis (ICA) and Wiener filter. These fusion techniques give high performance compared to the individual techniques listed. These can be used in recognizing the speech [22].

4 Mobile Audio Signal in High Noise Environment

This section focuses on the use of inter-microphone Posteriori SNR difference (PSNRD) for the estimation of the probability of speech presence, which is best than the previous used method power level difference (PLD) [17]. The noises in mobile applications are complex. The noise in mobile are divided into quasi-stationary and transient components. The first one has a spectrum comparative to engine noise and sensor noise. The estimation of the first noise can be done using single-channel estimators [23, 24]. However, the multichannel minimum-variance distortion-less response (MVDR) filter is best suited for reducing the noise in mobile audio[25].The MVDR performance is higher when the noise level is low [26]. However, when the noise is very high, it is not suitable to use. Because of the direct utilization of the noise signals, the distortion of the classical MVDR is lower [27]. The estimation of the correlation matrix of speech signals using MVDR is always an approximation [28].

5 Detection of the Image in the Presence of Noise

The detection of any climatic hazardous conditions is important in present scenario for avoiding accidents in the presence of rain, snow, or duct and also for ground traffic management [29, 30]. Consider a rainy image or snow image. The image may consist of useful information, but it may not be clear due to rain or snow. So in order to get the image clearly, the foreground and background must be separated using Gaussian mixture model [31]. The rain is detected in the foreground, and then rain drop diameter is measured [32]. Then a Kaman filter can be used to calculate the parameters of HoS model. Then by using this histogram of orientations of rain or snow streaks (HoS) method, two distributions are obtained, i.e., the Gaussian distribution [33] and the uniform distribution [34].The rain or snow orientation is indicated by Gaussian distribution, and the noise is indicated by the uniform distribution. These two distributions are separated by a technique called expectation maximization technique. Then Gaussian distribution is processed for smoothing, and its amplitude is used to indicate the presence of rain or snow [35]. Then, the HoS model detects the pixels of rain or snow, and then, the estimation of intensity of the pixels is done [36]. However, after making the image clear, if any additional noise is present in the image, by using median filter with Mel frequency cepstral coefficients, that noise can be removed. Hence, a clear image can be obtained [37]. By using these methods in video camera surveillance, accidents can be avoided to the great extent possible in winter or in rainy seasons [38]. In the image, noise can be reduced using PSNR and MSE values.

6 Detection of Biomedical Image Signal in the Presence of Noise

Biomedical signals are the electrical signals collected from the human body. In biomedical signal analysis, many times we need a perfect image to diagnose to locate the exact defect or presence of extra growth of cells. Many factors worsen the image quality while scanning. So, in order to obtain clear image [39], we need filters like median, Weiner filter, etc. Median filter has the capability of good denoising, and accuracy is high [40, 41]. The median filters reduce the noise in the image [42]. By application of this filter, then one can exactly know the location of tumor, etc., in the human body. Also when compared to the Gaussian filter and Weiner filter, median filter reduces the noise and gives high clarity [43]. This helps the doctors to carry out the treatment in a correct way and save the lives.

7 Applications

The applications of noise removal in different domains such as image, speech, and biomedical are listed in the following Table 2.

8 Summary and Conclusions

In summary, different filtering techniques are used for removing the noise in the image and speech. However, in the end, a high-quality signal is obtained. The signal processing methods used in this paper for reducing noise in speech, mobile audio, and image signal in presence of snow or rain, biomedical noisy signal are listed below in Table 3.

The performance results of various methods used in this paper for enhancing the signal in high noisy environment are listed in Table 4.

Table 2 Applications of noise removal methods in each domain

S.no	Type of the signal	Context	Application	Reference papers
1	Speech signal	Enhancing speech signal in high noisy environment	Conferences, political meetings, college functions	[21]
2	Audio signal	Enhancing mobile audio signal in high noise environment	Bus stations, railway stations, in heavy traffic conditions	[25]
3	Image	Detection of the image in the presence of noise like snow	Crop images in winter season, rain, in the presence of weeds, Vehicle images in winter season to avoid accidents, Detection of critical weather conditions	[35]
4	Biomedical image	Detection of the biomedical image signal	Detection of tumor images in presence of other biomedical noisy images	[43]

In Sect. 3, the usage of fusion techniques has given better performance in reducing noise in the speech compared to the other individual noise approaches. Also each signal processing method used for noise reduction has its own advantage. In Sect. 4, the proposed MVDR filter has achieved high performance. In Sect. 5, by using the HoS model, one can get a clear image irrespective of the weather conditions, and this can be used in video-surveillance systems. In images, by means of calculating the PSNR and MSE values, there are different signal processing methods for reducing noise in the image. In Sect. 6, by using the median filter and by converting biomedical signals into individual frequency components, a clear image of the biomedical image can be obtained. By using some more techniques or enhancing the existing methods, we can improve the signal accurateness in each of the domains. The reduction of noise in any of the image or signal can be done in a limited way, but we cannot make the entire noise to be zero in any of the case. But enhancing the image or signal to a possible extent by using different signal processing methods can be done and the method must be chosen according to the application needed.

Table 3 Signal processing methods used in different studies

S.no	Type of the signal to be enhanced	Signal processing methods used/Features used/ Classifiers used	Reference papers
1	Speech signal	Recursive least squares algorithm	[21]
		Gammatone frequency cepstral coefficients (GFCC)	[45]
		Adaptive LMS filter and ICA method, KNX technology, and LabView software tool	[46]
		Spectral enhancement and temporal enhancement of speech signal	[47]
		eigenspace normalization method	[48]
2	Mobile audio signal	Minimum-variance distortion-less response (MVDR) filter	[25]
3	Image	Gaussian mixture model, histogram of orientations of snow and rain	[35]
		Modified fuzzy set filter	[51]
		Decision-based adaptive median filter	[49]
		Collaborative Weiner filtering	[50]
		Improved median filter	[52]
4	Detection of the biomedical image	Band pass filters like Weiner filter, median filter with deep learning techniques	[43]

Table 4 Performance results of different signal processing methods

S.no	Type of the signal	Signal processing method used	Performance results	Reference papers
1	Speech signal	Recursive least squares method	Noise is reduced to a great extent	[21]
		Gammatone frequency cepstral coefficients (GFCC)	Better compared to MFCC	[45]
		Adaptive LMS filter and ICA method, KNX technology and LabView software tool	A success rate of 95% has been achieved	[46]
		Spectral enhancement and temporal enhancement of speech signal	55.7%	[47]
		eigenspace normalization method	Relative error rate reduction greater than 16%	[48]
2	Mobile audio signal	Minimum-variance distortion-less response (MVDR) filter	The mobile audio signal noise is reduced by using this filter	[25]
3	Image	Gaussian mixture model, histogram of orientations of snow and rain	HoS method is best suited for detection of rain, and then, Gaussian mixture model is used to reduce the noise	[35]
		Modified fuzzy set filter	High PSNR, low MSE	[51]
		Decision-based adaptive median filter	PSNR of 30.46 for noise density of 50% MSE	[49]
		Collaborative Weiner filtering	Based on quality and PSNR value	[50]
		Adaptive median filter	24.28% of PSNR for a noise density of 35%	[52]
4	Biomedical image	Band pass filters like Weiner filter, median filter with deep learning techniques	The median filter is best suited to reduce the noise	[43]

References

1. Hong, L.: Independent component analysis based single channel speech enhancement using wiener filter (2003)
2. Sharath R.K.: Improved iterative wiener filtering for non-stationary noise speech enhancement (2004)
3. Rosenblum, L.D.: In: Speech Perception as a Multimodal Phenomenon (2008)
4. Mahdi, J., Mehdi, Z.: Impulse noise detection and reduction using fuzzy logic and median heuristic filter. In: International Conference on Networking and Information Technology (2010)
5. Tonghan, W., Xingyi, L. (2011). An efficient impulse noise reduction algorithm; Mancuso, M., Poluzzi: filter architecture particularly for video applications (1992)
6. Kanagalakshmi, K., Chandra, E.: Performance evaluation of filters in noise removal of finger print image. 1. Doctoral Research Scholar, 2. Director, Dept. of Computer Science, DJ Academy Managerial for Excellence (2011)
7. Mohanta, A., Mittal, V.K.: Classifying speech of ASD affected and normal children using acoustic features. In: Proceedings 26th National Conference on Communications (NCC 2020), IIT Kharagpur, India, 21–23 Feb (2020)
8. John Phillip, B., Kalyan, S.S.S., Mittal, V.K.: Discriminating high arousal and low arousal emotional speech using mahalanobis distance among acoustic features. In: Proceedings 26th National Conference on Communications (NCC 2020), IIT Kharagpur, India, 21–23 Feb (2020)
9. Pietrzyk, U., Herholz, K., Schuster, A., von Stockhausen, H.M., Lucht, H., Heiss, W.D.: Clinical applications of registration and fusion of multimodality brain images from PET, SPECT, CT, and MRI. Eur J Radiol. **21**, 174–182 (1996)
10. Mittal, S., Mittal, V.K.: Biomedical requirements for human machine interface towards building a humanoid: a review. In: Proceedings 16th International IEEE India Conference (INDICON 2019), Marwadi University, Rajkot, Gujrat, India, 13–15 Dec (2019)
11. Zhang, W., Li, R., Deng, H., Wang, L., Lin, W., et al.: Deep convolutional neural networks for multimodality isointense infant brain image segmentation. Neuroimage **108**, 214–224 (2015)
12. Guo, Z., Li, X., Huang, H., Guo, N., Li, Q.: Deep learning-based image segmentation on multimodal medical imaging. IEEE Trans. Radiat. Plasma Med. Sci. **3**(2), 162–169 (2019)
13. Ramírez, J.: Voice activity detection. Fundamentals and Speech Recogn. Syst. Robustness (2007)
14. Shrawankar, U., Thakare, V.: Noise estimation and noise removal techniques for speech recognition in adverse environment international federation for information processing (2010)
15. Sundarrajan: A noise-estimation algorithm for highly non-stationary environments (2005)
16. Boll, S.F.: Suppression of acoustic noise in speech using spectral subtraction. IEEE Trans. Acoust. Speech Signal Process. **27**(2), 113–120 (1979)
17. Hu, Y., Loizou, P.C.: A comparative intelligibility study of single-microphone noise reduction algorithms (2007)
18. Hasan, T.: Suppression of residual noise from speech signals using empirical mode decomposition (2009)
19. Benesty, J.: Noise reduction algorithms in a generalized transform domain (2009)
20. Górriz, J.M.: A novel LMS algorithm applied to adaptive noise cancellation (2009)
21. Rahman, S.A.: Robust speech recognition using fusion techniques and adaptive filtering (2009)
22. Droppo, J., Acero, A.: Noise robust speech recognition with a switching linear dynamic model (2004)
23. Martin, R.: Noise power spectral density estimation based on optimal smoothing and minimum statistics. IEEE Trans. Speech Audio Process. **9**(5), 504–512 (2001)
24. Hendriks, R.C., Heusdens, R., Jensen, J.: MMSE based noise psd tracking with low complexity. In: IEEE International Conference on Acoustic, Speech, and Signal Processing (ICASSP), pp. 4266–4269. (2010)
25. Fu, Z.-H., Fan, F., Huang, J.-D.: Dual-microphone noise reduction for a mobile phone application, IEEE (2013)

26. Benesty, J., Huang, Y.: A single-channel noise reduction MVDR filter. In: IEEE International Conference on Acoustic, Speech, and Signal Processing (ICASSP), pp. 273–276. (2011)
27. Jeub, M., Herglotz, C., Nelke, C., Beaugeant, C., Vary, P.: Noise reduction for dual-microphone mobile phones exploiting power level differences. In: IEEE International Conference on Acoustic, Speech, and Signal Processing (ICASSP), pp. 1693–1696. (2012)
28. Habets, E.A.P., Benesty, J., Chen, J.: Multimicrophone noise reduction using interchannel and interframe correlations. In: IEEE International Conference on Acoustic, Speech, and Signal Processing (ICASSP), pp. 305–308 (2012)
29. Garg, K., Nayar, S.: Vision and rain. Int. J. Comput. Vision **75**(1), 3–27 (2007)
30. Hase, H., Miyake, K., Yoneda, M.: Real-time snowfall noise elimination. In: IEEE International Conference on Image Processing, vol. 2, pp. 406–409 (1999)
31. Boncelet, C.: Image noise models. In Bovik, A.C. (ed.)Handbook of Image and Video Processing (2005)
32. Hauser, D., Amayenc, P., Nutten, B., Waldteufel, P.: A new optical instrument for simultaneous measurement of raindrop diameter and fall speed distributions. J. Atmos. Oceanic Tech. **1**, 256–269 (1984)
33. Kalman, R.E., Bucy, R.S.: New results in linear filtering and prediction theory. Trans. ASME-J. Basic Eng. **83**, 95–107 (1961)
34. Bossu, J., Hautière, N., Tarel, J.-P.: Rain or snow detection in image sequences through use of a histogram of orientation of streaks. Int. J. Comput. Vision (2011)
35. Barnum, P., Narasimhan, S., Kanade, T.: Analysis of rain and snow in frequency space. Int. J. Comput. Vision **86**(2–3), 256–274 (2010)
36. Dallal, N., Triggs, B.: Histograms of oriented gradients for human detection. In: IEEE Conference on Computer Vision and Pattern Recognition, vol. 1, pp. 886–893. (2005)
37. Weyori, Benjamin, A.: improved median filtering algorithm for the reduction of impulse noise in corrupted 2d grey scale images, a thesis submitted to the Department Of Computer Engineering Kwame Nkrumah University of Science and Technology in Master of Philosophy (2011)
38. Gallen, R., Hautière, N., Glaser, S.: Advisory speed for intelligent speed adaptation in adverse conditions. In: IEEE Intelligent Vehicles Symposium, pp. 107–114. (2010)
39. Shruthi, B., Renukalatha, S., Siddappa, M.: Speckle noise reduction in ultrasound images- a review. Dept. of Computer Science and Engg. Sri (2015)
40. Chun-yu, N.[1,2], Shu-fen', L., Ming, Q.: Research on removing noise in medical image based on median filter siddhartha institute of technology, Tumkur, Karnataka, India. Method 1. Department of Computer Science and Technology, Jilin University, Changchun, 130012, China; 2. School of Life Science and Technology, Changchun University of Science and Technology, Changchun, 130022, China (2009)
41. Kumar, N., Nachamai, M.: Noise removal and filtering techniques used in medical images (2017)
42. Ismaila, S.M., Radwan, A.G.[b,c], Madian, A.H.[c,d], Abu-ElYazee, M.F.[d,e]: Comparative study of fractional filters for Alzheimer disease detection on MRI images. A Faculty of IET, German University in Cairo (GUC), Egypt. b Dept. of Engineering Mathematics and Physics, Cairo University, Egypt. c NISC Research Center, Nile University, Cairo, Egypt. d Radiation Engineering Dept., NCRRT, Egyptian Atomic Energy Authority. e Electronics and comm. Eng. Dept., Cairo University, Egypt (2016)
43. Shinde, B.: Study of noise detection and noise removal techniques in medical images. Department of Computer Science, R.B.N.B. College, Shrirampur. Affiliated to Pune University Maharashtra (2012)
44. Radha Krishna, G., Krishnan, R., Mittal, V.K.: An automated system for regional nativity identification of indian speakers from english speech. In: Proceedings 16th International IEEE India Conference (INDICON 2019), Marwadi University, Rajkot, Gujrat, India, 13–15 Dec (2019)
45. Zhao, X., Wang, D.L.: Analyzing noise robustness of MFCC and GFCC features in speaker identification. In: ICASSP (2013)

46. Martinek, R., Vanus, J.: Voice communication in noisy environments in a smart house using hybrid LMS+ICA algorithm. J. (2020)
47. Arai, T., Yasu, K., Hodoshima, N.: Effective speech processing for various impaired listeners. ICA, June (2014)
48. Yao, K., Visser, E., Kwon, O.-W., Lee, T.-W.: A speech processing front-end with eigenspace normalization for robust speech recognition in noisy automobile environments. In: Conference, Jan (2003)
49. Gao, Z.: An adaptive median filtering of salt and pepper noise based on local pixel distribution. In: International Conference on Transportation and Logistics, Information and Communication, Smart City (TLICSC 2018)
50. Shrestha, S.: Image denoising using new adaptive based median filter. Signal Image Process. : An Int. J. (SIPIJ) **5**(4), (2014)
51. Suneetha, A., Srinivasa Reddy, E.: Robust gaussian noise detection and removal in color images using modified fuzzy set filter. J. Intell. Syst. **30**(1), (2020)
52. Kaur, J., Garg, M.: An improved weighted median filter for the image processing application. Int. J.Sci. Res. (IJSR) (2013)
53. Kadali, D.B., Mittal, V.K.: Studies on paralinguistic sounds, emotional speech and expressive voices. In: Proceedings SMM20, Workshop on Speech, Music and Mind, ISCA, pp. 11–15. (2020)
54. Devi Bhavani, K., Mittal, V.K.: Studies on paralinguistic speech sounds. In: Proceedings 17th International IEEE India Conference (INDICON 2020), pp. 1–6. IEEE (2020)
55. Mohanta, A., Mukherjee, P., Mittal, V.K.: Prosody features characterization of autism speech for automated detection and classification. In: Proceedings 26th National Conference on Communications (NCC 2020),IIT Kharagpur, India, 21–23 Feb (2020)
56. Mohanta, A., Mittal, V.K.: Autism speech analysis using acoustic features. In: Proceedings 16th International Conference on Natural Language Processing (ICON 2019),IIIT Hyderabad, India, 18–21 Dec (2019)
57. Mohanta, A., Mittal, V.K.: Acoustic features for characterizing speech of children affected with ASD. In: Proceedings 16th International IEEE India Conference (INDICON 2019), Marwadi University, Rajkot, Gujrat, India, 13–15 Dec (2019)
58. Radha Krishna, G., Krishnan, R., Mittal, V.K.: Non-native accent partitioning for speakers of indian regional languages. In: Proceedings 16th International Conference on Natural Language Processing (ICON 2019),IIIT Hyderabad, India, 18–21 Dec (2019)
59. Radha Krishna, G., Krishnan, R., Mittal, V.K.: Foreign accent recogntion with south indian spoken English. In: 17th IEEE India International Conference (INDICON 2020), December, (2020)

An Implementation of IRTBS—for Observing Cardiac Patients

Bharat Prajapati, Satyen Parikh, and Jignesh Patel

Abstract Researchers have been endlessly contributing to discover cardiovascular diseases (CVDs) in an early stage using various intelligent practices when a report published by the WHO—World Health Organization—shows internationally CVDs are the primary reason for human deaths in 2017. They have been trying to control and prevent through various Global Hearts Initiatives. WHO also mentioned major risk factors for cardiovascular disease. These factors might be the use of tobacco, physical activity, consumption of high salt in daily food, consumption of foods with TRANS-FATTY acids, and high blood pressure ((2017). Cardiovascular diseases (CVDs). Available: https://www.who.int/news-room/fact-sheets/detail/cardiovascular-diseases-(cvds)). As per number published by the WHO, around 17.9 million (179 lakh) deaths were responsible due to cardiovascular in 2016. Heart attacks and strokes were the topmost reasons. Internet of Things (IoT) empowers people to get an advanced level of automating by evolving a system with the help of sensors, interconnected devices, and the Internet. In the healthcare segment, patient monitoring is a very critical and most vital activity since a small delay in decision related to patients' treatment may cause permanent disability or maybe death. Most ICU devices are furnished with various sensors to measure health parameters but to monitor it all the time is still a challenging job. In our research, we have proposed an IOT based intelligent model, which captures the various body parameters using the bedside monitor and discover the early stage CVDs disease using supervised machine learning classification algorithms. To improve the accuracy of the model, we have also used multiple algorithms to achieve better accuracy.

B. Prajapati (✉) · S. Parikh
AMPICS, Ganpat Vidhyanagar, Gujarat 384002, India
e-mail: bbp03@ganpatuniversity.ac.in

S. Parikh
e-mail: satyen.parikh@ganpatuniversity.ac.in

J. Patel
DCS, Ganpat Vidhyanagar, Gujarat 384002, India
e-mail: jmp03@ganpatuniversity.ac.in

T. Senjyu et al. (eds.), *ICT with Intelligent Applications*, Smart Innovation, Systems and Technologies 248, https://doi.org/10.1007/978-981-16-4177-0_21

1 Introduction

In the last decade, the Internet of Thing helped various areas such as manufacturing, construction, transport, agriculture, health care, and education to improve the quality of the services using automated capturing data from the various sensors. Economy growth of the country also depends on the health of citizens. Few of the nations are already facing healthcare problems such as expensive medical treatment, scarcity of the doctors, and lack of central healthcare records. Data facts published by the WHO showed that the primary cause of death was heart disease. The aim of the carried out research work is to develop the automated model which initially generates the patient profile of medical history and captures the various automated body parameters including heart rate, blood pressure, etc., using bedside monitor. In the initial phase, the rich libraries of Python helped to upload the taken parameters on a real-time firebase database. Finally, we have implemented the hybrid machine learning to supervise classifiers to detect heart disease in an early state. We have used the data set of the Framingham heart study dataset of around 4240 patients to develop the model [2]. Furthermore, we have also collected around 400 samples from the local hospitals and implemented our proposed model. Earlier, the highest accuracy of the model was 88.14%. We have extended this accuracy up to 88.28% with the help of ensemble methods. We have also considered the same features, so that we can compare our model with other researchers.

2 Related Work

Dataset is a primary source of research to find new truths. Different sensors have been producing massive data depending upon the time interval. It may be second, minute, hour, and days. These datasets provide input to different statistical methods to discover the facts. The group of the researcher will have the chance to study it and find the hidden facts in various sectors [3]. Different sensors attached to the Raspberry Pi board helps to capture different human body parameters such as pressure in heart, heartbeats, the heat of the body, and breathing [4] . In 2018, the WHO mentioned that every year around 41 million people died when they suffered from non-communicable diseases. The heart attacks, cancers, and diabetes were topmost caused [5]. The eBPlatform assisted the doctors to treat the patients very effectively [6]. Smart healthcare treatment enabled the group of peoples such as doctors, patients, researchers, and caretakers of the patient from the remote place. They have been also tracking the physical activities and monitor the reading even though they are located far away from the patient [7, 8]. S. M. White and J. Pateman developed a method of transmitting the data from the bedside monitor to digital records with the help of low-cost equipment. These devices are computer, Keyspan RS-232/ DB9 (male) USB PC/Mac serial adapter, StarTech DB9 RS232 serial null modem adapter (female/female), and an ATEN UC232A USB to serial converter (male/USB) [9].

Exclusive features of the smartphone such as video calling, portability, easy charge, applications, and network connectivity are beneficial to send and receive the notification, and report immediate [8, 10, 11]. Automation enhancement has been improving in healthcare sectors and expanding rapidly. Manual records are very hard to preserve since after passing several years. In medical, for a doctor, it becomes necessary to know the medical history of the patient for proper diagnoses of the disease. It becomes necessary to store these data on the cloud. The platform of cloud computing has useful features such as easy access, cross-platform, permanent storage, and scalability [12, 13]. Implementation of machine learning supervised and unsupervised becomes easy and powerful due to the Python Scikit-learn module and Python module. Different modules of Python offer a broad range of analytical methods to execute intelligent algorithms [14].

3 System Design

In this section, we have described the design of capturing real-time body parameters using the bedside monitor and uploaded it on the cloud for further data analytics. The following were the various component to achieve the goal.

Caretaker: He/she was responsible for creating the patient's profile before connecting bedside monitors. Essential information included in profile such as Name, Gender, Date of Birth, medical history—the habit of smoking, diabetes, medicine of blood pressure (Yes / No), cholesterol, body mass index, and level of glucose of the concerned patient. The digitalization of these records is easily possible with the help of Google form.

Bedside Monitor: It is the one of the most popular device useful to know the real-time body parameters. Due to easy availability, it is also available from small town to metro city. Bedside monitors help to capture the following healthcare parameters of human body [15].

- Arterial Carbon Dioxide Partial Pressure (PaCO2)
- Arterial Oxygen Partial Pressure (PaO2)
- Arterial Oxygen Saturation (SaO2)
- Central Venous Pressure (CVP)
- Central Venous Oxygen Saturation (ScvO2)
- End Tidal Carbon Dioxide (ETCO2)
- Mean Arterial Pressure (MAP)
- Mixed Venous Oxygen Saturation (SvO2)
- Oxygen Saturation At The Peak Of Blood Pulsation (SpO2)
- Positive End Expiratory Pressure (PEEP)
- Pulmonary Artery Catheter (PAC)
- Systemic Vascular Resistance (SVR)
- Others Parameters.

Vital Signs Capture: SourceForge is a Web-based platform that serves to share the open-source and freely available source code to the community. It creates online common space and extends the contributions of the developers to the end-users. Vital Sign Capture is also one of the available source code written in C#.NET. It helps to obtain various body parameters from the numerous medical devices such as Philips Intellivue, Spacelabs patient monitors, and BBraun Spacecom syringe. It can capture the real-time parameters and export it into different digital format, i.e., CSV, excel. Interval time also can be modified as per the necessities. It needs RS232 (COM) Port / USB to Serial cable connected to Laptop/PC and a regular null modem serial cable attached to the monitor for Datex S/5.

Firebase: It is one of the most adapted real-time database services. It allows you to keep a 24 h services such as an online server, writing backend code, managing real-time database, protection with authentication, and cloud functions. It is also improving assistance regularly. It served to manage electronic health records for further use.

Patients, Care taker, and Doctors: There were different stakeholders of the system who were participated directly or indirectly in the execution system. The caretaker person had helped to attach bedside monitors to capture real-time patient's records when patients admitted to the hospital for treatment of the different diseases. Doctors diagnosed the disease based on the gathered history. The developed model generated notifications with the help of a trained healthcare algorithm.

For a healthy life, it becomes necessary to keep heart health. A healthy heart helps to fight different heart diseases. A non-profitable organization American Heart Association—AHA—suggested ideal heartbeats per minute for the different age groups. Tachycardia (fast heart rate) and bradycardia (slow heart rate) conditions can occur if heartbeats per minute violate the boundary. World Health Organization—WHO—also recommended preferable systolic and diastolic standards for different age and gender wise. Abnormal flow of blood in vain is responsible for different conditions such as elevated, the primary stage of hypertension, and the secondary stage of hypertension. Patients need immediate admission to the hospital in the worst conditions because death is also possible. A real-time patient observation helped to minimize human error while treating the patients. As a human being, still it is challenging task. Various notifications (heart rate, blood pressure, oxygen-level pressure in the brain, etc.) based on standard healthcare parameters helped to make quick attention with fewer errors.

4 System Implementation

The following were the different stages of the model implementation, which were contributing to executing the intelligent automated healthcare system.

This was the primary stage of the system implementation when a patient came to the hospital for medical treatment. The Nurses/technician created a patient's history profile with the help of Google doc (see Fig. 1). It was easy to collect digital records

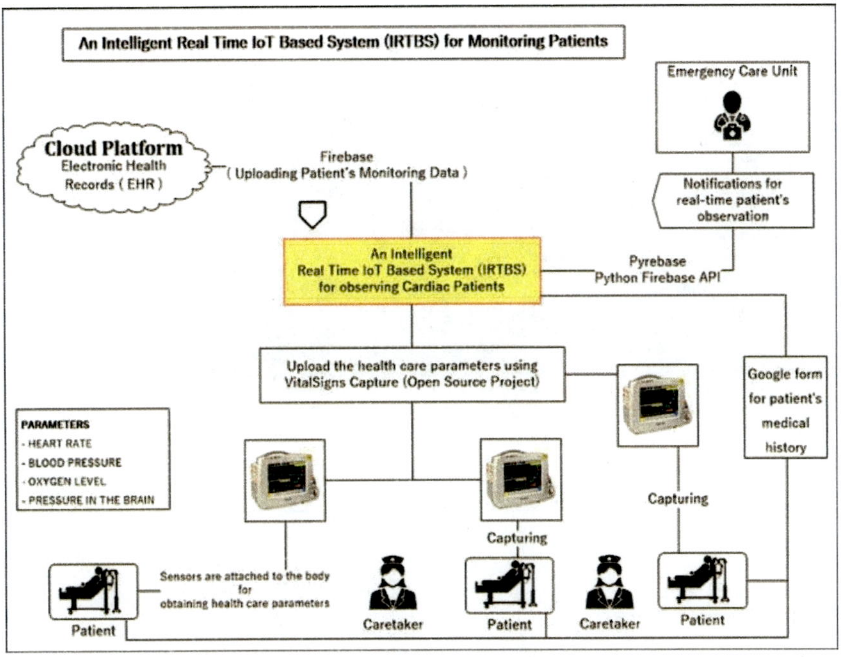

Fig. 1 Design for data capturing using bedside monitor and uploading on the cloud platform

for further process. We have manually collected a few parameters. This medical history was helpful during system implementation.

VitalSignsCapture—an open-source project helped to capture the real-time body parameter from the patient using your bedside monitor (Fig. 2).

Later on, these parameters uploaded on the cloud using the firebase library for further analysis (see Fig. 1).

Python module served to execute the model. Sklearn has been used to implement several machine learning algorithms. Panda was used to view the result graphically while Numpy handled to upload large multidimensional array for performing the matrix operations (Fig. 3).

In machine learning, concept of the ensemble is used to combine more than one algorithm to deal with the difficulties such as variance, bias, and improve the ratio of prediction. Ensemble methods are also known as bagging, boosting, and stacking. In the first group, the derive learner depended on base learner, whereas in the second group, both learners are independent. We have improved the accuracy of finding heart desires in the early stage which was up to 88.42% with the help of the IRTBS system (Fig. 4).

The developing countries are unable to speed up the process of healthcare treatment and reduce manual workloads. Moreover, it is also increasing the issues like expensive medical costs, human errors, delays in treatment, etc. The manual method

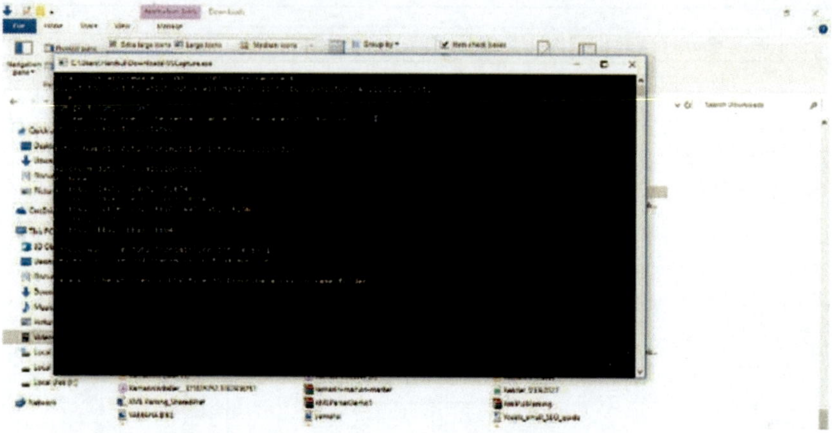

Fig. 2 Capture the real-time body parameters using bedside monitor

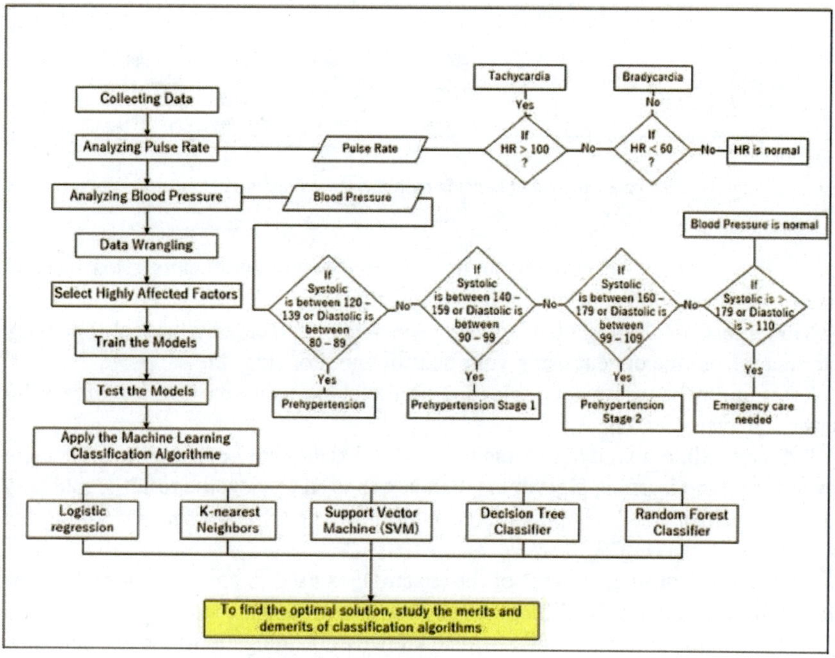

Fig. 3 Architecture of IRTBS for potions monitoring

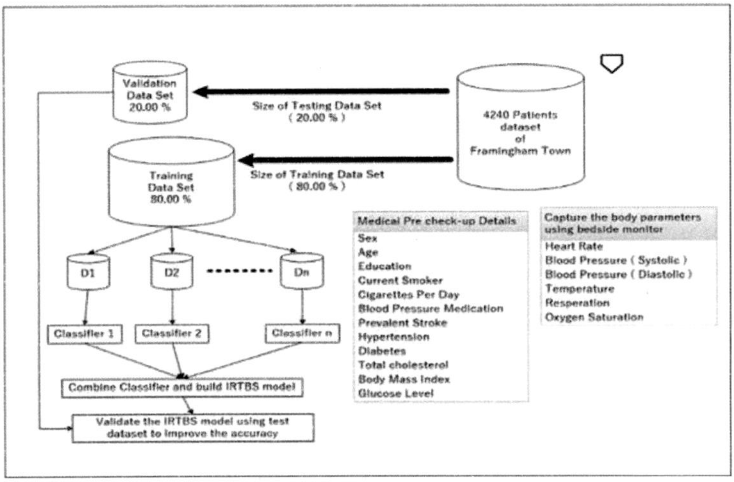

Fig. 4 Combination of the different classifier to achieve the best accuracy

of treating the disease is a significant key factor. The IRTBS model can catch complex patterns that would have been inspected during the human investigation for detecting heart disease. The model is also freeing up human workers to concentrate on tasks like precaution steps to prevent it. It will also support trade with more patients and recognize patterns much more quickly.

5 Conclusion and Future Work

The hybrid versions of the IoT, cloud computing, and machine learning can transform the world. It can deal with problems such as fulfill the gap of a scarcity of medical treatment, reduce the cost, instant medical treatment, improve the effectiveness of the drug, customize the diet chart, and many more. Scholars have been the opportunity to study the numerous body parameters and find more relevant patents to deal with various diseases. IRTBS system is a new approach for real-time patient monitoring using bedside monitors. It expands the capability of captured parameters using a bedside monitor. In our research, we have considered only selected parameters related to heart disease, which were highly correlated to detect the chances of heart disease. It can be expanded by adding further features in the dataset.

Acknowledgements We would like to thank Dr. Amol Agarwal, Dr. Pritesh Shah, and Dr. Mukul Shah for giving their worthwhile guidance to understand different medical parameters. We would also like to appreciate Mr. Aman Ajmera for storing a Framingham heart study dataset to examine the pattern [2]. Last but not least those patients who have actively engaged in research and willing to share their medical healthcare report with us. We acknowledged the hospitals that give us consent to access their medical devices to test our model. We are highly thankful to all the below-mentioned researchers in reference, who has encouraged us by their study and publications.

References

1. Cardiovascular Diseases (CVDs) (2017). Available https://www.who.int/news-room/fact-sheets/detail/cardiovascular-diseases-(cvds)
2. Ajmera, A.: In: Framingham Heart Study Dataset. 1st edn (2017)
3. Chen, Y.-K.:Eenges and opportunities of internet of things. In: 17th Asia and South Pacific design Automation Conference, pp. 383–388. IEEE (2012)
4. Kumar, R., Rajasekaran, M.P.: An IoT based patient monitoring system using raspberry Pi. In: 2016 International Conference on Computing Technologies and Intelligent Data Engineering (ICCTIDE'16), pp. 1–4. IEEE (2016)
5. Noncommunicable diseases (2018). Available https://www.who.int/news-room/fact-sheets/detail/noncommunicable-diseases
6. Liu, Y., Niu, J., Yang, L., Shu, L.: eBPlatform: an IoT-based system for NCD patients homecare in China. In: 2014 IEEE Global Communications Conference, pp. 2448–2453. IEEE (2014)
7. Archip, A., Botezatu, N., Şerban, E., Herghelegiu, P.-C., Zală, A.: An IoT based system for remote patient monitoring. In: 2016 17th International Carpathian Control Conference (ICCC), pp. 1–6. IEEE. (2016)
8. Al-Adhab, A., Altmimi, H., Alhawashi, M., Alabduljabbar, H., Harrathi, F., ALmubarek, H.: IoT for remote elderly patient care based on fuzzy logic. In: 2016 International Symposium on Networks, Computers and Communications (ISNCC), pp. 1–5. IEEE. (2016)
9. White, S., Pateman, J.J.A.: A method of recording electronic anaesthetic monitor data for research. **72**(2), 267–269 (2017)
10. Stutzel, M.C., Fillipo, M., Sztajnberg, A., Brittes, A., da Motta, L.B.: SMAI-mobile system for elderly monitoring. In: 2016 IEEE International Conference on Serious Games and Applications for Health (SeGAH), pp. 1–8. IEEE. (2016)
11. Blumrosen, G., Avisdris, N., Kupfer, R., Rubinsky, B.: C-SMART: efficient seamless cellular phone based patient monitoring system. In: 2011 IEEE International Symposium on a World of Wireless, Mobile and Multimedia Networks, pp. 1–6. IEEE. (2011)
12. Wu, J., Ping, L., Ge, X., Wang, Y., Fu, J.: Cloud storage as the infrastructure of cloud computing. In: 2010 International Conference on Intelligent Computing and Cognitive Informatics, pp. 380—383. (2010)
13. Tyagi, S., Agarwal, A., Maheshwari, P.: A conceptual framework for IoT-based healthcare system using cloud computing. In: 2016 6th International Conference-Cloud System and Big Data Engineering (Confluence), pp. 503–507. IEEE (2016)
14. Pedregosa, F. et al.: Scikit-learn: machine learning in Python. **12,** 2825–2830 (2011)
15. Prajapati, B., Parikh, S., Patel, J.: An intelligent real time IoT based system (IRTBS) for monitoring ICU patient. In: International Conference on Information and Communication Technology for Intelligent Systems, pp. 390–396. Springer (2017)

Early Detection of Parkinson's Disease Through Speech Features and Machine Learning: A Review

Ajay Sankar Gullapalli and Vinay Kumar Mittal

Abstract Parkinson's Disease (PD) is a kind of neurodegenerative disorder. The disease causes communication impairment based on its progression. In general, identification of PD carried out based on medical images of brain. But it was recently identified that voice is acting as biomarkers for several neurological disorders. A review of speech features and machine learning algorithms is presented. This might be helpful for development of a non-invasive signal processing techniques for early detection of PD. Several models developed for disease detection is discussed, which are developed based on features like acoustic, phonation, articulation, dysphonia, etc. Machine learning algorithms like Logistic Regression (LG), Support Vector Machine (SVM), Boosting Regression Tree, bagging Regression, etc., and their performance accuracies in classification of Patient with PD (PWP) and Healthy Controls (HC) are reviewed. All these classification algorithms are trained and tested on several repository corpuses and customized datasets. The Spontaneous Speech (SS) is an efficient tool for the early detection of diseases like Parkinson's, Alzheimer's, Autism and several other dementia types in elderly people.

1 Introduction

Most common disease in the elderly people next to Alzheimer's disease (AD) is Parkinson's Disease (PD). The PD is a neurological disease caused due to the loss of dopamine-producing brain cells [1]. It mainly effects the motor system of the central nervous system [2]. The Patient with PD (PWP) have symptoms like hypokinetic dysarthria, which causes reduction in phonation time, decreased breath support, harsh breathy voice, and increased acoustic noise [3]. PD manifests in two main vocal impairments, first is dysphonia (lack of ability to generate normal vocal sound) and the other is dysarthria (difficulty in pronouncing the words). The disease also shows

A. S. Gullapalli (✉)
Aditya Engineering College, Surampalem, India

V. K. Mittal
KL Deemed to be University, Vijayawada, India

T. Senjyu et al. (eds.), *ICT with Intelligent Applications*, Smart Innovation, Systems and Technologies 248, https://doi.org/10.1007/978-981-16-4177-0_22

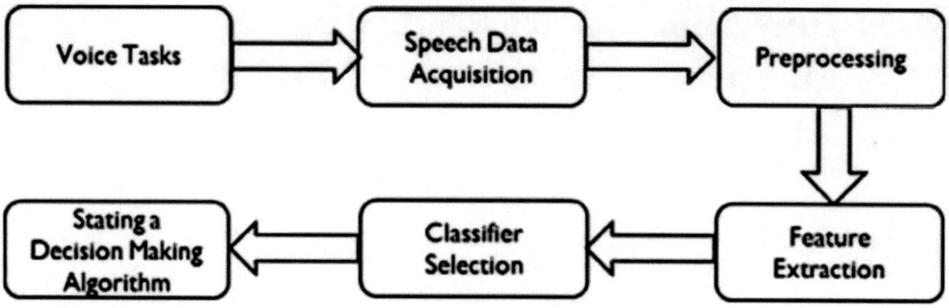

Fig. 1 Flow of disease detection mechanism based of spontaneous speech

impairments in speech, posture, and locomotion. The patient's Spontaneous Speech (SS) and extracted features might be used for the early detection and progression of the diseases like Alzheimer's, Parkinson's, Huntington's and Autism. Due to PD, the muscles used in the speech production are affected and thereby altering the various features of speech [4]. So, SS and extracted features may be treated as sophisticated biomarker for identification of disease and its progression [5]. This work might be supporting the development of machine-assistive technologies, which needs human–machine interface [6].

The general approach for discriminating the PWP and healthy control (HC) in a given dataset is a step-by-step process shown in Fig. 1. The approach involves voice tasks, speech data acquisition, preprocessing, feature extraction, classifier selection, and stating a decision-making algorithm [7].

Voice tasks means voicing of vowel and pronunciation of sentence. Voice tasks may include Storytelling, prosodic sentence reading, general conversations, vowel phonation, syllables repetitions etc. [8]. Speech data acquisition can be carried out by various devices like acoustic cardioid (AC) and smart phone (SP). In preprocessing step, segregation of voiced data (vocal cords vibrate) and unvoiced data (No vibration of vocal cords), standardization, application of randomization, normalization, and removal of outliers are carried out. Feature extraction block uses certain signal processing method and extracts the crucial features through which the impairment in speech is clearly marked.

Classifier selection means choosing a best fit classifier from various machine learning-based classifier algorithms. This might be useful for classifying an individual between the HC and PD. But when a classifier is applied alone, accuracy in discrimination might be very low. So, the classifiers are trained along with speech features for better accuracy in disease detection. Stating a decision-making algorithm means designing a best combination of speech processing models, features, and classifiers for disease detection.

The paper is organized by following chapters. Section 2 reviews various database corpuses used. Section 3 gives a brief idea of signal processing methods used for preprocessing/feature extraction. Section 4 describes various features used in the literature. Section 5 deals with classifiers. Accuracy comparisons can be seen in the tabular forms given in each sections.

2 Databases

Several databases are available in the form of corpuses for the speech processing of PD patients. Few are like UCI (University of California at Irvine) machine learning database [2], National Center for Voice and Speech (NCVS) [9], mPower [10], Neurovoz, GITA, CzechPD, Albayzin, and Fisher SP [11]. Neurovoz dataset is collected at otorhinoaringology and neurology services of the Gregorio Mara˜nón hospital in Madrid, Spain. Other than corpuses, clinically recorded speech data of PD patients are also described in Table 1. PD_Dataset-1 is a customized dataset collected at Parkinson disease movement disorder society's (PDMDS) centers in Mumbai, India. PD_Dataset-2 is by using Intel At-Home Testing Device for remote symptom progression telemonitoring. Accurate telemonitoring of PD is explained in [12]. PD_Dataset-3 is customized at Department of Neurology in Cerrahpaşa Faculty of Medicine, Istanbul University. Data definition, understanding and modelling for PD detection is explained in [13].

Table 1 Summary of databases for Parkinson's Disease

Database	Place/ Institution	Size	Ref	Remarks
UCI ML DB	Irvine	40 HC, 40 PWP	[2]	OGA for sampling
NCVS	Salt Lake City, Utah	10 HC, 33 PWP	[9]	263 phonations
mPower	USA	500 PWP and 500 HC	[10]	Mobile app. based
Neurovoz	Madrid, Spain	47 PWP and 32 HC	[11]	6 sentences and picture description
GITA	Colombian corpus	50 PWP and 50 HC	[11]	DDK task (/pa-ta-ka/), Monolog task and 6 read sentences
CzechPD	Czech	20 PWP and 14 HC	[11]	DDK task
Albayzin	Spain	136 Spanish speakers	[11]	UBM training
FisherSP	Linguistic Data Consortium	136 Spanish speakers	[11]	FAM training
PD_Dataset-1	PDMDS, Mumbai, India	45 PWP and 45 HC	[4]	Speaking a vowel /a/
PD_Dataset-2	Intel At-Home Testing Device for remote symptom progression telemonitoring	PD detection: 23 PWP, 8 HC PD telemonitoring: 42 PWP	[14]	195 vowel phonations
PD_Dataset-3	Istanbul University	188 PWP, 64 HC	[15]	vowel /a/ phonated
PD_Dataset-4	UCI ML repository	25 PWP, 20 HC	[16]	vowels 'a', 'o' phonated
PD_Dataset-5	PC-GITA	25 PWP, 20 HC	[16]	Randomly chosen
PD_Dataset-6	–	99 Subjects	[17]	vowel /a/ voicing
PD_Dataset-7	NCVS	147 PWP, 48 HC	[18]	195 vowel phonations

Table 2 Signal processing methods/tools

Signal processing method/tool	References
One against all based data sampling (OGABDS)	[2]
Mel frequency cepstral coefficients (MFCC)	[4, 7, 9, 11]
Joint time–frequency analysis (JTFA)	[10]
Praat and multidimensional voice program (MDVP)	[14]
Tunable Q-factor wavelet transform (TQWT)	[15]
Intrinsic mode function cepstral coefficient (IMFCC)	[16]

3 Signal Processing Methods

The speech data derived from the databases are further applied with various signal processing methods for preprocessing [19] and feature extraction. The speech processing is exemplified in [20]. The features extracted can be further used in developing, training, and testing the algorithm for detection of PD. Signal processing methods/tools for detection of PD are listed in Table 2.

The signal processing methods/tools can be used in extracting the features either manually or automatically. In the further sections, a description for the features extracted using these signal processing methods can be observed. MFCC reads the normalized speech signal and then finds the energy level and zero crossing signals. JTFA represents the frequency component evolution over time. MDVP is used for evaluation of spectral energy, fundamental frequency, and amplitude. Using TQWT Q-factor and redundancy can be tuned, and also radix-2 FFT can be computed.

4 Feature Sets

The spontaneous speech is identified to be one of the best tools for identification of PD in elderly people in a non-invasive method. The sample databases explained in the previous section are undergone with various feature extraction methods. Using these extracted features, various machine learning-based algorithms are trained and tested for early detection of PD. The features like vocal, acoustic [21–24], phonation and articulation, linguistic, dysphonia/baseline features, voice print features, time frequency, wavelet transform-based features, tunable Q-factor wavelet transform (TQWT)-based features, vocal fold features, features based on empirical mode decomposition (EMD), Speech rate [25] etc., are extracted. The extracted features can be grouped as Amplitude Parameters, Pulse Parameters, Frequency Parameters, Voicing Parameters, Pitch Parameters, Harmonicity Parameters [26]. For autism disease, detection using acoustic features is explained in [27, 28]. The extraction process of paralinguistic features is demonstrated in [29]. The accuracy comparisons for various extracted features are delivered in Table 3. The characterization of the disease autism based on acoustic features is exemplified [30].

There are several algorithms trained on a set of features, formed as a combination of different types of features. Based on Table 3, various conclusions can be made depending on the accuracy. The selection of a feature set depends on its % accuracy. The feature set with highest accuracy is best fit for the detection of Alzheimer's disease (AD).

Dysphonia/baseline features such as F0, jitter, pitch period entropy (PPE), shimmer, detrended fluctuation analysis (DFA), noise-to-harmonics ratio (NHR), recurrence period density entropy (RPDE), correlation dimension (D2), and harmonics-to-noise ratio (HNR) are given importance while classifying the subjects with Parkinson's [15]. Also as per [31] robust response is observed for pitch compare to formant (frequency band that determine the phonetic quality of vowel) manipulations in patients with PD.

Phonation features are like a vocal fundamental frequency average (F0), period perturbation quotient (PPQ), minimum vocal fundamental frequency (Flo), maximum vocal fundamental frequency (Fhi), jitter (%), jitter absolute value (Abs), shimmer in decibels (db), relative amplitude perturbation (RAP), three-point amplitude perturbation quotient (APQ3), local shimmer, five-point amplitude perturbation quotient (APQ5), average absolute difference between consecutive differences between amplitudes of consecutive periods (DDA), noise-to-harmonics ratio (NHR),

Table 3 Features extracted and attained accuracy for Parkinson's Disease detection

Features	Signal processing methods/tools	Accuracy	Database	References
Acoustic	OGABD, MFCC, IMFCC	89.46%, 81%, 94%, 82%, 70%	UCI, GITA, CzechPD, PD_Dataset-4, PD_Dataset-5	[2, 11, 16]
Phonation and articulation	MFCC	98%	PD Dataset-1	[4]
Dysphonia	MFCC, MDVP, TQWT	99%, 98.84%, 79%	NCVS, PD_Dataset-2, PD_Dataset-3	[9, 14, 15]
Voiceprint	JTFA	90.45%	mPower	[10]
Time frequency	JTFA	77%	PD_Dataset-3	[15]
MFCC based	MFCC, IMFCC	84%, 100%	PD_Dataset-3, PD_Dataset-4, PD_Dataset-5	[15]
Wavelet transform based	TQWT	78%	PD_Dataset-3	[15]
TQWT based features	TQWT	77%	PD_Dataset-3	[15]
Phonation	MFCC	100%	UCI	[7]

amplitude perturbation quotient (APQ) harmonics-to-noise ratio (HNR), recurrence period density entropy (RPDE), two measures of spread (Spread 1, Spread 2), detrended fluctuation analysis-based fractal (DFA), pitch period entropy (PPE), and correlation dimension (D2) [18].

Few features like time frequency features, MFCCs, wavelet transform-based features, vocal fold features, and tunable Q-factor wavelet transform (TQWT)-based features are also showing a better accuracy in disease detection in combination with a sophisticated classifier. Emotional Speech also acts as a discriminator [32, 33]. All these features are individually producing qualitative simulation results, but even better results can be produced when utilized in combination with other features.

As per the review, it can be observed from Table 3 that 100% accuracy is achieved through MFCC based features and phonation features when extracted from PD_Dataset-5 and UCI datasets, respectively. Next to these, dysphonia-based features extracted from NCVS dataset can able to classify with 99% of accuracy.

5 Classifiers

The extracted feature sets are used to train the algorithms designed using machine learning. The algorithms may use a type of classifier between several groups like healthy controls and PWP [34]. Similar work is examined for autism detection [35]. Various classifiers like Logistic regression, Support Vector Machine [16, 36], K-NN [37], CNN, Deep Neural Network [38], Boosting, Bagging [39], Random Forest [14, 16, 40] are available for training the algorithms.

Table 4 shows various available classifiers, and also a comparison on their accuracies is illustrated. The performance of classifiers might vary based on the selected datasets and extracted features which are demonstrated earlier.

From Table 4, it was identified that random forest has achieved 100% of accuracy in classifying the patient with Parkinson's disease. Here, the database considered is PD_Dataset-4. At first, the dataset was extracted for acoustic, MFCCs, and EMD based features by using IMFCC, and then, the extracted features are used as training and testing data for SVM and random forest classifiers. Random forest, MLP, and MLP-LSVM have presented an accuracy of 100%, and next to it, SVM and LSVM have scored an accuracy of 99%.

6 Conclusions

Each section in the literature has given a review of individual blocks in speech signal-based machine learning algorithm development. At first, a review of speech signal databases available and various customized datasets developed are demonstrated. In later section, the features extracted from the datasets is reviewed. The features like vocal, acoustic, phonation, demographic, dysphonia, time frequency domain,

Table 4 Classifiers and attained accuracy for Parkinson's Disease detection

Classifiers	Accuracy	Database	References
Logistic regression (LR)	84.30%, 79%	UCI, PD_Dataset-3	[2, 15]
Support vector machine (SVM)	98%	PD Dataset-4	[16]
SVM with Gaussian	88.76%	UCI	[2]
SVM (tenfold cross-validation)	99%	NCVS	[9]
Support vector regression (SVR)	92.78%	PD Dataset-2	[14]
SVMs with linear and RBF kernels	85%	PD Dataset-3	[15]
SVM + BO (Bayesian optimization)	92%	PD_Dataset-7	[18]
Lagrangian support vector machine (LSVM),	99.29%	UCI	[2]
MLP-LSVM	100%	UCI	[2]
Weighted k-NN	89.46%	UCI	[2]
1-nearest neighbor (K1)	94.55%	PD_Dataset-6	[17]
Neural network classifier	98%	PD Dataset-1	[4]
Convolutional neural network (CNN)	90.45%	mPower	[10]
Boosting regression tree	93.36%	PD Dataset-2	[14]
Bagging regression tree	90.78%	PD Dataset-2	[14]
Random forest	90.7%, 100%	PD Dataset-2, PD Dataset-4	[14, 16]
Multilayer perceptron	85%	PD Dataset-3	[10]
Gaussian mixture model—forced universal background model (GMM-fUBM)	94%	CzechPD	[11]
Forced Gaussian mixture model—forced universal background model (fGMM-fUBM)	81%	GITA	[11]
Optimized forced Gaussian mixture model—forced universal background model (opt-fGMM-fUBM)	81%	GITA	[11]
Multilayer perceptron (MLP)	100%	UCI	[2]

etc., can be extracted. The accuracy comparisons for various extracted features are explained. These accuracies are for classifying whether a patient is affected with PD or HC. The extracted features alone might not classify the subject, so a best classifier is needed for training the datasets. In the further section, a literature on various classifiers used for detection of PD is given. In the section of classifiers, the accuracy attained by various classifiers is explained. Several classifiers with best accuracies

(>90%) are observed. A combination of accurate features and best classifier is crucial for design of a reliable algorithm.

References

1. Hariharan, M., Polat, K., Sindhu, R.: A new hybrid intelligent system for accurate detection of Parkinson's disease. Comput. Methods Program. Biomed. **113**(3), 904–913 (2014). ISSN 0169–2607. https://doi.org/10.1016/j.cmpb.2014.01.004
2. Polat, K., Nour, M.: Parkinson disease classification using one against all based data sampling with the acoustic features from the speech signals. Med. Hypotheses **140**, 109678 (2020). ISSN 0306-9877. https://doi.org/10.1016/j.mehy.2020.109678
3. Harel, B.T., Cannizzaro, M.S., Cohen, H., Reilly, N., Snyder, P.J.: Acoustic characteristics of Parkinsonian speech: a potential biomarker of early disease progression and treatment. J. Neurolinguistics **17**(6) 439–453 (2004). ISSN 0911–6044. https://doi.org/10.1016/j.jneuroling.2004.06.001
4. Upadhya, S.S., Cheeran, A.N.: Discriminating Parkinson and healthy people using phonation and cepstral features of speech. Proc. Comput. Sci. **143**, 197–202 (2018). ISSN 1877–0509. https://doi.org/10.1016/j.procs.2018.10.376
5. Tracy, J.M., Özkanca, Y., Atkins, D.C., Ghomi, R.H.: Investigating voice as a biomarker: deep phenotyping methods for early detection of Parkinson's disease. J. Biomed. Info. **104**,103362 (2020). ISSN 1532-0464. https://doi.org/10.1016/j.jbi.2019.103362
6. Mittal, S., Mittal, V.K.: Biomedical requirements for human machine interface towards building a humanoid: a review. In: Proceedings 16th International IEEE India Conference (INDICON 2019), Marwadi University, Rajkot, Gujrat, India, 13–15 Dec (2019)
7. Parisi, L., RaviChandran, N., Manaog, M.L.: Feature-driven machine learning to improve early diagnosis of Parkinson's disease. Expert Syst. Appl. **110**, 182–190. (2018). ISSN 0957–4174. https://doi.org/10.1016/j.eswa.2018.06.003
8. Pompili, A. et al.: Automatic detection of parkinson's disease: an experimental analysis of common speech production tasks used for diagnosis. In: Ekštein, K., Matoušek, V (eds) Text, Speech, and Dialogue. TSD 2017. Lecture Notes in Computer Science, vol 10415. Springer, Cham. (2017). https://doi.org/10.1007/978-3-319-64206-2_46
9. Tsanas, A., Little, M.A., McSharry, P.E., Spielman, J., Ramig, L.O.: Novel speech signal processing algorithms for high-accuracy classification of Parkinson's disease. IEEE Trans. Biomed. Eng. **59**(5), 1264–1271 (2012). https://doi.org/10.1109/TBME.2012.2183367
10. Zhang, H., Wang, A., Li, D., Xu, W.: DeepVoice: a voiceprint-based mobile health framework for Parkinson's disease identification. In: 2018 IEEE EMBS International Conference on Biomedical and Health Informatics (BHI), Las Vegas, NV, pp. 214–217. (2018). https://doi.org/10.1109/BHI.2018.8333407
11. Moro-Velazquez, L., Gomez-Garcia, J.A., Godino-Llorente, J.I., Villalba, J., Rusz, J., Shattuck-Hufnagel, S., Dehak, N.: A forced gaussians based methodology for the differential evaluation of Parkinson's disease by means of speech processing. Biomed. Signal Process. Control **48**, 205–220 (2019). ISSN 1746–8094, https://doi.org/10.1016/j.bspc.2018.10.020
12. Tsanas, A., Little, M.A., McSharry, P.E., Ramig, L.O.: Accurate telemonitoring of Parkinson's disease progression by noninvasive speech tests. IEEE Trans. Biomed. Eng. **57**(4), 884–893 (2010). https://doi.org/10.1109/TBME.2009.2036000
13. Vadovský, M., Paralič, J.: Parkinson's disease patients classification based on the speech signals. In: 2017 IEEE 15th International Symposium on Applied Machine Intelligence and Informatics (SAMI), Herl'any, pp. 000321-000326 (2017). https://doi.org/10.1109/SAMI.2017.7880326
14. Despotovic, V., Skovranek, T., Schommer, C.: Speech based estimation of Parkinson's disease using gaussian processes and automatic relevance determination. Neurocomputing **401**, 173–181 (2020). ISSN 0925–2312. https://doi.org/10.1016/j.neucom.2020.03.058

15. Okan Sakar, C., Serbes, G., Gunduz, A., Tunc, H.C., Nizam, H., Sakar, B.E., Tutuncu, M., Aydin, T., Erdem Isenkul, M., Apaydin, H.: A comparative analysis of speech signal processing algorithms for Parkinson's disease classification and the use of the tunable Q-factor wavelet transform. Appl. Soft Comput. **74**, 255–263 (2019). ISSN 1568–4946. https://doi.org/10.1016/j.asoc.2018.10.022
16. Karan, B., Sahu, S.S., Mahto, K.: Parkinson disease prediction using intrinsic mode function based features from speech signal. Biocybernet. Biomed. Eng. **40**(1), 249–264 (2020). ISSN 0208–5216. https://doi.org/10.1016/j.bbe.2019.05.005
17. Almeida, J.S., Rebouças Filho, P.P., Carneiro, T., Wei, W., Damaševičius, R., Maskeliūnas, R., de Albuquerque, V.H.C.: Detecting Parkinson's disease with sustained phonation and speech signals using machine learning techniques. Pattern Recogn. Lett. **125**, 55–62 (2019). ISSN 0167–8655. https://doi.org/10.1016/j.patrec.2019.04.005
18. Lahmiri, S., Shmuel, A.: Detection of Parkinson's disease based on voice patterns ranking and optimized support vector machine. Biomed. Signal Process. Control **49**, 427–433 (2019). ISSN 1746–8094. https://doi.org/10.1016/j.bspc.2018.08.029
19. Orozco-Arroyave, J.R., Hönig, F., Arias-Londoño, J.D., Vargas-Bonilla, J.F., Daqrouq, K., Skodda, S., Rusz, J., Nöth, E.: Automatic detection of Parkinson's disease in running speech spoken in three different languages. J. Acoust. Soc. America 481–500, V139 (2016). .https://asa.scitation.org/doi/abs//doi.org/10.1121/1.4939739
20. Asgari, M., Shafran, I.: Predicting severity of Parkinson's disease from speech. In: 2010 Annual International Conference of the IEEE Engineering in Medicine and Biology, Buenos Aires, pp. 5201—5204 (2010).https://doi.org/10.1109/IEMBS.2010.5626104
21. John Phillip, B., Kalyan, S.S.S., Mittal, V.K.: Discriminating high arousal and low arousal emotional speech using mahalanobis distance among acoustic features. In: Proceedings 26th National Conference on Communications (NCC 2020), IIT Kharagpur, India, 21–23 Feb (2020)
22. Braga, D., Madureira, A.M., Coelho, L., Ajith, R.: Automatic detection of Parkinson's disease based on acoustic analysis of speech. Eng. Appl. Artif. Intell. **77**, 148–158 (2019). ISSN 0952–1976. https://doi.org/10.1016/j.engappai.2018.09.018
23. Wu, K., Zhang, D., Lu, G., Guo, Z.: Learning acoustic features to detect Parkinson's disease. Neurocomputing **318**, 102–108 (2018). ISSN 0925–2312. https://doi.org/10.1016/j.neucom.2018.08.036
24. Hsu, S.-C., Jiao, Y., McAuliffe, M.J., Berisha, V., Wu, R.-M., Levy, E.S.: Acoustic and perceptual speech characteristics of native Mandarin speakers with Parkinson's disease. J. Acoust. Soc. America **141**(3) (2017). https://doi.org/10.1121/1.4978342
25. Martínez-Sánchez, F., Meilán, J.J.G., Carro, J., Gómez Íñiguez, C., Millian-Morell, L., Pujante Valverde, I.M., López-Alburquerque, T., López, D.E.: Speech rate in Parkinson's disease: a controlled study. Neurología (English Edition) **31**(7), 466–472 (2016). ISSN 2173–5808. https://doi.org/10.1016/j.nrleng.2014.12.014
26. Agarwal, A., Chandrayan, S., Sahu, S.S.: Prediction of Parkinson's disease using speech signal with extreme learning machine. In: 2016 International Conference on Electrical, Electronics, and Optimization Techniques (ICEEOT), Chennai, pp. 3776–3779. (2016). https://doi.org/10.1109/ICEEOT.2016.7755419
27. Mohanta, A., Mittal, V.K.: Autism speech analysis using acoustic features. In: Proceedings 16th International Conference on Natural Language Processing (ICON 2019), IIIT Hyderabad, India, 18–21 Dec (2019)
28. Mohanta, A., Mittal, V.K.: Classifying speech of ASD affected and normal children using acoustic features. In: Proceedings 26th National Conference on Communications (NCC 2020), IIT Kharagpur, India, 21–23 Feb (2020)
29. Devi Bhavani, K., Mittal, V.K. : Studies on paralinguistic speech sounds. In: Proceedings 17th International IEEE India Conference (INDICON 2020), IEEE, pp. 1–6. (2020)
30. Mohanta, A., Mittal, V.K.: Acoustic features for characterizing speech of children affected with ASD. In: Proceedings 16th International IEEE India Conference (INDICON 2019), Marwadi University, Rajkot, Gujrat, India, 13–15 Dec (2019)

31. Mollaei, F., Shiller, D.M., Baum, S.R., Gracco, V.L.: The relationship between speech perceptual discrimination and speech production in Parkinson's disease. J. Speech Lang. Hear. Res. **62**(12), 4256–4268 (2019). https://doi.org/10.1044/2019_JSLHR-S-18-0425
32. Christine Schröder, M.D., Möbes, J., Martin Schütze, M.D., Szymanowski, F., Wido Nager, M.D., Bangert, M., Thomas Frank Münte, M.D., Reinhard Dengler, M.D.: Perception of emotional speech in Parkinson's disease. Movement Disorders **21**(10), 1774–1778 (2006).https://doi.org/10.1002/mds.21038
33. Möbes, J., Joppich, G., Stiebritz, F., Dengler, R., Schröder, C.: Emotional speech in Parkinson's disease. Movement Disorders **23**(6), 824–829 (2008). https://doi.org/10.1002/mds.21940
34. Gunduz, H.: Deep learning-based Parkinson's disease classification using vocal feature sets. IEEE Access **7**, 115540–115551 (2019). https://doi.org/10.1109/ACCESS.2019.2936564
35. Mohanta, A., Mukherjee, P., Mittal, V.K.: Prosody features characterization of autism speech for automated detection and classification. In: Proceedings 26th National Conference on Communications (NCC 2020), IIT Kharagpur, India, 21–23 Feb (2020)
36. Shahbakhi, M., Far, D., Tahami, E.: Speech analysis for diagnosis of parkinson's disease using genetic algorithm and support vector machine. J. Biomed. Sci. Eng. **7**, 147–156 (2014). https://doi.org/10.4236/jbise.2014.74019
37. Shirvan, R.A., Tahami, E.: Voice analysis for detecting Parkinson's disease using genetic algorithm and KNN classification method. In: 2011 18th Iranian Conference of Biomedical Engineering (ICBME), Tehran, pp. 278—283. (2011). https://doi.org/10.1109/ICBME.2011.6168572
38. Grover, S., Bhartia, S., Akshama, Yadav, A., Seeja, K.R.: Predicting severity of parkinson's disease using deep learning. Proc. Comput. Sci. **132**, 1788–1794 (2018). ISSN 1877–0509. https://doi.org/10.1016/j.procs.2018.05.154
39. Fayyazifar, N., Samadiani, N.: Parkinson's disease detection using ensemble techniques and genetic algorithm. In: 2017 Artificial Intelligence and Signal Processing Conference (AISP), Shiraz, pp. 162–165 (2017).https://doi.org/10.1109/AISP.2017.8324074
40. Aich, S., Younga, K., Hui, K.L., Al-Absi, A.A., Sain, M.: A nonlinear decision tree based classification approach to predict the Parkinson's disease using different feature sets of voice data. In: 2018 20th International Conference on Advanced Communication Technology (ICACT), Chuncheon-si Gangwon-do, Korea (South), pp. 638–642. (2018). https://doi.org/10.23919/ICACT.2018.8323864
41. Orozco-Arroyave, J.R., Hönig, F., Arias-Londoño, J.D., Vargas-Bonilla, J.F., Skodda, S., Rusz, J., Nöth, E.: Voiced/Unvoiced transitions in speech as a potential bio-marker to detect Parkinson's disease. In: INTERSPEECH 2015, 16th Annual Conference of the International Speech Communication Association, September 6–10, pp. 95–99 (2015)

Weight-Based Binary Channel Coding—A Machine Learning Approach

Piratla Srihari

Abstract Channel encoding provides the ability of having a reliable error-free reception in connection with digital data transmission. The channel noise affects some of the information bits transmitted (i.e., the received and the transmitted will not be the same), and such affected bits are referred to as errors in the received. Channel coding attends such influences of additive channel noise on the information transmitted, through the redundancy provided for the information prior to transmission. This redundancy provides overheads for transmission, and these are responsible for error-detecting/-correcting features of the communication system. The error-detecting/-correcting ability of a particular channel coding scheme depends on its Hamming distance. In the present discussion, the encoding and decoding (machine learning approach) of single error-correcting binary channel coding scheme referred to as weight-based code is proposed. An $'n'$ bit code word will be transmitted in the place of $'k'$ bit Information word, where $n = k + r$, (modulo-2 sum) is obtained by appending $'r'$ number of parity bits to $'k'$ bit Information word. Code efficiency is a measure of the useful number of bits in the total number of bits transmitted. Weight-based code is such (n, k) binary coding scheme with a code efficiency of $\frac{k}{n} \times 100\% = 50\%$ and is a systematic single error-correcting code.

1 Introduction

In an (n, k) weight-based binary coding scheme, each k-bit binary information word $I = \left[I_1\ I_2\ -\ -\ I_k \right]$ is transformed into an n-bit binary code word $C = \left[c_1\ c_2\ -\ -\ c_n \right]$. Each C of the code is constituted by $'k'$ bit information word appended with an $'r'$ bit binary parity word $P = \left[P_1\ P_2\ -\ -\ P_{n-k} \right]$ resulting in an $'n'$ bit word. A systematic weight-based code word is given by $C = \left[c_1\ c_2\ -\ -\ c_n \right]$

P. Srihari (✉)
Geethanjali College of Engineering and Technology, Hyderabad, India

T. Senjyu et al. (eds.), *ICT with Intelligent Applications*, Smart Innovation, Systems and Technologies 248, https://doi.org/10.1007/978-981-16-4177-0_23

$= \left[I_1\ I_2\ --\ I_k\ P_1\ P_2\ --\ P_{n-k} \right]$ or $\left[P_1\ P_2\ --\ P_{n-k}\ I_1\ I_2\ --\ I_k \right]$, i.e., either the first or the last $'k'$ bits of the code word are the Information bits.

In an (n, k) weight-based code, each of the 2^k information words (each of length $'k'$ bits) will be uniquely mapped to one of the 2^k code words (each of length $'n'$ bits) such as $C_1, C_2, \ldots . C_{2^k}$. The Hamming distance ($d_{\min}$) of an (n, k) weight-based code is the minimum number of bit positions in which any two code words of the code differ. An (n, k) weight-based code [1] with Hamming distance $d_{\min}$ can correct $\frac{d_{\min}-1}{2}$ number of errors [can detect $(d_{\min} - 1)$ number of errors].

In forward error correction schemes, the design of systematic linear (n, k) binary block codes also is on similar lines. In such codes, the parity bits are expressed as a linear combination of information bits. The parity bit structure is according to a predefined mathematical structure, and all the code words of the code will have the same parity bit structure, which is according to the already defined modulo-2 sum of the specific information bits. Unlike these linear systematic binary block code words, in a weight-based code word, the structure of the parity word is non-uniform across the code words of the code, i.e., all the code words of the code will not have the same parity word structure [1], as it is weight (of the Information word) dependent. Weight of an Information word is the number of 1 s it possesses.

2 Encoding

The code word of a weight-based code is expressed as $C = \left[I_1\ I_2\ --\ I_k\ P_1\ P_2\ --\ P_{n-k} \right]$, where $I = \left[I_1\ I_2\ --\ I_k \right]$ is the 'k' bit Information word and $P = \left[P_1\ P_2\ --\ P_{n-k} \right]$ is the parity word of length $(n - k)$ bits with $n = 2k$.

The encoding process of a weight-based code is as follows in Table.1:

Thus, the proposed weight-based codes are also referred to as data negation codes or data inverting codes [1]. The code words of a (6, 3) weight-based code are given below (Table 2):

The message word of any length can be mapped uniquely to a code word in a similar manner. The weight-based codes (under different information lengths) are single error-correcting codes. All these possible code words of the (n, k) code will be the training dataset for the decoder.

Table 1 Parity bit structure

Weight of I	Parity bits
Zero or even	$P_j = I_j$
Odd	$P_j = \overline{I_j}$(complement)

Table 2 Code words

Information word I			Code word C						
I_1	I_2	I_3	c_1	c_2	c_3	c_4	c_5	c_6	
			I_1	I_2	I_3	P_1	P_2	P_3	
0	0	0	0	0	0	0	0	0	C_1
0	0	1	0	0	1	1	1	0	C_2
0	1	0	0	1	0	1	0	1	C_3
0	1	1	0	1	1	0	1	1	C_4
1	0	0	1	0	0	0	1	1	C_5
1	0	1	1	0	1	1	0	1	C_6
1	1	0	1	1	0	1	1	0	C_7
1	1	1	1	1	1	0	0	0	C_8

3 Conventional Method of Locating the Error

Let the received code word be $R = \left[I_1\ I_2\ --\ I_k\ P_1\ P_2\ --\ P_k \right]$.

The decoding involves the following:

- Identifying the structure of the parity word;
- Estimating the possible locations of the errors in the received code word.
  - Identifying of the structure of parity bits:

    Under error-free reception, in R, modulo-2 addition of I and Pis

$$\begin{cases} \text{an all zero vector of size } 1 \times k \text{ if } P_j = I_j \\ \text{i.e. parity bits are same as Information bits} \\ \text{a vector of all 1s of size } 1 \times k \text{ if } P_j = \overline{I_j} \\ \text{i.e. parity bits are the complements of information bits} \end{cases}$$

    Under the reception with error, the above are, respectively,

$$\begin{cases} \text{vector of size } 1 \times k, \text{ with} (k-1) \text{zeros} \\ \text{vector of size } 1 \times k \text{ with} (k-1) \text{1s} \end{cases}$$

  - Estimating the possible error location

    The pair (I_j, P_j) of R whose modulo-2 sum differs from the error-free condition is an indication of the error location, i.e., either of this pair of elements is the bit in error.

After identifying the pair (I_j, P_j) of R where either element of the pair is in error, the exact element in error will be located as

- $2k$ summations, with each sum consisting of$k$ terms (either information bits or parity bits or both) is computed.
- In each sum, the number of information bits and the parity bits will differ by an integer from the set $\{0, 1, 2, ..k\}$.
  For example, in a (6, 3) data inverting code, the summations are

$$I_1 + I_2 + I_3;\ I_2 + I_3 + p_1$$

$$I_3 + p_1 + p_2;\ p_1 + p_2 + p_3;$$
$$p_2 + p_3 + I_1;\ p_3 + I_1 + I_2$$

- Under error-free reception, each of these summations results in (i) 0 when the parity bits of the code word are same as information bits (ii) alternate 1/0 and when the parity bits of the code word are the complements of information bits.
- These $2k$ summations are partitioned into two equal halves, such that the k summations in one of the partitions contain I_j, and the summations of other partition contain p_j.
- The partition in which the all the k summations differ from the error-free conditions is considered to contain the error and either I_j or p_j that is available in all those k summations of that partition is the bit in error.

4 Decoding (Machine Learning Approach)

4.1 Principle of Decoding

The decoding is based on the principle of Naïve Bayes classifier [2], under supervised learning [3]. Knowing the received, the probability for the transmitted to be, i.e., a posteriori probability will be calculated.

- The pair (I_j, P_j) results in two possible code words, among which one of them can be the transmitted.
- For a given R, if the possible code words transmitted are C_i and C_j, then the decision is as per MAP criterion, i.e., "knowing the received i.e. R, if the probability of transmitting C_i > probability of transmitting C_j, decision is in favor of C_i as the transmitted one.
- Otheriwse, decision is in favor of C_j [4, 5]

$$\text{i.e. } p(C_i/R) \underset{C_j}{\overset{C_i}{\gtrless}} p(C_j/R)$$

  – Computation of probabilities

- All the possible code words of the (n, k) weight-based code are with equal likelihood of transmission [6].
- Bit error probability is p_e.
- The probability for having the reception with $'m'$ number of errors is $p(X = m) = n_{c_m}(p_e)^m(1 - p_e)^{n-m}, m = 0, 1, \ldots n$; X is the random variable representing the no. of errors in the received [4].

– The two possible transmitted code words involving the pair (I_j, P_j) are C_α and C_β of the code.
– The training dataset enables the computation of a posteriori probabilities, and the decoder decides in favor of either C_α or C_β[7, 8].

5 Output Case Study

- Let the cross-over probability or transition probability [9] in the channel matrix be

$$p_e = p(1/0) = p(0/1) = \frac{1}{4}$$

- Let the received code word of a (6, 3) weight-based code be $R = 101011$
- The modulo-2 addition of the information bits and the parity bits of the R results in

$$I_1 + P_1 = 1 + 0 = 1; I_2 + P_2 = 0 + 1 = 1;$$
$$I_3 + P_3 = 1 + 1 = 0$$

- Since the sum 110 is consisting of two 1 s (k–1 = 3–1) only, it can be concluded that the parity bits are the complements of the information bits.
- Either I_3 or P_3 of the received code word can be in error.
- Considering P_3 is in error, the corrected code word is 101010.
- From the training dataset, the code word for the information word 101 is 101101(=C_5).
- Since the corrected is 101010, it can be treated that the received is with three errors.
- From the corresponding a posteriori probability of knowing the received R, the probability for the transmitted to be C_6 is $p(C_6/R)$ is same the probability for three errors to occur in the received, i.e., $p(X = 3) = n_{c_m}(p_e)^m(1 - p_e)^{n-m} = 6_{c_3}\left(\frac{1}{4}\right)^3\left(\frac{3}{4}\right)^3 = 0.13$
- Considering I_3 is in error, the corrected code word is 100011.
- From the training dataset, the code word for the information word 100 is 100011(=C_5).
- Since the corrected is 101010, it can be treated that the received is with 1 error.

- From the corresponding a posteriori probability of knowing the received R, the probability for the transmitted to be C_5 is $p(C_5/R)$ is same the probability for 1 error to occur in the received, i.e., $p(X = 3) = n_{c_m}(p_e)^m(1 - p_e)^{n-m} = 6_{c_1}\left(\frac{1}{4}\right)^1\left(\frac{3}{4}\right)^5 = 0.35$
- Since $p(C_5/R) > p(C_6/R)$, as per MAP criterion [10], it can be concluded that the transmitted code word is C_5.

6 Conclusion

The complexity of the conventional method of decoding can be reduced to a large extent through the concept of classification. Using the principle of Naïve Bayes classifier, the received code word can be classified to a group of its nearest equivalents among the transmitted code words, and using MAP criterion, the closest transmitted code word can be identified and thus the information or message word can be retrieved from the code word. Thus, error detection and error correction process can be made feasible. The similar approach can be implemented for multiple error correction also. But, under the case of reception with multiple errors, some more additional constraints are to be incorporated in the code word structure itself. A rectangular array (similar to product codes) as the code word structure instead of [1 × 2k] array can address the multiple errors in the received code word.

References

1. Sri Hari, P., Jinaga, B.C.: Data Inverting Codes. Ph.D Thesis, Jawaharlal Nehru Technological University, Hyderabad, India (2007)
2. Kay, S.M.: In: Intuitive Probability and Random Processes using MATLAB. pp. 215–237. Springer (2006)
3. Paluszek, M., Thomas, S.: In: MATLAB Machine Learning. A Press, 1.2.3.1 (2017)
4. Murphy, K.P.: In: Machine Learning-A Probabilistic Perspective. pp. 149–158. The MIT Press Cambridge, Massachusetts London, England (2012)
5. Unpingco, J.: In: Python for Probability, Statistics, and Machine Learning, pp. 111–113. Springer (2016)
6. Liese, F., Miescke, K.-J.: In: Statistical Decision Theory-Estimation, Testing and Selection-Springer Series in Statisitcs. pp. 293–247. Springer (2008)
7. He, R., Ding, Z.: (Eds) In: Applications of machine learning in wireless communications. IET Telecommunication Series, vol. 81, pp. 135–140. IET The institution of Engineering and Technology (2019)
8. Levy, B.C.: In: Principles of Signal Detection and parameter Estimation–Springer Texts in Statistics, pp. 15–62. (2008)
9. Shalev-Shwartz, S., Ben-David, S.: In: Understanding Machine Learning-From Theory to Algorithms, pp. 342–353. Cambridge University Press (2014)
10. Hoff, P.D.: In: A First course in Bayesian Statistical Methods- Springer Texts in Statistics, pp. 13–26 (2009)

Li-Fi: An Emerging Technology

Divya Singh

Abstract To handle the quickly developing number of cell phones and their extending requests for Internet administrations, network intermingling is conceived to coordinate distinctive innovation spaces. An as of late proposed and promising way to deal with indoor remote interchanges is coordinating light fidelity (Li-Fi). The notion and hypotheses of Li-Fi are very much clarified in this paper. Obvious protected light is used for remote entrance in such influenced conditions which additionally gives enlightenment. Moreover, with the confusions about Li-Fi are uncovered, and furthermore, the extent of Li-Fi in businesses and innovation is delineated. Light fidelity (Li-Fi) gives a successful answer to limit network traffic of remote communication. Lights turn out to be Li-Fi attocells due to controlled working tasks bringing about improved remote limit with respect to the 5G, Internet of Things, and past. The engineering of Li-Fi passage that executes OFDM is researched. The point by point concentrate on Li-Fi can show that this innovation can be the inescapable organization in future and can make an upset in media transmission.

1 Introduction

The present world significantly relies on the use of Wi-Fi in advanced mobile cells, and their utilization continues expanding because of interest on higher information rates. So as to get an answer, the organization associations grow new advancements similar to 5G, etc. The issues with RF range and social well-being are fretful, optical wireless communication assumes a crucial function in present- day correspondence [1] which utilizes its noticeable light range in visible light communication (VLC). It is anticipated that 44 zettabytes of information will be created by 2020 in accordance with explorers. Previously, remote information spread has been accomplished over RF range in the scope of millimeter wave spectrum. Power line communication (PLC) is joined with VLC in the innovation of home automation to communicate fast information and permit remote association among gadgets. For applying this

D. Singh (✉)
Department of Electronics and Communication, GLA University, Mathura, India
e-mail: divya.singh@gla.ac.in

T. Senjyu et al. (eds.), *ICT with Intelligent Applications*, Smart Innovation, Systems and Technologies 248, https://doi.org/10.1007/978-981-16-4177-0_24

innovation, LEDs are utilized as the passages, and VLC innovation is notorious as Li-Fi. LEDs are utilized in the innovation of Li-Fi for widespread remote communication over light range. It is reached out to noticeable light correspondence (VLC) that accomplishes fast of information broadcast, bidirectional, safety of information, and remote organizing [2–4].

Li-Fi-light commitment resembles Wi-Fi development that come for distant correspondence headways. The essential segment of this development consolidates totally organized, bidirectional, and quick distant. Nowadays, the most moving space in far off correspondence is Wi-Fi, and Web customers are similarly being extended year by year. For improving pace, adequacy, move speed, Li-Fi development has progressed [5–8]. The use of light is conceivable in the advancement of data transmission due to change in light force snappier than the regular eye for getting. This technology has snappier data transmission comparable to Wi-Fi on numerous occasions [9].

2 Working Principle of Li-Fi

Li-Fi scheme on a very basic level joins two segments to be explicit the source and recipient. The data signal can be offset at the transmitter zone with a precise interval of time then send the data in 0's and 1's structure using LED bulbs. Here, the glints of LED bulbs are implied with 0's and 1's. A photodiode is used at the receiver end to get the LED streaks invigorates the sign and gives the yield. The Li-Fi structure is shown in the block diagram as follows, and the transmitter zone joins the data, clock circuit, and a LED bulb. The commitment of the transmitter can be such a data like substance, voice, etc. The check circuit in this part is used to give the fundamental periods of time among each piece, and these are imparted to the authority end as LED streaks (Fig. 1).

The beneficiary area joins the photodiode in the same way as the intensifier. Here, the photodiode detects LED bulb bulbs that immediately turn the explosion into electrical signals. Eventually, the speaker receives signals from the photodiode and rises to give the output. Li-Fi is basically VLC that has high speed of this system. Li-Fi uses standard LEDs to move and accelerate the data at 224 Gigabits / sec [10–12]. Data transfer for this development should be considered structurally. The key features of this system are the diodes that illuminate brightly. The development of on / off LEDs allows for such data transfer as integrated codes; however, the feature's eye cannot detect this change and the bulbs appear with unchanging power [13].

3 Characteristics of Li-Fi

- High capacity or data transfer rate nearly to 1Gbps
- High efficiency
- Low cost

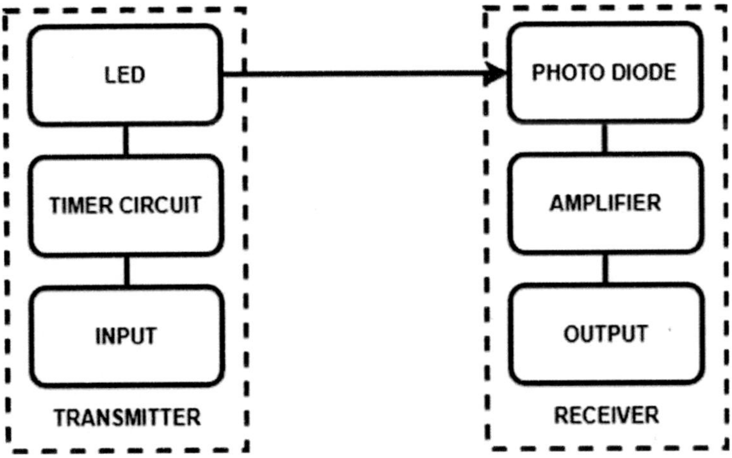

Fig. 1 Block diagram of Li-Fi

- High security
- High safety.

4 Advantages of Li- Fi

- Improved data density nearly to 1000 times
- Enhanced data security
- Improved energy efficiency nearly to 1000 times
- Used in RF banned places like oil platforms and petrochemical plants.
- The present data network infrastructures are used for making the intermediary connections in between the different light sources with Internet and its integrated Li-Fi modem.
- No electromagnetic radiation due to light as a medium that causes safe health environment
- Interference can be neglected among the signals for the safety of data because Li-Fi does not allow penetration through walls.

5 Disadvantages of Li-Fi

- Without light source, Web cannot be utilized. This may restrict the establishment of Li-Fi in a few circumstances and areas.
- Because it utilizes noticeable light, and light cannot infiltrate opaque items, the sign reach is restricted by actual boundaries.
- A totally different framework for Li-Fi would should be built.

- Deployment of new equipment for Li-Fi [14] organizing is needed to be introduced in existing PC, advanced mobile phones, and tablet PCs since they actually use equipment for Wi-Fi organizing. Li-Fi is not promptly in reverse viable.
- Since the LED lights stay on, it prompts light contamination because of its brilliant introduction to make up for conceivable impedance with other light sources.
- Manufacturing cost of Li-Fi is not that much as a result of the use of LED lights [15] which are a lot of less in cost. However, establishment of this innovation is costlier.

6 Scope of Li-Fi

Li-Fi is a new invention that hugy affects a very large number of businesses. It can open the IoT, improve future industrialization, take a more important part of the 5G cellular framework, improve road safety where there are more non-motorized vehicles, create new digital remote organizations, and provide new solutions for computer development. Li-Fi will cover two key businesses: (i) the telecommunications industry, and (ii) the lighting business. In the future, LED lighting is used for a large number of applications and prepares for the development of sharp urban environments. LaaS will be used extensively in the lighting industry [16, 17].

7 Applications of Li-Fi

- Airlines: Wi-Fi usage is maintained for the majority of carriers. As a Li-Fi substitute, it can be used with the necessary connectivity agencies that bring less effort, less weight, and higher technology.
- Indoor apparatuses: To reduce the power consumption and power management, Li-Fi frame can be imported into home appliances, for example, coolers, fixed heating frames, TVs, etc.
- LiBNet: Li-Fi inflatable organization is called LiBNet. Each inflatable, using remote domains, forms a remote connected network. This framework is used in disaster management where transparent light is used for the transmission of information at higher levels and in addition provides enlightenment on the affected situations.
- Li-Fi for street lights: The downlink framework is defined by a streetlamp as a transmitter that receives information about power light communication (PLC) and controls information [18]. The mobile camera as a beneficiary receives data through a small heartbeat with an LED bulb. For uplinks, Li-Fi is controlled by new features such as Bluetooth, ZigBee, or Wi-Fi.
- Atomic power plants: Since Li-Fi does not cause any electrical blockages, it can be best used to replace Wi-Fi in affected power stations such as atomic power plants.

- Efficiency: In hospitals, Li-Fi can be used to transmit rapid data to other critical types of gear clinic [19–21]. By wearing LED wrist watches, bands, lead, etc., presented by Li-Fi transmitters, we can check the well-being of the human body and report to a family specialist immediately if anything goes wrong with the web [1].
- Transmission of seawater information: The tendency to light can be seen when immersed when radio waves cannot increase in view of the harsh environment and the high flow of seawater; however, seawater reduces the quality of radio waves. Since connectors pose a risk to underwater writing, they are often installed by Li-Fi transmitters that can send data from submarines to the surface, another way around just like jumpers with their headlights [3, 18, 22].

8 Conclusion

Li-Fi brands are slowly growing. In the future, these innovations are further enhanced by the rapid arrival of information rates achieved in research institutions using the OFDM hypothesis. As the data is transmitted via LED, it poses no health risk and ensures a secure connection. It is common for Li-Fi to be the majority with the current remote organization that makes media transfer improvements. It can be seen that gadgets like laptops, advanced cells, and tablet laptops will be upgraded by Li-Fi. It has examined in this paper about the Li-Fi, improvement in this innovation when contrasted and the current framework, features of Li-Fi, network demonstrating, useful utilizations, upsides and downsides of this cutting edge innovation, etc. On the off chance that the VLC standard is upgraded to the enhancements of Li-Fi, it will act a substitute for Wi-Fi in numerous spots and circumstances and makes a brain gleaming world.

References

1. Tsonev, D., Chun, H., Rajbhandari, S., McKendry, J., Videv, S., Gu, E., Haji, M., Watson, S., Kelly, A., Faulkner, G., Dawson, M., Haas, H., O'Brien, D.: A 3-Gb/s single- LED OFDM-based wireless vlc link using a gallium nitride LED. IEEE Photon. Technol. Lett. **26**(7), 637–640 (2014)
2. Cheung, W.C., Quek, T., Kountouris, M.: Throughput optimization, spectrum allocation, and access control in two-tier femtocell networks. IEEE J. Sel. Areas Commun. **30**(3), 561–574 (2012)
3. Dimitrov, S., Haas, H.: Information rate of OFDM based optical wireless communication systems with nonlinear distortion. J. Lightw. Technol. **31**(6), 918–929 (2013)
4. Gibson, D.V., Kozmetsky, G., Smilor, R.W: In: Rowman and Littlefield (eds.) The technopolis Phenomenon: Smart Cities, Fast Systems, Global Networks (1992)
5. Tsonev, D., Videv, S., Haas, H.: Unlocking spectral efficiency in intensity modulation and direct detection systems. IEEE J. Sel. Areas Commun. **33**(9), 1758–1770 (2015)

6. Wang, Z., Tsonev, D., Videv, S., Haas, H.: On the design of a solar-panel receiver for optical wireless communications with simultaneous energy harvesting. IEEE J. Sel. Areas Commun. **33**(8), 1612–1623 (2015)
7. Chandrasekhar, V., Andrews, J., Gatherer, A.: Femtocell networks: a survey. IEEE Commun. Mag. **46**(9), 59–67 (2008)
8. Haas, H.: Wireless data from every bulb. TED Global. Edinburg, Scotland (2011)
9. Chaterjee, S., Agarwal, S., Nath, A.: Scope and challenges in light fidelity(Li-Fi) technology in wireless data communication. Int. J. Innov. Res. Adv. Eng. (IJIRAE) **2**(6), 6 (2015). ISSN: 2349–2163
10. Amrutha, S., Mathew, A., Rajasree, R., Sugathan, S., Aravind, S.: A visible light communication system for indoor application. Int. J. Eng. Innov. Technol. (IJEIT) **3**(12), (2014)
11. Shah, A.: Novel architecture for future Li-Fi wireless technology. Int. J. Eng. Res. Technol. (IJERT) **2**(11), (2014)
12. Zeng, L., O'Brien, D.C., Minh, H.L., Faulkner, G.E., Lee, K., Jung, D., Oh, Y.J., Won, E.T.: High data rate multiple input multiple output(MIMO) optical wireless communications using white LED lighting. IEEE J. Sel. Areas Commun. **27**(9), (2009)
13. Vucic, J., Kottke, C., Nerreter, S., Langer, K., Walewski, J.: 513 Mbit/s visible light communications link based on DMT modulation of a white LED. IEEE/OSA J. Light wave Technol. **28**(24), 3512–3518 (2010)
14. Haas, H., Yin, L., Wang, Y., Chen, C.: What is Li-Fi? J. Light wave Technol. **34**(6), 1533–1544 (2016)
15. Tsonev, D., et al.: A 3-Gb/s single-LED OFDM-based wireless VLC link using a gallium nitride μ LED. IEEE Photon. Technol. Lett. **26**(7), 637–640 (Apr. 2014)
16. Zanella, A., Bui, N., Castellani, A., Vangelista, L., Zorzi, M.: Internet of things for smart cities. IEEE J. Internet of Things **1**, 22–32 (2014)
17. Rani, J., Chauhan, P., Tripathi, R.: Li-Fi (Light Fidelity)-the future technology in wireless communication. Int. J. Appl. Eng. Res. ISSN **7**(11), 0973–4562 (2012)
18. Haas, H., Yin, L., Wang, Y., Chen, C.: What is Li-Fi? J. Lightwave Technol. **34**(6), 1533–1544 (2016)
19. Scuderi, A., Ragonese, E., Biondi, T., Palmisano, G.: In: Integrated inductors and transformers, CRC Press (2011)
20. Chang, M.H., Wu, J.Y., Hsieh, W.C., Lin, S.Y., Liang, Y.W., Wei, H.: High efficiency power management system for solar energy harvesting applications. In: 2010 IEEE Asia Pacific Conference on Circuits and Systems, Kuala Lumpur, pp. 879–882. (2010)
21. Wang, C.X., et al.: Cellular architecture and key technologies for 5G wireless communication networks. IEEE Commun. Mag. **52**(2), 122–130 (2014)
22. Haas, H.: Wireless data from every light bulb. TED Website, August (2011)

Identifying Learning Standards of Artificial Neural Networks: A Study

Eshetu Gusare Desisa, Shaik Janbhasha, and Gebreigziabher Abadi Meru

Abstract Artificial neural networks (ANNs) are now omnipresent, and there is continuous growth in excellent research to create new ones. ANN is capable of adaptive, dynamic learning. In ANN learning, we need to change the weights to enhance the input/output behavior. Therefore, with the help of which the weights can be adjusted, a method is required. These strategies are called learning laws, which are simply formulas or algorithms. Rules of learning are algorithms or mathematical logic that guides modifications is the weight of the network links. By employing the disparity between both the expected output and the real outcome to update its weights during training, they incorporate an error reduction mechanism. Learning laws improve the effectiveness of the artificial neural network and extend this rule to the network. Usually, the learning rule is repeatedly applied to the same set of training inputs over a large number of cycles, with error steadily decreasing over epochs. They are fine-tuned as the weights are. The present research strives, however, to assess the objective of the artificial neural network and the learning principles. In this, we analyzed ten different regulations studied by leading researchers that include rules based on supervised learning (perceptron, memory-based, delta, error correction, correlation, out star, supervised Hebbian) and unsupervised rules based on learning (competitive, competitive Hebbian, Hebbian); this defines how to adjust the weight of the nodes of a network.

1 Introduction

A neural network is an interconnected group of basic processing components, units or nodes that are loosely based on the functionality of the animal neuron. The network's processing capacity is stored in the assets or weights of the inter-unit connection, attained through a process of adaptability to a set of training shapes or learning from them. The other part of the art' of using neural networks spins around the various ways in which these distinct neurons can be assembled. In the human mind,

E. G. Desisa (✉) · S. Janbhasha · G. A. Meru
Assosa University, Assosa, Ethiopia

T. Senjyu et al. (eds.), *ICT with Intelligent Applications*, Smart Innovation, Systems and Technologies 248, https://doi.org/10.1007/978-981-16-4177-0_25

this clustering happens in such a way that data can be interpreted in a complex, collaborative, and auto-organizing way. Neural networks are biologically formed from microscopic components in a three-dimensional universe. These neurons seem capable of interconnections that are almost uncontrolled. In any planned, or actual, manmade network, that is not valid. Using recent technology, integrated circuits are two-dimensional structures with a small number of layers for intercommunication. This physical one, that Fact of the matter limits the types and scale of silicon artificial neural networks that can be developed. Neural networks are currently the fundamental clustering of primitive artificial neurons. By building layers that are then linked to each other, this clustering occurs. The other aspect of the 'art' of engineering networks to solve serious issues is how these layers communicate [1].

Basically, as shown in Fig. 1, all artificial neural networks have a striking resemblance or topography. Any of the neurons communicate with the natural world in that structure to obtain their inputs. Other neurons provide the natural universe with the outputs of the network. This performance may be the specific character that the network assumes it has inspected or the specific picture it feels is being displayed. The remainders of the neurons are all concealed from view. But more than a bunch of receptors are a neural network. Some early researchers attempted, without much success, to engage in similar neurons in a random way. Now, it is recognized that standardized devices are also the brains of snails. Developing layers of elements has become one of the fastest ways to construct a structure. It is the separation of these neurons into layers, the ties between these layers, and the functions of summing up and transfer that comprise a coherent neural network. For all networks, the specific terms used to define these features are similar.

While networks containing only this layer, or perhaps even one part, are useful, most applications need networks containing at least three normal layer types: source, secret, and output. In optimization problems, the input neuron layer collects data from either source code or directly from sensing devices. The output layer delivers data directly with the outside environment, to a secondary computing device, or to other machinery, such as a physical control system. There could be several hidden layers

Fig. 1 Basic neural network diagram with three layers

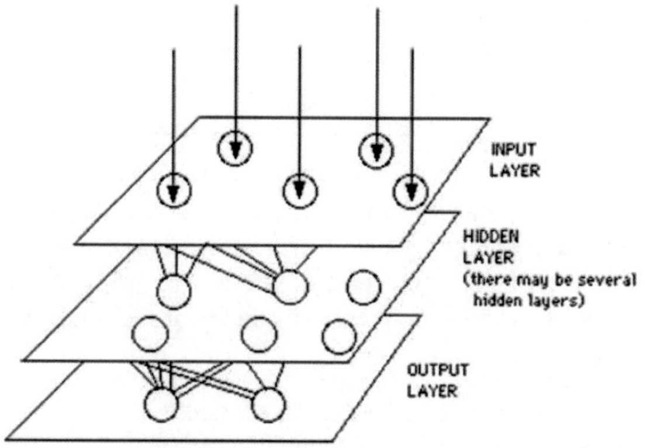

between those two layers. Many of the neurons in multiple interconnected systems produces these internal layers. Both of these secret neurons' inputs and outputs basically move to different neurons [2].

Each neuron in a secret layer receives the signals from all the neurons in the layer above it in most networks usually an input layer. It transmits its output to all of the neurons in the layer beneath it after a neuron performs its role, providing a feedforward path to the output. The essential aspects of neural networks are these channels of communication from one neuron to some other. The key to the device is them. They are the interconnections that give an input a variable power. There have been two kinds of these relationships. One enables the very next neuron's summing process to add, whereas the other stimulates it to deduct. One impresses, while the other prevents, in more human terms.

A few other networks want the other neurons in the very same layer to be hindered by a neuron. Lateral inhibition is called this. In the output layer, the most famous use of this is "For example, if the probability of a letter being a "P" is 0.85 in text recognition, and the probability of the letter being a "F" is 0.65 The network needs to pick and inhibit all the others with the greatest likelihood. With innervation, it can do that. Competition is also called this principle. Feedback is another form of link. This is where one layer routes production back to an earlier layer. Figure 2 demonstrates an example of this.

The manner neurons are related to one another has a huge effect on the network's activity. The user is required to add, remove, and monitor these links at will in the bigger, more professional application development packages. By "adjusting" parameters, it is possible either to inspire or impede these ties.

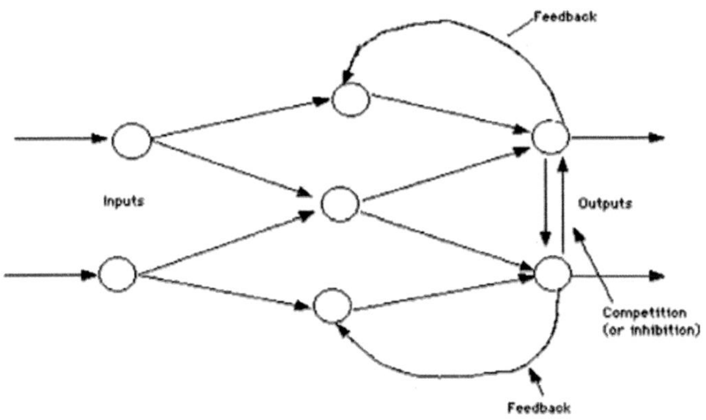

Fig. 2 Easy input and competition network

1.1 Learning in ANN

In essence, learning involves creating and adapting the adjustment in itself as and when the situation turns. Neural organization is a convoluted framework, and maybe more accurately, I can say it is a troublesome undertaking to accomplish, that relying upon the knowledge that courses through every last bit of it, might modify its inner form [3].

1.2 Importance of Learning

Learning in neural networks, being a complex adaptive system, means that is a computing machine able to adjust its actions at input data or output data because of the evolving world. Due to the defined role for activation and output or input function, when building a particular system, the significance of learning in ANN increases. We need to adjust the weights now to alter the input/output behavior.

2 Learning Rules for the Artificial Neural Network

A method or mathematical logic is a learning law or a learning mechanism. It enhances the efficiency of the artificial neural network as well as imposes this rule above the network. Therefore, learning rules adjust the weights as well as bias levels of a network when designed to simulate a network in a particular data setting. While ANN training, we have to modify the weights to alter the input/output conduct. Therefore, with the help of which the weights can be adjusted, a method is required. These strategies are called training laws, which are merely formulas or algorithms. The execution of the theory of learning is an iterative process. It allows an artificial neural network to learn and enhance its performance from the current conditions. Different learning principles are based on supervised and unsupervised learning in ANN.

2.1 Perceptron Learning Rule

This principle is an error implemented by Rosenblatt to improve the managed learning calculation of solitary feedforward neural organizations with a direct actuation work. The fundamental principle of this rule is that in order to quantify the error, a contrast must be made in between the ideal/target yield and the real yield. If any difference is detected, an adjustment must be made to the relation weights [4].

To illustrate its mathematical model, there "n" number of sets with limited input, X_n, alongside the ideal yield vector t_n, where $n = 1$ for N. As explained earlier, the yield 'y' would now be able to be determined based on the total info and the enactment work functional over that whole information can be communicated as follows.

$$Y = f(y_{\text{in}})\begin{cases} 1, & y_{\text{in}} > \theta \\ 0, & y_{\text{in}} \leq \theta \end{cases} \quad (1)$$

where the threshold is θ.

In the following two cases, the weight update can be performed,

Step i, if $t \neq Y$, then $W(\text{new}) = W(\text{old}) + t_x$.

Step ii, if $t = Y$, then no transformation in weight.

Every link under a neural network, as we know, has a related weight, which adjusts in the course of learning. The network begins its training by assigning a random number to every weight, as an example of supervised learning. Measure the output value in terms of a set of records for which we should know the predicted output value. This is the model of learning which indicates the whole definition. As a consequence, it is considered a sample of learning. The system then associates the value of the output measured with the value predicted. Next an error function $\in$, which can be the amount of squares of the errors that occur in the learning sample for each individual, is calculated as follows:

$$\sum_i \sum_j \left(E_{ij} - O_{ij}\right)^2 \quad (2)$$

Perform the first description on the learning set's individuals and perform the second summation on the output units. The anticipated and obtained values of the jth element for the ith distinct are E_{ij} and O_{ij}. The network then changes the weights of the various components, checking whether the error feature has gone up or down each time. This is a matter of overcoming a problem of least squares, as in a standard regression. Because the weights of nodes are allocated as indicated by clients, it is an illustration of supervised learning.

2.2 Rule for Memory-Based Learning

Everything previous involvements are directly deposited in a broad recollection of accurately labeled instances of yield inputs $\{(x_i, d_i)\}$ $N_i = 1$ in memory-based learning, where xi signifies an information vector and di indicates the ideal answer. "The calculation reacts by recovering and examining the preparation information in a nearby area" of X_{test} when arrangement of a test vector X_{test} (not seen previously) is required [5]. Two variables are involved in all memory-based learning algorithms, i.e., the criteria used to identify the X_{test} local neighbor and the training law functional to preparing models in the X_{test} nearby area. The vector $X_N \in \{X_1, X_2, \ldots X_N\}$ is the closest neighbor of X_{test} if min d $(X_i, X_{\text{test}}) = d\ (XN, X_{\text{test}})$ where XN is the X_{test} class which is the nearest neighbor rule (NNR). The variant of the NNR is the k-nearest neighbors law, which distinguishes the k ordered examples closest to X_{test} for any number k [6].

2.3 Delta Learning Rule (Widrow—Hoff Rule)

To reduce the error across all training patterns, Widrow and Hoff introduced the least mean square technique. It is somewhat directed learning calculation with having ceaseless enactment work. The rudimentary principle is that the gradient-descent approach, which continues indefinitely, is the basis of this law. In order to limit the whole contribution to the yield unit and the objective worth, the delta rule updates the synaptic weights [7].

The mathematical equation is given by the delta rule to update the synaptic weights.

$$\Delta W_i = \alpha \cdot X_i \cdot e_j \tag{3}$$

Here ΔW_i is a shift of weight for the pattern;

$\alpha =$ the positive and continuous rate of learning;

$x_i =$ Pre-synaptic neuron input value;

$e_j = (t\text{-}y_{\text{in}})$ is the contrast between the ideal/goal yield and the genuine yield yin.

The delta rule above is just for a solitary yield unit. The following two instances, the weight update can be performed.

Instance 1, if $\boldsymbol{t \neq y}$, then

$$W(\text{new}) = W(\text{old}) + \Delta W \tag{4}$$

Instance 2, when $t = y$, then no variation in weight.

One of the most famous learning rules is the delta rule. This standard expresses that the sympatric weight change of a node is equivalent to the error multiplication and the data. The delta rule in mathematical form is also as follows:

$$\Delta W = \eta(t - y)Xi \tag{5}$$

For a provided input vector, the correct response is to compare the output vector. No learning takes place if the difference is zero; otherwise, it changes its weights to decrease this difference. The transition from u_i to u_j in weight is:

$$dw_{ij} = r^* a_i^* e_j. \tag{6}$$

a_i represents the activation of u_i, where r is the learning rate, and e_j is the difference between the predicted output and u_j's actual output. If an arbitrary set forms the set of input patterns, then learn specific associations using the delta rule [8].

It was shown that there were no secret units for networks with linear activation functions. A paraboloid in n-space is the error squared vs. the graph of weight. The graph of such a function is curved in upward and has the slightest value, as the proportionality constant is negative. This paraboloid's vertex point that reflects where the error is minimized. The ideal vector of weight is then the weight vector corresponding to this point. For both one output unit and multiple output units, we should use the delta learning rule. When smearing the delta rule, accept that the error can be calculated directly. The purpose of smearing the delta rule is to diminish the contrast among the genuine and expected performance is the error.

2.4 *Error-Correction Learning Rule*

The method of associating the machine output to the desired output value as well as using that error to straight the training is Error-Correction Learning, used in supervised learning. The error values can be used on the most direct route to change the tap weights directly, using backpropagation algorithm. If the system output is y and x is known to be the desired system output, the blunder sign can be characterized as:

$$\mathrm{e} = x - y \tag{7}$$

At each training iteration, blunder adjustment learning calculations mean to limit this mistake signal. The backpropagation calculation is the well-known learning calculation for use in mistake rectification learning [9].

2.5 *Correlation Learning Rule*

The rule of correlation learning is based on a theory close to the rule of Hebbian learning. It implies that weights should be more favorable between reacting neurons, and weights should be more negative between neurons with opposite responses. The correlation law, in contrast to the Hebbian rule, is supervised learning. As an alternative of a genuine one, the reaction, o_j, the desired answer, d_j, is used to measure the weight shift. The correlation learning rule is as follows in mathematical form:

$$\Delta W_{ij} = \eta X_i d_j \tag{8}$$

where the desired output signal value is d_j. Typically, this training algorithm begins with initializing weights to zero. The correlation learning rule is an example of supervised learning as the desired weight is assigned by users [10].

2.6 *Outstar Learning Rule (Grossberg Learning)*

Supervised learning is concerned with this law, implemented by Grossberg, since the expected yields are set up. It is likewise called Grossberg training. This laws refers to neurons that are organized in a layer. It is expressly meant to achieve the ideal d-yield of the p-neuron layer. When we think nodes or neurons in a network organized in a layer, we use the out star learning law. Here, the weights linked to a certain node should be identical to the ideal neuron outputs linked by those weights. For a layer of n nodes, the output start rule produces a desired answer *t*. Apply this kind of learning to all nodes in a specific layer. As in Kohonen neural networks, weights for nodes are modified [11].

Express the out star learning in mathematical form as follows:

$$Wik = \begin{cases} \eta(y_k - W_{jk}) & \text{if node } j \text{ successes the competition} \\ 0 & \text{when node } j \text{ losses the competition} \end{cases} \tag{9}$$

In outstar learning rule, weight adjustment computations are as follows:

$$\Delta W_j = \alpha(d - w_j) \tag{10}$$

In the above, d is necessary performance of the neuron and α is the rate of learning.

2.7 *Supervised Hebbian Learning Rule*

The algorithm is used by taking each training data point (X_{train}, Y_{train}) to prepare a neural organization utilizing the preparation informational index and run the accompanying:

Feedforward sign utilizes a corrected form of the exaggerated digression enactment $Y = \tanh_{\text{rec}}\ (\Sigma X_i\ W_i - b_i)$ by measuring the activation of each neuron in the network. This $\tanh_{\text{rec}}$ function takes an activation of the hyperconstraint c which is the coefficient of x, which regulates tanh curve [12]. For each layer, the layer activation is attached to the list of activations. The desired output vector is strongly affixed to the activation list at the last layer so we can guarantee regulated pattern training. ReLU activation function is the function of the latest output layer to allow distinction of values.

The $\tanh_{\text{rec}}$ has the accompanying definition:

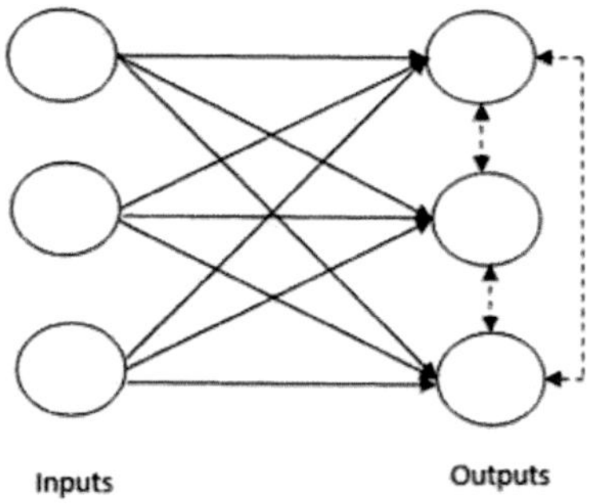

Fig. 3 General architecture

$$\text{Tanh}_{\text{rec}}(x, c) = \begin{cases} \frac{e^{ex} - e^{-ex}}{e^{ex} + e^{-ex}} & \text{if } x > 0 \\ 0 & \text{if } x < 0 \end{cases} \tag{11}$$

After receiving all data point activations, the updated Hebbian weight function is evaluated simultaneously for each weight, beginning from the initial layer:

$$\Delta W(x, y) = \begin{cases} +\eta_{\text{ltp}}xy, & \text{if } x^{*}y \geq \text{TH and } W\#0 \\ -\eta_{\text{ltp}}xy, & \text{if } x^{*}y < \text{TH and } W\#0 \\ 0.50, & \text{if } W = 0 \end{cases} \tag{12}$$

In the above updated weight function, TH is the positive integer indicating a threshold at which the mass should shift for apiece neuron. The values of x and y are in the middle of 0 and 1. Often to manage possible cases, we have to amend the weight update function. A dense rendition that assembles abundant data for the neuron to get the hang of utilizing the drawn out potentiation and long haul impedance measures is the weight update highlight above. If we try to model more biological processes/scenarios, we will extend the weight update functionality to include more situations. An extended version of the function is given below (Fig. 3):

$$\Delta W(x, y) = \begin{cases} +\eta_{\text{ltp}}xy, & \text{if } x > 0,\ y > 0 \text{ and } W > 0 \\ -\eta_{\text{ltp}}xy, & \text{if } x > 0,\ y > 0 \text{ and } W < 0 \\ -\eta_{\text{ltd}}x, & \text{if } x > 0,\ y = 0 \text{ and } W > 0 \\ +\eta_{\text{ltd}}x, & \text{if } x > 0,\ y = 0 \text{ and } W < 0 \\ +\eta_{\text{ltp2}}y, & \text{if } x = 0,\ y > 0 \text{ and } W > 0 \\ -\eta_{\text{ltp2}}y, & \text{if } x = 0,\ y > 0 \text{ and } W < 0 \\ +0.50 & \text{if } x^{*}y \geq \text{TH and } W = 0 \\ 0, & \text{Otherwise} \end{cases} \tag{13}$$

Use rectified linear unit function ReLU(x) to measure the newly modified weights as follows:

$$W_{\text{new}} = \text{ReLU}(W_{\text{old}} + \Delta W) \tag{14}$$

If $W_{\text{old}} > 0$ but $W_{\text{new}} < 0$, reset w to zero and vice versa, respectively.

The logic here is that a neuron does not alter its shape from being truly beneficial to receptor binding and the other way round. The ReLU(x) is described as follows:

$$\text{ReLU(x)} = \begin{cases} x, \text{ if } x > 0 \\ 0, \text{ if } x \leq 0 \end{cases} \tag{15}$$

Notice that the above laws derive since hypotheses as well as tries to design certain renowned processes for the functioning of plasticity. When determine extra knowledge about how versatility, long haul potentiation, and long haul melancholy work, we can still change these laws.

2.8 *Law of Competitive Learning*

It is concerned to unaided preparing in this law, the yield hubs endeavor to rival each other to mirror the example of information. We should comprehend the serious organization, which is given in this manner, to comprehend this learning rule. This organization resembles a solitary layer feedforward network with criticism joins between yields. The basic concept of the competitive network is the relationship between yields are inhibitory, appeared by spotted lines, meaning that rivals not ever endorse their own [13].

The basic principle of the competitive learning rule is that rivalry between the output nodes can take place. Therefore, the key concept is that the winner will be declared during preparation, the yield entity through the peak stimulation toward a certain effort decoration. Since only winning neuron is revised and the majority of the neurons are away untouched, this rule is often termed winner-takes-all. The succeeding three significant dynamics in the scientific construction of competitive learning law:

Situation to become a champion—Presume that there would be the following condition uncertainty a neuron Y_k would like to be the champion:

$$Y_k = \begin{cases} 1 \text{ if } Y_k > Y_j, \text{ for every } j \text{ and } j \neq k \\ 0 \text{ else} \end{cases} \tag{16}$$

This one implies that uncertainty somewhat neuron, roughly Y_k, desires to success, at that time the summary entity output, about Y_k, essential to be principal output of entire additional neurons in the system in its induced local area [14].

Situation for summation of over-all weight—One more drawback on theory of competitive training is the cumulative aggregate weights for given yield neuron would be 1. For instance, on the off chance that we think about neuron k, at that point,

$$\sum_j W_{kj} = 1 \quad \text{for every } k$$

Weight Shift for the Champion: On the off chance of a neuron not responding to the example of input, and, after that, there is no learning in that neuron. Be that as it may, the subsequent loads become altered as follows, if a solitary neuron succeeds.

$$\Delta w_{kj} = \begin{cases} -\alpha(xj - wkj), & \text{when neuron } k \text{ succeeds} \\ 0, & \text{when neuron } k \text{ failures} \end{cases} \quad (17)$$

From the above equation, α is rate of learning.

This specifically sets out the by changing its weight, we support the winning neuron What's more, if a misfortune happens in the neuron, at that point we try not to re-transform of its weight.

2.9 *Competitive Hebbian Learning*

In artificial neural networks, it is a type of unsupervised learning in which nodes vie for the option to react to a subsection of input data. Competitive learning, a form of Hebbian learning, works by enhancing the specialization of each network node. It is ideal for identifying clusters inside information. Vector quantization and self-organizing maps are models and algorithms based on the competitive learning theory (Kohonen maps) [15]. A competitive learning rule has three basic elements: A collection of neurons that are all the same apart from certain synaptic weights that are randomly distributed and thus reacting distinctively to a specified arrangement of information designs and cap forced on every neuron's "strong point." The process makes neurons battle for the option to react to a specified info subsection, with the goal that just one yield neuron is active at a time. A "winner-take-all" neuron is considered the neuron that successes the competition [16]. Consequently, the network's individual neurons learn to specialized in an array of identical patterns and thus develop 'include identifiers' for various categories of info designs. The fact that competitive networks record correlated input sets to one of a few output neurons effectively eliminates the representation redundancy that is an integral part of biological sensory system processing. Neural networks that include a hidden layer that is widely known as a "competitive layer" are typically implemented. Any competitive neuron i remains signified through the variable of weight. The calculation of correlation among the incorporating information and weigh vector are defined and quantified. The modest neurons " contend" by one another for any input vector to see which of them is the most similar to that specific variable of input and winning neuron "m" determines the production $i = 1$ and any remaining serious neurons set yield $i = 0$, i= 1... M and $i\#m$. The inverse of the Euclidean distance is generally used to calculate comparability among the vector of information and the vector of weight.

2.9.1 Hebbian Learning Rule

The first learning law was the Hebbian rule. It was developed by Donald Hebb in 1949 as a learning algorithm for the feedforward, unmonitored neural network. We may use it to define how to increase the weight of a network's nodes. A rule in neuroscience that offers an illustration for the adaptation of areas of the brain during the learning process is the Hebbian learning rule. It explains a fundamental synaptic plasticity process where an improvement in synaptic efficacy occurs from the regular and sustained stimulation of the post-synaptic cell by the pre-synaptic cell. It is also called the law of Hebb. Hebb's concept can be defined from the perspective of neurons and neural organizations as a technique for concluding in what way to adjust the loads among classic neurons. If both of these neurons activate together the weight between the two nerve cells goes up, then down in the event that they actuate independently. There are high positive weights in hubs that seem, by all accounts, to be either certain or both adverse simultaneously, where as those that are normally reversed have pessimistic loads.

Hebbian learning law can be expressed as:

$$\Delta W = \eta z(k, w)k \tag{18}$$

where k is the input signal vector arranged the neuron synapses, W is the neuron-related load variable, $\boldsymbol{\eta}$ is coefficient of learning rate, $\boldsymbol{\Delta W}$ is the vector for weight update, and z (k, w) is neuron's initiation of post-synaptic; a function of the input as well as loads is believed to be positive [17].

If two neighbor neurons were simultaneously activated and deactivated, which implies the Hebb learning rule. It should then increase the weight linking these neurons. The weight between them should reduce for neurons operating in the opposite process. The weight does not shift if there is no signal association. If all nodes' inputs are either favorable or unfavorable, there is a clear positive weight between the nodes. A heavy negative weight occurs among the nodes if the input of a node is optimistic as well as undesirable for one of the other node. Values of all weights are set to "0" at the beginning. For both soft- and tough functions, this rule of learning can be set to zero. Since desired neuronal responses are not included in the training process, this is the law of unsupervised learning. Typically, the outright estimations of the loads are corresponding to the undesired training period [18]. It is also possible to define the Hebbian learning rule with the following formula:

$$W_{ij} = K_i^* K_j \tag{19}$$

where W_{ij} is the heaviness of the association from neuron j to neuron i and K_i the contribution for neuron i. Note that this is design learning. In a Hopfield organization, associations W_{ij} are set to zero on the off chance that $i = j$. With twofold neurons, associations would be set to 1 if the associated neurons have a similar initiation. The fundamental definition of this norm is "The point at which an action potential of cell A is close to sufficient for energize a phone B and more than once or industriously participates in terminating it, certain development cycle or neuronal alteration happens with one or the two cells with the end goal that A's effectiveness, as one of the phones terminating B, is expanded.

From the above hypothesize, we can presume that the associations between two neurons may be reinforced if the neurons fire simultaneously and might debilitate in the event that they fire at various occasions [19].

According to the Hebbian learning law, the mathematical formulation is the formula for increasing the relation weight at each point of time.

$$\Delta W_{ji}(t) = \alpha x_i(t) \cdot y_j(t) \tag{20}$$

Here, $\Delta W_{ji}(t)$ = Significantly raise by which the link weight rises at the time stage t.

α = the non-negative and constant learning rate

$X_i(t)$ = The pre-synaptic neuron input value at the time point t

$Y_i(t)$ = The yield of time t of the pre-synaptic neuron at the identical time.

3 Conclusion

The present study tried to define the target of the artificial neural network and the rules of learning. In this, we examined ten different rules studied by various researchers that include supervised learning rules (perceptron, memory-based, delta, error correction, correlation, outstar, supervised Hebbian) and unsupervised learning rules (competitive, competitive Hebbian, Hebbian) that describe how to change the weights of a network's nodes. The supervised learning rule is a simpler system, extremely detailed, and reliable rules, offline learning method. Input and output variables will be given in a supervised learning model. However, the rule of unsupervised learning is computationally intensive, quite precise, and reliable, and the process of learning takes place in real time, and the model of learning is unsupervised, and only input data will be produced. Based on the type of neural network design and type of training, each technique has its own advantages and drawbacks. Improve these approaches and test for real data in the future. In addition, these learning rules are algorithms or mathematical logic that guide adjustments in the weight of a network's ties. It enhances the efficiency of the artificial neural network and imposes this rule over the network.

Consequently, training laws update the loads and predisposition levels of an organization when simulating a network in a particular data setting. By using the distinction between the desired output and an actual output to change its weights during training, they implement an error reduction mechanism. Learning laws boost the efficiency of the artificial neural network and extend this rule to the network. The learning rule is usually applied frequently over a huge number of epochs to the same set of training inputs with error steadily decreasing over epochs as the weights are fine-tuned.

References

1. Abi, O.I., Jan, A., Om, A.E., Dada, K.V., AbdElatif Mohamed, N., Arshad, H.: State-of-the-art in neural network techniques: a survey. **4**(11) (2018)
2. Alhama, R.G., Zuidema, W.: Pre-wiring and pre-training: what does a neural network need to learn truly general identity rules (2016)
3. Altmann, G.T.: Learning and development in neural networks—the importance of prior experience. Cognition **85**(2), B43–B50 (2002)
4. Gkanogiannis, Kalamboukis, T.: A modified and fast perceptron learning rule and its use for tag recommendations in social bookmarking systems. ECML PKDD Discovery Challenge 2009 (DC09) **497**(71), 1–13 (2009)
5. Mee, M., Rou, É., Hill, N.: Optimization of the largest annotated Tibetan corpus combining rule-based, memory-based and deep-learning methods. Trans. Asian Low-Reso. Lang. Info. Process. (2021)
6. Tzi, N., Tsak, N., Benr, E., Theo, J., Zal, G.: A memory-based learning approach utilizing combined spectral sources and geographical proximity for improved VIS-NIR-SWIR soil properties estimation. Geoderma (2019)
7. Chakraverty, S.: Training law of delta and backpropagation law for multi-tier perceptron. Chakr Concepts of Soft Computing. Springer Nature Singapore Pte Ltd (2019)
8. Jumah Al, O.Q.: Speech recognition based radon-discrete cosine transforms by delta ANN training law. IEEE (2016)
9. Sma, B., Goru, F.: Error-correction training for neural networks using the Bayesian paradigm. Techn. Autom. Med. Diagn. **52**, (2014)
10. Tao, Wa, C, Wang, J., Cheng, J.J., Chen, X.: Particle-swarm optimization of ensemble neural networks with unfavorable correlation training for forecasting short-term wind speed of wind farms in western China. Inform. Sci. (2019)
11. Snider, G.: Instar and outstar learning with memristive nanodevices. Nanotechnology **22**(1), 015201 (2011)
12. Ra: A supervised modified hebbian learning method on feed-forward neural networks. 11 Dec (2019)
13. Grossberg, S.: Adaptive pattern classification and universal recoding: parallel development and coding of neural feature detectors. Biol. Cybern. **23**(3), 121–134 (1976)
14. Lobov, S.A., Chernyshov, A.V., Krilova, N.P., Shamshin, M.O., Kazantsev, V.B.: Competitive learning in a spiking neural network: towards an intelligent pattern classifier. Sensors **20**, 500 (2020a). https://doi.org/10.3390/s20020500
15. Mar, T., Gielen, S., Kappen, B.: (eds) Competitive hebbian, training law forms perfectly topology preserving maps. In: ICANN '93. ICANN. Springer, London (1993)
16. Tim, M., Edda, S., Ham, A.: Mode region detection using improved Competitive Hebbian Learning for unsupervised clustering. Eng. Sci. Technol. **7**(4), 26–35 (2017)
17. Goltsev, A., Gritsenko, V.: Modular neural networks with Hebbian learning rule. Neurocomputing **72**(10), 2477–2482 (2009)
18. Gerstner, W., Ritz, R.: van Hemmen JL "Why spikes? Hebbianlearning and retrieval of time-resolved excitation patterns." BiolCybern **69**, 503–515 (1993)

19. Herz, A.V.M., Sulzer, B., Kühn, R., van Hemmen, J.L.: Hebbian learning reconsidered: representation of static and dynamic objects in associative neural nets. Biol. Cybern. **60**, 457–467 (1989)

Smart UAV Framework for Multi-Assistance

Vaishnavi Patil, Vaishnavi Potphode, Utkarsha Potdukhe, Vishal Badgujar, and Kaushiki Upadhyaya

Abstract Unmanned aerial vehicles, which are also familiar as drones, play an important role in military and national emergency. Our motive in this project is to present the real-time possibilities of using UAVs in rescue operations. UAVs can be practically used to transport goods on demand, provide medical support in urban areas, save people stuck during floods, analyze the scale of damages, monitor large performance inspection activities, human gatherings, ORS, deliver food and other necessary material, supply automated external defibrillators, support rescue operations and for emergency transport activities.

1 Introduction

Artificial intelligence and drones are pair made in technology. Pairing technology of the machine learning with AI along with the art of UAV drones allows ground-level operators an overview of entire area. Since a long time, drones play key diagnostic roles in a very form of sectors—including defense, natural disaster relief, agriculture [10], security and construction. With their capability to extend potency and enhanced safety, drones became one of the important tools for everybody from fire—fighters to farmers.

As we all know these days the unmanned aerial vehicles (UAVs) have become a popular technology with the ability to upgrade the commercial production industries and also the property right outside of the military. UAVs would be able to accelerate recovery operations and rescue from natural disasters and may be used for autonomous delivery [6] systems. Unmanned aerial vehicles (UAVs) are developing technology with firm resolution for fixing many common public and personal

V. Patil (✉) · V. Potphode · U. Potdukhe · V. Badgujar · K. Upadhyaya
Department of Information Technology, A.P. Shah Institute of Technology, Thane, India

V. Badgujar
e-mail: vsbadgujar@apsit.edu.in

K. Upadhyaya
e-mail: ksupadhyaya@apsit.edu.in

T. Senjyu et al. (eds.), *ICT with Intelligent Applications*, Smart Innovation, Systems and Technologies 248, https://doi.org/10.1007/978-981-16-4177-0_26

operations. Public departments, such as police department, public safety and transportation management and public health care, are beginning to use UAVs to convey prompt warnings regarding the catastrophe and improve the performance of rescue and resumption operations when a telecommunication infrastructure in an exceeding area is broken or otherwise not available. Natural disasters have been increased at an distressing rate in the world. Every now and then, we hear in the news about a locality full of catastrophe, tsunami or a hurricane, etc. There is a necessity to look at such a catastrophe stricken areas before tasking the required rescue and help operations.

Smart UAVs are so popular, in fact, that the method of getting medical attention in remote regions in India should be improved to extend wellness management accessible. The healthcare system should try and aim to supply the simplest and quick necessary medical help to all any one far and near the country while enhancing for time and cost efficiency. However, India's current healthcare and wellness system is disorganized and lacks the accessibility of medical services in various parts of country. AI in health care is used for complex algorithms and software to emulate human cognition with analysis of medical data.

The primary thought of AI applications related to health is to analyze relationship between treatment techniques and patient health. Drones are used as a security camera for monitoring outside activities like violence of rules or theft and protest to acknowledge the people or monitor their activities. Even for street surveillance [9], concerned authorities are using drones to transmit messages and spread awareness about the lockdown measures that the government has undertaken, particularly in areas that lack open communication channels for health insights and information. Drones equipped with loudspeakers are conventional to make public announcements to stay back indoors, take all the necessary precautions, maintain social distancing and wear a mask when stepping outside from home. Doctors and hospitals need medical supplies and laboratory testing quite often, and drones are the fastest and safest ways to deliver medication supplies and transport required samples from hospitals to laboratories.

The use of drones for medical reasons brings about numerous advantages, such as instant help, less time for traveling to the destination of patient, reduction of complications within the injured due to a brief time to attend for rescue, support and enhancement of basic functioning of medical necessity and their emergency teams, and therefore, the room to succeed in places which are otherwise inaccessible for basic means of medical support.

AI drones used in construction companies can scan and cover a larger area of buildings within some minutes than any human would have required to complete. In defense and military areas, drones have become suitable to develop for the unmanned tools to attack or bombard on the enemies during the conflicts. Military officers often should petrol areas so as to go looking for any potential threat, illegal activity or any intrusion within the borders of the city that may put the lives of the civilians in threat. Such areas involve very high risk to human life.

The UAV system can be effectively used to get the required job done without the loss of human life, also the rate at which the operation will be done will definitely going to be quicker with help of the drones. Until recently, though, drones were only ready to display what their cameras captured. Now, because of computer science software, they will perceive their surroundings, which enables them to map location, track objects and provide analytical feedback in real time.

2 Literature Review

2.1 Nelakuditi et al. [1] proposed an architecture stating IoT-based unmanned aerial vehicle system for agriculture applications. The pesticides and fertilizers are helpful to retain the bloom of the crops and avert the suffering caused by contagion that live on these crops. Using unmanned aerial vehicles (UAV) for spraying, one can spray pesticides all over the crops and can cover larger area in very limited time also managing the use of chemicals and water at the same time. A sprayer mechanism was enabled to the drone. With this mechanism, one can easily spray and prevent the crop from getting damaged with very less efforts.

2.2 Kataev et al. [2] presented a model. According to the model, the monitoring of the research processes predicting crop yields, supervising the growth stages of the plant, using unmanned aerial vehicles are feasible for farming. One of the effortless solution for this problem is to process the received data (images) from the unmanned aerial vehicle (UAV). An essential segment of image processing is linked with application of the clustering method to isolate the different varieties of plants in the field, such as weeds. The given report presents us the examples of an application of the standard K-means method for clustering images of agricultural fields.

2.3 Elloumi et al. [3] have presented their design in 2018. In this paper contains design for monitoring the traffic on road using a UAV-based system. This UAV is capable of monitoring the traffic situation on a road, and they collect and send the real-time data of information about the vehicle to a traffic processing center to regulate the movement of vehicles. This paper states that the performance of their system is finer than the performance of those UAV which are fixed for trajectory traffic monitoring system in terms of coverage and events detection rates.

2.4 Motlagh et al. [4] proposed a very clear and precise description about UAV-based IoT platform A crowd surveillance use case. This paper consists of a high-level perspective of a UAV-based integrative IoT platform for delivering the IoT assistance from greater height, in addition to the system orchestrator all together. As a visualized use case of the platform, the paper demonstrates how one can use UAVs to monitor the crowd depending upon the face recognition technology. They have designed a testbed consisting of a local processing node and one MEC node.

2.5 Noguchi and Komiya [5] proposed an architecture where in the disaster prone area can be remotely monitored in real-time using collective UAVs. Here, they have used a network of UAVs where each UAV will pass on the collected information to its subsequent UAV and then to the GCS at last. Every UAV maintains its connections with other UAVs and yet fly autonomously. The received information from the monitoring UAV is passed to other UAVs in sequence as to transfer the data to GCS. The proposed system replaces the UAVs in-flight with standby UAVs according to the remaining energy in order to prevent the monitoring disruption caused by their absence from the network.

3 Existing System Architecture/Working

The following Fig. 1 shows the existing system architecture of Arduino-based UAV. This system has Arduino as a core component and can be controlled through mobile GUI using Bluetooth.

Working: Usually, the drones are controlled with the help of a radio frequency (RF) remote or in case using an autopilot enabled with a GPS module for automatic controlling by directing the proper path to drone. Here in this system, a drone is constructed using an Arduino Uno.

The major motive of this project is to gather the data that is the video information from a particular known public area. So as to get the video information, we have to designate the rate of maximum dimensions that the drone is capable of covering by using Arduino programming. The system is enabled with autopilot functioning, which means the system need not have to be controlled by any user as it will be controlled by itself, and the flight controller will be responsible to handle the controlling actions of the drone with the help of built-in sensors that balances the drone. Arduino Uno is the main controller of the system, which gives the appropriate signal to flight controller. To maintain the feasibility and stability, it consists of an Open Pilot CC3D microcontroller (or any flight controller), as well as a camera to take the real-time data. Finally, the system includes a Bluetooth module to turn the drone on/off and display the live data using an Android mobile device or Bluetooth low energy module.

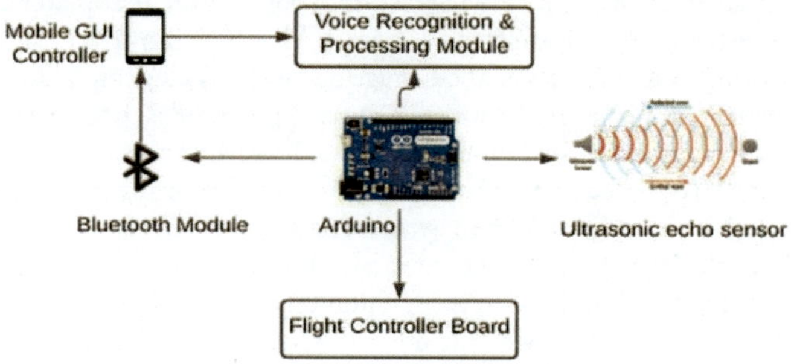

Fig. 1 Existing system architecture

4 Objective

(1) To make use of UAV for spraying disinfectant in public areas like airports, railway station, park and roads.
(2) To reduce human efforts as drones can disinfects approximately 50 times more area in comparison to any other traditional methods and at the same time keeps human operators out of harm.
(3) To make people aware, apart from spraying disinfectants, these drone have public monitoring and warning system equipped with camera.
(4) To monitor high risk places by real-time analysis for crowd gathering and keeping an eye on the crowd gathered in a particular area and give relevant instructions to the gathered crowd using the fitted speakers.
(5) Besides, this to trace the real-time location of UAV flying in a particular area, and the data is stored which can be used when necessary.

5 Proposed System Architecture/Working

The following Fig. 2 represents the system architecture of UAV drone armed with surveillance cameras and speaker for making the public briefing. The capability of UAV for aerial inspection is emphatically strengthening the police's capability to contest and manage the spread of the pandemic. By conveying circumstantial

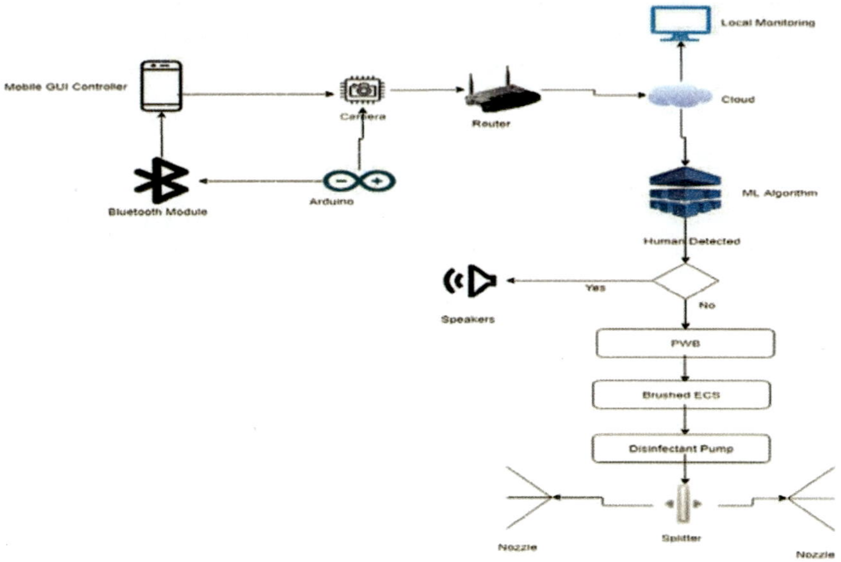

Fig. 2 Proposed system architecture

awareness on a real-time basis, this technology gives the police the means to interpret and set up resources very quickly and easily to manage emerging situations.

Working: Here we have used Arduino as the core component and all the other components are interfaced with Arduino. The DC motors are connected to the Arduino. These motors are mainly used for motion controlling and positioning the drone. The DC motor creates its own magnetic field which is an input to the motor's rotor which will help the drone to take off from the ground. The brushless DC motors are attached to the propellers these DC motors produce the necessary thrust that helps propellers to cause the circular motion. These components are covered with drone frame that gives shape to the drone and helps to hold all the components together. The strength of the drone depends largely on the quality of frame used. The IP camera is connected which will capture all the real-time images/video of a particular area and pass the data to cloud over the Internet. From cloud, the data is accessed and here machine learning [7] comes into picture. The main purpose of using machine learning is to make sure that any object (humans particularly) is detected, and then depending upon the output, the further action is taken.

For object detection, we have used YOLO algorithm, i.e., you only look once (YOLO) [8]. This algorithm is used for real-time object detection, and it is the most effective object detection algorithm. It has convolutional neural network (CNN) that uses deep learning algorithm for object applies one single neural network to an entire image and then divides the whole image into boundaries and predicts the bounding boxes. For this, it uses pre-trained data set called COCO data set. YOLO algorithm detection. It classifies the objects within the images frame. It is prominent because it brings about high efficiency thereby also being able to run it in real time. The meaning of algorithm "only looks once" is that the algorithm scans the given image or video at once and requires one forward propagation to pass through neural network to predict the object in the given image or video. Now depending upon the results we get from Machine Learning, the upcoming task will be scheduled like if the area is crowd-free then the system will spray the disinfectant with the help of fitted nozzle and splitter. We have made use of splitter here as it will divide the flow of disinfectant in two direction thereby covering a larger area. And if the output shows the presence of some crowd, then it will take the action of making the necessary announcement with speakers as to maintain the social distance to control the spread of disease. For communication between user and the drone we made use of transmitter and receiver. A radio control system consists of two elements that are the transmitter and receiver. The antenna on the transmitter will help transmitter to send the control signal to receiver, and the receiver in turn passes on the signal to flight controller with which the drone can be moved. The drone is enabled with flight controller, and it directs the RPM of each DC motor depending upon the input received. The command is fed to the flight controller, and it in turn manipulates the DC motor accordingly. The flight of the drone is controlled using Bluetooth low energy module which makes use of Web Bluetooth. Web Bluetooth is a technology mainly used of making two Bluetooth devices communicate with each other. All these components are powered using lithium polymer or LiPo batteries.

The above image is the snippet of the output showing the objects that are detected using YOLO algorithm. This output is then stored in cloud and can be used when needed.

The above image shows the output when YOLO algorithm is applied to the video, and it detects particularly the humans as our project concentrates not just the detection of objects but specifically on detecting the crowd in the particular area.

6 Conclusion

The UAV can provide the best and quick possible medical attention to all needy people over a specified region along with optimizing for time and cost efficiency. The drone along with GPS can also be used as a function of surveillance and counting a crowd in a specific location which can be treated as future advancement of our UAV. Also drone can be made more efficient by increasing its weight capacity which can be then also used for emergency delivery system in remote regions as to bring health care at convenience. The change in direction of wind flow, increase in distance between controller and the drone, or any other environmental changes like heavy rain, fog can limit the efficiency of proposed UAV.

Acknowledgements We would like to express our genuine obligation to A. P. Shah Institute of Technology Thane for letting us such an opportunity to work under this model. With this design has given us a chance to grasp the detailed knowledge and about the practical working of model. We are grateful to our college faculties helping and encouraging us by making us available with the necessary components and supervising us in the entire process. We would also thank our Head of the Department Mr. Kiran Deshpande for referring us such an ingenious concept to work upon. Lastly, we would like to express our sincere gratitude to our parents for their immense support and motivation.

References

1. Nelakuditi, U.R., Manoj Vihari, M., Purna Teja, M.: IoT based unmanned aerial vehicle system for agriculture applications. In: 2018 International Conference on Smart Systems and Inventive Technology (ICSSIT) (2018). https://doi.org/10.1109/icssit.2018.8748794
2. Kataev, M.Y., Qadeer, M.M.D.: In: International Multi- Conference on Engineering, Computer and Information Sciences (SIBIRCON) (2019)
3. Elloumi, M., Dhaou, R., Escrig, B., Idoudi, H., Saidane, L. A.: Monitoring road traffic with a UAV-based system. In: 2018 IEEE Wireless Communications and Networking Conference (WCNC) (2018). https://doi.org/10.1109/wcnc.2018.8377077
4. Motlagh, N.H., Bagaa, M., Taleb, T.: UAV-based IoT platform: a crowd surveillance use case. IEEE Commun. Magaz. **55**(2), 128–134 (2017)
5. Noguchi, T., Komiya, Y.: Persistent cooperative monitoring system of disaster areas using UAV networks. In: 2019 IEEE Smart World, Ubiquitous Intelligence and Computing, Advanced and Trusted Computing, Scalable Computing and Communications, Cloud and Big Data Computing, Internet of People and Smart City Innovation (2019)
6. Oniniwu, G., Onojo, O., Ozioko,O., Nosiri, O.: Quadcopter design for payload delivery. J. Comput. Commun. **4**, 1–12 (2016)
7. Alpaydin, E.: Introduction to Machine Learning. MIT Press, Cambridge, MA, USA (2014)

8. Hu, Y., Wu, X., Zheng, G., Liu, X.: Object detection of UAV for anti-UAV based on improved YOLO v3. In: 2019 Chinese Control Conference (CCC) (2019)
9. Wada, A., et al.: A surveillance system using small unmanned aerial vehicle (UAV) related technologies. NEC Tech. J. **8**(1), 68–72 (2017)
10. Erdelj, M., Król, M., Natalizio, E.: Wireless sensor networks and multi-UAV systems for natural disaster management. Comput. Netw. **124**, 72–86 (2019)

Tool for Assessing Intellectual Quotient Among Children

B. Vinitha, K. Rishita, K. Pranavi, R. Pallavi Reddy, and N. Kalyani

Abstract Cognitive assessment (or intelligence testing) is used to evaluate the general thinking and reasoning abilities of a person, often referred to as intellectual functioning or IQ. Higher IQ scores represent higher cognitive performance and lower scores reflect poor cognitive performance. Nevertheless, individual domain scores that provide a more accurate representation of an individual's cognitive ability than the overall IQ score because the scores between domains differ greatly. Intelligence testing can determine various aspects of a child's cognitive capacity. Such intellectual skills include verbal awareness, visual and spatial perception, logical reasoning, working memory and speed of processing. A cognitive assessment can also help recognize particular areas of strength or weakness in children using psychometric tests, so that effective approaches can be built to improve learning and growth through the implementation of different machine learning models. Recommendations should be provided with a deeper understanding of how a child thinks to improve weak skills and reinforce areas of strength. Sometimes, an assessment can also help to explain certain behaviors that a child may exhibit.

1 Introduction

Intelligence is the ability to solve problems that are respected or produce goods within one or more cultural settings. A global notion that includes an individual's ability to behave consciously, think rationally and cope effectively with the environment. Intelligence is the ability to inhibit an instinctive adjustment, the ability to redefine the inhibited instinctive adjustment in light of imaginably observed trial and error and the ability to realize the changed instinctive adjustment in open action as a social creature for the good of the individual. Intelligence is also the capacity to perform activities marked by (1) difficulty, (2) complexity, (3) abstractness, (4) economy, (5) adaption to a goal, (6) social meaning, and (7) emergence of originals, and to sustain these activities under conditions requiring concentration of

B. Vinitha (✉) · K. Rishita · K. Pranavi · R. P. Reddy · N. Kalyani
Department of Computer Science Engineering, G. Narayanamma Institute of Technology and Science (For Women), Shaikpet, Hyderabad, Telangana, India

T. Senjyu et al. (eds.), *ICT with Intelligent Applications*, Smart Innovation, Systems and Technologies 248, https://doi.org/10.1007/978-981-16-4177-0_27

energy and emotional resistance. The eight factors that influence intelligence are: linguistic, logical mathematics, music, spatial, bodily kin esthetic, interpersonal and intrapersonal.

Human intelligence quantification has been a challenging and contentious issue, because intelligence is viewed differently by different people. Intelligence is calculated by non-physical metrics that are officially referred to as cognitive tests or 'intellectual quotient' (IQ) tests; IQ tests come in several types. They can help diagnose intellectual disorders or assess the intellectual capacity of someone. The development of many IQ tests was worked on by psychologists. The most common types of IQ tests are: (a) Stanford–Binet intelligence scale, (b) universal nonverbal intelligence, (c) differential ability scales, (d) Peabody individual achievement test, (e) Wechsler individual achievement test (WIAT), (f) Woodcock Johnson III tests of cognitive disabilities.

The importance of IQ testing is most obvious in educational or clinical environments. Children who have learning disabilities or extreme behavioral problems will be tested to assess if the difficulties of the child can be related in part to an IQ score that is substantially different from the mean for the age group. Intelligence can change within time. But to a rather significant degree, an IQ test gives the same answer even over a span of a year. The older the test score, the more stable it would be. In IQ scores, the most variance is in childhood, mainly in adolescence. Therefore, children needing additional assistance cannot be detected effectively without IQ testing or some other indicator of intelligence, and they cannot be taught how to use what they have, more effectively and better. IQ tests can show certain disabilities in kids that could occur. Three of the standard IQ tests used today include: 1. Wechsler intelligence scale for children (wisc-v), 2. Stanford–Binet intelligence scale and 3. differential ability scales(das-ii).

Cognitive skill assessments or intellect or IQ tests measure the skills involved in learning (e.g., reasoning, comprehension, memory, verbal and mathematical capacity and problem-solving). These assessments pose questions designed to assess the ability of candidates to use mental mechanisms to solve work-related challenges or gain new knowledge about employment.

The main objective is to solve the above-mentioned challenges and to further develop a decision-making support system for the assessment of intellectual skills through a Web application which implements:

- Assessing children's thinking abilities based on various intellectual factors.
- To compute an individual's intellectual quotient (IQ).
- Create a performance profile of a child in IQ testing.
- To generate a study on child performance improvement in comparison with specific abilities.
- To minimize human effort.

These objectives can be achieved by automating the process of evaluating children's intellectual skills which can reveal cognitive strengths and weaknesses.

Kids usually feel discomfort in the cognitive assessments, as these assessments are monotonous and long. Evaluations are mediocre, more often than not. There are more creative and stimulating ways of evaluating a child than the traditional assessment. Assessments are changing, and gamification is constantly being emphasized. Cognitive assessments based on games can easily captivate children with their fun nature. Automated tests greatly reduce the amount of time spent by teachers or administrators, particularly in high-volume testing. Assessments provide meaningful insight into the functioning of the child. Overall, automated appraisals save time and money, making the process more effective.

The tool can be classified into six modules: verbal comprehension task (VCT), processing speed task (PST), working memory task (WMT), general intelligence task (GIT), profile generation (PG) and statistical analysis (SA). The VCT module evaluates the child's verbal understanding or vocabulary. The child's response stimuli are calculated by PST. WMT checks an individual's short-term memory (STM). GIT tests the child's overall performance in verbal understanding, processing speed and working memory. The profile module compares the results of the first three modules with those of the GIT tests. The statistical analysis includes documentation of the child's success in bell curves through various modules.

- **VCT:** The verbal comprehension challenge tests a child's capacity to apply word knowledge to thinking skills to assist in understanding his environment.
- **PST:** Processing speed task is a way of describing how the brain receives, understands and responds to information.
- **WMT:** Short-term memory is also called working memory. To concentrate on its functionality rather than its length, it was redefined.
- **GIT:** General intelligence, also referred to as the g factor, refers to the presence of a large mental capacity that affects the output of measures of cognitive ability.
- **PG:** The data accumulated across the VCT, PST, WMT, and GIT are made available to the user through a document.
- **SA:** Statistical analysis performs analysis of the data acquired from the profile generation module and projects its visual representation.

Further organization of the paper is described as follows. Section 2 discusses the literature survey. Sect. 3 highlights the methodology for assessing the intellectual abilities. Implementation of the results is focused in Sect. 4. The conclusion and future enhancements are discussed in Sect. 5.

2 Literature Survey

This segment examines current research work and provides insight into the issues, and attempts to identify the gaps in established approach. The abbreviation 'IQ' derives from the expression intelligence quotient, first coined by the German psychologist William Stern, in the early 1900s (from the German Intelligenz-Quotient). Alfred Binet and Theodore Simon, who published the Binet–Simon intelligence scale, the first modern intelligence metric, later used this notion in 1905. In several other nations, the Binet–Simon scale was modified for use, because it was simple to administer. IQ assessments aim to calculate and offer a quotient of intelligence, a score derived from a standardized test designed to access human intelligence. These tests now have several variations that have built on the original test designed to identify children in need of remedial education and extended it. Currently, the IQ tests are used to study distributions of scoring across various populations. Over time, variations in other variables, such as actions, success, and well-being, have been correlated with these scores; they vary according to cultural norms. It has been shown that scores on IQ tests have high margins of error and can vary greatly depending on the day, the circumstances of the test and even the relationship with the person being tested.

The reasoning for most cognitive tests is loosely based on Spearman's early observation that success is positively correlated on a wide variety of cognitive tasks (i.e., if you perform well on one, you appear to perform well on others) and as such can be reduced to a single index of aggregate results across a battery of diverse tests [1]. The thesis illustrates the introduction to the field of intelligence, demonstrating exceptional objectivity and lack of partiality [2]. Long back in 1997, in Osawa and Yuba, the concept of human cognitive capacity was used for output debugging of parallel programs in order to distinguish coordination and computational ability in terms of shapes, positions and other related characteristics [3]. The relationship between the construct of mental focus and 'giftness' is not well known. Gifted people may make good use of their executive functions, and this could be due to their capacity for mental focus [4]. The relationship between the intelligence quotient (IQ) and cognitive control mechanisms has been thoroughly developed. Several studies have shown that IQ correlates with executive function capacities, such as disturbance suppression, as tested with laboratory tasks such as Stroop and Flanker tasks [5]. This [6] is intended to discuss the main concerns related to the formal diagnosis of intellectual disability, with an emphasis on youth.

IQ testing or cognitive testing or psychometric testing is performed based on theories of intelligence listed as follows:

- Charles Spearman: General intelligence
- Howard Gardner: Multiple intelligences
- Robert Sternberg: Triarchic theory of intelligence
- Louis L. Thurstone: Primary mental abilities

The benefits of the proposed study include:

1. It is a way of recognizing 'gifted' people. 2. It provides useful insights into diagnosis. 3. It allows one to understand how the brain functions. 4. It can be used as a means of improving intelligence. 5. It may recognize students at risk of slipping through the cracks.

3 Methodology

This section contains the framework that explains the comparative analysis approach used to perform the study.

Figure 1 depicts the architecture of the assessment of intellectual skills, which comprises eight elements, namely the user, assess verbal comprehension, assess processing speed, assess working memory, the cumulative test, calculate individual test score, analysis phase, and profile generation. To complete all the modules, the user must enter the details. The user takes all the individual tasks such as verbal comprehension, processing speed, working memory and general intelligence. All individual scores of the tasks are calculated and stored in the database. User profile is depicted in the profile generation module based on the values stored in the database. The user's analysis is displayed in the analysis module in the form of normal distribution curves, based on the values stored in the database. The Web application has six modules which are elaborated in the section below.

A .**Implementation**

- **General Intelligence Task:**

All questions of the above modules such as verbal comprehension, processing speed and working memory are collectively presented in this module in a random manner. Based on the question, students must choose the nearest alternative available. Student scores one point for every correct answer.

- **Verbal Comprehension Task:**

It is the ability to grasp and express general knowledge and explain concepts using a variety of vocabulary. It is the capacity to recognize the significance of verbal stimuli. This can include understanding of listening, understanding of reading, as well as understanding of sentences. Intelligence verbal comprehension assessments

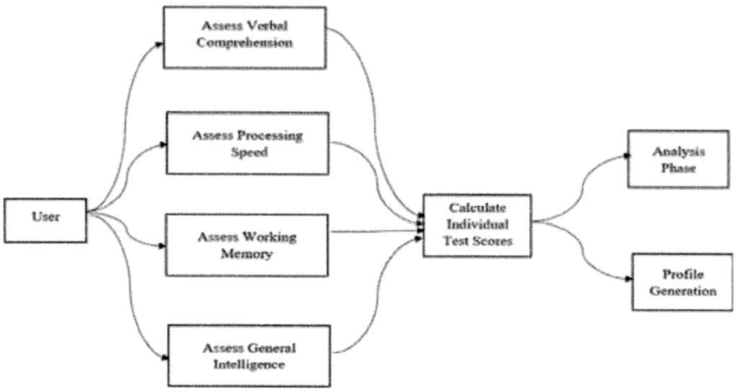

Fig. 1 Architecture of assessing intellectual abilities

provide an evaluation of the capacity of a person to think, reason and solve problems in various ways.

The module of verbal comprehension contains two sub-modules, namely odd one out and antonyms. An item that is different from or held apart from others in odd one out will form a category or package. The objective of odd one out is to guess which one of four objects does not belong to group. Having successfully found the odd one out, he/she will score one point for selecting the correct object.

The second sub-module is to find antonym for a term. An antonym is a phrase that has a contrary meaning to another word. The root words for the word 'anonymous' are 'anti,' meaning 'against' or 'opposite,' and 'anonymous,' meaning 'name.' Antonym is the relation between concepts that are meaninglessly opposite to the meaning that exists. The child must choose an antonym from the options for a given term in this module. A student scores one point for each correct answer.

- **Processing Speed Task:**

Processing speed is one of the main components of the cognitive process and is one of the most important competencies in learning academic success, mental development, reasoning and experience. Speed of processing is a cognitive skill that could be described as the amount of time it takes a person to perform a mental task. This is related to the speed at which a person, whether visual (letters and numbers), audible (language) or motion, can understand and respond to the information they are receiving; i.e., the time between receiving a stimulus and responding is processing speed.

Slow processing speed means some preferred tasks are harder than others, such as reading, doing math, listening and taking notes or holding conversations. As a person with slow processing speed will have tougher schedule times, set goals, make decisions, start projects, pay attention, etc., this can also interfere with executive functions.

Each module has two additional sub-modules, namely compatibility and image patterns. An arrow appears on the screen in compatibility module. Student should press opposite or the same arrow button as quickly as possible, depending on the direction of the arrow. Student scores one point for every correct answer. Pattern recognition happens when environmental information is acquired and inserted into short-term memory, which triggers automatic activation of a similar long-term memory material. Pattern recognition helps one to foresee and anticipate what is to come. The pattern recognition process involves comparing the acquired information with the information already stored in the brain. Making the relation between memories and perceived knowledge is a pattern recognition phase called identification. Pattern identification requires the replication of interactions. Student must predict the next possible pattern for the given picture correctly. Student scores one point for every correct answer.

- **Working Memory Task:**

Working memory is a limited-capability cognitive device that can momentarily retain information. Working memory is essential to reason and direct decisions and behavior. Working memory is often used synonymously with short-term memory, but some scholars find the two forms of memory to be different, suggesting that working memory enables the retrieval of stored information, whereas short-term memory only applies to the short-term storage of information. Working memory is fundamental to cognitive psychology, neuropsychology and neuroscience as a scientific term.

Working memory module contains two sub-modules, number series and memory game, for example. Number series can be defined as an arrangement of numbers in a certain order, where some numbers are mistakenly placed into the number series and some numbers are missing in that sequence, we need to observe and find the exact number to the number series. This module consists of a number sequence to which the student must guess the number on the basis of the problem. In memory game, for the first few seconds, a picture is shown to the student and then disappears, and a question is shown based on the picture shown. Students have to pick the most acceptable response based on the question. Student gains one point for every correct answer.

- **Profile Generation:**

The results obtained from the above tests are shown in this section. The results of individual experiments are stored in the database. These results are extracted from the database, and a report is produced based on the results obtained.

- **Analysis Phase**

Results gathered from all modules in this series are represented as normal distribution curves or bell curves. For each user's individual test results, the normal distribution curves are drawn.

The normal density curve represents a normal distribution: Two numbers are defined to describe any normal distribution: mean(μ) and standard deviation (σ). At the middle of the symmetric normal curve that is taken as 100 in this work is the mean of a normal distribution. The standard deviation is the distance, taken as 16, from the center to the change of curvature points on either side.

For a normal distribution with a mean (or expected value) of μ, the probability density function and the standard deviation of σ is given by $\boldsymbol{(x|\mu,\sigma) = [-(x-\mu)^{2}/2\sigma^{2}] \div \sqrt{(2\pi\sigma^{2})}}$.

Properties of normal distribution

- Mean = Median = Mode.
- The curve is center-symmetric; i.e., 50% of the values are lower than the mean (μ) and 50% of the values are higher than the mean (μ).The total area under the curve is 1.
- 68% of values are within 1 standard deviation of the mean, $\mu \pm \sigma$.
- 95% of values are within 2 standard deviations of the mean, $\mu \pm 2\sigma$.
- 99.7% of values are within 3 standard deviations of the mean, $\mu \pm 3\sigma$.

To find the probability of a value occurring within a range in normal distribution, integrate the function, which is nothing but area under the curve.

Integration: $[-2, 2] \int f(x).\, dx$ **where** $f(x) = exp[-(x-\mu)^2/2\sigma^2] \div \sqrt{(2\pi\sigma^2)}$; $\mu = 100$, $\sigma = 16$.

Test it by integrating the area under the curve (using the quad function in Python) for $\pm\ 2\sigma$, on either side of the mean (μ). For example, consider $\mu = 100$ and $\sigma = 16$, which are best known as standard normal distribution. The output is 0.9544997361036417 ~ 95.4% which means that 95% of the values are between 2 standard deviations of the mean.

4 Results

Children whose age ranges between 4 and 10 years take the test. Student logins into the program by presenting information of him/her. These details are stored in the database. If the student has already registered with the system, and the system will redirect to a page saying that the user has registered with the system already. The student redirects to the home page after successful login to the program where he/she will undergo a few cognitive assessments related to verbal understanding, processing speed, working memory and a composite examination. The system is totally composed of six modules in which the cognitive assessments apply to four modules. All these tests have sub-modules. The student is permitted to attempt every test once. The results of the tests are stored in the database after completion of the tests as shown in Fig. 2.

Questions related to all three modules such as verbal comprehension test, processing speed test, and working memory test appear randomly in the cumulative test module. The test results are also stored in the database. All the results are retrieved from the database in the profile module, and a profile is created. For each test, normal distribution curves are drawn for all the students who attempted the tests in the analysis module.

Figure 3 depicts the page with user information where it requires users to register with the application, including credentials such as name, age and ID. Both these records will be stored and validated against the database.

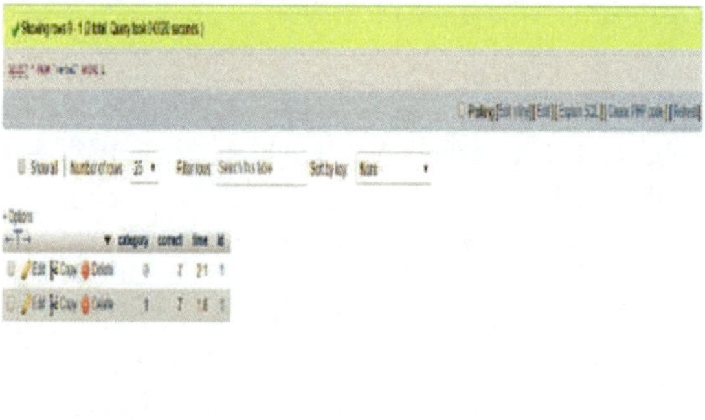

Fig. 2 Tests results stored in database

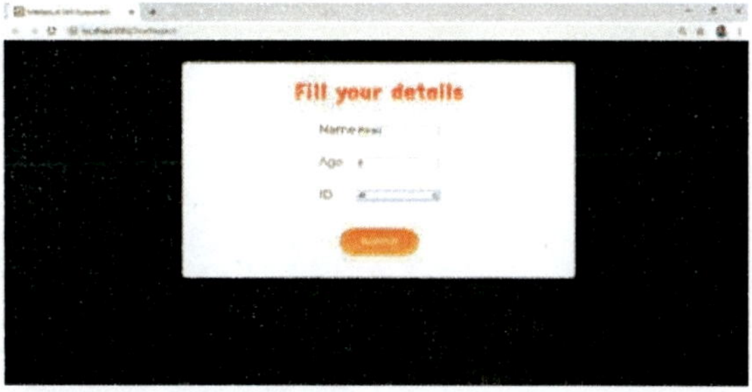

Fig. 3 User details page in the application

Figure 4 displays the home page which allows users to access all modules in the application. The user can start any task like task of verbal comprehension, task of processing speed and task of working memory. Figure 5 explains the question page in verbal comprehension module. Once the user selects the option, the answer is evaluated and indicates whether the choice selected is correct or not. In this figure, the option selected is incorrect.

Figure 6 explains the question page in verbal comprehension module. After the user selects the option, the answer is evaluated and shows whether the selected choice is correct or incorrect and displays the response summary. Figure 7 depicts the time up query page in processing speed module. If the user is unable to answer the question

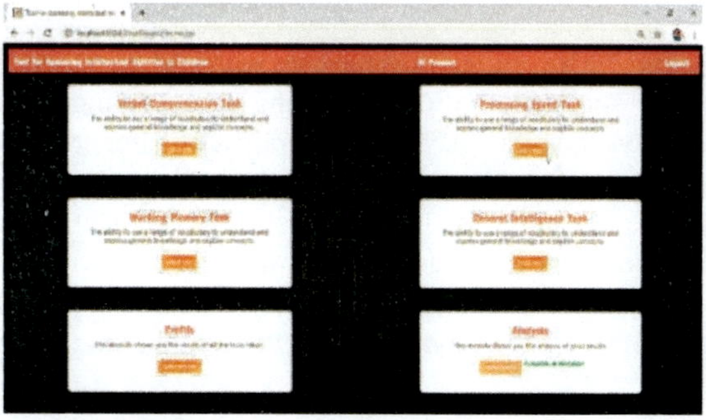

Fig. 4 Home page in the application

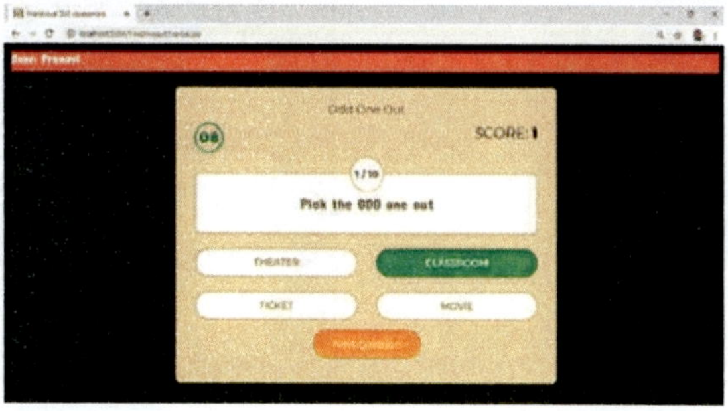

Fig. 5 Verbal comprehension module question page

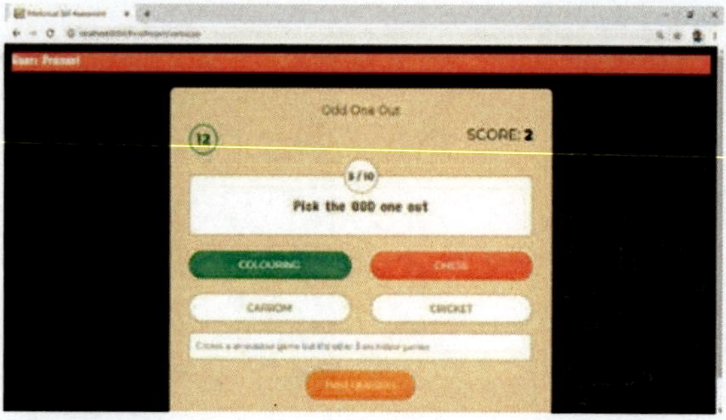

Fig. 6 Verbal comprehension module question page

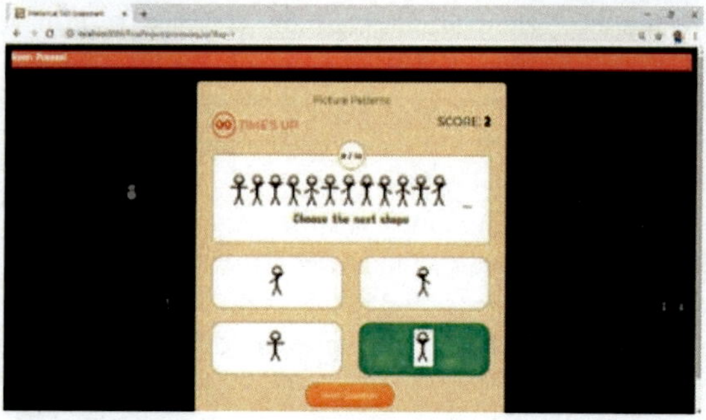

Fig. 7 Processing speed module time up question page

within the time limit, then the user will be shown the answer directly. The result page in the application after completing each sub-module is shown in Fig. 8. Figure 9 displays the home page in the application after completing all the tasks.

Figure 10 displays the profile page in the application after completing all the tasks. Figure 11 shows the profile page in the application after completing all the tasks. Downloading PDF button allows the user to download the results of all tasks.

Figure 12 shows the application's analytics page after all the tasks have been completed. The graph shown in the figure is the normal distribution curve for the

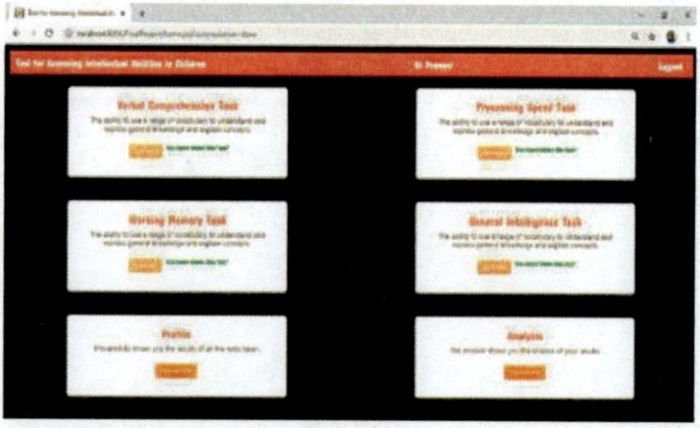

Fig. 8 Result page in the application

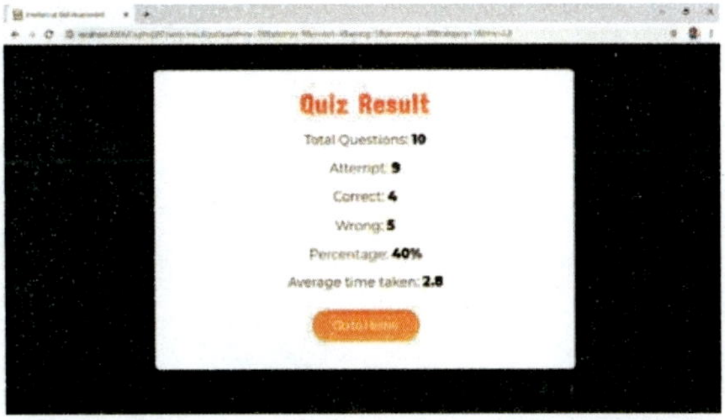

Fig. 9 Home page in the application

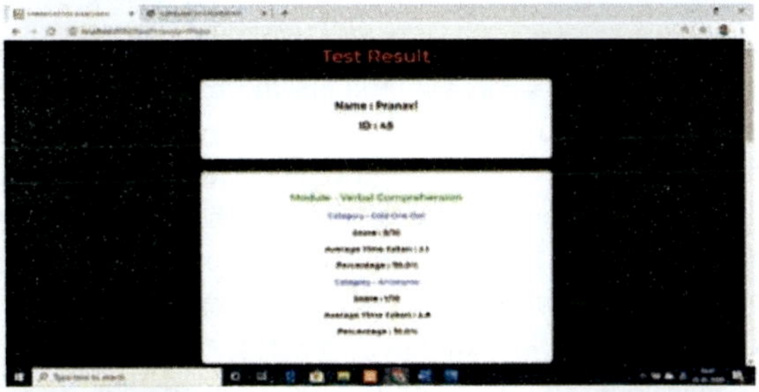

Fig. 10 Profile page after completion of all tasks

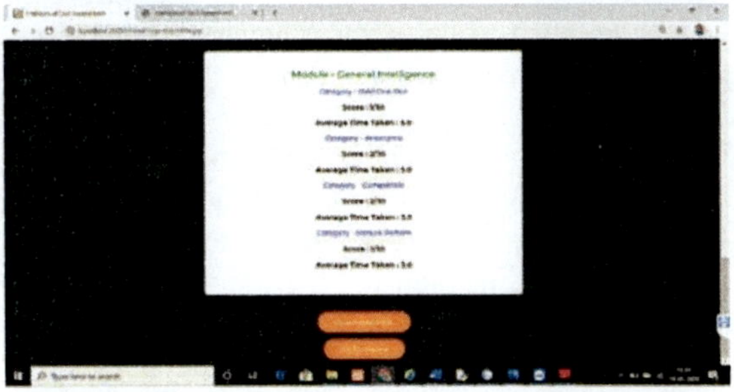

Fig. 11 Profile page after completion of all tasks

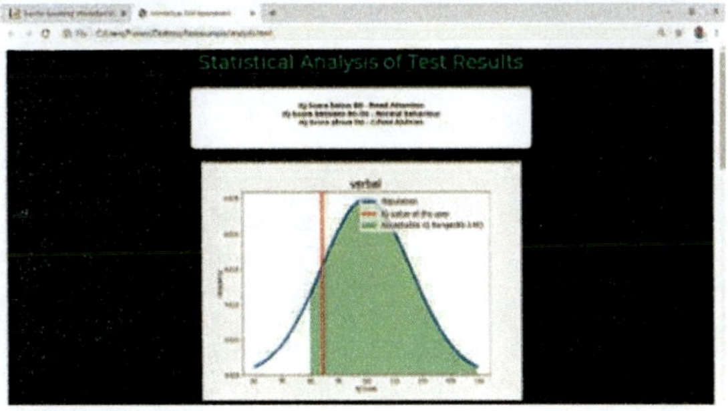

Fig. 12 Analysis page for the verbal comprehension module

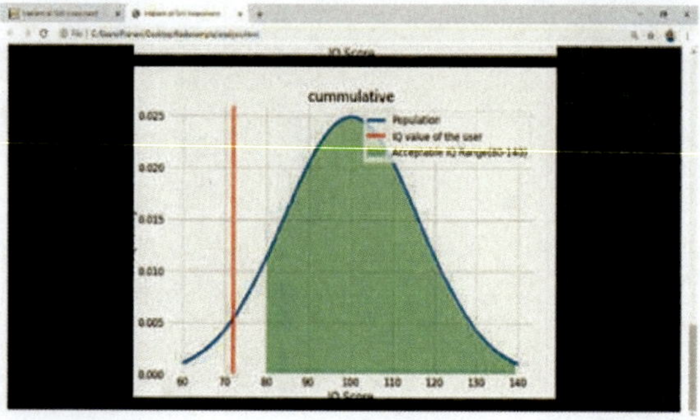

Fig. 13 Analysis page for general intelligence module

verbal comprehension module. A person's acceptable IQ value is shown in the green region. In that particular module, the red line indicates the user's IQ score. If the IQ score is below 80, the user needs some care, if the IQ score is above 110, the user has gifted abilities, and if the user has IQ score between 80 and 110, then the user is a normal person. The graph shown indicates that the user is a normal person. Figure 13 shows the application's analytics page after all the tasks have been completed. The graph shown represents the normal distribution curve for the general intelligence module. A person's acceptable IQ value is the green region. In this particular module, the red line indicates the user's IQ score. If the IQ score is below 80, it means that the user needs some care, if the IQ score is above 110, then the user has talented skills, and if the user has IQ score between 80 and 110, then the user is a normal person.

The graph depicted indicates that this module requires the user's attention. Figure 14 shows the application's analytics page after all the tasks have been completed. The graph shown in the figure is the normal distribution curve for the working memory module. A person's acceptable IQ value is the green region. In this module, the red line indicates the user's IQ score. If the IQ score is below 80, the user needs some care, if the IQ score is above 110, the user has gifted abilities, and if the user has IQ score between 80 and 110, then the user is a normal person. The graph depicts that the user is a talented person. Figure 15 is the application's profile document after all tasks have been completed. Downloading the PDF button on the

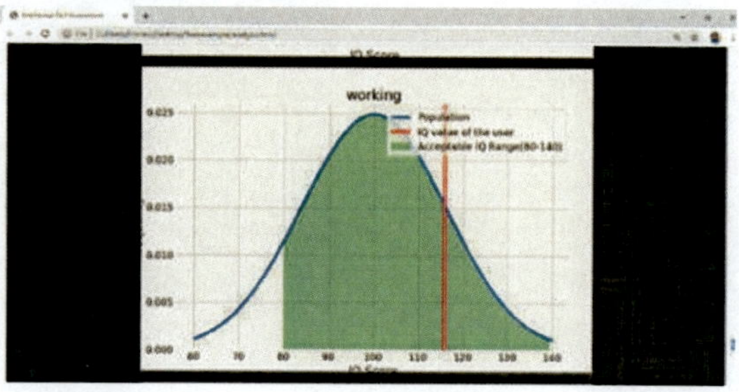

Fig. 14 Analysis page for the working memory module

Fig. 15 Profile document after completing all tasks

profile page allows the user to download all task results. The figure shows profile page downloaded text.

5 Conclusion and Scope for Future Enhancements

It is here to remain the notion of intelligence and so are the IQ measures that index intelligence. However, there is a case for a more detailed psychological analysis of what IQ scores represent. Examining the functioning of thought, a move in that direction is a factor that underlies the performance of the IQ tests. An accurate account of what the IQ scores reflect will allow these numbers to be used more carefully to assess access to opportunities that shape individual life outcomes in modern meritocratic societies. Growing evidence indicates that the cognitive performance of a standard test taker is energized and motivated by motives. Similarly, characteristics such as attitude to development, openness to learning and the need for cognition, modulate the willingness to seek and process information that affects a person's performance on an IQ test. There are implications for understanding and using IQ tests for evidence about the impact of motivational and affective processes. Differences in outcomes on IQ tests are conventionally thought to represent only differences in the cognitive functioning of the test takers' accuracy. However, with the new proof on the anvil, variance in results on the IQ tests also seems to be a function of the sophistication and intensity of motives experienced during the test by the test takers. This study calls for a change in the perception of IQ scores in order to make real-world decisions. Therefore, in the context of the IQ tests, more concerted efforts are required to unravel the effect on cognitive functioning of motivational and affective processes.

References

1. Matzel, L.D., Sauce, B.: Intelligence Quotient(IQ), Department of Psychology, Rutgers University, Piscataway, Springer International Publishing AG, NJ, USA (2017)
2. Mackintosh, N.: Human Intelligence. Oxford University Press, IQ and human intelligence, Oxford (1998)

3. Sinha, G.R., Srujan Raju, K., Kumar Patra, R.: Research studies on human cognitive ability. Int. J. Intell. Defense Support Syst. **5**(4), 298–304 (2018)
4. Navarro, J.I., Ramiro, P., Lopez, J.M., Aguilar, M.: Mental attention in gifted and non-gifted children. Europ. J. Psychol. Educ. **21**, 401–411(2006)
5. Checa, P., Fernǻndez-Berrocal, P.: The role of intelligence quotient and emotional intelligence in cognitive control processes. Front. Psychol. **01,** (2015)
6. Colmar, S.: Assessing intellectual disability in children: are IQ *measures sufficient, or even necessary*?. Austral. J. Guid. Counsel. **16**(n2), 177–188 (2006)
7. Mackintosh, N.J.: IQ and Human Intelligence. Oxford University Press, Oxford (1998)
8. Baron, J.: In: Rationality and Intelligence, University of Pennsylvania, Cambridge University Press (2005)
9. Richardson, J.T.E., Engle, R.W., Hasher, L., Logie, R.H., Stoltzfus, E.R., Zacks, R.T.: In: Working Memory and Human Cognition, Oxford University Press (1996)
10. Carroll, J.B.: In: Human Cognitive Abilities—A Survey of Factor Analytic Studies, University of North Coralina at Charpel Hill
11. Louis, D.: Matael and Bruno Sauce. Rutgers University, Tiscataway, NJ, USA, IQ (The Intelligent Quotient), Research Gate, Dept. of Psychology (2017)
12. Lincoln, A.J., Allen, M.H., Lman, A.K.: In: The Assessment and Interpretation of Intellectual Abilities in People with Autism. Springer (1995)
13. Sinha, G.R., Srujan Raju, K., Kumar Patra, R.: Research studies on human cognitive ability. Res. Gate (2018)
14. Navarro, J.I., Ramiro, P., Lopez, J.M., Aguilar, M.: In: Mental Attention in Gifted and Nongifted Children. Springer (2006)
15. Checa, P., Fernǻndez-Berrocal, P.: The role of intelligence quotient and emotional intelligence in cognitive control processes. Front. Psychol. (2015)
16. Webb, C.A., DelDonno, S., Killgore, W.D.S.: The role of cognitive versus emotional intelligence in iowa gambling task performance: what's emotion got to do with it? Intelligence **44**, 112–119 (2014). https://doi.org/10.1016/j.intell.2014.03.008
17. Rothbart, M.K., Derryberry, D., Hershey, K.: Stability of temperament in childhood: laboratory infant assessment to parent report at seven years. In: Molfese, V.J., Molfese, D.L. (eds.) Temperament and Personality Development Across The Life Span pp. 85–119. New York, NY, Routledge (2000)
18. Lozano, J.H., Gordillo, J., Pérez, M.A.: Impulsivity, intelligence, and academic performance: testing the interaction hypothesis. Pers. Individ. Dif. **61–62**, 63–68 (2014). https://doi.org/10.1016/j.paid.2014.01.013
19. https://www.talentlyft.com/en/resources/what-is-cognitive-ability-testing
20. https://brandongaille.com/15-intelligence-testing-pros-and-cons
21. https://www.oxfordlearning.com/whats-the-score-problems-with-iq-tests-for-kids
22. https://theconversation.com/the-iq-test-wars-why-screening-for-intelligence-is-still-so-controversial-81428
23. http://www.brainkart.com/article/A-Hierarchical-Model-of-Intelligence_29401
24. http://www.psychologydiscussion.net/psychology/theories-of-intelligence-top-6-theories-psychology/1502
25. https://www.verywellmind.com/theories-of-intelligence-2795035
26. https://www.verywellmind.com/what-is-general-intelligence-2795210
27. https://towardsdatascience.com/understanding-the-normal-distribution-with-python-e70bb855b027
28. https://medium.com/@soniknitr/an-introduction-to-normal-distribution-using-python-549d2285ea09

A Framework and Survey of Digital Learning Tools for Online Teaching

Y. Swathi

Abstract Over the past two decades, there has been a paradigm change in higher education with the rise of online education. The quality of online teaching and learning may be difficult to define because "quality" appears to be evaluated based on a relative experience or the expectation level of a person. Technical tool incorporation can allow teachers to use teaching and learning resources based on technology rather than traditional teaching methods. This paper provides a detailed analysis of some of the most common teaching methodologies and frameworks along with various technical tools used to assess the effectiveness of instruction. A framework is proposed for effective learning and usage of online tools, as demonstrated for the machine learning course. It is analyzed through student's survey that the usage of online tools helps to measure learning efficiently in this paper.

1 Introduction

The technology serves as a catalyst for many improvements in the learning methods and teaching methodologies in the work environment and the use of information and knowledge. Helping students rather than using traditional techniques to learn the use of technological tools has become more common in the twenty-first century in the education field. Students have different analytical capacities and types of learning that favor the acquisition of knowledge. As a consequence instructors are interested in ways that enable students to better understand and learn effectively. We are residing in an ever-changing environment where innovations are often taking place particularly in areas of information and technology communication. The position of technology in education is very important since the use of technology tools increases the efficiency of education.

In this age of technology, there are various forms and techniques available for integrating technology into the learning process. Teachers have to look for more efficient teaching methods to meet the educational standards of students. An institution's

Y. Swathi (✉)
Asst. Professor, Dept. of ISE, CMR Institute of Technology, Bengaluru, India
e-mail: swathi.y@cmrit.ac.in

T. Senjyu et al. (eds.), *ICT with Intelligent Applications*, Smart Innovation, Systems and Technologies 248, https://doi.org/10.1007/978-981-16-4177-0_28

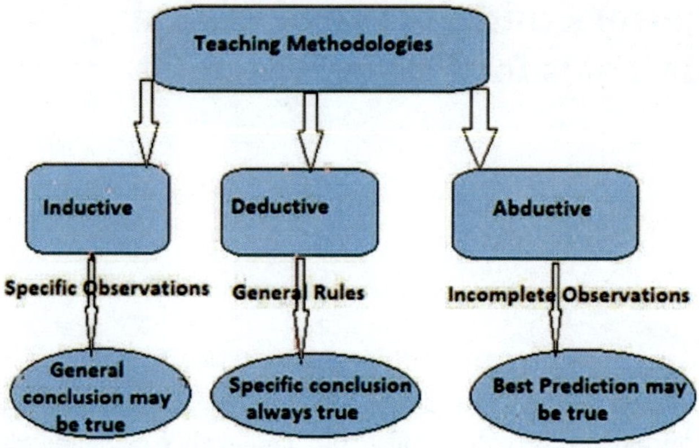

Fig.1 Teaching methodologies used

goal of achieving its better learning results can be supported by improved teaching methods as in Fig. 1.

Methods of teaching may be inductive deductive or abductive. The method of inductive teaching goes from particular to general and can be focused on experiential learning exercises. Emphasis on the deductive teaching method from general conception to particular usage. Typically abductive reasoning starts with an incomplete set of results and progresses to the most possible explanation for the set. This article discusses the literature that explains the essential variables for teachers and students in the online mode. It also introduces different frameworks used for teaching.

The rest of the paper is as follows. Section 2 provides a thorough analysis of the various skills required for students and teachers along with various existing frameworks and tools for effective teaching–learning. The proposed framework was discussed in Sect. 3, and some tools used for the machine learning course were summarized in Sect. 4.

2 Literature Survey

In this survey, we have started with some of the important skills teachers and students should have in online teaching followed by various existing frameworks used for teaching. Later, we have summarized various popular digital educational tools available for online teaching. Some of the most important skills of online teachers resulting from the above studies include:

- Communication expertise;
- Technology skills;
- Provision insightful feedback;
- Administrative skills; abilities;

- Reactivity;
- Provision of student care support.

A variety of variables and circumstances can act as barriers to student involvement in online studies, and it has long been recognized that online environments experience higher attrition rates than on-campus study modes.

- Technical complications;
- Separation perceived;
- Challenging juggling research;
- Job and relatives obligations;
- Confusion with material content;
- Bad educational performance; or
- Lack of absence of inspiration.

2.1 Frameworks for Teaching Learning

2.1.1 TPACK Model

The TPACK model [1] as in Fig. 2 promotes meaningful integration of technology content information and pedagogy. In this model, there is an equal overlap between all three information styles leading to the execution of all three aspects in the classroom. This is considered the ultimate aim for a twenty-first-century instructor but takes time to develop and relaxed with. We can apply this model to consider teacher skills in intensive online environments.

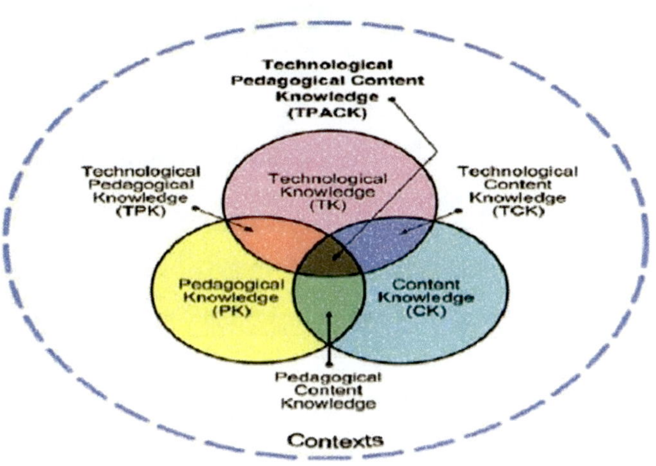

Fig. 2 TPACK Model

This system helps teachers to overcome the problems involved in incorporating teaching and learning technology.

TPACK is not meant to be a complete view of technology integration; for example, each approach to technology, integration would look different from undergraduate to graduate. Often if the teacher has not done his due diligence in preparing or evaluating the technology, the simple act of incorporating technology would not necessarily increase the course or lesson.

2.1.2 SAMR Model

The model SAMR [2] in Fig. 3 categorizes four distinct degrees of classroom integration of technology. The letters' SAMR' stands for substitution augmentation modification and redefinition. The model SAMR was created to exchange a common language across disciplines as teachers strive to assist students to imagine comp model.

In order of their complexity and transformative potential, the SAMR model [3] sets out four stages of online learning described roughly: replacement augmentation change and redefinition. Teachers also concentrate on the first two stages when transitioning to an online format replacing conventional materials with digital ones: translating lessons and worksheets into PDFs and posting them online or recording video lectures and making them accessible for example for asynchronous learning.

The key weakness with this model is getting caught up in the simple act of adding in technology which will not necessarily enhance the course or lesson if the teacher has not done their due diligence in their preparing or checking the technology.

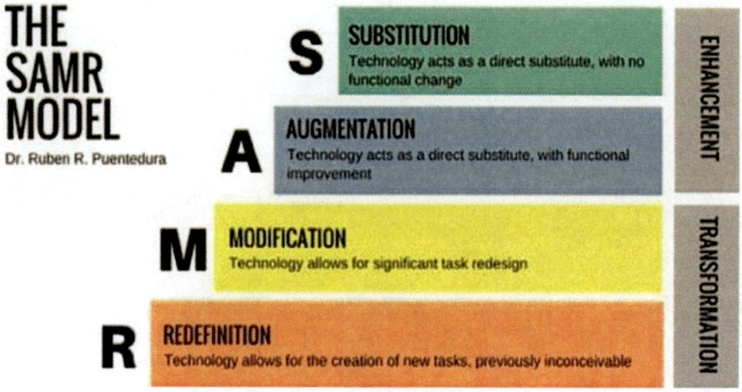

Fig. 3 SAMR model

2.2 The Most Popular Digital Education Resources for Teachers

To encourage collaboration and promote communication between teachers and learners, we present the 13 most common tools [4] used for teaching online.

2.2.1 Edmodo

Edmodo is an educational Web site that connects and is incorporated into a social network for educators and learners. In this one, teachers will create online collaborative groups, manage and provide instructional content, analyze student achievement and connect with parents.

2.2.2 Socrative

Socrative is a system that enables teachers to develop exercise or educational games that students can solve using smartphones laptops or tablets.

2.2.3 Project

Project is a tool that helps with interactive slides to create multimedia presentations. Among other choices, we can embed interactive map links to online quizzes, Twitter timelines and videos. During a class session, teachers will exchange academic presentations with students that are visually adapted to different devices.

2.2.4 Thinglink

With music, sound texts, and images, Thinglink allows educators to create interactive videos. These can be shared on other Web sites or Twitter and Facebook, for example, on social networks.

2.2.5 TED-Ed

TEDEd is an educational forum that, with the cooperation of teachers, students, and animators who typically want to broaden information and good ideas, enables the development of educational lessons. This Web site enables both teachers and students to democratize access to information.

2.2.6 ClassDojo

to enhance student behavior: teachers give instant feedback to their students so that points are 'rewarded' for good disposition in class. The knowledge gathered about student actions will later be shared via the internet with parents and administrators.

2.2.7 EduCipper

This forum helps educators and students to exchange references and instructional materials and to discuss them. Data found on the Internet can be compiled and then exchanged with members of previously formed communities which allows to more efficiently manage online academic content.

2.2.8 Storybird

Teachers can build immersive and creative books online with this tool via a simple and easy-to-use GUI.The created stories can be embedded, among other choices, in blogs sent by email and printed. In Storybird, teachers can also create projects with students that provide continuous input and coordinate classes and grades.

2.2.9 Animoto

This allows you to make high-quality videos in a fast time and from any smartphone device encouraging students and helping improve academic lessons. The Animoto GUI is friendly and realistic allowing teachers to develop audiovisual material that adapts to educational needs.

2.2.10 Kahoot!

It is a teaching tool focused on games and questions. Teachers may build questionnaires or surveys that supplement academic lessons through this method. In the classroom, content is projected and questions are answered at the same time by students while playing and studying. It encourages game-based learning that enhances student engagement and creates a diverse atmosphere for a social and enjoyable education.

2.2.11 Quizlet/Quizlet Live

Quizlet provides a Web site for students and teachers to create and share their learning resources, including flashcards and diagrams. Students must all contribute to remain focused and communicate well to win in this entertaining and interactive game.

Quizlet provides excellent reviews daily and is a fantastic way to carry study notes into the twenty-first century.

2.2.12 Khan Academy

With Khan Academy you can learn something literally all for free. Lessons are delivered by immersive exercises and challenges using videos. In line with their successes, learners also win badges and can save their progress on their profile.

2.2.13 Seesaw

Seesaw is an easy-to-use framework for a learning portfolio that helps students to track and reflect on what they are learning at school. The job can also be made available to parents.

3 Proposed Framework

First, we will discuss the challenges and opportunities with offline teaching vs online teaching [5] as per the below Table 1 and then describe the proposed framework (Table 1).

Building a digital infrastructure would take a lot of time, maybe years. Till then, we need to focus on the current situation. For that, let us divide the population into three major categories [6]. One set with access to good infrastructure and data. Here, we can test with any number of existing tools and also innovate with new ones. The second, where there is limited accessibility, i.e., the phone might be available but not for more than 1 h a day with low bandwidth. Here, the teachers can focus on a low-instruction, high practice pedagogical model. The third is where there is no access at all. In such cases, we will have to focus on home delivery of resources, IVR channels for sending instructions and reminders, community radio for longer sessions coupled with doubt clearing slots. To cope up with the current pandemic situation and to make online teaching effective, we have proposed a framework as mentioned below. Here, the usage of ICT tools plays a major role to make learning more effective.

The main aim of this study was to establish a conceptual structure for teaching and assessing online education courses. This research, therefore, began with a review of comprehensive literature on best practices in online education concentrating on certain mechanisms for quality criteria used [7]. The existing frameworks discussed do not address the learning outcomes of the students, and it focuses on technology and the proposed framework is given as in Fig. 4.

In this framework, the quality of learning is measured through certain aspects discussed below [10].

Table 1 Online versus offline teaching: challenges, opportunities, advantages and drawbacks

Distinctiveness	Offline teaching	Online teaching
The location of the process of learning	Offline education implies that a student must go to a school, a classroom, and attend a class face to face with a teacher	Online learning enables you, wherever you want, to study anywhere you want
Student–teacher interaction	Students listen to long lectures and take notes, get bored sometimes quickly, and can doze off a little	New technical teaching strategies such as interactive whiteboards and videos can be used for successful interaction with students through online classes
Tools of learning	Whereas there is only one form of teaching, i.e., one method of learning for everyone	Online learning offers many platforms, including email, online chat, and video conferences, through which students and teachers communicate with each other
Flexibility	Only one time class by the instructor and student cannot view again if they miss	With many things like time, recording lectures, and personal space, online courses provide a lot of versatility
Challenges		
Curriculum preparation	Teachers will have the option of preparing the course curriculum class-wise and modify accordingly	Before the course even starts, online instructors must also have all class material planned and ready for learners
Grading and assignment completion	Teachers can manually correct the assignments and provide feedback to students	Using some online tools teachers can set the grades to evaluate the assignments and provide the feedback the same as offline. But without using tools it is difficult to do this
Facilitating class discussions	Students can interact with teachers directly and clear their doubts	In online courses, On forums and message boards, teachers may require students to share their views
Learning outcomes	Teachers can find directly whether students have understood the topic and actual learning happened or not through direct interaction	To support the classroom-based student's experience certain usage of tools plays a major role here
Advantages and disadvantages		

(continued)

Table 1 (continued)

Distinctiveness	Offline teaching	Online teaching
	Classroom teaching also has some disadvantages 1. Inflexible 2. Big class sizes 3. Old equipment and technology	Online teaching has several advantages, such as: • Without the need for physical travel, you can conduct courses with students from different time zones • Access to online materials anytime • Usage of tools to check the learning outcomes immediately

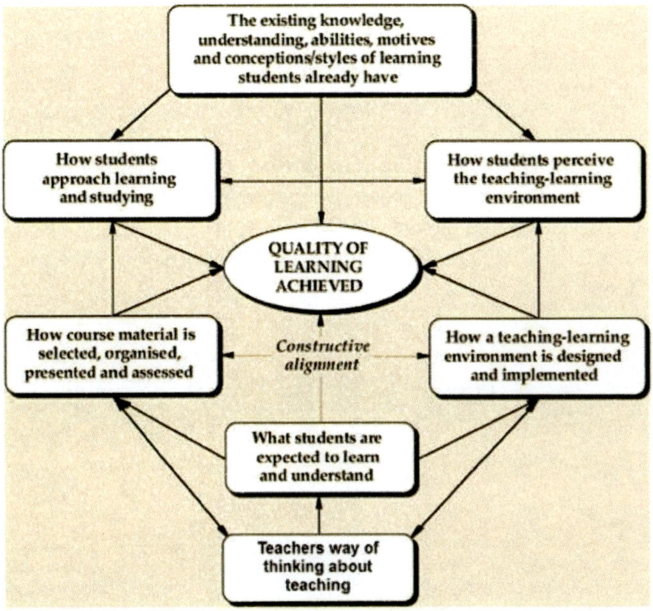

Fig. 4 Framework for effective teaching–learning

- The existing knowledge abilities students already have
- Students approach to learning
- Preparation of course material
- Teaching–learning environment design through various ICT tools
- Expectations from students to learn
- Teachers perspective on teaching

Here, first we focus mainly on the existing skills students have and what expectations and approaches they are looking for learning [8]. Based on these aspects, teachers prepare course material keeping in mind all the categories of students and ICT tools will play a major role here to make students feel classes more interactive and learn effectively. Below tools are used as part of effective delivery for machine learning course to make online learning more effective.

3.1 Graphical Tablet

Graphic tablet can be used for problematic subjects to provide a blackboard experience to students. Step-by-step explanation of problems/numerical is possible using the tablet. Students also find this easy to understand than writing using a mouse.

3.2 OpenBoard Software

This provides options to write using different pen tips and colors to make classes interesting as in the Fig. 5. Slide materials created can be duplicated (for multiple sections) and also saved and exported in PDF format. This will be like sharing the live class lecture notes with students. It also allows to import of PDF materials and write over them.

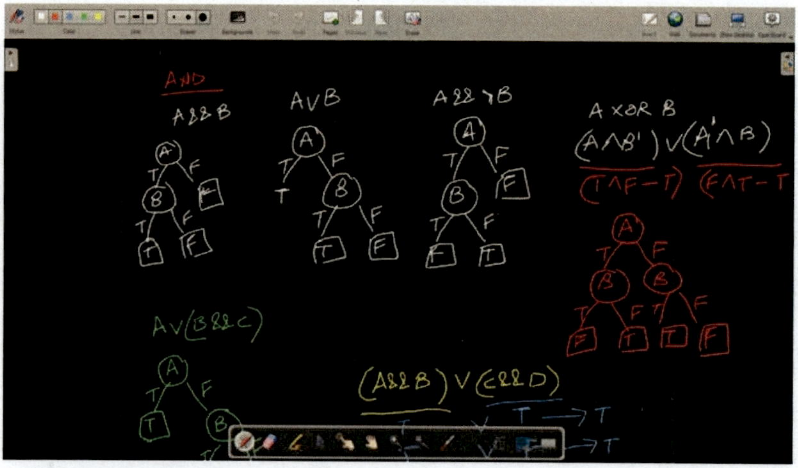

Fig. 5 Graphic tablet with open board

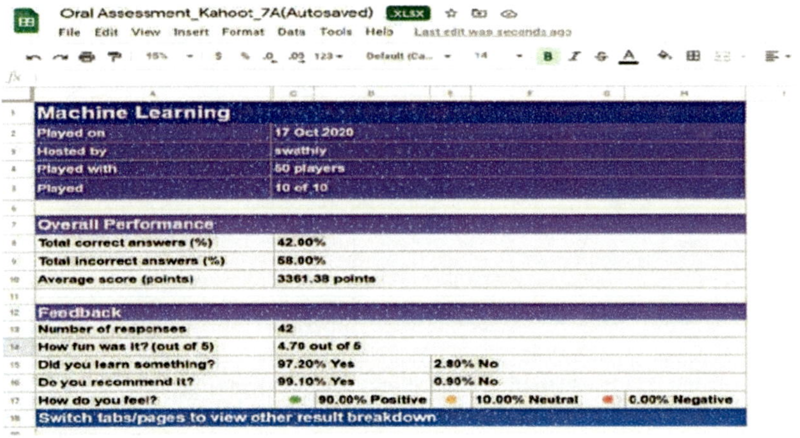

Fig. 6 Kahoot quiz demonstration

3.3 *Kami Browser Extension*

This allows you to write over PDFs. Internal assessment tests and assignment submissions can be evaluated using this extension. A graphical tablet is also used in combination with this browser extension for a better writing experience. This allows giving feedback in a better manner to each student for each answer instead of providing a consolidated comment in the comments section.

3.4 *Kahoot Quiz*

This allows to test the knowledge of students in a live manner as we can frame time-bound questions in a Kahoot quiz which needs to be taken by the students either live during a Google Meet session or within a limited time (for those with network issues). Another advantage of this is that students do not get the time to switch tabs or check answers in google or any other platform without compromising on time. Figure 6 shows that more than 90% of students feel positive about it and more than 97% say that they have learned something and will recommend this tool.

4 Results

An online survey of the online resources used for machine learning course was given to 101 students and Fig. 7 displays the instructional resources the students used in their online courses.

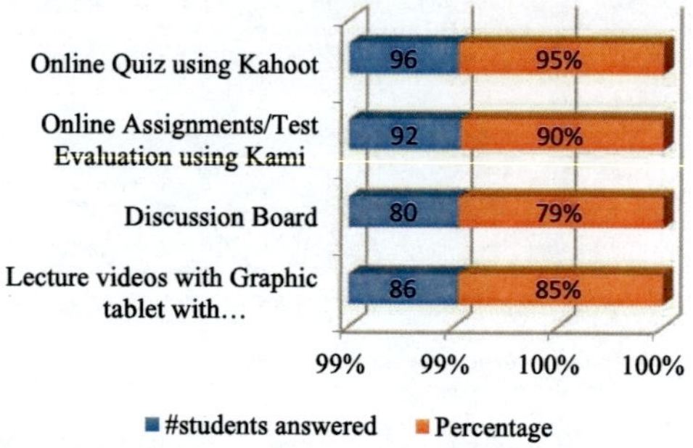

Fig. 7 Survey of online tools used for ML course

79% of the students replied that they used a learning discussion board and 85% liked graphic tablet lecture videos for online instruction. 90% liked the Kami tool for reviewing online assignments and internal assessments, and 95% liked Kahoot online quiz.

5 Conclusion

This paper established a conceptual structure for online course teaching and evaluation in higher education based on a comprehensive literature review [9]. This paper provides proof that students like the usage of online tools for effective learning. Persistence can be more difficult in an online world than in other classes of public administration. Besides, engagement can be less daunting, and online classes can increase the quality and quantity of interaction. This study demonstrates that custom-designed, mixed programs aiming to boost students' information retention capacity and learning skills. To boost their learning curve, interactive learning courses can be clubbed with other learning mediums, such as videos, podcasts and even immersive courses.

References

1. https://www.emergingedtech.com/2015/04/examples-of-transforming-lessons-through-samr/
2. https://www.edutopia.org/articles/model-understanding-good-tech-integration
3. https://creately.com/blog/tools/online-teaching-tools
4. Allen, I., Seaman, J.: In: Grade Change: In the United States Monitoring Online Learning. Wellesley MA, Babson College/Sloan Foundation (2014)
5. Barnard, R.: Identifying bias in distance education study meta-analysis: broad photos that can be focused on distance education. **35**(3) (2014)
6. Farrar, D.: Flexible learning: September 2014 Update Flexible Learning, the University of British Columbia (accessed 20 February 2015, but no longer available) (2014)
7. Firmin, R. et al.: Case study: using MOOCs for conventional college coursework. Distance Educ. **35**(2) (2014)

8. Muilenburg, L.Y., Berge, Z.L.: Digital badges in education: Trends, issues, and cases. Routledge, Taylor & Francis, New York (2016)
9. Colosimo, A.L.: E-learning technologies. Issues Sci. Technol. Librarianship **92** (2019)
10. Premlatha, K.R., Geetha, T.V.: Learning content design and learner adaptation for adaptive e-learning environment: a survey. Artif. Intell. Rev. **44**(4), 443–465 (2015)

Water Quality Analysis and Prediction Techniques Using Artificial Intelligence

Jitendra Pandey and Seema Verma

Abstract Water is one of the most valuable natural resources and the most important commodity for human consumption, with a valuable role to play in determining the portability of drinking water. The everlasting elixir is water. Clean and hygienic drinking water is the need of the civilizations, and the trend is showing that the water quality and availability is becoming a big issue for many societies. In this study, water quality prediction model used in various studies was analyzed. The proposed article uses the past researches on water quality prediction and after deep study proposes the work on future improvements. This research explores the applications of artificial intelligence (AI), which help achieve water quality parameters (WQPs). Current AI applications in the water sector include (a) predictive water infrastructure management, (b) water demand and usage forecasting, (c) water reservoir and dam monitoring, (d) water quality tracking, and (e) water-related disaster monitoring and prediction. The literature review shows: (a) The rate of adoption of AI-based solutions in water infrastructure predictive maintenance has increased as AI becomes increasingly available, and data analytics and smart sensors are becoming more effective and affordable. (b) Deep learning technology has created a new generation of water management systems, capable of producing short (everyday) and long (yearly) forecasts. C) Countries worldwide are witnessing a rise in water reservoir and dam building, and AI-based techniques are being successfully applied in the production and operation of reservoirs. D) Control of water quality has become the most dramatically influenced by AI in comparison to other activities. AI is used in test-based water quality control, large body of water, and real-time monitoring of water quality. (E) AI could be used to predict water-related hazards with greater precision, frequency, and lead time, enabling focused post-disaster recovery. The research ends by illustrating the difficulties of achieving water-related WQPs by implementing AI.

J. Pandey (✉) · S. Verma
Banasthali Vidyapith, Vanasthali, India

T. Senjyu et al. (eds.), *ICT with Intelligent Applications*, Smart Innovation, Systems and Technologies 248, https://doi.org/10.1007/978-981-16-4177-0_29

1 Introduction

An adequate supply of clean drinking water is completely essential for all human beings on earth and is a basic requirement. The consistency of water relates to its physical, chemical, and biological characteristics. In relation to the needs of one or more biotic organisms and to every human need or intent, it is an indicator of the water's state. This is most commonly used to refer to a series of standards that can be checked to conform with. The most common criteria used to assess the quality of water applies to environmental sustainability, the safety of human touch and drinking water. Drinking water quality criteria describe the standards for collection of drinking water quality. A broad variety of natural and human activities affects the water quality. Water quality can be described in terms of the concentration and condition (dissolved or particulate) of, along with other physical characteristics of water, some or all of the organic and inorganic matter present in the water. Geographic information is knowledge about earth's surface features and phenomena. Spatial information systems are information systems that are expanded to manage geographic data which is also sometimes referred to as space data. Which means that in addition to the IS features, a GIS provides different types of data, methods for data access, and methods for spatial data analysis. Spatial analysis is a method of analyzing the location, attributes, and relationship of spatial data characteristics through overlay and other analytical techniques to answer a query or useful information provided. Water quality is an interdisciplinary subject, and geographers make use of it for various purposes as well.

The data is collected from various research articles, summarized and then gap is identified for the proposed research. The paper focusses on common issues in drinking water, in terms of impurities and methods used by the organizations to eliminate the impurities. An algorithm and framework is proposed here to showcase the best method which will be used to differentiate between drinkable and non-drinkable water. Algorithm will be formed based on the artificial intelligence tools, and testing will be done using the appropriate data and programs (Fig. 1).

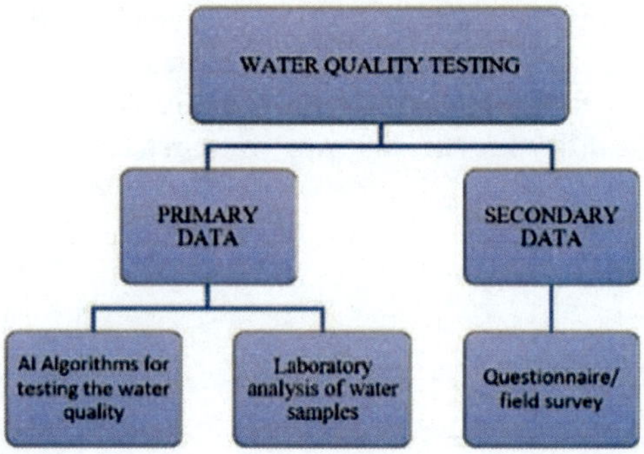

Fig. 1 Chart showing the data collection methods

Technology is being used today in every aspect, and with the advent of new devices and techniques, sensor networks may help detect any water quality problem in the water supply or treatment (e.g., liquid AI) at an early stage. This may also be extended to a drink company during its manufacturing cycle or on its watershed. Besides such instances, in order to optimize AI desalination plants in Chile and Israel for both drinking water and industrial use, AI instruments are already in operation. Industries typically have strong water flow databases, which means data can be easily accessed to gain insights using AI. This generates chances like: facilitating combined sewer discharges during rainy season to prevent combined sewer overflow. AI will help ensure compliance with local laws, maximize energy consumption, and lower sewage discharge rates. Optimize in-house flows, pumping speeds, and leakage inside the building [1].

The algorithm to find the impurities and drinkable quality of the water will be developed in the advanced stages of this research. Data will be tested using the algorithm to test the effectiveness of the same. Artificial intelligence tools will be used to test the data.

2 Problem Statement

Today, pollution of water is the greatest human problem that deteriorates water quality. For both drinking and household use, various human activities make water unhealthy. Chemical fertilizers and pesticides that enter rivers and streams into untreated waste water and industrial effluents that flow near towns and lowlands are the main sources of water pollution. A drop in the prevalence of some waterborne communicable diseases by the use of polluted water is causing some dangerous diseases.

3 Literature Review

The Kaithal City (Haryana) India Research on Physiochemical Analysis of Ground Water in Selected Region. The authors obtained groundwater samples in the 25 km area from different locations of the city of Kaithal, Haryana, India. Those water samples were tested for their physico-chemical characteristics from 20 Kaithal sampling points. Laboratory checks were conducted for pH, color, odor, hardness, chloride, alkalinity, TDS, etc. Considering the conclusions set by the Indian Medical Research Council (ICMR) and the World Health Organisation (WHO) against standards for the consistency of drinking water, it is observed that due to the high concentration of one or the other parameter, some water samples are non-potable to humans. The utility of these parameters has been explored in predicting properties of ground water quality. So an attempt was made to find the drinking water quality in and around the town of Kaithal Town [2].

Investigation into Deoli Bhorus Dam Water Physico-Chemical Parameter Determination. Water samples from two separate locations of Chalisgaon Tahsil, Dist Deoli Bhorus Dam, Jalgaon were collected for physico-chemical research in the State of Maharashtra, India, in the research review. The methods used by APHA, WHO, ICMR for the study according to standard methods are suggested. According to the standard limits, the values obtained are compared. The findings of this research suggest that the physico-chemical parameters of some parameters are below the maximum allowable WHO limit, with some small variations. For household, irrigation, and drinking purposes, water is therefore safe and sufficient [3].

Assessment is conducted on physico-chemical groundwater quality in rural areas near Sagar City by the authors [4]. In their analysis, groundwater is one of the main drinking water supplies in the city of Sagar (M.P.). In their groundwater quality analysis of the selected 02 Villages near Sagar Town, forty groundwater samples gathered from entire neighborhoods were taken for study and tested for their appropriateness for human consumption. The multiple regression analysis and regression equation showed that Gambhira and Baheria's reduced water quality is caused by anthropogenic activities and inadequate rural water management plan of action.

AI innovations can make water supply and wastewater schemes even easier and safer and to ensure compliance with drinking and waste water management requirements. This is extremely beneficial considering climate change. This can be a valuable method for creating stable water supplies that can overcome the consequences of extreme weather incidents and allow us to quickly reorganize and rebuild from a catastrophe. As the uses vary from emergency water and sanitation to the maximization of desalination facilities, including beverages, drainage, infrastructure, canals, and many others, there are many possibilities in the water market. Deceit is the limit. But it needs details. And there is data inside. Just a couple like that [5].

The disintegration of underground water due to saltwater contamination is considered to be a serious limit to the utilization of marine water supplies. Groundwater salinity projection models can also be used to create and solve computer-feasible models of sustainable coastal aquifer management as proxy models within a virtual simulation optimization process. Two machine learning algorithms are currently available to simulate the density-based approach to saltwater and concentrations of salt in illustrative coastal aquifers, namely the genetic programming (GP) and the Gaussians phase regression (GPR) models. The findings show that both GP and GPR models can be used when trained and validated to measure salinity at chosen sampling sites of the model aquifer in complex groundwater injection conditions.

In the creation of models for forecasting complex saltwater infiltration processes in a coastal aquifer environment, GP and GPR machines learn algorithms were used [6]. Many of the machine learning algorithms have demonstrated robust saltwater intrusion prediction capability. Nevertheless, all performance measuring metrics (RMSE, r, and N) were used in GPR models more stable. As a result, the GPR models will estimate the dynamic 3D numeric coastal aquifer response to a numerical simulation model. It can also replace the complex numerical model in a similar S-O model with computationally viable and sustainable methods for coastal aquifer management.

There has been a significant growth in the use of machine learning algorithms over the last decade in the area of saltwater intrusion simulation science. Machine learning algorithms have models, which can be used easily, computationally, and cheaper to operate, as a reasonable approximation of the complex numerical model. Several algorithms were used to construct mathematical models that can estimate the dynamics of saltwater penetration in coastal aquifers correctly. Many include the artificial neural network [7] and the support for the regression of vectors [8, 9] and the adaptive regression of C-medium-clustered clusters [10].

The authors presented a system for two publicly available datasets for intelligent water management related to Tanzania and Nigeria, demonstrating that the proposed models surpass multiple benchmarks in terms of precision, retrieval, and F1, including logistic regression, SVMs, and multilayer perspectives. The authors show that the models can achieve superior predictive efficiency for forecasting the operating status of water pumps for different water extraction methods. Detailed feature analyses have been conducted to assess the effect on model efficiency of water pump operating status, quality of water, and the fine grain feature analysis in the different feature categories (e.g., space, management) to determine how actual features, and not just features, are impacted. In order to accurately forecast the various water problems, research is helpful to identify the forms of data that should be obtained in the future [11]. Smart water management prediction models for forecasting water pump operating state, water quality, and quantity in emerging regions have been developed in order to prove by studies that models function efficiently. The baseline SVM model is superior in terms of accuracy, recall, and F1 score for all three water prediction issues. As indicated, the models for the different methods of water extraction can reliably predict the operating status of the pump.

The study indicates that the LSTM NN method of water quality estimation is used. Input, secret layer and output layer were used by the authors in a simulation process. Secondly, historical calculations of water quality are used to train the model. Third, parameter collection and a sequence of simulations improve predictive precision. Lastly, the LSTM NN water quality prediction approach is contrasting with two methods: one is based on the BP NN and the other on the online extreme learning machine (OS-ELM). The results suggest that the proposed approach has been successful [12]. The data obtained in this article are data from Taihu Lake from 2000 to 2006, observed each week. Keras develops a deep education predictive model based on LSTM NN. The Taihu Lake model tests the dissolved (DO) and cumulative (TP) oxygen levels of Taihu Lake with 657 data. 9:1 is the evaluation set and training set mix. Lastly, the LSTM NN-based solution is contrasted with the RMSE BP NN and OS-ELM-based method.

4 Problem Formulation

4.1 Techniques Used

Naive Bayes algorithm.

It is a classification technique based on Bayes' theorem, presuming that predictors are independent of each other. A Naive Bayes classificator assumes, in the basic words, that there is no connection to any other feature in one class with a certain function. For starters, it is possible to call apples if the fruit has a red, circular, and approximately 3 inches of diameter. And when these features depend on one another or the presence of the other, these features relate separately to the possibility that the fruit is an apple. Therefore, it is referred to as "Naive" (Fig. 2).

- The posterior probability of the class (c, target) given predictor is $P(c|x)$ (x, attributes).
- prior likelihood of class's is $P(c)$.
- $P(x|c)$ is the likelihood of the class given by the predictor.
- $P(x)$ is the predictor's prior likelihood.
- How does the Naive Bayes algorithm function?

How about using one instance to understand it. Below, I have a set of weather data and the required "play" vector objective (suggesting replay possibilities). Now, depending on weather conditions, we must decide whether or not players will play. To execute it, let us go through the steps below.

Phase one: Turn data collection into a frequency table.

Phase two: The probability is 0.29, and the playing probability is 0.64, construct the likelihood table.

Phase three: The posterior probability for each class is now calculated using the Naive Bayesian method. The prediction results are the class with the greatest posterior probability. Problem: When the bacteria are absent, water is drinkable. Is that statement accurate? We will solve it using the form of posterior probability discussed above (Fig. 3).

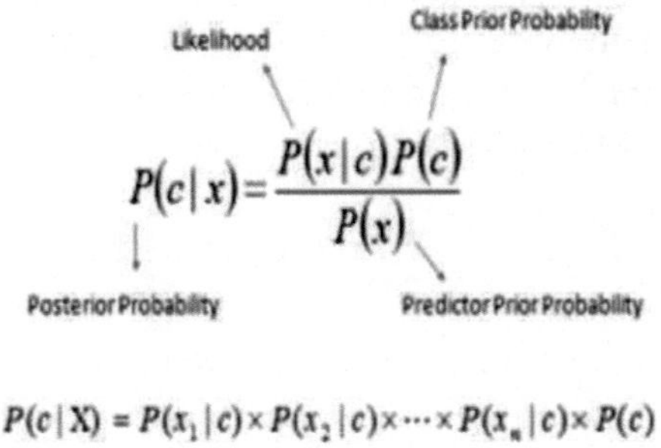

Fig. 2 Bayes theorem

Fig. 3 Parameters used for analysis

Parameter	Drinkable
More Sulphur	No
More Salt	No
Normal Sulphur	Yes
Normal Salt	Yes
Bacteria	No
No Bacteria	yes
More Sulphur	No
More Salt	No
Normal Sulphur	Yes
Normal Salt	Yes
Bacteria	No
No Bacteria	yes
More Salt	No
Normal Sulphur	Yes
Normal Salt	Yes
Bacteria	No
More Salt	No
Normal Sulphur	Yes
Normal Salt	Yes

Frequency Table		
Parameter	Yes	No
Bacteria	2	3
Salt	4	4
Sulphur	4	2

Likelyhood Table			
Paramete	Yes	No	
Bacteria	2	3	0.31
Salt	4	4	0.57
Sulphur	4	2	0.43
Total	10	9	
	0.53	0.47	

4.2 *Analysis*

Prob Bacteria = Prob Bacteria * Prob / Prob Bacteria.

Here, we have Prob Bacteria Yes = 2/9 = 0.22, Prob Bacteria = 5/19 = 0.26, Prob Yes = 14/19 = 0.74.

Then, P (Yes | Sunny) = 0.22 * 0.74 / 0.26 = 0.63, which is more probable.

This method uses a standard approach, based on various attributes, to estimate different class probabilities. This algorithm is used mainly to classify texts and problems in many schools.

Naïve Bayes Advantages.

- The type of test dataset is simple and fast to forecast. It also works well in forecasting various classes.
- Where independence is presumed, the Naive Bayes classification works well in comparison to other models such as logistic regression, as less training data are required.
- It performs better when opposed to the numerical variable for group input variables (s). The normal distribution (bell curve, explicitly assumed) is assumed for the number vector.

Usage in the analysis.

- Real-time prediction: Naive Bayes is an optimistic learning classifier and it will be easy. It can also be used for forecasting in real time.
- Multiclass prediction: The multiclass prediction function is also well known in this algorithm. We will measure the probability of a target variable in many groups here.

- Classification of the text/spam/emotional analysis: the classification of Naive Bayes is more efficient than the classification algorithms often used for text (due to better multiclass problems and independence rules). As a consequence, spam screening (identifying spam e-mail) and sentiment analysis are commonly used (identifying positive and negative user feelings in social media data).
- Together with the Naive Bayes classification and collective filtering a recommendation method can use deep learning and data mining to process unknown knowledge, to decide whether a user has a particular resource or not.

4.3 The Multilayer Perceptron—Artificial Neural Networks

From the perspective of the 1950s, the perceptron is a general strategy designed for binary detection, the neural network, whose name stipulates how the future looks; that is, it predicts whether or not knowledge corresponds to a certain category of concern: drinkable or not drinkable, clean or not clean.

In the past of neural networks and artificial intelligence, the perceptron plays a crucial role, as the initial excitement for its success led to the reaction of Minsky and Papert and more spreading outcry that for decades placed a pall over neural network science, a neural net winter that was only frozen with Geoff Hinton's study in the 2000s, the results of which have since swept the machine [13].

It was popularized as an instrument rather than an algorithm by the godfather of the perceptron, Frank Rosenblatt. As hardware, the perceptron first entered the universe. US funding was received by Rosenblatt, a psychologist who studied at Cornell University and later lectured. Naval Research Office to build a machine that is capable of understanding (Fig. 4).

Perceptrons are a straight classifier, i.e., they are an algorithm that categorizes inputs into two groups. Normally, the input is a Vector x feature which is applied by w to biases *b*: $y = w * x + b$. *b*.

A perceptron created one output based on several real-time inputs by using its input weights to create a linear combination (and also passing the output through a nonlinear activation function) (Fig. 5). This is how you can mathematically write that:

Fig. 4 Mark I perceptron

Fig. 5 Perceptron formula

$$y = \varphi(\sum_{i=1}^{n} w_i x_i + b) = \varphi(\mathbf{w}^T \mathbf{x} + b)$$

where *w* denotes *a* vector of weight: *x* is the vector of entry, *b* is the distance, and the nonlinear function is phi.

A single-blue perceptron was developed by Rosenblatt. In other words, his hardware algorithm did not have several layers to model feature hierarchy for neural networks. As Minsky and Papert have shown in their book, therefore, it was a shallow neural system which prevented its perceptive from performing nonlinear classes like XOR (a XOR operator is activated when the input reveals both, but not both; it stands for "special OR") (Fig. 6).

Multilayer Perceptrons (MLP)

Subsequent multilayer perceptrons experiments have shown a correlation between XOR and other nonlinear elements. Like the one that Rosenblatt, invented in 1943, based on a neuron of McCulloch-Pitts, impressions are basic constructs, which are useful in broader roles as multilayered perceptrons.

The multilayer perceptron is the deep learning "hi-world" a good place to begin when exploring deep learning.

A profound artificial neural network is a multilayered perceptron (MLP). The perceptron consists of a signal, an output layer for the judgment or forecast of data and an infinite number of hidden layers in between the two which are the real engine of the MLP calculations. Any continuous function of MLPs with a hidden layer can be approximated.

Multilayer perceptons are also used to solve supervised learning problems: they refer to a variety of input-output pairs and learn to model the relation between those inputs and outputs (or dependence). The teaching requires adjusting models parameters, weights, and biases to eliminate errors. Backpropagation is used to cause weight and partition to adjust relative to the error, and the error is calculated in a series of ways, including by root mean square error (RMSE) (Fig. 7).

Networks like basketball, tennis, or ping pong, like MLPs, are similar. They are primarily active in two movements, one repetitively back and forth. This tangle of suppositions and answers can be seen as a kind of rapid quest for what we think we

Fig. 6 XOR function

XOR

Input x_1	Input x_2	Output
0	0	0
0	1	1
1	0	1
1	1	0

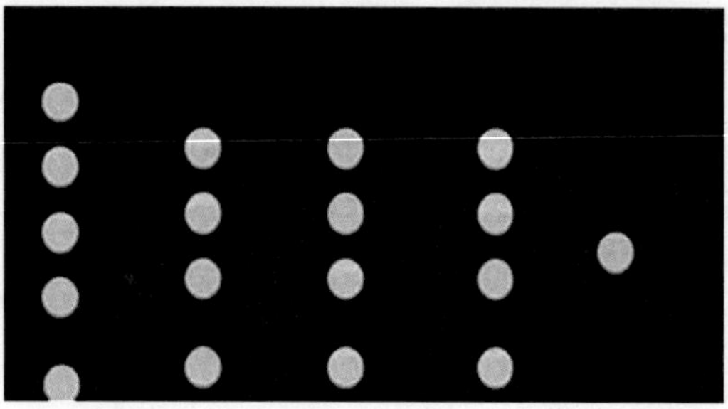

Fig. 7 Multilayer perceptrons model

know, for each guess is a test, and each answer is an input that tells us how poor we are!

In the forward step, the signal flow passes through the hidden layers from the input layer to the output layer and calculates the judgment of the output layer against ground reality labels. The defining act in which parameters can be adjusted as they put the MLP closer than the minimum error is given as a gradient or error landscape. This can be achieved for any algorithm for gradient-based optimization, such as stochastic descent. The association continues to play the tennis game until the error cannot go any lower. That condition is referred to as convergence. Similar concept can be used for the water quality testing, with multiple parameters. At what stage the water can be considered as drinkable and what should be best combination of parameters for the same.

5 Parameters and Methods

The layers consist of points. A node is just a position where measurement takes place, loosely patterned in the human brain on a neuron that fires when appropriate stimuli are found. A node combines data input with a set of coefficients or weights that either intensify or dampen the input, giving inputs value in terms of the purpose the algorithm is attempting to learn; for instance, which input is most useful for data classification that is error-free? These input weight objects are summarized and the total is then distributed to decide whether and to what degree the signal will begin to flow through the network to influence the final outcome, Say, A Classification Act, through the so-called activation mechanism of a node. If the signals move, the neuron is "activated."

Here is a plot of what seems like a single node (Fig. 8).

The proposed model shows the stages involved in the elimination and segregation process. The inputs are identified as P1, P2.., i.e., parameters to be used for the measurement of water quality. Weights A1, A2,.. are the intensity of the each parameters in the water sample taken for analysis. Basic cleansing is done with the data collected, as it is quite obvious that in huge datasets there are some values missing. The same can be done by replacing them with appropriate methods, so that it does

Fig. 8 Proposed model

not affect the results much. After cleansing the developed algorithms will be used to test. For each algorithm, the output will be generated and analyzed. The best output from the algorithm will be identified and proposed as the suitable method for analysis in the future.

6 Results and Discussion

The proposed research talks about the Bayes theorem and perceptrons model to identify the errors in the process. The logic can be used in the water impurities calculation also, thereby segregating the drinkable elements and non- drinkable elements.

7 Conclusion

This research reviews multiple articles in the field of water quality detection and the techniques used in the process. The identified models and techniques are found to be suitable for carrying out further research in the area. With the identified set of parameters and the new algorithm, it is assumed that the water quality can be determined for any source. The use of artificial intelligence techniques and algorithm will make the process efficient and faster in terms of decision making. The sample collection can be a challenge so that the accuracy is guaranteed. The proposed models would be able to accurately mimic the water quality parameters with a comparatively small prediction error, considering the extremely stochastic nature of the proposed water quality parameters. In forecasting the electrical conductivity, dissolved solids and turbidity parameters with different data input patterns, the proposed model prediction will demonstrate robustness and reliable efficiency.

References

1. Huneeus, C.: Digital water transformation: the promise of artificial intelligence. (online) https://medium.com/datadriveninvestor/digital-water-transformation-the-promise-of-artificial-intelligence-7d88fb07e79b (2019). Accessed on 14 July 2020
2. Gupta, D.P., Sunita, Saharan, J.P.: Physiochemical analysis of ground water of selected area of Kaithal City (Haryana) India Researcher **1**(2)

3. Kalwale, A.M., Savale, P.A.: Determination of Physico-Chemical parameters of Deoli Bhorus Dam water. Adv. Appl. Sci. Res. **3**(1), 273–279 (2012)
4. Pathak, H.: Assessment of physico-chemical quality of municipal water samples of Makronia sub-urban area of Bundel khand region, India. Analele University Nii din Oradea–Seria Geografie Year **XXII**(1), 122–127 (2012)
5. Huneeus, C.: Digital water transformation: the promise of artificial intelligence, data driven investor. (online) https://medium.com/datadriveninvestor/digital-water-transformation-the-promise-of-artificial-intelligence-7d88fb07e79b (2019). Accessed on 19 July 2020
6. Lal, A., Datta, B.: Genetic programming and gaussian process regression models for groundwater salinity prediction: machine learning for sustainable water resources management. In: 2018 IEEE Conference on Technologies for Sustainability (SusTech), Long Beach, CA, USA, pp. 1–7 (2018). https://doi.org/10.1109/SusTech.2018.8671343
7. Bhattacharjya, R.K., Datta, B., Satish, M.G.: Artificial neural networks approximation of density dependent saltwater intrusion process in coastal aquifers. J. Hydrol. Eng. **12**, 273–282 (2007)
8. Lal, A., Datta, B.: Development and implementation of support vector machine regression surrogate models for predicting groundwater pumping-induced saltwater intrusion into coastal aquifers. Water Resour. Manag. 1–15 (2018)
9. Lal, A., Datta, B.: Multiple objective management strategies for coastal aquifers utilizing new surrogate models. Int. J. Geomate **15**, 79–85 (2018)
10. Roy, D.K., Datta, B.: Fuzzy C-Mean clustering based inference system for saltwater intrusion processes prediction in coastal aquifers. Water Resour. Manag. 1–22 (2016)
11. Bejarano, G., Jain, M., Ramesh, A., Seetharam, A. Mishra, A.: Predictive analytics for smart water management in developing regions. In: 2018 IEEE International Conference on Smart Computing (SMARTCOMP), Taormina, pp. 464–469 (2018). https://doi.org/10.1109/SMARTCOMP.2018.00047
12. Wang, Y., Zhou, J., Chen, K., Wang, Y., Liu, L.: Water quality prediction method based on LSTM neural network. In: 2017 12th International Conference on Intelligent Systems and Knowledge Engineering (ISKE), Nanjing, pp. 1–5 (2017). https://doi.org/10.1109/ISKE.2017.8258814
13. Wiki A.I.: A beginner's guide to important topics in AI, machine learning, and deep learning, (online). https://pathmind.com/wiki/multilayer-perceptron#one. Accessed on 12-Jul-20

Super Bot Controlled by IoT and Android

P. Ajitha, S. Gowri, A. Sivasangari, Jithina Jose, Bevish Jinila, and J. Vimali

Abstract *The Android-based mobile phones are turning into each time all the more impressive and* outfitted with a few adornments that are helpful for robot. This venture depicts to control a robot utilizing versatile through Internet, a few highlights about Internet innovation, and segments of the portable along with robot. We show a survey of robots constrained by mobile phone by means of shifting the robot up, in reverse, right as well as left side by the Android application, for instance, NodeMCU. Web has transformed how people use advanced gadget at office or home as well as has moved customary wired computerized gadgets into remote gadgets. Proposed work utilizes IoT, Arduino programming to interface the NodeMCU microcontroller. As suggested by orders got from Android, the robot progress could be managed. We inferred straightforward answers for the given system to build robots with minimal effort yet with high calculation and detecting capacities given by the advanced cell that is utilized as a control gadget. People have ceaselessly been at the most elevated of the Animalia. it is a benefit we keep an eye on all appreciative for, anyway would not or not it's wonderful on the off chance that we will in general may substitute our errands for a couple of relaxation time we will start by sending minimal valuable self-sufficient vehicles. Self-sufficient vehicles are the basic strides of progress toward headway in IoT and processing. During this paper, we will in general bless the idea of a modest self-ruling vehicle, which can be constrained by voice orders, given by the client. The client could likewise be settled at some remote area, anyway as long as he/she is associated with the net, the vehicle can follow voice headings. The idea is authorized through NodeMCU ESP866, Adafruit, and IFTTT. The headings are taken care of to the vehicle through Google Assistant

1 Introduction

In the present time, PDAs are getting all the more remarkable with strong processors, bigger capacity limits, more extravagant diversion capacity, and more specialized

P. Ajitha (✉) · S. Gowri · A. Sivasangari · J. Jose · B. Jinila · J. Vimali
Sathyabama Institute of Science and Technology, Chennai, India

T. Senjyu et al. (eds.), *ICT with Intelligent Applications*, Smart Innovation, Systems and Technologies 248, https://doi.org/10.1007/978-981-16-4177-0_30

strategies. Speech recognition is considered as the main term when we talk about voice control, for example, creating the framework to comprehend human voice. Discourse acknowledgment is exactly where the framework realizes the phrases provided through discourse. At the end of the day, speech acknowledgment innovation is an incredible guide to concede the test as well as it is a conspicuous innovation for human–computer interaction (HCI) and human–robot interaction (HRI) for what is to come. Discourse is an exact strategy for automated regulation as well as correspondence. The disclosure recognition of circuit we will plot works independently from the robots fundamental vision. To regulation and order equipment by addressing it, this can allow it to be easier, while increasing the proficiency as well as adequacy of working with this gadget. At its most essential level discourse acknowledgment allows the user to execute equal assignments, while proceeding to work with the PC or machine. Disclosure is a perfect strategy designed for automated regulation as well as correspondence. The principle objective is to limit the expense included in the undertaking and force utilization fulfilling the need of the day simultaneously. Our objective stretches out in including our proposition to give effective, availability, and control of ordinary items. Presently day's PDAs are getting all the more impressive with strong processors, more extravagant amusement capacity, bigger capacity limits as well as more specialized and modern strategy. The main part useful for information trade is Bluetooth, insert new highlights to PDAs.

In 1994, telecom merchant Ericsson creates a Bluetooth innovation [1], which shows its preferred position from incorporating with PDAs. This task presents a survey of current robots constrained via cellular phone as well as a shut circle control framework which uses sound channels of cellular phones, for instance tablet PCs along with telephones. In this work, with the help of the Android application, we move the robot, for instance "BT Voice control for Arduino." This particular paper talks about Android application-based Bluetooth-controlled automated vehicle. Here in [2] principles are saying for our task to control the vehicle with the help of Android application. Presently, we utilize primarily Arduino Uno and Bluetooth module. People have ceaselessly been at the most noteworthy of the Animalia. It is a benefit we keep an eye on all appreciative for, anyway would not or not it is glorious in the event that we will in general may substitute our tasks for a couple of relaxation time we will start by conveying minimal helpful self-governing vehicles. Self-sufficient vehicles are the rudimentary strides of progress toward headway in IoT and processing. During this paper, we will in general bless the idea of a modest independent vehicle, which can be constrained by voice orders, given by the client. The client could likewise be settled at some remote area, anyway as long as he/she is associated with the net, the vehicle can follow voice headings. The idea is authorized through NodeMCU ESP866, Adafruit, and IFTTT. The headings took care of to the vehicle through Google Assistant

2 Related Work

Various sorts of inquiries have been made to create Android gadget-controlled mechanical framework. The application areas for these examines are considered as unique. Hence, specialists [3] have projected distinctive engineering system for automated framework as well as Android application. Winter et al. [4] have created Android-controlled versatile robot. Target of this venture was remote Internet information move between advanced cell robot and building up the robot and its correspondence framework under an ease as well as open-source reasoning. He has utilized 3D planning strategy to plan the automated arrangement through the assistance of parametrical displaying programming. This structure could be taken care to the 3D printer that will print the mechanical parts layer by layer as well as afterward utilizing those segment robots could be collected effectively. He have utilized Arduino miniaturized scale controller as well as Wi-Fi network right now. Selvam et al. [5] have proposed design to build up an automated framework connected with remote camera with the end goal of observation. He has utilized Bluetooth innovation for network among robot and PDA. He has utilized remote night vision camera for remote reconnaissance.

Through RF signal, videos caught using the camera are transferred to the TV unit. He has utilized 8051 miniaturized scale controllers for mechanical framework. Guardia et al. [6] have projected a structure of a Bluetooth innovation. Android application for microcontroller guided robot. The principle goal of this work is to prove that a solitary Android application is fit for working with various electronic gadgets regularly utilized inside the side interest as well as armature automated field. Pahuja [7] has built up a correspondence convention for Android cell phone as well as mechanical stage over a Bluetooth innovation network. Activity Bot Robot Kit [8] is utilized by the researcher which is produced by Parallax as well as a proper small scale controller board consist of eight free 32-piece processors along with RN-42 Bluetooth innovation connector by Parallax for the mechanical framework. Indira et al. [9] have created Android-controlled mechanical design for pick and holding of items. The principle target of their work was making a pick as well as hold a robot which could be utilized for a bomb dispersion from safe separation. Narayana [10] has utilized two engines for automated hand as well as two engines for robot wheels to control the development. Monson [11] has utilized LPC2148 microcontroller, remote camera for remote observation as well as Bluetooth module for availability. They include the built-up this undertaking by remembering its modern as well as military applications.

The WI fly module speaks with the Wi-Fi passageway which indicated in the modified arrangement. It is mounted on a Uno board. Arduino WI fly Mini sends the video feed through a 1.3 GHz 300 mW video transmitter. The Arduino starts a sequential association with the WI fly, and these two gadgets convey as well as trade data. The Arduino WI fly procedure traded data in User Datagram Protocol (UDP) bundle structure. Although UDP is not solid as TCP, however, its reaction will be quicker than Transmission Controller Protocol (TCP). A triple pivot accelerometer is mounted at the front of the vehicle. This utilizes to recognize the adjustment in quickening the robot speed C. Versatile Arduino Experimental Platform (MAEP) 2.0 MAEP is a microcontroller robot which fit to maintain a strategic distance from impediment encompassed by identifying them through an infrared sensor as well as ultrasonic worked by a specialist with a moniker General Geek.

The security devices and sensors are utilized for prospecting the houses, for example, shopping centers, organizations, and so forth. The security frameworks [8] like alarm and different contraptions are not present in all the houses. They are not reasonable by everyday citizens. Aside from utilizing these devices, they are absent in an equivalent machine. For alarm in [2], various devices are utilized and for security purposes, various contraptions are being utilized which makes the expense significantly more.

3 Proposed Work

Created so that it converts the voice order to content and move the content to the associated Bluetooth device. The Bluetooth associated on the Arduino board gets content from the Android application as characters and put away them as string to the doled out String. There are words pre-customized (forward, switch, right, left, and stop) to the Arduino, at whatever point they got content matches with the premodified words, the Arduino executes the order that doled out to the words. Arduino can associate with laptop to screen sequential correspondence and check the working procedure and the words got by the Bluetooth (Fig. 1).

The telephone-controlled robot that could be controlled through an application on your versatile. The control orders are sent through Bluetooth. The user, Bluetooth module, Arduino, robot vehicle, sensors. Client is for giving the headings to control

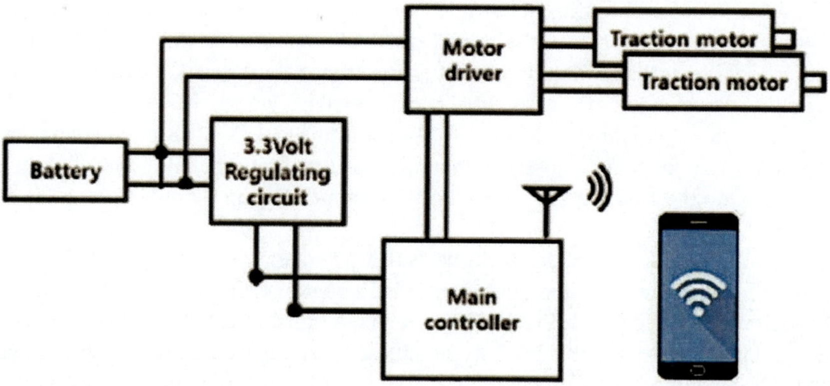

Fig. 1 Overview of system architecture

a robot vehicle dependent on the circumstances. Bluetooth module is for correspondence reason; through this client can offer orders to the vehicle. Arduino Uno board is for interfacing all the Bluetooth module, robot vehicle, and different sensors. Robot comprises of one engine driver and two DC engines; these two DC engines are associated with the wheels of the vehicle, through the DC engines just conceivable to change the vehicle bearings. We are giving charge to engines by 9V batteries. It additionally comprises of various kind of sensors like ultrasonic sensor to identify the impediment, temperature sensor and moistness sensor used to detect the temperature and stickiness of the outside condition separate Android telephone-controlled robot.

This model is built utilizing an Arduino Uno board, ultrasonic handset HC-SR04, Bluetooth module, engine driver L293D, direct momentum engines M1 as well as M2, along with few parts. The model utilizes 2, 9V batteries 1 battery to control the Arduino board as well as the another to control the engines separately, Managed 5V supply for the rest of the circuit" is feeded by the Arduino board itself. Focused on the board determine nearness of intensity supply. Bluetooth module JY MCU BT is utilized right now, that it can interface with any gadget, through implicit universal asynchronous receiver/transmitter (UART) interface, to convey. With other Bluetooth-empowered gadgets, for example, cell phones, handheld PCs as well as PCs, the module runs on a 3.6V to 6V supply. Ultrasonic handset module: ultrasonic handset module HC-SR04 utilizes sonar, similar to bats and dolphins, to decide separation to an item. It offers brilliant non-contact run discovery of 2cm to 400cm with high precision as well as stable readings in an easy-to-utilize bundle. It accompanies an ultrasonic recipient and a transmitter module.

3.1 Microcontroller Arduino

The Uno microcontroller is going to go about as the cerebrum of the robot. The robot progression is going to be picked by the microcontroller. Right now will use the microcontroller called Arduino Uno that consists of ATMEGA 328P microcontroller chip.

3.2 Motor Driver (L293d)

Engine driver IC is utilized to control the DC engines. It is in like method, interfaced with circuit affiliations as well as with the microcontroller. The Bluetooth component will go about as an interface among microcontroller as well as smartphone. We are going to use HC-05 Bluetooth module for the structure, which could be used as either transmitter or gatherer. Everything considered the transmitter is going to be moved mobile phone and authority is going to be Bluetooth module. Bluetooth module is going to give the solicitations given by PDA to the microcontroller.

3.3 Amplifier

A class-D intensifier or exchanging speaker is an electronic enhancer where the heightening gadgets fill in as electronic switches, rather than as quick addition gadgets as in different speakers. The sign to be elevated is a train of unsurprising abundancy beats, so the dynamic contraptions switch quickly forward and in reverse between an absolutely conductive and non-conductive state. The essential sign to be improved is changed over to a development of heartbeats by beat width adjustment, beat thickness balance, or other framework before being applied to the enhancer.

3.4 DC Motors

A DC engine is an electrical engine that unforeseen spikes looked for after every direct stream (DC) power. DC engines were used to run mechanical equipment which will gather together, as frequently as conceivable disposing of the need for a near to steam motor or internal turn over motor. DC engines can work direct from battery power which is filled with batteries, giving the objective a capacity to the fundamental electric vehicles.

3.5 L293D Motor Drivers

This gadget is a solid with high voltage, high to as well as for development 4 channel driver wanted to perceive standard DTL or TTL reason levels along with drive inductive loads, and exchanging power transistors. Each couple of channels is outfitted with various empower information. Another stock data has obliged the explanation, permitting development of a lower voltage as well as inside fasten different diodes are joined together. This gadget is reasonable for utilization in exchanging applications at frequencies up to 5 kHz.

3.6 IR Sensor

An infrared sensor is an electronically contraption to produce so as to distinguish two or three pieces of nature. An IR sensor can quantify the gleam of an article likewise as recognizing the development. These sorts of sensors check basically infrared radiation, instead of discharging it that is called as a limited IR sensor. This sensor resembles human's visionary recognition, which can be utilized to recognize obstructions and it is one of the essential applications in authentic time. This circuit incorporates the going with segments.

3.7 Speaker

Enhancer or speaker, gadget used to change over electrical criticalness into sound. It contains basically of a thin flexible sheet considered a stomach that is made to vibrate by an electric sign from a speaker. The vibrations make sound waves unmistakable all around the speaker. In a fascinating speaker, the most by and large saw kind, the stomach is cone molded and is related with a hover of wire suspended in an engaging field passed on by a suffering magnet.

4 Conclusion

This is in all actuality a shrewd and capable to achieve something. The interest lies in the manner that it is a monetarily Android endeavor with a direct and easy use of interface appeared differently in relation to existing ones. Like this manner the Bluetooth RC Controller application is also clear. In size the robot is little so it might be used for the reason of spying. With scarcely any increments and adjustments in the process, this robot can be utilized in furnished power for distinguishing as well as organizing the land mines. The robot could be utilized for the way in which something is understood. In future, we could interface sensors to this robot through the objective that it can screen a couple of parameters as well as we could improve the profitability using Internet of things. We could moreover incorporate remote camera, in order to join other security features.

References

1. Android Developers Guide Android Architecture [online] 2013. http://.android.com/about/versions/index.html.
2. Piyare, R. Tazil, M.: Bluetooth based home automation system using Android phones. In: IEEE 15TH International Symposium on Consumer Electronics (ISCE), 14–17 June 2011, Singapore (2011)
3. Saini, A.K., Sharma, G., Choure, K.K.: BluBO: bluetooth controlled robot. Int. J. Sci. Res. (IJSR). In: National Conference on Knowledge, Innovation in Technology and Engineering (NCKITE), 10–11 Apr (2015)
4. Winter, J.K.: Android controlled mobile robot (2013)
5. Selvam, M.: Smart phone based robotic control for surveillance applications. IJRET (2014)
6. Guardi, V.M.: Design of a bluetooth enabled android application for a microcontroller driven robot (2014)
7. Pahuja, R., Kumar, N.: Android mobile phone controlled bluetooth robot using 8051 microcontroller. Proc. IJSER **2**(7), 2347–3878 (2014)
8. Potts, J., Sukittanon, S.: Exploting bluetooth on android mobile mobile devices for home security application. In: Proceedings of Southeastcan, Florida, USA (2012)
9. Indira, K., Ajitha, P., Reshma, V., Tamizhselvi, A.: An efficient secured routing protocol for software defined internet of vehicles. In: 2nd International Conference on Computational Intelligence in Data Science, Proceedings (2019)

10. Narayana, M., Alishety, A., Chapala, H.: Voice control robot using android application. IJEIR **4**(2), 332–337 (2015)
11. Monson, H.: Bluetooth Technology and Implementations. Wiley (1999)
12. Sivasangari, A., Ajitha, P., Gomathi, R.M.: Light weight security scheme in wireless body area sensor network using logistic chaotic scheme. Int. J. Netw. Virtual Organ. (IJNVO) **22**(4), (2020)

Developing a Framework to Measure Cyber Resilience Behaviour of Indian Bank Employees

Tanvi Godbole, Saikat Gochhait, and Dilip Ghosh

Abstract Banks have technology-driven business model and are heavily dependent on the Internet. As banks continue to digitize, the attack surface is increasing. Cyber-crimes have the propensity to impact not only an organization's reputation but also its profitability and economic growth. The apex bank of the country, Reserve Bank of India (RBI), has detected 50,547 cyber-frauds worth a whopping INR 145.08 Crore in India's scheduled commercial banks the financial year 2018–19 (Ganjapure, V.: RBI reports over 50,000 cyber frauds in FY 2018–19. Times of India (2019). Retrieved from: https://timesofindia.indiatimes.com/city/nagpur/rbi-reports-over-50000-cyber-frauds-in-fy-2018-19/articleshow/71174635.cms#:~:text=NAGPUR%3A%20The%20Reserve%20Bank%20of,credit%20cards%20and%20internet%20banking) [6]. Hence, the need for banks to be cyber resilient is becoming more and more essential. Cyber resiliency of an organization depends on the IT infrastructure and cyber security measures undertaken; however, a prominent factor to achieve resilience depends on the employees. Confronted with this bleak realization, it is only recently that the need for cyber resilient culture has come into existence. While seminal contribution has been made in the field of cyber security measures in the banking sector, these have typically proposed and explored techniques to enhance infrastructure rather than cyber resilient behaviour. There is a need to look at the cyber resilient behaviour of employees as a potential source of cyber resilience of bank as a whole. This paper presents a novel framework to measure cyber resilience behaviour of Indian bank employees. Cyber-attacks are inevitable, but what will matter now is how fast a bank can recover from vulnerabilities, and this will heavily depend on cyber resiliency of the staff.

T. Godbole (✉) · S. Gochhait
Symbiosis Institute of Digital and Telecom Management, Constituent of Symbiosis International Deemed University, Pune, India

D. Ghosh
Banking Sector Expert, Mumbai, India

T. Senjyu et al. (eds.), *ICT with Intelligent Applications*, Smart Innovation, Systems and Technologies 248, https://doi.org/10.1007/978-981-16-4177-0_31

1 Introduction

Reference [1] Banks across the world are susceptible to damaging cyber-attacks. Due to rampant digitalization, it is expected that cost of cyber-crimes and associated losses incurred will only increase. With the quick digitalization of the banking industry, cyber-attacks have become more complex and pervasive. Unprecedented and major security challenges faced in the recent past have bought the issue of cyber security and resilience to the Indian banking sector like never before. Internet usage in India is growing rapidly. A recent report by NITI Aayog [2] claims that India has the third highest number of Internet users (730 million users) in the world after the USA and China. Further, India is also among the top 5 countries affected by cyber-crime worldwide and has been the second most cyber-attack-affected country during 2016–18 as per the Data Security Council of India (DSCI) report.

Indian banks are witnessing higher incidence of digital frauds in the rising age of online payments. As per a report by Websense Security Labs, average number of attacks aimed at financial institutions are four times more than that of companies in other industries [3]. Cyber-attacks cost banks over $1 trillion each year. Following the global surge in cyber-attacks, securing information has become vital for the Indian banking sector.

There are multiple reasons why cyber resilience for the Indian banking industry must be studied. Evolving nature of cyber-crimes and heavy dependence on third-party vendors are key challenges faced by the banking sector today. Although much research exists in cyber security assessment, what Indian banking environment lacks is an approach to measure cyber resilience. Problems pertaining digital security and privacy are on the rise. Cyber resilience assumes all the more importance for banking sector as the very foundation of banking lies in nurturing trust and credibility. There is a need that cyber security and resilience culture are ingrained in bank's culture. Measures undertaken to increase cyber resiliency are majorly focussed on enhancing the security infrastructure; however, unless the banking staff is trained adequately to behave in cyber resilient manner, banks will continue to be exposed to a wide range of cyber risks. While technological security infrastructure is important to prevent against cyber-attacks, it will be not be effective in the absence of a cyber resilient culture.

This research attempts to propose a framework for measurement of cyber resilient behaviour of employees in the Indian banking sector.

2 Review of Literature

Cyber-attacks are not new for the banking industry. For decades, banks around the world have been battling with numerous borderless cyber-crimes. With the quick digitalization of the banking industry, cyber-attacks have become more complex and pervasive. In the race to provide a superior customer experience, banks are embracing digital payment technologies. This has increased the attack surface for security risks in the banking sector. Hackers are also exploring new attack channels. Cyber-attacks on financial sector have grown from simple cyber-crime to attempts to dangerously cripple the economy. As per the RBI annual report, total cases of frauds reported by banks increased by 28 per cent by volume and 159 per cent by value during FY 2019–20. Banking industry remains the major target of cyber criminals, and notably, financial gain is the key motivation behind cyber-attacks. Rampant cyber security incidents have necessitated research into and discourses on cyber resilience [4]. Many corporate organizations and governments have realized that despite huge investments in cyber security defences, cyber-crime incidents are still increasing in frequency and in sophistication.

The etymology of **resilience** is rooted in material sciences and physics where resilience characterizes the property of any material to absorb energy when it is subjected to strain and to resume or maintain its original shape after being bent, stretched or compressed [5]. By extension, this concept was then applied to medical sciences, ecology, psychology and new emerging technologies. **Cyber resilience**, unlike general resilience, is a relatively new and growing research area. Due to the increasing number of sophisticated cyber-attacks, every organization needs to be cyber resilient. National Academics of Science (NAS) defines cyber resilience as the ability to prepare and plan for, absorb, recover from and more successfully adapt to adverse events, especially those associated with catastrophic cyber-attacks [6].

Seville [7] recognizes resilience as the ability of an organization to plan for and adapt to crises, in order to both survive and thrive in a world of uncertainty. The National Institute of Standard and Technology (NIST) defines cyber resilience as 'the ability to prepare for and adapt to changing conditions and withstand and recover rapidly from disruptions'. Resilience includes the ability to withstand and recover from deliberate attacks, accidents or naturally occurring threats or incidents. Cyber security and cyber resilience are related but different concepts. Both cyber security and cyber resilience are forms of defence mechanisms against cyber-crimes. Cyber security refers to measures implemented to 'protect' the occurrence of cyber-attack, while cyber resilience is the ability of an organization to withstand or quickly recover from cyber events that disrupt usual business operations. Cyber resilience recognizes the fact that security measures may not work and enables an organization to operate effectively and deliver intended outcomes despite adverse cyber-attacks. Cyber resilience is also often confused with robustness. Institute of Electrical and Electronics Engineers (IEEE) standard 610.12.1990 defines robustness as the degree to which a system operates correctly in the presence of exceptional inputs or stressful environmental conditions. Resiliency, on the other hand, refers to the ability to recover or regenerate after an unexpected impact [8]. Popular dictionary Merriam-Webster defines resiliency as the ability to recover from or easily adjust to unpredictable changes.

2.1 Need for Cyber Resilience in Banking Sector

The banking sector heavily depends on complex IT systems so as to conduct the business. Digitization of sensitive data and banking procedures has become the foremost agenda of banks at board meetings. This urgency has led to an increase in use of digital technologies, such as artificial intelligence (AI), cloud computing and analytics, the Internet of things (IoT) and machine learning (ML). With the transformation of technology, sensitive information is stored on remote servers. Higher digitization and remote operations have resulted in increased vulnerabilities that open up doors for cybercriminals, exposing cyber risks to banks.

Many studies have suggested that the banking sector is a lucrative target for cyber criminals owing to the critical environment they work in. Evolving nature of cyber-crimes and heavy dependence on third-party vendors are key challenges faced by the banking sector today. Devastating cyber-attacks on banking sector in recent past with outcomes like loss of job and reputation, heavy fines imposed by regulatory authorities have bought in the realization that no amount of security measures will ensure prevention from cyber-attack; rather, cyber incidents are inevitable and focus should be on reducing the impact of cyber-attacks. As per a recent report by Deloitte [9], in just the last week of June 2020, 40,000 cyber-attacks were attempted by hackers on India's IT and banking industry. No matter how advanced the platform might be, cybercriminals are able to adapt quickly to any virtual operating environment. Moreover, unlike independent fraud, which presents risks to individual organizations, cyber-crimes on bank can have very serious implications (compounded by financial and technical relations among financial institutions and non-financial institutions), resulting in significant economic losses. As per the recent International Monetary Fund (IMF) report (2020), cyber-crimes have the potential to be an emerging source of a macro-critical risk. To understand why cyber resilience has become a pressing necessity for Indian banking sector, it will be helpful to have a look at recent disruptive cyber-attacks that have targeted the financial sector. Table 1 highlights major cyber-attacks that targeted Indian banks in recent past.

Employees are the weakest link in an organization's cyber security resilience. This has been highlighted by several authors in their efforts to identify information security threats and their resultant impact on banks. As opined by Bogale et al. [10], employee awareness is by far the best way to protect company's data. North et al. [11] suggest that cyber security and resilience are all about governance and culture requiring everyone in the organization to participate. While technological security infrastructure is important to prevent against cyber-attacks, it will not be efficient in lack of a cyber resilient culture. As per a report by Wyman [12], a **cyber resilient culture** is a state of maturity in which employees make conscious efforts to behave in a manner that can help protect the organization against cyber threats and in which the staff is supported by appropriate mechanisms to inculcate the required behavioural changes. The cyber resiliency of an organization depends on the IT infrastructure and cyber security measures undertaken, but a prominent factor to achieve resilience depends on the employees. Cyber-attacks are inevitable, but what will matter now is

Table 1 Disruptive cyber incidents on Indian banks in recent past

Year	Victim	Incident details	Cost
2013	Cosmos Bank Pune	A malware was deployed on switch through which V.ISA and RuPay Debit cards operate. Details of many cardholders were compromised, and money was immediately withdrawn by hacker gangs from ATMs in 23 countries	INR 94.42 Crore
2013	Canara Bank	ATM servers were targeted, and skimming devices were used to steal information of debit card holders	INR 20 Lakh +
2016	Multiple Banks	3.2 million debit cards of multiple banks were compromised. Data breach was due to a malware that was injected into payment gateway network of Hitachi payment systems	INR 1.3 Crore
2013	Multiple Banks	Hackers used SIM swap systems and blocked individual SIM cards, post which they transacted money through online banking	INR 4 Crore
2017	Union Bank of India	The attack was initiated by a malware code in email attachment that forged a central bank email. This helped hackers get access to Society for Worldwide Interbank Financial Telecommunication (SWIFT) codes	$ 170 M (attempted)

how fast a bank could recover from vulnerabilities, and this will heavily depend on cyber resiliency of the staff.

2.1.1 Research GAP

Studies on cyber resiliency till date have majorly focussed on improving security infrastructure. It is imperative to realize that cyber risk management requires the commitment of the entire organization to create an environment free from cyber-crimes. Many studies have recognized cyber defence in terms of advanced techniques and tools that leads to cyber security and resilience. A critical open question remains that despite investing so much in cyber security measures, even the most advanced and tech-savvy banks are also not immune to cyber threats leading to catastrophic results. This highlights the lack of resilience level against ever-increasing risks. Banks will continue to be exposed to a wide variety of cyber threats until the banking staff is appropriately trained to operate and behave in a cyber resilient manner.

As opined by Parsons et al. [13], most studies in the domain of cyber resilience have focussed on technical issues of cyber resilience. Any number of security tools will not help protect the banking sector from cyber-attacks; rather, it is the people who can reduce the occurrence and impact of cyber-crimes through responsible behaviours. Employees need to be appropriately motivated, adaptive and have the right mindset to operate in a cyber resilient manner. Even though employees can be seen as the first line of defence against cyber-crimes, this approach to focus on cyber resilience has remained briefly addressed in the existing literature. This study thus provides a model to measure cyber resilience and cyber security preparedness of Indian banks through its employees rather than just with technologies.

3 Conceptual Framework for Measurement of Cyber Resilience Behaviour

It is imperative to be able to quantify the cyber resilience behaviour of the bank employees. Many industry frameworks have attempted to formalize the cyber resilience concept. Ponemon Institute has constructed an instrument to measure cyber resilience of an organization using 7 factors, namely preparedness, agility, strong security posture, knowledgeable or expert staff, leadership, preparedness, planned redundancies and ample resources [14]. In the year 2018, NIST came up with a flexible framework to improve critical infrastructure cyber security. This widely recognized framework groups the constructs into broader categories of identify, detect, protect, respond and recover. Each of them is broken down into twenty-one sub-categories. From a theoretical standpoint, the idea of modelling frameworks to address cyber safety and cyber resilience has been documented previously.

Proposing a framework to measure cyber resilient behaviour of organizations in general can be formidable. Hence, this study narrowed the scope to a more regulated and structured banking sector. The study attempts to look at contributions of the 8 identified factors to measure cyber resilient behaviour of employees in the banking sector. These 8 factors were then operationalized into the proposed framework as independent variables, while the cyber resilient behaviour of bank employees was operationalized as the effect of changing individual independent variables. Hence, cyber resilient behaviour is treated as the dependent variable for the study. Refer Fig. 1 for the proposed model. Each of the factors considered for the study is explained below:

3.1 Factor 1: Employee Expertise

This factor deals with the educational background, years of work experience in the banking domain and possessing knowledge regarding cyber security. Employees with more years of experience and related background in computer studies may be more sensitive towards cyber incidents. Even with all the industry best practice approaches, cyber resilience will eventually rely on cognitive abilities of the

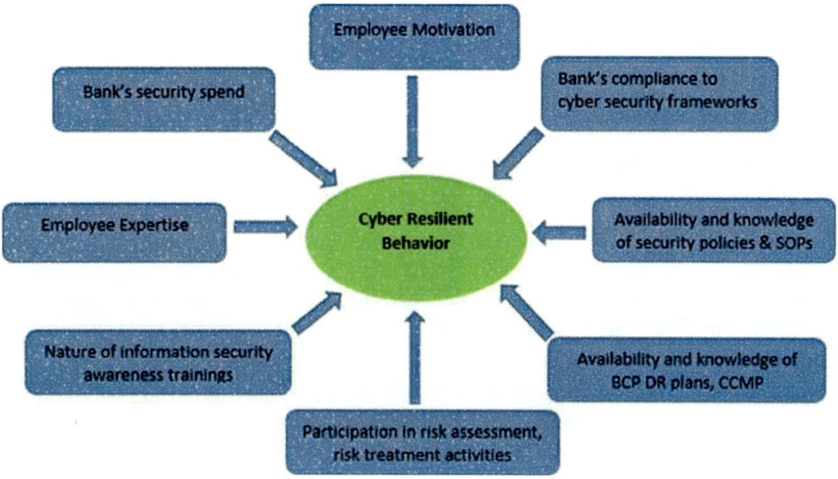

Fig. 1 Proposed conceptual model displaying factors affecting cyber resilient behaviour of bank employees

employees. Expert and skilful employees ultimately can contribute to cyber resilient organization culture. With this factor, it is planned to study the effect of employee expertise on cyber resilience behaviour.

3.2 Factor 2: Bank's Security Spend on Employees

Lack of an adequate budget has historically been a key challenge in addressing cyber security challenges. In the wake of rise in data breach numbers, banks have started investing in cyber security efforts. As per a joint report developed by Data Security Council of India (DSCI) and PricewaterhouseCoopers (PwC) published in December 2019, India's BFSI industry accounts for 26% of the total expenditure in cyber security market. It is expected that the sector will increase its investment from USD 518 m in 2019 to USD 810 m in 2022. The forecasted growth can be attributed to factors like tightened regulations and rapid adoption of digital payment technologies.

While Indian banks do spend a significant amount on improving the cyber defence stand, it is attempted to investigate whether banks invest sufficiently towards the development of cyber resilient culture and thus this factor tries to understand bank's security spend on its employees. This can be measured by looking at investments like having robust incident management tools, conducting periodic mock drills for cyber crisis situations, security workshops and town halls organized by the banks for their employees. Organizations that make a strong investment in cyber security have a higher level of confidence in their defences against cyber threats. These organizations enjoy a strong cyber security commitment from top management, which enables them to protect and prevent themselves from the devastating impacts of cyber-attacks.

3.3 Factor 3: Employee Motivation

Employees must be adequately inspired, adaptive and have the right mindset to function in a cyber resilient way. This factor tries to analyse whether employees of the bank consider the capability to behave and respond to cyber-crime as something important. Employee motivation can also depend on role or level of responsibility handled by employees. Relevant functions like information security, information technology and internal audit may be more committed to function in a cyber resilient manner. Motivation can be described as all the processes that may energize and guide behaviour of employees. Lack of appropriate motivation may hinder the resilient functioning of banks.

3.4 Factor 4: Bank's Compliance Towards Cyber Security Frameworks

Meeting industry standards like the International Organization for Standardization (ISO) 27,001 standard, NIST cyber security framework equips banks with the ability to deal with unexpected cyber incidents. Although the ISO 27,000 standard does not specifically include resilience as one of its goals, it recommends several measures that can effectively contribute to cyber resilience. Apart from generic cyber security standards developed by ISO/IEC and NIST, a more specific framework like that of CERT-Resilience Management Model (CERT-RMM) has been developed by a division of the Software Engineering Institute at Carnegie Mellon University. It is believed that ensuring compliance with such industry standards can have positive impact on cyber resilient behaviour of the employees. This factor attempts to analyse impact of bank's compliance towards industry standards on cyber resilience of bank employees.

3.5 Factor 5: Availability and Knowledge of Security Policies and Standard Operating Procedures

As pointed by Killmeyer (2016), effective cyber security management can be achieved by effective policies, standards, procedures and action plans that contribute to ensure confidentiality, integrity and availability of data and information in organizations. Employees require policies and procedures to guide them so as to behave in a responsible manner during cyber crisis situation. Further, as per the circular released by RBI in the year 2016, it is mandatory for all scheduled commercial banks in India to have a distinct cyber security policy which is separate than the broader information security policy.

This factor tries to investigate the relationship between adequate documentation of security policies and procedure with a responsible cyber ready behaviour. It further seeks to assess whether these policies are updated and communicated on a timely basis and are accessible to employees when required.

3.6 Factor 6: Availability and Knowledge of BCP-DR Plans and CCMP

As per RBI Circular (2016), all scheduled commercial banks operational in India must have a Cyber Crisis Management Plan (CCMP) in place. Recognizing that cyber risk varies from other digital threats, conventional strategies for business continuity plan and disaster recovery (BCP-DR) may not be appropriate and must therefore be revisited in view of the complexities of cyber risks. As banks are employee-driven organizations, it is imperative for employees to be aware of the BCP-DR and CCMP. It is believed that increased understanding of BCP-DR and CCMP will have a positive impact on cyber resilience behaviour.

3.7 Factor 7: Participation in Risk Assessment and Risk Treatment Activities

It is imperative to identify and implement preventive measures to mitigate cyber risk. Cyber security risk assessment and risk treatment are important processes to understand, manage, control and mitigate cyber risk across organization. This process assists to identify all possible risks associated with information assets and helps organizations to address them. Risk assessments help to detect threats and vulnerabilities. A good practice to conduct effective risk assessments is to set up communication across the organization. All employees in every department of the organization should have a comprehensive and common understanding of the vulnerabilities of their workplace and security systems. Hence, this factor attempts to understand whether bank employees participate in risk assessment and risk treatment activities. It is believed that greater participation in risk management activities will have positive impact on cyber resilience behaviour.

3.8 Factor 8: Nature of Information Security Trainings

According to NIST Special Publication 800–16 Revision 1, security awareness and role-based training applies to all employees regardless of whether they use or interact with information systems or data. This factor deals with information security awareness training of employees. It tries to understand the nature (mandatory or optional) of the security trainings. An effective security awareness programme should be enriched with the latest security information and inculcated from induction and ongoing basis. This factor attempts to seek the impact of effective information security trainings on cyber resilient behaviour of employees.

4 Scope and Limitation

The study is focussed at bank employees working in Indian banks. It is planned to restrict respondents from participating banks to subject matter experts (SMEs) who are part of cyber security and related functions operational in the bank. Due to the ever-changing cyber security landscape, the scope is limited to the year 2021–22 so as to ensure that all outcomes are referenced to point in time. Cyber resilience in any organization is dynamic in nature. The dynamics of a framework measuring cyber resilience delves into technological factors that may become obsolete very soon.

5 Conclusion

Given the ever-increasing rate of cyber-crimes in the banking sector, it is the need of the hour that banks become cyber resilient. A prominent factor to achieve effective cyber security is the cyber resilience of employees of the bank. Building a cyber resilient culture should be seen as a continual effort. This study is novel in terms of measuring cyber resilience behaviour basis the identified factors. The study is intended not only to investigate cyber resilience but also to help the Indian banks understand the current state of cyber security preparedness and improvement areas. It is believed that this study will help to address the identified research gaps and contribute in whatever small way towards the existing body of knowledge.

References

1. Ganjapure, V.: RBI reports over 50,000 cyber frauds in FY 2018–19. *Times of India* (2019). Retrieved from: https://timesofindia.indiatimes.com/city/nagpur/rbi-reports-over-50000-cyber-frauds-in-fy-2018-19/articleshow/71174635.cms#:~:text=NAGPUR%3A%20The%20Reserve%20Bank%20of,credit%20cards%20and%20internet%20banking

2. Saraswat, V.: Niti Aayog report on cyber security (2019). Retrieved from: http://niti.gov.in/sites/default/files/201907/CyberSecurityConclaveAtVigyanBhavanDelhi_1.pdf
3. Korolov, M.: Banks get attacked four times more than other industries (Online) (2015). Available: http://www.csoonline.com/article/2938767/advancedpersistent-threats/report-banks-get-attacked-four-timesmore-than-other-industries.html
4. Obura, F.: Kenya worst hit in East Africa by cyber crime (2017). Retrieved 16 July 2019 from https://www.standardmedia.co.ke/article/2001235820/kenya-worst-hit-in-eastafrica-by-cyber-crime
5. OED (Oxford English Dictionary) Online. Resilience, OED Online 2018. www.oed.com/view/Entry/163619. Retrieved 14 Mar 2020
6. Kott, A., Linkov, I.: Fundamental concepts of cyber resilience: introduction and overview. In: Kott, A., Linkov, I. (eds.), Cyber Resilience of Systems and Networks. Springer, Switzerland (2019). Retrieved from: https://link.springer.com/chapter/https://doi.org/10.1007/978-3-319-77492-3_1
7. Seville, E., Opstal, V., Vargo, J.: A primer in resiliency: seven principles for managing the unexpected. In: Global Business and Organizational Excellence (2015). Retrieved from: https://onlinelibrary.wiley.com/doi/abs/https://doi.org/10.1002/joe.21600
8. Dalziell, E.P., McManus, S.T.: Resilience, vulnerability, and adaptive capacity: implications for system performance (2008). Retrieved from https://www.researchgate.net/publication/29489371
9. Deloitte, T.: Cyber security in Indian banking industry: part 1 (2020). Retrieved from: https://www2.deloitte.com/content/dam/Deloitte/in/Documents/risk/in-ra-cybersecurity-in-the-indian-banking-industry-noexp.pdf
10. Bogale, M., Lessa L., Negash, S.: Building an information security awareness program for a bank. In: 25th Americas Conference on Information Systems. Cacun, Mexico (2019). Retrieved from: https://www.researchgate.net/publication/336133212_Building_an_Information_Security_Awareness_Program_for_a_Bank_Case_from_Ethiopia
11. North, P., James, R., Pascoe, R.: Cyber security and resilience It's all about governance. Governance Direct **68**(3), 146–151 (2016). Retrieved from: https://search.informit.com.au/documentSummary;dn=003855469276044;res=IELBUS;type=pdf
12. Wyman, O.: Building a cyber resilient culture (2019). Retrieved from: https://www.oliverwyman.com/content/dam/oliver-wyman/v2/publications/2019/October/building-a-cyber-resilient-culture.pdf
13. Parsons, K.M., Young, E., Butaviciu, M.A., Cormac, A., Pattinson, M.R., Jerram, C.: The influence of organizational information security culture on information security decision making. **9**(2), 117–129 (2015). Retrieved from: https://journals.sagepub.com/doi/full/https://doi.org/10.1177/1555343415575152
14. Ponemon Institute.: The Third Annual Study on the Cyber Resilient Organization (2018). Retrieved 31 Dec 2018 from: https://public.dhe.ibm.com/common/ssi/ecm/55/en/55015655usen/the-3rdannual-study-on-the-cyber-resilient-organization-whitepaper_55015655USEN.pdf
15. Caralli, R., Allen, J., White, D.: CERT resilience management model, version 1.2. Pittsburgh: Carnegie Mellon University (2016). Retrieved from: https://resources.sei.cmu.edu/library/asset-view.cfm?assetid=508084
16. Joseph, G., Heedon, K., Zhi, L., Cheng, W., Galen, S., Chris, Y.: Cyber risk surveillance: a case study of Singapore. In: IMF working paper, Monetary and Capital Markets Department (2020). Retrieved from: https://www.imf.org/en/Publications/WP/Issues/2020/02/10/Cyber-Risk-Surveillance-A-Case-Study-of-Singapore-48947
17. NIST: Framework for improving critical infrastructure cybersecurity. (V1.1) (2018). Retrieved 25 Jan 2019 from https://nvlpubs.nist.gov/nistpubs/CSWP/NIST.CSWP.04162018.pdf
18. Reserve Bank of India: Cyber Security Framework in Banks. Retrieved from Reserve Bank of India Circular Number: DBS.CO/CSITE/BC.11/33.01.001/2015–16 (2016)
19. Vedashree, R., Vishwanath, S.: Cyber Security India Market (2019). (Joint report published by PwC and DSCI) Retrieved from: https://www.pwc.in/assets/pdfs/consulting/cyber-security/cyber-security-india-market.pdf

IoT-SDWD: Internet of Things-Based Security Device for Women in Danger

V. Kanchana Devi, Alluru Kavya, E. Umamaheswari, A. Karmel, D. Kavitha, and D. Shruthi

Abstract Security for women has become an issue nowadays. This paper aims to defend the women in danger by using an electronic gadget to recognize the problem and caution the surroundings, saving them on time. This paper recommends another perspective to utilize innovation to ensure ladies assistance with smartphones with an incorporated component that alarms and gives location-based data. This paper presents a modeled gadget with Global Positioning System (GPS) and Global System for Mobile Communication (GSM)-based "Women Security Device," which provides the blend of GPS gadgets just as a handheld device and "I AM IN TROUBLE, PLEASE HELP ME!!!" message is sent when a button is triggered. This application assistance is created in android with Graphical User Interface (GUI); it gives the degree of quality, accessibility, and similarity. The results provided in this paper show the customized message sent when the button is pressed with the exact location.

1 Introduction

This paper aims to produce safety for ladies, and therefore, the device ought to be a transferrable one. It also creates awareness on the time of emergency scenario for ladies. First of all, user is ought to click the emergency button embedded on

V. Kanchana Devi (✉) · A. Kavya · E. Umamaheswari · A. Karmel · D. Shruthi
Vellore Institute of Technology-Chennai, Chennai, Tamilnadu, India
e-mail: kanchanadevi@vit.ac.in

A. Kavya
e-mail: kavya.alluru@vit.ac.in

E. Umamaheswari
e-mail: umamaheswari.e@vit.ac.in

A. Karmel
e-mail: karmel.a@vit.ac.in

D. Kavitha
Sri Sairam College of Engineering, Chennai, Tamilnadu, India
e-mail: kavithad.cse@sairam.edu.in

T. Senjyu et al. (eds.), *ICT with Intelligent Applications*, Smart Innovation, Systems and Technologies 248, https://doi.org/10.1007/978-981-16-4177-0_32

the device. When the button gets pressed, the device gets activated and sends short message service (SMS) to those contacts saved during registration process. The SMS carries the customized message and the current location fetched using the embedded GPS module in safety device. This paper provides a deactivation service to be used in safety places. Its fundamental aim is to design and develop a cost-effective IoT devices which includes GPS and GSM units, mostly beneficial for ladies called "Women Safety System (WSS)." The objective is to trace the user present location using GPS unit in the device, which extracts the user longitude and latitude and indicates the predefined contacts about the current situation. WSS is designed in such a way, stress-free while carrying and integrated with many functionalities. The use of smartphones has increased a lot in the world. Many mobile applications and smart gadgets have been developed to be used whenever women faced a problem or in danger. Regardless of several functionalities and features, various devices have been developed but did not reduce the rate of sexual offenses in applications. Women's safety devices or applications should be designed carefully with many features that helps in everyday life situations and real emergency scenarios.

2 Literature Survey

Hyndavi in [1] developed a smart wearable device safety using IoT. Crime against women has increased significantly and often heard about harassment, jokes, and rape in public places in society. This article proposes a smart women's safety device that uses pressure sensors, pulse rate sensors, and temperature sensors to automate emergency alert systems and anomaly detection. An innovative security device is designed which automatically uses outlier detection technology is suggested. The system uses the woman's position coordinates to detect and send alerts to loved ones, without the woman interacting during critical times. Automatically send emergency messages to relatives and nearby police stations. Three sensors, namely pressure, temperature, and pulse rate sensors, are used to detect outrage automatically. Pressure sensors, to determine if a woman is facing pressure beyond acceptable limits, are used. Temperature sensors, to detect temperature deviations, are used. Pulse rate sensors, to identify abnormalities in pulse rate, are used. The measured values of these three sensors are used to identify critical conditions. The device also offers a button that must be pressed when the user feels insecure. When any one of such abnormalities occurs, the alarm in the device gets activated to alert individuals nearby conveying that the user is in a risky condition. Then, the users' corresponding location is automatically identified and conveyed to the registered list.

In [2], authors have developed the IoT-based smart security women's safety. The primary approach is to send a warning message too close to the immediate location and call the registration numbers of parents, friends, media, police, etc., to avoid this annoying incident. The strap is incorporated with several sensors such as heart rate monitor, GPS for location capture, GSM for messaging, motion sensor, analog-to-digital converter, and alarming buzzer. These sensor modules are connected to a

Raspberry-Pi controller is used. The strap has an emergency call button that allows women to use this smart strap when they are in danger. When the button is pressed during the dangerous situation, and the specified threshold is reached, an alarm buzzer sounds, a loud beep sounds to attract the attention of people near the victim, an urgent message indicating the location is generated, and a call to that number is sent to the number. It is already pre-programmed in GSM unit of the designed device. This device gets activated when the motion sensor and heart rate sensor reach the specified threshold value. All these sensor modules are programmed using the Python language. The notion here is to design a device and create a fully portable, self-defending universal IoT security system for women to help women in order to escape in crises. One can also use the IoT platform to track the whereabouts of women with problems remotely.

Ramachandiran in [3] proposed a comprehensive survey on safety device for women using IoT components. In current scenes, women's safety in urban and rural regions is considered the primary problem. It is not easy to change the mind-set of society as a whole. However, one can take various safety precautions for women exposed to sexual harassment, acid attack, harassment, etc. Various intelligent devices and applications have been developed to ensure security. There are also many smart devices and applications on the market. However, this is not a practical solution. Specialists from various fields have exposed a safety device for women that works in both manual mode and automatic mode. This article inspects multiple techniques related to women's safety. The disadvantages and possibilities of safety devices for women are also discussed. A smart device has been designed and developed to make women safe. Safety equipment for women is particularly designed for women or user in a hazardous situation and crisis. Safety devices for women should be simple, easy to carry, easy to handle, and capable of integrating with many other features. Smartphone usage has increased dramatically worldwide. Many mobile apps and smart devices are being designed that help women by the government and people when they are in trouble. Even though they have developed various tools and applications, the rate of sexual offenses has not decreased. Female safety equipment or applications should combine with many features used in day-to-day activities and actual danger scenarios.

3 Proposed Work

The system was developed with GPS and GSM's help and offers a combination of alarm and message options with an emergency call button. In case of trouble, the message "I AM IN TROUBLE PLEASE HELP ME" is sent to the registered contacts when the button is pressed. This block diagram explains the connections of the components fixed inside the locket; the lithium polymer battery is connected with a charging circuit and Bluetooth (HC-05), the Bluetooth is associated with the emergency button, the button is combined with GPS and GMS module to send a message via the controller, and an alarm is set to alert the nearby people (Fig. 1).

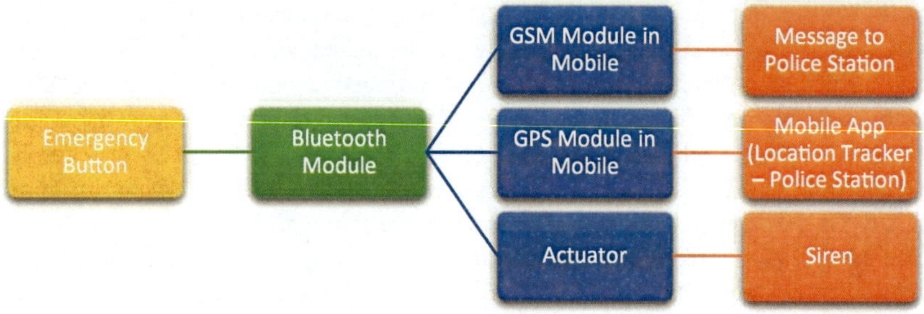

Fig. 1 Flow diagram of safety locket for women

4 Application Development Specifications

This application is designed to protect women. The purpose of designing this application is to spread alertness about women's safety around the world. The locket gets activated, when the emergency button is pressed, the Bluetooth HC05 module is placed inside helps in transmitting the signal to the smart for sending the message of help and support is sent to the predefined contacts stating that the person is in danger.

- *This application is used for the safety of women.*
- *This application can be used as an evidence of a crime with prompt location details for prosecution.*
- *This application exhibits a useful device for safety of user.*
- *This application is feasible and productive.*
- *The application is simpler to utilize by all women.*
- *The application is a portable device.*
- *For women, there is no need for outside equipment or knowledge of programming to utilize this application. This application is free for women, which does not need any money.*

5 System Architecture

In Fig. 2, the device developed has been provided, in which a button can be seen, which can be used in the time of emergency. Several separate components were used, and these components are combined on a single device to function as per the customization of the structure. The processor chip is incorporated to control the other components such as GPS and GSM modules for the specified operations like location tracking and sending a message. This small handheld device is designed in such a way mainly to be embedded into a small locket kind of structure and can be worn by any human being and used in a time of emergency.

Detailed Design of the System: Data Flow Diagram (DFD) (Fig. 3).

Fig. 2 Design and implementation locket for women

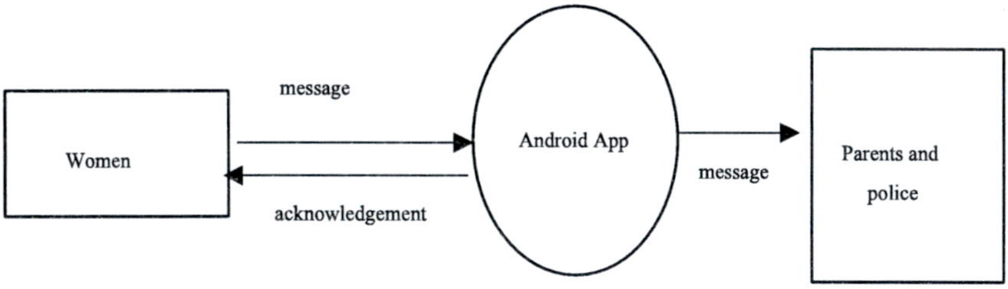

Fig. 3 First level: data flow diagram (DFD)

This first-level DFD explains how the message gets transferred from the android app to parents/police (Fig. 4).

This second-level DFD explains how the message, call, and the location are sent to the parents and police.

Technical Specification:

- Bluetooth connection is a must—HC05 Bluetooth module
- Lipo battery −3.7 V battery
- Lipo battery charging circuit
- The button is connected to the charging circuit to enable the trigger to the Bluetooth module (Table 1).

The sample output after implementation of the designed android application has been provided below which shows various tabs such as stop application, add a

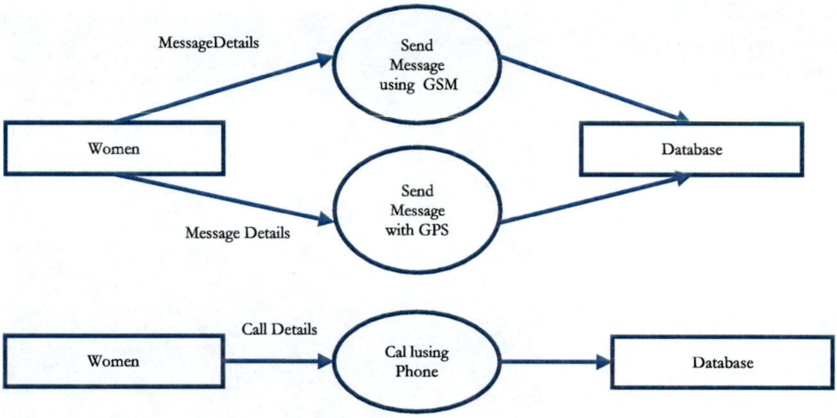

Fig. 4 Second level: data flow diagram (DFD)

Table 1 Results obtained from test cases

Test case Id	Test case description	Expected result	Actual result
1	About us	Description of the women's safety system locket	PASS
2	Add contact	Predefined emergency names and contacts	PASS
3	Select the device	The pairing of the hardware and software via Bluetooth	PASS
4	Add message	The help message is included	PASS
5	Deactivate the device	The pairing of the device via Bluetooth is switched off	PASS

message, deactivate, update contact, about us, and about the application. The "about us" tab holds the details regarding the development, software, and hardware used in the device. "Stop application" tab is used to terminate the application. "Update contact" tab is for updating the contact details as per the user's desire. It can be any relatives' number, friends' phone number, emergency contact number, or the police number. The deactivate tab is for keeping the application in silence while using it in a safer place (Fig. 5).

6 Conclusion

The issue of women's safety in this environment is growing rapidly. This is an android application designed to avoid certain suspicious activities and informed the authorities using android mobile phones that helps prevent them. This system helps in overcoming the fear that makes every woman in the country fear about the safety

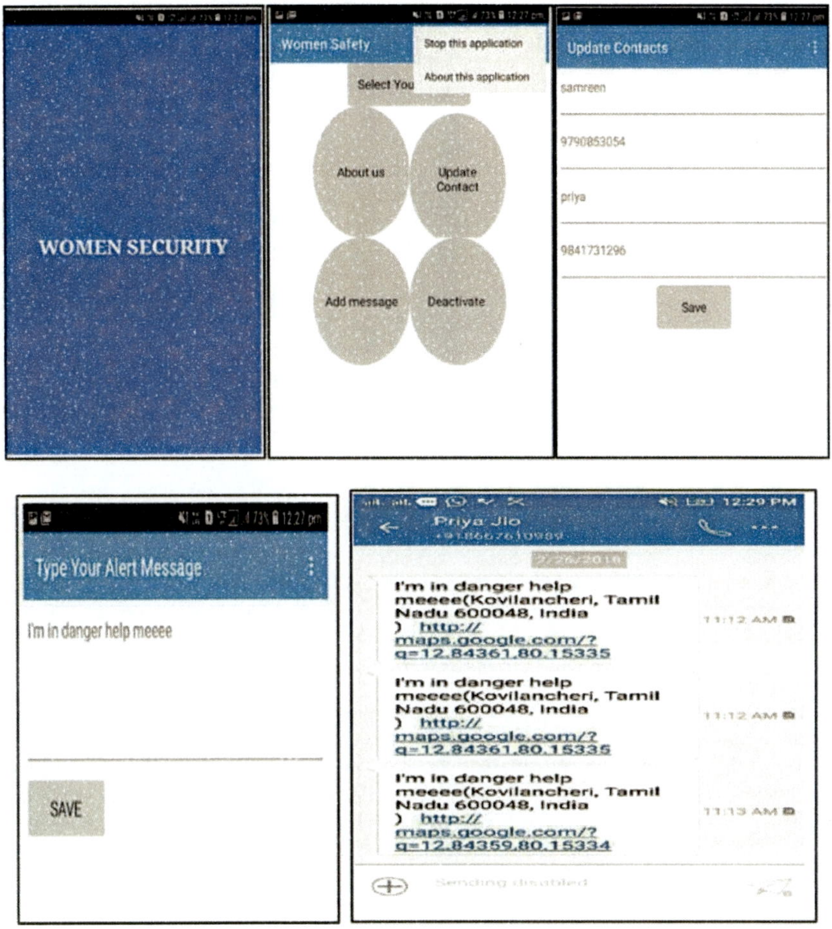

Fig. 5 Various output screenshots of the implemented application

and security. An app sends online and offline SMS to a particular emergency contact number. In this paper, a system for the protection of women was proposed. Using a wireless transmission, an alert message is sent to a predefined number like police, relatives, emergency contact. Upon pressing the push button embedded in the device, information about the user is collected by the sensor, and then it is sent to listed numbers with the voice call. This particular system helps in monitoring of the safety of women using GPS tracking engine. Predefined message is sent as SMS to stored numbers, this saves time, and the sufferer gains help without wasting time. The paper aims to assist any women in danger by sending message with the stored relatives contact and the nearest police station contact using the recent concept Internet of things, GSM, and GPS and ensure the safety of women by detecting problems and threats and automatically sending messages about the victim's condition.

References

1. Hyndavi, V., Nikhita, N.S., Rakesh, S.: A smart wearable device for women's safety using IoT. In: 2020 5th International Conference on Communication and Electronics Systems (ICCES), pp. 459–463. IEEE (2020)
2. Nagamma, H.: IoT based smart security gadget for women's safety. In: 2019 1st International Conference on Advances in Information Technology (ICAIT), pp. 348–352. IEEE (2019)
3. Ramachandiran, R., Dhanya, L., Shalini, M.: A survey on women safety device using IoT. In: 2019 IEEE International Conference on System, Computation, Automation, and Networking (ICSCAN), pp. 1–6. IEEE (2019)
4. More, S.A., Borate, R.D., Dardige, S.T., Salekar, S.S., Gogawale, D.S.: Smart band for women security based on internet of things (IoT). IJARSE **6**(11), 840–845 (2017)
5. Murphy, M.L.: Android programming tutorials. CommonsWare, New York (2009)
6. Bhilare, P., Mohite, A., Kamble, D., Makode, S., Kahane, R.: Women employee security system using GPS and GSM based vehicle tracking. Int. J. Res. Emerging Sci. Technol. **2**(1), 65–71 (2015)
7. Gadhave, S.N., Kale, S.D., Shinde, S.N. Bhosale, A.C.: Electronic jacket for women safety. IRJET (2017)
8. Sharma, S., Ayaz, F., Sharma, R., Jain, D., Student, B.E.: IoT based women safety device using ARM7. IJESC **7**(5), 11465–11466 (2017)
9. Bhavale, D.M., Bhawale, M.P.S., Sasane, M.T., Bhawale, M.A.S.: IoT based unified approach for women and children security using wireless and GPS. Int. J. Adv. Res. Comput. Eng. Technol. (IJARCET) **5**(8) (2016)
10. Jain, R.A., Patil, A., Nikam, P., More, S., Totewar, S.: Women's safety using IoT. Int. Res. J. Eng. Technol. (IRJET) **4**(05), 2336–2338 (2017)
11. Zainuddin, S., Yadav, M.S.: Women security system using IOT (2019)
12. Prathap, P., Reddy, K.V., Sreelatha, T.: Security and privacy protection methods for women's safety based on IoT. Indian J. Publ. Health Res. Dev. 1069–1072 (2018)
13. Thavil, J.K., Durdhawale, V.P., Elake, P.S.: Study on smart security technology for women based on IoT. Int. Res. J. Eng. Technol. (IRJET) **4**(02) (2017)
14. Arias, O., Wurm, J., Hoang, K., Jin, Y.: privacy and security on the internet of things and wearable devices. IEEE Trans. Multi-Scale Comput. Syst. **1**(2), 99–109 (2015)
15. Vani, A., Purohit, A., Tiwary, D.: A Smart technique for women and children's security system with location tracking

Stock Prediction Using Machine Learning Algorithms with Special Reference to Technical Indicators

Chetan Gondaliya, Ajay Patel, and Tirthank Shah

Abstract Machine learning algorithms have increasingly become chosen tools for stock price prediction. Using a variety of financial news as an input to compare various algorithms for accuracy level has been extensively studied. However, taking some of the prominent technical indicators as an input to test the algorithms' prediction accuracy for a stock price has remained less explored. This study focuses on using chosen seventeen technical indicators to compare selected algorithms to test the prediction accuracy for six Indian stocks as a sample. This study covers the critical time period of the outbreak of the Covid-19 pandemic and attempts to capture the impact on accuracy levels of algorithms. Three algorithms are tested, and among them random forest algorithm has demonstrated superior results. Based on these results, this study proposes a framework to create a platform for further application.

1 Introduction

In the world of finance, the stock market has been center of attraction for researchers as well as practitioners due to its ability to remain elusive about its prediction. The stock market forecast has remained one of the most researched domains for many decades across various fields. It was considered that the accurate forecast of the stock market price movement is a futile exercise. To substantiate this point, the studies from the finance field such as [1, 2] have established that markets are efficient in nature, and hence, the stock market behaves randomly leading to no possibility to predict it. Hence, the forecast of the stock market has been perceived as a paramount challenge. Stock markets are considered to be very dynamic, and many times traders who trade

C. Gondaliya (✉) · A. Patel
AMPICS, Ganpat University, Mehsana, Gujarat, India
e-mail: cpg01@ganpatuniversity.ac.in

A. Patel
e-mail: ajaykumar.patel@ganpatuniversity.ac.in

T. Shah
Institute of Management, Nirma University, Ahmedabad, Gujarat, India

T. Senjyu et al. (eds.), *ICT with Intelligent Applications*, Smart Innovation, Systems and Technologies 248, https://doi.org/10.1007/978-981-16-4177-0_33

in the market based on applying various indicators or techniques of technical analysis are a major force.

The field of using various data feeds and technical indicators as a feature set to build or train predictive models have generated significant research attention from researchers. This study focuses on training the model by using the features set based on a sizable variety of technical indicators. The main objective of this study is twofold; the first objective is to analyze if the features set created based on most used technical indicators are having any valuable contribution to predicting the stock market movement as compared to other financial news feeds, the second objective is to compare K-nearest neighbors (KNN), neural network (NN), and random forest algorithms for their predictive accuracy. Studies such as [3–6] have used technical indicators to build features set for training the model. Random forest classifier is applied by [3] for testing the forecasting ability of the algorithms, while [5] has used ML classifiers and classifiers based on deep learning to test the accuracy of predictive models. Basak et al. [6] focused on tree-based classifiers to compare the accuracy rate of algorithms.

Multivariate analysis and time-series modeling were earlier more famous as a statistical tool for forecasting of a stock price. Prediction of stock price movement applying ML models is still a recent phenomenon. Different algorithms are used for the forecasting of stock market movement such as artificial neural networks, Naive Bayesian classifier, support vector machine, linear discriminant analysis, KNN, linear regression methods, and so on. Most of such research work such as [7] has used financial news from various sources as feed for building the models. However, very limited work has been done toward focusing on the output of technical indicators to train the models to test accuracy.

Technical analysis consists of a full range of technical indicators that are used to analyze the past price and volume of the stock to forecast the future movement of the stock. Study conducted by [5] has used indicators such as stochastic oscillator, relative strength index, Williams %R, money flow index, exponential moving average, and moving average convergence and divergence which are applied to develop features set.

This paper is segregated into subsections where Sect. 2 elaborates the review of literature, Sect. 3 discusses the data sets, methodology, algorithms, and framework proposed, Sect. 4 deliberates on the empirical findings based on testing, and the last Sect. 5 presents the closing remarks and future scope of the research work.

2 Literature Review

An extensive literature review has provided a platform to better comprehend the existing research work and identify the research gap. Article [3] has used a random forest classifier to predict the direction of the stock market. Some of the technical indicators such as stochastic oscillator and relative strength index (RSI) were used as inputs to develop a model. Hartono and Sulistiawan [3] have used an ensemble of multiple decision trees as a learning model. This algorithm has shown better results when technical indicators were used as inputs. Predicting stock market movement based on machine learning algorithms where information is used as an input was discussed in [7]. Financial news data and social media data were used to run algorithms to measure the accuracy of stock market prediction for the next ten days in [7]. The study [7] has identified the random forest classifier as the most accurate and consistent classifier where its ensemble was instrumental in achieving a higher level of accuracy. The study done by [4] has focused on analyzing the market quality's impact on the profitability of technical analysis with the special case of inter-country data. The study [4] confirmed that market quality which was measured as market capitalization was affecting the performance of technical analysis based on data of 50 companies selected across from 21 countries.

Adam optimizer was used for the optimization of multilayer deep learning neural networks by [8] to predict the stock price of the stock. Different firms' six years of data have been used to develop the model to test the model's robustness to a variety of data set from varied sectors. The study [8] has observed minimal prediction error score pointing toward the strength of the model. Linear regression technique was compared with support vector machines (SVMs) under machine learning technology for prediction of the stock market in the study conducted by [9]. Methodologies such as environment, time-series forecasting, and sliding-window method were used for comparison of two algorithms to test for better accuracy in [9]. This study has observed that support vector machines (SVMs) performed better than linear regression based on accuracy. Using data feed from various indicators under technical analysis to predict the movement of the stock market has been studied by [5] with the purpose of investigating the efficiency of using technical analysis to predict the stock price movements. This study has applied stock technical indicators as well as optimal long short-term memory deep learning to predict the trend of the stock prices. A classifier based on deep learning and ML classifiers was compared to measure mean prediction accuracy of the proposed models in the study undertaken by [5]. The study [10] has proposed a new prediction algorithm that leveraged temporal correlation among various international stock markets and financial instruments to forecast the trend of the stock the next day. Support vector machines as an algorithm was used for the analysis purpose alongside different regression algorithms that were used to observe the actual change in the market trend. Prediction accuracy in the range of 70–80% was observed in [10]. This study has also proposed a stock trading model to test the power of prediction for the algorithms which were analyzed.

Tree-based classifiers such as XGBoosted trees and random forest were used for stock market prediction purpose where technical indicators such as stochastic oscillator and relative strength index were used as features to develop the model in the study of [6]. This study has indicated that tree-based classifiers were having better performance than the other algorithms. The study [6] has taken the problem of prediction of the stock price as a classification problem instead of looking at it as a problem of the traditional forecasting model. The formulation of a trading strategy was also suggested as the results of tree-based classifiers were superior.

3 Data, Research Methodology and Proposed Framework

3.1 Data

For any model building exercise, data as input plays a very essential part in it. Time period which has been under focus for this study is from January 01, 2018, to September 30, 2020. This time period crucially covers the impact of the Covid-19 pandemic. A total of six stocks from very active sectors such as pharmaceuticals and banking were identified for the testing purpose. These stocks are Torrent Pharma, Sun Pharma, Biocon, State Bank of India (SBI), Axis Bank, and HDFC Bank. The source of the data collection is the 'fix_yahoo_finance' library of Yahoo finance. The sample code file as an image is shown in Fig. 1.

A list of technical indicators' inputs is used as a feature set in order to train the model. A total of 17 technical indicators that are frequently used by traders for prediction of stock price movements are included in this study purpose which is as listed in Table 1.

Some of the sample formulas for some technical indicators are exhibited as below;

1. Relative strength index (RSI): Relative strength index is measured by applying below formula;

$$\text{RSI} = 100 - \left[\frac{100}{1 + \frac{n_{\text{up}}}{n_{\text{down}}}}\right]$$

where N up = average of n-days up close and N down = average of n-days down close.

2. Simple moving average: Simple moving average is calculated by applying below formula;

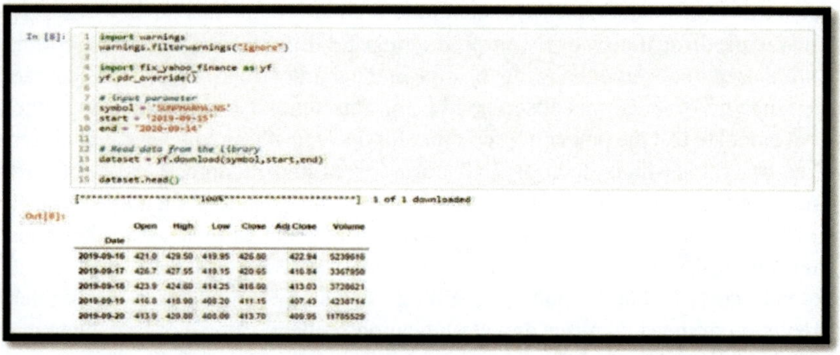

Fig. 1 Sample code for data collection from Yahoo library

Table 1 List of technical indicators

Technical indicators	
1. Increase/Decrease (Volume) 2. Return (Based on close price) 3. Up_Down (Based on return) 4. Buy_Sell 5. MA (Moving average) 6. EMA (Exponential moving average) 7. RSI (Relative strength index) 8. ROC (Rate of change) 9. BB (Bollinger bands)	10. MACD (Moving average convergence divergence) 11. KST (Know sure thing) 12. OBV (On balance volume) 13. CCI (Commodity channel index) 14. VWAP (Volume weighted average price) 15. STDDEV (Standard deviation) 16. EOM (Ease of movement) 17. SO (Stochastic oscillator)

$$\text{SMA} = \frac{A_1 + A_2 + \cdots + A_n}{n}$$

where A_n = Refers to closing price at period n and n = total number of period.

3. Stochastic oscillator (SO): Stochastic oscillator is measured by applying below formula;

$$\%\text{K} = 100 * (\text{CP} - \text{L14})/(\text{H14}–\text{L14})$$

where CP = close price, L14 = The lowest price of last 14 previous trading sessions and H14 = The highest price of last 14 previous trading sessions

3.2 Research Methodology

Wider choice of technical indicators as an input was essential to build a robust prediction model. A total of 17 technical indicators as given in Table 1 were included as a part of the study to measure the impact. Features selection methods such as filter method, wrapper method, and embedded method are used initially to check which method has generated better feature selection. Filter method has demonstrated better feature selection, and hence, it is used as a feature selection method in the proposed frame work. Three machine learning algorithms namely the k-nearest neighbors (KNN), neural network (NN), and random forest were identified to measure the prediction accuracy.

1. The k-nearest neighbors (KNN) algorithm: The regression problem and classification problem can be solved using the simple method called "KNN." The k-nearest neighbors method is based on a supervised learning technique. KNN is considered to a nonparametric algorithm hence making no assumptions on

underlying data. At the heart of the "KNN" algorithm is the classification of data based on similarity. Storage of all available data is created by KNN algorithm first, then as the new data appears based on the similarity of this new data set to existing stored data, it is categorized appropriately to most similar feature category. Hence, the new data points are easily classified based on similarity by applying the "KNN" algorithm. However, the KNN algorithm is a slow learner as it just stores the data and does the classification of data based on similarity of new data with existing data set.

2. Neural network (NN) algorithm: Neural network algorithm is inspired from the functionality of human brain which consists of billions of neurons. Neural networks are modeled based on the architecture of neurons in the brain. A neural network algorithm will have many units which are artificial neurons arranged in the series of layers. These layers are connected with other layers. Inputs units are designed in such a way that it receives information in various forms from outer world. On this sets of information, the network will attempt to learn or identify the pattern. The output units are on the other side of the network signals the responses to the information it has learnt. Hence, different layers are part of neural networks that are used for learning and analyzing the data. Neural networks are used in this study with the primary purpose of prediction of stock price movement.
3. Random forest algorithm: The random forest algorithm is a flexible supervised learning algorithm. The random forest algorithm is utilized to solve both the regression problem as well as a classification problem. The "bagging" method is used to an ensemble of decision trees known as a forest. Random forest can increase the accuracy and stability of prediction as multiple decision trees are built and merged together. Random forest classifier consists of number of decision trees which are based on various subsets of identified data set. Then, random forest classifier takes into consideration the prediction from each decision tree, and depending on majority of predictions it shows the prediction for final output. Hence, higher number of decision trees in this algorithm leads to higher accuracy of the results.

The process starting from data collection to applying the three mentioned algorithms leading to the generation of accuracy rate for prediction of stock prices is delineated under the proposed framework.

3.3 Proposed Model Framework

The study proposed the framework for stock price prediction based on 17 technical indicators which are used as input in Fig. 2.

For the empirical testing purpose, first, the data pertaining to Open-High-Low-Close (OHLC) for six stocks namely Torrent Pharma, Sun Pharma, Biocon, State Bank of India (SBI), Axis Bank, and HDFC Bank were extracted from the library

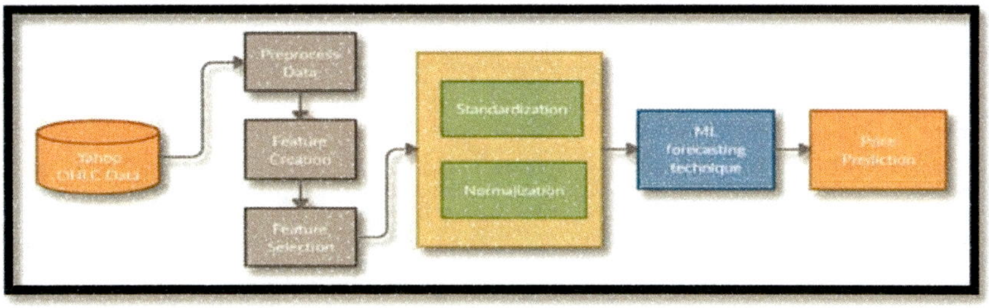

Fig. 2 Proposed framework

of Yahoo finance. Once the data is acquired, the processing of the data is done. Afterward, the feature creating task is performed. As a preprocessing step, various features selection methods are applied. Among three methods which are applied, filter method is used for the features selection purpose due to better result. In this method, statistical tests are applied to identify those features which have demonstrated strong relationship with the output variables. Consequently, right variables are identified as a part of feature selection process. Henceforth, standardization and normalization of the data are done. Three machine learning algorithms are used to generate accuracy results for the price prediction of the six stocks. The accuracy results are discussed in the findings and discussion section.

4 Finding and Discussion

Six stocks were identified from the Indian stock market to measure the price predication accuracy of three algorithms where 17 technical indicators were used to train the feature set of the model. Among the chosen machine learning algorithms, the random forest has exhibited superior prediction accuracy for five out of six stocks. The random forest algorithm has demonstrated more than 80% accuracy level for stock price prediction of Torrent Pharma, Sun Pharma, Biocon, and Axis Bank. Table 2 depicts the accuracy level of three algorithms tested for selected six stocks.

The study has indicated the higher accuracy level of the random forest algorithm in comparison with KNN and NN algorithm, and this has led to understanding that random forest may be more desirable while using technical indicators as inputs to train the feature sets with the aim of prediction of the stock price.

5 Conclusion

The covering time period of the spread of the Covid-19 pandemic for the empirical testing in this study has established a crucial background for testing algorithms. A

Table 2 Accuracy level of three algorithms for selected stocks

Stock name	ML technique	Accuracy
Torrent Pharma	KNN	72
	NN	79
	Random forest	82
Sun Pharma	KNN	78
	NN	75
	Random forest	84
Biocon	KNN	74
	NN	80
	Random forest	84
SBI	KNN	74
	NN	81
	Random forest	79
Axis Bank	KNN	78
	NN	77
	Random forest	87
HDFC Bank	KNN	70
	NN	70
	Random forest	76

total of 17 technical indicators which are prominently used by traders and investors to predict the stock price movement have been used for input purpose. Among the three algorithms such as KNN, NN, and random forest which are subject for testing in this study, the random forest algorithm has demonstrated a higher prediction accuracy level in five out of six selected stocks. The study observed that random forest may turn out to be a more robust algorithm for stock price prediction using technical indicators as inputs especially covering the volatile time period of the year 2020 due to the impact of Covid-19. Noticeably, this study has cross-validated favorably the results generated based on testing with the chosen machine learning algorithms with actual results.

This study may provide motivation to researchers to apply proposed framework to test repeatability of such results as well as extend the research work by expanding scope to other competing algorithms and comparing the accuracy level. Research work can also be developed toward building prediction models using the random forest as an algorithm of choice.

References

1. Malkiel, B.G., Fama, E.F.: Efficient capital markets: a review of theory and empirical work. J. Financ. **25**, 383–417 (1970)
2. Malkiel, B.G.: The efficient market hypothesis and its critics. J. Econ. Persp. **17**(1), 59–82 (2003)
3. Khaidem, L., Saha, S., Dey, S.R.: Predicting the direction of stock market prices using random forest (2016). arXiv abs/1605.00003
4. Hartono, J., Sulistiawan, D.: The market quality to technical analysis performance: intercountry analysis. Gadjah Mada Int. J. Bus. **16**(3), 243 (2014). https://doi.org/10.22146/gamaijb.5658
5. Agarwal, M., Khan, A., Shukla, P.: Stock price prediction using technical indicators: a predictive model using optimal deep learning. Int. J. Rec. Technol. Eng. **8**(2) (2019)
6. Basak, S., Kar, S., Saha, S., Khaidem, L., Dey, S.: Predicting the direction of stock market prices using tree-based classifiers. North Am. J. Econ. Finan. **47**, 552–567 (2019)
7. Khan, W., Ghazanfar, M.A., Azam, M.A., Karami, A., Alyoubi, K.H., Alfakeeh, A.S.: Stock market prediction using machine learning classifiers and social media, news. J. Ambient Intell. Humaniz. Comput. (2020). https://doi.org/10.1007/s12652-020-01839-w
8. Das, S., Mishra, S.: Advanced deep learning framework for stock value prediction. Int. J. Innov. Technol. Explor. Eng. **8**(10) (2019)
9. Gururaj, V., V.R.S, K.A: Stock market prediction using linear regression and support vector machines. Int. J. Appl. Eng. Res. **14**(8), 1931–1934 (2019)
10. Shen, S., Jiang, H., Zhang, T.: Stock market forecasting using machine learning algorithms. Comput. Sci. (2012)

A Brief Review on Multiple-Valued Logic-based Digital Circuits

Nagarathna and S. Jamuna

Abstract Traditional digital circuits use binary logic for implementation which requires more number of data lines to carry information. In contrast to binary, multiple-valued logic (MVL), digital circuits require less number of data lines to represent information. MVL uses more than two levels to represent a number; thereby reducing the number of interconnects. One of its kind is ternary logic, (known as radix-3 system) and uses three different voltage levels, whereas the other quaternary logic (radix-4 system) uses four different voltage levels to represent a number. Various digital circuits are implemented using MVL which potentially reduces the chip area, power consumption, and propagation delay. This article discusses about fundamental concepts of ternary, quaternary system and reviews some of arithmetic and logic circuits using ternary and quaternary logic. This paper also presents different methodologies used to implement arithmetic and logic operations using MVL. It is found that, due to the effective utilization of interconnects, MVL digital circuits provide better performance over binary digital circuits in terms of area, power, and delay.

1 Introduction

In recent years, there have been major advances in digital circuits which employ more than two discrete levels of signal. Such circuits are called as multiple-valued logic (MVL) circuits, which offers several potential opportunities for the improvement of present VLSI circuit designs [1, 2]. Multiple-valued logic uses radix-n number system to represent a number with $n > 2$ [2, 3]. Balanced and unbalanced are the two conventions used in MVL to represent signed and unsigned numbers, respectively. The number system with radix-3 is known as ternary logic, and radix-4 is known as quaternary logic [2, 5, 17]. The number of digits used to represent a number depends

Nagarathna (✉) · S. Jamuna
Dayananda Sagar College of Engineering, K S Layout, Bangalore, Karnataka, India
e-mail: nagarathna-tce@dayanandasagar.edu

S. Jamuna
e-mail: jamuna-ece@dayanandasagar.edu

T. Senjyu et al. (eds.), *ICT with Intelligent Applications*, Smart Innovation, Systems and Technologies 248, https://doi.org/10.1007/978-981-16-4177-0_34

on the radix of the number system. A large-valued number can be represented with comparatively less number of digits if the radix is more. Since MVL uses $n > 2$, number of digits required to represent a value is less compared to that of conventional binary system [2, 3].

In VLSI, 30–40% of the chip area is consumed by interconnects and 60–70% of latency because of path delay is due to interconnects [10, 17]. Higher number of interconnects leads to more chip area and performance degradation. As MVL uses less number of lines to represent large amount of information, the number of interconnects can be reduced if MVL is used for designing the digital circuits. MVL also increases the data processing capability per unit chip area. These MVL digital circuits are implemented using MOSFETs and carbon nanotubes (CNT) FETs, of variable threshold voltages [4, 16].

In this paper, two techniques of MVL ternary and quaternary logics are discussed with its features. It also discusses about design and implementation of digital circuits like adders, subtractors, multipliers using VLSI and compares the timing delay and chip area of the designs.

2 Ternary and Quaternary Logic

2.1 Ternary Logic

There are two types of MVL frequently used to construct the circuits: ternary and quaternary. As against to binary, ternary being one of the logic in MVL which uses three switching levels to represent a digital signal. This logic uses 0(Low level logic), 1(High level logic), and Z(Medium level logic) as switching levels. Each of these symbols is called "Trit" (Ternary digit) used to represent information in ternary. With all the advantages of MVL, ternary has additional advantages because of its radix-3. For effective representation of a value, radix $e(=2.718)$ is the optimum logic [2–4, 18]. As ternary uses three states, next integer is close to e and hence can be considered for representation of information in digital circuits over existing binary circuits. Using of three states reduces the number of digits used to represent a number by 36.90% compared to binary [3]. For example, to represent a number 50,000, it requires 10 trits in ternary over 16 bits in binary. The balanced ternary logic uses 0, 1, -1 states, whereas the unbalanced ternary logic uses 0, 1, 2 states to realize the digital circuits [4, 5, 18]. The conventional binary logic experiences a problem of consuming additional bit in signed representation. The major advantage of balanced ternary over conventional binary is that it uses same number of bits for representation of both positive and negative numbers because of its symmetry. In addition to all these, approximation of a value results in less loss of information in ternary as compared to that of binary [5].

2.2 *Quaternary Logic*

The quaternary logic uses four voltage levels and each of them is called as "quaternary digit." The number of digits used to represent a value reduces by 50%, when four states are used for representation [6]. For example, to represent a number 50,000, it requires only 8 bits. The ordinary quaternary logic uses four states 0, 1, 2, and 3 to represent a number. All arithmetic operations are carried out using modulo 4. In signed digit representation, digits can be attached with −(minus) for indicating negative numbers. In quaternary signed digit (QSD) system, both positive and negative numbers are represented by the digits from the set [−3, −2, −1, 0, 1, 2, 3]. In QSD, any decimal value D can be represented using n quaternary signed digits using Eq. 1 [3, 4, 13, 20].

$$D = \sum_{i=0}^{n-1} X_i 4^i \tag{1}$$

where $X_i \in [-3, -2, -1, 0, 1, 2, 3]$. QSD provides carry-free arithmetic operations [13, 14]. In contrast to ternary logic, quaternary logic is compatible with binary, as conversions from binary 2's compliment to QSD can be easily made as the radix 4 can be expressed as 2^2 [20].

Arithmetic logic unit (ALU) is a primary block of central processing unit (CPU) and floating point units (FPU). This paper reviews scholarly work carried out on different MVL digital circuits used in implementing ALU.

3 Ternary Logic Circuits

Da et al. have analyzed ternary number system and their benefits. They have also proposed new algorithms for arithmetic operations on ternary numbers. This paper highlights on advantages of ternary logic system such as reduction in the path delay due to the reduction in number of interconnections, which can be reduced by $1/\log_2 3$ and reduction in power dissipation [7, 17]. It presents algorithms for addition, subtraction, shifting, multiplication, and division of balanced ternary numbers using 3's complement method.

For example, if $A = 64$ in decimal, it is represented as 02,101 in ternary by using Eq. 2 [4].

$$0 \times 3^4 + 2 \times 3^3 + 1 \times 3^2 + 0 \times 3^1 + 1 \times 3^0 = 02101_{(3)} \tag{2}$$

If $B = 48$ in decimal, it is represented as 01,210 in ternary by using Eq. 3.

$$0 \times 3^4 + 1 \times 3^3 + 2 \times 3^2 + 1 \times 3^1 + 0 \times 3^0 = 01210_{(3)} \tag{3}$$

Table 1 Operations on two consecutive trits to obtain product [7]

Combination	Operation
22 or 00	Right shift the partial product
20 or 21	Subtract the multiplicand from partial product followed by Arithmetic right shift
00 or 01	Simply right shift the partial product
02	Add the multiplicand to partial product followed by Arithmetic right shift
10 or 11	Add multiplicand to partial product followed by simple right shift
12	Add twice of the partial product and multiplicand

The operations $A + B$ and $A - B$ are carried out as follows:

A+B	A-B
02101(64)	02101(64)
+01210(48)	+21020(3'scomplement of 48)
$11011_{(3)} = 112_{(10)}$	$100121 = 16_{(10)}$ (discarding carry)

The multiplication is carried out by considering each trit of the multiplier as MR[0] and the preceding trit as MR[1]. The partial product is initialized as 0, and the operations are performed based on the combination of 2 consecutive trits as given in Table 1.

Similarly, it proposes the algorithm for division as shown in Fig. 1 The performance of these algorithms is analyzed. The time complexity of addition, subtraction is $O(n)$ and multiplication, division is $O(n^2)$.

Dhande et al. have implemented 2 bit ternary ALU, which is capable of performing arithmetic operations on 2 bits of unbalanced ternary numbers. The different operations include addition, subtraction, multiplication, comparison, as well as logical operations ternary NAND, NOR, AND OR, ExOR, etc. The ALU is implemented using CMOS transmission gates. Each of the input (say X) is converted to X^0, X^1, X^2 by a decoder for ternary implementation. W and Z function lines are used to select the appropriate modules of ALU to carry out operations. Figure 2 shows the module of 2 bit ternary ALU (TALU). Each module of operation is designed using equations that are obtained by k-map technique for ternary function [8]. Comparative analysis of number of binary gates and ternary gates required for half adder, half subtractor, and full adder is carried out in terms of average propagation delay of MOSFET(τ) and as shown in Fig. 3.

Authors A. P. Dhande, et al. have implemented 1 bit adder, multiplier, Ternary logic gates using VHDL. The operations of these digital circuits are simulated [9]. 1-Trit balanced ALU is designed and developed using double-pole double-throw (DPDT) relays by Malachy Eaton, Member, IEEE. The device designed uses two pins bicolor LEDs to represent ternary output for $+1/-1$. Better visualization of TALU operations are reflected through DPDTs and LEDs. These circuits are energy efficient compared to CMOS circuits [10]. It finds application in cybernetic systems, in representing individual states of the system.

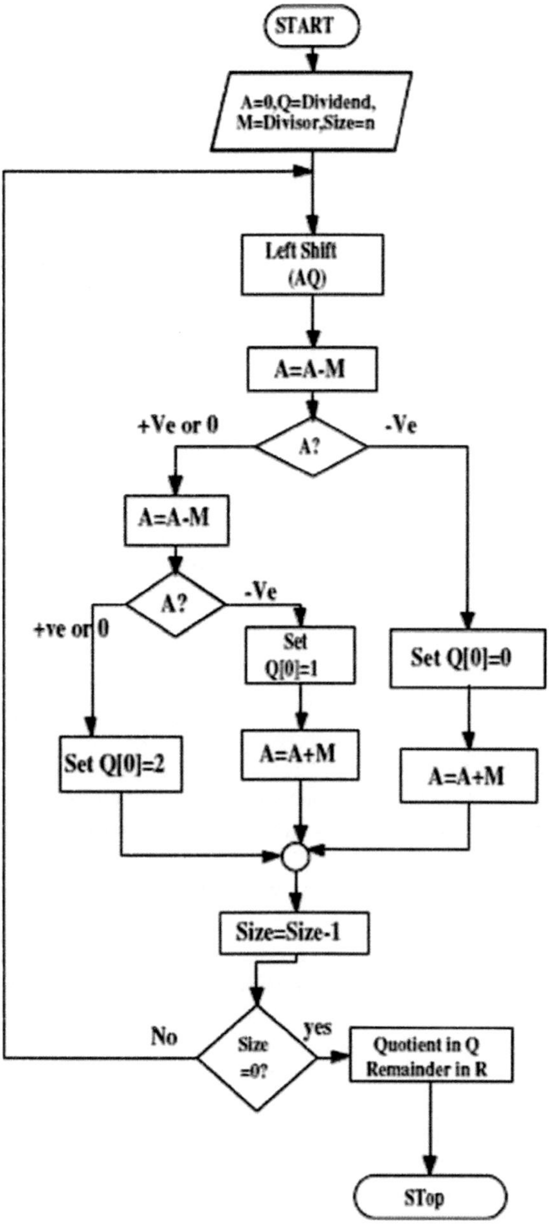

Fig. 1 Flowchart for division [7]

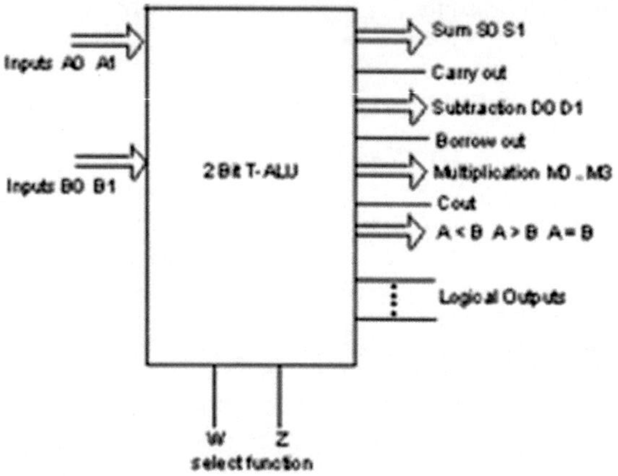

Fig. 2 Block diagram of T-ALU [8]

Fig. 3 Comparison table in terms of τ

Adder/Subtractor	Propagation delay (Scheme based on T gates)	Propagation delay (Scheme based on binary gates)
Half adder	6τ	7τ
Full adder	7τ	8τ
Full subtractor	6τ	-

Carbon nanotubes (CNT) are cylindrical molecules that consists of cylindrical structure [11, 16]. These are classified as single-wall CNT and multiwall CNT based on the diameters. Carbon nanotube field-effect transistors (CNTFETs) uses a single or an array of carbon nanotube(s) as a channel in place of silicon in conventional MOSFETs. CNTFETs have higher carrier mobility and low leakage current. Unlike CMOS, both *n*-type and p-type CNTFETs have same carrier mobility. When ternary circuits are implemented using CMOS, it requires multiple supply voltages. This can be avoided by deploying CNTFETs which has the feature of threshold voltage adjustments. Sneh Lata Murotiya et al. have implemented circuits for arithmetic and logic operations using CNTFETs. The circuits designed are analyzed and evaluated using Synopsys Hspice Simulator with the Stanford model of 32 nm CNTFET. These circuits exhibit less power delay product (PDP) over its CMOS counterparts [11].

Satish Narkhede et al. have presented designing of a 4 trit (ternary digit) processor using VHDL, which consists of data path unit and control unit. An instruction set of 21 instructions supporting various addressing modes and arithmetic and logic operations is developed. Figure 4 shows the block diagram of ternary processor, which has instruction decoder, array of registers, TALU as building blocks of data path unit. The necessary control signals to control the operations of these units are generated by ternary timing and control unit [12].

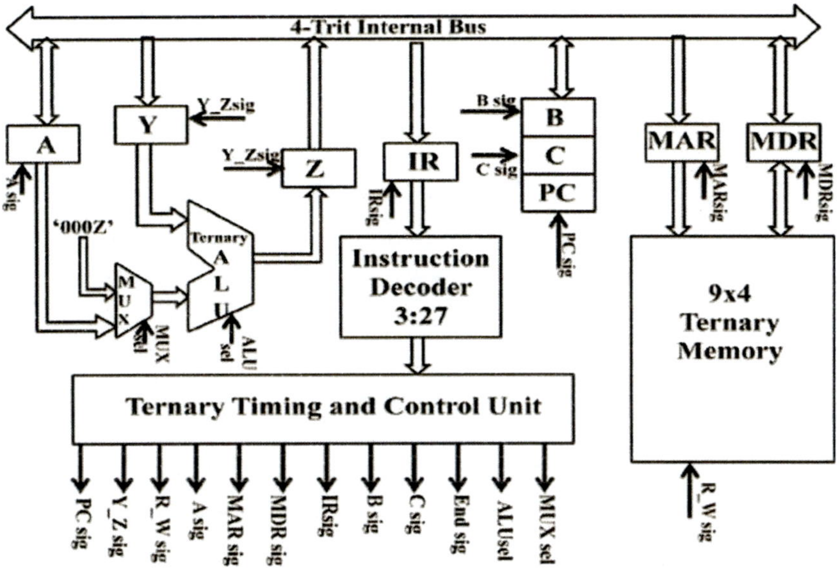

Fig. 4 Architecture of 4 trit ternary processor [12]

4 Quaternary Logic Circuits

Songpol Ongwattanakul et al. presented design of ALU to carry out addition and multiplication operations on quaternary signed Digits (QSD). The design is implemented on Altera FLEX10K using VHDL. The timing delay of *n* bit binary ripple carry and Altera megafunction adders are compared with *n*/2 bit quaternary adder. QSD adders provide consistent timing delay irrespective of input size and proved to be faster for larger values of *n* compared to other adders. The number of logic cells used by these adders is compared (Fig. 5) which shows the number of cells consumed by quaternary adders is relatively more [13]. The work presented by Nagamani A. N. et al. proposes ALU which includes QSD adder, multiplier, inverter, and comparator circuits. ALU is implemented using VHDL and synthesized using SPARTAN 3. Timing delay of QSD adder is compared with various types of adders as shown in Fig. 6. QSD adder is faster and exhibits constant time delay with respect to increase in input size. It also proposes novel algorithms for converting binary to QSD and vice versa [14].

A Galois field is a set of elements on which the operations of multiplication, addition, subtraction, and division are defined to satisfy certain basic rules. In the case of quaternary, it can be represented as GF(22) = GF(4) [19]. Modulo 4 adders and multipliers are implemented and simulated using HSPICE by Vasundara Patel et al. [15]. Fazel Sharifi et al. have implemented 2 bit ALU using CNTFETs, which is capable of performing operations on quaternary inputs. ALU is simulated, and the results are analyzed using Synopsys HSPICE simulator [16].

Number of bits	Ripple carry adder	Altera megafunction adder	QSD adder
4	125.00	125.00	55.55
8	62.89	125.00	45.24
16	51.02	125.00	45.04
32	21.97	78.74	39.84
64	12.46	31.94	34.60
128	5.89	16.52	41.49

Fig. 5 Timing delay comparison of various adders in [13, 14]

N BIT	RCA	CLA	QSD	CSA
8	17.650ns	17.044ns	15.729ns	17.585ns
16	28.806ns	27.312ns	15.729ns	28.741ns
32	51.117ns	47.847ns	15.729ns	51.052ns
64	95.739ns	88.918ns	15.729ns	95.675ns
128	184.984ns	171.058ns	15.729ns	184.92ns

Fig. 6 Timing delay comparison of various adders in [13, 14]

5 Conclusion

Ternary and quaternary logics are most widely used multiple-valued logic (MVL) to design digital circuits. A literature survey is carried out to study different methodologies used for implementing arithmetic and logic operations using MVL. It is observed that in contrast to binary, MVL circuits are faster especially when the size of the input is high. These arithmetic and logic circuits implemented using ternary and quaternary logic consume less area compared to binary circuits. CNTFET based implementation of MVL arithmetic circuits provides lower power consumption compared to CMOS. MVL arithmetic circuits can be considered for high-speed and low-power applications over conventional binary circuits. These circuits find potential application in handling big data.

References

1. Dubrova, E.: Multiple-valued logic in VLSI: challenges and opportunities. In: Proceedings of NORCHIP'99, pp. 340–350
2. Smith, K.C.: Multiple valued logic: a tutorial & appreciation. IEEE Trans. Comput. **21**(4), 17–21 (1988)
3. Hurst: Multiple-valued logic—its status and its future. IEEE Trans. Comput. **C-33**(12), 1160–1179 (1984)
4. Porat, D.I.: Three-valued digital systems. Proc. Inst. Electr. Eng. **116**(6), 947–954 (1969)
5. Stakhov: Brousentsov's ternary principle, Bergman's number system and ternary mirror-symmetrical arithmetic. Comput. J. **45**(2), 221–236 (2002)
6. Gupte, A., Goel, A.: Study of quaternary logic versus binary logic. In: Proceedings First Great Lakes Symposium on VLSI, Kalamazoo, MI, USA, pp. 336, 337 (1991)
7. Das, S., Dasgupta, P.S., Sensarma, S.: Arithmetic algorithms for ternary number system. Prog. VLSI Des. Test **7373** (2012)
8. Dhande, A.P., Ingole, V.T.: Design and implementation of 2 bit ternary ALU slice. In: 3rd International Conference: Sciences of Electronic, Technologies of Information and Telecommunications, March 17–21, 2005, Tunisia
9. Dhande, A.P., Jaiswal, R.C., Dudam, S.S.: Ternary logic simulator using VHDL. In: 4th International Conference: Sciences of Electronic, Technologies of Information and Telecommunications, 25–27 March 2007, Tunisia
10. Eaton, M.: Design and construction of a balanced ternary ALU with potential future cybernetic intelligent systems applications. In: 2012 IEEE 11th International Conference on Cybernetic Intelligent Systems (CIS), Limerick, pp. 30–35 (2012)
11. Murotiya, S.L., Gupta, A.: Design of CNTFET-based 2-bit ternary ALU for nanoelectronics. Int. J. Electron. 1244–1257 (2014)
12. Narkhede, S., Kharate, G., Chaudhari, B.: Design and implementation of an efficient instruction set for ternary processor. IJCSA **83**(16) (2013)
13. Ongwattanakul, S., Chewputtanagul, P., Jackson, D.J., Ricks, K.G.: Quaternary arithmetic logic unit on a programmable logic device. In: Proceedings of IEEE Conference (2001)
14. Nagamani, A., Nishchai, S.: Quaternary high performance arithmetic logic unit design. In: 14th Euromicro Conference on Digital System Design. Architectures, Methods and Tools, pp. 148–153 (2011)
15. Vasundhara Patel. K.S., Gurumurthy, K.S.: Multi-valued logic addition and multiplication in galois field. In: International Conference on Advances in Computing, Control, and Telecommunication Technologies, Trivandrum, Kerala, pp. 752–755 (2009)
16. Sharifi, F., Moaiyeri, M.H., Sharifi, H., Navi, K., Thapliyal, H.: On the design of quaternary arithmetic logic unit based on CNTFETs. Int. J. Electron. **7**(1), 1–13
17. Ozer, E., Sendag, R., Gregg, D.: Multiple-valued logic buses for reducing bus energy in low-power systems. In: IEE Proceedings—Computers and Digital Techniques, vol. 153(4), pp. 270–282, 3 July 2006
18. Dhande, A.P., Ingole, V.T., Ghiye, V.R.: Ternary Digital System: Concepts and Applications. SM Online Publishers LLC Copyright © (2014)
19. da Silva, R.C.G., Boudinov, H., Carro, L.: A novel Voltage-mode CMOS quaternary logic design. IEEE Trans. Electron Dev. **53**(6), 1480–1483 (2006)
20. Chattopadhyay, T., Sarkar, T.: Logical design of quaternary signed digit conversion circuit and its effectuation using operational amplifier. Bonfring Int. J. Power Syst. Integr. Circ. **2**(3) (2012)

An Approach of Enhanced PNCC for Resident Identification Applications

D. Duy, N. H. Dat, H. T. M. Tram, N. H. Son, and N. M. Son

Abstract The performance of applying voice control in home automation can significantly drop under multi-resident situations and noisy environments. It therefore requires some appropriate approaches for smart home applications to address the problem of resident identification. Voice recognition, which explores characteristics of voice, is a potential biometric modality for such problem in smart home. In this paper, the power-normalized cepstral coefficient (PNCC) of voice biometrics is applied to identify individuals in smart home. A new technique of power-law nonlinearity and an algorithm of noise suppression based on asymmetric filtering are used to enhance feature extraction and reduce environmental noise. This proposed approach extremely reduces error rate and achieves high performance on different data sets.

1 Introduction

Determining who is performing what action is important and useful for monitoring smart homes. Based on such current resident's behavior, the control system can make proper decisions and control end devices to give suitable feedback. However, taking the personalized actions in situation of multiple residents is really a challenge to researchers, especially in case of no individual identification. There have been a number of researches investigating the multi-occupancy in ambient intelligent

D. Duy (✉) · N. H. Dat · H. T. M. Tram · N. H. Son · N. M. Son
VNUHCM -University of Information Technology, Ho Chi Minh City 70000, Vietnam
e-mail: duyd@uit.edu.vn

H. T. M. Tram
e-mail: tamhtm.12@grad.uit.edu.vn

N. M. Son
e-mail: sonnm@uit.edu.vn

T. Senjyu et al. (eds.), *ICT with Intelligent Applications*, Smart Innovation, Systems and Technologies 248, https://doi.org/10.1007/978-981-16-4177-0_35

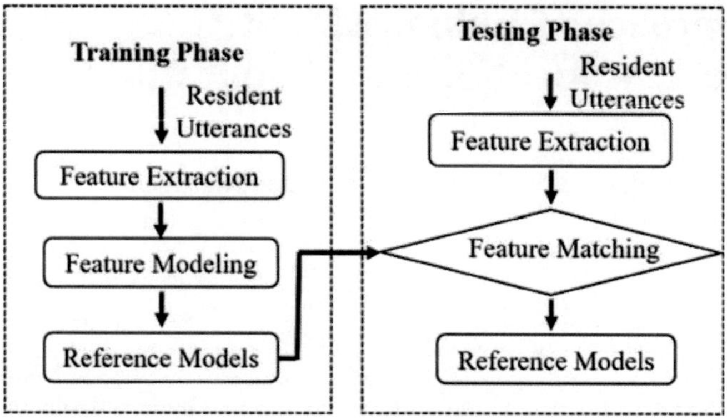

Fig. 1 Basic resident identifying process using voice biometrics

environments. In order to recognize multi-residents' activity, sequence models are employed to perform prediction over time, which is known as temporal modeling. As showed in [1], artificial intelligence algorithms in machine learning such as Bayesian networks and recurrent neural networks are encountered. Typical models are found in [2–5] in which alternative forms of hidden Markov model [6] and conditional random field [7] are implemented.

On the other hand, non-temporal models are also introduced to identify modeled activities of multi-residents [3, 5]. This approach explores data sets of photographs collected from smart homes during a certain period to imply activity prediction. Both temporal and non-temporal models are using temporal features as their input. However, since these models are examined using different data sets, it is difficult to achieve comparative results in effectiveness. In 2019, Son et al. [8] tried to adopt these models onto the same data sets and then investigate the effectiveness and efficiency of each candidate.

Although achieving remarkable results, existing solutions are using records (photographs) over time of residents in the considered environment, which rises uneasy problems regarding the personal privacy. Therefore, this research tries to use voice biometrics as an identifying mean for smart home applications. Specifically, the DARPA Resource Management (RM1) database is used for human voice identifying with the training set of 1600 clean speech of utterances and the testing set of 600 clean or degraded speech of utterances. In addition for the testing on word, we use the DARPA Wall Street Journal (WSJ) 5000-word database in which the WSJ0SI-84 is applied to the training phase and the WSJ0 5 K is applied to the testing phase.

The created voice recognition system is then constructed in two phases: training and testing. Figure 1 depicts the process of the resident identifying system in smart home using voice biometrics. In the training phase, each utterance of given residents will get through a feature extraction algorithm to extract the voice features which is modeled as person identification for the successive phase of testing. In the testing phase, a certain utterance will be featured and then compared with the reference identifications obtained in the training phase to make a proper decision. Several classification techniques (normally used for pattern matching) such as dynamic time warping (DTW) [17], neural networks [18, 19], and hidden Markov models (HMMs) [20–22] can be applied for the feature matching step.

In this paper, the power-normalized cepstral coefficient (PNCC) of voice biometrics is applied to identify individuals in smart home. A new technique of power-law nonlinearity and an algorithm of noise suppression based on asymmetric filtering are proposed to enhance feature extraction and reduce environmental noise. This contribution is appropriate to apply to smart home applications with noise.

The rest of the paper is organized as follows. Section 2 describes the proposed approach for feature extraction process. Section 3 introduces the implementation of the voice identifying system using the proposed feature extraction. Finally, the experimental results come in Sect. 4.

2 Feature Extraction Process

Toward the resident identifying problem in smart homes, the main concern of this study is to deal with noisy situations. Multiple speakers giving speech at the same time, white noise in background sound, and music interference are considered, for example, of noise environments. For this purpose, we employ the PNCC for feature extraction. The introduction of PNCC [9] is to obtain a set of practical features for the problem of speech recognition. Because of its acoustical variability, PNCC approach achieved stronger robustness without reducing performance on undistorted speech and computational complexity comparable with MFCC [10, 11] and PLP coefficients.

PNCC provides several innovations for speech processing as follows:

- PNCC introduces a effective combination of medium-time and short-time processes for speech recognition. Specifically, the "medium-time" processing takes frames in a duration of 50–120 ms and the traditional short-time Fourier analysis processes frames of 20–30 ms. This combination enables analysis of the environmental degradation parameters and is helpful in conventional speech recognition systems.
- For every time frame and frequency bin, using "asymmetric nonlinear filtering" form allows to estimate the level of the acoustical background noise efficiently.
- Introduction of signal processing block is effective to realize temporal masking with a similar approach.
- That the log nonlinearity of MFCC is replaced with a power-law nonlinearity approximates the nonlinear relation between signal intensity and auditory nerve firing rate. This is considered as short-time signal intensity measuring over frequency.
- PNCC introduces a concept of computationally efficient realization of the algorithm to allow "online" real-time processing. There is no considerable non-causal look-ahead of the input signal required for PNCC calculation.

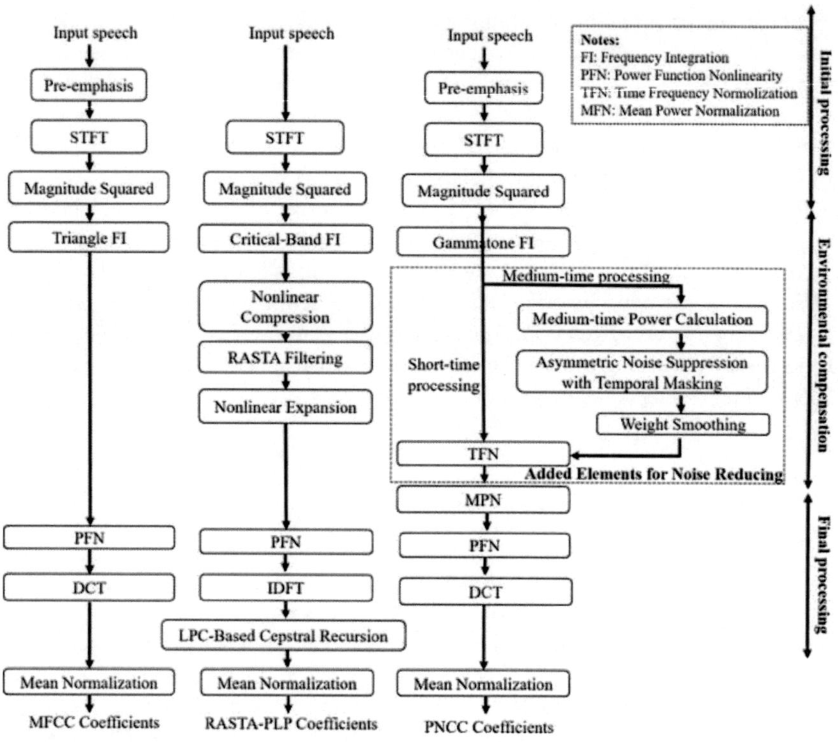

Fig. 2 Structure of MFCC, PLP, and PNCC algorithms

Figure 2 shows the structure of the proposed PNCC algorithm in comparison with MFCC processing [12] and PLP processing [13]. As one of the major improvements, the proposed PNCC processing consists of a function of redesigned nonlinear rate intensity and series of extra mechanisms to eliminate impacts of acoustical background noise based on medium-time analysis.

In detail, PNCC, MFCC, and PLP processes are almost in alignment at initial processing stages, except the difference at the frequency analysis stage where Gammatone filters are performed in PNCC rather than triangular and critical and filters in the other approaches. Then, series of nonlinear time-varying operations are conducted for environmental compensation. Different from MFCC and PLP, the proposed PNCC processing suggests a longer duration temporal analysis in order to reduce noise and increase robustness for reverberation. Certainly, in the medium-time processing, we suggest operations of calculating medium-time power, suppressing asymmetric noise, and smoothing weights to deal with noise before the stage of time–frequency normalization. It is important that the technique of temporal masking is exploited to reduce noisy elements. Finally, the three algorithms finish with similar final stages. Note that in our analysis the power-law nonlinearity is carefully selected with exponent of 1/15.

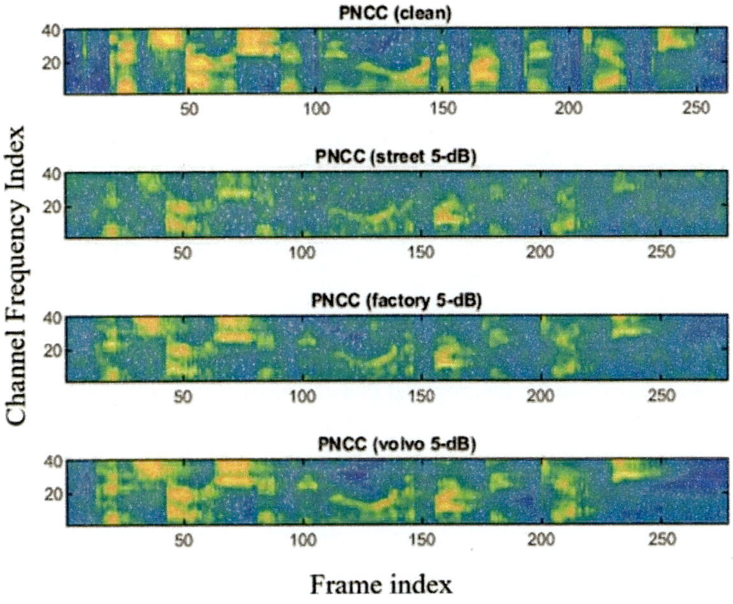

Fig. 3 Spectrograms of utterances composed of 9 words in clean environment, street noise, factory noise, volvo car noise added

Figure 3 illustrates an example of spectrograms of utterance recorded in different environments. The yellow spectrum represents the filtered sound, and the blue one represents the environmental noise of the input signals. Depending on the different filter media and levels, different results are obtained. According to the assessment, volvo noise environment is recorded with the highest accuracy followed by factory noise and street noise. However, the differences are lightly small and they are quite stable compared to other methods. This is a very important point for the modeling process afterward.

3 Voice Detection System

Voice detection system is implemented for ambient intelligent environments such as smart home. In the voice detection system, voice recognition is a critical operation which leads to successive control operations. Basically, voice recognition consists of two phases: verification and identification [14]. The verification phase decides whether an identity claim of speaker should be accepted for detection or rejected. The identification phase processes the given utterance and determines the proper speaker using a database (references) of the registered voices. In the database, the

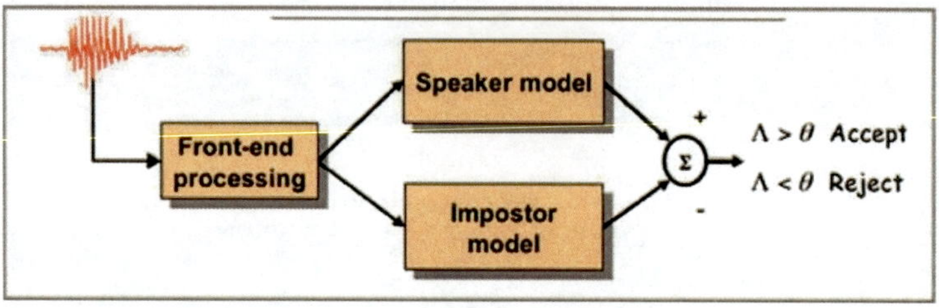

Fig. 4 Speaker verification

voice of registered people like the home owner and family members is modeled based on the feature extracted by the proposed PNCC processing.

3.1 *Voice Verification*

The voice verification phase is first taking the targeted person's voice sample [15] for verification. This phase applies a model of 1:1 matching. Specifically, one input voice is going to be matched to a voice model prepared in the database. When an utterance comes and needs verification, it will be processed for feature extraction and modeling at the front-end processing phase. The model of the coming utterance is then compared with the models of both speaker and imposter in the database using the pattern matching scheme depicted in Fig. 4. The acceptance decision is released according to the comparison between the matching rate Λ and a selected threshold θ. If the match rate is greater than the selected threshold, the identity claim is accepted. Determining an appropriate threshold is therefore substantial task toward the recognition accuracy of this system. Based on this approach, it is apparent that the system is safer from the impostors with a higher threshold, whereas if the selected threshold is relatively low, it causes the risk of misidentification. That is, the system may reject the correct person and accept the unverified people to enter the system.

A basic model of speaker verification is shown in Fig. 4. Accordingly, a given utterance is processed with feature extraction and modeling at the front-end processing step before compared with the registered models of claimed speakers. The models of imposter speakers are also included into the reference for comparison step to detect unverified cases. The voice acceptance is decided by comparing the matching ratio of speaker and imposter, which is indicated by the likelihood ratio statistic (Λ), with the selected threshold (θ). If $\Lambda > \theta$, the utterance is accepted. Otherwise, if $\Lambda < \theta$, then the utterance is rejected.

3.2 Voice Identification

The second phase of the voice detection process is voice identification whether a unknown voice is detected for a certain person. This action is done by comparing the featured model of the coming voice with the registered voices in the database [16]. The comparison model is mapping one to many (1:N model), which is illustrated in Fig. 5.

The problem of voice detection is classified into several situations. On the one hand related to the database of registered voices, there exist closed-set and open-set situations. In the situation of closed set, the person requiring the detection has his voice registered into the database before, whereas in the case of open set, the person has not registered yet at the detection moment. On the other hand, in accordance with characteristic of registered voices, the detection is considered as text-independent and text-dependent paradigms. Namely, the sentences spoken by the person at detection time and at registration time are fixed or unfixed. If the sentence is fixed, it is recognized for text-dependent paradigm. On contrary, the detection system is a text-independent one when the involved sentence is unfixed. Due to the different situations, the rate of successful detection may be varied a lot.

In the identification process (in Fig. 5), at the front-end processing step, the coming voice is processed for feature extraction, modeling, and verifying. If verified, the model of the involved voice is then processed for detection step. The detection process is a bit different for the closed-set and open-set cases. For identification of the closed set, the identified speaker is decided as the model with the highest scoring matching. For identification of the open set, the registration state of the speaker is determined by comparing the model with the highest scoring matching and a predefined threshold. In the implementation of this work, the paradigm of open-set text-independent identification is considered for the smart home environment.

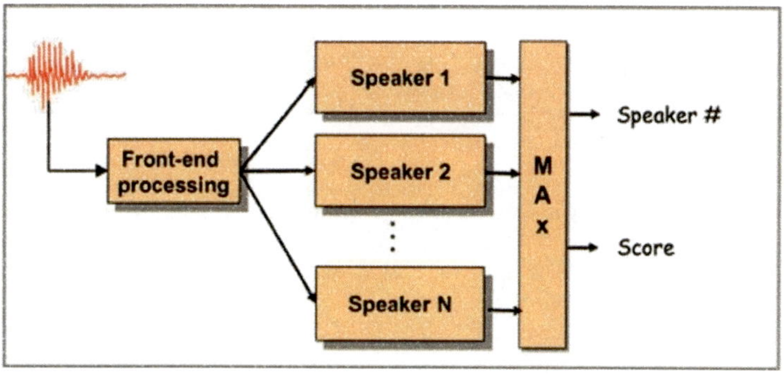

Fig. 5 Voice identification

4 Experimental Results

4.1 Accuracy

The recognition accuracy of the proposed PNCC is assessed in comparison with MFCC feature and RASTA-PLP feature processes. Table 1 shows the recognition accuracy of candidates in different noise environments at high level of noise. The results show that the proposed PNCC is significantly better in recognition accuracy than both MFCC process and RASTA-PLP process for all scenarios under evaluation. Similar results are obtained in cases of medium and low level of noise as shown in Tables 2 and 3, respectively. Apparently, by applying temporal masking techniques to the process of PNCC, noise is effectively suppressed and therefore stably high recognition accuracy is achieved.

4.2 Computational Complexity

Computational complexity of feature extraction process is estimated for MFCC, RASTA-PLP, and PNCC algorithms. In this estimation, it is assumed that the window length is equal to 25.6 ms and the interval between successive windows is 10ms. Additionally, the sampling rate is decided at 16 KHz and the 1024-pt FFT is applied to each frame. The estimation results are shown in Table 4.

Table 1 Recognition accuracy in different noise environments at high level of noise

Comparison criteria	PNCC (%)	MFCC (%)	RASTA-PLP (%)
White noise (SNR = 5 db)	78.7	17.2	18.3
Street noise (SNR = 5 db)	82	62	48.4
Music noise (SNR = 5 db)	68.3	42	32.6
Interfering speaker (SNR = 5 db)	40	31	18.5
Reverberation (time = 0.9 s)	39.5	30.2	18.4

Table 2 Recognition accuracy in different noise environments at medium level of noise

Comparison criteria	PNCC (%)	MFCC (%)	RASTA-PLP (%)
White noise (SNR = 10 db)	82.7	37.2	38.3
Street noise (SNR = 10 db)	83	81.8	69.5
Music noise (SNR = 10 db)	82.3	68.2	52.6
Interfering speaker (SNR = 10 db)	62.2	48.1	40.5
Reverberation (time = 0.5 s)	50.5	41.2	20.4

Table 3 Recognition accuracy in different noise environments at low level of noise

Comparison criteria	PNCC (%)	MFCC (%)	RASTA-PLP (%)
White noise (SNR = 15 db)	90.7	62.2	62
Street noise (SNR = 15 db)	88.5	83.8	80.5
Music noise (SNR = 15 db)	82.3	81.3	74.6
Interfering speaker (SNR = 15 db)	80	72.9	61.8
Reverberation (time =1.2 s)	30.5	24.2	17.4

Table 4 Computational complexity in different noise environments at the low level of noise

Item	MFCC	PLP	PNNC
Pre-emphasis	410		410
Windowing	410	410	410
FFT	10240	10240	10240
Magnitude squared	512	512	512
Medium-time power calculation	N/A	N/A	40
Spectral integration	958	4955	4984
ANS filtering	N/A	N/A	200
Equal loudness pre-emphasis	N/A	512	
Temporal masking	N/A	N/A	120
Weight averaging	N/A	N/A	120
IDFT		504	
LPC and cepstral recursion	N/A	156	
DCT	480		480
Sum	13010	17289	17516

Overall, with a very simple stage of environmental compensation (see Fig. 2), MFCC processing has lowest computational complexity. The proposed PNCC process introduces a longer duration analysis, which results in higher time complexity. Certainly, PNCC processing costs about 34.6 and 1.31% more than MFCC and PLP, respectively. Besides, PLP processing is about 32.9% more costly than the MFCC one.

4.3 Running Time

In addition to the computational complexity, candidates are also assessed by the running time on a typical hardware configuration. The hardware platform is set up with normal configurations. The recorded results show that PNCC process spends around 80 ms for feature extraction on the configured hardware platform. It is believed that the execution of PNCC algorithm could be improved on specific hardware configuration such as embedded system.

5 Conclusion

We have introduced an improved approach of PNCC for voice identification in multi-resident situations. The proposed approach combines different attributes to deal with the problem of noise in different noisy environments. The considered attributes include auditory processing including the rate-level nonlinearity, temporal and spectral integration, and temporal masking. Especially, the implement of temporal masking technique effectively reduces extra noise and reverberation. As a result, the proposed PNCC processing achieved a significantly higher recognition accuracy compared with MFCC and PLP processes in trade-off with only about 33% computation more increasing. However, many advanced algorithms of speech recognition have been introduced recently to solve noisy problem. Most of solutions investigate the behavior models of the human auditory system in which noisy speech is injected. The future works are expected to investigate deep learning algorithms to model the features taken from PNCC together with ours Vietnamese voice database. The proposed method of voice biometrics is then probably applied to practical applications with variety of noise sources.

Acknowledgements This work is supported by the University of Information Technology - Vietnam National University Ho Chi Minh City under grant No. D1-2020-09.

References

1. Benmansour, A., Bouchachia, A., Feham, M.: Multioccupant activity recognition in pervasive smart home environments. ACM Comput. Surv. **48**(3), 34:1–34:36 (2015)
2. Cook, D.J.: Learning setting-generalized activity models for smart spaces. IEEE Intell. Syst. **27**, 32–38 (2012)
3. Alemdar, H., Ertan, H., Incel, O.D., Ersoy, C.: Aras human activity datasets in multiple homes with multiple residents. PervasiveHealth'13. ICST, Brussels, Belgium, pp. 232–235 (2013)
4. Chen, R., Tong, Y.: A two-stage method for solving multi-resident activity recognition in smart environments. Entropy **16**(4), 2184 (2014)
5. Alemdar, H., Ersoy, C.: Multi-resident activity tracking and recognition in smart environments. J. Ambient Intell. Hum. Comput. **8** (2017)
6. Rabiner, L.R.: Readings in speech recognition. A tutorial on hidden markov models and selected applications in speech recognition. Morgan Kaufmann Publishers Inc., San Francisco, pp. 267–296 (1990)
7. Lafferty, J., McCallum, A., Pereira, F.: Conditional random fields: probabilistic models for segmenting and labeling sequence data. In: Proceedings of 18th International Conference on Machine Learning. Morgan Kaufmann. pp. 282–289 (2001)
8. Son, N.T., Dung, N., Tung, S.N., Son, V.X., Long, H., Qing, Z., Mohan, K.: On multi-resident activity recognition in ambient smart-homes. In: Artificial Intelligence Review, Springer (2019)

9. Kim, C., Stern, R.M.: Power-normalized cepstral coefficients (PNCC) for robust speech recognition. IEEE/ACM Trans. Audio Speech Lang. **24**(7), 1315–1329 (2016)
10. Cornaz, C., Hunkeler, U., Velisavljevic, V.: An Automatic Speaker Recognition System. Lausanne, Switzerland (2003)
11. Kumar, P., Vardhan, K., Krishna, K.: Performance evaluation of MLP for speech recognition in noise environments using MFCC & wavelets. Int. J. Comput. Sci. Commun. (IJCSC) **1**(2), 41–45 (2010)
12. Muda, L., Began, M., Elamvazuthi, M.: Voice recognition algorithms using mel frequency cepstral coefficient (MFCC) and dynamic time warping (DTW) techniques. J. Comput, 139–140 (2010)
13. Dave, N.: Feature extraction methods LPC, PLP and MFCC in speech recognition. Int. J. Adv. Res. Eng. Technol. 1–4 (2013)
14. Reynolds, D.A.: An overview of automatic speaker recognition technology. In: Proceedings of IEEE International Conference on Acoustics, Speech, and Signal Processing (ICASSP) (2002)
15. Fazel, A., Chakrabartty, S.: An overview of statistical pattern recognition techniques for speaker verification. IEEE Circ. Syst. Mag. 62–81 (2011)
16. Togneri, R., Pullella, D.: An overview of speaker identification: accuracy and robustness issues. IEEE Circ. Syst. Mag. **11**(2), 23–61 (2011)
17. Xihao, S., Miyanaga, Y.: Dynamic time warping for speech recognition with training part to reduce the computation. in: International Symposium on Signals, Circuits and Systems ISSCS2013, pp. 1–4 (2013)
18. Pawar, R.V, Kajave, P.P., Mali, S.N: Speaker identification using neural networks. World Acad. Sci. Eng. Technol. (2005)
19. Bhushan, C.K.: Speech recognition using artificial neural network. A Rev. Int. J. Comput. Commun. Instrum. Eng. **3**(1) (2016)
20. Shahin, I., Botros, N.: Text-dependent speaker identification using hidden Markov model with stress compensation technique. in: Proceedings IEEE Southeastcon 98 Engineering for a New Era, Orlando, pp. 61–64 (1998)
21. Maesa, A., Garzia, F., Scarpiniti, M., Cusani, R.: Text independent automatic speaker recognition system using mel-frequency cepstrum coefficient and gaussian mixture models. J. Inf. Secur. 335–340 (2012)
22. Campbell, G.P.: Speaker recognition: a tutorial. Proc. IEEE **85**(9), 1437–1462 (1997)

A Study of Mathematical Backgrounds Behind the Digitally Communicative Switches Design

Man Singh, Sanyam Gupta, and Laxminarayan Das

Abstract Dial tone frequency modulation, demodulation process through a microcontroller sound card and suitable codec program are a conventional telecommunication technique used in telephone service systems. Practically, the dial tone frequency modulation, demodulation tasks are performed through a finite sequence obtained from the Fourier transformation of a sine(nx) or cosine(nx) sequence functions. The paper considered mathematical algorithms are applied to encode and decode the linked matrix element switch positioning row and column sampling frequencies to form the finite Fourier sine and cosine transforms with the help of MATLAB program to experiment a sound beat formed sample bit image spectra. The obtained tuning sound forms are audible from the computer microphone speaker if the sine transform spectra images are processed through the computer memory processor input signal formatted sound encoder decoder card. The paper also emphasizes the numerical technique's brief concept to establish the congruence modulo relations to represent switch matrix number and symbols.

1 Introduction

Digitally communicative switches are applied in automation systems. The computer network applied digital TV channel regulation, asynchronous transmission system in petrol pump operation and banking system are some of the instances where digitally communicative switches are designed. In this article, we have studied the mathematical transformation, functions and relations behind the digitally communicative switches design.

M. Singh (✉) · S. Gupta · L. Das
Delhi Technological University, Delhi 110042, India
e-mail: maan@spm.du.ac.in

S. Gupta
e-mail: sanyam_phd2k18@dtu.ac.in

L. Das
e-mail: lndas@dce.ac.in

T. Senjyu et al. (eds.), *ICT with Intelligent Applications*, Smart Innovation, Systems and Technologies 248, https://doi.org/10.1007/978-981-16-4177-0_36

1.1 The Fast Fourier Transform

A sound sample can be represented in a computing system's sound card device by encoding, decoding operation of the finite numbered sinusoidal signal's binary-coded decimal number sequences. A sinusoidal signal computation device sound codec determines $\cos(nx)$ and $\sin(nx)$ formal function values and represents it as a relational form in the circle circumference arc $2n\pi$ and a unit radius. In the process, the circle $(\cos(nx), \sin(nx))$ circumference arc is traversed one cycle if $n = 1$ and memorises the functional value coordinates $\cos(nx)$ and $\sin(nx)$ are in circle circumference form with the unit radius. Thus for every *n*th value traverse, the unit circle's circumference arc routes one point value, and the *n*th value repeats at $nx\pi$ position values. If the arc traversed path values are calculated frequently, the finite Fourier transformation algorithm is applied. Discrete time frequency function relating Fourier series and the finite discrete Fourier transform synthetic notation of the Fourier series, and Euler's identity complex form are mentioned in the works of Cooley–Tukey [1] and Frigo–Johnson [2].

$$f(x) = \frac{A_0}{2} + \sum_{n=1}^{\infty}[A_n\cos(nx) + B_n\sin(nx)] \tag{1}$$

The equivalent form in relations in terms of complex exponentials by using the Euler's identity,

$$e^{ijx} = \cos(jx) + i\sin(jx)$$

We can write equation (1),

$$\begin{aligned} f(x) &= \sum_{j=0}^{\infty}\left(c_j e^{ijx} + c_{-j}e^{-ijx}\right) \\ &= 2c_0 + \sum_{j=1}^{\infty}\left[\left(c_j + c_{-j}\right)\cos(jx) + i\left(c_j - c_{-j}\right)\sin(jx)\right]. \end{aligned} \tag{2}$$

From Eqs. (1) and (2), we have

$$c_j = \frac{A_j - iB_j}{2},\ c_{-j} = \frac{A_j + iB_j}{2}$$

When $f(x)$ is real valued, it is easy to show that $c_0 = \underline{c}_0$ and $c_j = \underline{c}_{-j}$, where the bar represents complex conjugate.

For integers *j* and *k*, it is true that

$$\int_0^{2\pi}\left(e^{ikx}e^{ijx}\right)\mathrm{d}x = \int_0^{2\pi}\left(e^{i(k+j)x}\right)\mathrm{d}x = 0, \quad \text{for } k \neq -j$$

and

$$\int_0^{2\pi}\left(e^{ikx}e^{ijx}\right)\mathrm{d}x = \int_0^{2\pi}\left(e^{i(k+j)x}\right)\mathrm{d}x = 2\pi, \quad \text{for } k = -j$$

1.2 *Finite (Discrete) Fourier Transform*

The sequence of N complex numbers $x_0, x_1, x_2, \ldots, x_{N-1}$ is transformed into the sequence of N complex numbers $X_0, X_1, X_2, \ldots, X_{N-1}$ by the microprocessor using discrete Fourier transform (DFT), and the expression is

$$X_k = \sum_{n=0}^{N-1} x_n e^{\frac{-2\pi ikn}{N}}, \quad k = 0, 1, 2, \ldots, N-1$$

$$x_n = \frac{1}{N}\sum_{k=0}^{N-1} X_k e^{\frac{2\pi ikn}{N}}, \quad n = 0, 1, 2, \ldots, N-1$$

1.3 *FFT Decomposition*

- The N-point time domain signal decomposed to N time frequency domain signal and each composed of a single point.
- Now calculate DFT of the N single points for single point signals, the frequency spectrum is equal to the time domain signal.
- Combine the N frequency spectra into one N-point frequency spectrum.

  Example: $N = 8$

Decomposition of 8 Symbolic Switches

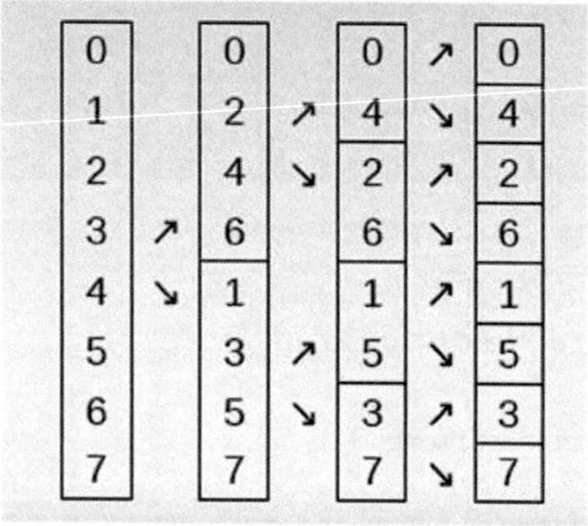

1.4 Touch—Tone Dialling

When we talk about the technological development of the telephone set, the rotary dial was used in the initial stages. Slower dialling was one major disadvantage associated with the rotary dial. It took 12 s to dial a 7-digit number on a rotary dial. The step-by-step switching elements of the Strowger switching system cannot respond to rates higher than 10–12 pulses per second.

It uses the DTMF technology, prior to which the pulse dialling technique was used. In the pulse dialling technique which is also called a **loop disconnect** technique, repeated connecting and disconnecting of the lines are done, like clicks of a switch; this is interpreted by the exchange as the dialled tone, according to the number of clicks with the introduction of the common control subsystems into switching exchanges, and there came the feasibility for higher rates of dialling. Hence, a new system called the **touch-tone dialling** was developed in telephony to replace the rotary dial; this was considered to benefit the customer with higher speed. This also removed the disadvantages of limited usage and limited signalling capacity along with lower speed. Touch-tone dial is a transform and synthesis of the dual tone multi-frequency system (DTMF).

1.5 Modulation, Demodulation of Dial Tone

The dial tone time frequency **modulation** is the process of encoding information into a transmitted signal form. The **demodulation** is the process of extracting information from the encrypted transmittance signal. A modem at the far end demodulates the audio signal to recover the data stream from encoded or modulated signal.

2 Scheme

The press button or touch screen switches are commonly used in telephonic conversion sound wave or typed message's electromagnetic impulse wave transmission. The alphanumeric symbolic group switches in telephone telegraphy instrument transact signal formed voice or editable text message electromagnetic impulse form by using a finite Fourier sine transformation. In this paper we explain array formed digital symbolic matrix switches those are used in digital equipment touch button dialling. Each digit or chosen symbols touch after visual spectra can be traced by using the Fourier sine or cosine transforms which are described in the paper.

2.1 *Matrix Representation of Switches*

The press button telephone dialling pad acts as either a 4 by 3 or 4 by 4 matrix associated with alphanumeric switches, and each switch position has a row and column frequency allotment. In case there are 12 press button switches in the telephone dialling pad, the switch matrix row frequencies are f-row $\{800, 830, 982, 1071\}$ and column frequencies are f-col $\{1339, 1466, 1607\}$. The labels of each press button switch can be 1, 2, 3, 4, 5, 6, 7, 8, 9, 0, *, #. The labels position corresponding row index and column index can be denoted with the following format, which is mentioned in the works of Cleve Moler [6]

Switch label '0' has row index $k = 4$ and column index $j = 1$;
Switch label '*' has row index $k = 4$ and column index $j = 2$;
Switch label '#' has row index $k = 4$ and column index $j = 3$;

Otherwise, if d is the digit within 1, 2, ..., 9, the column index is $j = \text{mod}(d - 1, 3) + 1$ and row index is relating with the rule $k = (d - j)/s + 1$.

In case there are 16 push button switches in the dialling pad, the matrix row frequencies are f-row $= \{800, 830, 982, 1071\}$ and matrix column frequencies are f-col $= \{1339, 1466, 1607, 1763\}$. There are sixteen labelled push button switches, and switch labels can be 1, 2, 3, 4, 5, 6, 7, 8, 9, A, B, C, D, 0, *, #, where the labels A, B, C and D positional values are 10, 11, 12 and 13, respectively. In general, if s is a character that labels one of the buttons on the keypad, the corresponding row index and column index can be found with the following forms.

switch label s,	row index,	column index
case '0',	$k = 4$,	$j = 2$
case' $*$ ',	$k = 4$,	$j = 3$
case '#',	$k = 4$,	$j = 4$

Otherwise,

$$d = s - `0', \quad j = \mathrm{mod}(d-1,4), \quad k = \frac{(d-j)}{4} + 1$$

End

3 MATLAB Code for Finite Fourier Transformation

The MATLAB code concerning finite Fourier transformation of the sine sequence is given below:

```
% Touch Tone Dialling Algorithm
clc
clear
fr = [800 830 982 1071];
fc = [1339 1466 1607 1763];
s =
menu('Input','1','2','3','4','5','6','7','8','9','A','B','C','D','0','*','#
')
switch s
% k is row
% j is column
    case '0', k = 4; j = 2
    case '*', k = 4; j = 3
    case '#', k = 4; j = 4
    otherwise,
        d = s-0
        j = mod(d-1,4)+1
        k = (d-j)/4+1
end
Fs = 32898
t = 0:1/Fs:0.015;
y1 = sin(2*pi*fr(k)*t);
y2 = sin(2*pi*fc(j)*t);
y = (y1 + y2)/2;
plot(t,y,'r')
xlabel('Time')
ylabel('Frequency')
title('Sine Spectra of Switch 1')
sound(y,Fs)
```

16 Symbolic Switches

OUTPUT

s =
1
d =
1
j =
1
k =
1
Fs =
32898

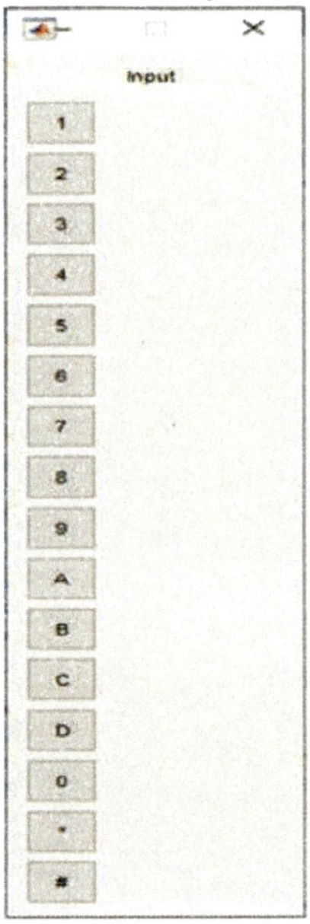

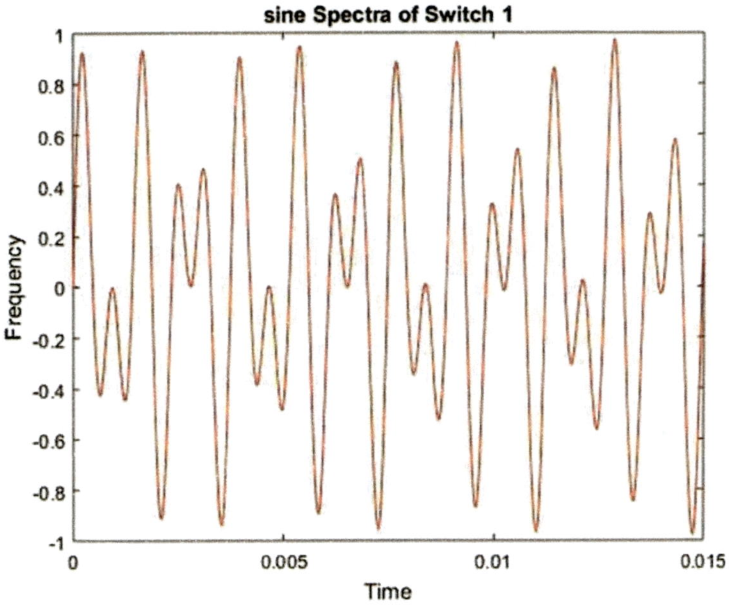

3.1 The MATLAB Code Concerning Finite Fourier Transformation of the Cosine Sequence is Given Below

```
% Touch Tone Dialling Algorithm
clc
clear
fr = [800 830 982 1071];
fc = [1339 1466 1607 1763];
s =
menu('Input','1','2','3','4','5','6','7','8','9','A','B','C','D','0','*','#
')
switch s
% k is row
% j is column
    case  '0', k = 4; j = 2
    case  '*', k = 4; j = 3
    case  '#', k = 4; j = 4
    otherwise,
        d = s-0
        j = mod(d-1,4)+1
        k = (d-j)/4+1
end
Fs = 32898
t = 0:1/Fs:0.015;
y1 = cos(2*pi*fr(k)*t);
y2 = cos(2*pi*fc(j)*t);
y = (y1 + y2)/2;
plot(t,y,'g')
xlabel('Time')
ylabel('Frequency')
title('Cosine Spectra of Switch 1')
sound(y,Fs)
```

OUTPUT

s =

1

d =

1

j =

1

k =

1

Fs =

32898

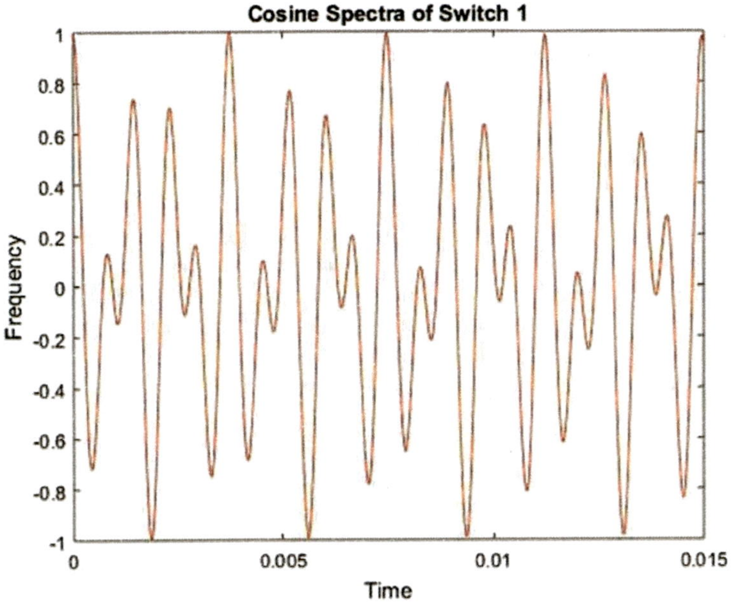

4 Observation

Numerical analysis of the quadratic spline and the discrete finite Fourier sine transform is observed from the obtained sine transform spectra lines by utilizing MATLAB program. The interpolating quadratic spline curves are the polynomial ordinate's functional values fitted to make a connected smooth continuous spectra line. The coordinates of the abscissas and subintervals' functional values are lying in the spectra maps. The nodes are the points where values of the interpolating spline arcs are specified and connected to interval endpoint ordinates functional values to form a continuous waveform. In the quadratic spline functions, there are $n + 2$ interpolation conditions and $2n - 2$ conditions from the continuity of the quadratic spline and its derivative, the independent variable t is chosen from the first and the last subinterval endpoints of the domain knots, and also, the midpoint between the knot to knot lies in the interval.

We executed the MATLAB-coded program using an electronic computer and the output from the program displayed dial-able switch button patterns as shown in the figure mentioned above. While we touched the switch button 1 with the cursor, the dial tone rang and was heard from the computer speaker. The sine and cosine spectra of each dialable switch button are also resulted from the program output. We have considered only switch button '1' for Sine and Cosine spectra but there are sixteen switch button on the electronic keypad. If we considered all sixteen switches button then the size of the paper will be increasing drastically. So for precise the size of the paper, we used only switch '1' for sine and cosine spectra.

5 Conclusion

In our calculation, we determine the dial tone by press button switches from a set of sixteen symbolic switches, and each switch matrix bottom position's row and column frequencies finite discrete Fourier sine, cosine transform spectra mappings is obtained from the discussed numerical techniques and the outcome from the computer MATLAB coding and discrete Fourier sine, cosine transforms output mappings. The sine transform spectra line control points' numeric domain decompositions are made by fitting the B spline and Bezier curves. In the numerical mathematics and computing book [4], the description of spline and Bezier curve fittings is found. The discrete time signal processing [5] text may be a good reference to decide the control points in the sine transform spectra line decomposition.

The numerical analysis involved in these studies is relating to the number decomposition procedure of the symbolic switching system. The decomposed numbers are represented as line graph forms.

Acknowledgements The authors acknowledge gratitude to the unknown reviewers who have reviewed the paper with patience. We have incorporated their comments in the revision of the paper.

References

1. Cooley, J.W., Tukey, J.W.: An algorithm for the machine calculation of complex Fourier series. Math. Comput. **19**, 297–301 (1965)
2. Frigo, M., Johnson, S.G.: FFTW: an adaptive software architecture for the FFT. In: Proceedings of the 1998 IEEE International Conference on Acoustics Speech and Signal Processing, vol. 3, pp. 1381–1384 (1998). http://www.fftw.org
3. Van Loan, C.: Computational frameworks for the fast fourier transform. SIAM, Philadelphia (1992)
4. Cheney, W., Kincaid, D.: Numerical Mathematics and Computing. Books Cole publishing Company
5. Oppenheim, A.V., Schater, R.W., Buck, J.R.: Discrete—Time Signal Processing, 2nd ed. Prentice Hall, Upper Saddle River, New Jersey, 07458 (1998)
6. Moler, C.: Numerical Computing with Matlab, 2nd Repriting, pp. 257–262. SIAM, 15 February 2008

Stay Home and Stay Safe with the Budgeting Tool a One-Stop Finance Tracker

R. Radhika, Anagha Praveen, G. Gokul Krishna, Adithya Anand, and T. Anjali

Abstract COVID-19 has devastated people's lives across the globe. Declining economy, job losses and market uncertainty have created restlessness in public. These unprecedented times have led to concerns about the future and fear of restricted income. To overcome these strenuous and tough times, individuals and families need to take certain vital steps in order to manage and monitor their budgets judiciously. Though COVID-19 will continue to have its adverse consequences globally, a few smart decisions may help common people maintain healthy and stable finances. So it's essential that we keep a track of the transactions every day. To do so, we've proposed a web application with the help of HTML and CSS to make it visually appealing. It is capable of tracking the minute transactions that we make. The main objective of this application is for the user to monitor their expenses. The user can store his income and expenses based on the transaction, date of the income or expense and the category of items. This could be achieved by linking the database (MySql) with the PHP script. We have developed a web based application which is used to manage a users daily expense in a more structured and organized manner. Thus, we can minimize the hand-operated computations which would've been involved otherwise and also track expenditure. This application asks the user to provide his income which is further split across various categories in order to calculate his total expenses per day, and these results will be stored for each user separately. We also included a feature to scan and store the images of the bill in the database which could be downloaded at any given point of time. In addition to this, we have integrated a calculator which is an optional feature for the user to do the extra calculations within the website rather than going for an external calculator.

1 Introduction

"A budget is telling your money where to go instead of wondering where it went" goes the proverb. The economic and social disruption caused by the pandemic is

---

R. Radhika (✉) · A. Praveen · G. G. Krishna · A. Anand · T. Anjali
Department of Computer Science and Engineering, Amrita School of Engineering, Amrita Vishwa Vidyapeetham, Amritapuri, India

T. Senjyu et al. (eds.), *ICT with Intelligent Applications*, Smart Innovation, Systems and Technologies 248, https://doi.org/10.1007/978-981-16-4177-0_37

devastating. covid has caused a lot of disruption on account of which many businesses couldn't operate in the usual way. As a consequence of which numerous people have lost their job or part of their income. It is likely that there will be more job losses in the months to come. Hence, keeping a regular track of our expenses has become all the more crucial in current circumstances. Every penny has to be monitored and spent judiciously. It is the responsibility of every individual to plan out their expenses especially in these uncertain times when we all have just begun to adapt to the new normal, in order to ensure that their money is being spent wisely and also that they end up saving money. Since the world is now progressing towards the digital era, automation is being implemented almost everywhere. Expenses are an inevitable part of our everyday life and with the enormous attention being given to automation, we have developed a web based expense tracking system to help make managing expenses a lot easier. The principal objective of this application is that it allows the user to monitor all the expenses made by them, based on the transaction date of the expenses or category of items. We are developing a web based application which is used to track the user's daily expenses in a more efficient and organized manner [1]. Thus, we can minimize the hand-operated computations and also keep a track of the expenditure. A user is only allowed to spend a specific amount of money per day, and this threshold is calculated on the basis of the income of that particular user. These results will be unique for all users depending on their income amount.

2 Related Works

In the field of applications related to expense managing, the ratio of web application is less compared to Android Applications. S. Chandini1, T. Poojitha2, D. Ranjith3, V.J. Mohammed Akram4, M.S. Vani5, and V. Rajyalakshmi of Department of Computer Science and Engineering, Mother Theresa Institute of Engineering and Technology, Palamaner, Andhra Pradesh has developed an Tracking system which maintains daily,weekly and monthly income and expenses in a specific way [2]. Smart expense management model developed by S. Yadav, R. Malhotra and J. Tripathi which can monitor the cash outflow. This model can capture the incomes and outcomes of a household owner. They have also integrated Smart home with a Smart expense management system [3]. E-Expense Android app developed by Shahed Anzarus Sabab1, Sadman Saumik Islam2, Md. Jewel Rana and Monir Hossain of Department of Computer Science and Engineering, Northern University Bangladesh is an automated system that can capture the expense bills using OCR techniques and can track the regular expenses of the user [4], Time Scheduling and Finance management is an app developed by J.L. Yeo, P.S. JosephN, K.A. Alezabi and H.C. Eaw FBM, UCSI University, Malaysia, allows students to efficiently manage their financial funds by adding various income and expenses from transportation, food, etc. It also has notes featured along with it [5]. P. Woodson and Z. J. Wang designed an online clothing store using php and MySql, separate administrative and admin logins are present in this application [6].

3 Proposed System

Our project is designed in a utilitarian manner which comes handy with all the necessary tools available all at one place. The application checks if the user is legitimate by cross checking the credentials entered in the login page with the one stored in the database. If someone is new to this application platform, they will have to sign in, in order to create their account and get going. Once the account has been created, they can login to this platform in the similar manner as an existing user. Once the user successfully logs in, they will find four broad categories available which include:

- Income
- Expense
- Edit Profile
- Calculator
- Reminder

The income category consists of various options which include salary, rent (in case the user owns a house), investments, etc. Thereby giving a lucid image as to where the amount comes from. Once you fill these details and click the enter button, they will get stored in the database showing us the date we had received it on and also the time at which we had entered these details. The expense category has another set of options including investments, rent, education, food, bills, etc. The unique feature available here is that we can scan the bills and upload them so as to refer to them in the near future. All the expenses will be stored in the database corresponding to each category. The edit profile button here helps you assist with the changes of your personal details including change in your mobile number, updation of password or e-mail id and so on. There is also a provision of an inbuilt calculator which is a vital tool to perform calculations and also for the user to keep a check on his/her expenses. The reminder tab here enables the user to set reminders of paying their bills and clearing their dues if any on the dates set by them [7]. Since it is a web based project and is prone to a lot of attacks like DOS (Denial of Service), SQL Injection, etc., which are inevitable penetration and automation testing should be applied (Figs. 1 and 2).

4 Methodology

- This application provides a password-based secured access to the user thereby ensuring authentication.
- There is a dashboard providing an easy and simple view of all the functionalities in one place.
- The user is required to add income to different categories provided.
- The user is also required to add his/her expenses corresponding to different categories provided such as food, clothing, etc.

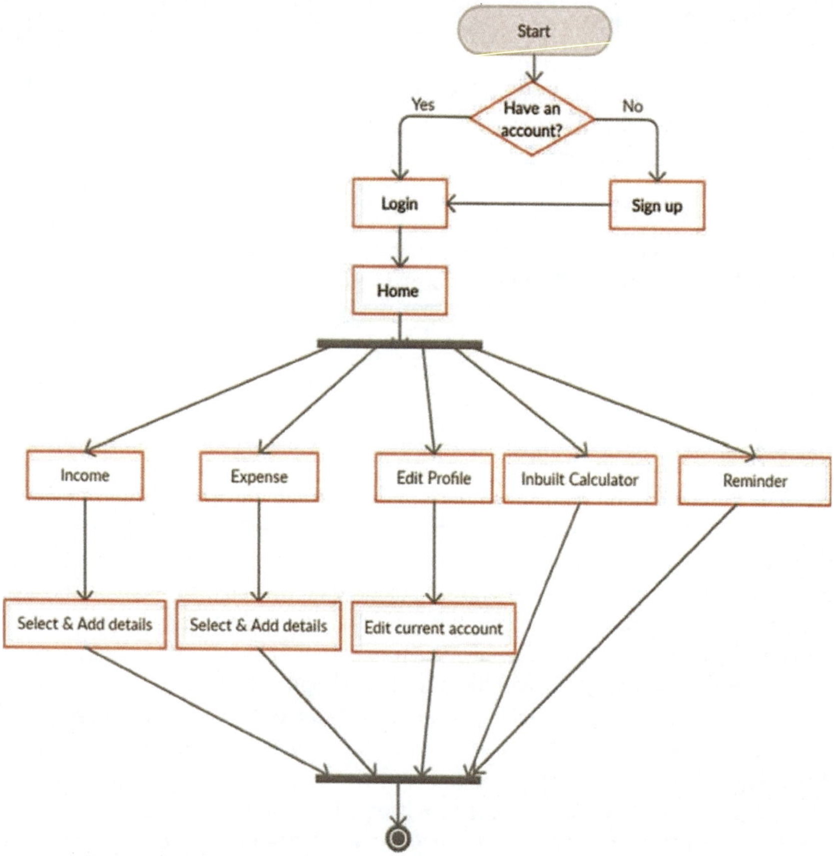

Fig. 1 Activity diagram

- The user may add/remove their categories of income or expenses as per their requirement.
- This application will also allow the users to set their monthly budget in order to monitor and control their budget.
- It also provides the facility to set alerts on budgets regarding expenses (eg: paying electricity bill).
- There is also an inbuilt calculator provided so that the calculations become a lot easier and the user has everything available at one place. Thereby, making this application a one-stop destination for all the requirements of the user (Fig. 3).

Here are the steps of the process functioning briefly explained:

STEP 1: The user enters the URL of the website they wish to visit.
STEP 2: The web-browser then requests for those dynamic pages.

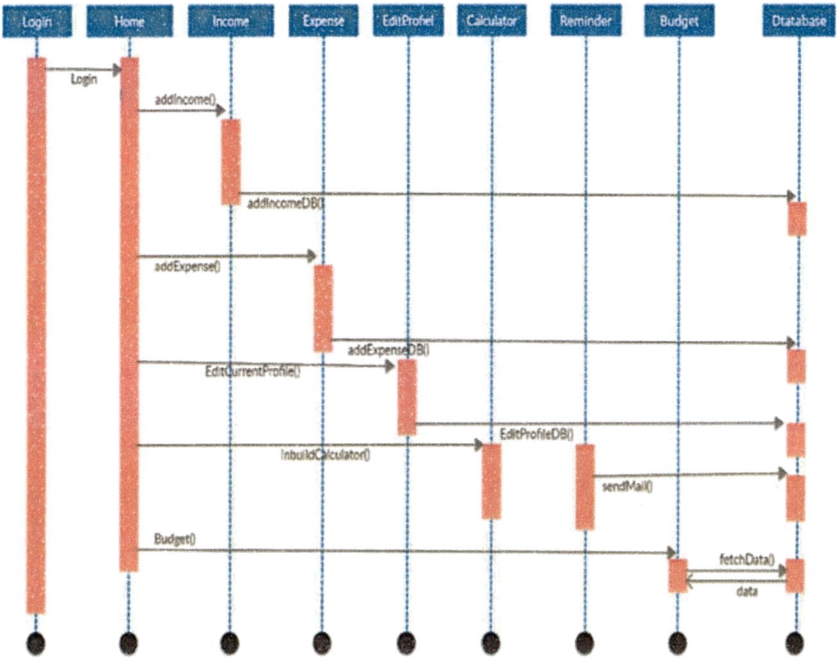

Fig. 2 Sequence diagram

STEP 3: The web server finds those pages and forwards them to the application server
STEP 4: Application server then scans those pages for instructions.
STEP 5: The driver executes the query against the database.
STEP 6: The record-set is returned to the driver.
STEP 7: Driver passes this record-set to the application server.
STEP 8: Application server inserts data into the pages and passes these pages to the web server.
STEP 9: Web server then sends these finished web pages to the web-browser and the user is able to view it.

5 Advantages

- This application allows the user to record their monthly expenses in an efficient and organized format.
- The application helps us minimize the hand-operated computations by letting us segregate our income and expenditure into different categories and thereby gives us a lucid idea to keep a track of the expenditure.

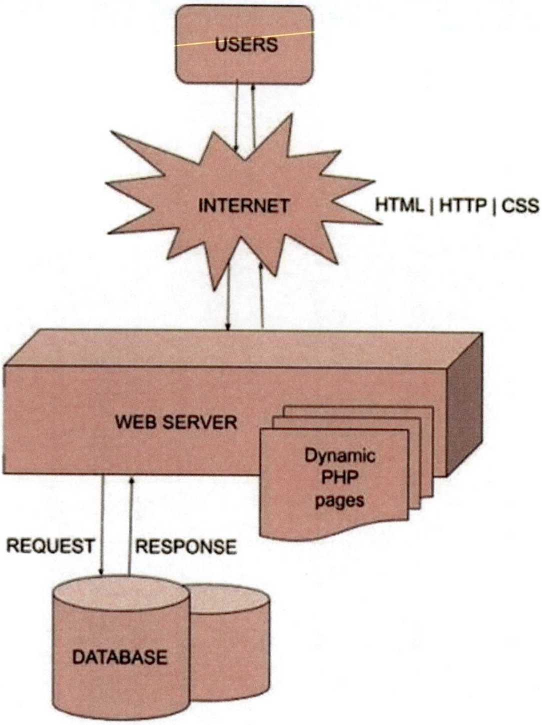

Fig. 3 Overall view of the process functioning

- Users can come up with their income to figure out the total expenses (allowed) per day, and these results will be unique for each user.
- Users may add/remove their categories of income and expenses as per their motive.
- Users can also add the bills by attaching the image of the bill so as to refer to it in the near future.
- Users can also get to know about their daily threshold amount.
- There is also an inbuilt calculator provided so that the calculations become a lot easier and the user has everything available at one place. Thereby, making this application an all-purpose destination to formulate a user's expense.

6 Result

This expense managing system makes it easier for the users to efficiently compute and manage their expenses on a daily basis. There is also an inbuilt calculator provided for the user to use. There is also an intriguing feature added through which we can scan our bills and refer to them in the future if needed, thereby keeping you sorted. Another vital and eye catching feature is a mechanism through which the application sends an e-mail to the mentioned id reminding the user to pay their bills before the due date approaches in order to stay on track and avoid any sort of late fee payment. Also there is a daily threshold amount being set which the user shouldn't exceed in order to manage and save the expenses as desired. The different categories available in the expense and income webpage make it easier for a user to track their expenses and how were they spent. This web application is definitely a one-stop destination for any user (Figs. 4, 5, 6 and 7).

We connected to the database using the mysqli_connect method. We have the fields named $usname and $Paassword for storing the user credentials, i.e the username and the password entered by the user and posting it into the database. Next, we have the sql queries to insert the values into the database (Fig. 8).

Here, we have introduced a couple of categories and features with the help of two separate php files namely expense.php and expenseimg.php for storing numerical values and bill images, respectively, into the database (Fig. 9).

Here, we have an option to scan and insert images of the bills and store it in the database for later use which could also be downloaded anytime, here we insert the bill images into a table named tbl_billings and finally with an if condition it shows the updation of status of image inserted (Figs. 10 and 11).

The captivating feature in this application is setting reminders. The user can select a date and an auto-generated e-mail reminding the user to pay his bills or clear his dues will be set in charge thereby helping the users to be on track and up to date with their expenses.

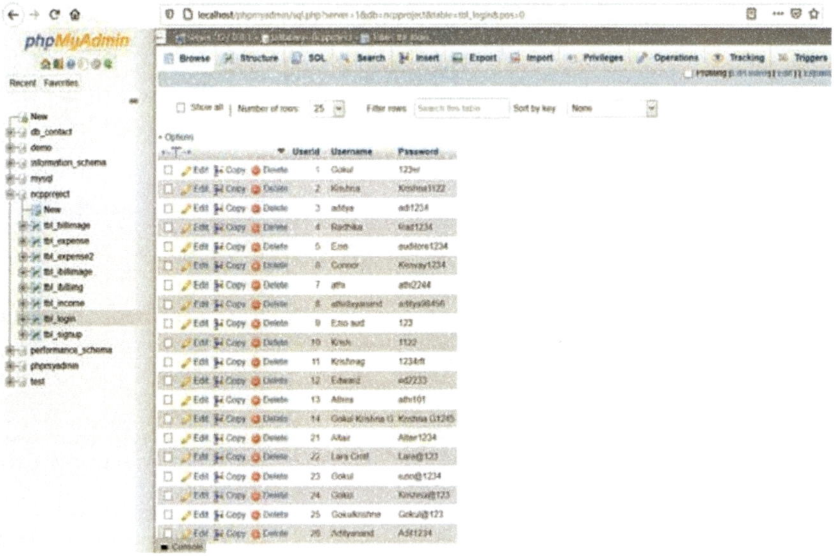

Fig. 4 Backend database for storing user credentials

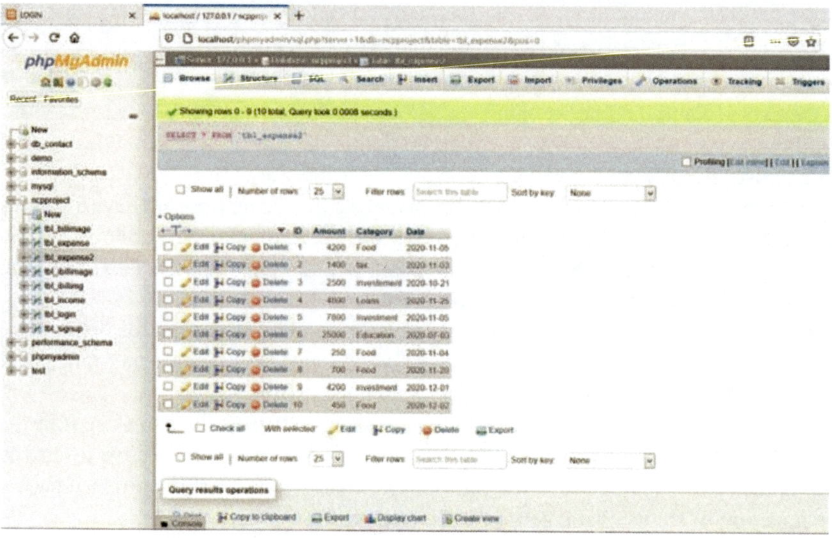

Fig. 5 Backend database for storing user expenses

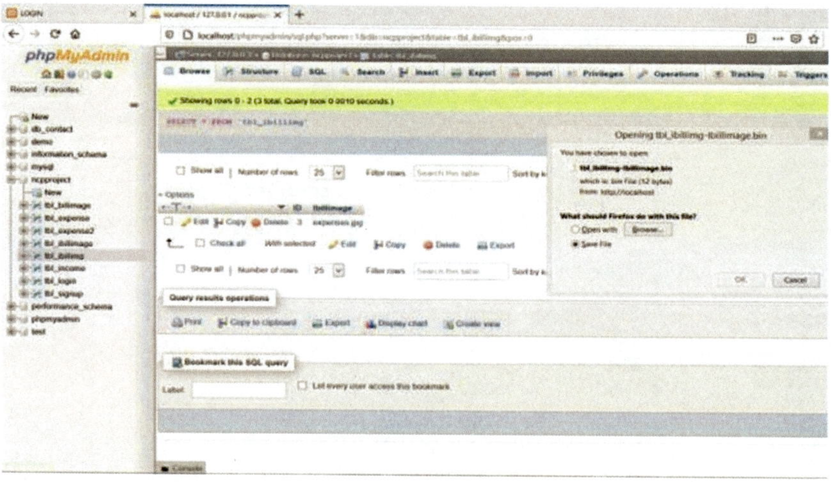

Fig. 6 Backend database bill storage with download option

7 Conclusion

Save For The Rainy Day. This we all need to ensure and for which we need to re-evaluate our budget. Families need to re-evaluate their budgets and sort expenses. A clear demarcation between necessary and extravagance spend needs to be done

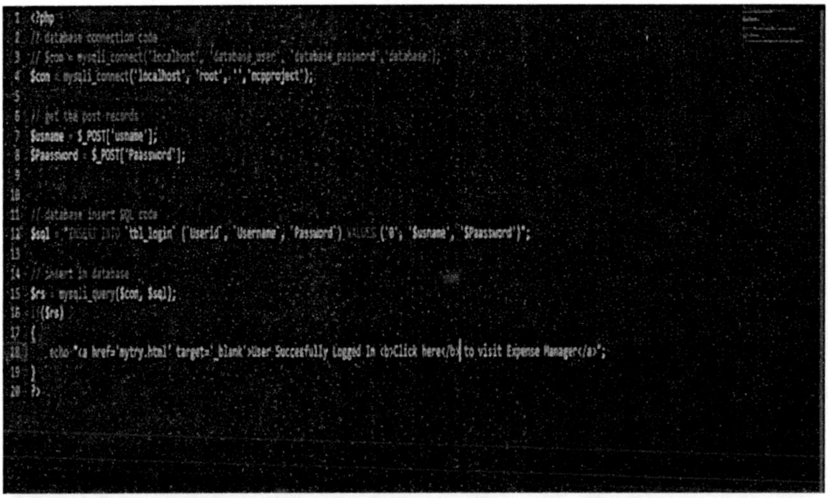

Fig. 7 Database connection for login page using php

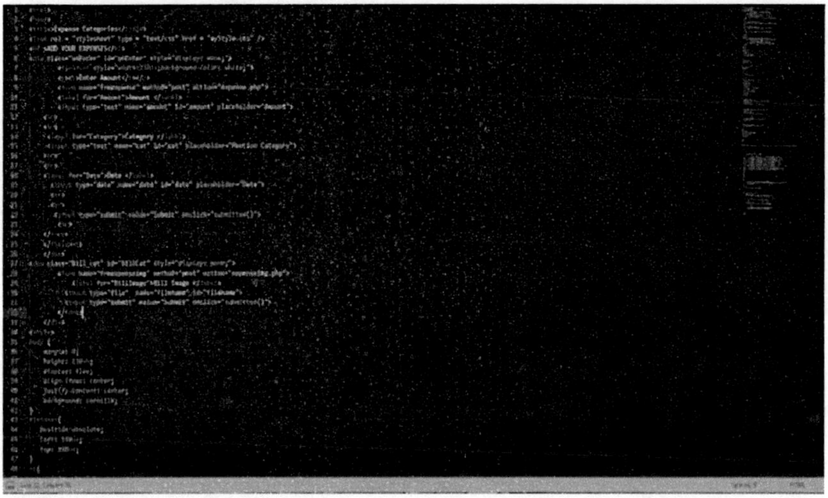

Fig. 8 Html code for expense page

and needless to say the latter needs to be curtailed and to achieve this, our web application comes to the rescue. As long as you are able to monitor and supervise your earnings and detail each expense, you will be able to gain a better perspective of your financial status. That means it provides a lot of lucidity and also helps making more informed choices and avoids any additional financial stress whenever possible. A sound budget is a major part of overall security for you and your family. A useful

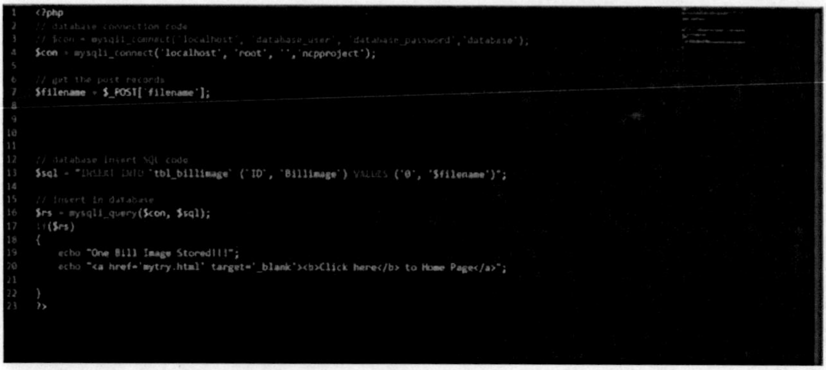

```php
<?php
// database connection code
// $con = mysqli_connect('localhost', 'database_user', 'database_password','database');
$con = mysqli_connect('localhost', 'root', '','ncpproject');

// get the post records
$filename = $_POST['filename'];




// database insert SQL code
$sql = "INSERT INTO `tbl_billimage` (`ID`, `Billimage`) VALUES ('0', '$filename')";

// insert in database
$rs = mysqli_query($con, $sql);
if($rs)
{
    echo "One Bill Image Stored!!!";
    echo "<a href='mytry.html' target='_blank'><b>Click here</b> to Home Page</a>";

}
?>
```

Fig. 9 PHP code for storing bill images

```html
<!DOCTYPE html>
<html>
<head>
    <meta charset="utf-8">
    <meta http-equiv="X-UA-Compatible" content="IE=edge">
    <title>Send you a reminder</title>


<body>
    <div class="main">
        <div class="info">Send You a Reminder!</div>
        <form action="mail_handler.php" method="post" name="form" class="form-box">
            <label for="name">Name</label><br>
            <input type="text" name="name" class="inp" placeholder="Enter Your Name" required><br>
            <label for="email">Email ID</label><br>
            <input type="email" name="email" class="inp" placeholder="Enter Your Email" required><br>
            <label for="date">Date</label><br>
            <input type="date" name="date" class="inp" placeholder="Enter Date" required><br>
            <label for="message">Message</label><br>
            <textarea name="msg" class="msg-box" placeholder="Enter Your Message Here..." required></textarea><br>
            <input type="submit" name="submit" value="Send" class="sub-btn">
        </form>
    </div>
</body>
</head>
</html>
```

Fig. 10 HTML code for setting reminders

project is developed using the web application. After entering the expense, there is a threshold amount set for the user to spend on a daily basis. Also there is a reminder which can be set and a mail will be received by the user on the set date in order to pay the bills before the due date in order to avoid any sort of fine and also the bills can be scanned and uploaded for reference in the future. Hence, by this project, we can keep our expenses organized.

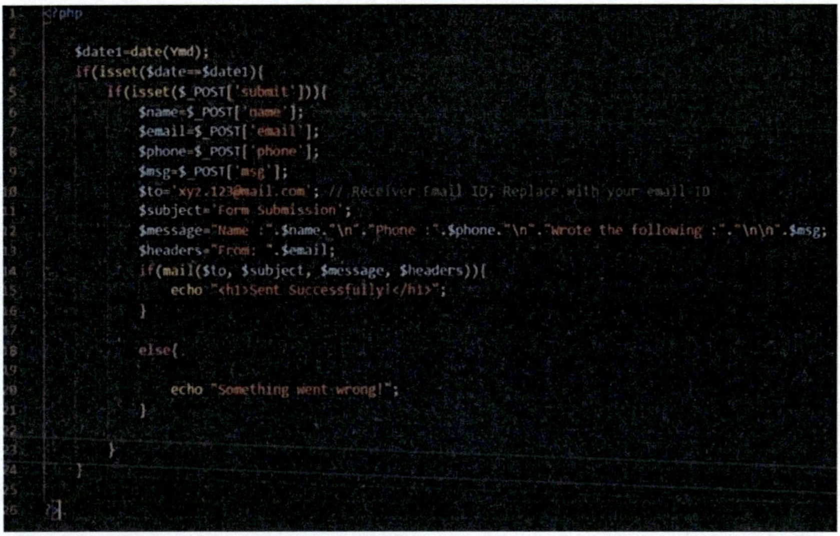

```php
<?php

    $date1=date(Ymd);
    if(isset($date==$date1){
        if(isset($_POST['submit'])){
            $name=$_POST['name'];
            $email=$_POST['email'];
            $phone=$_POST['phone'];
            $msg=$_POST['msg'];
            $to='xyz.123@mail.com'; // Receiver Email ID, Replace with your email ID
            $subject='Form Submission';
            $message="Name :".$name."\n"."Phone :".$phone."\n"."Wrote the following :"."\n\n".$msg;
            $headers="From: ".$email;
            if(mail($to, $subject, $message, $headers)){
                echo "<h1>Sent Successfully!</h1>";
            }

            else{

                echo "Something went wrong!";
            }

        }
    }

?>
```

Fig. 11 PHP code for sending mail as reminders

8 Future Scope

Since we are adapting to the new normal during these pandemic situations, it is highly important as to how efficiently we manage our expenditures and savings. Online transactions play a vital role in the future [8], slowly the world is adapting to a digital build of online transactions, so in the near future, we could plan to build an environment where online transaction history of Money transfer apps using UPIs gets linked to our website so that we could directly extract the payment details of the online payment apps to keep track of the income and expenses for a particular user. We could plan to improve the cross-platform compatibility to make it more user-friendly UI and also easy to track [9] the finances and budget for all kind of devices [10].

References

1. Mandal, S., Roy, S., Amar Raju, G.: Exploring the role of website attractiveness in travel and tourism: empirical evidence from the tourism industry in India. Tourism Plann. Dev. **14**(1), 110–134 (2017)
2. Chandini, S., Poojitha, T., Ranjith, D., Mohammed Akram, V.J., Vani, M.S., Rajyalakshmi, V.: Online income and expense tracker (2019)
3. Yadav, S., Malhotra, R., Tripathi, J.: Smart expense management model for smart homes. In: 2016 International Conference on Computational Techniques in Information and Communication Technologies (ICCTICT), pp. 544–551. IEEE (2016)

4. Sabab, S.A., Islam, S.S., Rana, M.J., Hossain, M.: eExpense: a smart approach to track everyday expense. In: 2018 4th International Conference on Electrical Engineering and Information and Communication Technology (iCEEiCT), Dhaka, Bangladesh, pp. 136–141 (2018). https://doi.org/10.1109/CEEICT.2018.8628070
5. Yeo, J.L., JosephNg, P.S., Alezabi, K.A., Eaw, H.C., Phan, K.Y.: Time scheduling and finance management: university student survival kit. In: 2020 IEEE Student Conference on Research and Development (SCOReD), Batu Pahat, Johor, Malaysia, pp. 1–6 (2020). https://doi.org/10.1109/SCOReD50371.2020.9250969
6. Woodson, P., Wang, Z.J.: A shopping store online database system. In: 2012 International Conference on Computing, Measurement, Control and Sensor Network, Taiyuan, pp. 56–59 (2012). https://doi.org/10.1109/CMCSN.2012.15
7. Mouli, V.R., Jevitha, K.P.: Web services attacks and security—a systematic literature review. Procedia Comput. Sci. **93**, 870–877 (2016)
8. Jayakumar, H.A.M., Gupta, D.: Indians' choice of payment mode(s): a study based on the impact of demonetization. In: Information and Communication Technology for Intelligent Systems, pp. 83–93. Springer, Singapore (2019)
9. Chaitanya, Ch., Gupta, D.: Factors influencing customer satisfaction with usage of shopping apps in India. In: 2017 2nd IEEE International Conference on Recent Trends in Electronics, Information and Communication Technology (RTEICT), pp. 1483–1486. IEEE (2017)[10] P. Das, Jisha R. C., and Dr. Sajeev G. P., "Adaptive web personalization system using splay tree", in 2017 International Conference on Advances in Computing, Communications and Informatics (ICACCI), 2017.
10. Das, P., Jisha, R.C., Sajeev, G.P.: Adaptive web personalization system using splay tree. In: 2017 International Conference on Advances in Computing, Communications and Informatics (ICACCI) (2017)

A Novel Encryption Algorithm Using Random Number Generation for Ciphertext Expansion

Eeshan Mundhe and Amit Birajdar

Abstract Information security has a major role to play in this ever-increasing digital world. One of the many important aspects in information security is cryptography, which is used to encrypt and decrypt information thereby protecting it from eavesdroppers. The modern cryptographic techniques make it difficult to decipher a message for even the most powerful computers. With advances in supercomputing, it is getting easier to break down these complex codes within minutes. This paper introduces a new algorithm for symmetric key encryption which will increase the length of the ciphertext as per the requirement and make it harder to decode back the original plaintext without knowing the secret key. The proposed algorithm in this paper generates a large random private key by using two relatively small random numbers: A natural number and a binary number, which is then used to encrypt the plaintext message.

1 Introduction

1.1 Cryptography

Cryptography has become an inevitable tool to secure communication between two parties so as to avoid a third party from intercepting the communication. The main aim is to maintain the confidentiality of data, be it data related to communication or storage. It refers to the techniques used for protecting information on computing systems. It secures the information by encoding or encrypting it. This encrypted information can then be accessed only using an appropriate key.

E. Mundhe (✉)
K. J. Somaiya College of Engineering, Mumbai, India
e-mail: eeshan.m@somaiya.edu

A. Birajdar
University of Southern California, Los Angeles, USA
e-mail: abirajda@usc.edu

T. Senjyu et al. (eds.), *ICT with Intelligent Applications*, Smart Innovation, Systems and Technologies 248, https://doi.org/10.1007/978-981-16-4177-0_38

Cryptography mainly performs three checks, (i) authenticity of the sender, (ii) authenticity of the message/data integrity check, to make sure if the data is not being modified during transmission, and (iii) non-repudiation, where a sender cannot deny the authenticity of its signature on a message that it sent.

The original message, also known as the plaintext, is encrypted using an encoding technique. After encryption, this encrypted message is known as the ciphertext. The ciphertext is communicated over an untrusted communication medium. The receiver can decrypt this encoded message (ciphertext) to get back the plaintext using a key. This process will be discussed in detail later in this paper.

1.2 Encryption and Decryption

Encryption refers to the process of converting a meaningful message (plaintext) into a meaningless one (ciphertext). This is done at the sender's side to ensure the data sent over the communication channel is not intercepted by a third party. It aims to protect the integrity of data. The plaintext is encrypted using a key and the resulting ciphertext is sent across the communication channel.

Decryption on the other hand is exactly opposite. It involves converting the ciphertext (encrypted message) back to plaintext (original message) using the key (Fig. 1).

Public Key A public key is generally a numerical value that is used in encrypting data. As the name suggests, it is publicly available within a defined repository or

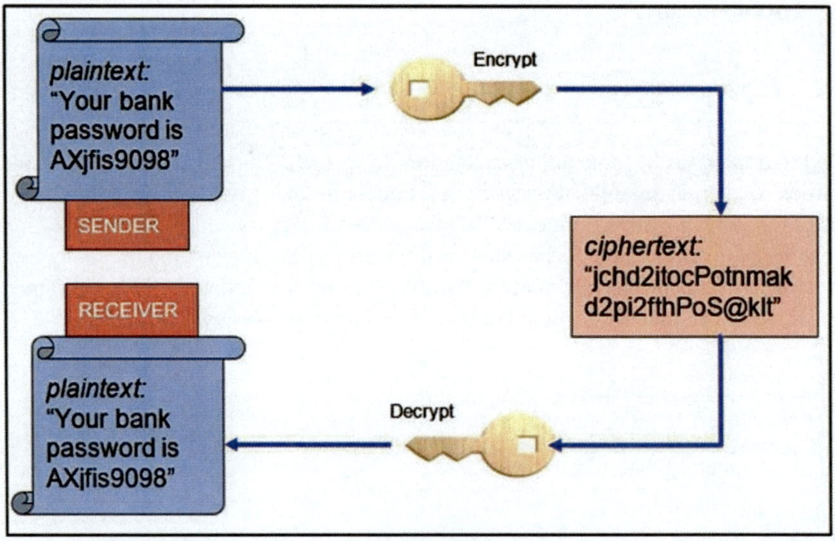

Fig. 1 Shows the process of encryption and decryption

directory. Public key cryptography, also known as asymmetric encryption, two keys are used. One is used for encryption, and the other key is used for decryption. The public key is used to encrypt the plain text, converting it into cipher text and a private key is used by the receiver to decrypt the cipher text and get back the plain text.

Private Key A private key is again a numerical value which is in some way related to the public key. In private key cryptography, only a single secret (private) key is used for encryption and decryption. This is also known as symmetric encryption as the same key is used to encrypt and decrypt a message. It is known to be faster than the public key cryptography.

Difference between Public Key and Private Key The public and private key are mainly large numbers that are mathematically related to each other but are not the same. The relation between the public and the private key is such that whatever is encrypted by the public key can only be decrypted by the related private key. A person cannot guess the private key based on knowing the public key and because of this, a public key can be freely shared. However, the private key only belongs to one person.

1.3 Benefits of Symmetric Key Encryption

Symmetric keys are generated and distributed safely to the sender and receiver and are unknown to any other entity. The recipient can be assured of the confidentiality, integrity and authenticity of the data by using symmetric keys for encryption and decryption.

Confidentiality is ensured as the content which is encrypted by the sender with the private key, can only be decrypted by the receiver with the same private key. Since, the private key is only available to the sender and the receiver, the message cannot be decrypted by anyone else, thereby ensuring confidentiality. This ensures that only the intended recipient can decipher the contents of the message.

Integrity is not provided by symmetric encryption on its own. Although, there are certain methods like including message authentication codes (MAC) as MAC is generated and validated by the same key. Using MAC also ensures authenticity.

There is another reason for using symmetric encryption over asymmetric encryption in our proposed algorithm. Symmetric encryption is faster and more efficient than asymmetric encryption. Asymmetric encryption affects the networks negatively due to performance issues with data size and heavy CPU usage. Symmetric cryptography, on the other hand, is typically used for bulk encryption or encrypting large amounts of data owing to its faster processing compared to asymmetric encryption.

1.4 Ciphertext Expansion

Ciphertext expansion takes place when there is an increase in the length of the encoded message as compared to the original message.

2 Proposed Algorithm

2.1 Key Generation Algorithm with Example

The proposed algorithm works by generating a private key (K), also known as secret key, using two numbers, one of which is a randomly selected natural number 'A' and other is a randomly selected binary number 'B'. The key is generated as follows:

- We match the total number of bits with value = 1 in the binary number 'B' with the total number of digits of the natural number 'A' from left to right.
- If the number of bits in 'B' with value = 1 is greater than the number of digits in 'A', then discard the extra bits from left to right in the binary number starting from the first extra bit with value = 1 to obtain the modified value of 'B'.
- If the number of bits in 'B' with value = 1 is less than the number of digits in 'A', then concatenate the same binary number to itself bit-by-bit from left to right at the right end until the number of bits with value = 1 are equal to the number of digits in 'A'. Thus, the modified value of 'B' is obtained.
- Now, we make the length of 'A' equal to the total number of bits in 'B' by inserting 0's in between 'A' wherever the corresponding bit in 'B' is 0. Thus, the modified value of 'A' is obtained.
- Now, subtract the original value of 'A' from the modified value of 'A' and divide the result by 9 to obtain the key (K).

Let the Plaintext be 'M'. The Ciphertext or Encrypted Text (E) is obtained as follows: $E = M + K$. We can obtain the Plaintext back from the ciphertext as follows: $M = E - K$ (Table 1).

2.2 Math Used in the Algorithm with Proof and Examples

We use the following two properties in the algorithm.

1. $k \times (10^n - 1)$ is always divisible by 9 for every natural number n and k.
2. If we insert any number of zeros between or at the end of a number and subtract the original number from it, the resultant is always a multiple of 9.

Table 1 Depicts the examples of encryption and decryption using the proposed algorithm

S. No.	Plaintext (M)	Random integer (*A*)	Random binary value (*B*)	Modified *B*	Modified *A*	Key (*K*)	Ciphertext (E)
1	6234	1475	10,010	10,010 – 1001	100,407,005	11,156,170	11,162,404
2	276	95,781	101,100	101,100 – 101	905,700,801	100,622,780	100,623,056
3	52,576	438	10,010,010	10,010,010	40,030,080	4,447,738	4,500,314
4	87	65	1,011,010	101	605	60	147
5	1592	136	1,001,010,100	100,101	100,306	11,130	12,722

Proof for property 1

In the first property, $10^n - 1$ is a number with n 9's. For example,

$$10^3 - 1 = 999 \text{ or,}$$
$$10^7 - 1 = 9999999.$$

Hence, the $10^n - 1$ is always divisible by 9.

Now, let k be any natural number. As $10^n - 1$ is divisible by 9, $k \times (10^n - 1)$ is also divisible by 9.

For example, let k be 7 and $n = 2$, then

$$k \times \left(10^n - 1\right) = 7 \times 99 = 693$$

693 is divisible by 9.

Proof for property 2

For proving the second property, we make use of the first property.

Let's take an example of a three-digit number

$$\text{`}N' = abc,$$

where a, b, c are the first 10 whole numbers.

'N' can be represented as:

$$\left(a \times 10^2\right) + \left(b \times 10^1\right) + c.$$

If we add 'k' 0's at the right end of abc, and 'l' 0's in between b and c, and 'm' 0's between a and b, then the new value obtained for 'N' is N':

$$N' = \left(a \times 10^{2+k+l+m}\right) + \left(b \times 10^{1+k+l}\right) + \left(c \times 10^k\right)$$

Now, we subtract N from N' and obtain:

Table 2 Shows the examples of second property

Random integer (N)	Modified integer (N')	$N' - N$	$(N' - N)/9$	$(N' - N)\%9$
39,787	3,090,078,070	3,090,038,283	343,337,587	0
213	201,003,000	201,002,787	22,333,643	0
13,896	10,038,906	10,025,010	1,113,890	0
45	4005	3960	440	0
8,624,729	862,004,072,009	861,995,447,280	95,777,271,920	0

$$a \times \left(10^{2+k+l+m} - 10^2\right) + b \times \left(10^{1+k+l} - 10^1\right) + c \times \left(10^k - 1\right)$$

This can be expressed as:

$$a \times 10^2 \times \left(10^{k+l+m} - 1\right) + b \times 10^1 \times \left(10^{k+l} - 1\right) + c \times \left(10^k - 1\right).$$

Now all the terms with coefficient a, b, c are in the form of $k \times (10^n - 1)$. Therefore, this value of $N' - N$ is divisible by 9. Hence, the second property is also true.

Let us take a numerical example for the same with:
$N = 379, k = 3, l = 1, m = 2.$

So, the new value N' becomes 300,070,900. Now, we subtract N from N' and obtain 300,070,521 which is 33,341,169 times 9 (Table 2).

3 Results and Discussion

The proposed algorithm generates a secret key with the help of randomly generated values. Due to this randomness, the entropy increases which results in better security. Using the proposed algorithm, the length of the ciphertext can be controlled by the sender and receivers. Longer than secret key, longer is the ciphertext. Thus, enabling ciphertext Expansion.

3.1 Advantages of the Proposed Algorithm

- No need to use modular arithmetic for generating the key, making it less complex.
- The length of the secret key is dependent on the number of bits in the binary number with value 0, hence it is easier to generate a large key by selecting a binary number with many zeroes.
- Harder to decode the original message as the value of the key gets larger.

Table 3 Shows the comparison between AES, DES and the proposed algorithm

Parameter	DES	AES	Proposed algorithm
Maximum length of key	56 bits	128, 192, or 256 bits	No restriction
Maximum length of plaintext	64 bits	128 bits	No restriction
Processes involved in encryption	Expansion, XOR Operation, Substitution and Permutation	Byte Substitution, Shifting, Mixing Columns and Key Addition	Random Number Generation, Number Expansion, Division, Key Addition

3.2 *Steps to Increase the Length of Secret Key for Ciphertext Expansion*

Select a binary number where the distance between bits with value 1 is longer. This will ensure that the resultant value of the key is longer as a greater number of 0's will get inserted into the selected integer. Larger the length of the key, difficult it is to guess.

3.3 *Possible Ways to Find Out the Secret Key by an Attacker*

There are two ways in which an attacker can find the secret key:

1. By guessing all the integers up to the value of the key.
2. By guessing the values of the two random numbers used to generate the key.

3.4 *Comparison with Existing Symmetric Encryption Algorithms*

See Table 3.

4 Conclusion

This paper has introduced a new symmetric key encryption algorithm which not only increases the randomness of the private key, but also gives the users control over the length of their encrypted message. Using the proposed algorithm, a very long encrypted message can be generated for even a very small original message. This makes it almost impossible for an attacker to decipher the original message without knowing the value of the key, or the values used to generate the key which are randomly selected. This method will be suitable in encryption of very small files as the resulting encrypted file can be as large as possible.

References

1. Rachmawati, D., Budiman, M.A.: New approach toward data hiding by using affine cipher and least significant bit algorithm. In: 2017 4th International Conference on Computer Applications and Information Processing Technology (CAIPT), Kuta Bali, pp. 1–6 (2017). https://doi.org/10.1109/CAIPT.2017.8320737
2. Babu, S.A.: Modification affine ciphers algorithm for cryptography password. Int. J. Res. Sci. Eng. **3**(2) (2017)
3. Chandra, S., Paira, S., Alam, Sk.S., Sanyal, G.: A comparative survey of symmetric and asymmetric cryptography. In: 2014 International Conference on Electronics, Communication and Computational Engineering (ICE CCE)
4. Nath, A., Ghosh, S., Mallick, M.A.: Symmetric key cryptography using Random key generator. In: Proceedings of the 2010 International Conference on Security and Management, SAM 2010, 12–15 July 2010, Las Vegas Nevada, USA
5. Chatterjee, D., Nath, J., Dasgupta, S., Nath, A.: A new symmetric key cryptography algorithm using extended MSA method: DJSA symmetric key algorithm. In: 2011 International Conference on Communication Systems and Network Technologies, Katra, Jammu, pp. 89–94 (2011). https://doi.org/10.1109/CSNT.2011.25
6. Srikantaswamy, S.G., Phaneendra, D.H.: Improved Caesar cipher with random number generation technique and multistage encryption. Int. J. Cryptogr. Inform. Sec. (IJCIS) **2**(4), 39–49 (2012)
7. Thakur, J., Kumar, N.: DES, AES and Blowfish: symmetric key cryptography algorithms simulation based performance analysis. Int. J. Emerg. Technol. Adv. Eng. **1**(2), 6–12 (2011)
8. Zibideh, W.Y., Matalgah, M.M.: Modified-DES encryption algorithm with improved BER performance in wireless communication. In: 2011 IEEE Radio and Wireless Symposium, pp. 219–222. IEEE (2011)
9. Churi, P.P., Ghate, V., Ghag, K.: Jumbling-salting: an improvised approach for password encryption. In: 2015 International Conference on Science and Technology (TICST), Pathum Thani, pp. 236–242 (2015). https://doi.org/10.1109/TICST.2015.7369364
10. Stalings, W.: Cryptography and network security data encryption method and network security. PHI
11. Kansal, S., Mittal, M.: Performance evaluation of various symmetric encryption algorithms. In: 2014 International Conference on Parallel, Distributed and Grid Computing, Solan, pp. 105–109 (2014). https://doi.org/10.1109/PDGC.2014.7030724
12. De Cannière, C.: Analysis and design of symmetric encryption algorithms. Doctoral Dissertaion, KULeuven (2007)
13. Agrawal, M., Mishra, P.: A comparative survey on symmetric key encryption techniques. Int. J. Comput. Sci. Eng. **4**(5), 877 (2012)
14. Yassein, M.B., Aljawarneh, S., Qawasmeh, E., Mardini, W., Khamayseh, Y.: Comprehensive study of symmetric key and asymmetric key encryption algorithms. In: 2017 international conference on engineering and technology (ICET). pp. 1–7. IEEE (2017)
15. Murillo-Escobar, M.A., Abundiz-Pérez, F., Cruz-Hernández, C., López-Gutiérrez, R.M.: A novel symmetric text encryption algorithm based on logistic map. In: Proceedings of the International Conference on Communications, Signal Processing and Computers, vol. 4953 (2014)
16. Singh, G.: A study of encryption algorithms (RSA, DES, 3DES and AES) for information security. Int. J. Comput. Appl. **67**(19) (2013)

17. Patel, J.R., Bansode, R.S., Kaul, V.: Hybrid security algorithms for data transmission using AES-DES. Int. J. Appl. Inform. Syst. (IJAIS) **2**(2), 15–21 (2012)
18. Mousa, A., Hamad, A.: Evaluation of the RC4 algorithm for data encryption. IJCSA **3**(2), 44–56 (2006)
19. Abd Elminaam, D.S., Abdual-Kader, H.M., Hadhoud, M.M.: Evaluating the performance of symmetric encryption algorithms. IJ Netw. Sec. **10**(3), 216–222 (2010)
20. Islam, M., Shah, M., Khan, Z., Mahmood, T., Khan, M.J.: A new symmetric key encryption algorithm using images as secret keys. In: 2015 13th International Conference on Frontiers of Information Technology (FIT), Islamabad, pp. 1–5 (2015). https://doi.org/10.1109/FIT.2015.12

Segmentation of Lung Region: Hybrid Approach

H. M. Naveen, C. Naveena, and V. N. Manjunath Aradhya

Abstract The separation of tumor region from normal tissue in the lung is a challenging task. Separating the tumor region from the normal lung region in the CT image is called nodule Segmentation. There are several methods available for nodule segmentation. Segmenting the nodules is a challenging task since some of the nodules may be attached to the other parts of the lungs. Correct and reliable segmentation of the lung nodule is required to study the characteristics of the nodules is essential, using computed tomography. In this work, we have proposed the hybrid approach using random walk algorithm with weights and watershed algorithm to segment the CT image of lungs. To measure the performance, we have used the segmentation which is measured using and boundary displacement error (BDE), rand index, global consistency error and variation of information. Results are promising and encouraging.

1 Introduction

Nodules, cancerous tissues that are to be examined, are situated within the lungs parenchyma, wherein the lungs occupy not more than half of the region of computed tomography image slice, hence a lot of computational burden can be saved if the segmentation routines execute only on the inner part of its lung region. Therefore, the count for false detection of lesions found in a segmented image is drastically lesser than that of designated in the same image without segmentation, since no signals will appear exterior of the lungs. Lung extraction is basic pre-processing step of the automated lung diagnosis systems. Edge detection algorithms play an

H. M. Naveen (✉)
Department of CS and E, BIET, Affiliated to VTU Belgavi, Davanagere, India

H. M. Naveen · C. Naveena
Department of CS and E, SJBIT, Affiliated to VTU Belgavi, Banglore, India

V. N. M. Aradhya
Department of Computer Applications, JSS Science and Technology University, Mysuru, India
e-mail: aradhya@sjce.ac.in

T. Senjyu et al. (eds.), *ICT with Intelligent Applications*, Smart Innovation, Systems and Technologies 248, https://doi.org/10.1007/978-981-16-4177-0_39

important role in lung segmentation. Edge detection is a method of segmenting the image into multiple regions. Some of the edge detection algorithms are Canny, Prewitt, Sobel, Robert [1]. Computed tomography data sets consist of more images, hence manually segmenting the lungs is a tiresome job and is prone to inter and intra observer variability. The radiology department across the globe produces lakhs of images per day. Hence it demands an automatic segmentation procedure for lung partition from computed tomography images [2].

2 Related Work

Hu et al. [3] proposed an iterative thresholding algorithm for segmentation of lung, followed by opening and closing operations using morphological operations for refining the segmentation output. Yim et al. [4] used region growing method with connected component analysis for extracting lung fields. Pu et al. [5] used an initial threshold for segmenting the lung region and for improving the segmentation process a border marching algorithm is used. Gao et al. [6] used a threshold based method for separating the pulmonary vessels, lung airways, left and right lungs separately. Jibi et al. [7] introduced a method that uses various levels of thresholding to segment the lung from other parts of the CT image.

Itai et al. [8] used the deformable models to fit the lung boundaries. Silveria et al. [9] proposed the level set active contour model to separate the left and right lungs. Annangi et al. [10] used this shape-based deformable model with the prior information of the shape of the lung for segmenting the lungs. Shi et al. [11] introduced a technique that uses the prior knowledge of the shape of the lungs and principal component analysis for extracting the prime feature points. Sun et al. [12] suggested a lung segmentation process in which active shape models are first used to extract rough lung borders, later global optimal surface finding method is used for refining the lung segmentation [13] and used graph based shape models to segment the lung fields. Sofka et al. [14] developed a CAD system that used graph based shape for automatic detection of anatomical landmarks and refined through an iterative surface deformation approach. Hua [15] proposed a graph based search algorithm. The image is treated as weighted directed graph. The algorithm operates on a cost function which incorporates intensity, gradient and boundary smoothness that minimizes the cost to segment the lung region [16]. A user interface framework is introduced for lung segmentation, which corrects the results obtained from K-nearest neighborhood approach [17]. A method is proposed in which the initial lung segmentation is done using Bayesian classification and the refinement through Markov Gibbs Random Field (MGRF) method. Campedelli et al. [18] used first order derivative for initial segmentation and Laplacian of Gaussian (LoG) operator for detecting the edges sharply. Korfiatis et al. [19] for highlighting the lung borders selected an optimal threshold for refining the segmentation. Shaungfeng Dai et al. [20] introduced a novel algorithm that uses improved graph cuts with Gaussian mixture model (GMM) and expectation maximization (EM) algorithm. Xu et al. [21] adopted 2D contour optimization, and the boundaries are detected by converting the image polar coordinate system. In Wang et al. [22] approach, sequence of 2D dynamic programming is applied to each slice separately. In this method, the average overlap obtained is 0.66. Wang et al. [23] proposed another system in which dynamic programming is applied by converting the 3D spherical structure to 2D polar coordinates. The mean overlap is 0.66 as against the desired value of 1.

Sema Candemir et al. [24] have proposed lung segmentation method using image retrieval for patient data. Here, the detect lung boundaries surpasses with the performance. Badura et al. [25] have introduced a multilevel method that segments various types of pulmonary nodules in computed tomography studies. The paper describes with following sections. Section 3 describes proposed method, and Sect. 4 discusses about segmentation algorithms. Further, Sect. 5 presents results of proposed method.

3 Proposed Method

This work is focused on an enhanced hybrid method for segmenting the nodules with irregular boundaries interactively with the random walk algorithm by modifying the weights and watershed algorithm. Leo Grady used random walk algorithm in which two pixels are picked by the user as labels for nodule and lung region. For each of the unlabeled pixels, the probability of arriving nodule and lung pixels is computed. The anticipated technique finds a 2-tuple array of probabilities for every unlabeled pixel and the nodule segmentation is done by recognizing the nodule pixels as the pixels with higher probability to reach the nodule pixel. The resulting image generated is segmented with the watershed algorithm for more precise result.

4 Segmentation

The general architecture diagram for the hybrid system is given in Fig. 1. The general architecture diagram provides the overall design of the process of segmentation of

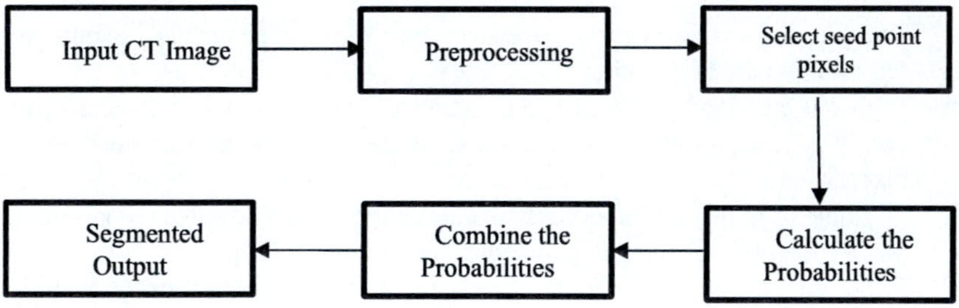

Fig. 1 Proposed flow diagram

lung tumors from CT images. The input to the pre-processing subsystem is a DICOM image of a chest CT scan of size 512×512 pixels. The source of noise in CT image is random noise and Gaussian noise. They are removed by pre-processing the images. In the first step, quality of image is improved by removing noise using median filter. It is a statistical approach that is applied to remove "salt and pepper" noise by preserving edges. The segmentation and extortion of the lung region is done by global thresholding.

Procedure for nodule segmentation using proposed method are as follows:

- The CT image is pre-processed to eliminate the noise and to enhance the image.
- Select input seed points for the nodule and lung regions.
- Using random walk algorithm with improved weights, calculate the probability of all the pixels in the image.
- For each pixel image, two probabilities are defined one for nodule and the other for lung region.
- The estimated couple of likelihoods for each unlabeled pixel is united using the watershed technique, and they are assigned as nodules if it has maximum probability to obtain nodule segmentation.

4.1 Random Walk Algorithm

The most significant phase in a computer aided design system for lung cancer detection is to segment the nodule region. Several automatic segmentation methods are available. Yet, they do not give exact results permanently, as the nodule dimensions as well as location is different by diverse pixels range . The other way is segmenting the nodules interactively with the help of the radiologist's assistance. Thus, applications involving medical image processing use interactive segmentation. The region-based segmentation is one of the method that is most widely used interactive segmentation technique.

Steps in Random Walk Algorithm

1. Input image is considered as a graph with a set of vertices and edges where vertices and edges represent the pixels and the relationship between pixels in the image, respectively.
2. Two user defined seed points are read, one for marking the nodule and other for representing the lung region.
3. With 2 pre-defined pixels as labels, compute two probabilities Pn and Pl for each pixel the probability of walking to reach the nodule and the lung seeded pixels used.
4. A couple of probabilities for each pixel is defined with two values representing one for the nodule region and the other for the lung region.
5. From the vector of probabilities, select the pixels with maximum probability ($pn > pl$) for nodule pixel and label that pixel as a nodule.
6. Segment the nodule labeled pixels as separate region representing the nodule.

4.2 Random Walk Algorithm with Weights

In the proposed method, some modifications to the existing automated systems have been made to improve the accuracy and image visibility.

Steps to calculate weights:

The following procedure is used to calculate the weights of the pixels in the image.

1. Compute the intensity differences between the unlabeled pixel and the seed point using Euclidean distance method.
2. Normalize the resultant value.
3. Calculate the geometric distance if required
4. Compute Gaussian weights using Equation.
5. Use random walk algorithm for segmenting the nodules.

Where the h_{ij} represents the Euclidean space among nearby pixels i and j and g_i and g_j represent the intensity at the position i, j. In the initial method, the probability depends only on the inclination between pixels, but not directly on their intensity.

4.3 Watershed Algorithm

Watershed algorithm may be used for segmenting the images using mathematical morphology. The notions of watersheds and catchment basins are finely identified with landscape images. Watershed shape separates all catchment basins. Image details are inferred as a topographical plane in which gradient image with gray-levels denote elevations.

In nodule segmentation process, the watershed able to produce a complete division of the image in separated regions so nodules with irregular boundaries can be extracted using the watershed algorithm as the boundaries of the nodules resemble the catchment basins and watershed lines. Using random walk algorithm with improved weights, calculate the probabilities of each pixel. The vector of probabilities is united into a single value by multiplying all likelihoods in the path to get the resulting image R. It can be represented as

$$R = \prod_{j} x_u^j$$

In the areas with same likelihood, the resulting image tends to have greatest reading. If an unlabeled pixel has identical possibility to arrive at all labeled pixel, the area will have highest probability. A limited edge development is noted in the region of every labeled pixel in the resulting image R since the likelihood of an unlabeled pixel to arrive at labeled pixel declines if it goes out of it. The labeled pixel area is extended with every possible order in the image R till its equivalent point position is arrived. This ends up with extension of seed points to ROI. At last, the resulting image R is reversed, and an indicator controlled watershed transform is executed on the image in which the markers are the labeled pixel region. Result is segmentation of the tumor area with more precise difference between the tumor and its boundaries.

Similarity Measure

The segmentation precision is calculated with dice similarity coefficient. DSC is a statistical confirmation measure to evaluate the dissimilarity among physical segmentation and automatic segmentation of images. The dice similarity coefficient is calculated as: With M as segmenting the nodule manually and A represents nodule segmentation using random walk. Images with high irregularity have the greatest of DSC values using random walk is 0.92. The followings are the rewards of proposed method • Nodules with irregular shapes are segmented precisely. • Free parameter is replaced by a constant β. • With small amount of seed points accuracy is improved.

5 Results

The proposed method is tested using 100 patient's dataset taken from LIDC-IDRI dataset. The random walk with improved weight (RWIW) is used to segment the nodules. Segmentation precision is measured based on the boundary descriptors. Some of the evaluators are rank index, variation of information, global consistency error and boundary displacement error. The average displacement error of boundary pixels between two segmented images is measured using the boundary displacement error (BDE).

Results obtained from the data sets are discussed in Fig. 3. From Fig. 2, we observe that the segmentation measures good results. The suggested method is applied to spot out lung nodules from CT images having asymmetrical boundaries. The principle of random walk watershed algorithm that is used for tumor detection is to combining the probabilities of each unlabeled pixel to get a resulting image. The resulting image is then subjected to segmentation using watershed method. The use of watershed method is to define and segment the region of interest using given seed point. The random walk algorithm considers is Euclidean distance among neighboring pixels, which implies that weight function is not only the gradient of the pixel, hence random walk algorithm is upgraded. The random walk algorithm is further improved with Watershed algorithm for segmenting irregular boundaries. Some of the successful results obtained from the proposed method is shown in Fig. 3.

6 Conclusion

In this work, we have proposed a using hybrid approach for lung segmentation. The random walk and watershed algorithms are used in this approach. Proposed lung

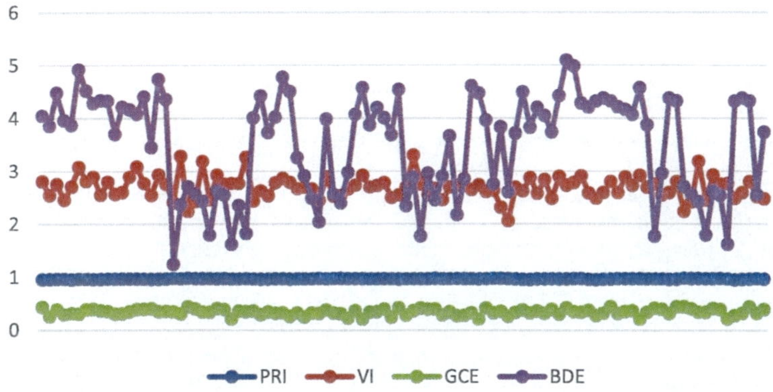

Fig. 2 Results on all the 100 samples considered

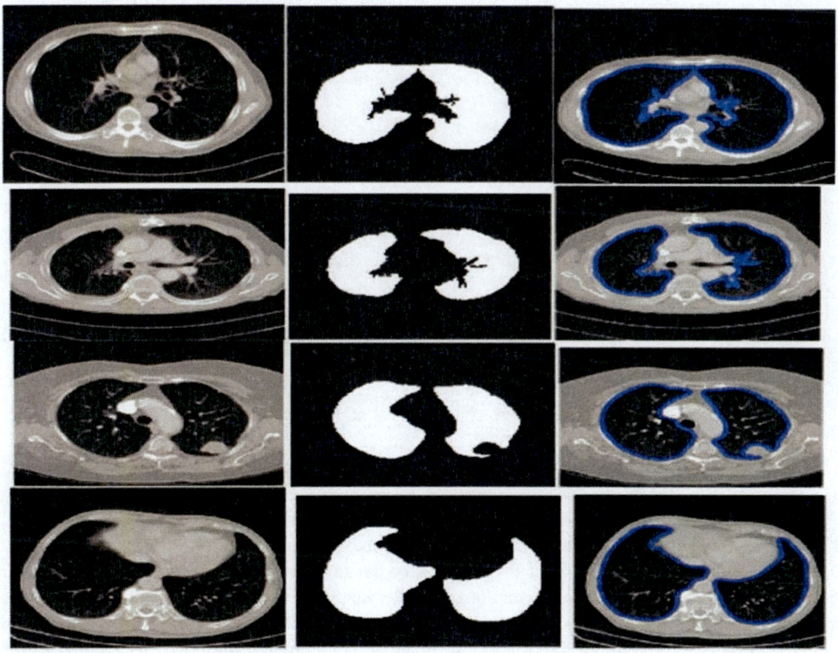

Fig. 3 Proposed segmentation results

segmentation algorithm performs well with irregular shape of boundary of the lung and the type of the lung nodule. To measure the performance, we have used global consistency error, boundary displacement error (BDE), rand index and variation of information. Experiment shows the proposed method performs well and promising.

References

1. Rashmi, M.K., Saxena, R.: Algorithm and technique on various edge detection: a survey. Int. J. (SIPIJ) **4**(3) (2013)
2. Sun, S., Bauer, C., Beichel, R.: Robust active shape model based lung segmentation in CT scans. Int. J. Eng. Sci. Technol. (2012)
3. Armato, S.G., Giger, M.L., Moran, C.J., Blackburn, J.T., Doi, K., MacMahon, H.: Computerized detection of pulmonary nodules on CT scans. Radiographics **19**(5), 1303–1311 (1999)
4. Yim, Y., Hong, H., Shin, Y.G.: Hybrid lung segmentation in chest CT images for computer-aided diagnosis. In: 7th International Workshop on Enterprise Networking and Computing in Healthcare Industry, HEALTHCOM 2005, pp. 378–383, kor, June 2005
5. Pu, J., Roos, J., Yi, C.A., Napel, S., Rubin, G.D., Paik, D.S.: Adaptive border marching algorithm: automatic lung segmentation on chest CT images. Comput. Med. Imag. Graph. **32**(6), 452–462 (2008)
6. Gao, Q., Wang, S., Zhao, D., Liu, J.: Accurate lung segmentation for X-ray CT images. In: Proceedings of the 3rd International Conference on Natural Computation (ICNC '07), vol. 2, pp. 275–279 (2007)
7. John, J., Mini, M.G.: Multilevel thresholding based segmentation and feature extraction for pulmonary nodule detection. Procedia Technol. **24**, 957–963 (2016)
8. Itai, Y., Kim, H., Ishikawa, S., et al.: Automatic segmentation of lung areas based on SNAKES and extraction of abnormal areas. In: Proceedings of the 17th IEEE International Conference on Tools with Artificial Intelligence (ICTAI '05), pp. 377–381, November 2005
9. Silveira, M., Nascimento, J., Marques, J.: Automatic segmentation of the lungs using robust level sets. In: Proceedings of the 29th IEEE Annual International Conference of Medicine and Biology Society (EMBS '07), pp. 4414–4417 (2007)
10. Annangi, P., Thiruvenkadam, S., Raja, A., Xu, H., Sun, X., Mao, L.: Region based active contour method for X-ray lung segmentation using prior shape and low level features. In: Proceedings of the 7th IEEE International Symposium on Biomedical Imaging: from Nano to Macro (ISBI '10), pp. 892–895, April 2010
11. Shi, Y., Qi, F., Xue, Z., et al.: Segmenting lung fields in serial chest radiographs using both population-based and patient-specific shape statistics. IEEE Trans. Med. Imag. **27**(4), 481–494 (2008)
12. Sun, S., Bauer, C., Beichel, R.: Automated 3-D segmentation of lungs with lung cancer in CT data using a novel robust active shape model approach. IEEE Trans. Med. Imag. **31**(2), 449–460 (2012)
13. Besbes, A., Paragios, N.: Landmark-based segmentation of lungs while handling partial correspondences using sparse graph-based priors. In: Proceedings of the International Symposium on Biomedical Imaging (ISBI '11), pp. 989–995 (2011)
14. Sofka, M., Wetzl, J., Birkbeck, N., et al.: Multi-stage learning for robust lung segmentation in challenging CT volumes. In: Proceedings of the International Conference on Medical 27 Imaging Computing and Computer Assisted Intervention (MICCAI '11), pp. 667–674 (2011)
15. Hua, P., Song, Q., Sonka, M., Hoffman, E.A., Reinhardt, J.M.: Segmentation of pathological and diseased lung tissue in CT images using a graph-search algorithm. In: Proceedings of the International Symposium on Biomedical Imaging (ISBI '11), pp. 2072–2075 (2011)
16. Kockelkorn, T.T.J.P., Van Rikxoort, E.M., Grutters, J.C., Van Ginneken, B.: Interactive lung segmentation in CT scans with severe abnormalities. In: Proceedings of the 7th IEEE International Symposium on Biomedical Imaging: From Nano to Macro (ISBI '10), pp. 564–567 (2010)
17. El-Baz, A., Gimel'farb, G., Falk, R., Abou El-Ghar, M., Holland, T., Shaffer, T.: A new stochastic framework for accurate lung segmentation. In: Proceedings of the International Conference on Medical Imaging Computing and Computer-Assisted Intervention (MICCAI '08), pp. 322–330, New York, NY, USA, September 2008

18. El-Baz, A., Gimel'farb, G., Falk, R., Holland, T., Shaffer, T.: A framework for unsupervised segmentation of lung tissues from low dose computed tomography images. In: Proceedings of the British Machine Vision Conference, pp. 855–865, University of Leeds, Leeds, UK, September 2008
19. Campadelli, P., Casiraghi, E., Artioli, D.: A fully automated method for lung nodule detection from posterior-anterior chest radiographs. IEEE Trans. Med. Imag. **25**(12), 1588–1603 (2006)
20. Dai, S., Lu, K., Dong, J., Zhang, Y., Chen, Y.: A novel approach of lung segmentation on chest CT images using graph cuts. Neurocomputing **168**, 799–807 (2015)
21. Yankelevitz, D.F., Reeves, A.P., Kostis, W.J., Zhao, B., Henschke, C.I.: Small pulmonary nodules: volumetrically determined growth rates based on CT evaluation. Radiology **217**(1), 251–256 (2000)
22. Xu, N., Ahuja, N., Bansal, R.: Automated lung nodule segmentation using dynamic programming and EM based classification. In: Medical Imaging: Image Processing, vol. 4684 of Proceedings of SPIE, pp. 666–676, February 2002
23. Wang, J., Engelmann, R., Li, Q.: Segmentation of pulmonary nodules in three-dimensional CT images by use of a spiral scanning technique. Med. Phys. **34**(12), 4678–4689 (2007)
24. Wang, Q., Song, E., Jin, R., et al.: Segmentation of lung nodules in computed tomography images using dynamic programming and multi direction fusion techniques. Acad. Radiol. **16**(6), 678–688 (2009)
25. Candemir, S., Jaeger, S., Palaniappan, K., Musco, J.P., Singh, R.K., Xue, Z., Karargyris, A., Antani, S., Thoma, G., McDonald, C.J.: Lung segmentation in chest radiographs using anatomical atlases with non rigid registration. IEEE Trans. Med. Imag. **33**(2) (2014)

Next Generation 5G Wireless Technologies in Healthcare

Apoorva Gupta and Yasha Hasija

Abstract With vastly developing technology regime, healthcare sector is witnessing next generation upgrades that are showing a lot of promises with impetus on assistance to healthcare infrastructure. Coincidently, telecommunication advancements like 5G will emphatically revolutionize the due course in this domain with super low latency and highly precise information transfer. With IoT devices available to retrieve personal medical data (MIoT), the paradigm shift toward personalized medicine is leveling itself. This paper in detail covers each aspect of 5G technology, and how it could act as a path breaker in recovering from present challenges in healthcare. Critical evaluation of present applications is also being made to contrast the past practices with hypothesized futuristic advantages. Quantitative data transfer at broad scale is a next generation requirement to facilitate efficient transfer of information and healthcare sector pan globe with integrated and well-devised policies could be the real beholder of these beneficiaries

1 Introduction

In 2016, the United Nations (UN) underwent serious deliberations for the establishment of world order in the spirit of sustainable development and the umbrella organization came with the policy of sustainable development goals (SDGs). These are a series of issue-specific propositions intended to guide people and the planet toward a better future in a comprehensively devised integrated manner. Under this very directive goal, 3 for better health and well-being of every citizen is proposed. But since the advent of these measures primarily devised under the continuation of millennium development goals, the progress has been very little. Despite bilateral and multilateral policies centered toward healthcare and large sums of budgets involved, the beneficiaries are very less thus making the rest of the people devoid of basic healthcare facilities resulting in increased mortality rates. The deaths due to non-communicable diseases are considerably high which otherwise could be less

A. Gupta · Y. Hasija (✉)
Delhi Technological University, Delhi, India

T. Senjyu et al. (eds.), *ICT with Intelligent Applications*, Smart Innovation, Systems and Technologies 248, https://doi.org/10.1007/978-981-16-4177-0_40

as compared to present trends. Inaccessibility to immediate healthcare response in case of emergency is one of the major contributors to such soaring numbers. Inaccessibility here means both in terms of distance from patient to the nearest center (due to remoteness) and the potential of nearest primary health care center to consult specialists in case of exceptional situations. In the classical context, to solve such problems, infrastructure-centric policies would be the best solution but it requires larger capital investments which become difficult for developing countries at times. This paper in detail tries to introduce an alternative. Section 2 deals with basic definition of 5G, followed by Sect. 3 highlighting pitfalls in modern healthcare system, and Sect. 4 briefing on certain additional benefits from applications in 5G. Sections 5 and 6 are based on technical throughputs and applications, respectively. In the last, Sect. 7 concludes the work with assertions in 5G medicated healthcare.

2 Defining 5G

The 5G is an acronym for fifth generation which is an upgrade in wireless data transmission technology. It is upgraded from 4G which is used in major parts of the world with better network speeds, capacity, and scalability. As concluded in a recently conducted project by ITU (International Telecommunication Union) this advanced networking technology has some key features [1]:

1. It is more efficient both in terms of energy and spectrum efficiency with 100 × and 3 × efficiency, respectively
2. Up to 500 km/h mobility can be achieved
3. Peak data should reach 10 Gbps but with better logistical implementations and technological feedback analysis with comprehensive implementations, it can go to 20 Gbps enabling large machine communications.

Unlike 4G and 3G, 5G is an integration of 4G with Wi-Fi and millimeter-wave technology with various other technical wireless communication methods resulting in formation of a transformative ecosystem. This comprises of cloud-based network architecture, virtualized network core, smart edge services with state-of-the-art attributes, and dispersed computing model that derives insights from the data generated by billions of devices which includes biosensors, system-to-system communications (applications in telesurgery) and advanced digital services (in diagnostics). Algorithms and machine design that enables 5G confers end-to-end encrypted communication on a cloud platform. This is an evolved version of point-to-point channel in which it gathers data from millions of sources simultaneously and relays that bunch of user specific information to the directed device with utmost precision over a suitable cloud enabled operational platform [2]. So, to make 5G assist the healthcare sector in every technical and computational manner, following features are either developed or are in the phase of development:

1. It allows multiple connections at the same time which enables all the biosensors to function at the same time with very little delay in relay information transmission. Scalability must be given additional attention in m-heath (mobile-health) enabled technologies because better scalability execution can accommodate node fluctuations without any impact on network processing [3].
2. The best advantage of having 5G is low latency. Latency is the time lapse between the time at which the demand for the execution of a command is made to the time at which that command appeared.
3. It supports larger bandwidth thus supporting higher frequency information transmission rates with broader spectrum utilization. This mainly enables distant healthcare providers to see high-resolution images for diagnosis and exchange of ultra-high-definition medical content.

3 Present Shortcomings in Healthcare Sector

3.1 Logistical and Manpower

As per the global standard set by WHO, the minimum threshold required of health workers required in India to facilitate optimum delivery of healthcare services is 22.8 per 10,000 people but the present estimates are much lower than the desired ones banked at only 16 per 10,000 people [4]. Out of the top 5 reasons for mortality in India, preterm birth complications (first), ischemic heart disease (third), and chronic pulmonary infections (fifth) are non-communicable, and their mortality tally can be reduced by the immediate availability of emergency medical services. But this can be done by more health care centric recruitments in remote rural areas which further requires fiscal investments and comprehensive policymaking which is a time-taking process. A better alternative is to make utilization of innovations made under 5G wireless communication technology which are efficient, require less manpower, and can provide an immediate response that too from a specialist from urban areas. The same process will not only reduce the fiscal deficit burden from the government but also assist the logistics sector present at ground level as better networking will lead to faster production in supply chains and delivery more swiftly. Thus, making use of proposed innovations will not only assist stakeholders in disseminating technology in rural areas but also make them by-pass what otherwise would be a mountainous task especially in developing countries like India (Fig. 1).

3.2 Diagnostic

Diagnosis of medical conditions plays a very crucial role in the initiation of treatment. Higher precision rates can guarantee the prompt treatment of infirmity thus reducing the chances of fatality, i.e., targeted disposal of medicinal services could reduce the

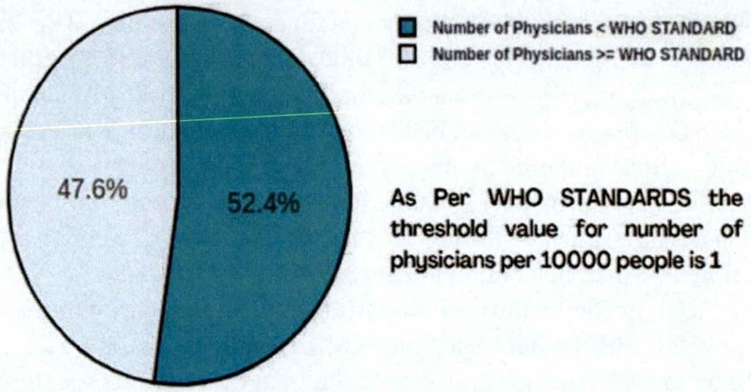

Fig. 1 Nation wise distribution of physicians

chances of death. Since the basic nature of medicinal diagnosis is based on trial-and-error approach, the time delays become more evident resulting thus contributing as a risk factor which could be reduced by better availability of data (both personal and community) and efficient machine learning algorithms which have trained on a lot of data related to fixed medical conditions. 5G wireless technology thus can serve as a medium to propagate and analyze data available to the cloud with highly rated computational efficiency Not only this, but it will also deliver the required specialist consultations with high-definition image resolution features in a very crucial narrow duration of time.

3.3 Economic

As per 2019 public health expenditure report, the at present Indian government invests at an average 1.3% of its GDP in healthcare which as per CAG report under national health program it must be 2.5% [5]. Thus, a double amount of GDP spending per capita is required to provide equal and efficient medical service to every citizen with rapid fluctuations in demand in case of health emergencies like COVID. Jeopardy arises when this high demand for investment comes with a relative rise in population growth rate thus enhancing the burden on the existing healthcare structure. With immediate investments required, the task has become unbearable but phasing out secure digital policies can qualitatively rescue us. With very less investment (especially when compared with the grade of development of physical infrastructure) major challenges can be resolved promptly. Also as stated in the above three points this innovation has overlapping benefits, reduced costs, better treatment, and patient-specific are some of the key features. With a far reach of technology in rural areas, the accessibility of city-based healthcare centers will also increase thus reducing the amount of logistical and transportation costs therefore higher benefit–cost ratio with all shortcomings handled via a single set-up (Figs. 2 and 3).

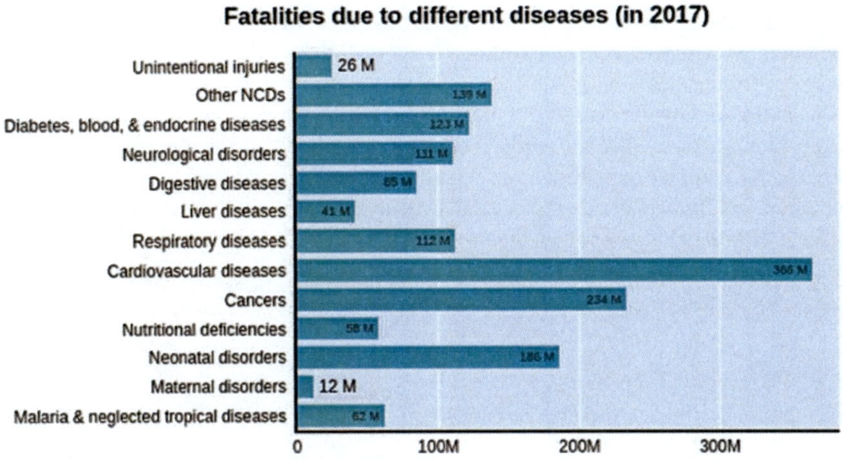

Fig. 2 Fatalities due to different diseases (worldwide)

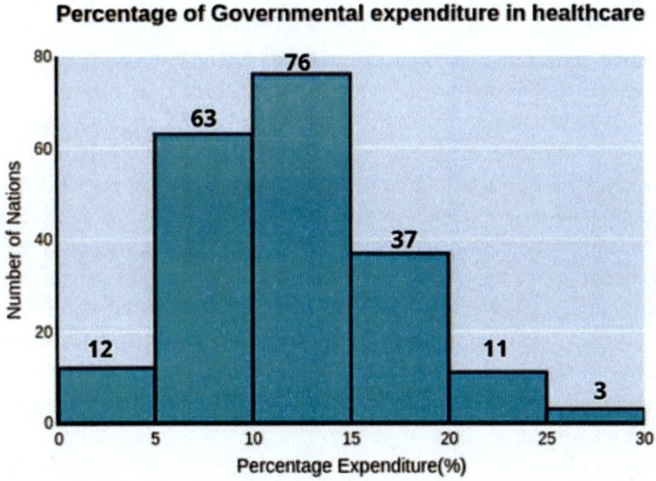

Fig. 3 Nation wise Expenditure in Healthcare

4 Fringe Benefits

4.1 5G Cognitive Systems [6]

Unlike the prospects of availability of basic healthcare facilities via long-range low latency, transmission 5G can also be used for treating psychological conditions. The 5G—Cognitive system is a way ahead. It consists of a data cognitive engine and resource cognitive engine. The latter covers the architectural and technical aspects including learning of data networks, very low latency, and very high reliability, whereas former deals with analytics of big data over cloud therefore in the diagnosis and treatment. It perceives the user's emotions via environmental, emotional, and behavioral data. To attain medical cognitive intelligence with algorithms that can process the available data from three categories to achieve real-time emotion communications. It further will be used in rendering drug recommendations for patients. To achieve low-data latency within core network functionalities like network fusion and content delivery are developed.

4.2 Wireless Tele-Surgery

As the name suggests, this is innovation is subjected to facilitate surgical/operational services in a remote area from a specialist distant apart via robots. With ultra-low latency levels, this task seems achievable. The specialist can either be appointed to conduct the surgery or be consulted any time in between from a well-equipped decentralized medical center. Low latency should come with stable haptic feedback and improvised data rates for high accuracy and efficient visual communication. Cyber-sickness (a condition in which there is a time delay in perception of movement by our vestibular system and observation made by eyes) of surgeons should be taken care of [7]. A reaction time less than or equal to 100 ms (auditory), 10 ms (visual), and 1 ms (manual) [8]. Robots will be in control of haptic gloves; they will thus enable transferring of tactile data back and forth to the specialist. Since it is logistically intensive with high-data connectivity demands with the simultaneous requirement of several connections on the cloud, which is required for high reliability of tactile applications, some applications like network slicing are very helpful in this context as the operator develops customized relay networks for enhanced solutions. Mobile edge computing (MEC) [9] is another domain that guarantees faster relay and low latency to hosted software's in processor trusted domain in proximity of retrieval point for robot [10].

4.3 Wireless Service Robots

It is based on creating an AI-enabled edge cloud platform to command robots in far away from healthcare centers. These robots hence can be used to help patients, especially older ones in hospitals, hospices, homes, and nearby areas. IoT enabled sensors will keep a track of their circadian rhythms with special emphasis on the blood-sugar level, blood pressure, and oxygen level as these are most vital physiological parameters which in general practice show deviations from threshold values in case of emergencies. If any of these deviates from the expected physiological range, robots linked via 5G enabled wireless transmissions can assist based on data retrieved from cloud storage thus reducing the chances of fatal complications in case of absence of human assistance. Also, for fully autonomous robots to meet the required performance, it is very important to decide where and what to compute, and how to transmit data and signals, at a given bandwidth [11].

5 Advent of 5G Wireless Technology

The onset of 5G technology specifically in healthcare will transform the architecture and features of m with major advancements in cloud data analytics and haptic feedback.

5.1 Tactile Internet

An integral part associated with this upgrade in modern-day internet is its potential to transfer real-time point-to-point touch feedback back and forth from users. Therefore, its development is inevitable for further enhancements in technologies like telesurgery. Suitable robotics and haptics equipment are required along with a swift and stable communication network. Simultaneous relay of physical tactile experiences over the web will shift the role of the internet from content delivery to skill-set delivery. For deployment of telemedical technologies at basic level, tactile internet-based servers and resource centers have to be developed as high fidelity and extreme precision are the prerequisites. Additional benefits will come with telediagnosis services which can become available anywhere anytime without physicians' actual presence [12]. The latency of not more than 1 ms is required to prevent fitful oscillatory movements of multi-connected robots and to ensure real-time synchronous and visual-haptic feedback.

5.2 Big Data Analytics Over Cloud

As mentioned in the diagnosis section, larger availability of aggregated data with faster applications to transfer the concluded information will help healthcare facilitators in precision medicine unlike prescriptions based on average populations, while also considering an individual's medical and genetic history. But to make it a reality, primarily we require electronic health records (EHRs). It is a repository that stores patient's digital data [13] in case of IoT sensors utilized by the patient for regular tracking of several physiological variables (blood pressure, pulse rate, heartbeat, etc.) it will store those values as well. Thus, an established reliable channel for error-proof transfer of data is required which 5G can provide. In case the availability of such large data related to a cohort is studied, community medicinal approaches can be devised, and population-based health patterns can be developed which become crucial, while dealing with epidemics and pandemics. Big data provide the necessary support for developing a patient-centric personalized healthcare system in which the right health intervention for a given person and the health problem can be identified in an evidence-based manner. The collaborative cancer cloud is an analytics platform that integrates patient information from a variety of organizations. It allows participating institutions to "securely share patient genomic, imaging, and clinical data for potentially lifesaving discoveries. It will enable large amounts of data from sites all around the world to be analyzed in a distributed way, while preserving the privacy and security of patient data at each site" [14]. Thus, cloud computing enabled big data analytics could support every aspect of 5G enabled healthcare by rendering accuracy.

In addition to these benefits, this technology could safeguard the quality of treatment as well. Some examples of this are:

1. Undermined communities will get better access to state-driven healthcare facilities as people living in remote/isolated locations could get better medical services.
2. For older/physically disabled people it becomes painful to travel distant medical centers even within cities but with the advent of 5G communications home-health therapies will come into play which could transfer vital medical data regularly thus enabling distant physicians/specialists in therapeutic prescriptions and diagnosis.
3. In a study conducted in Indiana, USA on patients of congestive heart ailments it was deduced that remote assistance reduced hospital readmissions. There was a contrast of 12%, in the number of patients that did not render with remote services (15%) readmitted to the ones that rendered the service (3%) [15].

6 Related Studies

The advent of 5G communication technology has come with several promising advancements in the healthcare sector. Governments and private stakeholders are investing capitals in research and development of state-of-the-art tools and applications based on 5G transmissions. The thrust area of research includes algorithm design and analysis for better data aggregation and estimation with minimum time complexity and achieving low-latency rates via architectural and systematic computational amendments. Here are some projects and their outcomes, and they could enable a head start for future developments.

1. To develop 5G enabled healthcare, they need a resilient architecture system. This system utilizes transparent interconnection of lots of links (TRILL) which is a protocol used for delivery of data packages and management of mobile applications. Moreover, it renders bridge routing functions to every eNB. Unlike 4G which has limited bandwidth and does not allow multiple IoT devices connected

over the same network. For stable multimedia services, a high-speed network is a prerequisite. Users can access critical medical information as it provides pragmatic roaming services. To make 5G more efficient, some technical aspects of 4G are also used in this system for better mobility control. Link layers in the routing bridge incorporated in system architecture also helped in minimizing overhead tunneling [16].

2. Under the domain of M-health solutions with 5G enabled machine-to-machine (M2M) communication [3] an M2M healthcare gateway is considered. These mechanized instruments enable bidirectional communication between healthcare sensors and back-end servers which have enabled better screening facilities. In the case of chronic infections and age-related complications, it can provide better monitoring applications with immediate treatment in case of an emergency. Linked with EHR reports and prescriptions can be directly sent to the patient's guardians via M2M gateway by the doctor on duty without any restriction and time delay.

7 Conclusion

After a comprehensive analysis of 5G wireless technology with a briefing about economic, social, logistical, and technical advantages which it can provide one thing becomes natural to understand that if not today, this is the innovation for tomorrow. It has the potential to modify existing medicinal regimes with far-reaching benefits and more accurate and swifter disposal of medical service. We are living in an era wherein a time span of 20 years the world has faced regional epidemics in the form of SARS (2003) and MERS (2013) and the global epidemic of COVID-19 with continental outbreaks in the form of Zika virus and Ebola. Global antibiotic resistance is also a much-awaited phenomenon that will check our technological advancements in medical science. Despite being breakthroughs in medicine, we still lag in the dissemination of universal healthcare guarantees and with the count of community transmittable diseases increasing fourfold with sidelining existing personal ailments the burden on modern-day services is more than ever. The situation worsens in case of highly transmissible infections like COVID, thus making those with regular exposure more susceptible like our healthcare practitioners, for the first-time concepts like social distancing came into play just to mitigate the flow of virus and break the chain but still many medical service providers got infected due to direct exposure. Such instances should make governments realize the need for alternative mechanisms that can replace humans thus reducing the chances of exposure. Several international organizations like ITU and private players have determined the potential of 5G, and states are undergoing deliberations to remodel them for better results so that after proper standardization process, they can become a part of national healthcare regimes.

References

1. Recommendation ITU-R M.2083-0. IMT Vision—Frame-work and overall objectives of the future development of IMT for 2020 and beyond. In: International Telecommunication Union-Radiocommunication Sector (2015)
2. West, D.M.: How 5G technology enables the health internet of things
3. M-Health Solutions Using 5G Networks and M2M Communications IT Pro, May/June 2016
4. Chen, M., Yang, J., Hao, Y., Mao, S., Hwang, K.: A 5G cognitive system for healthcare
5. Jay, C., Glencross, M., Hubbold, R.: Modelling the effects of delayed haptic and visual feedback in a collaborative virtual environment. ACM TOCHI **14** Article N. 8
6. Simsek, M., Aijaz, A., Dohler, M., Sachs, J., Fettweis, G.: 5G-enabled tactile internet. IEEE J. Sel. Areas Com. **34**(3), 460-473 (2016)
7. 3GPP TR 23.799: Study on architecture for next generation system. Release **14**
8. Cavallo, F.: Toward cloud service robotics in active and healthy ageing applications. IEEE MMTC E-Lett. **10**(4), 31–34 (2015)
9. Din, S., Paul, A., Ahmad, A., Rho, S.: Emerging mobile communication technologies for healthcare system in 5G network
10. Arata, J., et. al.: A remote surgery experiment between Japan and Thailand over internet using a low latency CODEC system. In: Proceedings of the IEEE International Conference on Robotics and Automation (ICRA), April 2007, pp. 953–959
11. Ali, A., Qadir, J., ur Rasool, R., Sathiaseelan, A., Zwitter, A., Crowcroft, J.: Big data for development: applications and techniques. Big Data Anal. 1, 2 (2016)
12. Vanian, J.: Intel's cancer cloud gets new recruits. Fortune, 31 March 2016
13. Vashist, S., Luppa, P., Yeo, L., Ozcan, A., Luong, J.: Emerging technologies for next-generation point-of-care testing. Trends Biotechno.
14. Snell, A., Smalley, J.: Beacon community research study: reducing hospital readmissions via remote patient management. Indiana Health Inform. Exchange (2013)
15. Care Innovations: How Mississippi is leading the way in innovation (2015)
16. Din, S., Paul, A., Ahmad, A., Rho, S.: Emerging mobile communication technologies for healthcare system in 5G network. In: 2016 IEEE 14th Intl Conf on Dependable, Autonomic and Secure Computing, 14th Intl Conf on Pervasive Intelligence and Computing, 2nd Intl Conf on Big Data Intelligence and Computing and Cyber Science and Technology Congress(DASC/PiCom/DataCom/CyberSciTech), pp. 47–54 (2016). https://doi.org/10.1109/DASC-PICom-DataCom-CyberSciTec.2016.23

Manipuri Meitei Mayek Numeral Classification by Using HOG-Assisted Deep Learned Features

Palungbam Roji Chanu and Oinam Nickson

Abstract In this paper, a histogram of oriented gradient (HOG)-assisted deep learning algorithm is proposed for classification of Manipuri Meitei Mayek (MMM) numerals. In HOG, gradient computes the magnitude and direction of largest change in intensity among small neighborhood of every pixel. The extracted HOG feature vectors are trained and recognized by using multilayer perceptron (MLP) artificial neural network. The stacking method is used to gain robust features. After that, it is applied to as Softmax classifier and aims for the recognition of MMM's character. To prove the effectiveness of our work, various comparisons are made on existing state of the art of Manipuri optical character recognition (OCR) algorithms and it shows that our proposed algorithm is superior comparing to the other existing Manipuri OCR work.

1 Introduction

Optical character recognition (OCR) has become a very interesting research area among many scholars and business firms. It has become one of the most promising and helpful strategies for human-computer interaction. Text data stored in the form of handwritten documents is very beneficial as they can be used for future reference while keeping records for events, culture, literature, history, and many other events. Automatic handwritten OCR has entered the era of large-scale application with the growing interest in mobile information gadgets like cell phones, tablets, and handheld computers. The challenge in automatic recognition of written scripts is that one has to encounter with large variety of variations in handwritten styles. Characters from various writers are very much different in terms of shapes and sizes. This makes the task of recognition more difficult.

Machine learning algorithms have been used for feature extraction and classification of images in the literature [1–3]. Artificial neural network (ANN) is a common technique for classification and recognition purposes. This is because of

P. R. Chanu (✉) · O. Nickson
National Institute of Technology Nagaland, Dimapur, India

T. Senjyu et al. (eds.), *ICT with Intelligent Applications*, Smart Innovation, Systems and Technologies 248, https://doi.org/10.1007/978-981-16-4177-0_41

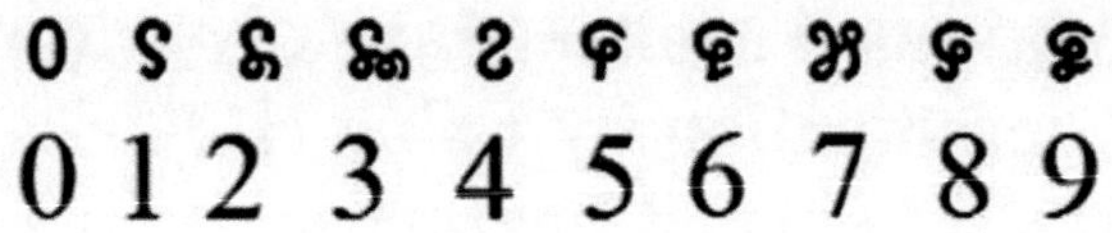

Fig. 1 Meitei Mayek numerals and English digits

their good learning capability. They can be used in solving task like MMM classification and recognition by applying multilayer perceptron neural network (MLPNN) data recognition through a learning approach.

Meitei Mayek is the original script of Manipuri and Manipur's official language, a northeast Indian state. It is spoken in Manipur's valley region, mostly by ethnic Meitei group. It is also spoken in some parts of northeastern Indian states like Assam and Tripura, and countries like Myanmar and Bangladesh. A detailed description on the evolution of Meitei Mayek is depicted in the literature [4–6]. Meitei Mayek consists of Mapung Iyek/Iyek Ipee and has 27 alphabets (18 alphabets are original, and 9 letters are derived from these 18 alphabets and known as Lom Iyek), Cheitek Iyek (8 symbols), Khudam Iyek (3 letters), Lonsum Iyek (8 letters), and Cheising Iyek (10 digits). Six additional vowel letters are also included. All the Meitei Mayek alphabets are named after different human body parts [5]. The Meitei Mayek numerals for which classification and recognition are carried out are shown in Fig. 1.

In the early days from the eleventh century to the eighteenth century, Meitei Mayek was used. But in the early eighteenth century, Vaishnavism was introduced, and all the documents written in Meitei Mayek were burnt and destroyed in the kingdom of Manipur. This incident is known as "Puya Meithaba" which has led to the substitution of original Manipuri script by Bengali script. This fact shows the importance of storing documents in the digital form to preserve its survival. Many efforts were made to revive this script, and Manipur government gave approval in 1980. However, the Manipuri government included this script in the academic curriculum in 2005 which was 25 years after its approval. Since then, research on Manipuri Meitei Mayek (MMM) becomes an important area. In this paper, we have proposed an algorithm which can recognize MMM numerals efficiently by using HOG and MLP.

Literature review is given in Sect. 2. Section 3 describes the proposed work. The experimental results are discussed in Sect. 4 followed by conclusions in the last section.

2 Literature Review

Recognition of numerals system has been designed for different languages globally in different countries. In case of Indian scenario, a recognition system for offline handwritten Gujarati numerals is developed in [7] by using neural network approach. The results are computed for three different image sizes with overall accuracy of 87.29%, 88.52%, and 88.76%. In [8], a classification algorithm for handwritten Devanagari digit is developed based on neural network. Different features and multistage classifiers are used to achieve a recognition accuracy of 97.87%. A handwritten Gurmukhi numeral recognition method is found in [9]. It relies on zone-based hybrid approach for feature extraction. In [10], a recognition algorithm for handwritten Assamese numerals is proposed based on mathematical morphology. It is shown that the algorithm can achieve an average recognition rate of 80% for handwritten Assamese digits. Limited works have been carried out in case of Manipuri Meitei Mayek numerals recognition due to its complex nature. A detection method of Meitei Mayek numerals is designed in [11] which is based on the multilayer feedforward neural network approach. It uses backpropagation learning method and achieves an overall performance of 85%. Another recognition system of Meitei Mayek digit is designed in [12]. It uses Gabor filter for feature extraction and classification by support vector machine (SVM) and has achieved an overall performance of 89.58%. Another SVM-based handwritten Manipuri Meitei Mayek numeral digits is presented in [13]. In this work, various methods for feature extraction are used such as background directional distribution (BDD), histogram-oriented features, zone-based diagonal, and projection histograms. The accuracy for recognition rate is 95%.

3 Proposed Work

Feature detection of MMM is a tedious process because of the similarity found in every character. Therefore, one needs a feature detector which is robust. The detector must be able to discriminate the shape or structure of the images clearly. In our current work, a handwritten classification of Manipuri Meitei Mayek numerals is presented based on HOG-assisted deep learned features and trained by multilayer perceptron (MLP). Features extracted from HOG are given as input to MLP which are trained and used for recognition purposes. In the beginning, all the input images undergo preprocessing stage for removing noise and extract them individually. Then, the images are passed into HOG descriptor to extract the feature vectors. Finally, the feature vectors extracted from HOG are trained and recognized by MLP.

3.1 Multilayer Perceptron

A multilayer perceptron (MLP) is the form of artificial neural network (ANN) which is feedforward in nature. It represents a network of fully connected neurons having at least three layers of nodes. These layers are known as input, hidden, and output layers. The neurons in first layer are connected to neurons in the second layer, and the chain continues. For training purposes, MLP uses a supervised learning method called backpropagation. A detailed description on MLP is presented in [14].

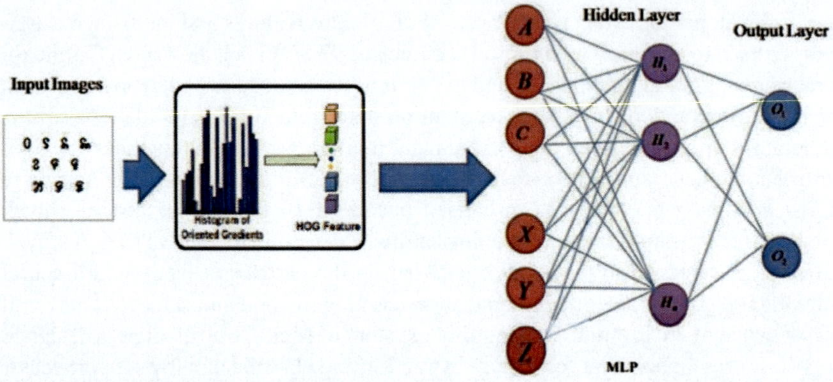

Fig. 2 Illustration of the proposed model of HOG-assisted deep features using MLP

3.2 *MMM Numerals Classification Using MLP with HOG Feature Descriptor*

In Fig. 2, the proposed HOG-assisted deep feature learning using MLP for classification of MMM numerals is shown. It can be seen that the proposed model starts with a feature extraction of the input images using HOG followed by deep learned feature extraction using MLP. Histogram of oriented gradient (HOG) is a feature descriptor, which is commonly used in the field of image processing for object detection. HOG was first used by N. Dalal and B. Triggs for human detection [15]. In our work for classifying handwritten MMM, we deployed HOG feature vector to assist deep feature extraction which can discriminate the shape or structure of given image so as to differentiate every character clearly. HOG deals with the image gradients, which are indicated by a directional change in the intensity or color in an image. It gives a count of the occurrence of orientation of gradient in some localized parts of an image. To compute HOG, the whole image is divided into smaller patches of cells. For each cell, the horizontal and vertical gradients of every pixel are computed as follows:

$$g_h(u, v) = I(u + 1, v) - I(u - 1, v) \tag{1}$$

$$g_v(u, v) = I(u, v + 1) - I(u, v - 1) \tag{2}$$

where $I(u, v)$ denotes the value of pixel at coordinate (u, v) of the input image, and $g_h(u, v)$ and $g_v(u, v)$ represent the horizontal and vertical gradient. Then, the gradient magnitude $g(u, v)$ and gradient orientation $\alpha(u, v)$ are computed as follows:

$$\alpha(u, v) = \arctan \frac{g_v(u, v)}{g_h(u, v)} \tag{3}$$

$$\alpha(u, v) = \arctan \frac{g_v(u, v)}{g_h(u, v)} \tag{4}$$

Generation of final HOG feature vector: As shown in Fig. 2, the original input image of size 80 × 80 is divided into 25 different blocks each of size 60 × 60; to improve the local correlation, we used overlapping strategy with 5-pixel stride. Further, each block is divided into 36 small cells of 10 × 10 pixels.

4 Experimental Results and Discussion

The proposed algorithm is tested over a dataset of 10,000 samples. These datasets are collected from different people belonging to different age groups and professions. Figure 3 depicts a sample image from the dataset of handwritten Manipuri Meitei Mayek numerals used for the experiment.

Training set: Training dataset consists of the sample images used for learning the neural network. We have collected 1000 samples for each 10 digits, out of which 999 samples are used for training the neural network and the remaining 100 samples are used for testing purposes. The test dataset is used for the evaluation of the generalization of the proposed system to the unseen data.

After training the model, we have created two charts:

1. Accuracy plot as shown in Fig. 4.
2. Loss plot as shown in Fig. 5.

The history for the validation dataset is labeled test by convention as it is indeed a test dataset for the model. From the plot of accuracy, we can see that the training model

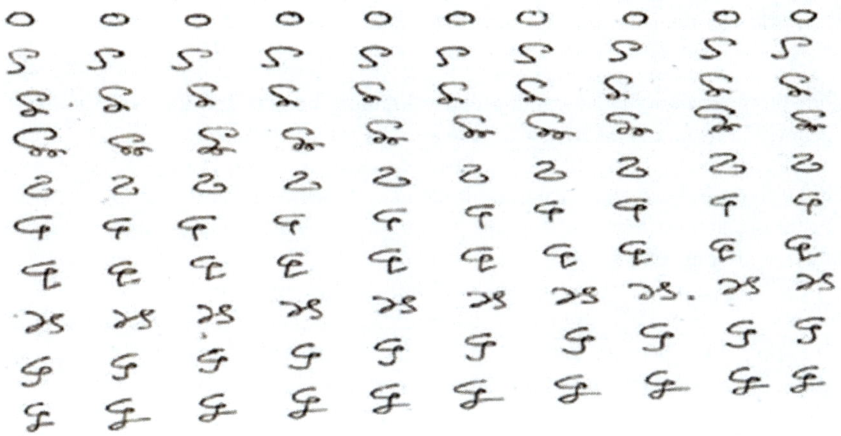

Fig. 3 An image sample of MMM numeral dataset

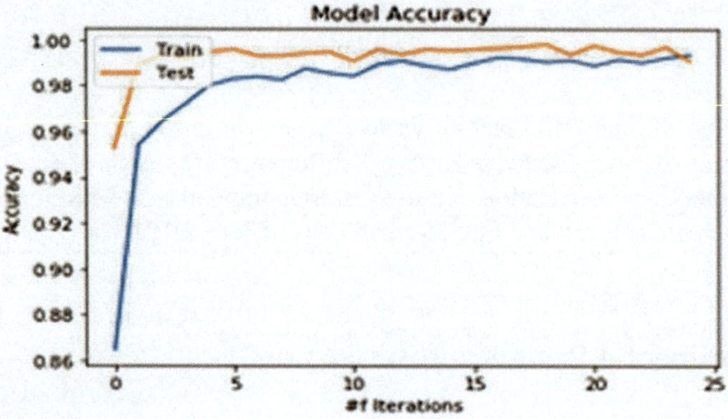

Fig. 4 Plot of accuracy

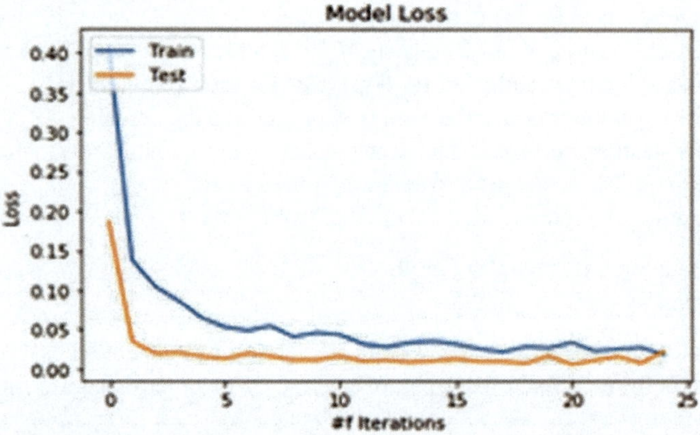

Fig. 5 Plot of loss

does not overlearn on the training and validation datasets and the best validation accuracy is found at epoch 18 after that accuracy does not increase.

From the plot of loss, we can see that the model has comparable performance on both trained and validation datasets. It indicates that we can further train our model but for our convenience we stop our training because the model accuracy does not increase after epoch 18.

The confusion matrix is shown in Table 1. The testing is performed on 1000 images, in which each digit has 100 images. This matrix shows the number of times each digit is recognized correctly and the number of times each digit is recognized falsely. The elements in the diagonal represent the number of digits which are correctly recognized, whereas the elements in the remaining rows show the number of digits which are misclassified against the digit given in first column. Consider digit 0, 99 digits out of 100 are correctly recognized, while it is wrongly classified

Table 1 Confusion matrix of Manipuri Meitei Mayek 0 (zero) to 9 (nine)

	0	1	2	3	4	5	6	7	8	9
0	99	0	0	0	0	0	0	0	1	0
1	0	100	0	0	0	0	0	0	0	0
2	0	0	100	0	0	0	0	0	0	0
3	0	0	1	99	0	0	0	0	0	0
4	0	0	0	0	100	0	0	0	0	0
5	0	0	0	0	0	99	0	0	1	0
6	0	0	0	0	0	0	100	0	0	0
7	0	0	0	0	0	0	0	100	0	0
8	0	0	0	0	2	0	0	0	98	0
9	0	0	0	3	0	0	0	0	0	97

as 8 once. It is observed that digits 1, 2, 4, 6, and 7 are classified 100% correctly, whereas digit 0 is predicted 99% correctly and 1% wrongly as digit 8, digit 3 as 99% correctly and 1% wrongly as digit 2, digit 8 as 98% correctly and 2% wrongly as digit 4, class 0 predict 98% correctly and 2% wrongly as class 4, digit 9 as 97% correctly and 3% wrongly as digit 3. This confusion matrix arose due to the similarity in the shapes of the digits. The total number of true predictions is 992 out of 1000, while false prediction is 8 out of 1000. The final accuracy obtained is computed as Accuracy = 992/1000 = 0.992 = 99.2%. Table 2 shows the performance of the proposed model for all testing digits. For each digit, 100 images are used for recognition and the output is shown along with the accuracy of each digit. CR stands for correct recognition, and FR stands for false recognition.

It was observed that the proposed method using HOG features integrated with MLP shows better results than method using MLP only. Table 3 shows the comparison of method using HOG+MLP and method using MLP only.

Table 2 Percentage accuracy for recognition of Manipuri Meitei Mayek digits

Digits	Attempts	CR	FR	Accuracy (%)
0	100	99	1	99
1	100	100	0	100
2	100	100	0	100
3	100	99	1	99
4	100	100	0	100
5	100	99	1	99
6	100	100	0	100
7	100	100	0	100
8	100	98	2	98
9	100	97	3	97

Table 3 Comparison between methods using HOG+MLP and MLP only

	HOG+MLP (%)	MLP only (%)
Test loss	16.1	17.7
Test accuracy	99.2	98.2

Table 4 Comparison of proposed method with other methods

Methods	No. of classes	Class types	Accuracy
Romesh et al. [11]	10	Manipuri digits	85%
Maring and Dhir [12]	10	Manipuri digits	89.58%
Chandan and Sanjib [13]	10	Manipuri digits	95%
Proposed method	10	Manipuri digits	99.2%

Many techniques have been developed for numeral digit recognition system for different languages. Certain research have been carried out in Indian languages such as offline handwritten Gujarati system, Devanagari digit recognition system, handwritten Gurmukhi numeral recognition system, Assamese numerals classification method, etc. However, when it comes to Manipuri numerals recognition, only limited works can be found. Table 4 gives a comparison of our proposed method with some existing methods used for recognition of Manipuri digits. In [11], a neural network approach is used for simulation and recognition of handwritten Meitei Mayek numerals which achieved an accuracy of 85%. In [12], another method is developed for recognition of Manipuri digits using support vector machine and an accuracy of 89% is achieved, while our proposed method which uses both HOG and MLP can achieve a higher accuracy of 99.2%.

Main limitation of our work lies in the fact that we need to find an optimal size of HOG block feature for recognizing MMM characters efficiently. Accuracies obtained after considering different cell sizes such as 6×6, 7×7, and 8×8 are compared, and the cell size with largest accuracy is chosen for recognition. Different cell sizes take different times for training purposes; 6×6 is 155 s, 7×7 is 318 s, and 8×8 is 69 s.

5 Conclusions

In our work, we have developed a novel method for recognizing Manipuri Meitei Mayek digit efficiently using HOG integrated with MLP. About 10,000 handwritten samples of digits are collected from different groups of people with different professions and age group. The maximum accuracy which can be achieved is 99.2% which is higher than other methods developed for Manipuri digit recognition. The key novelty of the proposed method lies in the design of a recognition system which explore the deep learned feature of MLP and HOG-assisted feature extraction. Our method can be extended to recognition of degraded text also.

References

1. Inunganbi, S., Prakash, C., Khumanthem, M.S.: Meitei Mayek handwritten dataset: compilation, segmentation, and character recognition. Visual Comput. 1–15 (2020). https://doi.org/10.1007/s00371-020-01799-4
2. Inunganbi, S., Prakash, C., Khumanthem, M.S.: Local texture descriptors and projection histogram based handwritten Meitei Mayek character recognition. Multimedia Tools Appl. 2813–2836 (2020). https://doi.org/10.1007/s11042-019-08482-4
3. Nongmeikapam, K. et al.: Handwritten Manipuri Meetei-Mayek classification using convolutional neural network. In: ACM Transactions on Asian and Low-Resource Language Information Processing (TALLIP) vol. 18, no. 4, pp. 1–23 (2019). https://doi.org/10.1145/3309497
4. Wangkhemcha C.: A short history of Kangleipak (Manipur) Part-II. Sagolband Thangjam Leirak, Imphal, India: Kangleipak Historical and Cultural Research Centre (2007)
5. Kangjia, M.: Revival of a closed account, a brief history of kanglei script and the Birth of phoon (zero) in the world of arithmetic and astrology. Sanmahi Laining Amasung Punshiron Khupham (SalaiPunshipham), Lamshang,Imphal (2003)
6. Hodson, T.C.: The Meitheis. Low Price Publications, Delhi (1908)
7. Avani, R.V., Sandeep, R.V., Kulkarani, G.R.: Performance evaluation of different image sizes for recognizing offline handwritten gujarati digits using neural network approach. In: International Conference on Communication Systems and Network Technologies, pp. 270–273, (2012). https://doi.org/10.1109/CSNT.2012.66
8. Rajiv, K., Kiran, K.R.: Handwritten Devnagari digit recognition: benchmarking on new dataset. J. Theoritical Appl. Inf. Technol. **60**(3), 543–555 (2014)
9. Gita, S., Rajneesh, R., Renu, D.: Handwritten Gurmukhi numeral recognition using zonebased hybrid feature extraction techniques. Int. J. Comput. Appl. **47**(21), 24–29 (2012)
10. Medhi, K., Kalita, S.K.: Recognition of assamese handwritten numerals using mathematical morphology. In: Advance Computing Conference (IACC), IEEE International, pp. 1076–1080, (2014). https://doi.org/10.1109/IAdCC.2014.6779475
11. Romesh, L., Angom, U.S., Chandrakumar, N.S., Suresh, A.S., James, H.: Simulation and modeling of handwritten Meitei Mayek digits using neural network approach. In: Proceedings of the International Conference on Advances in Electronics, Electrical and Computer Science Engineering-EEC (2012)
12. Maring, K.A., Renu, D.: Recognition of cheising iyek/eeyek-manipuri digits using support vector machines. IJCSIT **1**(2), 1694–2345 (2014)
13. Chandan, J.K., Sanjib, K.K.: Recognition of handwritten numerals of Manipuri script. Int. J. Comput. Appl. **84**(17), 1–5 (2013). https://doi.org/10.5120/14674-2835
14. Wilson, E., Donald, W.T.: Multilayer perceptron design algorithm. In: Proceedings of IEEE Workshop on Neural Networks for Signal Processing. IEEE (1994) https://doi.org/10.1109/NNSP.1994.366063
15. Dalal, N., Bill, T.: Histograms of oriented gradients for human detection. In: IEEE Computer Society Conference on Computer Vision and Pattern Recognition (CVPR'05*)*, vol. 1. IEEE (2005) https://doi.org/10.1109/CVPR.2005.177

Effectiveness of Artificial Intelligence in Cognitive Behavioral Therapy

Aakriti Mathur, Harsha Munshi, Srishti Varma, Ankita Arora, and Archana Singh

Abstract The socio-economic conditions around us have been significantly deteriorating taking a toll on the mental health of each contributing to an escalation in the number of people agonizing mental health issues such as anxiety, stress, etc. Due to the corona virus outbreak, many individuals suffered the feeling of helplessness and depression during this time, although the action and reaction vary from person to person, some may be panicking or some may be suffering from collective hysteria, the feeling of helplessness is also emerging due to the desperation and hopelessness caused by the disease. These circumstances demand compelling contemplation in order to eradicate or control the shoot up in the number of cases worldwide. In this paper, we introduce Cognitive Behavioral Therapy (CBT) and its importance in today's time looking at the cur- rent scenario. Throughout the years, there has been an evolution in the way CBT treatment is done. Nowadays, CBT has become more easily accessible as compared to before because sometimes a person's condition might not be that serious and so they might feel awkward going for a therapy session. It will help to treat a lot of people at one time and that too according to their convenience. As of now, we can combine CBT and Artificial Intelligence (AI) to help and reach out to larger audiences keeping in mind to provide convenience and ease, while having the therapy. The motive is to make them appreciate every part of life and look for a positive reason behind everything.

1 Introduction

The corona virus disease 2019 abbreviated as COVID-19 is caused due to the outbreak of a Severe Acute Respiratory Syndrome 2 (SARS-CoV-2) which is an infection that has led to major economic and social crisis worldwide. Along with the social and

A. Mathur (✉) · H. Munshi · S. Varma · A. Arora · A. Singh
Information Technology, Amity University, Uttar Pradesh, Noida, India

A. Singh
e-mail: asingh27@amity.edu

T. Senjyu et al. (eds.), *ICT with Intelligent Applications*, Smart Innovation, Systems and Technologies 248, https://doi.org/10.1007/978-981-16-4177-0_42

economic factors, one of the most alarming and sensitive factors is the psychological distress caused by COVID-19. Due to the vast outreach of the pandemic, all non essential movements for an individual have been prohibited, in order to reduce the spread of the disease. Psychological problems involving depression, frustration, anxiety, stress and even uncertainty of what will happen next has been emerging sequentially. Mass quarantine was implemented in order to eradicate the COVID-19 spread which has led to a feeling of fear and anxiety with the pandemic outbreak and also the escalation of new cases in the number of thousands each day [1].

Most people are suffering from regular negative thinking which has a large number of negative outcomes the worst being suicidal behavior. The World Health Organization (WHO) [2] has been releasing articles about coping up with stress and anxiety in this pandemic which shows that statistically people are suffering from mental ailment worldwide. With this rate of the number of cases regarding stress and anxiety increasing extensively, there need to be self help measures in a way which can lead an individual out of such situations by showing them a positive perspective and teaching them how to look at the greener side of the grass themselves. Such an approach is called Cognitive Behavioral Therapy (CBT). We aim to introduce CBT integrated with Artificial Intelligence (AI) in a way which can be helpful in establishing the importance of CBT in today's time of COVID-19 pandemic and in future for elevating mental health of an individual. CBT has become popular in today's era as it has garnered more attention from people because of how effective it looks for people that are under the treatment of anxiety, depression, etc. (Fig. 1). The first wave

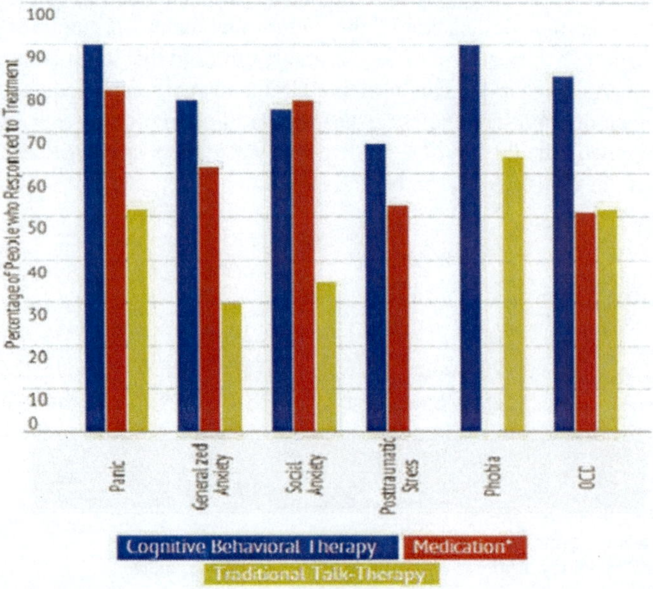

Fig. 1 Representing CBT as the most useful method in battling signs of anxiety

of CBT originated around 1940, and it was for the veterans returning from WWII to treat their short-term depression [3]. During the second wave which happened to be around 1960 Beck, who is a psychiatrist, discovered the link between feelings and thoughts that people go through. He learned that people might not always understand their emotions but they can identify and report them through this treatment. Although these methods were effective it still wasn't that helpful for people under depression so, for the third phase, they merged the cognitive (how you think) and behavior (what you do) to help people facing depression, anxiety related to a number of reasons in their daily lives [4]. Nowadays, CBT has become more easily accessible as compared to before because sometimes a person's condition might not be that serious and so they might feel awkward going for a therapy session for example they might just slightly feel nervous or have an adrenaline rush before a big event so by keeping this in mind and due to its popularity people came up with CBT apps that are just one touch away and people can have a private therapy session for whichever reason they like and help themselves accordingly. The patients can do the assessment provided online anytime. They have to follow a structured course and complete the homework given to them according to their situation. It can be done independently, and it is also affordable [5].

Moreover, it will help to treat a lot of people at one time and that too according to their convenience. Usually, they are given tasks following which they can tend to have a positive outlook and count their blessings. The motive is to make them appreciate every part of life and look for a positive reason behind everything.

2 Literature Review

Alsharif et al. (2015) talked about the psychological approach known as CBT that has been widely used in the smoking cessation method. Most of the smoking cessation apps out there lack commitment and follow-up which is needed for people to quit smoking. The proposed system smoke mind has been used to help track the smoking intake of the users and the goal is to help them quit smoking completely [6]. Michelle et al. (2014) proposed a CBT Assistant app used to help patients diagnosed with a social anxiety disorder (SAD). It has been designed to provide an end-to-end support to the therapist and patient. After the sessions, the patients can do their homework on the app in a more efficient way with an interactive interface [7]. Jarzabek et al. (2018) proposed a CBT assistant app, a combination of web and mobile apps. Therapists can customize treatments through the web app and the patients can view that through the mobile app, while minimizing the time of therapy and serving more clients [8]. Addepally and Purkayastha (2017) introduced an application called MoodTrainer which is built on Rapid Prototyping taking continuous inputs and feedback from the customers. It replaces the old log diary application with a real-time application which is able to manage the user's behavior including sensors which are used to detect if the user is going to an isolated location and inform their emergency contacts for the same. It tracks the user's voice pitch to identify the mood of the user. In order to enlighten the user's mood, the application also plays motivational quotes and the user's proffered songs according to the surveys held earlier [9]. Marzuki et al. (2018) studied about the attrition rate which is the amount of time a user is focused on the CBT based technology and studied various modes of technology integrated with CBT and their attrition rates leading to the conclusion that game based applications have the highest attrition rate among others [10]. Xu et al. (2011) aims at reducing the number of mental health cases in china due to the cases being more than the number of doctors to treat them and the procedure being extensively expensive by introducing a pervasive online Metal Coach which targets anxiety and depression by applying machine learning algorithms and having the provisions of self help as well as mutual help which incorporated the help of medical experts or therapists [11]. Butler et al. (2006) discuss the empirical status of cognitive behavioral therapy by reviewing 16 meta-analyses such as anxiety disorder, panic disorder, social phobia and many more. Based on the study, the authors conclude that CBT provided them promising results for all the 16 meta-analyses. They also concluded that CBT is a better option for treatment in adults than any anti-depressant [12]. Barrera (2017) discusses that although a small portion of the human population; adults do suffer from intellectual disabilities also known as cognitive deficits in which a person is not able to perform his daily functions properly. So, based on systematic review, the author finds that adults with borderline and mild intellectual disability suffering from anger, depression, tensions give out positive results for CBT, making it a person-centered approach [13]. Bekirogullari (2018) discusses that people who are getting old suffer from learning disabilities because of several reasons but this can also lead to mental problems in them.

So, the author reviews that CBT plays an important role in treating people with learning disabilities based on evidence study [14]. Fletcher et al. (2011) discuss the use of wearable sensor systems which can monitor a patient's physiology and is able to deliver therapeutic sessions with a patient's mobile phone based on specific autonomic nervous system activity. Another key feature of the system is the ability to customize the sensor classification algorithm for every individual [15]. Algorri (2003) discussed the need for 3D graphic simulators in therapy to make the fears like height or closed spaces and disorders like eating be visualized in children. Five different simulators are used for the treatment of five different behavior therapies. The simulators can be used by the therapist in the office, in a safe, effective manner. The simulators consist of interactive, virtual worlds where the therapist can introduce progressively stressing situations and help the child deal with them [16]. Patel et al. (2019) discussed the computer program which conducts a conversation with humans either in voice or textual method commonly called a chatbot. It passes the famous Turing Test. In order to identify emotion of user chat text, three deep learning algorithms namely, CNN, RNN, and HAN were deployed. With the help of a label of emotion, it also identifies the mental state of the user such as stressed or depressed. Tokenization of the chats is used for classification and analysis [17].

3 Methodology

3.1 Cognitive Behavioral Therapy (CBT)

Cognitive behavioral therapy or CBT is a psychological technique which involves guiding people to a different perspective which enables them to look at a brighter side of a situation that brings about a change in their thought pattern or procedure [18]. It is based on the fact that a person's thought and belief system is the root of their behavior and their feelings. Basically, it focuses on the thoughts and beliefs of a person which is the cause of the person's outlook or perspective toward a situation and talks the person out enabling them to see things from a better perspective. CBT is a co-operative therapy in which the therapist and the individual have to work in partnership, eventually as stated by the American Psychological Association (APA), the individual eventually becomes his own therapist by having a better perspective and guiding himself through any situation. Therapists believe that CBT is based on the hypothesis that issues emerge from the implications or meanings individuals provide for a situation, just as the situation themselves. Unhelpful implications can make it hard for an individual to work positively in various circumstances. It aims at having a positive influence on how an individual feels about a situation, along with this CBT enables the individual to act and engage in strategies which will help one cope from a tough situation. In order to cater to the various aspects of life, there are different types of CBT techniques like emotional and social challenges to show an individual better perspective of a situation.

3.2 Artificial Intelligence (AI)

The process of making the machines think and act like humans by programming is known as Artificial Intelligence. Any machine which exhibits the human traits like problem solving or ability to learn itself will still have artificial intelligence in the machine. With more transparency in algorithms studied by the researchers, there has been a great increase in the use of Artificial intelligence in every technology. Artificial intelligence has built its roots in almost every field of study but the most recent focus of the researchers lies in the human interaction, cognitive sciences, psychology, and social sciences (Fig. 2). From ages, there is a hope that one day humans can interact with machines with complete simulation and machines will be able to learn and mimic humans. Artificial intelligence can be used to either make innovations in the existing technologies or develop new ones. Making machines smart comes with some difficulties. For example—If artificial intelligence is being applied to the machines which are responsible for the social and mental well being of humans then there should not be any mis-happening because it will ultimately cause destruction [19].

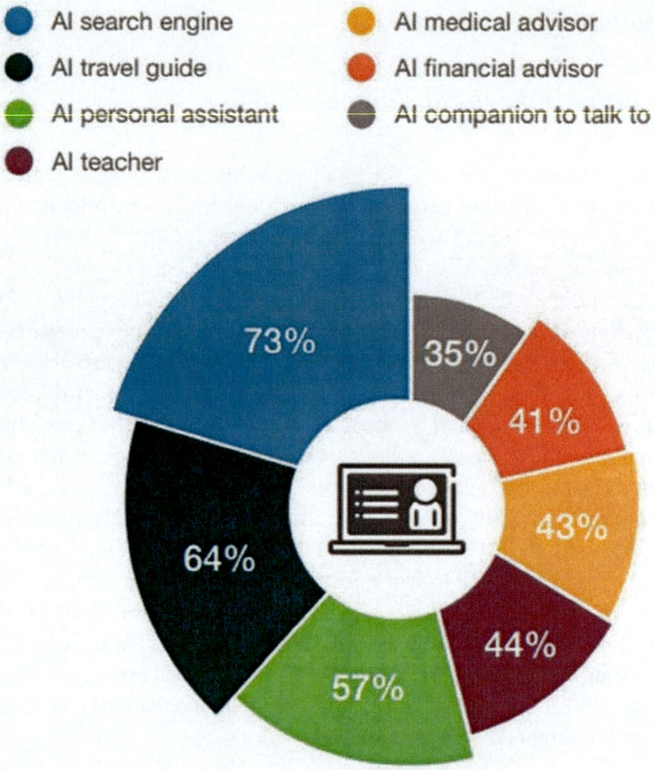

Fig. 2 Representing the consumers who think AI is the best solution

3.3 *Cognitive Behavioral Therapy with AI*

The expanded adoption of artificial intelligence and its progressions are opening new doors for a huge number of people who are affected by common health problems. Artificial Intelligence Therapy overcomes social anxiety, grief, substance abuse, career, etc.; which are based upon a carefully gathered database from different users. This data is utilized to naturally recognize the user's particular thoughts and practices. Because of Artificial Intelligence applications such as recommendation systems, symptom monitoring, etc.; use of CBT has become widely popular to cope with mental stress/ illness, everyday challenges and many more. Such applications use Natural Language Processing (NLP), mood inputs, validated measures and active symptom monitoring to deliver the right mediation to an individual at the right time [20]. As we know that AI is an effort to build machines that can actually learn from the environment, from mistakes and people; one such application of AI is Machine learning (ML). Additionally, AI takes what ML has gained from its current circumstance and utilizes it to make intelligent decisions. The machine learns from the data and starts its decision-making training. After this training, when a user gives an input to the machine, it gives back a logical output or a solution such that it is adapted according to every person's emotional state and in the moment experience.

3.4 Datasets in CBT

To assemble correct data about issues is an important way to overcome them. CBT approach is a treatment that is based on evidence which means that the therapists follow the most effective scientific research. Each person's therapy depends on what is best for them. The type of data that is gathered during the process between therapists and clients is the details about the dilemma, and its symptoms that are causing any kind of stress.

The symptoms or complications that they look for in the client are their emotions, thoughts, memories, behaviors, body sensations, compulsions, etc. The treatment is effective when therapists focus on what matters such as by focusing on some events that took place, basically situations like "Yesterday I drove by this place where the accident took place." The motive behind why CBT focuses on events is because those moments grouped together affect our life. We can track how strong the symptom was and how frequently it happened. Also, record the situations you were in like what you were doing during those symptoms. Sometimes the trigger that caused the symptom is unclear that is why it is necessary to know about the context. As one of the issues for depression is a low level of activity, and the effective solution to it will be increasing the relevant activity level. There are various tools through which therapists can collect data such as thought records, memory records, diaries, online surveys, and some tasks given by therapists to clients like behavior assessment [21].

4 CBT Integrated AI Frameworks

4.1 Existing Frameworks

It would not be wrong if we say that chatbot are taking over the world. One such chatbot is Woebot which provides talk therapy. It was developed by Stanford's psychologists and AI Experts. Woebot utilizes brief day-by-day visit discussions, tracking state of mind, curetted recordings, and word games to assist individuals with overseeing emotional well-being. Patients are urged to discuss their emotional reactions to life occasions, and afterward stop to distinguish the mental snares that cause their pressure, tension, and depression [22]. Another application that is available for everyone on the web is CBT Companion which was developed by Swasth Inc. It is one of the best companion applications to learn CBT Technique with easy to follow visuals tools. CBT-i Coach is a mental health app created by a collaborative effort between the VA's National Center for PTSD, Stanford School of Medicine, and DoD's National Center for Telehealth and Technology. It depends on the treatment manual, Cognitive behavioral therapy for Insomnia in old people [23]. EndeavorRx is the first FDA-cleared video game based treatment delivering application. It is used to improve the attention of children from ages 8–12 years for inattentive or combined-type ADHD. It is used as part of the therapeutic program and not as stand-alone therapy [24]. SuperBetter is a free app that gamifies positive psychology and helps to change behavior helping users to battle anxiety, depression, and post-traumatic stress. The basic idea is to set the mental health goals using the games and naturally build up the psychological strengths which are the best reflected, while playing games [25]. The Quest modular app is a game based application used to learn six evidence-based skills. The game includes different islands and oceans with different avatars, the user travels between the islands to possess different skills. Though gamifying, the therapy concept is very useful but the main challenge lies in the individuality and personalization for every individual [26].

4.2 *Proposed Framework*

After researching and learning about the broad scope of Cognitive behavioral therapy, it can be concluded that the best outcome lies in the therapy by applications using Artificial Intelligence. The aim is to show the effectiveness of CBT with AI by gamifying the mental health concept and designing a personalized application to provide therapy on the skills that need to be considered and also combining it with a conversational agent to be more precise (Fig. 3).

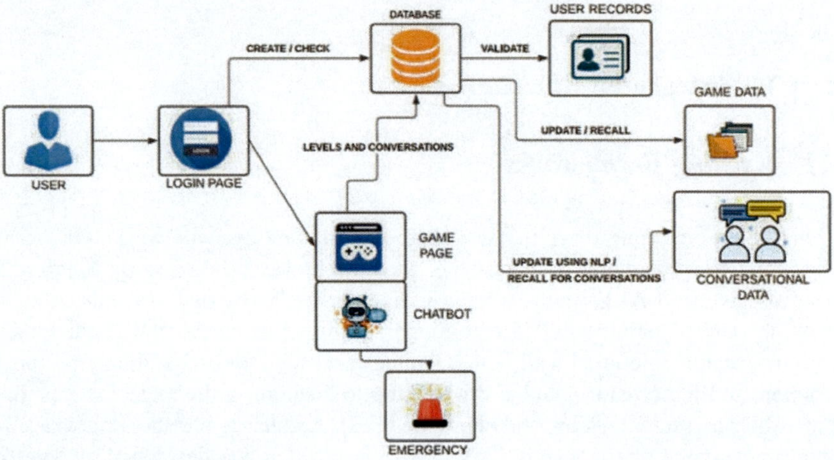

Fig. 3 Proposed framework

A well-defined user interface can be created with characters for better involvement and giving baits to achieve good state of mind by making changes in the gaming life and ultimately instilling the change in the real life. The structure of the interface can be designed to work like fuzzy cognitive systems so that the decisions can be more accurate and will be knowledge-based to exclude the assumptions and treat with real solutions. The interface will be able to interact with every individual according to their needs and the skill set which needs to be improved. Also, including an agent who can extract the features and target skills will be an advantage for making the application more efficient. The AI simulated agent can have conversation with the user to get insight of the user's state of mind and to provide the personalized solutions for the therapy. An emergency feature must be available in the application for situations where the indulging interface can't help to contact acquaintances for support. This framework aims at targeting the individuals who are dealing with depression, anxiety or any kind of negative feeling and are unable to overcome them, in order to showcase to them that there is a path of happiness which they can tread by achieving a positive outlook in life through means of Cognitive behavioral therapy. There can be many advantages of this application like it gives a personalized solution to every problem as every individual has different targets, it also provides a solution to the people who has a mindset of not approaching therapist for the negativity as it gives a friendly atmosphere to share the thoughts and to take responsibility of changing negative ones to positive ones. The application can have a wide future perspective as the personalization can take place according to the advanced cognitive methods and different skill sets could be added as per discoveries. Artificial intelligence is a technology which can show many future advances which could be very beneficial to this application.

5 Conclusion

Our research work aimed at gathering information about Cognitive behavioral therapy and how we can make it more effective in practice with the help of assistance in the form of Artificial Intelligence. Based on our study and our vision for an interface with AI assistance to CBT, we have proposed a framework for our project which we will be carrying forward. The proposed framework aims at bridging the gaps which we found, while studying the existing frameworks along with this our framework aims at enhancing the experience of the user with eradicating their stress and anxiety, thus, leading to a path of positive mental health.

6 Future Scope

In the long run, the idea is to incorporate all of the features that were talked about in this paper and make an intelligent bot with the help of artificial intelligence and

Cognitive behavior therapy. The motive is to help people attain therapy even if they are not financially stable to have direct interaction with the therapist. A fuzzy cognitive method will be used to make decisions by the intelligent bot which will be integrated with voice assistant to make it more alluring and productive. Users going through different situations won't be given therapy in the same way but it would rather be more skill set based means they will be given solutions according to the category they fall in. With the successful incorporation of all these features and testing the intelligent bot, it would also come under the guidance of a psychologist so that the therapy would be legitimate.

References

1. https://academic.oup.com/qjmed/article/113/8/531/5860841
2. https://www.who.int/docs/default-source/coronaviruse/coping-with-stress.pdf?sfvrsn=9845bc3a_2
3. https://www.klearminds.com/blog/history-cognitive-behavioural-therapy-cbt/
4. https://journals.sagepub.com/doi/full/https://doi.org/10.1177/1755738012471029
5. https://thiswayup.org.au/learning-hub/cognitive-behavioural-therapy-explained/#:~:text=Cognitive%20Behavioural%20Therapy%20(CBT)%20is,more%20positive%20and%20fulfilling%20solutions.
6. Alsharif, H., Philip, N.: Cognitive behavioural therapy embedding smoking cessation program using smart phone technologies. In: 2015 5th World Congress on Information and Communication Technologies (WICT), Marrakech, pp. 134–139, (2015).https://doi.org/10.1109/WICT.2015.7489660
7. Michelle, T.Q.Y., Jarzabek, S., Wadhwa, B.: CBT Assistant: MHealth App for psychotherapy. In: 2014 IEEE Global Humanitarian Technology Conference—South Asia Satellite (GHTC-SAS), Trivandrum, pp. 135–140, (2014). https://doi.org/10.1109/GHTC-SAS.2014.6967572
8. Jarzabek, S., Cheong, K., Lim, Y., Wong, J., Kayanoth, R., Teng, J.: CBT assistant platform: web/mobile co-design solution for cognitive behavioural therapy. J. Hosp. Manag. Health Policy. **2**, 34–34 (2018). https://doi.org/10.21037/jhmhp.2018.06.02.
9. Addepally, S., Purkayastha, S. (2017). Mobile-Application based cognitive behavior therapy (CBT) for identifying and managing depression and anxiety. 3–12 (2017). https://doi.org/10.1007/978-3-319-58466-9_1
10. Marzuki, Noor, S.F.M., T. Siti, T., Wook, M.T., Ismail, W.S.W.: Review of computerized cognitive behavioral therapy for depression in adolescents. In: 2017 6th International Conference on Electrical Engineering and Informatics (ICEEI), Langkawi, pp. 1–6 (2017). https://doi.org/10.1109/ICEEI.2017.8312463.
11. Xu, X., Zhu, T., Zhang, R., Lit, L., Li, A., Kang, W., Fang, Z., Ning, Y., Wang, Y.: Pervasive mental health self-help based on cognitive-behavior therapy and machine learning. In: Proceedings—2011 6th International Conference on Pervasive Computing and Applications, ICPCA (2011). https://doi.org/10.1109/ICPCA.2011.6106507.
12. Butler, A.C., Chapman, J.E., Forman, E.M., Beck, A.T.: The empirical status of cognitive-behavioral therapy: a review of meta-analyses. Clin Psychol. Rev. **26**(1), 17–31 (2006). https://doi.org/10.1016/j.cpr.2005.07.003. Epub 2005 Sep 30. PMID: 16199119
13. Crystal, B.: Cognitive behavior therapy with adults with intellectual disabilities: a systematic review, 5-2017, Clinical Research Papers School of Social Work.
14. Zafer, B.: Cognitive behavioral therapy in treating persons with learning disabilities. J. Educ. Sci. Psychol. VIII(LXX), 2/2018, 31–39 (2018)

15. Fletcher, RR, Tam, S, Omojola, O, Redemske, R., Kwan, J.:Wearable sensor platform and mobile application for use in cognitive behavioral therapy for drug addiction and PTSD. In: 2011 Annual International Conference of the IEEE Engineering in Medicine and Biology Society, Boston, MA, 2011, pp. 1802–1805 (2011).https://doi.org/10.1109/IEMBS.2011.6090513
16. Algorri, M.E..: Interactive virtual environments for behavioral therapy. 1, 537–540 (2003). https://doi.org/10.1109/IEMBS.2003.1279792
17. Patel, F., Thakore, R., Nandwani, I., Bharti, S.K.: Combating depression in students using an intelligent chatbot: a cognitive behavioral therapy. In: 2019 IEEE 16th India Council International Conference (INDICON), Rajkot, India, pp. 1–4 (2019).https://doi.org/10.1109/INDICON47234.2019.9030346
18. https://www.apa.org/ptsd-guideline/patients-and-families/cognitive-behavioral
19. https://www.investopedia.com/terms/a/artificial-intelligence-ai.asp#:~:text=Artificial%20intelligence%20(AI)%20refers%20to,as%20learning%20and%20problem%2Dsolving.
20. https://www.ncbi.nlm.nih.gov/books/NBK279297/
21. https://www.psychologytools.com/self-help/data-gathering/
22. https://www.wired.com/2017/06/facebook-messenger-woebot-chatbot-therapist/
23. https://emerj.com/ai-application-comparisons/chatbots-mental-health-therapy-comparing-5-current-apps-use-cases/
24. https://www.endeavorrx.com/
25. https://www.superbetter.com/
26. https://questapp.in/

Discrete Wavelet Transform Based Non-blind Video Authentication Using Principal Component Analysis

Canavoy Narahari Sujatha, Gudipalli Abhishek, and Jeevan Reddy Koya

Abstract This article introduces a non-blind video watermarking scheme embedded with QR code using Principal Component Analysis on Wavelet Transform Coefficients. The proposed method segregates the color video into individual frames. Each frame in video sequence is converted into YUV color space. Luminance component of each frame is decomposed using Discrete Wavelet Transform. Block based Principal Component Analysis (PCA) is applied to selective band of coefficients to get high invisibility. The binary image is chosen to be inserted in principal components of selected coefficients. In this proposal, embedding takes place in mid-frequency coefficients of DWT with least degradation of the perceptual quality of video. Hidden watermark is detected from the marked video by following the inverse processing steps. Strength of the algorithm is tested against various attacks like median filtering, Gaussian noise, salt and pepper noise and rotation, etc. Experimental results exhibit the imperceptibility of video and robustness of watermark image from the proposed scheme.

1 Introduction

With expeditious development of networks like internet, intranet and multimedia schemes, digital content is extensively used. Because of illegal copying of digital data transmission through internet, protection of digital data became popular. Watermarking is one of the digital data protecting methods like cryptography and steganography. Visual transparency and robustness are the two important characteristics of

C. N. Sujatha (✉) · J. R. Koya
Department of ECE, Sreenidhi Institute of Science and Technology, Hyderabad, Telangana, India
e-mail: cnsujatha@sreenidhi.edu.in

J. R. Koya
e-mail: jeevanreddyk@sreenidhi.edu.in

G. Abhishek
Department of Electrical Engineering, Vellore Institute of Technology, Vellore, Tamilnadu, India
e-mail: abhishek.g@vit.ac.in

T. Senjyu et al. (eds.), *ICT with Intelligent Applications*, Smart Innovation, Systems and Technologies 248, https://doi.org/10.1007/978-981-16-4177-0_43

watermarking. Watermarking takes place by inserting logo into host image in such a way that it causes less damage to marked image and high invisibility [1]. Watermark is embedded either in spatial or transform domain. In the present work, embedding is done in spatial transformation of transform coefficients [2–4]. Broadly, transform coefficients are obtained by using Discrete Fourier Transform (DFT), Discrete Cosine Transform (DCT) or Discrete Wavelet Transform (DWT) and so on. To attain spatial transformation, Singular Value Decomposition (SVD), Principal Component Analysis (PCA), Independent Component Analysis (ICA) like techniques are used [5–8].

PCA is a powerful technique to find linear spatial transformation that reduces the image dimensions. The principal components contain most of the image information and also constitute the characteristics of an image [9]. PCA is also known as KL Transform. As DWT is computationally more efficient and have an excellent spatio-frequency localization property, it is highly suitable to identify location in the video frame for inserting the secret imperceptibly. The wavelet decomposition of the video frame still keeps some amount of correlation between the wavelet coefficients. PCA is mainly used to form hybridized DWT algorithm as it has the inherent property of removing the correlation between the wavelet coefficients [5, 7, 10]. PCA helps in distributing the watermark data bits in the sub band used for insertion. Thus, result in a more robust watermarking scheme that is resistant to almost all possible attacks. Earlier few researchers attempted to embed in individual components of YUV color map [11]. After comparing the results of individual components, researchers find that Y component is more suitable for implanting with less human eye sensitivity. The watermark data is implanted into the luminance component of the individual frames as it has least sensitivity to the human visual system (HVS). Many of the video authentication algorithms are based on the image authentication schemes [10].

Singular values based authentication provides not only invisibility but also dimension reduction based on the requirement [12]. Blind video authentication with scrambling and scene change technique helps to get more robustness in addition with quality [13]. Using entropy analysis, a block with low entropy will be chosen to decompose into channels. According to tradeoff between quality and robustness, channels are picked to apply singular value decomposition to insert the logo with high invisibility [14]. Similarly, to improvement in quality was observed by embedding Hilbert coefficients obtained from gray image into video frames with integer wavelet filters using PCA. This type of mechanism is employed to generate digital signature [15].

Ant colony optimization scheme along with DWT and PCA is used for copyright protection, and here threshold value is calculated using Ant colony optimization algorithm to adopt best frame to watermark [16]. Dimensionality reduction scheme with wavelet transform is used to insert the binary information in LL band without affecting the human vision. But this hybridization is not much resistant to different geometrical and noise attacks [17]. Assorted techniques for blind and non-blind watermarking process are available, but enormous improvement in imperceptibility and similarity were achieved with two level DWT and SVD [18]. PCA can be used to embed the secret information as well as to reduce the dimension of feature set, while sorting the embedding coefficients [19, 20].

An innovative scheme with Tchebichef transform is introduced for image watermarking and compared with robust PCA and SVD schemes in terms of imperceptibility and robustness [21].

In the present work, a non-blind invisible watermarking for QR code embedding in a video sequence is proposed where the information about original multimedia is used to extract the watermark. The main objective is to verify the imperceptibility of present marked algorithm and robustness against disparate assaults. The rest of the paper is structured as follows. Section 2 describes the methodology for hybridized watermarking algorithm. Section 3 presents the experimental results, and Sect. 4 narrates the conclusion.

2 Methodology

In this section, two basic methods of watermarking operation are presented: Implanting method and Detection method for multimedia. Prior to implanting the watermark, location is selected which is suitable for watermarking efficiently with less visual degradation. The following sub sections give the details about embedding and detection methods.

2.1 *QR Code Implanting Method*

A binary QR code image of size n × n is selected as a watermark, where each pixel is represented with 0's or 1's. Size of frames in video sequence is taken as 2nx2n.

1. Host video is isolated into N number of frames
2. Frames are converted from RGB color space to YUV color space
3. Y component is decomposed into sub-bands by applying DWT
4. LH band is partitioned into blocks of size same as watermark size n × n
5. Each block is converted into a row vector
6. Calculate the mean and standard deviation of row vector
7. Centered values of row vector are computed by dividing difference between elements of row vector and mean with standard deviation
8. PCA is applied on centered values vector to acquire principal component coefficient matrix
9. Find the vector component score by multiplying the centered values vector with coefficient matrix
10. Binary bits of the watermark are scaled with scaling factor and get added with component score to produce modified principal component score.
11. Step 10 is repeated for each block for embedding watermark bits
12. Modified wavelet coefficients are obtained by applying inverse PCA

13. Marked luminance component is acquired by applying inverse DWT on LL, modified LH, HL and HH bands
14. Thus marked frame is converted back to RGB space
15. Watermarked video is generated by combining all the frames
16. Imperceptibility is computed between host and marked videos.

2.2 QR Code Detection Method

QR code is the trade mark of the two-dimensional barcode which is machine readable label attached to an item to record the information about the item. QR code consists of black square dots on white background that is read by any imaging device like camera or mobile and processed until the image is properly interpreted. Then, data is identified from the vertical and horizontal patterns of the image. The steps involved in watermark detection are follows.

1. Watermarked video is divided into N number of frames
2. Each frame is converted from RGB format to YUV color space
3. Y component is subjected to DWT to get four sub bands
4. LH sub band is divided into blocks of size nxn
5. Steps 6 to 9 of embedding method are repeated to form principal components score from modified LH band
6. With the help of original frame principal component score, watermark bits are obtained from that of each block of modified LH band at chosen scaling factor
7. Watermark is produced by reshaping the watermark bits vector into nxn
8. Steps 2 to 6 are repeated for each frame to detect the watermark even under various subjected attacks
9. Similarity factor is calculated between original and detected watermarks.

3 Results and Discussion

A suzie video of 150 frames of size 256 × 256 is taken as host media for estimating the efficiency of present algorithm. Simulations are performed using Matlab. QR code image of size 32 × 32 has taken as watermark logo. In this algorithm, video is partitioned into frames from which each frame is converted into YUV color map. Then, Y component is decomposed into four resolution bands of size 128 × 128 using Haar wavelet. LH band is chosen for embedding by subjecting PCA on LH band using statistical parameters. Thus, watermark bits are embedded into principal component scores of LH band coefficients.

Due to the insertion of same watermark in all frames, detection of watermark is easy even with single frame after frame swapping and frame dropping. But with frame averaging, it is observed that retrieving is difficult. Figure 1a–d shows original video snap shot, input watermark, marked video shot and extracted watermark. PSNR

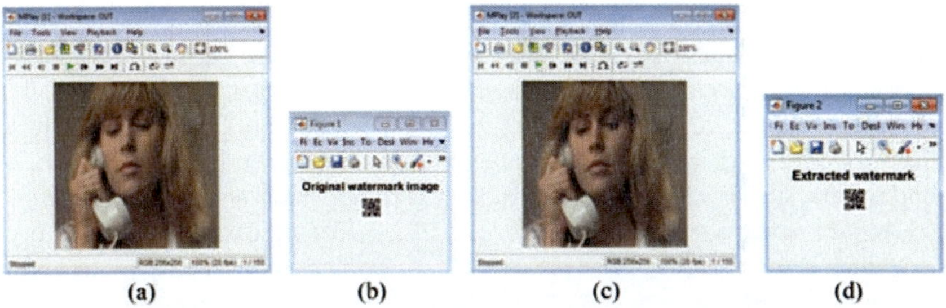

Fig. 1 **a** Snap shot of original video, **b** input watermark image, **c** snap shot of watermarked video, **d** extracted watermark

measured without any distortion is 54.15 dB. Similarity factor is unity under no attack and against frame dropping and swapping. Marked video is subjected to various attacks as displayed in Fig. 2a–i with respective detected watermarks. Figure 2a shows a watermarked frame with Gaussian noise attack with variance 0.02 and zero mean. If the variance increases, similarity factor drastically falls down with this algorithm. Similarly, salt and pepper noise attacked video with noise density 0.02 is

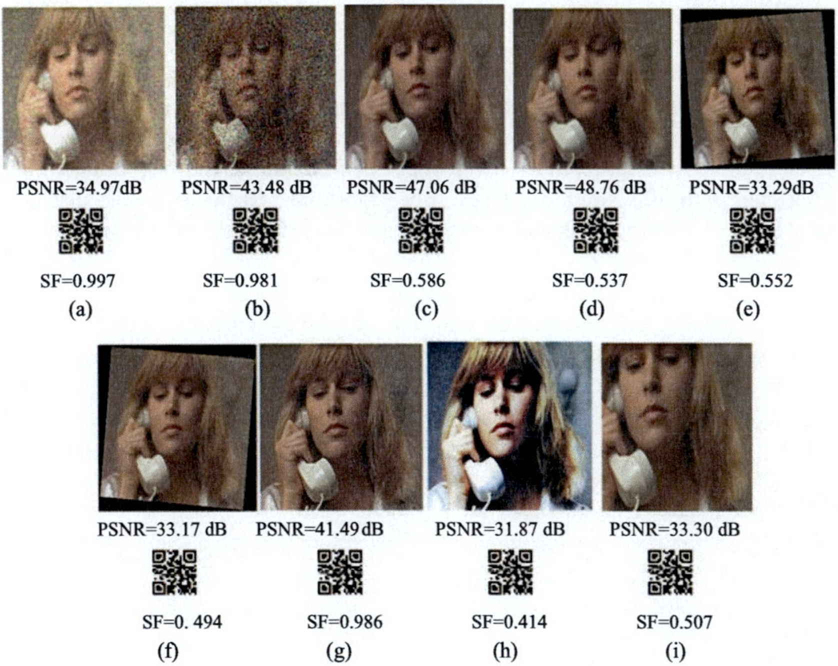

Fig. 2 Watermarked video frame with detected watermarks under **a** Gaussian noise, **b** Salt and pepper noise, **c** average filtering, **d** median filtering, **e** rotation in anti-clock wise direction, **f** rotation in clock wise direction, **g** sharpening, **h** histogram equalization, **i** cropping

shown in Fig. 2b and found that this algorithm is more resistant to Gaussian and salt and pepper noise attacks at respective low variance and noise density values. The algorithm is weak to recognize the watermark at large noise density values.

From the results, it is observed that the algorithm is least resistant to rotation either in clock wise or anti-clock wise direction, filtering, histogram equalization and cropping attacks. Though the imperceptibility is preserved in marked video against median and average filtering with a mark of size 3×3, identified watermark similarity is not much better to characterize the strength of the algorithm. The values of SF in Fig. 2 show that the proposed algorithm is stubborn to sharpening attack compared to rest of all distortions. The results show that the algorithm is successful in identifying the embedded watermark under all distortions but poor in maintaining the similarity factor between embedded and detected watermark. So there is scope to improve the performance of the algorithm by combining with other suitable transforms in order to maintain the high robustness. Like in singular value decomposition (SVD) based algorithm, scaling factor plays a vital role in maintaining transparency of host media in PCA based algorithm. In the proposed work, scaling factor is chosen for watermarking process after making several observations.

4 Conclusion

A PCA based non-blind invisible video watermarking algorithm in DWT domain is presented in this paper. The frames of video are mapped into YUV color space to adopt luminance component for embedding. Binary watermark bits are implanted into principal component scores of LH band in each frame of video. The hidden watermark is detected with high-similarity factor. The algorithm resistance against different possible attacks is described, and it ensures a transparent watermark. In the proposed algorithm, only LH band is selected for embedding, instead remaining HL and HH bands can be chosen as embedding location to improve the embedding capacity, while maintaining invisibility. The selection of both middle- and high-frequency bands HL and HH leads to implant more number of secrete logo bits to transmit huge information in concealed way compared to single band. It has been understood that the present algorithm is lagging in resisting rotation, histogram, filtering and cropping assaults. In future, this algorithm can be further enhanced to prove to be more robust to conquer these limitations and used to embed color logo in middle- and high-frequency coefficients of each frame of video. It can also be implemented with scene change analysis technique to achieve more secured authentication.

References

1. Chang, C.-C., Chan, C.-S.: A watermarking scheme based on principal component analysis technique. Informatica **14**(4), 431–444 (2003)

2. Kaarna, A., Toivanen, P.: Digital watermarking of spectral images in PCA/Wavelet-transform Domain, IEEE, pp. 3564–3567 (2003)
3. Hien, T.D., Chen, Y.-W., Nakao, Z.: Robust digital watermarking based on principal component analysis. Int J Comput Intell Appl 4(2), 183–192 (2004)
4. Mirza, H.H., Thai, H.D., Nagata, Y., Nakao, Z.: Digital video watermarking based on principal component analysis. IEEE (2007)
5. Mostafa, S.A.K., Tolba, A.S., Abdelkader, F.M., Elhindy, H.M..: Video watermarking scheme based on principal component analysis and wavelet transform. Int. J. Comput. Sci. Netw. Secur. (IJCSNS) **9**(8), (2009)
6. Amini, M., Yaghmaie, K., Sadreazami, H.: A new scheme for dual watermarking using DWT-PCA technique. Int. Conf. Imaging Theory Appl. (IMAGAPP) 43–46 (2010)
7. Yassin, N.I., Salem, N.M., El Adawy, M.I.: Block based video watermarking scheme using wavelet transform and principle component analysis. Int. J. Comput. Sci. Issues (IJCSI) **9**(1, 3), 296–301 (2012)
8. Liu, T., Xu, X., Wang, X.: M-band wavelet and cosine transform based watermark algorithm using randomization and principal component analysis. Int J. Sci. Eng. Invest. **2**(13), 1–4 (2013)
9. Chimanna, M.A., Khot, S.R.: Digital video watermarking techniques for secure multimedia creation and delivery. Int. J. Eng. Res. Appl. (IJERA) 3(2),839–844 (2013)
10. Saurabh, P., Pooja, D., Sharayu, B., Nishigandha, J.: Video watermarking scheme based on DWT and PCA for copyright protection, IOSR J. Comput. Eng. (IOSR-JCE) 9(4), 18–24 (2013)
11. Saboori, A., Hosseini, S.A.: Color image watermarking in YUV color space based on combination of DCT and PCA. In: 23rd Iranian Conference on Electrical Engineering (ICEE), pp. 308–313. IEEE (2015)
12. Sujatha, C.N.: Invisible color image authentication in SVD domain. In: International Conference on Information and Communication Technology for Intelligent Systems (ICTIS 2017), 25th–26th Mar 2017, Ahmedabad, Gujarat, pp. 262–269, Smart Innovation, Systems and Technologies, Vol. 83. Springer (2017)
13. Sujatha, C.N., Satyanarayana, P.: DWT based blind video watermarking using image scrambling technique. In: International Conference on Information and Communication Technology for Intelligent Systems (ICTIS 2018), 6th–7th April 2018, Smart Innovation, Systems and Technologies 106, Ahmedabad, Gujarat, 2018
14. Rasti, P., Samiei, S., Agoyi, M., Escalera, S., Anbarjafari, G.: Robust non-blind color video watermarking using QR decomposition and entropy analysis. J. Vis. Commun. Image R. **38**, 838–847 (2016). https://doi.org/10.1016/j.jvcir.2016.05.001
15. Agilandeeswari, L., Ganesan, K.: An efficient hilbert and integer wavelet transform based video watermarking. J. Eng. Sci. Technol. **11**(3), 327–345 (2016)
16. Sake, A., Tirumala, R.: Hybrid discrete wavelet transform-methods of principal components based video watermarking for copyright protection using ant colony optimization techniques. Global J. Pure Appl. Math. **13**(2), 795–815 (2017). ISSN 0973–1768
17. Sinha, S., Bardhan, P., Pramanick, S, Jagatramka, A., Kole, D.K., Chakraborty, A.: Digital video watermarking using discrete wavelet transform and principal component analysis. Int. J. Wisdom Based Comput. 1(2), (2011)
18. Shanmugam, M., Chokkalingam, A.: Performance analysis of 2 level DWT-SVD based non blind and blind video watermarking using range conversion method. Microsyst. Technol. (2018). https://doi.org/10.1007/s00542-018-3870-x
19. Rathour, G., Singh, S.: Principal component analysis based digital video watermarking. Int. J. Res. Eng. Sci. Manag. 1(9), (2018). ISSN: 2581–5782
20. Verma, V.S., Bhardwaj, A., Jha, R.K.: A new scheme for watermark extraction using combined noise-induced resonance and support vector machine with PCA based feature reduction. Multimedia Tools Appl. (2019) https://doi.org/10.1007/s11042-019-7599-z
21. Noor, R., Khan, A., Sarfaraz, A., Mehmood, Z., Cheema, A.M.: Highly robust hybrid image watermarking approach using Tchebichef transform with secured PCA and CAT encryption. Soft. Comput. (2019). https://doi.org/10.1007/s00500-019-03838-2

Automation of MC/DC Coverage Test Case Suite Deploying the Optimal Strategic Choice for Tool Development

Sonal Shekhawat, Arshad Iqbal, Usha Srinivsan, and Pratibha Menon

Abstract The avionics standard for software development, DO-178B, has mandated the modified condition/decision coverage (MC/DC) test criterion for avionics software testing. MC/DC is a structural coverage decisive factor which requires the specific test cases to be executed where each condition is shown to be independently affecting the decision outcome. There can be various methods possible to achieve the same. This paper discusses two distinct approaches to automate the test case generation process for 100% MC/DC coverage achievement. Software tool pertaining to the chosen methodology has been carried out deploying python as the programming language. Further, this paper has demonstrated the implemented tool by embedding the preferred approach using python language.

1 Introduction

DO-178B, "Software Considerations in Airborne Systems and Equipment Certification", guideline for software production for airborne system, classifies system software as per safety assessment and allied risk. Multiple objectives are mandated by DO-178B as per the assigned design assurance levels. When software failure results in catastrophe events, it is categorised as level A software and is considered to be the most critical and crucial software. The obligatory objectives for level A ensure rigorous testing to detect and avoid any kind of off-nominal conditions.

As per DO-178B, Level A software should ensure 100% MC/DC in addition to the decision coverage and statement coverage [1]. MC/DC coverage subsumes all other types of coverage by attaining the defined objectives for them. To assure 100% MC/DC coverage, there are multiple test cases that are required to be executed based on the number of inputs and type of logical operator used. This paper has proposed two different approaches to automate the test case suite generation for achieving 100% MC/DC coverage. Subsequently, one algorithm will be chosen based on the apt functionality, least number of test cases and authentication by compliance to the

S. Shekhawat (✉) · A. Iqbal · U. Srinivsan · P. Menon
Gas Turbine Research Establishment, DRDO, Bangalore, India

T. Senjyu et al. (eds.), *ICT with Intelligent Applications*, Smart Innovation, Systems and Technologies 248, https://doi.org/10.1007/978-981-16-4177-0_44

definition of MC/DC defined by Chelenski and Miller [2]. This paper starts off with briefing of MC/DC followed by the algorithms proposed to automate the test case suite generation process. Further, one of the proposed algorithms will be chosen to be implemented as part of the automation process based on the comparative study analysis mentioned in Sect. 4. Finally, Sect. 5 demonstrates the generated tool.

2 What is MC/ DC?

To ensure that the software is free from all the bugs, there is a test criterion that is to be carried out on the software as a part of Independent Verification & Validation (IV&V). The metric to measure the extent to which one activity has achieved its objective is termed as coverage [1]. The coverage analysis percentage gives the degree measure up to which the given software has been verified [3].

Requirement coverage and structural coverage are two types of coverage measures. Requirement coverage assesses the software with respect to the functionality of the software. It authenticates the software with respect to the requirements but, may not have executed the full code, resulting in oversight of the unexecuted code structure. So, to assure full code coverage, the code is subjected to an additional check called structural coverage analysis. Control flow analysis and Data flow analysis are two methodologies which are covered under structural coverage analysis [4]. Focus of data flow analysis is to analyse and inspect the usage and flow of data in the software, whereas control flow analysis examines the flow of control by wielding the control constructs, inspecting the execution of statements and evaluating the decision outcomes. Statement coverage (SC), decision coverage (DC), condition coverage (CC), condition/decision coverage (C/DC), multiple conditions coverage (MCC) and modified condition /decision coverage (MC/DC) are the control flow testing techniques [5].

In aerospace domain, to realise the intricate functionality for safety critical software, lengthy and complex boolean expressions are used. Test case execution with permutations and combinations of multiple inputs is essential to verify and validate the functionality of the software and to rule out the possibility of presence of any error.

Adequacy of test case suite is assessed by the percentage of coverage it provides to test the boolean expressions [6]. Statement coverage analyses the coverage for each statement present in the code. As depicted in Table 1, decision coverage mainly focuses on the decision part of the source code. Every decision must take all possible outcomes. Unlike decision coverage, focal point of condition coverage is the individual conditions. It says that all the conditions in a decision should take all possible values at least once. C/DC requires all the conditions in a decision to take all possible values in such a way that the decision also results in all possible values at least once. C/DC covers both decision and condition coverage, but does not ensure exhaustive testing. To ensure exhaustive and in-depth testing, multiple condition coverage

Table 1 Coverage requirements

S. No.	A	B	A OR B	DC	CC	C/DC	MCC	MC/DC
1	False	False	False	•		•	•	•
2	False	True	True		•		•	•
3	True	False	True	•	•		•	•
4	True	True	True			•	•	

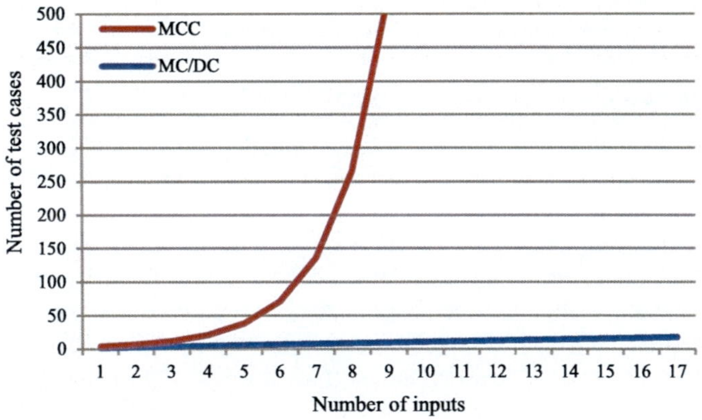

Fig. 1 Variation in quantum of test cases with respect to number of inputs

(MCC) analysis is carried out [7]. It requires execution of all possible input combinations at least once. The shortcoming of this coverage analysis is the exponential increase in number of test cases with linear increase in number of inputs. For n inputs, the number of test cases to ensure 100% MCC coverage will be 2^n, which is not feasible as depicted in Fig. 1.

Here comes the MC/DC into the picture, which finds the middle ground by retaining the linear increase in the number of test cases while ensuring the profound and thorough testing at the same time [6]. MC/DC thoroughly tests the independent effect of each condition on the decision outcome. Effect of each condition is analysed by varying that condition exclusively while keeping remaining conditions constant [8].

The manual approach to achieve 100% MC/DC requires expertise in proper test case selection, and it becomes more time-consuming and tedious with increase in number of inputs. So, for better time management and accurate results, process automation is proposed. Automated solution of any problem requires a well-defined algorithm. This paper discusses two different algorithms as per the definition of MC/DC to generate the test case suite for 100% MC/DC coverage achievement and will further implement one of the algorithms to automate the complete process of test case suite generation.

3 Algorithms for MC/DC

The foremost and vital step in the process of automation is the design of the algorithm. One can seek multiple courses of action in order to derive the dynamic algorithm for MC/DC test case suite generation. Proposed approaches to derive the test case suite are as follows:

- Minimal Case Analysis Approach (MCAA)
- Duo Case Analysis Approach (DCAA)

3.1 Minimal Case Analysis Approach (MCAA)

This approach analyses the test case requirement for basic decision statements consisting of one logical operator as per the definition of MC/DC and examines them further to determine the test cases for complex decision statements expression. It starts by inspecting the independent effect of each condition on the basic decision statement and will further use the obtained results to generate the test cases for the derived decision statements. The AND and OR operator test case suite for MC/DC forms the baseline for MCAA. Further complex Boolean expressions are evaluated using results obtained from the base lined test cases.

AND Operator. The truth table of AND operator with two inputs is shown in Table 2. Going by the definition of MC/DC, the independent effect of each input on the decision outcome has to be shown in the final test case suite. For two inputs, four test case combinations are possible.

Test case 1 and 3 clearly depicts the output variations, by varying X input alone, keeping the other input Y constant. Thus, the independent effect of X is evident as mandated by MC/DC definition. Similarly, test case 1 and 2 show the independent effect of condition Y on the decision outcome. So, to ensure 100% MC/DC coverage, test case 1, 2 and 3 are sufficient for two input AND operator.

OR Operator. For OR operator, Table 3 shows all the possible test cases. Test case 2 and 4 show the independent impact of X, while test case 3 and 4 are depicting the effect of input Y on the outcome. So in case of a two input OR operator, test case 2, 3 and 4 are adequate to ensure 100% MC/DC coverage.

Table 2 Test case required for AND operator

S. No.	X	Y	OUTPUT (X AND Y)
1	TRUE	TRUE	TRUE
2	TRUE	FALSE	FALSE
3	FALSE	TRUE	FALSE
4	FALSE	FALSE	FALSE

Table 3 Test case required for OR operator

S. NO.	X	Y	OUTPUT (X OR Y)
1	TRUE	TRUE	TRUE
2	TRUE	FALSE	TRUE
3	FALSE	TRUE	TRUE
4	FALSE	FALSE	FALSE

Evaluation of complex expression using MCAA. The specific test cases identified for AND and OR operator shall be used to evaluate complex Boolean expressions involving a combination of multiple logical operators. Considering the example X = ((P AND Q) OR S) = C OR S, where C = (P AND Q), the sequence of steps associated with MCAA is as follows:

Step 1: Tabulate all the possible 2^n combinations with n inputs as shown in Table 4.

Step 2: Considering base lined test case suite generated Tables, namely Table 2 and 3, map the corresponding test cases of Table 4 and tabulate the results in Table 5.

Step 4: Choose the test cases in such a way that minimum number of test cases fulfil the requirement given in Table 5 to accomplish 100% coverage (Table 6).

Table 4 Truth table for expression X

S. No.	P	Q	C (P AND Q)	S	X = ((P AND Q) OR S)
1	FALSE	FALSE	FALSE	FALSE	FALSE
2	FALSE	FALSE	FALSE	TRUE	TRUE
3	FALSE	TRUE	FALSE	FALSE	FALSE
4	FALSE	TRUE	FALSE	TRUE	TRUE
5	TRUE	FALSE	FALSE	FALSE	FALSE
6	TRUE	FALSE	FALSE	TRUE	TRUE
7	TRUE	TRUE	TRUE	FALSE	TRUE
8	TRUE	TRUE	TRUE	TRUE	TRUE

Table 5 Mapping of baseline test cases with complex expression from Table 4

Condition	Requirement	Test case Sl no
(P AND Q)	(TRUE TRUE)	7,8
	(TRUE FALSE)	5,6
	(FALSE TRUE)	3,4
(C OR S)	(FALSE FALSE)	1,3,5
	(TRUE FALSE)	7
	(FALSE TRUE)	2,4,6

Table 6 MC/DC test case suite

S. No.	P	Q	C (P AND Q)	S
3	FALSE	TRUE	FALSE	FALSE
6	TRUE	FALSE	FALSE	TRUE
7	TRUE	TRUE	TRUE	FALSE

The base test case requirements for AND operator [(True, True),(True, False),(False, True)] are realised by test cases number 7, 6 and 3, respectively. Similarly [(False, False), (True, False), (False, True)] , test case 3, 7 and 6 are sufficient, and hence, test case 3, 6 and 7 will ensure 100% MC/DC coverage for expression X. The point to be observed here is that the final test case suite does not explicitly show the independent impact of each condition as per the definition of MC/DC.

3.2 *Duo Case Analysis Approach (DCAA)*

This approach contemplates the definition of MC/DC (the independent effect of each condition on the decision outcome). The focal point of this idea is to decide on those test cases, wherein changing one condition alone keeping the other conditions constant, the decision of the expression toggles.

Considering the example Y = ((P AND Q) OR (R AND S)), the sequence of steps associated with DCAA is as follows:

Step 1: Tabulate all the possible 2^n conditions with n inputs in Table 7 same as MCAA.

Step 2: For each input, select the test case pairs in such a way that on varying that specific input while keeping the others constant, the outcome of the expression toggles and record the test case pair data as shown in Table 8. For example: test case 5 and 13 show the individual effect of input P on the outcome.

Step 3: Generate pair wise mapping of the test case pairs corresponding to each input for the complete set of test cases as shown in Table 9. This quantifies the contribution of each test case to all the inputs and this way, the minimal number of test cases can be identified.

Step 4: This step is the most crucial step in this strategy where the test cases will be selected in a minimal way while ensuring the visible independent effect of each condition's contribution.

The tool developed uses the correlation matrix to provide the optimal derived test case suite. One such example is depicted in Fig. 2, wherein the independent effect of P is indicated by test case 6 and 14, Q is indicated by test case 10 and 14, R is indicated by test case 10 and 12, and S is indicated by test case 11 and 12.

Table 7 Truth table for expression Y

S. No.	P	Q	R	S	Y = ((P AND Q) OR (R AND S))
1	FALSE	FALSE	FALSE	FALSE	FALSE
2	FALSE	FALSE	FALSE	TRUE	FALSE
3	FALSE	FALSE	TRUE	FALSE	FALSE
4	FALSE	FALSE	TRUE	TRUE	TRUE
5	FALSE	TRUE	FALSE	FALSE	FALSE
6	FALSE	TRUE	FALSE	TRUE	FALSE
7	FALSE	TRUE	TRUE	FALSE	FALSE
8	FALSE	TRUE	TRUE	TRUE	TRUE
9	TRUE	FALSE	FALSE	FALSE	FALSE
10	TRUE	FALSE	FALSE	TRUE	FALSE
11	TRUE	FALSE	TRUE	FALSE	FALSE
12	TRUE	FALSE	TRUE	TRUE	TRUE
13	TRUE	TRUE	FALSE	FALSE	TRUE
14	TRUE	TRUE	FALSE	TRUE	TRUE
15	TRUE	TRUE	TRUE	FALSE	TRUE
16	TRUE	TRUE	TRUE	TRUE	TRUE

Table 8 Test case pairs for toggled inputs

Toggled input	Test case pair no
P	5-13,6-14,7-15
Q	9-13,10-14,11-15
R	2-4, 6-8,10-12
S	3-4,7-8,11-12

Table 9 Input correlation matrix with test cases

	1	2	3	4	5	6	7	8	9	10	11	12	13	14	15	16
p					1	2	3						1	2	3	
q									1	2	3		1	2	3	
r		1		1		2		2		3		3				
s			1	1			2	2			3	3				

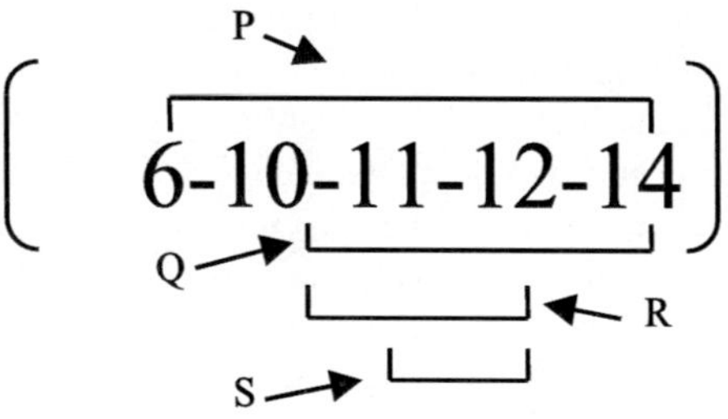

Fig. 2 MC/DC test case suite

4 Comparative Study of Algorithms

Since the quantum of test cases increases with every additional input, MCC is not practically feasible. This major constraint became basis for proceeding with MC/DC. Moving on to MC/DC implementation as indicated in section 3, both MCAA and DCAA can be used to achieve 100% MC/DC individually. Evaluation of both the approaches with respect to the significant parameters of interest enabled the apt selection of the meritorious approach for MC/DC test case suite generation. Further, to identify the most favourable strategy among these two, a comparative study has been carried out as follows:

- Both algorithms assure 100% MC/DC coverage.
- In terms of number of test cases required, both MCAA and DCAA show linear rate of growth in number of test cases with respect to the increase in number of inputs.
- MCAA uses the baseline cases of logical operations to get minimal number of test cases for the complex derived decision statements. On the contrary, focal point of DCAA is to determine the impact of each individual condition on the decision outcome in the final test case suite.
- The tabulated results in Table 10 reveals that the number of test cases generated by MCAA is less than or equal to the number of test cases generated by DCAA.
- The final test case compilation of MCAA does not evidently show the independent effect of each condition on the overall result, whereas results of DCAA are absolutely consistent with the definition of MC/DC.

This comparative study and analysis yielded:

- No major variation is observed is number of test cases required for MCAA and DCAA. The rate of growth of number of test cases with respect to the input is equivalent.
- Considering the authentication of the strategies to the definition of MC/DC, DCAA emerges as the valid choice for automatic MC/DC tool generation.

5 Design of MC/DC Tool Using Python

Python is a high-level general purpose programming language [9]. It is interpreted, strongly typed and highly structured language. The reason to choose python as the programming language to automate the tool generation process is the vast and robust library it offers [10]. The tool for generation of test case suite for MC/DC has been implemented in python using DCAA. On execution, the tool prompts for the entry of the expression for which the MC/DC test cases need to be generated. Once the required expression is keyed in, the tool applies the DCAA and start analysing all the 2^n test case combinations possible with n inputs. It will then compare each test case combination with the other combinations to see the individual impact of each

Table 10 Comparative analysis of MCAA and DCAA

Expression	No. of test cases using MCAA	No. of test cases using DCAA
(A OR B)	3	3
(A AND B)	3	3
(A OR (B OR C))	3	4
(A AND (B OR C))	3	4
((A AND B) OR C)	3	4
(A OR (B AND C))	3	4
((A OR B) AND C))	3	4
(A AND (B AND C))	3	4
((A AND B) OR (C AND D))	3	5
((A OR B) OR (C AND D))	4	5
((A AND B) OR (C OR D))	4	5
((A OR B)) OR (C OR D))	4	5
((A AND B) AND (C AND D))	3	5
((A OR B) AND (C AND D))	4	5
((A AND B) AND (C OR D))	4	5
((A OR B)) AND (C OR D))	3	5
(((A AND B) OR (C OR D)) OR E)	5	6

condition, and if the same is observed then, it will be stored as a possible pair against that particular condition. The same is repeated for all the test cases, and possible test case pairs will be noted down against each condition as shown in Fig. 3 which will further be used to choose the least number of test cases for MC/DC.

6 Conclusions

Structural code coverage qualitatively assesses the aptitude of the testing process by executing each statement of the source code. MCC examines all possible combinations of the inputs of the decision statement to ensure exhaustive testing. However, with the increase in number of inputs, the exponential increase in number of test

```
Enter expression: (a and b) or (c and d)
+-----+-----+-----+-----+-------------------------+
|  a  |  b  |  c  |  d  |  (a and b) or (c and d)  |
|-----+-----+-----+-----+-------------------------|
|  1  |  1  |  1  |  1  |            1            |
|  1  |  1  |  1  |  0  |            1            |
|  1  |  1  |  0  |  1  |            1            |
|  1  |  1  |  0  |  0  |            1            |
|  1  |  0  |  1  |  1  |            1            |
|  1  |  0  |  1  |  0  |            0            |
|  1  |  0  |  0  |  1  |            0            |
|  1  |  0  |  0  |  0  |            0            |
|  0  |  1  |  1  |  1  |            1            |
|  0  |  1  |  1  |  0  |            0            |
|  0  |  1  |  0  |  1  |            0            |
|  0  |  1  |  0  |  0  |            0            |
|  0  |  0  |  1  |  1  |            1            |
|  0  |  0  |  1  |  0  |            0            |
|  0  |  0  |  0  |  1  |            0            |
|  0  |  0  |  0  |  0  |            0            |
+-----+-----+-----+-----+-------------------------+
Test Case pairs for input  a value:  [(1, 9), (2, 10), (3, 11)]
Test Case pairs for input  b value:  [(1, 5), (2, 6), (3, 7)]
Test Case pairs for input  c value:  [(4, 6), (8, 10), (12, 14)]
Test Case pairs for input  d value:  [(4, 5), (8, 9), (12, 13)]
Solution
1 : (True, True, True, False)
9 : (False, True, True, False)
8 : (False, True, True, True)
10 : (False, True, False, True)
5 : (True, False, True, False)
```

Fig. 3 Tool generated MC/DC test case suite

cases makes it practically unfeasible. As discussed earlier, the linear increase in the number of test cases with respect to the input and the ability to sustain a high error-detection probability makes MC/DC feasible. The independent effect of the inputs on the outcome is the basis of MC/DC. Achievement of MC/DC for level A safety critical software is essential to accomplish DO-178B standard. There can be multiple approaches to achieve the same and generate the test case suite for 100% MC/DC coverage. This paper has discussed two distinct strategies to attain 100% MC/DC coverage and has identified the best to automate the process of test case suite generation.

Both MCAA and DCAA provide 100% MC/DC coverage with the generated test case suite. MCAA is more of a constructive approach with focus on the base case scenarios for the logical operators. Having basic test cases in hand with the independent effect of each condition shown, test cases for the overall expressions will be generated. However, DCAA shows the independent effect of each input in the final test case suite generated. In terms of number of test cases and coverage, both MCAA and DCAA qualify to be implemented for automatic MC/DC test case suite generation tool. However, the definition of MC/DC entailing the individual effect of each input is being accomplished only by DCAA. So, to align with the definition of MC/DC while ensuring the linear growth rate in number of test cases, DCAA strategy has been deployed to develop the MC/DC test case suite generation tool in python. This automation enabled apt test case generation, overcoming the laborious time-consuming manual approach for MC/DC.

Acknowledgements The author would like to thank Director, GTRE, for permitting to present this research activity. The author is grateful to all the team members of Software Quality Assurance Group for their continual support and motivation throughout the development process.

References

1. DO-178B: software considerations in airborne systems and equipment certification. Requirements and Technical Concepts for Aviation, Washington, DC, December 1992.
2. Chelenski, J.J., Miller, S.P.: Applicability of modified condition/decision coverage to software testing. Softw. Eng. J. **9**(5), 193–200 (1994)
3. Shahid, M., Ibrahim, S.: An evaluation of test coverage tools in software testing. In: 2011 International Conference on Telecommunication Technology and Applications: Proceeding of CSIT vol. 5 (2011) © (2011) IACSIT Press, Singapore
4. Khurana, H.: Software Testing, Pearson India Education Services Pvt. Ltd. (2016)
5. Hayhurst, K.J., Veerhusen, D.S., Chilenski, J.J., Rierson, L.K.: A practical approach to modified condition/decision coverage. In: Proceedings of the 20th Digital Avionics Systems Conference (DASC), pp. 1B2/1-1B2/10, Daytona Beach, Florida, USA, Oct 2001
6. Sundmark, D., Carlson, J., Punnekkat, S., Ermedahl, A.: Structural testing of component-based systems. https://doi.org/10.1007/978-3-540-87891-9_12
7. http://www.westfallteam.com/sites/default/files/papers/Testing%20-%20Systematic%20Code%20Coverage%20Teschniques.pdf
8. Salem, M.A, Eder, K.I.: Novel MC/DC coverage test sets generation algorithm, and MC/DC design fault detection strength insights. In: 2015 16th International Workshop on Microprocessor and SOC Test and Verification (MTV 2015): Proceedings of a Meeting held 3–4 Dec 2015, Austin, Texas, USA (pp. 32–37). [7548935] (Proceedings of the International Workshop on Microprocessor and SOC Test and Verification (MTV)). Institute of Electrical and Electronics Engineers (IEEE). https://doi.org/10.1109/MTV.2015.15
9. Kuhlman, D.: A Python Book: Beginning Python, Advanced Python, and Python Exercises, 12–14,
10. https://docs.python.org/3/library/

Underwater Robotics

Arpit Gothi, Priyanka Patel, and Mrudang Pandya

Abstract In the age of technology, the world is slowly moving toward the automated system and self-traveling vehicles in all the fields. The emerging development of the underwater security are in defense and other exploration. The underwater robot is a type of bionic robot, which has the locomotion-link living thing. The type of resemblance of the prototype that creates the virtual habitat of marine life for ocean environment researches. This bionic robot can also gather much information about corals by going deep inside the corals where humans cannot possibly go. This model can also help in estimating the amount of oil spills in oceans. This robotic model can also be used for locating leakages in pipelines deep inside the ocean by giving exact coordinates of that location. And, the most important thing is this robot goes inside the water (sea), measures and detects the level of pollution done due to leakage of the pipeline, as well as, it can detect the level of natural pollution cause due to human activates.

1 Introduction

Investigation of submerged indigenous habitats by perceptions of marine life is valuable in contemplating their practices, swim examples and cooperation inside their living spaces. A biomimetic submerged observatory for considering further comprehension of marine life, particularly their social practices and how ecological changes influence the fragile parity inside the marine world [1]. One of the best approaches to accomplish this is utilizing submerged vehicles that can swim alongside marine life for profound perceptions. The underwater robots can work for this, and the robot works in two modes. One is remotely operated vehicles (ROVs), and the other is autonomous underwater vehicles (AUVs) in the sea; situations use big propellers or a stream-based drive framework. This propulsion system generates an amount of turbulence and sound which scares marine life and does not make observations

A. Gothi (✉) · P. Patel · M. Pandya
Smt. Kundaben Dinsha Patel Department of Information Technology, Chandubhai S. Patel Institute of Technology, Charotar University of Science and Technology, Changa, Anand, Gujarat 388421, India

T. Senjyu et al. (eds.), *ICT with Intelligent Applications*, Smart Innovation, Systems and Technologies 248, https://doi.org/10.1007/978-981-16-4177-0_45

successful. A submarine does not coordinate well with marine conditions. The most developed ROVs engineering is mind boggling to the point that it requires exorbitant manufacture and perplexing control techniques and the most significant upkeep cost.

Water assets and sea-going environments are confronting a different physical, concoction, and organic dangers from environmental change, modern contamination, ill-advised waste removal, infection-causing organisms, and even fear-based oppressors' assaults. Detecting oceanic conditions is of expanding significance to general well-being, port, and stream security, environment manageability, sea life science, and the aquaculture industry. Manual inspecting, by means of pontoon/transport or with handheld gadgets, is yet a typical practice in the checking of sea-going situations. The aim of this research is to create an underwater robot that, by adjusting the direction of its rotational forces, can freely change its attitude. While going across the sea, we would want the robot to be robust. We adopted a water propeller for the propulsion system.

2 Literature Review

The project itself is a challenge to us. Below are some of the challenges for an underwater robot. There is a challenge to make this model to make it complete water resistance; as we know, salt water is a good conductor of electricity; so, if there is no isolation, then it leads us to an electric short circuit. And, the model will be malfunctioned. The second challenge to make this is to control the buoyancy of the model. There is one more challenge, and this challenge is about the pressure. If our model goes to an extreme deep ocean area, then it will be crushed by extremely high pressure. Communication under aquatic area is the most challenging because if we want live readings from the model then it is difficult to have. An unconfined environment remains one of the open challenges. The solution to the communication problem is one can put a network of multiple nodes in specific positions to receive the data from far in the ocean [2, 3].

3 Prototype Design

Various types of ROVs, with different structural design, dimensions, size, and also with different materials, have been planned and created [4]. In a significant field of research, we try biomimetic underwater robots, which mimic the behavior of living organisms. In conventional aquatic vessels and robots, steering techniques include adjusting rudder angles or utilizing the differential propulsive powers of three or more boosters [5, 6].

The successful working model of underwater robot has three propellers; Two propellers as the main thrust/boost and another single propeller are used to change the vertical height of the model itself. The result of this design lets us make our model robust and vigorous. The mentioned underwater robot design is chosen after taking many trials with different designs.

The Arduino UNO and the whole connections shown in Fig. 1b can be implemented on **Raspberry Pi** also, but due to budget constraints, we implement it on the Arduino UNO. Here, Arduino's work is for handling the propeller of this model. It can easily handle the model's propeller, in which we gave instructions. Figure 2

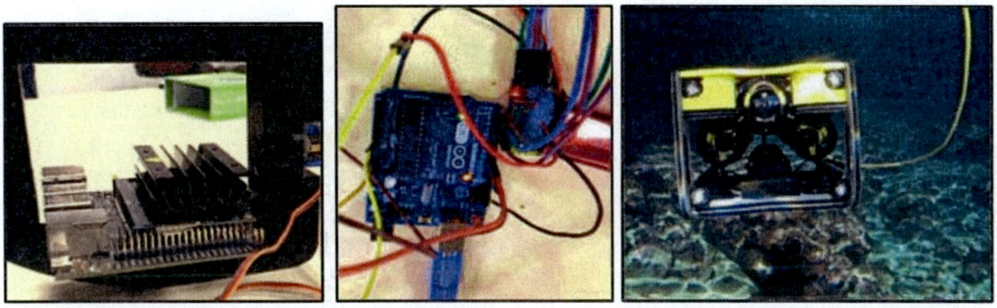

a) Initial Design b) Arduino UNO c) Underwater robot [2]

Fig. 1 Model of underwater robot

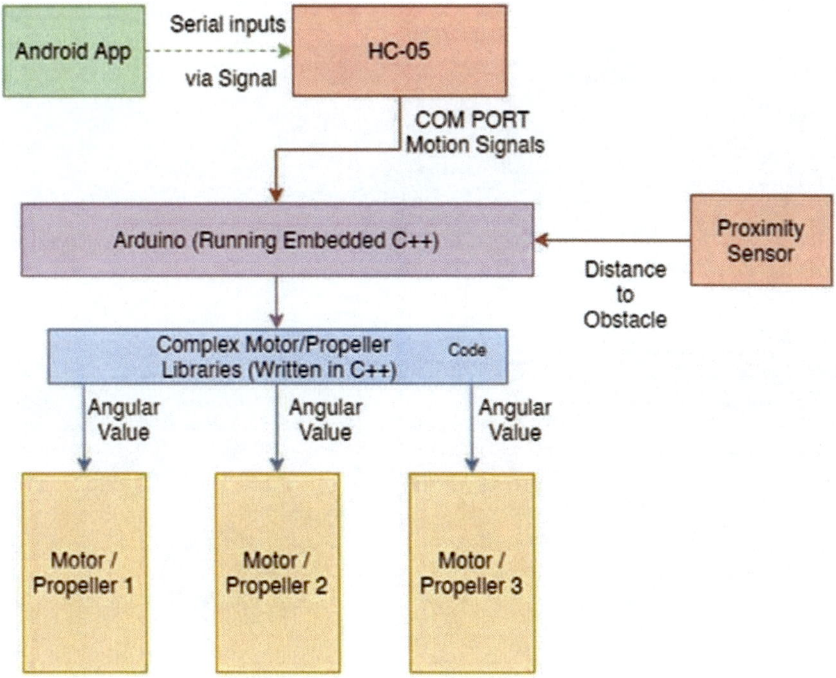

Fig. 2 General control flow diagram

shows the flow of this model. The code is too complex to explain, so this diagram will help to know the flow of the model.

The above diagram shows the information about how this robot model can go inside the water and moves forward and backward, up and down, and left and right.

3.1 The Flow of Sensor Processing:

Figure 3 shows the core sensor process of the mentioned underwater robot model. The Raspberry Pi got sufficient data from a different sensor like a chemical sensor, pressure sensor, temperature sensor, and many other different sensors. Through the camera, Pi will capture the environment footage. Two cameras are installed; one is a normal camera, and another is a nightscape camera. Pi will work according to the instructions given by the operator.

This model is sensing the different parameters of pollution in the aquatic environment with the help of different sensors.

- **Temperature Sensor**: The temperature-detecting sensor detects the temperature and changes over the simple temperature esteem into computerized esteem. It

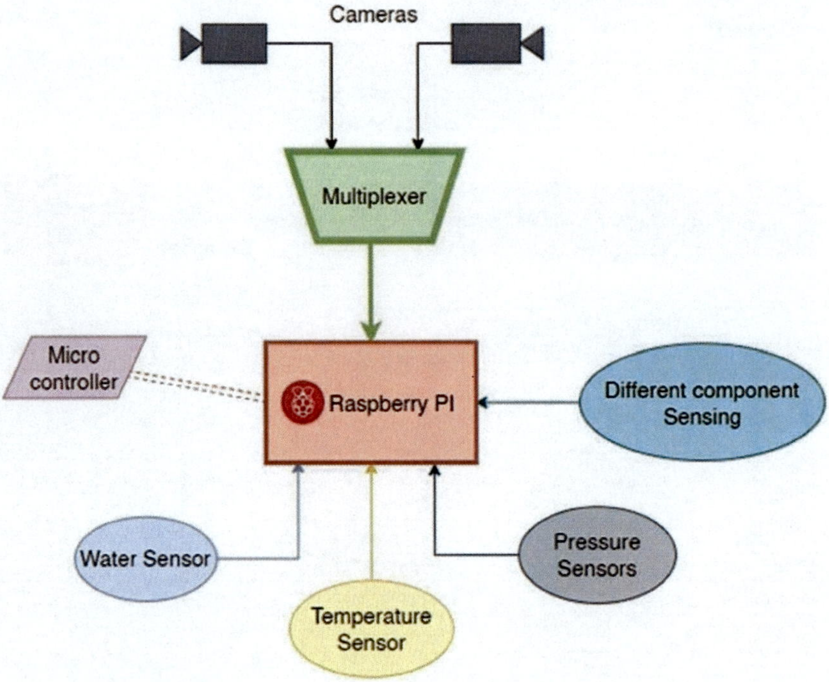

Fig. 3 Flow of sensor processing

likewise fits for detecting the information in the wet condition with the variable piece fragment of temperature readings [7]. All the information is locally formed by a microcontroller before sending it to the sensor information authority.
- **Pressure Sensor**: A pressure sensor is an instrument composed of a pressure sensitive part that is used to evaluate the real pressure being applied to the sensor and certain components that are used to translate the input into an output signal.
- **Chemical Sensor**: The toxic concoction substance nearness in the water ought to be detected in a persistent way, and the errand was completed by these sensors. It likewise quantifies the different components of mass and compound recommendations in it. The detected information is refined into a helpful configuration for examination in the pre-preparing stage [7].
- **pH Sensor**: This model has a pH sensor to convey the pH level of different zones in the sea and compares it. This sensor gets pH measurements at ± 0.1 pH. For the most specialists, this extraordinary precision range and its minimal cost make this an incredible device for bio-robotics and venture this way!
- **DO Sensor**: DO means dissolved oxygen. This is used to measure the dissolved oxygen in water to reflect the water quality.

This model can have many other sensors as per the need. And, the data of these sensors will be stored on Raspberry Pi/Nvidia Jetson Nano as per the interval which is given by the user.

3.2 Machine Learning Model

(1) The input image phase: The first step is to acquire videos from deep water through the underwater robot and convert them into the image.
(2) The preprocessing phase: Convert an image into color scaling and also do histogram equalization for the correct feature detection of image.
(3) The feature extraction phase: Image features are crafted with a pre-trained VGG16 neural network.
(4) The classification phase: Classify the image (Fig. 4).

4 Logical Steps and Working of Model

4.1 Basic Movement for Working Model

1. The electrical connections are made as shown in Fritzing diagram (Fig. 5).
2. Once the connections are made and checked (to prevent short-circuiting underwater), the batteries are connected to the battery headers provided in the circuit.

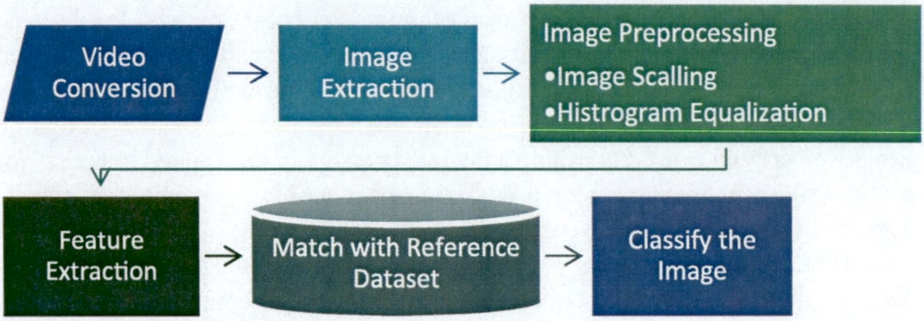

Fig. 4 Flow of machine learning model

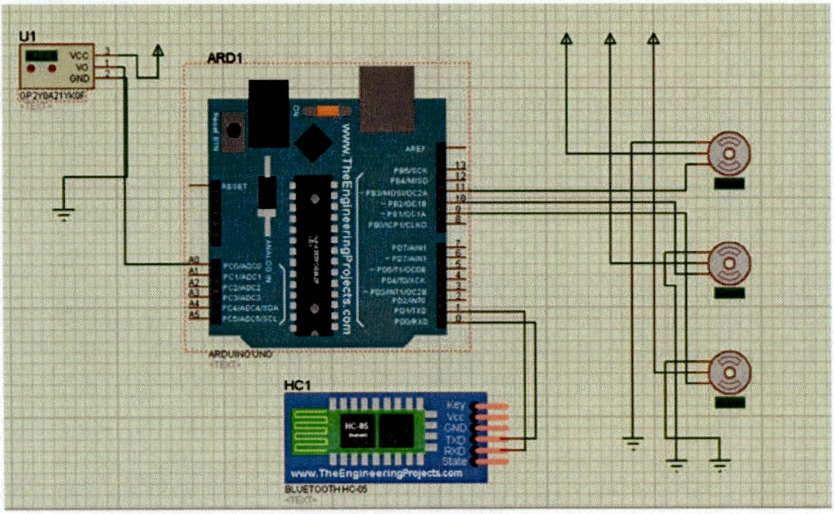

Fig. 5 CORAL-SDN graphical user interface (dashboard)

3. Once the Bluetooth has been connected, wait for a while as the code script executes and flushes the EEPROM onboard the UNO to prevent interference from previously stored values.
4. The robot can be maneuverer as required using the buttons on the android app. Six movements are possible:

Forward: Robot's two propellers move in a clockwise motion for movement. **Backward**: Robot's two propellers move in anticlockwise motion for movement. **Upside**: Robot's propellers (direction on topside) move in a clockwise motion for movement. **Downside:** Robot's propellers (direction on topside) move in a clockwise motion for movement. **Left:** Robot's right side propeller from the main propeller runs fast compared to the left side to deflect toward the left. **Right:** Robot's left-side propeller from the main propeller runs fast compared to the right side to deflect toward the right.

This model has two functional features. (1) This model runs on Bluetooth in the range of 1 meter, and if we want to go deeper in water, then there is another facility of locomotion. (2) By locomotion facility, the model goes through the predefined path and observes the situation of underwater, saves that data, and gives us that data.

4.2 Working of Fully Assembled Model

First of all, if we choose a predefined path then according to the user's predefined path, our model goes through the path. In between destination and source, our model gets some information regarding pollution. And, the information got by our model is precious. Firstly, the model's camera captures the footage, stores it and sends it to processing unit to understand the environment and also to get different info about ocean environment, fishes, and marine culture, and then scans for the pollution that causes due to leakage in underwater pipelines [8]. It also captures the different values got by the pre-installed sensors like the above mentioned. The VGG16 architecture is a convolutional neural network. The VGG16 model is used to detect steel destruction and to detect cracks in the pipeline. This detection is necessary to detect pollution. The U-Net architecture is a fully convolutional network (FCN) that does image segmentation. These architectures are used to find different obstacles, and it can be also used to find fish and observe the pattern of marine life.

5 Results

5.1 Accuracy of Making System vs Existing System

Right now, identification exactness might be spoken in a graphical strategy. The technique shows the precision level co-relation of contamination discovery between the canny sensor-based submerged observing the framework and the past strategy which can be utilized for checking the angle in submerged contamination; here, the chart shows the scope of high exactness levels in the astute sensor-based submerged checking framework. The unit model is used for the detection of fish inside the water, and the VGG16 model [9] is used for the detection of a defect in steel to find pollution. The following are the results (Figs. 6 and 7).

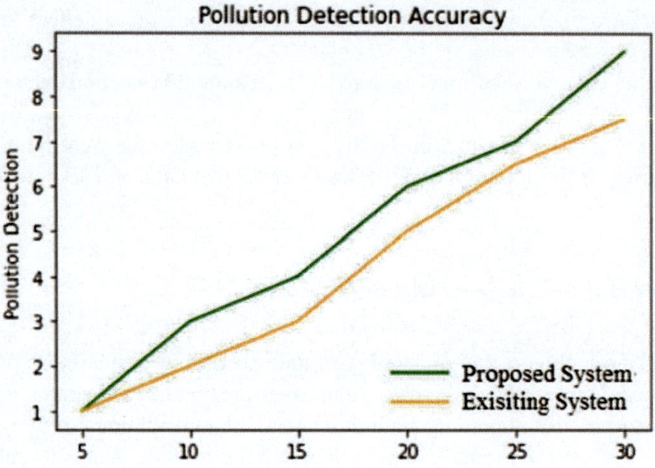

Fig. 6 Graph of pollution detection vs accuracy

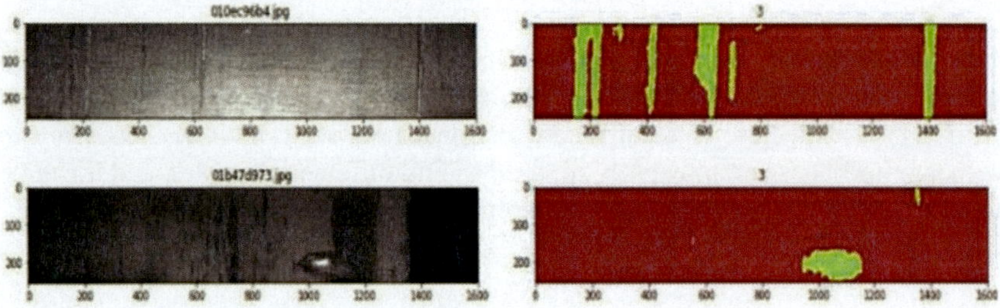

Fig. 7 Detected mask of steel that causes pollution

6 Discussion

By this model, we can get information about the level of pollution in underwater (sea). This data can be used to prevent the pollution by doing some proper actions. Like if the main pipeline (underwater) got a crack and leaking toxic liquids, then it causes damage to the natural environment of the sea as well as harms the marine life of the sea. In the above example, if we do not have any acknowledgment about the leakage, then a supplier of that particular toxic or nontoxic liquids has a loss due to leakage.

7 Conclusion

Robotic technology allows aquatic applications to achieve access to the underwater world regularly and indefinitely. The creation and introduction of more sophisticated embedded tools that can increase the autonomy levels of the platform require challenges relevant to this area. This robot is used as an underwater observatory. This robotic robot can also detect new organisms that are not in the database. We can use this robot for detecting the defects in the oil pipeline for leakages and measuring the level of pollution by leakage. Measuring pollution is important because due to this leakage the whole marine environment is affected. The underwater robot can capture images and acquire data from different sensors in different conditions of the surrounding underwater.

Future Scope

In the future work, this project can be extended to use it as ROV/AUV to control the pollution under the sea areas. And, this ROV/AUV can surveillance the sub-aquatic area for spying purposes.

References

1. Katzschmann, R.K., et al.: Exploration of underwater life with an acoustically controlled soft robotic fish. Sci. Robot. **3**(16), (2018). eaar3449
2. Tan, X.: Autonomous robotic fish as mobile sensor platforms: challenges and potential solutions. Mar. Technol. Soc. J. **45**(4), 31–40 (2011)
3. Hu, Y., Zhao, W., Wang, L.: Vision-based target tracking and collision avoidance for two autonomous robotic fish. IEEE Trans. Industr. Electron. **56**(5), 1401–1410 (2009)
4. Liu, J., Dukes, I., Hu, H.: Novel mechatronics design for a robotic fish. In: 2005 IEEE/RSJ International Conference on Intelligent Robots and Systems. IEEE (2005)
5. Liu, J., Hu, H.: Biological inspiration: from carangiform fish to multi-joint robotic fish. J. Bionic Eng. **7**(1), 35–48 (2010)
6. Raj, A., Thakur, A.: Fish-inspired robots: design, sensing, actuation, and autonomy—a review of research. Bioinspiration & Biomimetics **11**(3), 031001 (2016)
7. Premkumardeepak, S., Krishnan, M.B.M.: Intelligent sensor-based monitoring system for underwater pollution. In: 2017 International Conference on IoT and Application (ICIOT). IEEE (2017)
8. Zakikhani, K., Nasiri, F., Zayed, T.: A failure prediction model for corrosion in gas transmission pipelines. Proc. Inst. Mech. Eng. Part O: J. Risk Reliab. (2020). 1748006X20976802
9. Nash, W., Drummond, T., Birbilis, N.: A review of deep learning in the study of materials degradation. npj Mater. Degrad **2**(1), 1–12 (2018)
10. Erena, M., et al.: Use of drones for the topo-bathymetric monitoring of the reservoirs of the segura river basin. Water **11**(3), 445 (2019)

Iris Recognition System Based on Convolutional Neural Network

Manisha A. Nirgude and Sachin R. Gengaje

Abstract With an ever-increasing emphasis on security, the requirement of reliable methods for user authentication has amplified drastically and biometric methods for authentication are gaining wide acceptability. In contrast to the existing biometric features of human beings, iris, a part of human eye, has several interesting and unique characteristics. We solely address the iris characteristics and traits for human identification, explicitly for iris recognition. We have proposed convolutional neural network (CNN) model to develop robust, scale and rotation invariance, and scalable iris recognition system. Pre-trained network model "AlexNet" and data augmentation techniques, viz. clockwise and anticlockwise rotation, increased contrast, histogram equalization, addition of noise, and dilation, are used to create different datasets to train the CNN model. The performance of the proposed CNN model has been validated on five databases. We have achieved almost 100% iris recognition accuracy with reduced ERR as small as 0.46% as compared to existing iris recognition systems handling the issues of scale and rotation invariance and scalability using complete eye images.

1 Introduction

Biometric authentication has been considered as a reliable and the most secure mechanism replacing conventional verification mechanisms and is being widely used everywhere in the world at banks, airports, and documents like passports, identity cards, and Aadhar cards. Among these types of biometrics currently being used, iris patterns are unlike and unique for each person and even different for left and right eye of the same person and statically proven to remain stable and unchanged almost at all times. Iris recognition system is a biometric authentication based on iris pattern. Typical

M. A. Nirgude (✉) · S. R. Gengaje
Walchand Institute of Technology, Solapur, Maharashtra, India
e-mail: manirgude@witsolapur.org

S. R. Gengaje
e-mail: srgengaje@witsolapur.org

T. Senjyu et al. (eds.), *ICT with Intelligent Applications*, Smart Innovation, Systems and Technologies 248, https://doi.org/10.1007/978-981-16-4177-0_46

iris recognition systems often include image acquisition, preprocessing localization, normalization and enhancement, and feature extraction and matching.

1.1 Iris Image Database

To evaluate the performance of the developed recognition system, many free and public databases are available. In most of the research work, standard databases like CASIA, IITD, UBIRIS, BATH, MMU, UPOL, and WVU are used. But in real-time scenario, there are chances that these eye images may be rotated or scaled or both while capturing. Apart from these, there are other issues such as occlusion in eye images, blur images, poor illumination, and distance between eye and camera which make segmentation a challenging task that affects feature extraction. Thus, the desirable property that a classifier should have is invariance to rotation and scale, and also it should be scalable, i.e., perform well with increasing number of persons. Thus, the solution is adoption of convolutional neural network (CNN) model which integrates feature extraction and classification in a single module. As CNN model takes an image as an input, eye images can be used directly instead of segmenting iris and finding features which solves the issues of segmentation and also saves the computation cost of preprocessing.

Hence, we have proposed CNN model to develop robust and scale and rotation-invariant iris recognition system. We have used two different databases CASIA and IITD including images of different sizes to evaluate scale invariance performance. We have generated three databases by applying data augmentation techniques to CASIA images to evaluate rotation invariance performance and robustness of the proposed approach.

Section 2 discusses the first related work, and in Sect. 3, we provide the details of convolutional neural network (CNN). In Sect. 4, we discuss about proposed iris recognition system. The results are discussed, and its subsequent discussion in Sect. 5 followed conclusion in Sect. 6.

2 Related Work

Biometrics is one of the most powerful methods for identification and verification of people today than the existing methods. The first iris recognition system [1] employed 2D Gabor wavelet for feature extraction and hamming distance for finding matched identity for a large database. Many researchers did study of different phases of iris recognition to improve performance of the system. Since feature extraction is a challenging task, various researchers focused their study on feature extraction. Most of these features are evaluated using algorithms for a given database. We have also proposed iris recognition where wavelet transform was used for feature extraction and support vector machine for classification [2] before this CNN approach. Thus, before feature extraction, preprocessing is required. Also, there are issues like guaranteed system performance for different databases and scalability. To overcome these issues, one solution is used called convolutional neural network (CNN) model [3–5]. In deep learning, instead of features, images are passed as an input directly and network finds best features automatically. Different CNNs have been used for classification, localization, detection, segmentation, etc. Convolutional overfit network was trained for object classification [4]. Features extracted from this network are used for various recognition tasks which are parts of object image classification such as scene, fine-grained recognition, and attribute detection. A CNN model is called as AlexNet which is trained using 1.2 million for 1000 classes and has got state-of-the-art results [6]. Features of different layers (varying from 6 to 13) are extracted and PCA is applied to restrict number of features to 256 at each layer. Complete eye images are used instead of segmenting iris from eye images. SVM classifier is used for classification and got the accuracy 99.4% for IITD database. How features are transferred from one layer to next is studied, and this transferability of features is inversely proportional to distance between base task and target task [7]. Transfer learning is used to exploit deep representations when there is limited or few data. A feature extraction method is proposed, in which instead of storing activation values at the last layer, the activation values generated on every convolutional and fully connected layer of the model are stored to generate a full-network embedding [8]. This full-network embedding is used on three datasets for image annotation resulting in high performance. Thus, many researchers claimed that pretrained models can be used to tune any other dataset. The idea is to use pretrained models for feature extraction, in which weights of the pretrained model can be used for training new dataset. Therefore, only the last layer, i.e., fully connected layer, needs to be modified as per number of classes.

3 Convolutional Neural Network

Convolutional networks are inspired by biological processes showing excellent performance in hand-written digit classification and face detection as LeNet architecture [3]. Much deep learning architectures are derived from LeNet. CNN became extremely popular in 2012, when a CNN model called as AlexNet was introduced [6]. Convolutional neural network is a deep feed-forward neural network with images as input. CNN is an architecture comprising of set of layers, namely convolution, pooling, ReLU, softmax, and fully connected layers. One can design a new CNN architecture for image classification defining number of layers, weights, filers, etc. Accurate training of the model will require large volume of data which will take large time for training. An alternative solution is to use pre-trained network models which are already trained on other large datasets. The only issue with the use of pre-trained CNN models in other application is that already trained CNN models had used millions of images for training, whereas other classification applications may have small size dataset. This issue can be solved by applying transfer learning. A large dataset can be created be increasing number of samples from existing dataset by applying data augmentation techniques such as vertical and horizontal flipping, cropping, rotation, and translations. Thus, data augmentation is applied for twofold reasons: to evaluate the performance of iris recognition-based CNN for scaling and rotation invariance and to create a large dataset for avoiding the problem of overfitting. We have created three new datasets from existing CASIA database by applying different augmentation techniques.

The First dataset (CASIA-ROT I) is created by applying rotation, i.e., 1^0, 2^0, 3^0, 4^0, and 5^0, in clockwise and anticlockwise directions to the images from CASIA resulting in dataset consisting 7140 images. The second dataset (CASIA-ROT II) is created by applying higher angle of rotation, i.e., 5^0, 10^0, 15^0, 20^0, and 25^0 in clockwise and anticlockwise directions to the images from CASIA database resulting in dataset consisting 7140 images. The third dataset (CASIA-augmented) is created by applying five operations: increase contrast of the image, histogram equalization, addition of noise "salt-and-pepper" with intensity 0.02, dilation of an image, and rotation of image by angle 10^0 in clockwise direction producing additional 5 images resulting in dataset containing 4284 images to check robustness of the proposed system.

4 Development of Proposed CNN-Based Iris Recognition System

Since AlexNet has obtained state-of-the-art results on highly challenging dataset using supervised learning, we have used AlexNet as a pre-trained CNN model.

Procedure to train and test CNN model based on AlexNet is as follows:

4.1 Procedure to Train CNN Model

- Resize all images to 227 × 227 × 3 size for making input images of equal size.
- Store all the resized images in "ImageDatastore" object which will store these images with their respective classes.
- Split stored dataset in "ImageDatastore" object randomly into training and testing dataset.
- Load pre-trained AlexNet convolutional neural network containing 25 layers.
- Since number of classes in our dataset are different than AlexNet dataset (1000 classes), we need to configure the last three layers for of the model. Therefore, we have configured the last three layers, i.e., fully connected layer, a softmax layer, and a classification output layer in proposed CNN architecture.
- Specify the parameters for training CNN model as follows:
- Solver name: "sgdm" (stochastic gradient descent with momentum) [12]

- Initial learning rate: Learning rate is used to determine how the weights is updated. We have set learning rate to 0.0001 empirically.
- Maximum number of epochs: To find the best recognition accuracy, we have trained the network for different number of epochs varying from 5 to 50.
- Batch size: Size of batch used for each iteration is kept 10 or 15 experimentally.
- Number of iteration per epochs:

$$\text{No. of iteration per epochs} = \frac{\text{No. of training images}}{\text{Batch Size}} \tag{1}$$

- Execution environment: In CNN, if graphics processing unit (GPU) is available, then it will be used else CPU will be used. There are two important requirements for CNN: large dataset and high computation facility.
- Once all training parameters are set and layers are configured, the network model is trained using training dataset containing images and label associated with it.
- Store trained network model for further performance evaluation of CNN.

Thus, CNN model is configured and trained separately for different datasets.

4.2 *Feature Extraction Using Proposed CNN Model*

CNN learns features itself during the training process. The features are learned at every convolutional layer. Fully connected layers (FC) give higher level of features. Thus, as we increase number of convolutional layers and filters, more detailed information or higher-level features are obtained. This trained network is used for extracting features of images from our database. The performance of network model is evaluated using testing dataset based on linear discriminant analysis function. Probability of test sample x is calculated as defined in Eq. 2.

$$f(x|\pi_i) = \frac{1}{(2\pi)^{p/2}\Sigma^{1/2}}\exp\left[-\frac{1}{2}(x-\mu_i)'\Sigma^{-1}(x-\mu_i)\right] \tag{2}$$

where

π_i—class I, μ_i—mean vector for class I, Σ—variance–covariance matrix.

Decision is based on linear score function, a function of the class means for each of our n classes, as well as the pooled variance–covariance matrix.

5 Experiment Results

The performance of proposed CNN model is evaluated on five different datasets. The details are given in Table 1. The CNN model is trained separately for five different databases tuning various training parameters to achieve the optimum results.

For training CNN model, we have split every dataset into 90% images as genuine cases and remaining 10% images are used as imposters. From the genuine cases, 60% images are selected randomly for training the model and remaining 40% images are used for evaluating the performance of the model. Further, images from other database are combined with 10% imposter images to create imposters' dataset for calculating equal error rate (EER) and receiver operating characteristics (ROC) curve. EER is the point where FAR and FRR are equal, and smaller the ERR, better the approach is. By varying threshold, false acceptance rate (FAR) and false rejection rate (FRR) are calculated. The experiments are performed using desktop Core i5 with 3.20 GHZ processor, 8 GB RAM with NVIDIA GeForce GTX 1050 Ti 4 GB graphics card/graphics processing unit and MATLAB R2017a (64 bit), and neural network toolbox(TM) model for AlexNet.

The CNN models are trained and their performance is evaluated for five datasets. The recognition accuracy obtained for these datasets for different number of epochs is shown in Fig. 1. The recognition accuracy and EER values for datasets are as shown in Table 2. Further, we trained CNN model for different number of training samples

Table 1 Datasets used for evaluation

Dataset no.	Database	Input images to CNN model	Size of images in pixels	No. of persons/classes	No. of images
Dataset 4	Left CASIA: left eye images	Complete eye images	320 X 280	102	714
Dataset 5	IITD: left and right images	Complete eye images	320 X 240	224	2240
Dataset 6	CASIA-ROT I: left eye images rotated by 1^0, 2^0, 3^0, 4^0, and 5^0 clockwise and anticlockwise	Complete eye images	320 X 280	102	7140
Dataset 7	CASIA-ROT II: left eye images rotated by 5^0, 10^0, 15^0, 20^0, and 25^0 clockwise and anticlockwise	Complete eye images	320 X 280	102	7140
Dataset 8	CASIA-augmented: left eye images augmented by five techniques	Complete eye images	320 X 280	102	4284

Fig. 1 Recognition accuracy for different datasets

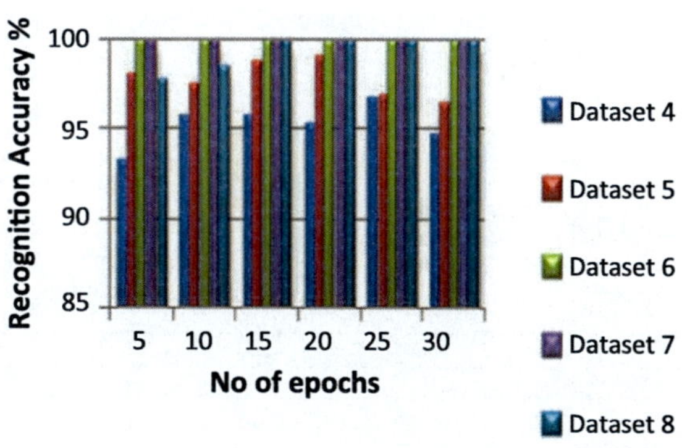

Table 2 Recognition accuracy and EER

Dataset	Recognition accuracy %	EER %
Dataset 4	96.77	8.25
Dataset 5	99.01	2.21
Dataset 6	100.00	1.16
Dataset 7	100.00	1.94
Dataset 8	99.84	5.35

varying from 10% to 60% for Dataset 6 and Dataset 7 to find the optimum performance of the system. From the experiments, it has been observed that recognition accuracy is at 100% and relatively constant from 30% training samples. Slight variation in EER is observed for different trained models. We have obtained minimum ERR value 0.46% at the cost of 30% training samples and 1.94% ERR for Dataset 7 for 60% training samples. The recognition accuracy and ERR are reported in Table 3 for Dataset 6 and Dataset 7, respectively. A comparison of proposed CNN model with existing iris recognition system on CASIA and IITD database is shown in Tables 4. We have achieved 100% recognition accuracy using CNN and significantly less ERR

Table 3 Recognition accuracy and EER for Dataset 6 and Dataset 7 varying number of training images

	Dataset 6		Dataset 7	
% of images used for training	Recognition accuracy %	EER %	Recognition accuracy %	EER %
10	97.70	3.43	96.76	11.83
20	98.79	6.15	99.58	4.82
30	100.00	0.46	99.78	3.33
40	100.00	1.21	99.82	2.77
50	100.00	1.22	100.00	3.53
60	100.00	1.16	100.00	1.94

Table 4 Performance comparison of CNN-based system with existing systems

CASIA database			IITD database		
Methodology proposer	Recognition accuracy %	EER %	Methodology	Recognition accuracy %	EER %
Daugman [1]	100.00	0.08	Swati et al. [12]	96.20	-
Zhao [10]	98.94	0.60	Umer et al. [13] (tenfold cross validation)	100.00	0.08
Ma et al. [11]	99.98	1.76	Bhateja et al. [14]	99.20	–
Nirgude et al. [2]	96.33	–	Minaee et al. [15]	99.40	–
Proposed CNN dataset 6	100.00	0.46	Proposed CNN	99.01	2.21
Proposed CNN dataset 7	100.00	1.94			
Proposed CNN dataset 8	99.84	5.34			

comparable to existing systems and 100% recognition accuracy for iris recognition system showing stable performance for rotation and scaling invariance.

6 Conclusion

We have proposed a CNN model which integrates feature extraction and classification in a single model. The performance of the proposed model has been validated on five databases containing complete eye images. We have achieved almost 100% iris recognition accuracy with reduced ERR 0.46% comparable to existing systems handling the issue of scale and rotation invariance and scalability using complete eye images. Thus, the use of complete eye images has laid to solve the issues in iris preprocessing such as variation in scale, camera, and distance between camera and eye and saved the computation cost of preprocessing of iris images. By applying data augmentation techniques, we have achieved high recognition accuracy for small dataset. Thus, this proposed CNN model can be used for iris recognition even with a relatively small database and also other recognition applications.

References

1. Daugman, J.: How iris recognition works. IEEE Trans. CSVT **14**(1), 21–30 (2004)
2. Nirgude, M., Gengaje, S.: Iris recognition system based on multi-resolution analysis and support vector machine. Int. J. Comput. Appl. **173**(7), 28–33 (2017)

3. LeCun, Y., Boser, B., Denker, J.S., Henderson, D., Howard, R.E., Hubbard, W., Jackel, L.D.: Backpropagation applied to handwritten zip code recognition. Neural Comput. **1**(4), 541–551 (1989)
4. Razavian, A., Azizpour, H., Sullivan, J. Carlsson, S.: CNN features off-the-shelf: an astounding baseline for recognition. In: IEEE Conference on Computer Vision and Pattern Recognition Workshops, pp. 806–813 (2014)
5. Karpathy, A.: Convolutional neural networks for visual recognition. Convolutional Neural Networks. http://cs231n.github.io/convolutional-networks/ (2012)
6. Krizhevsky, A., Sutskever, I., Hinton, G.E.: Imagenet classification with deep convolutional neural networks. Adv. Neural. Inf. Process. Syst. **1**, 1097–1105 (2012)
7. Yosinski, J., Clune, J., Bengio, Y., Lipson, H.: How transferable are features in deep neural networks? Adv. Neural Inf. Process. Syst. **27**, NIPS (2014)
8. Garcia-Gasulla, D., Vilalta, A., Pares, F., Moreno, J., Ayguadé, E., Labarta, J., Cortes, U., Suzumura, T.: An Out-of-the-box full-network embedding for convolutional neural networks. arXiv:1705.07706 [cs.LG] (2017)
9. Bishop, C.M.: Pattern Recognition and Machine Learning. Springer, New York, NY (2006)
10. Zhao, D., Luo, W.: Negative Iris Recognition. IEEE Trans. Dependable Secure Comput. **15**(1), (2018)
11. Ma, L., Tan, T.: Efficient Iris Recognition by characterizing key local variations. IEEE Trans. Image Proc. **13**(6), (2004)
12. Dhage, S.S., Hegde, S.S., Manikantan, K., Ramachandra, S.: DWT-based feature extraction and radon transform based contrast enhancement for improved Iris Recognition. Procedia Comput. Sci. **45**, 256–265 (2015)
13. Umer, S., Dhara, B.C., Chandaa, B.: A novel cancellable iris recognition system based on feature learning techniques. Inf. Sci., Volume **406–407**, 102–118 (2017)
14. Bhatejaa, A.K., Sharma, S., Chaudhury, S, Agrawal, N.: Iris recognition based on sparse representation and k-nearest subspace with genetic algorithm. Pattern Recogn. Lett. **73**, 13–18 (2016)
15. Minaee, S., Abdolrashidiy, A., Wang, Y.: An experimental study of deep convolutional features for iris recognition. In: IEEE Signal Processing in Medicine and Biology Symposium (SPMB) (2016)

Database Management Systems—An Efficient, Effective, and Augmented Approach for Organizations

Anushka Sharma, Aman Karamchandani, Devam Dave, Arush Patel, and Nishant Doshi

Abstract Big and small firms, organizations, hospitals, schools, and other commercial offices are generating moderate to huge amounts of data regularly and need to constantly update and manage these data. These data are not only used at that instance, but generally, the retrospective analysis of data helps tremendously to improve the business strategies and the marketing trends. With time, these data may grow and become unmanageable if handled conventionally, like the file system. These factors resulted in the introduction of the terms database and database management system. Hierarchical, network, relational, and object-oriented approaches of DBMS are discussed in this paper. A highlight of the new-generation database approach called NoSQL is also included in this paper along with an insight into augmented data management. A model based on the database design for the Study in India Program is discussed. It is followed by a graphical user interface developed in Java for the same which ensures the ease of access to the database.

1 Introduction

A database is a collection set of interrelated data that can be inserted, retrieved, and deleted from it. A database management system (DBMS) is an application that stores and manages the data. Some of the popular relational DBMS like MySQL, Oracle, PostgreSQL, and SQLite have been discussed in the paper. A DBMS can be used for personal as well as commercial use. For example, a university can have its own IT admin application, while the traffic control system, supermarkets, or telecommunication sector have applications that are used at a huge level. In the following paper, we have talked about the variety of ways in which a DBMS can be used in a corporate setting, as well as the plethora of significant advantages they provide over the traditional file system. The aim of this paper is to enlighten the readers on why DBMS is the way going forward, as well as giving a brief

A. Sharma · A. Karamchandani · D. Dave · A. Patel · N. Doshi (✉)
Department of Computer Science and Engineering, School of Technology, Pandit Deendayal Energy University, Gandhinagar, India
e-mail: Nishant.Doshi@sot.pdpu.ac.in

T. Senjyu et al. (eds.), *ICT with Intelligent Applications*, Smart Innovation, Systems and Technologies 248, https://doi.org/10.1007/978-981-16-4177-0_47

explanation to the future trends in addition to the future technologies related to database management systems.

Outline: The following is the structure of the paper. Section 2 gives us the literature survey, which tells us about the history of DBMS briefly. Section 3 showcases the various advantages database management systems have to offer over a typical file system. Various types of databases are highlighted in Sect. 4. Section 5 has highlighted some of the most used database software available in the market at present. Section 6 has provided some useful insights regarding the future trends of DBMS, and the various software are mentioned in Sect. 5. Section 7 gives a brief explanation of augmented data management, an up-and-coming technology closely ingrained to DBMS. In Sect. 8, a database that we designed for the Study in India Program is presented; with Sect. 9 showcasing the graphical user interface (GUI), we designed to go along with it. Finally, Sect. 10 discusses the entire paper and concludes the work.

2 Literature Survey

Before the evolution of the database, every information is used to be recorded on papers. The evolution of DBMS started from somewhere in the mid of the twentieth century by Charles Bachman who designed the first DBMS system. Further, in 1970, Edgar Codd, who was working with IBM's information management system (IMS), wanted to get a better search engine for his research. This resulted in the introduction of relational databases, popularly based on 12 Codd's rules on his name. In 1976, Peter Chen defined the entity-relationship model also known as the ER model very much useful to define relationships among entities. From 1980 onward, the relational model became a widely accepted database component. Object-oriented DBMS was developed in 1985, and gradually, Microsoft MS access and Internet database applications were introduced with rising demand to manage data. To give support to standard input data formats, in 1997, XML was integrated into DBMS products by many vendors. Relational databases are the best solutions to store structured data, but in this era, in response to the major challenges faced by the internet data and the need for faster search engines for unstructured data, NoSQL proved to be useful.

3 Advantages of Database

The file system manages data using files in the disk. In this, data are entered by the user in Excel sheets, Word documents, or PowerPoint presentations with specific names under certain files and folders. Hence, database proves itself to be better due to the following cases:

- Data redundancy: It refers to the duplication of the same data at different places. For example, if an employee's department gets changed then it has to get updated in all the places where this information is stored. This also shows that the same data are present in multiple locations or files. Data redundancy can increase the size of the database, thus making the search or access process time slower.
- Data inconsistency: In the example stated above, if an employee has been a part of two different projects then his details have been stored twice. Now, if he asks to change his email address, it is possible that out of the two, one instance of his email address gets changed and the other would stay intact. This would lead to data inconsistency due to the presence of different email addresses of the same person in the office.
- Data searching: In order to search for an employee, project or any specific attribute, it is much easier to carry it out in DBMS. On the other hand, in file system, entire search operations need to be made to carry out the same.
- Data isolation: In the file system, data will be scattered among multiple files of different formats such as Excel, PPT, and Word. Here, it would be difficult to create a program to carry out an operation. This is majorly faced by some offices where they do not use a database and are dependent on only the documents.
- Data integrity: In DBMS, the integrity of the data is maintained due to the integrity constraints. Data integrity makes sure that processes such as insertion, deletion, and updating of records are successful. For example, the phone number of any employee cannot exceed ten digits in India.
- Data security: The date of the employees and other confidential information remain more secure in the DBMS as compared to the file system. The security of data is of the utmost importance to an office.
- Atomicity of updates: If a system failure occurs while a process or update is being carried out in DBMS, this would lead to the database being in an inconsistent state. For example, if the increment or any other important reformation of some information of an employee is being carried out then due to some system failure or electricity cut-off, the system shuts down. No problem would occur in the database upon reopening.
- Concurrent access by multiple users: If a database does not support concurrent access then it would be impossible for the database to be in control. For example, the admin office of a company is handled by different users. So, if there is a concurrent by the users, the information would be updated simultaneously without any problem.

4 Types of Database

A database comprises a set of interrelated data. The linking of data is explained through different types of databases. These databases are discussed below:

4.1 Hierarchical Database

In the 1960s, IBM developed an hierarchical data model, the data model in which the data are stored in a tree structure called the hierarchical data model. Data are stored as records that are connected to other records through links. A record is a collection of fields that can store one value each by defining the type of record, and the type of the field can be defined. In the hierarchical data model, it is required by rule that each parent can have only one child whereas each parent record can have more than one child records. To get data from such a data structure, the whole tree needs to be traversed starting from the root. Shared-memory, shared-disk, and shared-nothing architectures are different type of architectures that are combined by the hierarchical database architecture. An interconnected network connects the nodes present in the system at the highest level, not sharing the memory with one another. Therefore, a shared-nothing architecture is present at the highest level, whereas each node of the system comes under the shared-memory architecture, including some processors [1].

4.2 Network Database

The network database system was the first system to use multi-parent structure. It was developed to replace the traditional-used hierarchical database model which in contrast had a single parent system. This structure, when visualized, forms a web-like network consisting of multiple parent–child relations. This allows the network database to have well-modeled relationships between its entities. The lack of flexibility given by the hierarchical database model was overcome; hence, in the early 1970s, it was highly implemented and used. It provided more flexibility in terms of parent–child relationships, as it offered multiple parents to a single record (entity). This made accessing the data fast and fairly simple. This was the first model to introduce many-to-many relationship between tables. This kind of many-to-many relation (having multiple links) applied in two ways: the schema and the database, which itself is a generalized graph representation of the records connected by relationships.

This model failed in the long run due to the introduction of relational database model. It was difficult to maintain and was not as flexible as the relational model. Also, it was only accessible by well-qualified engineers and programmers. Still, it needed the programmer to understand the data structure well in order to make it more efficient.

4.3 Relational Database

The most widely used and the most popular database is the relational database. In this kind of database, the data are stored in the form of rows and columns in a table with a specified table name. As the name suggests, the data in the tables are related to each other; hence, these tables are called relations. The rows and columns are the tuples and the attributes of the table, respectively. The information stored in the relations is called a record. Every table has an attribute that uniquely defines the records under it. This field or attribute is called the primary key. Since the data are organized in the form of rows and columns, the information can be retrieved easily since the data are stored in a comparatively systematic manner. Relational database management system has proved to be the most easily designed database. The open-source software such as MySQL, PostgreSQL, SQLite, and many more has made the database storage and management process easier, especially for start-ups and for small companies who are opting for RDBMS. The data in the database are accessible through the queries entered by the user which is obtained by the means of relational algebra and relational calculus. An advantage of RDBMS is that a user can view the data in whichever way or order he wants just by writing the query as per his requirements. All the advantages of DBMS are present in RDBMS making it ideal for office use. DBMS, as discussed above, includes hierarchical, network, and object-oriented. But, none of these DBMS stores data in a tabular manner. In addition to this, these systems do not apply any security when it comes to data manipulation. RDBMS follows ACID properties, i.e., atomicity, consistency, isolation, and durability. Also, RDBMS can store and handle a comparatively large amount of data and can support access to multiple users.

4.4 Object-Oriented Database

In an object-oriented database, information and data are represented by objects, which are instances of classes. An object-oriented database can be called a database having the capabilities of a relational database along with the capabilities of object-oriented programming. The object-oriented database management system has three main parts to it which are object structure, object classes, and object identity. The term object-oriented database management system (OODBMS) first came into play circa 1985. Several research projects have been done on the subject, with the most notable one being ORION [2].

The main advantage that object-oriented databases have over other databases is that for one, they handle complex data in a better and more effective manner as compared to other databases. As a result of object-oriented programming languages such as C++ and Java being used in these databases, less coding is required, and as a result, the coding used is more efficient. One other advantage is that data can be accessed quickly and the response time is comparatively less due to the use of pointers while searching for something. In the memory sector, OODBMS does wonderfully well, taking into account the use of single-level memory and automatic-method storage.

While there are several advantages of the object-oriented database, there are a handful of disadvantages as well. Perhaps, the most major drawback of using object-oriented databases is that they do provide adequate tools for querying and reporting the database. A feature of the object-oriented is like double-edged sword-object-oriented programming. On one hand, there are advantages to using the object-oriented programming languages, on the other hand, object-oriented databases are limited to these specific languages and other languages cannot be implemented. The major application of object-oriented databases is done in scientific fields as these fields require complex data to be handled in the most efficient as well as the most effective way possible such as engineering, high energy physics, and molecular biology. Examples of object-oriented databases are Smalltalk, Perst, etc. [3].

5 Database Software

Database software helps small and big firms and companies to manage the data efficiently. It provides a front end, simple interface through which, even people with a non-technical background can carry out the tasks. Database software can perform various tasks such as storing, managing, changing, searching, and extracting the data within a database. A database system as such, also secures the information in the database which is crucial for all companies. It also increases the overall accuracy as well as efficiency of using databases. Some of the most popular database software presently are Oracle, MySQL, PostgreSQL, MongoDB, etc. We have shown some statistics and qualities of these software (Fig. 1 and Table 1).

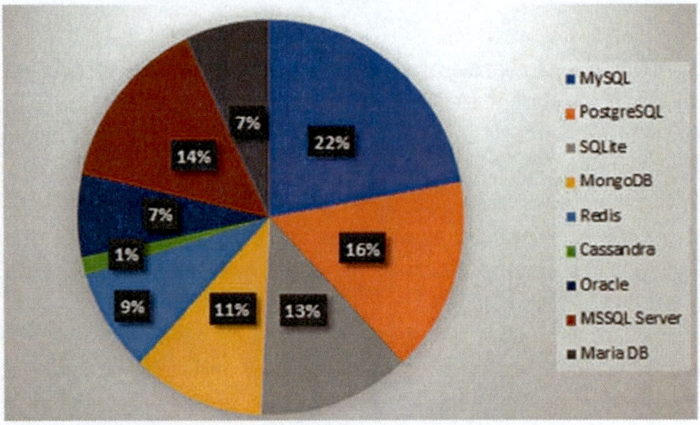

Fig. 1 Database comparison in terms of usage [6]

Table 1 Software comparison chart [13]

Software	Data structure	Scaling capacity	Usage ease
MySQL	SQL	Vertical, complex	Easy
MariaDB	SQL	Vertical	Easy
Oracle	SQL, multi-model	Vertical	Hard
PostgreSQL	SQL, object-oriented	Vertical	Hard
MSSQL server	T-SQL	Vertical, complex	Hard
MongoDB	NoSQL, document-oriented	Horizontal	Easy
Redis	NoSQL, key-value	Horizontal	Easy
Cassandra	NoSQL, column-oriented	Horizontal	Hard

6 Future Trends

The database software market continues to expand with the constant generation of data. With the emergence of big data, the amount of semi-structured and unstructured data has critically escalated. "IDC Technologies has predicted that in 2025, the world will create and replicate 163ZB of data, representing a tenfold increase from the amount of data created in 2016." [4] The relational database model (or RDBMS) is capable of handling only the structured type of data. NoSQL, known as Not Only SQL, is the solution for handling unstructured or semi-structured data and provides a mechanism of faster storage and retrieval of data. It is considered to be the next-generation database. Some NoSQL databases provide suggestively higher data throughput than traditional RDBMSs. MongoDB, Neo4j, and Amazon DynamoDB are some of the examples of NoSQL databases [5].

However, the latest trends include adopting a hybrid database structure instead of a single database structure. In this way, the users can utilize the capabilities of both, SQL and NoSQL. This bridges the relational database with the non-relational database (Table 2).

Table 2 Features comparison [6]

Feature	NoSQL	Relational database
Schema	Dynamic	Pre-defined
Properties	CAP (consistency, availability, and partition tolerance)	ACID
Scalability	Horizontal	Vertical
Consistency and reliability	Poor	Good
Availability	Good	Good
Data	High amount of unstructured data	Low to medium amount of structured data
Example	Cassandra and MongoDB	PostgreSQL and Oracle

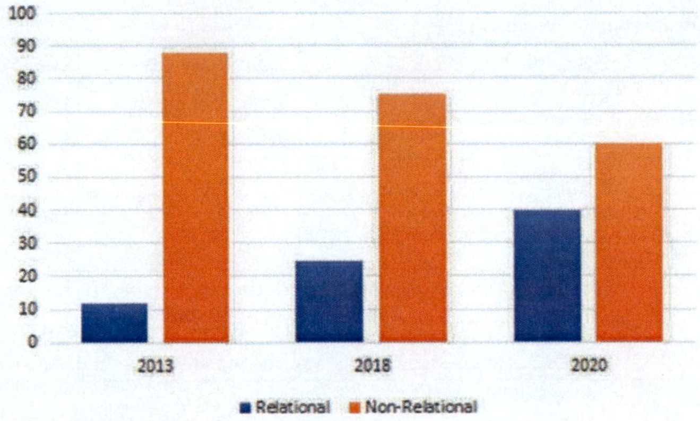

Fig. 2 Relational versus non-relational database [6]

Security is going to be one of the most prominent complications in the near future. IDC states the data to be managed has reached 40ZB in 2020, and only half of that is protected. For protection, data logging is necessary. Since the data are getting generated from many different places such as web servers, file servers, and credit card payment gateways, it is not structured in the same way for each of them, making it very difficult to design a relational database for the same [7]. Therefore, a hybrid model can be optimized here.

Management of database is a tedious and time-consuming task for the most administrators. Hence, automating database systems is an emerging trend. AI is used to simplify maintenance which includes handling, updating, and upgrading. This is known as augmented data management. Also, with cloud service providers (CSPs) offering more options, organizations (and DBAs) are embracing their current infrastructure with the cloud services. Some of the databases, also working on cloud platforms, are Microsoft Azure, Amazon Web. Thus, for small firms and organizations, the relational database has always been useful, but as and when the number of data increases, there is a shift from SQL to NoSQL [8, 9] (Fig. 2).

7 Augmented Data Management

Augmented data management is a newly emerging technology in the vast expanse of data management and data analytics. This process stems from the capabilities of two of the most popular and robust technologies that the world has to offer at this moment in time, artificial intelligence and machine learning. The main purpose of augmented data management is to increase the number of automated tasks, and thus, reducing manual tasks such as handling and configuring as well as tampering of huge amounts of data. This includes various processes which include copying, synchronizing, restoring, migrating, archiving, and backing up of data, and the most tedious ones being cleaning and preserving the quality of data. Gartner, a global research and advisory firm, has predicted that by 2022, 45% of the manual tasks of database management would be reduced due to the introduction of machine learning and automated service-level management and that owing to the introduction of this AI-enabled automation, and 20% of the specialized data management professionals will no longer be in need. According to the firm, the technology is expected to influence effectively every area of data management, including database management systems data integration, metadata management, and various others [10].

An immense drawback that is faced by data scientists and data analysts is the tremendous amount of time that is spent on unnecessary manual tasks such as the ones mentioned above. According to an analysis carried out by AIM, they spend 80% of their time in the maintenance and organization of data. The implementation of augmented data management hopes to reduce this time significantly and thus increasing the efficiency of not only the data of an organization or company, but also the time and productivity of its employees as well as the financial costs of operations by decreasing the data complexities. Another facet in the kaleidoscope of advantages that augmented data management provides to a company is ensuring authentic data as human errors can lead to distorted data which can, in some cases, lead to major consequences.

Augmented data management can be used in almost every sector where database management systems are in place. For example, if a pharmaceutical company or bank wants to take decisions based on machine learning, the necessity to make sure that the data are of the highest quality is of the utmost importance. In a financial firm, any specific trading portfolio is judged on metadata. Augmented data management assists in the management of such metadata. An industry that is going to benefit greatly from this technology is the media and entertainment industry. Machine learning and AI are the roots of new-age digital advertising, which is projected which is estimated to increase by 23% by 2023, according to Statista. It can also be used to monitor and control the security and privacy of online users who decide to upload photos and/or videos online. With the substantial increase in the number of media being uploaded to the net, augmented data management can also help with the relevant data storage in specific segments. Thus, augmented data management is one of the most exciting upcoming prospects to keep a keen eye on in the next couple of years [11].

8 Office Management Using Relational Database Approach for Study in India Program (SIP)

This database is designed for the Study in India Program. This database helps the current way of storing data to change into a smart and simpler way of storing data. This project emphasizes more on smart and efficient work rather than the tedious data entries and repeated documentation of every new Study in India Program. Since, the amount of data generated is not in TB or ZB, it is more feasible to develop a relational database for SIP.

8.1 About SIP

The Study in India Program is an integral part university which is known for establishing connection with foreign universities. Study in India Program gives the exposure of the Indian culture, business, psychology, and tradition to the international students from various countries [12].

8.2 SIP Requirements

SIP generates a lot of data related to the students, faculties, and university which is further used to prepare reports and post-analysis. To ease this activity, we have established an interface to generate an integrated database management system. This database will help SIP to ease the office activities documentation, reporting activities and to perform the retrospective analysis from past datasets.

From multiple discussions with SIP team members and our own experience and knowledge regarding SIP, following major requirements at the highest level of abstraction are identified:

- To store the itinerary/time table of SIP in database to keep the record for the future reference.
- To store student, faculty, and university details.
- To store the feedback by the students.
- To keep record of student and volunteer attendance.
- To store pictures or their links.
- To store and manage other useful information like documents/PPTs/lectures.
- To keep record and the information of the managers that the future programs might need.
- Stay details for SIP.

8.3 SIP Database Design

The entity–relationship diagram (ER diagram) is a model that shows the relationships between the tables. It is the initial step while designing a database. Here, there are four types of relationships, one to one, one to many, and many to one. We have not used many to many, as it was not required. The tables have primary keys and some have composite keys (their combination becomes the primary key for the table). Also, there are foreign keys, which are the primary keys of some other table and may or may not be the primary key of the table it is related to. The dotted lines represent non-identifying relationship, in which the primary key of the parent table is not the primary key of the child table. The solid lines represent identifying relationship, in which the parent table and the child table have the same primary key. (refer pg. 7 for ER diagram).

9 Graphical User Interface

Graphical user interface forms an interface between the user and the devices by using icons and other graphical notations. After the designing part, i.e., the creation of the database management system, the project moves ahead. The focus is now on making an application that is user-friendly and can be easily accessed by an office or company. Without a GUI, access to a database is quite cumbersome. A person needs to know the SQL statements to insert, delete, update, or search in the database. GUI makes this process easy, as the database gets linked to Java (GUI) and develops an application that is secure and can be modified as and when required. SWING graphical user interface is a part of the Java framework and has been used to develop a desktop application that can be used in any system and any platform.

Firstly, a code is written to establish a connection with the required database. Here, Java Eclipse IDE (integrated development environment) is used with SQLite (database). The first page of the application is the login page which makes sure that only the registered people or the people who have the correct username and password are the only ones who are allowed to access the menu page (Fig. 3).

The menu page is the main content page which provides a handle to work upon five main options given on the menu bar (see Fig. 4). International students, volunteers, and patrons tabs give four options—insert, delete, update, and search student. The insertion and deletion of data are done using PreparedStatement, in which the general SQL statement is passed. For the search and update operations, ResultSet along with PreparedStatement is used along with SQL update query.

Contact MenuItem is for the purpose of getting contact information about the main admin.

The SIPs MenuItem contains the information related to a specific SIP. As the name suggests, the Change Password button gives the user an option to change the password. The password can be changed only when a person enters the right username and old password. Register New gives an option to the admin or the other users to register for new users.

In conclusion, the application can perform all the SQL queries along with easy access to the database. This application also ensures the security of the database along with the atomicity of updates.

The most important factor is reusability. Just by changing the names of the label and buttons, and establishing a new database connection, other offices or companies can also use the same GUI. Other advantages such as data redundancy and searching are also there which make DBMS GUI more reliable than the trivial file system

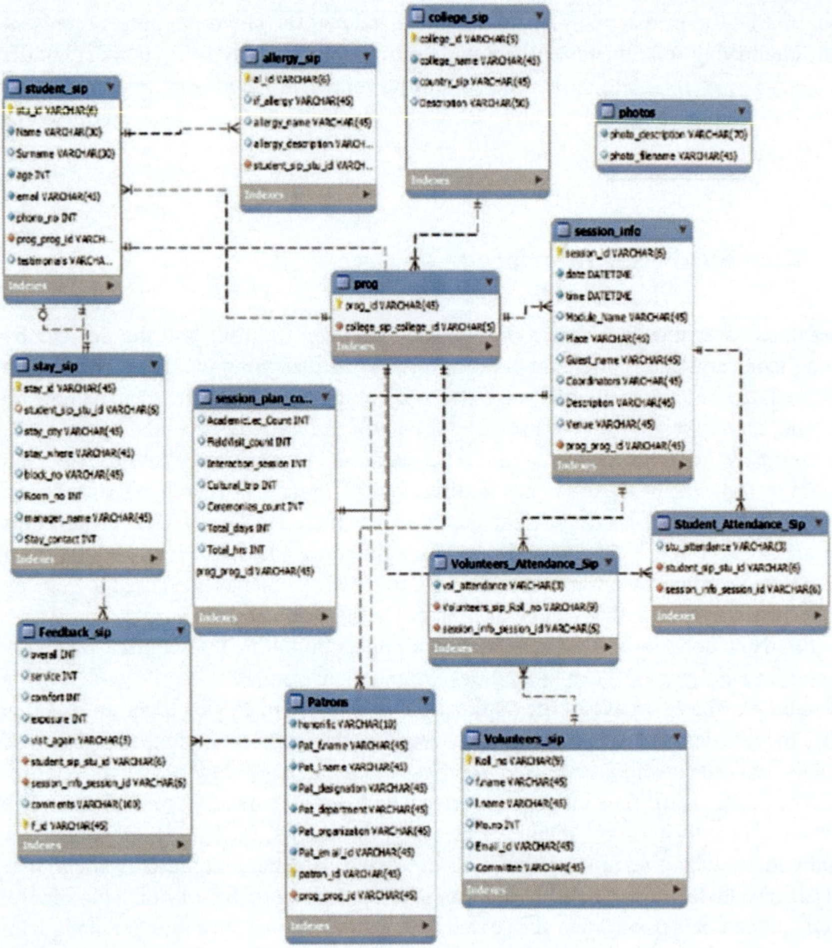

Fig. 3 ER diagram

method. This was just an example of a small office. The same GUI can be used for huge companies and organizations to store thousands of records. But, this also has disadvantages when it comes to a large level; i.e., companies like Google use NoSQL, because of the huge amount of data as the data are increasing at a fast rate and thus SQL might become a less reliable source.

Fig. 4 Menu page

10 Conclusion

The database provides the best solution for the data generated from big and small firms, businesses, schools, colleges, and other commercial offices. Post-analysis of data is required to improve the business strategies, future planning, and to know about the customer's liking and marketing trends. The database is proven to be more advantageous than old file-based systems. Network, hierarchical, relational, and object-oriented-based databases are the popular approaches to store and manage the database. In the era of big data where a huge amount of structured, semi-structured, and unstructured data needs to be managed, the NoSQL approach has emerged and has become popular day-by-day. A small database application for office use is discussed in this paper which is based on the relational database approach. Implementation to store and retrieve the actual data into the database has been done using Java—JDBC.

Acknowledgements We would like to extend our gratitude to Prof. Nigam Dave, Head of Office of International Relations, PDPU, and Dr. Ritu Sharma, Associate Professor, PDPU, for providing insight into SIP requirements. We are immensely grateful to them for guiding us through our project and providing us with information as and when required.

References

1. Silberschatz, S., Korth, H.F., Sudarshan, S.: Database system concepts
2. Object database page. https://en.wikipedia.org/wiki/Object_database

3. Ketabchi, M.A., Mathur, S., Risch, T., Chen, J.: Comparative analysis of RDBMS and OODBMS—a case study
4. Database Trends and Application. http://www.dbta.com/Editorial/News-Flashes/Key-Database-Trends-Now-and-for-the-Future-131888.aspx, last accessed 2019/5/21
5. ChristofStrauch, Prof. Walter Kriah: NoSQL Database
6. Dbzone webpage. https://www.dbzone.org (for figure 1,2)
7. NoSQL and hybrid databases. https://www.stratoscale.com/blog/dbaas/hybrid-databases-combining-relational-nosql/
8. Sethi, B., Mishra, S., Patnaik, P.K.: A study of NoSQL database. Int. J. Eng. Res. Technol. (IJERT) (2014)
9. Padhy, R.P., Patra, M.R., Satapathy, S.C.: RDBMS to NoSQL: reviewing some next-generation non-relational database's. (IJAEST) Int. J. Adv. Eng. Sci. Technol. (2011)
10. https://www.gartner.com/en/conferences/apac/data-analytics-india/gartner-insights/rn-top-10-data-analytics-trends/augmented-data-management
11. AnalyticsIndiaMag webpage. https://analyticsindiamag.com/how-startups-can-leverage-augmented-data-management-to-drive-business/,last accessed 2019/10/14
12. PDPU official site. https://www.pdpu.ac.in/exposure-program.html
13. Comparing Database Management Systems. https://www.altexsoft.com/blog/business/comparing-database-management-systems-mysql-postgresql-mssql-server-mongodb-elasticsearch-and-others. Last Accessed 20 June 2019

Parallel Computation of Hadamard Product for Dynamic Dataset on Google Colab

Sarjil Shariar and K. M. Azharul Hasan

Abstract Hadamard product, the element-wise product of tensors, is an important tool for many real-life applications such as machine LSTM, multimodal learning, object recognition, and segmentation. In this paper, we propose a faster processing of hadamard product by the parallel computation on graphics processing unit (GPU) for dynamic data. Most of the computations for real-life data are multidimensional and dynamic which expand during run time. In this scheme, we use dynamic memory allocation that allows further expansion of the dimension size and high memory utilization based on the notion of extendable array. Since high dimensionality is difficult to handle, we use a dimension conversion technique where the n-dimensional array structure is transformed into a 2-dimensional structure. Furthermore, to handle the future data, we apply the converted extendable array structure to get Scalable Hadamard Product (SHDP) for dynamic dataset. Then, we implement a parallel algorithm for the parallel processing of the SHDP that partitions and distributes the element-wise product during runtime. We utilize the maximum available GPU threads for the faster performance of the hadamard product. We found good speed-up for the proposed partitioning while applying to "Google Colab".

1 Introduction

Hadamard product is significant for many computational problems in different fields of physics, statistics, and computer science [1, 2]. This product indicates the element-wise multiplication between tensors and demands the equal number of row and column setup after tensor-to-matrix conversion. In this paper, we are conducting the parallel processing of the hadamard product with a transformed scalable data structure on GPU. Traditional multiway array (TMA) is mainly the carrier of the multidimensional models with n index variables. But, storing massive amount of data in this structure can be problematic because of the memory adjacency, fixed index

S. Shariar (✉) · K. M. A. Hasan
Department of Computer Science and Engineering, Khulna University of Engineering & Technology, Khulna, Bangladesh
e-mail: az@cse.kuet.ac.bd

T. Senjyu et al. (eds.), *ICT with Intelligent Applications*, Smart Innovation, Systems and Technologies 248, https://doi.org/10.1007/978-981-16-4177-0_48

size, and smaller stack memory. To work with the multidimensional model in the hadamard product processing, we introduce a scheme named as Scalable Hadamard Product (SHDP) which uses a two-dimensional indexing technique constructed by the dynamic allocation. This scheme uses the concept of dimension conversion where n-dimension is converted into two-dimension [3]. To compute the SHDP, we use Scalable Tensor Structure (STS) that can scale (extension or reduction) the dimension size during runtime for the inclusion of the new tensor elements in future [4]. The structure is chunked in such a way that the task parallel scheme maintains simplicity for the parallel tensor computation [5, 6]. To achieve the parallelism, we implement a parallel algorithm which distributes the elements of the tensors dynamically and runs the element-wise multiplication simultaneously using GPU threads. By the utilization of the maximum available threads, this parallel processing technique shows good speed-up than the sequential scheme. Also, we design our algorithm such a way that the increase of the tensor elements or the GPU threads will not have any complication regarding the load balance [7]. With the increase of the GPU threads in the system, we expect to attain more speed-up with our parallel algorithm. The scheme can be applied to dynamic computation for multimodal learning [8], parallel tensor computation [6], large-scale data analysis [9], etc.

2 Related Work

There has been some influential study to work with the properties of the hadamard product. In [10], a learnable hadamard product model is used for the low-light image and underwater image enhancement problems. A bilinear pooling is developed in [8] using the hadamard product for an efficient attention mechanism of the multimodal learning. The tensor computation for the hadamard product is analyzed in [11, 12]. But the above schemes use static data set and the maintenance for dynamic data during runtime is more difficult. Moreover, the hadamard operation can take large computation time with the increase of the elements in tensors. In this era of high-performance computing (HPC), the use of low-cost parallelism devices like GPU for the faster data processing is popular. For example, [5] explains the parallel computations of the dense-sparse matrices using various GPU models, [6] explains the parallel decomposition techniques such as canonical decomposition and Tucker decomposition, [9] uses a framework called FastQuery to distribute the subarrays and execute them concurrently on CPU cores and [13] describes the data parallelism on the relational database scheme. In addition, some other methods such as training-based linear regressions [7] and MapReduce model [14] can be applied to achieve the load balance in various systems. In our task-based parallel algorithm, load balance is maintained by the dynamic element-wise data distribution on the available GPU threads. Therefore, this scheme requires only the mapping-based parallel technique that can save the computation cost. In this paper, we are working with a scalable structure rather than the conventional TMA structure for high memory utilization in

SHDP. With this scheme, we do not need to worry about the big adjacent memory location to hold the tensor elements as we are converting the large array into smaller subarrays.

3 Scalable Hadamard Product (SHDP)

In SHDP, we transform the n-dimensional array into a two-dimensional array and allocate the memory dynamically as a segment called subarrays. Therefore, dynamic data is possible to handle even if the size of the tensor overflows. The dynamic space does not have the necessity to allocate consecutively in the memory. Hence, improved space utilization is achieved. SHDP is constructed with the converted two-dimensional dynamic arrays which are actually the row and column indexing of the matrix. The odd and even dimension of the n-dimensional indexing is used for the row and column indexing, respectively. We assume, an nth-order array A has dimension $N \in \mathbb{N}$ and dimension length $L \in \mathbb{N}$ where every dimension of N is denoted by d_i and every dimension length of L is denoted by l_i for the index position, $i = 1, 2, \ldots, n-1, n \in \mathbb{N}$. Figure 1 shows the dynamic data structure used in the scheme to calculate SHDP. Based on the memory space size, we can extend each dimension

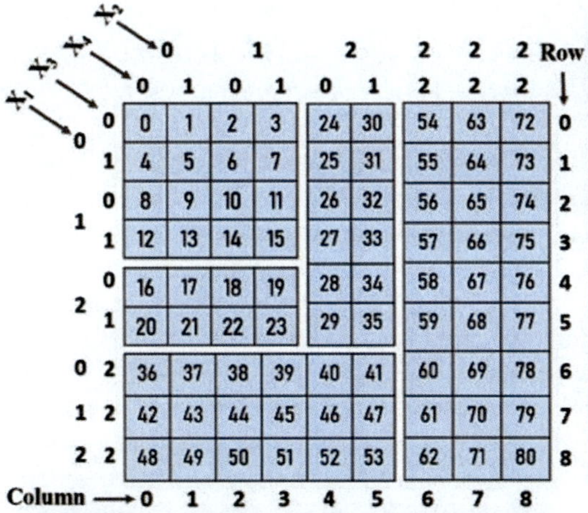

Length	IP	CP	EDP	ATP
2, 2, 2, 2	0	2, 1	0	0
3, 2, 2, 2	2	2, 1	1	1
3, 3, 3, 2	2	1, 3	3	2

Length	2, 2, 2, 2	3, 3, 2, 2	3, 3, 3, 3
IP	0	2	2
CP	2, 1	2, 1	1, 3
EDP	0	2	4
ATP	0	1	2

Fig. 1 Scalable array realization for SHDP

length L up to λ. Let A be an nth-order array with length $L(l_1, l_2, \ldots, l_{n-1}, l_n)$. Now, we obtain the row and column length of the subarrays following the Eqs. 1–4.

$$\text{If } n \text{ is odd, } row\_len = l_1 \times l_3 \times \ldots \times l_n \quad (1)$$

$$\text{If } n \text{ is even, } row\_len = l_1 \times l_3 \times \ldots \times l_{n-1} \quad (2)$$

$$\text{If } n \text{ is odd, } col\_len = l_2 \times l_4 \times \ldots \times l_{n-1} \quad (3)$$

$$\text{If } n \text{ is even, } col\_len = l_2 \times l_4 \times \ldots \times l_n \quad (4)$$

With the size of $row\_len \times col\_len$, we allocate the array in the memory. In addition, we have to form an auxiliary table (AT) with the following pointers for the future maintenance (See Fig. 1).

- **Index pointer**: **IP** indicates the increased index value, initially IP = 0.
- **Extended dimension pointer**: **EDP** indicates the extended dimension number, initially EDP = 0.
- **Auxiliary table pointer**: **ATP** maintains the AT index number in the row and column direction, initially ATP = 0.
- **Coefficient pointer**: For an nth-order array, the coefficients **CP** will be calculated following the Eqs. 5–8.

$$\text{If } n \text{ is odd, CP}[row] = \{l_3 \times l_5 \times \ldots \times l_n, \ldots, l_{n-2} \times l_n\} \quad (5)$$

$$\text{If } n \text{ is even, CP}[row] = \{l_3 \times l_5 \times \ldots \times l_{n-1}, \ldots, l_{n-3} \times l_{n-1}\} \quad (6)$$

$$\text{If } n \text{ is odd, CP}[col] = \{l_4 \times l_6 \times \ldots \times l_{n-1}, \ldots, l_{n-3} \times l_{n-1}\} \quad (7)$$

$$\text{If } n \text{ is even, CP}[col] = \{l_4 \times l_6 \times \times l_n, \ldots, l_{n-2} \times l_n\} \quad (8)$$

In Fig. 1, we have shown some converted subarrays obtained from an order-4 array. When we need to retrieve any index from those subarrays, we convert the n-dimensional index into 2-dimensional index using the above pointers. To retrieve any index such as (1, 0, 2, 1) from any four-dimensional array, we first search for the maximum row and column value, i.e., 2, 1. Comparing these values with IP, we then obtain EDP and comparing with EDP, we obtain ATP for the row and column dimension. Now, we multiply the CP values of the two ATPs with the four-dimensional row and column index values, i.e., (1, 2), (0, 1). The addition of these values forms the two-dimensional row and column index. With this two-dimensional index, we finally retrieve the tensor data. Figure 2 shows the SHDP scheme with the dynamic subarrays that can handle the dynamic data during runtime. Element-wise products are computed using the auxiliary tables. Therefore, dynamic data that occurs during runtime can be computed without intervening the existing data in the array.

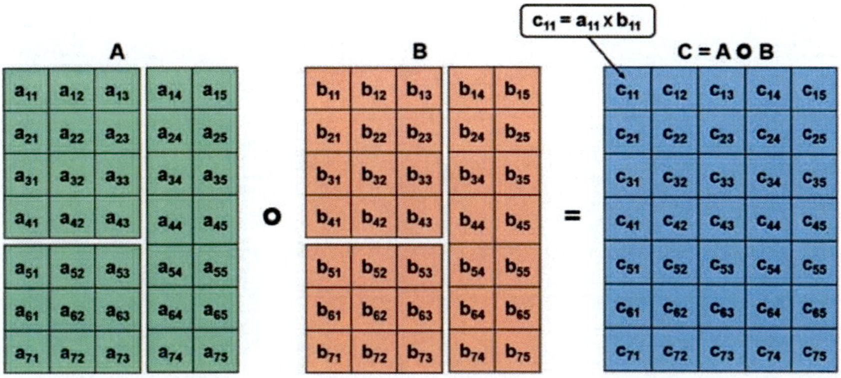

Fig. 2 Scalable Hadamard Product (SHDP)

4 Parallel Computation on SHDP

We implement the parallel algorithm in a way that the element-wise products are distributed on the GPU threads. It maintains the load balance dynamically with the increase of tensor elements and GPU threads. This parallel implementation has two parts namely CPU segment and GPU segment. CPU segment calculates the number of element-wise products ne per thread with the modulus operation between total product number and maximum threads (i.e., $ne = n\%th$, where n and th being total number of products and total number of threads, respectively). Also, the extra element-wise products ed are calculated after the equal division. Then, these two parameters are passed to the GPU segment with the kernel launch from CPU. Algorithm 1 shows the CPU segment of the algorithm.

In the GPU segment, the main job of the parallel GPU threads are converting the indices and processing the product operations simultaneously. We find the running thread number tn by the block indexing $blockIdx.x$ and thread indexing

Algorithm 1: Parallel Algorithm CPU Segment

let n be total number of element-wise products
$th \leftarrow 24000$ `// maximum parallel threads`
$tb \leftarrow 128$ `// threads per block`
$ne \leftarrow n\%th$ `// element-wise products per thread`
$ed \leftarrow n - ne$ `// extra element-wise products`
if $n < th$ **then**
  $bg \leftarrow ceil(n/tb)$ `// blocks per grid`
else
  $bg \leftarrow ceil(th/tb)$
kernel_execution $<<< bg, tb >>>$ (ne, ed) `// kernel launching`

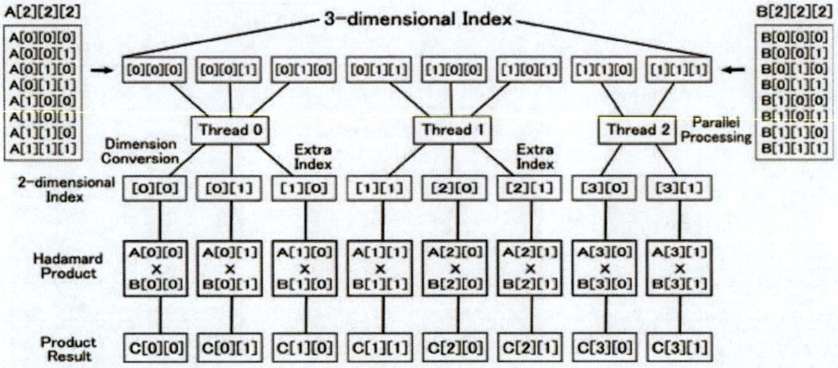

Fig. 3 Illustration of parallel scheme

$threadIdx.x$ as $tn = (blockIdx.x \times tb) + (threadIdx.x + 1) - 1$. With the running thread number and the previously computed ne and ed, we find the particular n-dimensional indices for the running thread. Then, we use the dimension conversion technique to convert the n-dimensional indices into two-dimensional indices. Finally, we retrieve the elements by the two-dimensional indices, execute the element-wise product operations, and store the results concurrently in a 2D array C. Figure 3 shows an illustration of 3D tensor operation with the parallel computation. In this system, the input is the three-dimensional elements of tensor A and B. After processing the elements with the proposed parallel algorithm on GPU threads, the output is the element-wise hadamard product with the 2D array C. This algorithm dynamically adjusts the workloads with the increase of GPU threads, therefore creates the path to extend the system with more GPUs. Algorithm 2 shows the GPU segment of

Algorithm 2: Parallel Algorithm GPU Segment

$tn \leftarrow (blockIdx.x \times tb) + (threadIdx.x + 1) - 1$ `// thread number`
if $tn + 1 \leq n$ **then**
  if $tn + 1 \leq ed$ **then**
    $x \leftarrow tn \times (ne + 1)$ `// indices computed in previous threads`
    $y \leftarrow ne + 1$ `// indices computation for running thread`
  else
    $x \leftarrow tn \times ne + ed$
    $y \leftarrow ne$
for $i \leftarrow x$ **to** $x + y - 1$ **do**
  $A[r][c] \leftarrow dim\_conv(ndA\_idx + i)$ `// dimension conversion`
  $B[r][c] \leftarrow dim\_conv(ndB\_idx + i)$
  $C[r][c] \leftarrow A[r][c] \times B[r][c]$ `// element-wise hadamard product`
return

the algorithm where the function $dim\_conv()$ performs the index conversion on the parallel threads to apply the parallelism on the SHDP scheme.

5 Experimental Results

A prototype of our proposed parallel algorithm for hadamard product has been simulated on Google Colab. This system is configured with Intel(R) Xeon(R) CPU @ 2.00GHz processor, 12 GB RAM, NVIDIA Tesla T4, GDDR6 16 GB VRAM, and 56320 KB cache memory. We indicate the sequential (non-parallel) computation using the term SEQ-HP and the parallel computation using the term PL-HP. We have used the number of dimensions for the tensors varying from 4 to 12 and the number of threads varying from 6K to 18K to show the performance of the proposed scheme.

Figure 4a shows the comparison between the PL-HP scheme and the SEQ-HP scheme varying number of dimensions. The execution time rises heavily with the increase of dimension number when the sequential processing happens. There is a

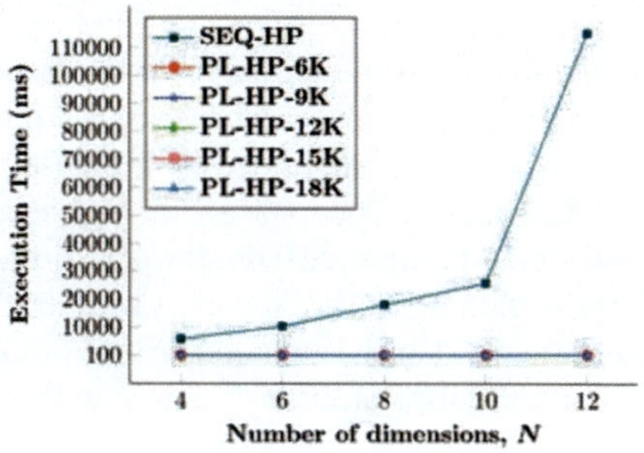

(a) Sequential vs parallel computation

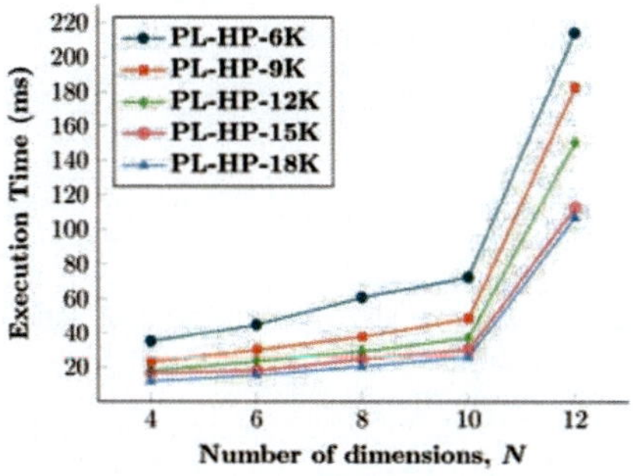

(b) Parallel computation for various dimension

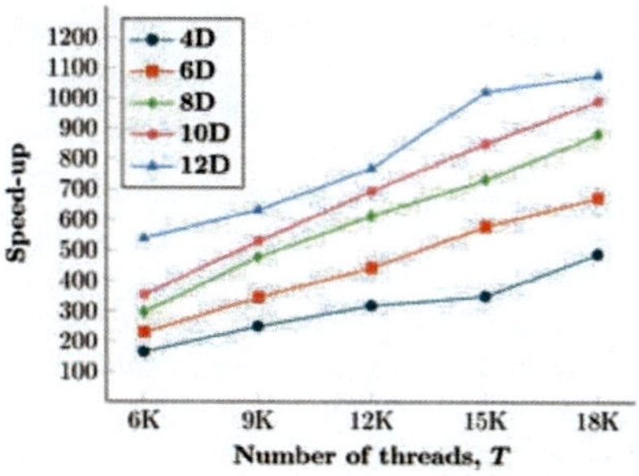

(c) Speed-up for various thread

Fig. 4 Computation comparison of hadamard product

clear improvement of PL-HP over SEQ-HP. With the same dimension and thread variation, we get very low execution time in the parallel processing on GPU. The comparison of execution time for different number of threads and varying number of dimensions is shown in Fig. 4b. The result shows that for increasing the number of threads the execution time decreases drastically.

In Fig. 4c, we have analyzed the speed-up for parallel processing with the same parameters. With our scheme, the more dimension we apply, the more efficiency we get. This is because when the number of dimensions increases, the data volume increases, therefore parallel algorithm works well. When the data volume increases, the processing time increases, respectively, but with more number of parallel threads, we can get better speed-up. For example, the speed-up of the 4D, 6D, 8D, 10D, and 12D tensors indicates the maximum speed-up for 12D in Fig. 4c. Here, the speed-up increases with the parallelism increase. The distribution of the element-wise products is conducted such a way in our scheme that the maximum parallel threads are used with dynamic load balance.

6 Conclusion

We have implemented a parallel scheme on GPU device for the faster processing of the hadamard product. In our scheme, we have used a dimension transformation technique for the dynamic dataset namely Scalable Hadamard Product (SHDP). To hold the elements of the hadamard operation with high memory utilization, we have used the two-dimensional structure. SHDP allows n-dimensional array to two-dimensional subarray conversion as well as the expansion of the dimension length for the inclusion of new elements. For the efficient parallel processing, we have integrated a parallel algorithm that distributes the element-wise products on the GPU threads. Our experimental analysis based on this parallel algorithm shows faster performance in the cloud environment of the Google Colab. We believe the scheme can be applied to the dynamic tensor computations effectively and integrated with other parallel devices for faster data analysis.

References

1. Dilip, A.S.A.: The controllability gramian, the hadamard product, and the optimal actuator/leader and sensor selection problem. IEEE Control Syst. Lett. **3**(4), 883–888 (2019)
2. Wang, J., Li, Y., Sun, H.: Lie-trotter formula for the hadamard product. Acta Mathe. Sci. **40**(3), 659–669 (2020)
3. Hasan, K.M.A., Shaikh, M.A.H.: Efficient representation of higher-dimensional arrays by dimension transformations. J. Supercomput. **73**(6), 2801–2822 (2017)
4. Omar, M.T., Hasan, K.M.A.: A scalable storage system for structured data based on higher order index array. In: Proceedings of the 3rd IEEE/ACM International Conference on Big Data Computing, Applications and Technologies, pp. 247–252 (2016)

5. Shi, S., Wang, Q., Chu, X.: Efficient Sparse-Dense Matrix-Matrix Multiplication on GPUS Using the Customized Sparse Storage Format (2020)
6. Austin, W., Ballard, G., Kolda, T.G.: Parallel tensor compression for large-scale scientific data. In: Proceedings of the 2016 IEEE International Parallel and Distributed Processing Symposium (IPDPS), pp. 912–925 (2016)
7. Lin, C.S., Hsieh, C.W., Chang, H.Y., Hsiung, P.A.: Efficient workload balancing on heterogeneous GPUS using mixed-integer non-linear programming. J. Appl. Res. Technol. **12**(6), 1176–1186 (2014)
8. Kim, J., On, K.W., Lim, W., Kim, J., Ha, J., Zhang, B.: Hadamard Product for Low-rank Bilinear Pooling. CoRR abs/1610.04325 (2016)
9. Chou, J., Wu, K., Rubel, O., Howison, M., Ji Qiang, Prabhat, Austin, B., Bethel, E.W., Ryne, R.D., Shoshani, A.: Parallel index and query for large scale data analysis. In: Proceedings of the 2011 International Conference for High Performance Computing, Networking, Storage and Analysis, pp. 1–11 (2011)
10. Liu, R., Li, S., Liu, J., Ma, L., Fan, X., Luo, Z.: Learning hadamard-product-propagation for image dehazing and beyond. IEEE Trans. Circuits Syst. Video Technol. 1-1 (2020)
11. Xu, Y., Li, Y., Li, Z.: Some results on the hadamard product of tensors. Bull. Iranian Math. Soc. **45**(4), 1193–1219 (2019)
12. Kressner, D., Perisa, L.: Recompression of hadamard products of tensors in tucker format. SIAM J. Sci. Comput. (SISC) **39**(5), A1879–A1902 (2017)
13. Hasan, K.M.A., Tsuji, T., Higuchi, K.: A parallel implementation scheme of relational tables based on multidimensional extendible array. Int. J. Data Warehousing Mining **2**(4), 66–85 (2006)
14. Li, J., Liu, Y., Pan, J., Zhang, P., Chen, W., Wang, L.: Map-balance-reduce: an improved parallel programming model for load balancing of mapreduce. Future Generation Comput. Syst. **105**, 993–1001 (2020)

Cooperative Hypothesis Building for Computer Security Incident Response

Shota Motoura, Ayako Hoshino, and Itaru Hosomi

Abstract We have been developing an AI-based system that automatically collects evidence from logs on computers within an intranet and infers the most plausible hypothesis of a cyber-attack. In this paper, we propose a user-interactive functionality which enables the system to incorporate knowledge and insight from the user in order to build the hypothesis. We show the system's practicality with example use cases and also confirm the necessity of such functionality for the system by interviewing with cyber security analysts.

1 Introduction

Recently, Computer Security Incident Response Teams (CSIRTs) in companies and organisations attract considerable attention, as cyber-attacks sometimes cause severe system failure in a company or an organisation and, in the worst case, put a critical infrastructure to a halt.

Under a cyber-attack, the CSIRT is required to recover as soon as possible from the ongoing attack and to prevent further damage. Because of this time demand, the security analysts at CSIRTs need to draw and utilise the whole picture of the attack, rather than to examine all alert logs one by one. More precisely, the analysts (1) hypothesise the whole attack process on the basis of the evidence they have already found, then (2) check the consistency of the evidence with some modification if needed, and (1') revise the hypothesis of the whole attack process.

S. Motoura · A. Hoshino · I. Hosomi
Data Science Research Laboratories, NEC Corporation, Kawasaki, Japan
e-mail: ayako.hoshino@nec.com

I. Hosomi
e-mail: i-hosomi@nec.com

S. Motoura (✉) · I. Hosomi
Cyber Security Strategy Division, NEC Corporation, Kawasaki, Japan
e-mail: motoura@nec.com

T. Senjyu et al. (eds.), *ICT with Intelligent Applications*, Smart Innovation, Systems and Technologies 248, https://doi.org/10.1007/978-981-16-4177-0_49

As the cyber-attacks are becoming more and more stealth and complex, companies' CSIRTs are required for more and more time and efforts for investigation. To reduce the time and efforts, our AI-based system supports such investigation by automating inference of the whole attack process in phase (1) above. In order that the user can judge whether the hypothesis is plausible or not, our system outputs self-explaining hypothesis built upon system's logs.

However, it is not sufficient that the system outputs a hypothesis only once in the investigation cycle (1), (2), (1'),$\cdots$. The system has to support the user's revision of the hypothesis.

To deal with this problem, we propose the concept of *cooperative hypothesis building* in this paper, which means that the system and the user, in cooperation, build and revise the hypothesis of the attack process as the analysts do in the aforementioned investigation cycle. We first briefly explain our automatic inference for (1) and (1'), and then describe our feedback functionality for (2) in detail with a concrete example of how our system incorporates the user's feedback on the consistency of evidence to revise the hypothesis.

The contributions of the study are as follows:

1. We implemented three kinds of feedback functions that accept complementary or corrective evidence from the user.
2. We confirmed the necessity of the feedback functions on use of such AI-based system via interviewing with the analysts in NEC Corporation.

2 Related Work

Automatic inference has been applied to various tasks in the cyber security domain [5, 7, 10]. Kandefer et al., with the motivation of identifying true attack traces among enormous IDS/IPS alerts, developed a rule-based system where the users are allowed to specify rules written in SNePSLOG logical expression to determine true positives based on multiple sources [5]. Zhong et al. have recorded the cyber security analysts' investigation processes in a form of their proposed action–observation–hypothesis tree [10], which can be used to support collaboration among multiple analysts investigating the same incident [9]. They worked towards automation of investigation process and constructed automata that detect true positives among alert logs [11]. While all the above studies eventually boil down to a determination of true attack or false alarm, our system can output a possible whole picture of an attack.

If we are to output a whole picture that is plausible to the user, it is natural to incorporate feedback and information given by the user. A framework of sharing information between a user and a system for proving one proposition is known as inquiry dialogue (cf. [2, 3]), and it is implemented by Hiraoka et al. [4]. However, it is not easy for analysts to use an inquiry dialogue system since its input has to be a logical formula or a sentence in a natural language that can be uniquely translated to a logical formula. In contrast, the interaction proposed in this paper mainly requires clicks on several buttons and is easier for the user to feedback to the system. For example, the user can deny or approve a part of the hypothesis as in the following section.

3 Cooperative Hypothesis Building

Corresponding to the cycle of cyber security investigation explained in Sect. 1, our system operates as follows:

1. The initial input to our system is a set of pieces of evidence automatically drawn from computer logs.
2. The system outputs the hypothesis of the most plausible whole picture which is expressed in a directed graph.
3. The user examines the output hypothesis to decide whether he continues his investigation for new pieces of evidence or trusts the hypothesis and breaks the loop of this operation.

1' The user investigates for new pieces of evidence, inputs them if any and goes to (2), or breaks the loop otherwise.
We explain (2) the hypothesis building part and (1') the feedback part below.

3.1 Hypothesis Building

Formally, a piece of evidence is expressed, in our system, as a *literal* in first-order logical language such as $is\_user\_of(Alice, PC1)$, and thus the input is a set of literals. In addition, our system also has a set of rules, which are Horn clauses such as $is\_user\_of(x, y) \rightarrow has\_access\_to(x, y)$ (if x is a user of y, then x has an access to y). The system obtains and outputs a *hypothesis* by chaining instances of rule, like $is\_user\_of(Alice, PC1) \rightarrow has\_access\_to(Alice, PC1) \rightarrow \cdots$.

We have implemented an actual hypothesis building module, using a general logical reasoning engine called David [1]. David performs an abductive reasoning; that is, for evidence and a rule set, it generates possible hypotheses in graph-based representation and outputs the most plausible hypothesis on the basis of its evaluation function.

Figure 1 shows an example of a hypothesis output by the system. A hypothesis has tree structure of three layers with a root labelled with 'END'. Here, 'ComplementaryFactorsForSceduledTask' is an auxiliary node. The taxonomy of the node labels of the top two layers follows MITRE ATT&CK[1] as we explain below.

[1] MITRE ATT&CK provides an organised taxonomy of cyber-attack processes and currently considered a global standard in describing cyber-attacks ([8]).

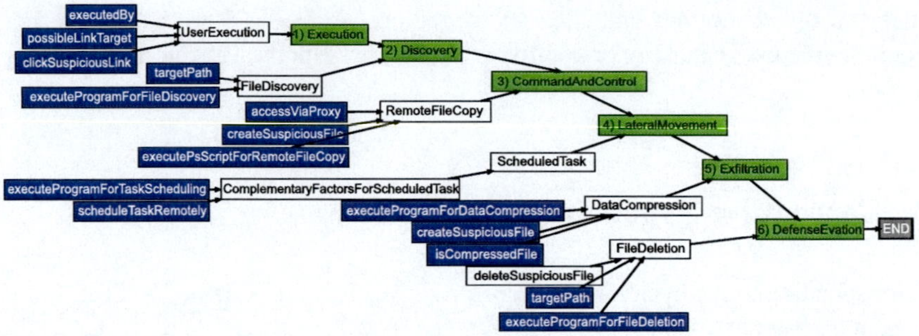

Fig. 1 An example of an inferred attack process. This figure is processed for readability

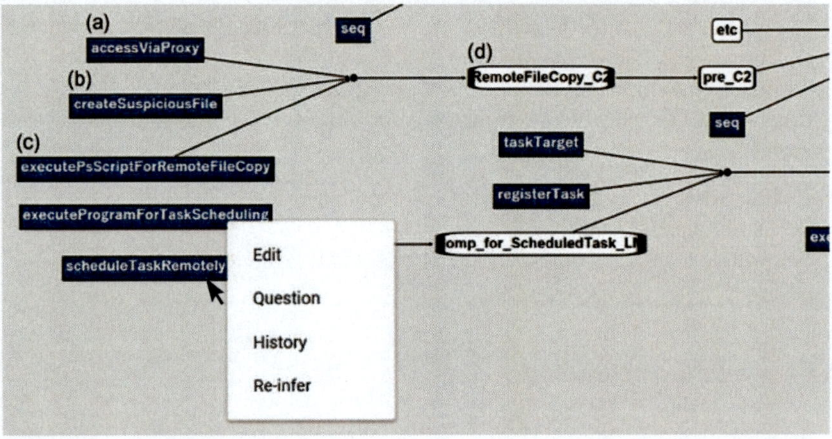

Fig. 2 A screenshot of a close view around the evidence of (3) Command And Control in Fig. 1 and a drop-down menu that appears when clicking on the node scheduleTaskRemotely

In the bottom layer, each leaf corresponds to a piece of evidence. Its colour indicates whether or not the piece of evidence has already been observed: it is blue if it has been *observed*, while it is white if it has not. In the latter case, the piece of evidence is called a *hypothesised* piece of evidence.

In the middle layer, a node corresponds to a *technique* in MITRE ATT&CK framework. A technique is inferred to be taken, on the basis of the pieces of evidence in the lower layer by a rule. For example, in Fig. 2, (a) Access via Proxy, (b) Create Suspicious File, and (c) Execute PS Script for Remote File Copy are related to (d) Remote File Copy. This expresses that it is inferred that a technique of 'Remote File Copy' has been taken, on the basis of the evidence that there was access from outside the network via a proxy server, a suspicious file has been created, and a PowerShell script for remote file copy has been executed.

The top layer chronologically describes the overall flow of the process, corresponding to a *tactics* in MITRE ATT&CK, and the chain of tactics is called a *cyber kill chain* among analysts. For example, in Figs. 1 and 2, a cyber kill chain starts with (1) Execution, goes through (2) Discovery, (3) Command and Control, (4) Lateral Movement, (5) Exfiltration, and arrives at (6) Defense Evasion. An edge from a technique to a tactic indicates that the tactic has been accomplished by the technique: e.g. as in Fig. 1, tactic (3) Command and Control was established by the means of technique (d) Remote File Copy, since nodes (a-c) are observed.

3.2 Types of Feedback

We consider three kinds of user's feedback to the system below:

(1) Confirming a hypothesised piece of evidence;
(2) Denying an (either hypothesised or observed) piece of evidence;
(3) Inputting a constant for a variable in a piece of evidence.

1. *Confirming a hypothesised piece of evidence* The output hypothesis may contain a hypothesised piece of evidence. It can be considered as a question from the system to the user. We expect the user to try to verify whether it is true or false. If it is true, the user answers to the system that he has confirmed the validity of the evidence.
2. *Denying an observed or hypothesised piece of evidence* If a piece of evidence is found to be false, the user answers to the system that the piece of evidence should be negated. In addition, feedback of this kind can be used to point out that a piece of evidence is irrelevant to the current case of a cyber-attack: after the user denies a piece of evidence, the system will not use it in the hypothesis.
3. *Inputting a constant for a variable in a hypothesised piece of evidence* A hypothesised piece of evidence may contain a variable, as our system can assume existence without specifying a constant value such as a person's name or a machine's IP address. In this case, the user can input a constant that satisfies the condition indicated by the hypothesised piece of evidence; after the feedback, the system will consider the piece of evidence to be *observed*. For example, $is\_user\_of(x, PC1)$ might appear in the hypothesis, meaning that there is a user x of $PC1$ but x is unknown. Then, if the user (of our system) knows the user of $PC1$, say $Alice$, he can input $Alice$ for x and the system considers $is\_user\_of(Alice, PC1)$ to be observed.

We have chosen the above kinds of feedback as a first step of our study on the basis of easiness. (i) It is easy to input the feedback because these kinds of feedback do not require to input literal or a sentence in a natural language that can be translated

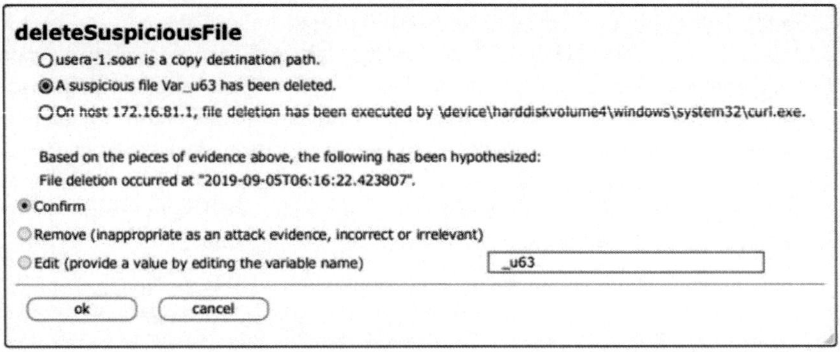

Fig. 3 A feedback window, translated from Japanese

into a literal. This is also the reason why we have not chosen feedback to type a new piece of evidence. (ii) It is also easy for the user to decide what to feedback because the content of feedback is whether the piece of evidence is true or false and thus objective. This is also one of the reasons why we have not chosen the feedback on the plausibility of an edge or a rule.

Figure 2 shows a drop-down menu to select the user's action on a node. A pop-up in Fig. 3 appears when the user clicks on a node in the bottom layer and chooses 'Edit'.

The pop-up displays radio-button-selectable options for the feedback target and those for the feedback action. The options for the feedback target show the contents of the clicked node, along with its sibling nodes and their contents, and the technique to which they lead are also displayed. A variable is displayed in the form of Var_u* such as Var_u63 in Fig. 3, where the prefix Var indicates that it is a variable. Selecting a radio button out of the displayed nodes, the user can choose the piece of evidence on which he gives feedback.

Selecting a radio button out of the lower three, the user can choose the type of the feedback:

1. 'Confirm' is for confirming the piece of evidence if it is a hypothesised one.
2. 'Remove' is for denying the piece of evidence.
3. 'Edit' is for inputting constants for the variables.

3.3 Example

The hypothesis in Fig. 1 is inferred from an input set of pieces of evidence reproduced and modified from the publicly available report on an actual cyber-attack against Japan Pension Service [6]. Let us consider giving feedback on this hypothesis. Suppose that we choose 'ScheduleTaskRemotely' in the hypothesis for our feedback

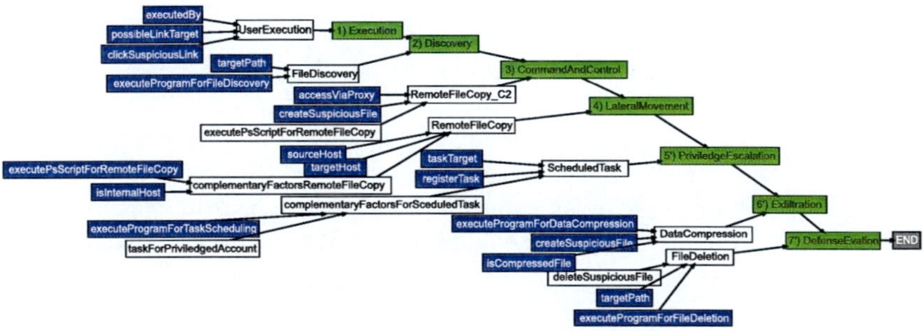

Fig. 4 A revised hypothesis

target and 'Remove' for our feedback action and have the system infer again to obtain the most plausible cyber kill chain reflecting this feedback. Then, the system outputs the revised hypothesis in Fig. 4. The revised cyber kill chain now consists of (1) Execution, (2) Discovery, (3) Command and Control, (4) Lateral Movement, (5') Privilege Escalation, (6') Exfiltration, and (7') Defense Evasion. (5') Privilege Escalation newly appears in the revised hypothesis. As exemplified above, the user can give fine-grained feedback, denying only a part of the whole attack process.

4 Interview

We had two interview sessions with two cyber security analysts in NEC Corporation on the above-explained user feedback function: during and after the implementation of the feedback function. To elicit comments from the analysts, we showed a mock-up scenario in the first interview and the example of Japan Pension Service in the second interview.

The following are comments from the analysts that are relevant to the use cases in their actual workplace. (The words in the parentheses are added by the authors.)

1. On hypothesis building
   - A hypothesis of the whole picture of a cyber kill chain could be useful because using it enables the analyst team to share and discuss the hypothesis on the attack process and narrow down the range of the investigation.
   - If the reliability (of the function that automatically collects evidence) of the system is low, the analysts would have to see whether the observed pieces of evidence (in an output hypothesis) are true as well as the hypothesised ones; if the reliability is high, they would not have to check the observed ones.
   - If the (observed) evidence supporting the hypothesis is insufficient, the analysts have to investigate to confirm whether the hypothesised pieces of evidence in the hypothesis are true.

2. On user feedback
    - Evidence correction function (such as our feedback function) is necessary, and the analysts would actually use it.
    - The analysts would like to know the hypothesis of a cyber kill chain after reflecting our feedback (beforehand).
    - Our system can be used as an automatically generated checklist (which tells the user which piece of evidence should be confirmed to complete the hypothesis).

5 Discussion

The positive comments according to concrete examples from the analysts ensure the necessity of hypothesis building for cyber kill chain inference. On the other hand, the negative comments suggest that the trustability of our system is the next issue that should be addressed.

6 Conclusion

In this paper, we have proposed the concept of cooperative hypothesis building for cyber kill chain inference and implemented functions that accept feedback on pieces of evidence in the hypothesis. On the basis of the interviews with cyber security analysts, our proposed feedback functions would be necessary in the investigation cycle, and the trustability has been revealed to be the next issue for our human AI cooperative system.

Compared to the practice as currently done by analysts which probably achieves a more trustable hypothesis, the coverage of the implemented feedback functionalities is still restricted. To address this issue, we shall

1. Expand the coverage of our feedback functionalities, adding the functionality that incorporates feedback on rules (edges) in the output hypothesis and the addition of a new node, which was not chosen to be implemented but referred to in the interviews.
2. Conduct a user or computer-simulated experiment to confirm the validity of our framework of the cooperative hypothesis building and to measure the coverage of our feedback functionalities.

References

1. Automated reasoning technology group, NEC: Open-David [GitHub]. https://github.com/aurtg/open-david. Accessed on 5 Jan 2021
2. Black, E., Hunter, A.: A generative inquiry dialogue system. In: Proceedings of the 6th International Joint Conference on Autonomous Agents and Multiagent Systems, AAMAS '07, pp. 1–8. Association for Computing Machinery, New York, NY, USA (2007). https://doi.org/10.1145/1329125.1329417.https://doi.org/10.1145/1329125.1329417
3. Black, E., Hunter, A.: An inquiry dialogue system. Auton. Agents Multi-Agent Syst. **19**(2), 173–209 (2009)
4. Hiraoka, T., Motoura, S., Sadamasa, K.: Detecticon: A prototype inquiry dialog system. In: 9th International Workshop on Spoken Dialogue System Technology, pp. 381–387. Springer (2019)
5. Kandefer, M., Shapiro, S., Stotz, A., Sudit, M.: Symbolic reasoning in the cyber security domain. In: Proceedings of MSS 2007 National Symposium on Sensor and Data Fusion McLean (2007)
6. Report on the Results of Investigation into the Incident of Information Leakage due to Unauthorised Access (in Japanese): Japan Pension Service. https://www.nenkin.go.jp/files/kuUK4cuR6MEN2.pdf. Accessed on 6 Jan 2021
7. Shepard, B., Matuszek, C., Fraser, C.B., Wechtenhiser, W., Crabbe, D., Güngördü, Z., Jantos, J., Hughes, T., Lefkowitz, L., Witbrock, M., Lenat, D., Larson, E.: A knowledge-based approach to network security: applying cyc in the domain of network risk assessment. In: Proceedings of the 17th Conference on Innovative Applications of Artificial Intelligence, IAAI'05, vol. 3, pp. 1563–1568. AAAI Press (2005)
8. The MITRE Corporation: MITRE ATT&CK. https://attack.mitre.org/. Accessed on 6 Jan 2021
9. Thomas, L., Vaughan, A., Courtney, Z., Zhong, C., Alnusair, A.: Supporting collaboration among cyber security analysts through visualizing their analytical reasoning processes. In: 2018 IEEE International Conference on Multimedia Expo Workshops (ICMEW), pp. 1–6 (2018). https://doi.org/10.1109/ICMEW.2018.8551497
10. Zhong, C., Kirubakaran, D.S., Yen, J., Liu, P., Hutchinson, S., Cam, H.: How to use experience in cyber analysis: An analytical reasoning support system. In: 2013 IEEE International Conference on Intelligence and Security Informatics, pp. 263–265 (2013). https://doi.org/10.1109/ISI.2013.6578832
11. Zhong, C., Yen, J., Liu, P., Erbacher, R.F.: Automate cybersecurity data triage by leveraging human analysts' cognitive process. In: 2016 IEEE 2nd International Conference on Big Data Security on Cloud (BigDataSecurity), IEEE International Conference on High Performance and Smart Computing (HPSC), and IEEE International Conference on Intelligent Data and Security (IDS), pp. 357–363 (2016). https://doi.org/10.1109/BigDataSecurity-HPSC-IDS.2016.41

Hierarchical Multi Label Classification of News Articles Using RNN, CNN and HAN

Pratik Sharma, Aman Shakya, Basanta Joshi, and Sanjeeb Prasad Panday

Abstract Labeling or tagging of any online news article is very important nowadays for quick information retrieval. In real world, news may belong to multiple categories. Moreover, those categories can be organized hierarchically. Such news classification is termed as hierarchical multi label classification. Hierarchical multi label classification in case of news domain has been very rarely explored. The base article for this paper created a pre-defined taxonomy of Bio-NLP labels, and articles were then categorized into multi labels based on pre-defined hierarchy. In this work, similar approach is used. The labels are placed in a pre-defined hierarchy, and then articles are classified in multi label manner with the pre-defined hierarchy in place. HAN (Hierarchical Attention Network) has rarely been used for hierarchical multi label classification. In this work, combination of CNN and GRU is implemented at first, and then HAN alone is implemented for the task. When CNN (Convolutional Neural Network) alone is used for classification as in the base paper, the average F1-score for different sets of experiments results in approximately 0.86. When CNN is combined with GRU (Gated Recurrent Unit), the average F1-score improvement is 1%.When HAN (Hierarchical Attention Network) is used then average F1-score improves by 3.35%. The models were trained and tested using news articles of the guardian newspaper published in 2014.

P. Sharma (✉) · A. Shakya · B. Joshi · S. P. Panday
Pulchowk Campus, Institute of Engineering, Tribhuvan University, Kirtipur, Nepal

A. Shakya
e-mail: aman.shakya@ioe.edu.np

B. Joshi
e-mail: basanta@ioe.edu.np

S. P. Panday
e-mail: sanjeeb@ioe.edu.np

T. Senjyu et al. (eds.), *ICT with Intelligent Applications*, Smart Innovation, Systems and Technologies 248, https://doi.org/10.1007/978-981-16-4177-0_50

1 Introduction

1.1 Background

Suppose there is a news article whose headline state "Lionel Messi is the God of football." It can be induced that this article falls under sports category of any newspaper. However, this same article may fall under football category for newspapers. Moreover, in some newspapers, this article may fall under "Lionel Messi" category as well. So, basically, the above piece of news may belong to any of the above-mentioned categories. In case of multi label, the article may belong to multiple categories at the same time. For example, this single article may be tagged with multiple labels like Messi, Football. Using multiple labels to represent, an article makes it easier to read, search interpret and gives clearer picture of what the news is about. An extension to multi label problem where the labels are inter related in a categorical hierarchy is termed as hierarchical multi label classification. The labels sports, Messi, football, etc., can be organized in a hierarchy. So say, there may be other news related to sports like cricket.

This can be pictorially represented as (Fig. 1):

2 Literature Review

Simon et al. [1] used the strength of hidden layer initialization for multi label classification taking advantage of co-occurrence between labels. This method used CNN architecture to perform hierarchical multi label classification of Bio-NLP research articles. Research articles unlike news articles are very much specific and concentrate toward single topic. You et al. [2] introduced AttentionXML which is used for extreme multi label classification. It uses a multi label attention mechanism whose input is raw text allowing to capture the most relevant parts of text to each label. It constructs PLT (Probabilistic Label Tree) which handles millions of child, i.e., tale labels. However, this method is really slow and cannot deal with large number of labels. You et al. [3] again proposed HAXMLNET (Hierarchical Attention Network for Extreme Multi Label Text Classification). HAXMLNET performs label wise attention and uses a probabilistic label tree for solving extreme-scale datasets. The probabilistic label tree consists of label hierarchy with parent label, intermediate

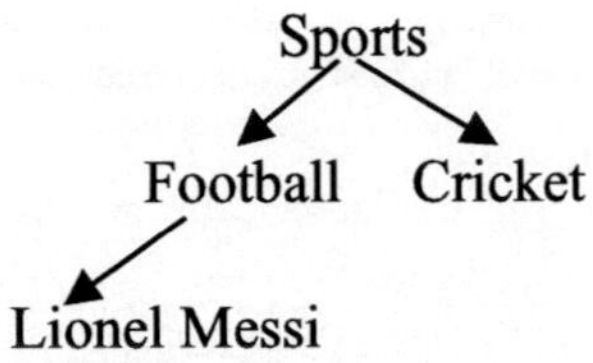

Fig. 1 Figure of hierarchically placed labels

label and child label. Here, two AttentionXML are trained, i.e., one for the dataset and another one for label. In this paper, three separate models were used on the same dataset and then compared. First, CNN is applied to the dataset to establish baseline performance by following methodology of Simon et al. [1]. The next methodology implemented combination of CNN and RNN. The final methodology implemented HAN. Combination of CNN and RNN has not been yet been applied for hierarchical multi label classification task. HAN has been applied for the task as mentioned in the previous paragraphs, but the methods are either slow or shows low performance even on the datasets it is implemented. This work focuses on improving Simon et al. [1] method for hierarchical multi label classification of news articles.

3 Methodology

3.1 Without HAN

The dataset was obtained from a forum named fastai. The dataset was extracted using scrapping tools from Guardian Newspaper as informed by its author. It was cross checked and verified. There were about 100,000 rows of data and its labels. However, the labels were random and were not placed in hierarchy. Data preprocessing was done, and labels were placed in hierarchy. Since only 2000 data was considered in the base paper, the experiment was done on 1200 data with the aim of expanding to 2000 based on results (Figs. 2, 3).

Hierarchical Multi Label Dataset

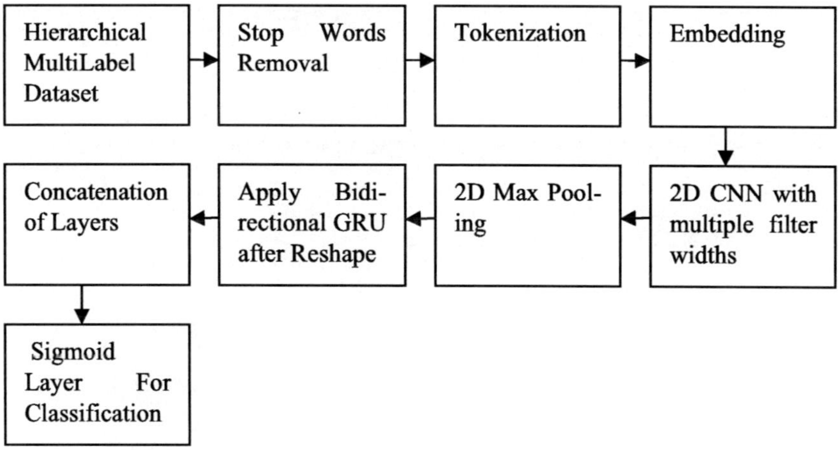

Fig. 2 Methodology without HAN

Hierarchical Multi Label Dataset

World	Sports	Politics	…..	Words	Tag_List	Lang	Type	Word_Count
1	0	1	…..	Anger o…	usnews| world	en	article	627

Fig. 3 Sample columns of the dataset

Label header columns as shown in table above have values 0 or 1. 0 indicates the news article doesn't belong to that label, whereas 1 indicates the news article belongs to that label. Say, the first article has value 1 in label world and 0 in label sports and politics. This means that the news article belongs to label world and not sports or politics. This can be verified by tag_list header of that corresponding article. There were 33 different labels in the original dataset. Only 16 labels are considered for this work. Here is the hierarchy of labels (Fig. 4).

Note that News is used in the above diagram just for representation. News label is not present in the dataset nor is it used for any calculation.

For each combination of parent node and child node, there are each 100 numbers of articles. So say, for combination of sport and each of its child nodes, i.e., tennis, cricket, football, rugby there are 100 articles each. Note that in case of society, there are 100 articles for each combination of society, lifestyle and its child nodes. Hierarchies were created manually taking labels into account. There are total 133,375 multi labeled articles in the original dataset. However, only those articles are taken which satisfied the hierarchy which resulted in total of 1200 articles. The ratio in

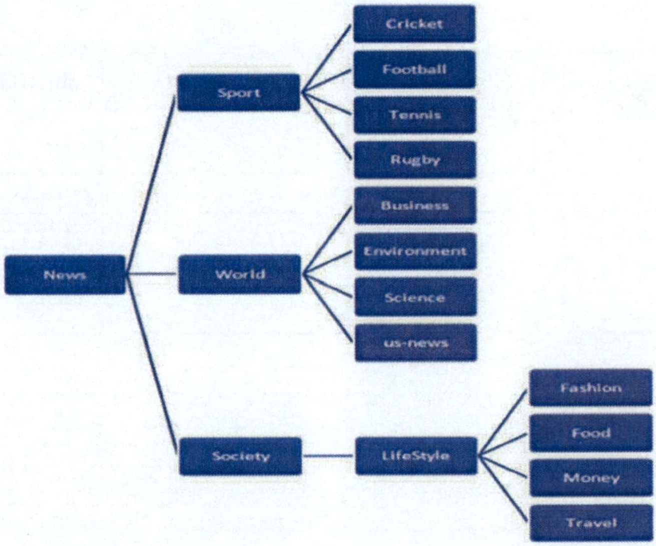

Fig. 4 Hierarchy of used labels

one	person	equals	one	vote	seems	simple	enough	eos	unfortunat	worked	m
tennis	week	joined	roster	major	sports	played	shadow	corruption	allegations	eos	et
like	bucketfuls	empty	ones	collected	younger	shells	sink	make	happy	eos	pa
louis	van	gaal	lack	sense	humour	eos	issuing	apology	journalists	eos	fi
images	new	zealand	show	pristine	environme	great	beauty	eos	therefore	comes	su
republican	chair	senate	foreign	relations	committee	said	friday	incoming	trump	administra	w
david	thorne	former	ceo	owner	rugby	union	side	wasps	one	parties	in
coral	reefs	probably	earth	life-packed	ecosystem	privilege	diving	tropics	know	reef	o
radio	today	presenters	never	particularl	happy	time	talking	cricket	relationshi	sport	th
saracens	signed	south	africa	world	cup	winner	schalk	burger	eos	flanker	jo

Fig. 5 Tokenized dataset snapshot

which the articles for this work is taken is 40 percent of the articles is taken from first 40,000 articles then next 40% is taken from next 40,000 articles, and finally last 20% is taken from the remaining articles of the original dataset.

Stop Words Removal

Stop words consists of words like the, is, are, etc., along with punctuations such as commas. Classification problems normally don't need stop words because it is possible to talk about the general idea of a text even if stop words is removed from it. Punctuation such as Full Stop is replaced with eos indicating end of sentence.

Tokenization

After removal of Stop words, the next step is to tokenize remaining text. At the end of tokenization each text document is denoted by a single row in a csv file. Another csv file is created which contains corresponding labels for each row of text, i.e., for each document. Hence, two csv files with rows of first one indicating documents and rows of second one indicating corresponding hierarchical labels of those documents are created. Here is the tokenized dataset snapshot (Fig. 5):-

Embedding

In this work, embedding layer of Keras is used. The input dimension of embedding layer is total number of unique words in the corpus plus one. The input length is taken as 500 as per Simon et al. [1] paper which implies every document will have length 500. If the length is less than 500 then padding will be added and if the length is greater than 500 then it will be truncated. The output dimension is taken as 200 because glove word embedding [4] is used which has a vocabulary of 6 billion words each of which is represented by 200 dimensional vectors.

Two Dimensional CNN with multiple filter widths

The output of embedding layer is in 2D format with the first dimension being maximum length of an article allowed, i.e., 500 and second dimension being 200 (since glove word embedding of dimension length 200 is being used as described in Embedding section). The input is actually reshaped to (None, 1, 500, 200). Here, None implies the batch size is None. For this work, 2D convolution is used. Filters of size 3, 4 and 5 are taken, and the number of filters taken is 600 with 200 of each size. For each different filter size, the number of feature vector obtained is 200.The convolution 2D operation is repeated over a loop where for each filter size, convolution operation is done. The activation function for this 2D convolution operation is RELU activation function.

Apply Two Dimensional Max Pooling

After the 2D convolution operation is done in the loop, 2D max pooling is performed.

Reshape and Apply Bidirectional GRU Layer

The output from Max pooling 2D layer is reshaped and then passed to a Bidirectional GRU layer in the loop. The reason for using Bidirectional GRU in place of normal GRU is because Bidirectional GRU allows for the use of information from both previous time steps and later time steps to make predictions about the current state unlike normal GRU. This feature of Bidirectional GRU allows it to consider future context as well. In this work, the activation function used is tanh and recurrent activation function used is hard_sigmoid for Bidirectional. In this work, combination of RNN and CNN is taken to take advantage of both the architectures.

Concatenation of Layers

At the end of each loop, flattening operation is performed. The flattened input sequence is then appended in an array at the end of each loop. All the elements of the array are concatenated at the termination of overall loop. Finally, a dropout of 0.5 is applied to the concatenated array.

Classification Using Sigmoid Function

Finally, a dense layer is used with activation of sigmoid function. Sigmoid activation is applied at the end of dense layer because the classification tasks are not mutually exclusive, and each class (label) is independent. In this work, sigmoid function basically gives percentage of how much an article can belong to each label. If the value is greater than 0.5 then it is assumed that the article belongs to that label. Sigmoid gives a value between 0 and 1 and those values are independent between different classes for a single input.

3.2 *With HAN*

All the other components are same as described in previous section (Fig. 6).

Apply HAN

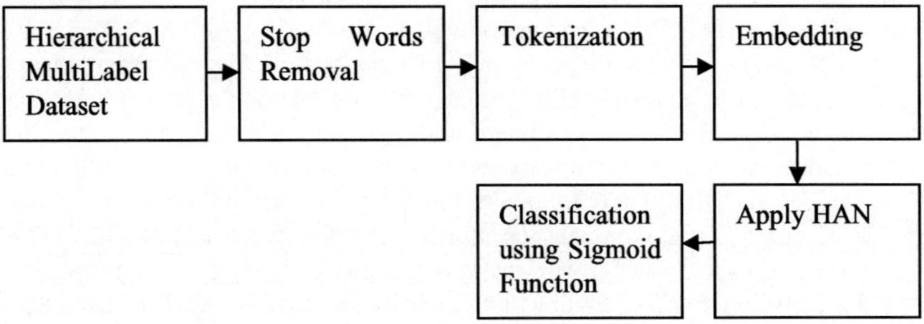

Fig. 6 Methodology with HAN

There are 4 different components in an HAN [5]. First is Word Encoder. In this component, each word is embedded to vectors through an embedding matrix. Bidirectional GRU is applied to the resultant vector which reads the sentence using forward GRU and backward GRU. Second is Word Attention. In this component, all the words which provide more meaning to a sentence are combined and then a sentence vector is formed for each sentence. Third is Sentence Encoder. Just like Word Encoder component, document vector now like word vector is calculated in a similar way. Fourth is Sentence Attention with the help of which those sentences which are more meaningful are identified.

4 Result and Discussion

Stratified sampling is done to the single csv file which splits dataset into train, test and validation csv file. Stratified sampling is done using python's built in functionality iterative train_test_split which uniformly distributes the multi label data into train, test and validation csv files. With stratified sampling, the data is divided into separate groups called strata. The stratified folding is done randomly 3 times over whole dataset to get different sets of data. The dataset is taken in split configuration such as 70 20–10 (70% training set, 20% validation set and 10% test set). Learning rate is taken as 0.001 (Tables 1, 2, 3).

Table 1 For CNN different metrics

Strata	Precision	Recall	Specificity	Sensitivity	F1-score	Jaccard Index
First strata	0.8996	0.8321	0.9841	0.8321	0.8645	0.8048
Second strata	0.8736	0.8642	0.9786	0.8642	0.8689	0.7902
Third strata	0.8621	0.8714	0.9762	0.8714	0.8667	0.7923

Table 2 For combination of CNN and RNN different metrics

Strata	Precision	Recall	Specificity	Sensitivity	F1-score	Jaccard index
First strata	0.9143	0.8392	0.9865	0.8392	0.8752	0.8159
Second strata	0.8892	0.8607	0.9817	0.8607	0.8747	0.8070
Third strata	0.9246	0.8321	0.9884	0.8321	0.8759	0.8127

Table 3 For HAN different metrics

Strata	Precision	Recall	Specificity	Sensitivity	F1-score	Jaccard index
First strata	0.9323	0.8857	0.9890	0.8857	0.9084	0.875
Second strata	0.9176	0.875	0.9865	0.875	0.8957	0.8606
Third strata	0.9047	0.8821	0.9841	0.8821	0.8933	0.87

5 Conclusion

In this work, hierarchical multi label classification is successfully improved. The base paper's [1] method only employed Convolutional Neural Network (CNN) for hierarchical multi-label classification of Bio-NLP articles. In this work, however, hierarchical multi label classification is done for news articles. The methodology in base paper, i.e., only Convolutional Neural Network (CNN) is implemented along with combination of Convolutional Neural Network (CNN) and Recurrent Neural Network (RNN) and finally Hierarchical Attention Network (HAN) is applied on the dataset. 1% improvement is achieved compared to CNN when combination of RNN and CNN is applied in general, whereas 3.35% improvement is achieved when HAN is applied.

There are few limitations. First, the dataset taken is small. Second, other deep learning architectures or combinations have not been explored. There is some room for improvements like hierarchical similarity can be taken into account for labels which can further improve the classification with related labels occurring more than unrelated labels. Other deep learning architectures can also be explored and implemented to improve classification.

References

1. Baker, S., Korhonen, A.: Initializing neural networks for hierarchical multi-label text classification. In: BioNLP 2017 August, pp. 307–315 (2017)
2. You, R., Dai, S., Zhang, Z., Mamitsuka, H., Zhu, S.: Attentionxml: extreme multi-label text classification with multi-label attention based recurrent neural networks (2018). *arXiv preprint* arXiv:1811.01727
3. You, R., Zhang, Z., Dai, S., Zhu, S.: HAXMLNet: hierarchical attention network for extreme multi-label text classification (2019). *arXiv preprint* arXiv:1904.12578
4. Pennington, J., Socher, R., Manning, C.D.: Glove: global vectors for word representation. EMNLP (2014)
5. Yang, Z., Yang, D., Dyer, C., He, X., Smola, A., Hovy, E.: Hierarchical attention networks for document classification. In: Proceedings of the 2016 conference of the North American Chapter of the Association for Computational Linguistics: Human Language Technologies June, pp. 1480–1489 (2016)

Enhancing the Date Rate by the Management of Bandwidth and Power Using Interference Temperature Constraint in Cognitive Radio Networks

Paurav Goel, Avtar Singh, and Ashok Goel

Abstract As the network nowadays is overcrowded, due to increase in demand of wireless applications requiring high data rates with full signal strength still there are parts in our overall wireless spectrum that are unutilized. Cognitive radio networks technology is a one step ahead in solving the problem of underutilization of certain parts of spectrum. Interference plays a negative role in the success of cognitive radio networks technology. Interference temperature gives the closest approximation of overall interference at a particular time and position. The proposed approach implements a feedback system in cognitive radio networks using interference temperature and achievable capacity as a constraint using back propagation neural network. The goal is to maximize data rate at a particular time at designated channel, and channel utilization is maximized.

1 Introduction

Due to enormous increase in number of wireless devices, increase in mobility and variation in RF transmitters administration of interference has become a difficult task. Also, users have been given greater flexibility in spectrum access. Interference is caused due to power produced by Radio frequency waves of transmitters and noise produced by RF sources that are nearby the receiver. It includes co-channel interference that is due emissions from other transmitting devices functioning on the identical frequency as the chosen transmitter and adjacent channel interference due to emissions from transmitting devices working on adjacent frequencies. Interference temperature is used as an appropriate scale to calculate and cope interference. It amounts to be the sum of the RF power produced by unwanted transmitters (I) and

P. Goel (✉)
Department of ECE, IKGPTU Campus, Jalandhar, India

A. Singh
IKGPTU Campus, Jalandhar, India

A. Goel
MRSPTU, Bhatinda, India

T. Senjyu et al. (eds.), *ICT with Intelligent Applications*, Smart Innovation, Systems and Technologies 248, https://doi.org/10.1007/978-981-16-4177-0_51

noise (N) sources that are existent in a receiver system ($I + N$) per unit of bandwidth. Interference temperature is measured in Kelvin, can be calculated as division of the input power at the antenna in watts and the related RF bandwidth measured in Hertz multiplied by a term known as Boltzmann constant K [1] as shown in Eq. 1 (where $K = 1.38 \times 10–23$ W sec/K, and 1 K = −272.15 °C).

$$\mathrm{Ti}(\mathrm{fc}, B) = \frac{\mathrm{Pi}(\mathrm{fc}, B)}{\mathrm{KB}} \tag{1}$$

Here, Ti denotes the interference temperature of licensed band with center frequency fc and bandwidth B and Pi denotes the power of the signal at same frequency and bandwidth. For a predefined geographic site, an interference temperature limit is fixed in order that system functions efficiently on an adaptive or real-time basis. In this cognitive radio system, the interference temperature within the band is calculated and knowledge is communicated to devices w.r.t to the limit. A feedback process is employed to confine the operation of devices so as to take care of the interference temperature at or below the extent of the limit $T\mathrm{l}(\mathrm{fc})$ [1] as given in Eq. 2.

There are two approaches to calculate interference temperature one is the ideal approach and other is generalized approach. In ideal approach, interference is calculated where the licensed signal is present, whereas in generalized approach interference temperature of entire frequency range is calculated. The generalized approach is used when we have no knowledge of signal environment, and it is difficult to distinguish licensed user from interference and noise.

$$\mathrm{Ti}(\mathrm{fc}, B) + \frac{\mathrm{MP}}{\mathrm{KB}} \leq \mathrm{Tl}(\mathrm{fc}) \tag{2}$$

Here, M denotes the combined effect of attenuation due to licensed and unlicensed users and P denotes the average power of unlicensed users in the entire frequency range. In this paper, we will use ideal approach. In the proposed approach, the unlicensed user will measure the interference temperature at its geographic site and add its own impact of RF energy and then will make a decision of whether to transmit or not transmit. If the resulting interference temperature is below the interference temperature limit set for that geographic site, then the device would send out [2]. This approach is beneficial for users of unlicensed devices in Cognitive Radio network in order to enhance the data rate [3]. In the bands where the interference temperature limit (IT limit) isn't surpassed, chances exist for bonus operation by "underlay" transmitters. The unlicensed users then can emit at higher power intensity in the same band since interference temperature bounds are not touched, leading to enhancement in the data rate. In such cases, the IT limit for the band would help in adding extra potential RF energy into the band. When primary user is not using the licensed spectrum, the interference temperature limit in that channel sufficiently decreases and now secondary users can transmit at higher data rate and power. By proper adaptation and reconfiguration of transmission parameters, the user needs of secondary network can be satisfied. This is called overlay approach.

In [4], power allocation strategy is studied in two different scenarios of terrestrial and satellite primary networks. Author tests the performance cognitive radio having a OFDM-based secondary network in presence of primary terrestrial and DVB-SH satellite system. In [5], a fast algorithm is used for power allocation. In this method, computational complexity is reduced from O(N3) to O(M2N). In [6], for power allocation, a non-orthogonal multiple access technique is employed. This method showed better results than tradition power control algorithms with linear complexity. In [7], power allocation is studied for full duplex networks in underlay scenario. This method gave high spectrum efficiency while guarantying the PU 's quality of service. In [8], author used genetic algorithm for power allocation considering both overlay and underlay spectrum mechanisms. The JOUSAM model is used which shows performance improvement. We have developed a new algorithm in which based on feedback from SU-Rx, SU-Tx in underlay will be dynamically allocated the transmit power [9] and bandwidth.

The rest of the paper is organized in Sect. 2 that describes the system model and proposed approach. Section 3 describes the implementation of proposed technique. Section 4 discusses the results. At last, conclusion is derived in Sect. 4.

2 System Model and Proposed Approach

2.1 System Model

A system model of cognitive radio (see Fig. 1) consisting of primary network and secondary network. Secondary network users can use the licensed spectrum that is

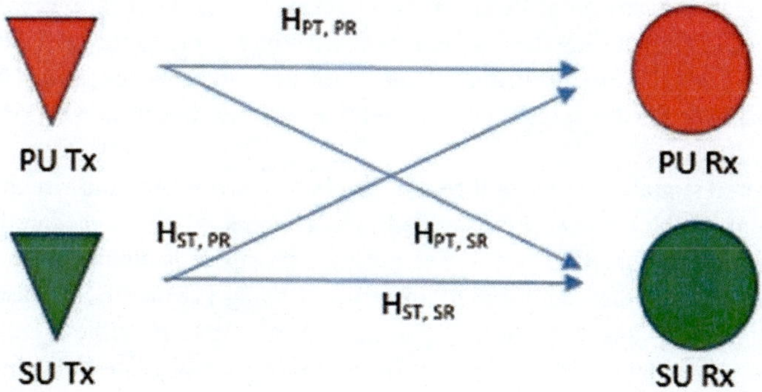

Fig. 1 System model

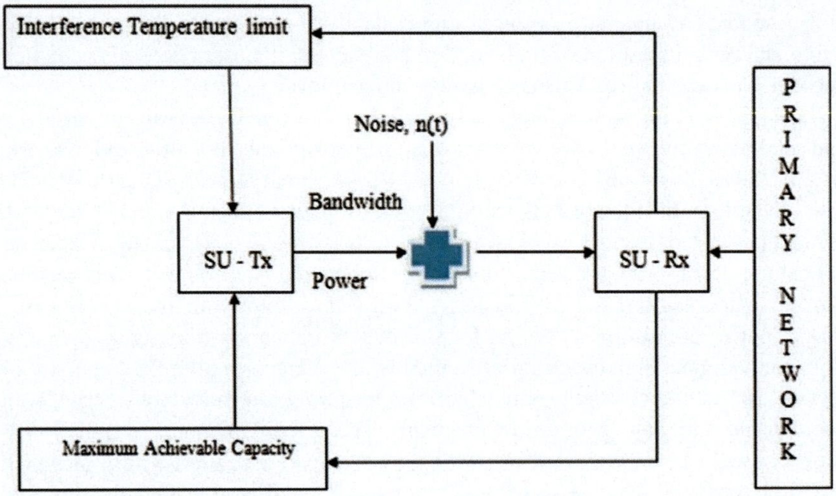

Fig. 2 Proposed feedback model

allocated to primary users in an opportunistic manner and can transfer data for several applications.

We have proposed an algorithm for enabling Secondary User to transmit in the licensed band in an opportunistic manner. An environment will be created which will have wireless primary and secondary users. Presence of PU's will be detected by the method of spectrum sensing. Receiving unit will determine the vacant spectra and intimate the SU transmitter to enable in its absence. For proficient utilization of spectrum, the transmitting parameter [10, 11] of SU-Tx will be adjusted according to availability of PU spectrum using feedback from the SU-Rx receiver based on algorithm up to the interference temperature limit and maximum achievable capacity (see Fig. 2). By assigning proper bandwidth and appropriate power to the SU transmitters, the demands of SU receiving units can be satisfied [12, 13].

Primary user transmitters PU-Tx is a WLAN [14] Transmitter with center frequency 5.15 GHz. Using the parameters of Orthogonal Frequency Division Multiplexing (OFDM) [15, 16], the primary user's message signals are generated (see Table 1). Binary Phase Shift Keying (BPSK) modulation is used over the OFDM message signals.

Modulated signals are passed through the channel with gain hi to which Additive White Gaussian Noise [17] $n(t)$ is added. The channel gain is considered constant as attenuation is assumed to be 0.9. Secondary User signal is obtained using OFDM parameters. Energy Detection-based spectrum sensing [18] method is used by the secondary units to detect the vacant spectrum pertaining to possible spectral access by the secondary device. Secondary Users are allowed to coexist with primary user for transmission using the same channel until their interference does not cross the allowable interference limit. They can coexist with the primary transmission using

Table 1 Primary user OFDM parameters [20]

S. No.	OFDM Parameters	Symbol	WLAN (802.11a)
1	FFT size	N	64
2	OFDM bandwidth	BW	20 MHz
3	Total number of subcarrier	Nst	52
4	Number of data subcarriers	Nsd	48
5	Number of pilot subcarriers	Nsp	4
6	Subcarrier frequency spacing	Fsp = BW/N	312.5 kHz
7	Ifft/fft period	Tfft = 1/Fsp	3.2 us
8	Guard interval time	Tgi = Tfft/4	0.8 us
9	Symbol interval time	Tsym = Tgi + Tfft	4 us
10	Number of symbols allocated to cyclic prefix	Ncp = N*Tgi/Tfft	16

underlay access method and can transmit in the vacant space using overlay access method. Both the scenarios for Secondary User's transmissions are considered.

In our approach, SU receiver will sense the channel for the presence and absence of primary user and the same will be communicated to SU transmitter to reconfigure its transmission parameters depending upon the presence or absence of primary user. This reconfiguration will be carried out using back propagation neural network. Initially, the SU transmitter may be allocated small bandwidth and power so that working of PU is not hampered. Then, slowly bandwidth and power are increased step by step while monitoring the constraints of interference temperature limit and maximum achievable data rate for that channel side by side.

2.2 Back Propagation Neural Network Model

To automate, the proposed approach back propagation neural network model is used which will give the feedback depending upon the desired output and actual output. Every neural network has three layers input layer, hidden layer and output layer. The inputs are Xi and the output is single Y; weights (Wi) are adjusted so that required output can be achieved. The back propagation model works on Delta rule that updates weights in order to reduce the error e between target Tg and output Y. The Bias b is chosen to give best fit values.

The objective is to minimize the gap between average interference temperature limit (IT_{lim}) chosen and interference temperature attained (IT_{att}) when SU is transmitting in underlay and overlay scenario, i.e., min (IT_{lim}–IT_{att}) and to minimize the difference between maximum achievable channel capacity and actual data rate at which secondary user is transmitting, i.e., min ($C^i{}_{av} - D_i$). Learning parameter β is chosen to be 0.5. At each step, interference temperature limit gap and achievable capacity gap are found, this error is used update the weights (W_i) which in turn

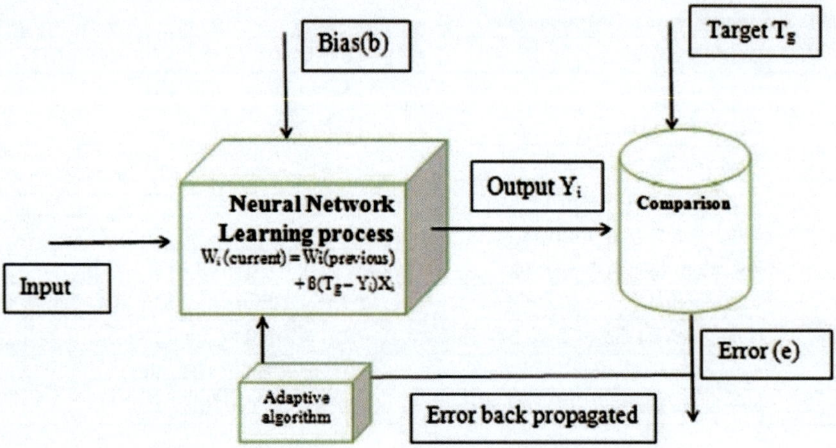

Fig. 3 Block diagram of back propagation neural network

update the input parameters of bandwidth and power. Firstly, the input parameter of bandwidth is increased step by step and after bandwidth synchronization transmit power is increased, and the desirable output, i.e., maximum transmission rate is achieved. This process continues till the error min ($IT_{lim} - IT_{att}$) and min($C^i{}_{av} - D_i$) are achieved (see Fig. 3).

Weight is updated in accordance to Eq. 3 given below

$$W_i(\text{current}) = W_i(\text{previous}) + \beta(T_g - Y_i)X_i \tag{3}$$

Maximum attainable Capacity for a user [19]

$$C^i_{av} = B\log_2(1 + S/N)\text{bits/s} \tag{4}$$

Data Rate achieved by each user [19]

$$D_i = \frac{\text{Bi}}{\text{Ni}}\log_2\left(1 + \frac{\text{Pi}^{\text{o}}|h(i)|^{2\text{o}}}{\text{No} + \frac{\text{Bi}}{\text{Ni}}}\right)\text{bits/sec} \tag{5}$$

Pi is the signal transmit power
Bi is the bandwidth of channel in MHz.

3 Implementation of Proposed Approach

The algorithm is simulated using MATLAB R2013b in which the code is written and results are obtained. The primary transmitter is a WLAN user operating at center

Table 2 Calculation of interference temperature

Total allowable interference temperature limit	465 dB K
IT of PU in dB K	223 dB K
IT difference	242 dB K
Data rate	54 Mbps
Achievable bandwidth	20 MHz
Achievable strength	60 Db
Achievable capacity	87 Mbps

frequency of 5.15 MHz and signal power of 50 dB m. Some assumptions are also made and then the simulation outputs for the PU transmitted signal, passing it through the channel and the received signal are obtained. This section demonstrates the proposed approach, and a relative investigation of our proposed feedback approach is made with that of a system lacking feedback. Equations 4 and 5 demonstrate the mathematical formulas used to calculate capacity and data rate. We have taken a channel with AWGN noise of 10 dB SNR in which PU transmission is taking place. Sampling frequency of 40 MHz is chosen so that 40,000 samples are created, and a varying subcarrier spacing is assigned to each OFDM user [13].

Energy Detection-based spectrum sensing is chosen to identify spectrum holes, i.e., the presence and absence of PU in the designated channel. The peak signifies the existence of the WLAN signal in the channel that is sensedI by the receiving front end of the SU using Energy Detection-based spectrum sensing. Interference temperature is the measure of RF power available at the receiving unit. It is calculated using formula given in Eq. 1. Table 2 calculates the IT caused by the PU transmission and the data rate achieved by it. The transmission of PU does not reach the maximal capacity and allowable IT limits. It is because subcarriers in PU that are less closely packed with spacing of 312.5 kHz, and its signal strength is also less.

As shown in Table 2, primary user is not transferring data at maximum capacity due to which spectrum is not used properly. As a result, there is gap of 242 dB K. An interference temperature limit of 465 dB K is chosen. Figure 4 shows that SU is transmitting in overlay at 40 dB power and 5 MHz BW using OFDM along with PU as the interference temperature is 440 dB which is less then interference temperature limit of 465 dB (refer Table 3).

As the PU gets quite as shown in Fig. 4, the IT limit gap increases and secondary user (SU) tries to readapt its bandwidth at every iteration up until it reaches BW equal to that of PU as shown in Table 4. Figures 5, 6, 7, 8, 9 and 10 show the iterations of increasing bandwidth, and transmit power is kept constant.

Figures 11, 12, 13, 14, 15 and 16 show power reconfiguration of SU-Tx. Table 5 shows the data rate achieved at different power reconfigurations as the limits of IT and available capacity are achieved. Figure 16 shows that power locking condition is attained once the SU attains the IT limit. No further reconfiguration is allowed after this Iteration 6, otherwise the interference limit will be crossed that would be harmful for the system. Figures 17 and 18 show the accomplishment of date rates during bandwidth adaptation and power reconfiguration. Figure 16 signifies that

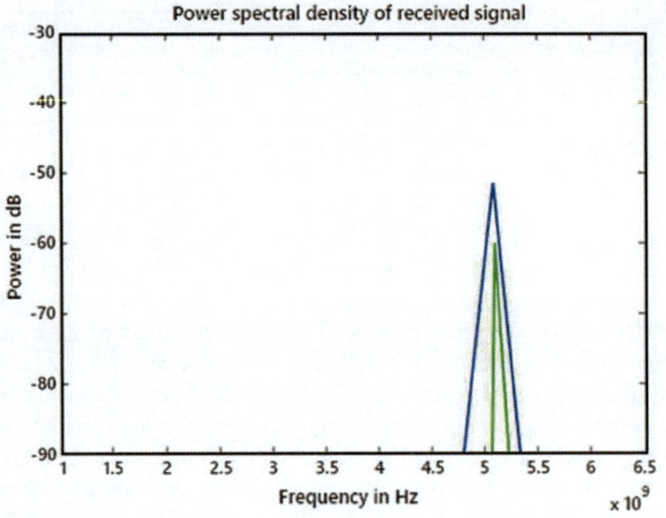

Fig. 4 PU and SU transmitting simultaneously

Table 3 Transmission of secondary user in underlay

Interference temperature by SU (R)	217 dB K
Interference temperature by (PU + SU)	440 dB K

Table 4 Bandwidth adaptation of secondary user in absence of primary user

Iteration no	BW adaptation (MHZ)	Power (dBm)	IT SU (K)	IT difference (K)	Data rate (Mbps)	Difference from achievable capacity
1	12.5	40	246	219	44	43
2	16.25	40	269	196	58.5	28.5
3	18.125	40	280	185	66	21
4	19.062	40	286	177	69	18
5	20	40	300	175	71	16

before IT limit was achieved the SU-Tx achieved a transmit power level of 60 dB m which was higher then what PU-Tx was transmitting at while achieving the full available Shannon capacity. In this way, cognitive radio system leads to better usage of available resources.

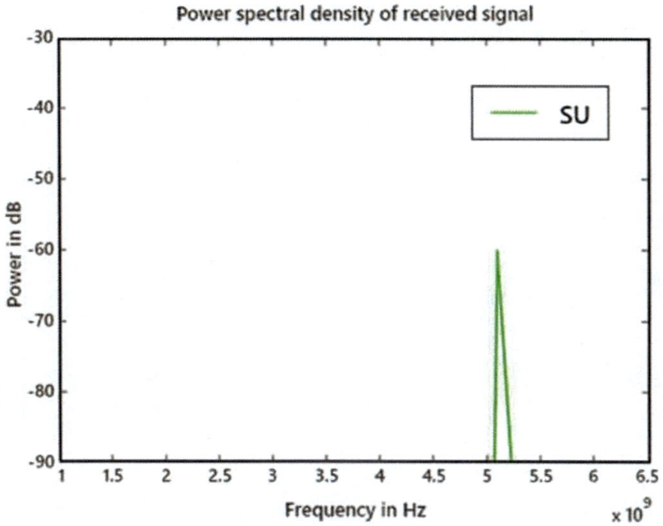

Fig. 5 SU transmitting in absence of PU

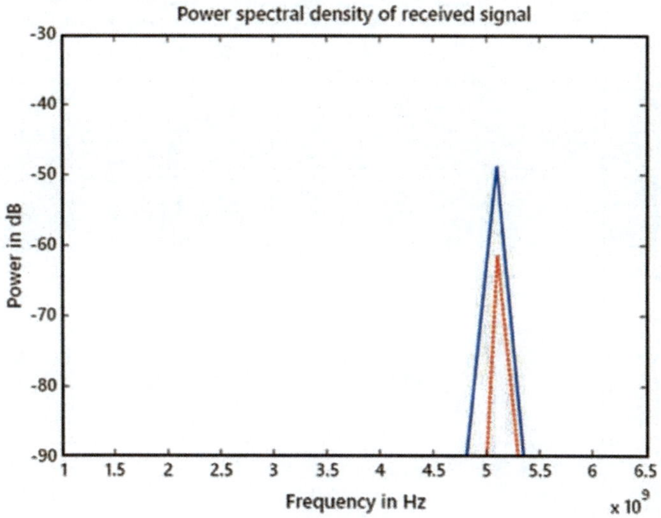

Fig. 6 Iteration-1 of bandwidth adjustment of SU-Tx

4 Results

Our feedback cognitive radio system with a better knowledge of Interference temperature limit and by proficient management of bandwidth and power provides better channel utilization from 43 to 95% as compared to existing systems without feedback

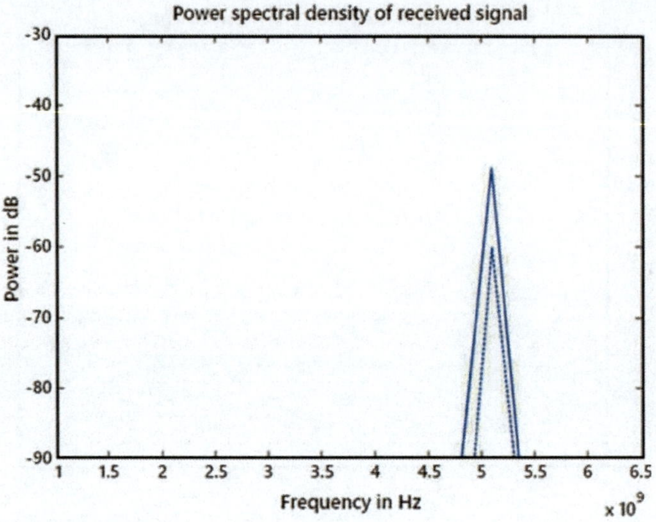

Fig. 7 Iteration-2 of bandwidth adjustment of SU-Tx

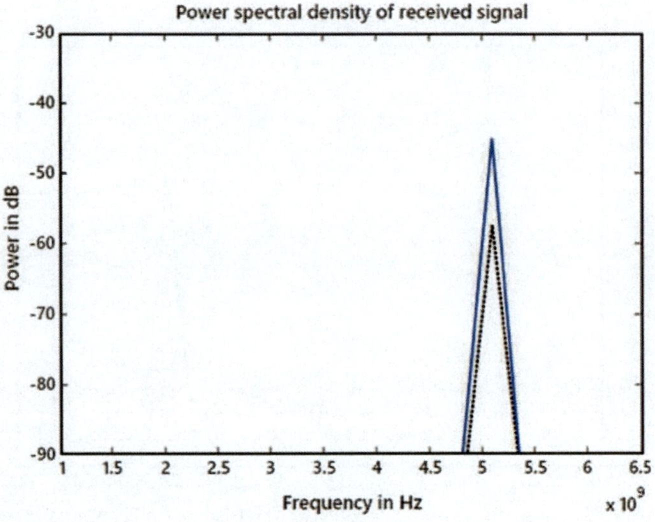

Fig. 8 Iteration-3 of bandwidth adjustment of SU = Tx

as illustrated by Fig. 19. Reconfiguration and adaptation of the bandwidth and signal strength of the Secondary Users helped to mitigate the gap between the achievable capacity and working capacity. The PU WLAN user was transmitting at an average data rate of 54 Mbps, but by our proposed method the SU used the same bandwidth of WLAN user but using higher transmit power enhanced the data rate up to

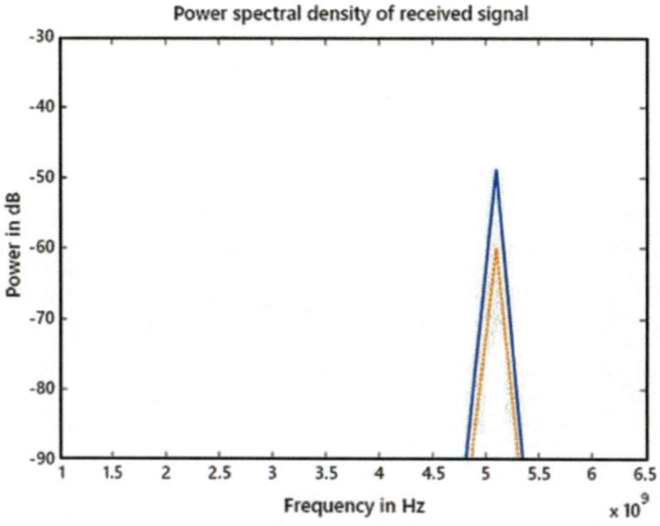

Fig. 9 Iteration-4 of bandwidth adjustment of SU = Tx

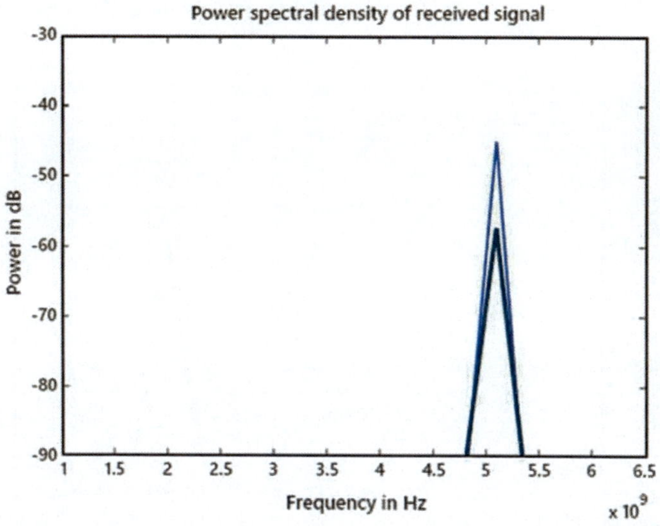

Fig. 10 Iteration-5 of bandwidth adjustment of SU

86.35 Mbps. With increase in transmit power the range of transmission may also be increased.

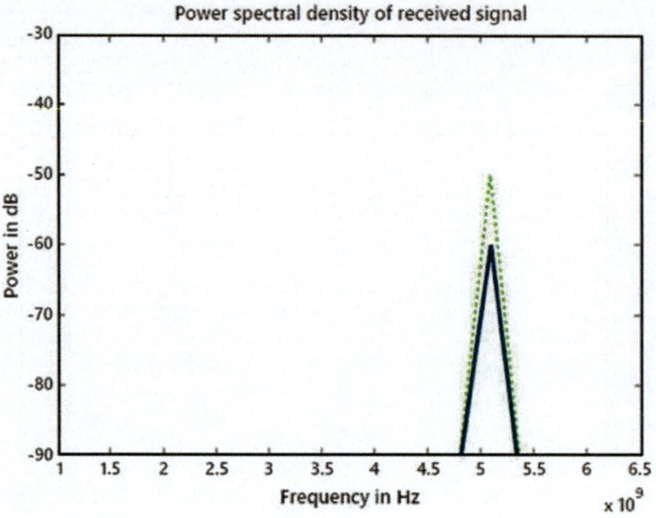

Fig. 11 Iteration-1 of power adjustment of SU-Tx

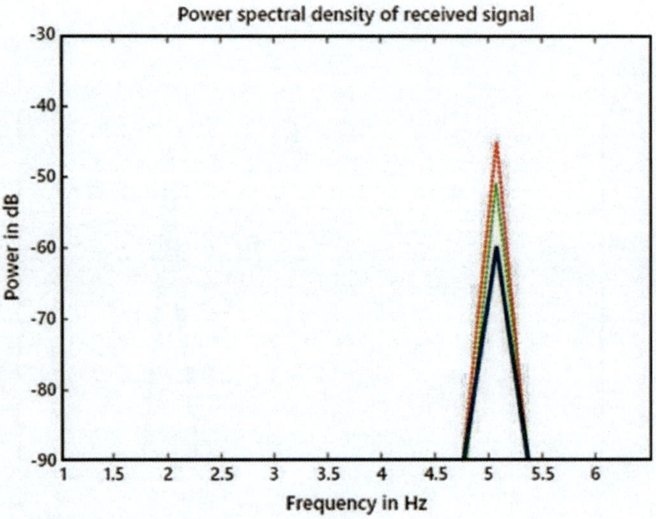

Fig. 12 Iteration-2 of power adjustment of SU-Tx

5 Conclusion

Cognitive radio is a promising technology that is solving the problem of network congestion and underutilization. The paper focused on data rate enhancement using the same channel as used by licensed user. Data rate has been enhancing due to full

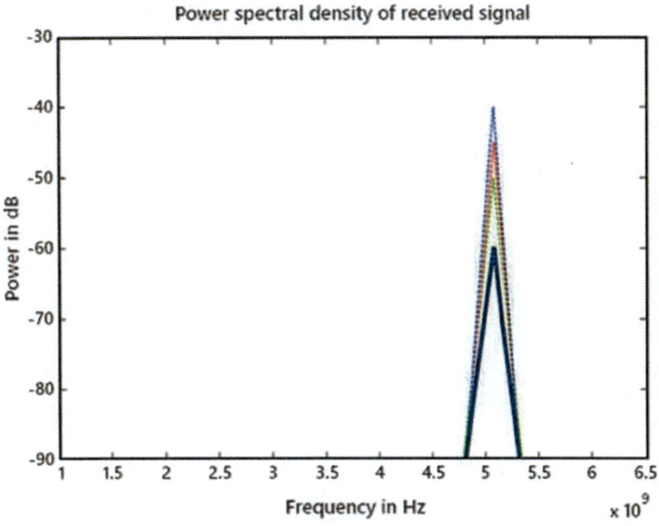

Fig. 13 Iteration-3 of power adjustment of SU-Tx

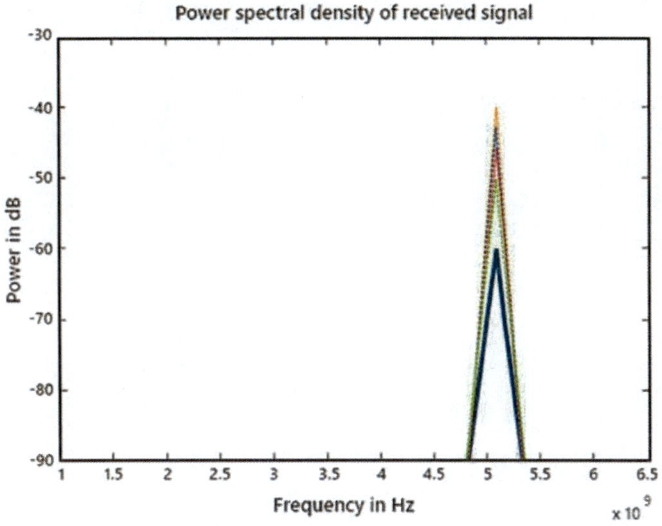

Fig.14 Iteration-4 of power adjustment of SU-Tx

utilization of channel. Adaptation and reconfiguration are the two pillars of cognitive radio technology. With gradual iterations in bandwidth and power not only the goal has been achieved and no harm was caused to activity of primary network. The interference temperature is the main constraint as it gives the close approximation of interference at particular channel and time. Further, enhancement in this method

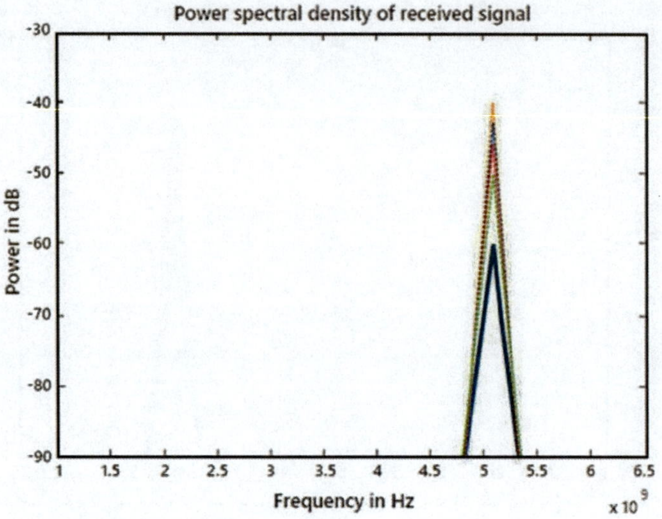

Fig. 15 Iteration-5 of Power adjustment of SU-Tx

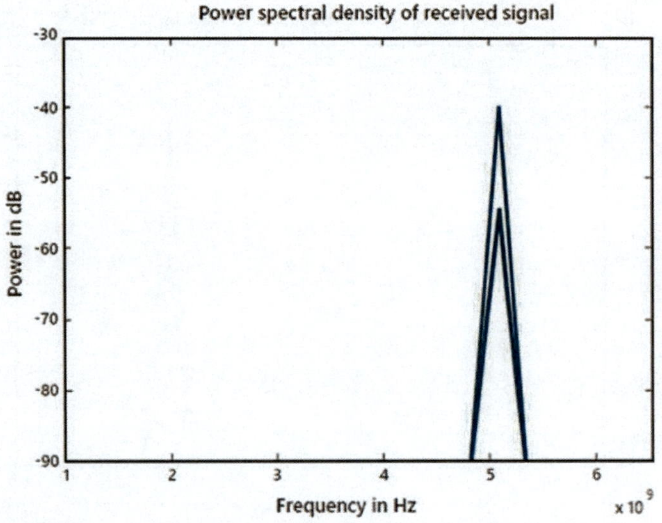

Fig. 16 Iteration-6 of power adjustment of SU-Tx

is going on with increasing the number of secondary users. Different scenarios are under consideration where all secondary users adapt their bandwidth and power simultaneously or one at a time. Not only data rate but various other parameters like S/N ratio and BER of secondary network would be analyzed. The current work has

Table 5 Power reconfiguration of secondary user in absence of primary user

Iteration no	Power reconfiguration (dBm)	BW (MHz)	IT SU (K)	IT difference (K)	Data rate (Mbps)	Difference from achievable capacity
1	50	20	386	79	79.7	7.3
2	55.6	20	420	45	82.9	4.1
3	57.5	20	448	17	84.8	2.1
4	58.75	20	457	8	85.7	1.3
5	59.075	20	461	4	85.8	1.5
6	60	20	464.5	0.5	86.35	0.65

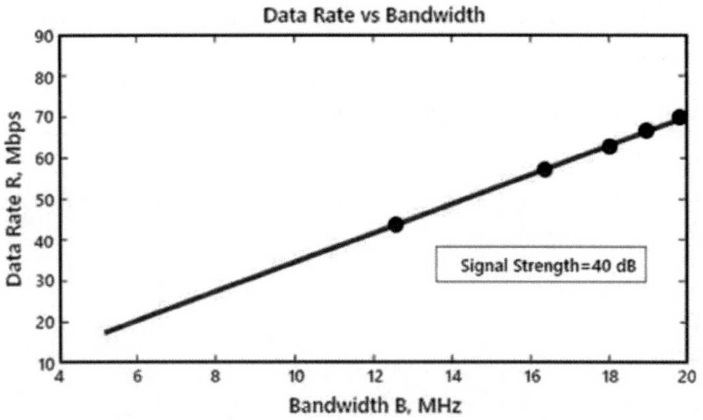

Fig. 17 Shows the curve for achieved data rate while bandwidth adaptation

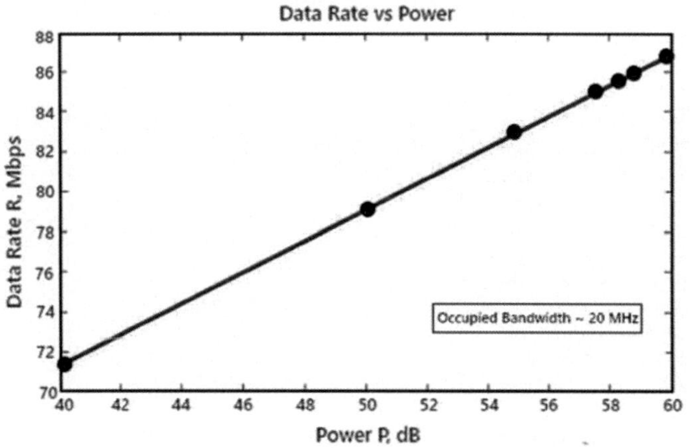

Fig. 18 Shows the curve for achieved data rate while power reconfiguration

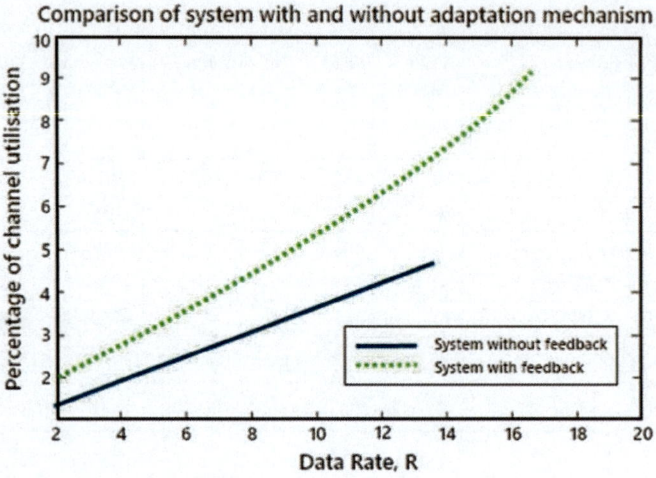

Fig. 19 Comparison of system with and without adaptation mechanism

been tested in AWGN channel conditions, but working of system can also be tested in different fading environments.

References

1. Measuring Interference Temperature. https://www.researchgate.net/publication/228862893
2. Spectrum Policy Task Force. Federal Communications Commission, Rep. ET Docket no. 02–135(2002)
3. Srinivasulu, S., Pagadala, P.: Efficient utilization of spectrum bandwidth using Cognitive Radio to improve throughput in wireless adhoc networks. Int. J. Eng. Sci. Res.-IJESR 1280–1284 (2013)
4. Del Re, F., Argenti, F., Ronga, L.S., Bianchi, T., Suffiritti, R.: Power allocation strategy for cognitive radio terminals. In: First International Workshop on Cognitive radio and Advanced Spectrum Management, pp. 1–5 (2008). https://doi.org/10.1109/COGART.2008.4509988
5. Wang, S., Huang, F., Zhou, Z.: Fast power allocation algorithm for cognitive radio networks. IEEE Commun. Lett. **15**, 8 (2011)
6. Zeng, M., Tsiropoulus, G.I., Dobre, O.A., Ahmed, M.A.: Power allocation for cognitive radio networks employing non-orthogonal multiple access. IEEE Glob. Commun. Conf. (2016). https://doi.org/10.1109/GLOCOM.2016.7842156
7. Lan, P., Zhang, G., Liu, L., Wang, B., Sun, F.: Power allocation for two way communication underlay full duplex cognitive radio systems with qos constraint of primary user. In: International Symposium on Intelligent Signal Processing and Communication Systems, pp. 130–134. China (2017). https://doi.org/10.1109/ISPACS.2017.8266459
8. Mishra, K., Trivedi, A., Gupta, R.: Power allocation in cognitive radio networks using Genetic Algorithm. Int. J. Innov. Technol. Explor. Eng. **8**, 11 (2019)
9. Zhou, Z., Jia, Y., Chen, F., Tsang, K., Liu, G., Han, Z.: Unlicensed spectrum sharing: from coexistence to convergence. IEEE Wireless Commun. **24**, 1536–1284 (2017)
10. Goel, P., Buttar, A.S.: Survey of methods for optimization of transmit power for cognitive radio networks. Int. J. Res. Eng. Appl. Manage. (IJREAM) **4**(7), 655–659 (2018)

11. Mitola, J.: Cognitive radios: making software radios more personal. IEEE Pers. Commun. **6**(4), 13–18 (1999)
12. Devi, M., Sarma, N., Deka, S.K.: Allocation and access mechanisms for spectrum sharing in CRNs—a brief review. In: International Conference on Accessibility to Digital World, pp. 117–120 (2016). https://doi.org/10.1109/ICADW.2016.7942524
13. Kaur, R., Buttar, A.S., Anand, J.: Spectrum sharing schemes in cognitive radio network: a survey. In: 2nd International Conference on Electronics, Communication and Aerospace Technology, Coimbatore, pp. 1279–1284 (2018)
14. Soldo, Malarić, M.: Wi-Fi parameter measurements and analysis. In: Proceedings of the 9th International Conference, pp. 339–342. Smolenice, Slovakia (2013)
15. Mahmoud, H.A., Yucek, T., Arslan, H.: OFDM for cognitive radio: merits and challenges. IEEE Wireless Commun. **16**(2), 6–15 (2009)
16. Yucek, T.,Arslan, H.: ofdm signal identification and transmission parameter estimation for cognitive radio applications. In: IEEE Global Telecommunications Conference, pp. 4056–4060. Washington, DC (2007).
17. Ujjinimatad, R., Patil, S.R.: Signal detection in cognitive radio networks over AWGN, Nakagami-m and Nakagami-q channels. In: Eleventh International Conference on Wireless and Optical Communications Networks (WOCN), pp 1–5. Vijayawada (2014)
18. Malhotra, M., Aulakh, I.K.,Vig, R.: A review on energy based spectrum sensing in Cognitive Radio Networks. In: International Conference on Futuristic Trends on Computational Analysis and Knowledge Management (ABLAZE), pp. 561–565. Noida (2015)
19. Rioul, O., Magossi, J.: Shannon's formula and Hartley's Rule. A Math. Coincidence **1641**, 105 (2015). https://doi.org/doi.org/10.1063/1.4905969
20. 802.11 OFDM Overview. http://rfmw.em.keysight.com

Dialogue Extraction and Translation from Stories on Thirukural Using Verb Cue Quote Content Source Identifiers

M. S. Karthika Devi, G. Umaa Mahesswari, and Baskaran Ramachandran

Abstract The term 'graphic novel' has been referring to a variety of narrative types published in the last twenty years. Recent researches in graphic novels focus on how the human brain processes the combination of images and text which as a result indicated that graphic novels are one among the excellent resources for advanced learners, dialogue analysis, conversion of image to graphical representation and many such related works. Thirukural (as always said to be as Ulaga Pothumurai (in Tamil), the common Veda), extracting content from the couplets and its existence in various articles and media elements is very complex. Researches in Thirukural include automatic couplet extraction from a database with keywords, bilingual language processing, rendering of couplets using different programming languages and works related to the same. Graphic novels are the most sought-after creative visualization tools that can be used to ensure a better way of approaching and teaching. In this work, quote content source identifier (QCSI) is used to automatically generate a storyboard containing seven panels with the constituents of a graphic novel in such a way that the moral behind a Thirukural is conveyed as a result of the sequence of dialogue exchanges between the characters. Couplets from Thirukural are chosen with stories on the same with dialogues between the characters are developed in the English language. Powerful natural language processing tools and libraries are used to extract the dialogues involved along with the source of each of these dialogues and to translate the extracted dialogues to Tamil. The techniques used upon evaluation of results are found to extract all the dialogues with appropriate characters involved in sublime translation results.

1 Introduction

In the field of visualization, graphic novels are believed to be an ideal medium for visual story telling. This is mainly because of the rich expressivity and high interoperability that they provide. When stories are expressed visually using the

M. S. Karthika Devi (✉) · G. Umaa Mahesswari · B. Ramachandran
Department of Computer Science and Engineering, Anna University, Chennai, Tamil Nadu, India
e-mail: mskarthikadevi2014@annauniv.edu

T. Senjyu et al. (eds.), *ICT with Intelligent Applications*, Smart Innovation, Systems and Technologies 248, https://doi.org/10.1007/978-981-16-4177-0_52

components of graphic novels that can be used to present information or narratives through characters, panels, gutters and word/speech balloons, the primary objective of understanding the context and the essence of the story are met. Powerful techniques in machine learning and natural language processing such as QCSI, along with the advantages of graphic novels are used to perform the story telling of Thirukurals, as Thirukural is indeed one of the biggest boons to the Tamil literature with a unique way of deliverance of code of conduct to the mankind to follow for all time to come. Given a story with dialogues between characters which aims at conveying the meaning of the Thirukural in its moral, the characters involved are identified using named entity recognition and the dialogues along with the source of each are extracted using a verb cue, quote content, source identifier that is developed using the Spacy and Textacy libraries. The character–dialogue pairs are extracted from the triplets and after pronoun replacement, the dialogues are translated to Tamil.

1.1 Overview

When stories are expressed visually using the components of graphic novels that can be used to present information or narratives through characters, frames, panels, gutters and thought/word/speech balloons, the impact that is created on the reader is durable. There is no doubt that the legacy of Thiruvalluvar's Thirukural has crossed geographical boundaries to teach moral values for life in depth. It is important that the ethics behind Thirukurals are conveyed to the young minds of the upcoming generations, with a deep understanding of the morals. Also, extracting content from the couplets and its existence in various articles and media elements is very complex compared to content extraction from a news article. Graphic novel makes it a less intimidating read for beginner and advanced readers by their 'image-based' aesthetic. It allows readers to understand 'words through pictures'. The components of a graphic novel, that are of our primary focus in this work, include distinct segments of graphic novel, containing images and text, called the panels. Frames are the lines and borders that contain the panels. Speech balloons are the components that enclose dialogues from a specific character/speaker's mouth and can vary in size, shape and layout. Characters are the image representations of those who are involved in rendering the dialogues that are the utterances or a conversation between two or more characters. A graphic organizer that contains all of these components displayed in a pre-determined sequence is called a storyboard. Regular expressions are used to extract dialogues from the stories. The stories are then fed to QCSI method developed using the Spacy library in Python in order to identify the dialogues, source of the dialogues and the corresponding verb cues involved. The sources identified by the verb cue quote content source identifier are then evaluated against the actual character sources. The Google Trans library is used to translate the extracted dialogues which are in turn placed in appropriate speech balloons. The extracted dialogue sources along with an analysis of the sentiment of the dialogues are used to obtain appropriate character images from the character database to be placed near the speech balloons containing the respective dialogues.

2 Related Work

2.1 Processing and Extraction of Thirukurals

Uma and Ilakiyaselvan [1] proposed a method for context-based searching of Thirukural using the techniques of natural language processing (NLP). A system [2] where the customers would get a multimedia Kural every day from the telecom service providers through a video call/link was proposed to develop Thirukural mobile—a cultural tool for all.

2.2 Automatic Comic Image Layout Preservation and Page Segmentation Systems.

Myodo et al. [3] proposed a framework for bringing together the low-level and high-level information in semi-structured and complex background documents such as comics and also proposed a conceptualization that allowed for semantic analysis and information deduction from automatically extracted visual elements. Li et al. [4] proposed image segmentation, polygon detection, line clustering, line segment detection techniques used to segment 2237 comic images from 12 different printed comic series. A novel neural storyboarding technique for visualizing stories with coherent image sequences was suggested [5, 6].

2.3 Comic Strips-Based Narrative and Story Telling Techniques

Takeda et al. [7] proposed a methodology that involved OCC-RDD (an acronym for objective, conflict, catastrophe—reaction, dilemma, decision), single-responsibility principle Bloom's taxonomy and UML techniques employed with a special focus of a system which converts temperatures from Fahrenheit to Celsius. References [8–11] implemented an optimized comics-based storytelling technique for temporal image sequences. Karthika Devi et al [12] proposed a system to retrieve a coherent flow of multiple sentences that use multimodal neural architecture and ranking-based summarization to generate the summarized description of possibly larger image streams.

2.4 Comic Role Plot Dialogue Recommendation

Tsai [13, 14] used an intelligent comic role plot dialogue recommendation system to select the role of dialogue, role selection and dialogue recommendations for different roles were used. Karthika Devi [15] proposed a system for generating synopsis of a comic book with unsupervised abstractive dialogues.

2.5 Quote Extraction and Analysis

Newell [16] proposed a method that used a CNN classifier trained with PARC annotations that represent a corpus with attribute relations, annotated for effective verb-cue classification. Luyuan [17] proposed unsupervised speech text localization for comics based on the training of a Bayesian classifier on aligned connected components and then detecting the rest of the text using the classifier for text or non-text separation.

2.6 Visualization of Large Image Collections

Semantic information extractor that employed an image captioning based on CNN used along with a visual layout generator to visualize large image collections [18–20]. Descriptive captions for images were transformed into semantic keywords.

3 System Design

3.1 Overall Design

The proposed architecture of the system is divided into three stages including collection of data, source–dialogue pair identification and translation of the dialogues. The first stage involves data collection. The input to the system is text file containing stories (that include the conversation between the characters) coined solely on the basis of the couplets chosen from Thirukural, as extracting content from the couplets and its existence in various articles and media elements is very complex. From the stories, using named entity recognition, the characters involved in conversations are identified. In this stage text tokenization, POS and IOB tagging, noun-phrase chunking and identification of the entities using the Spacy library are performed on the input text to perform character identification. The entities with the class PERSON are identified as a list of characters followed by elimination of the duplicates from the list. The second stage also involves the core methodology used in the proposal. A regular expression is used to extract the dialogues from quotes to generate a list

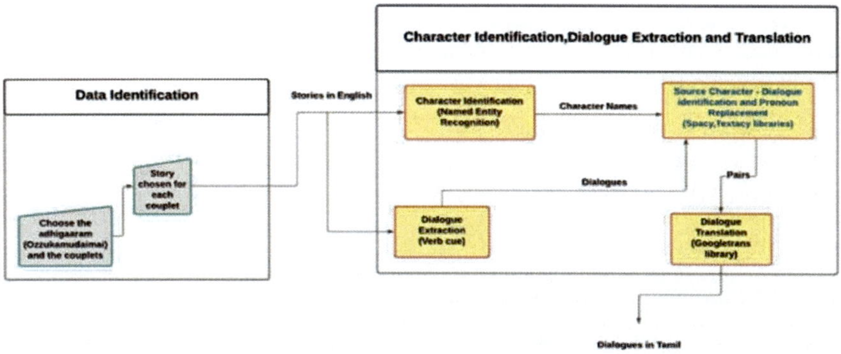

Fig. 1 Proposed architecture. *Source* Dialogue pairs identification

of dialogues involved. The Spacy and the Textacy libraries are used to process the text input in order to obtain direct quotations that are identified on the basis of the character names, verb cues and dialogues occurring in the stories. From the triplets, the source and dialogue pairs are obtained and replacements of pronouns referring to the characters are done using co-reference resolution techniques. The lists of sources are then used for evaluation and the dialogues are subjected to translation. In the final stage, the GoogleTrans library is used to translate the dialogues that are obtained from the source-dialogue pairs to Tamil as in Fig. 1. With the output of this module, the translated dialogues are obtained and sources of each of these dialogues are known.

3.2 Proposed Architecture

The architecture of the proposed system and the libraries used to implement them are stated below in Fig. 1.

3.3 Algorithms

Data Collection

With the goal of bringing about the importance of inculcating good mannerisms in a classroom environment among young minds, Thirukural couplets that extol the virtues of decorum are chosen, and stories are generated that would involve characters along with dialogues involved.

Step 1:**Character Identification**

In order to identify the characters involved in the conversation, the character identification module processes the input text to be able to identify the proper named entities of the class PERSON as the characters involved.

Algorithm*: Character Identification*
Input: The stories of couplets S
Output: A list of characters involved L.

```
T<- List of tokens obtained from S.
for i in T:
        pos_tags<- nltk.pos_tag(i)
pattern<-'NP: {<DT>?<JJ>*<NN>}
cp<- nltk.RegexpParser(pattern)
chunks<- cp.parse(T)
iob_tagged<- Tree2conlltags(chunks)
ifent.label<- "PERSON":
   L.append(ent.text)
while duplicates not in L:
   print(L)
```

As a result of POS and IOB tagging, the entities identified with the labelling named "PERSON" and extracted and displayed as characters involved in the stories. A list of characters are displayed as the final result of this module.

Step 2: Dialogue Source—Pair Identification

The most essential stage of QCSI involves identifying the dialogues in the story along with the character that has uttered that dialogue who is collectively termed as the source of the dialogue.

Algorithm: Pairs Extraction
Input: The stories of couplets S
Output: A list of Pairs P (character-source of dialogue, dialogue).

```
nlp<- spacy.load("en_core_web_sm").
Doc<- textacy.make_spacy_doc(raw_text_lines)
triplets<- direct_quotations(doc)
for t in triplets:
print(P)
```

***Step 3:* Replacement of pronouns**

In order to obtain a clarity on the source of each dialogue, the pronouns identified as sources in the previous module are replaced to the actual character names by making use of the Spacy library to perform co-reference resolution.

Algorithm: Pronouns Replacement
Input: Obtained source-dialogue pairs P with pronoun sources
Output: R—(sources, dialogues).

```
1. nlp<- en_coref_md.load()
      2. doc<- nlp(raw_text_lines)
      3. result_file<- doc._.coref_resolved
      4. for R in result_file:
      5.     print(R)
```

Dialogue Translation

The obtained dialogues are required to be translated to Tamil in order to make use of the information gained about the source of the dialogues to perform further operation on the translated dialogues. These operations include placement of the dialogues (in Tamil) within the speech balloons with a display of appropriate character images for storyboarding.

Algorithm: **Language Translation**
Input: Extracted dialogues in English D
Output: Dialogues in Tamil T.

```
 1. translator<- Translator()
2. translated<- translator.translate(D, src = "en",
dest = "ta")
     3. for trans in translated:
     4.    T.append(trans.text)
     5. print(T)
```

4 Implementation and Analysis of Results

4.1 Description About the Data Used

The data used involves a text file that include couplets from the adhigaaram "ozzuka-mudaimai" with their corresponding stories with dialogues between the characters involved.

4.2 Experimental Results

Translation of dialogues to Tamil

Figure 2 shows five instances of the dialogues from the source-dialogue pairs that are translated to Tamil using the GoogleTrans library. Thus, the dialogues in Tamil are obtained.

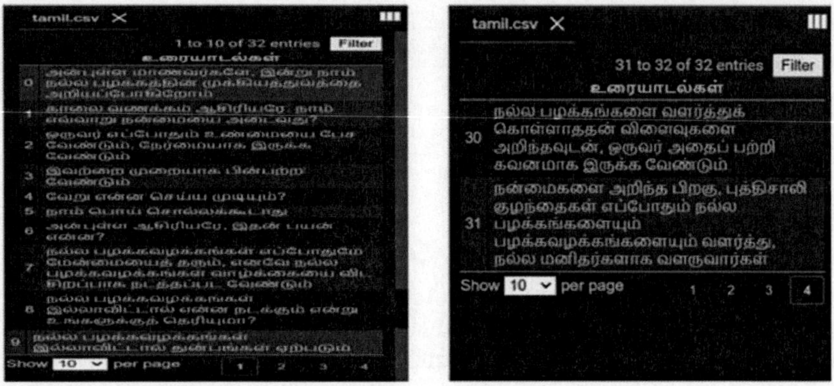

Fig. 2 Dialogue translation

4.3 Results of Sentiment Analysis on the Dialogues

Values between −1.0 and 0.0 indicate negative sentiments conveyed in the dialogues and that between 0.0 and + 1.0 indicate positive sentiments and 0.0 indicated neutral sentiment of dialogues as in Fig. 3. The same as estimated for all the dialogues involved. Figure 4 shows the trend of the subjectivity and polarity values the dialogues that will provide a means of analysis of the sentiment of dialogues in three categories: positive, neutral and negative. Figure 5 shows the plot of polarity values of each dialogue involved against the respective polarity score obtained. Subjective sentences generally refer to personal opinion, emotion or judgment and the subjectivity scores can be used to differentiate between the opinions that describe people's feelings

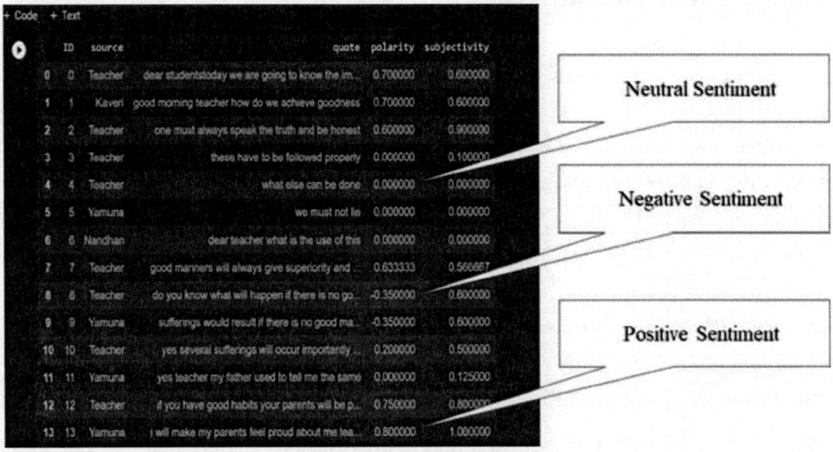

Fig. 3 Polarity and subjectivity scores of dialogues 1–13

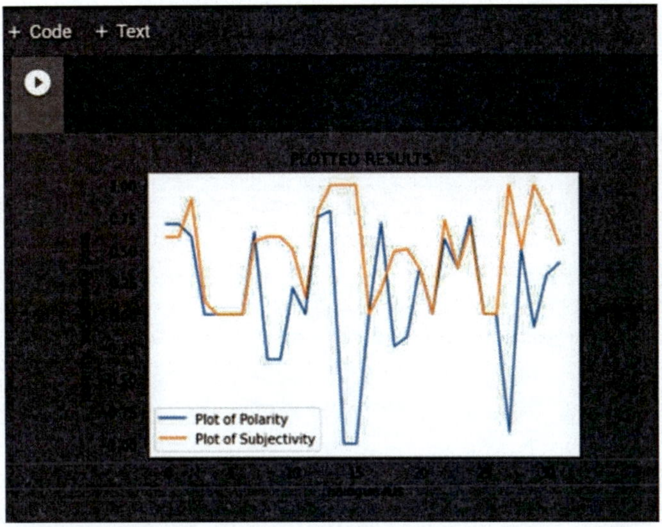

Fig. 4 Plot of polarity and subjectivity

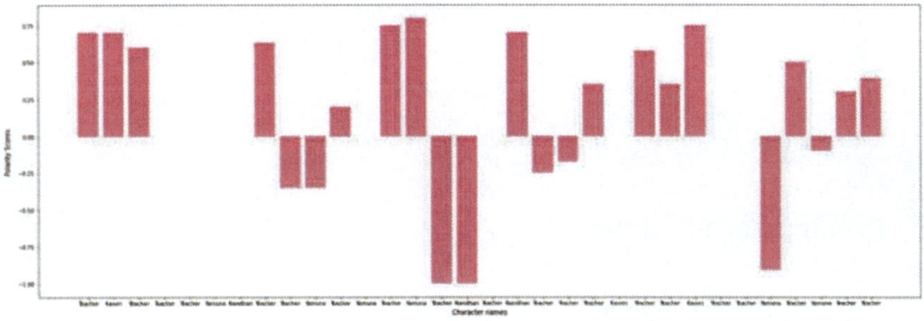

Fig. 5 Bar graph with polarity scores of character names for each dialogue

towards a specific subject or topic. Figure 6 shows the plot of subjectivity scores of characters involved in each dialogue. It is visible that subjectivity scores are either zero or positive in this case.

Figure 7 shows the count of each of the subjectivity and polarity values obtained. Dialogues with neutral polarity and subjectivity values are dominant.

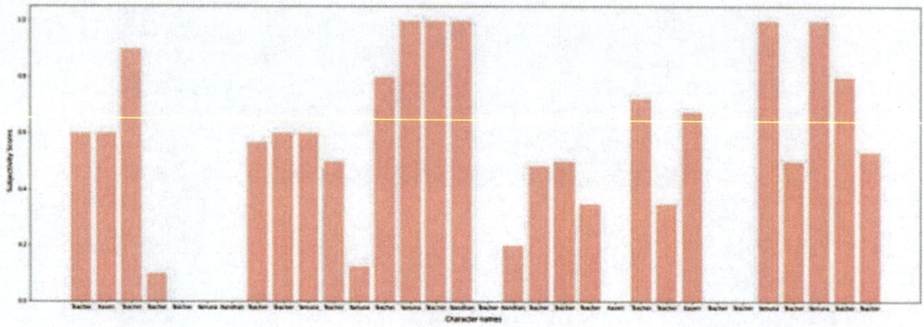

Fig. 6 Bar graph with polarity scores of character names for each dialogue

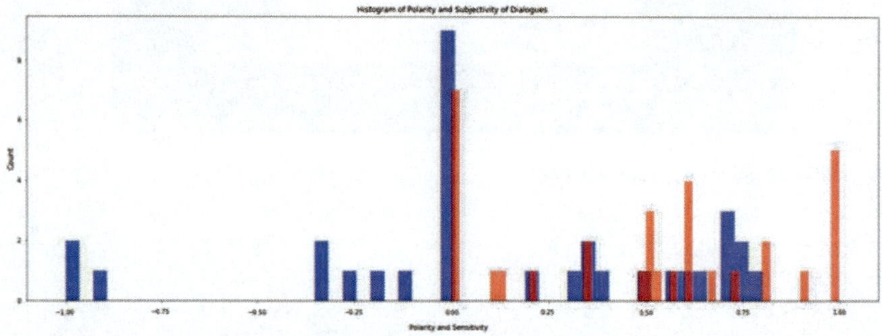

Fig. 7 Histogram on counts of subjectivity and polarity scores

5 Performance Metrics

5.1 Evaluation Metrics

The performance of the dialogue extraction process can be evaluated on the basis of comparison between the predicted character–dialogue pairs and the actual character–dialogue pairs. The metrics that are used for evaluation are as follows.

The confusion matrix as in Fig. 8 is found to provide a thorough understanding of the number of sources that are correctly identified to that of the actual number of sources of each dialogue by comparing them to the actual source names. The sources of each dialogue are matched to their corresponding dialogue utterances and are compared to the extracted sources to construct the confusion matrix. It is found that the sources of all the 32 dialogues are correctly identified by the process and therefore the precision, recall and F1-score values are optimal. Precision is the ratio of correctly predicted character names to those that are the total number of predicted character names. Recall is the ratio of correctly predicted character names to those that are the total number of actual character names. F1—Measure establishes a relation between the precision and recall values computed.

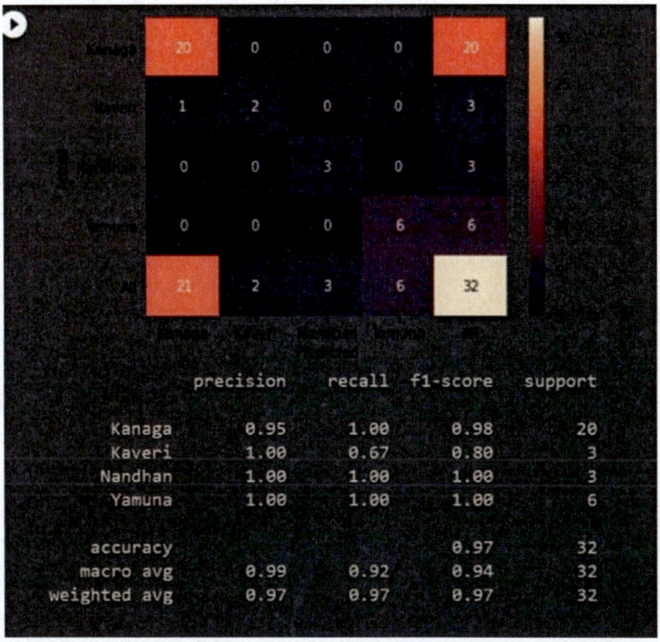

Fig. 8 Confusion matrix in the form of a heat map along with precision, recall and F1-score

Extracting content from the couplets and its existence in various articles and media elements is very complex when compared to content extraction from a news article. It can be seen from Table 1 that source identification of dialogues involved between characters in stories chosen for the couplets, using the QCSI system, seem to outperform the source identification system implemented using CRF and Spacy library that were used on BBC news articles that were attempted to be manually classified by the BBC journalists. The performance of the source identification system depends on the size, complexity of the dataset used in addition to the training and classification mechanisms used in case of large datasets.

Table 1 Comparison on Source Identification using CRF and QCSI

Methods→ Metrics↓	Source identification using conditional random fields and spacy (%)	Source identification using QCSI (%)
Precision	92.6	99
Recall	89.8	92
F1-score	91.2	94

6 Conclusion

It is obvious that graphic novels are indeed one of the powerful tools to convey information and ideas by making use of these components along with suitable domain knowledge. In this proposal, the domain chosen to be interpreted with is the facts and morals hidden behind the couplets by Thiruvalluvar. As initial steps towards developing graphic novel-based storyboards to perform story telling of Thirukurals, powerful natural language processing techniques are employed to extract dialogues and characters from the stories to match the dialogues with their sources and to translate the dialogues to Tamil. It is seen that the libraries and the methodologies used have served to provide the best performance by extracting all the dialogues involved along with their appropriate character sources. Coining stories with proper use of verb cues and proper involvement of characters and their dialogues is the most important requirement that could be tedious with an increase in the number of couplets chosen. Regular expression-based extraction of indirect dialogues requires appropriate algorithms. The placement of the dialogues in the speech balloons and the placement of the respective characters in suitable positions and steps to generate storyboards containing seven panels (for each couplet) with the generated elements using segmentation techniques are to be carried out to generate a graphic novel.

Acknowledgements This Publication is an outcome of the R&D work undertaken in the project under the Visvesvaraya Ph.D Scheme of Ministry of Electronics & Information Technology, Government of India, being implemented by Digital India Corporation (formerly Media Lab Asia).

References

1. Uma, G.V., Ilakiyaselvan, N.: Context based information search for Thirukural. In: Tamil Internet Conference. Coimbatore (2010)
2. Bhuvan Babu, G.: Thirukural Mobile A Cultural Tool for All, Tamil Internet Conference, Coimbatore (2010)
3. Myodo, E., Ueno, S., Takagi, K., Sakazawa, S.: Automatic comic like image layout system preserving image order and important regions. In Proceeding ACM MM, pp. 795–796 (2011)
4. Li, L., Wang, Y., Tang, Z., Gao, L.: Automatic comic page segmentation based on polygon detection, Springer Science. Multimed. Tools Appl. (2012). https://doi.org/10.1007/s11042-012-1241-7
5. Ravi, H., Wang, L., Muniz, C., Sigal, L., Metaxas, D., Kapadia, M.: Show me a story: towards coherent neural story illustration. In: Proceedings of the IEEE Conference on Computer Vision and Pattern Recognition, pp. 7613–7621 (2018)
6. Chen, S., Liu, B., Fu, J., Song, R., Jin, Q., Lin, P., Qi, X., Wang, C., Zhou, J.: Neural storyboard artist: visualizing stories with coherent image sequences In: 27th ACM International Conference on Multimedia (MM '19), p. 9. Nice, France. ACM, New York, NY, USA, 21–25 Oct 2019
7. Takeda, M.M., Marcondes, F.S., Vega, I.S., Gara, E.B.M., Gaspar, L.: A computational teaching approach through the use of a narrative technique and a comic strip. In: Ninth International Conference on Information Technology- New Generations (2012)
8. Kosara, R., Mackinlay, J.: Storytell.: Next Step Visual. Comput. **46**(5), 44–50 (2013)

9. Kosara, U.R., Mackinlay, J.: Even now you can draw! Introduction to 4-frame cartoon. Storytell.: Next Step Visual. Comput. 46(5):44–50 (2013)
10. Chu, W.-T., Member, IEEE, Yu, C.-H., Wang, H.-H.: Optimized comics-based storytelling for temporal image sequences. IEEE Trans. Multimedia **17**(2) (2015)
11. Wen, M.-H., Thawonmas, R., Chen, K.-T.: Pomics: a computer aided storytelling system with automatic picture-to-comics composition. In: Proceeding 2012 Conference Technology Application Artificial Intelligence, pp. 314–318 (2012)
12. Karthika Devi, M.S, Fathima, S., Baskaran, R.: SYNC—Short, Yet Novel Concise Natural language description: generating a short story sequence of album images using multhi-modal network, Nature Singapore Pte Ltd. 2020. In: Fond, S., et al. (eds.) ICT Analysis and Applications, Lecture Notes in Networks and Systems 93. Springer (2020)
13. Tsai, S.T.: Intelligent comic role plot dialogue recommendation system for comic script creation educational application. In: 1st IEEE International Conference on Knowledge Innovation and Invention (2018)
14. Tsai, S.: Design of intelligent comic make system for educational application based on comic script creation. Eurasia J. Math., Sci. Technol. Educ. **14**(3), 1023–1041 (2017)
15. Karthika Devi, M.S., Fathima, S., Baskaran, R.: CBCS—Comic VBook Cover Synopsis: generating synopsis of Comic Book with unsupervised abstractive dialogue. In: 9th World Engineering Education Forum. Elseiver Procedia Computer Science (2020)
16. Newell, C., Cowlishaw, T., Man, D.: Quote Extraction and Analysis for News. DSJM, London, UK (2018)
17. Luyuan, L., Wang, Y., Tang, Z., Lu, X., GaoL.: Unsupervised speech text localization in comic images. In proceedings of the IEEE International Conference on Document Analysis and Recognition, pp. 1190–1194 (2013)
18. Xie, X., Cai, X., Zhou, J., Cao, N., Wu, Y.: A semantic-based method for visualizing large image collections. IEEE Trans. Visual. Comput. Graph. **25**(7) (2019)
19. Huang, Y, Wu, Q., Song, C., Wang, L.: Learning semantic concepts and order for image and sentence matching. In: IEEE Conference on Computer Vision and Pattern Recognition, pp. 6163–6171 (2018)
20. Xu,T., Zhang, P., Huang, Q., Zhang, H., Gan, Z., Huang, X., Attngan, X.H.: Fine-grained text to image generation with attentional generative adversarial networks. In: Proceedings of the IEEE Conference on Computer Vision and Pattern Recognition, pp. 1316–1324 (2018)

A Survey on Using Machine Learning to Counter Hardware Trojan Challenges

Ravikant Khamitkar and R. R. Dube

Abstract In today's information-driven world where different things are connected together using various sophisticated embedded systems to collect the data for processing and decision making. These embedded systems are designed using various hardware components developed using several VLSI designs and technologies which focus on quality integrated circuits (IC). But these ICs during design and development phase have security concern of trigger-based hardware which can perform malicious activity or lead to destructive behavior of system. This malicious hardware is known as Hardware Trojan. There are lot of researchers who suggested various methods to detect the Hardware Trojans and to mitigate their effect. Research work is carried out for HT detection majorly from two perspectives: one is through ICs layout image analysis with physical parameter validation and second is during design phase where hardware description language along with assertion checkers are used to verify and validate the design to be implemented inside the IC or FPGA. In this paper, various research works carried for HT with their future perspective along with challenges in different approaches are discussed in detail. As hardware software co-design is widely used in embedded application development and also due to advancement in technology, it is possible to use the machine learning (ML) approach to identify the HT in SoC design. These various approaches are studied in this paper, therein tagging the security threats and requirements to provide the recommendation for future research in this field.

1 Introduction

With the development of various integrated circuits in the last decade, there is attached problem of malicious modifications in hardware, which is also known as Hardware Trojans (HTs). This modification has become serious concern about security in the IC

R. Khamitkar (✉)
NBNSCOE, Solapur, MH 413255, India

R. R. Dube
WIT, Solapur, MH 413006, India

T. Senjyu et al. (eds.), *ICT with Intelligent Applications*, Smart Innovation, Systems and Technologies 248, https://doi.org/10.1007/978-981-16-4177-0_53

development industry. ICs which have HTs can cause failure, leakage of information, or cause other devastating consequences. Thus, HT has been a problem or concern for the various stakeholders from industry, research and development teams [1–4] (Fig. 1).

From the early research on HT which was published by Agrawal et al. [5], a plenty of research work is carried out on detecting various forms of HTs. But, very little efforts are taken on the implementation of HTs in real-world scenario. To establish reliable HT finding and protection techniques, it is required to understand the practicability of inserting HTs in real implementations [6]. This has been the field that has received comparatively little consideration in the research domain where various HTs developed in the literature are circuits of small to medium size which are used in register transfer level (RTL) in the IC design flow [7]. However, the integration of HTs in design phase of system using field-programmable gate array (FPGA) device is not much explored at research level. This gives the opportunity to have new paradigms in this field for devising the efficient methods to detect the HTs. The HTs can be inserted in ICs at various vulnerable steps as shown below shaded portion in the Fig. 2.

It can be seen that the specification, package testing, deployment and monitor steps are trusted where no outsourcing is done by companies. But during design phase, the external intellectual property (IP) cores, standard definitions and libraries

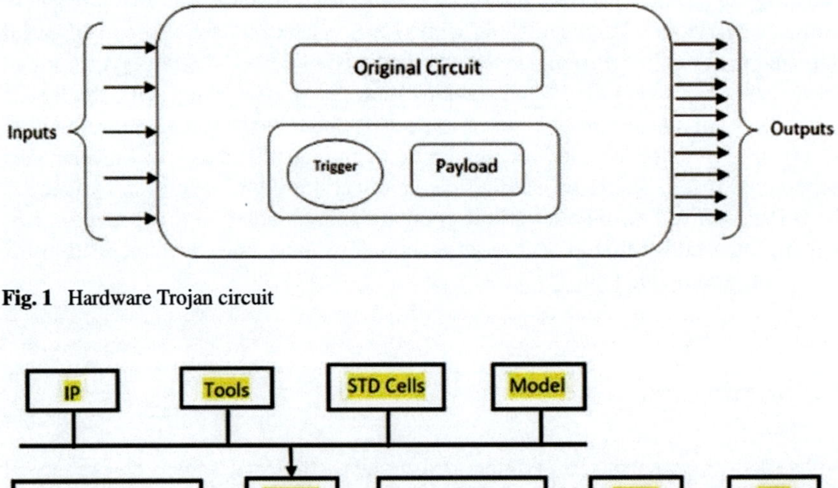

Fig. 1 Hardware Trojan circuit

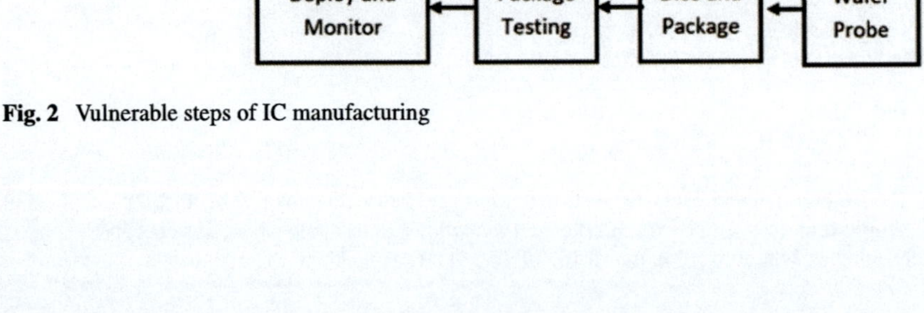

Fig. 2 Vulnerable steps of IC manufacturing

are taken from third party where possibility of HTs insertion is more. Also, during fabrication as these facilities are very limited in world, so it is also possible to get the HTs insertion during these steps [3]. As it is very difficult to integrate the HTs during fabrication and also various research studies shown that using electron scope, it is possible to detect the HTs insertion during fabrication process. So the major challenge remains in design step. In design step before fabrication, most of the companies use FPGAs to develop prototype of their system. So to efficiently check the HTs insertion in design phase, practical hardware implementation can be done using FPGA-based board to test the IP cores and other hardware design of system. As design and implementation of soft core-based systems on FPGAs require hardware plus software co-design, it is possible to incorporate the use of machine learning (ML) tools and integrate the artificial neural network (ANN) for efficiently detecting the HTs [8, 9].

1.1 HT Definition and Taxonomy

The malicious hardware added intentionally in the form of IP or bus control unit or soft core system to carry out adverse effect over normal operation of the intended hardware is called as Hardware Trojan. Initial studies carried out related to HT have classified it as combinational and sequential, but this classification was not sufficient to incorporate the HTs in analog hardware [10]. Further, the HTs are classified in three levels based on their action, activation and physical characteristics [11]. Still, this taxonomy is limited with wide variety of HTs examples.

Here, from above Table 1, it can be seen that the different taxonomies are used for studying the impact of HTs. We can present a new, simple and effective taxonomy for the HTs implemented in system on chip (SoC)-based integrated circuits. (1) User Logic (Glue Logic) Level HTs (2) Control Level HTs (3) SoC Level HTs as shown in Fig. 3.

User Logic (Glue Logic)-Level Trojans—this type of Trojans have scope of affecting the limited space of IC as like only single intellectual property (IP) core (as shown in Fig. 3) [15]

Table 1 Various HT taxonomies

Work	No. of classification	Features	Categories
[10]	2	Logical types	Combinational and sequential
[2, 11]	3	Trojan features	Physical, activation and action
[12]	2	Trojan impacts	Bug based and parasite based
[1]	2	Circuit features	Analog and digital
[13, 14]	5	Attributes	Design phase, abstraction level, activation, effects and location

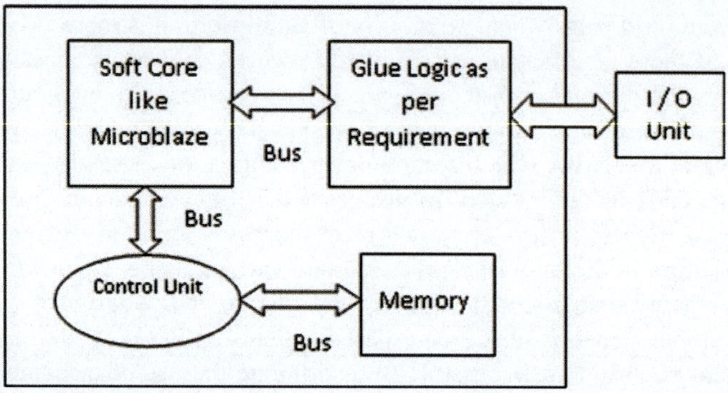

Fig. 3 Proposed taxonomy for the HTs

Control-level Trojans—These type of Trojans have capability to affect the bus level transactions to and from IC or inside the IC also can have scope to affect multiple IPs (as shown in Fig. 3) [16].

SoC-level Trojans—this type of Trojan have capability to affect at any place inside SoC or even it can affect outside performance of IC circuits. This type of Trojans have wide scope of affecting the performance (as shown in Fig. 3) [17].

1.2 Features or Properties of HT

There are several features or properties various researchers have considered for classification of HT threats is listed in following Table 2.

2 HT Detection Methods

Various researchers have adopted various methods for detecting HT at different levels of IC development cycle. Various HT detection methods can be classified as shown in Fig. 4.

2.1 Machine Learning in HT Detection

Figure 4 shows the different HT detection methods; from this, it can been seen that ML algorithms are used widely in side-channel analysis, as it is easy to test and categorize the various side channel parameters. The second frequently used is circuit feature analysis, but the reverse engineering method is having little scope for

Table 2 Various HT properties

Sr. No.	Property	Description
01	Location	This property gives idea about the scope of HT Attack, like if Control Level, HT will have impact over bus control units and data accessibility
02	Attack target	The target is which part of system can be affected. If the user logic-based HT will have impact over working simple logic circuit. Whereas, SoC level HT can impact on other cores or any part of the system
03	Trigger condition	Understanding triggering condition in case of HT is very important from designing defense against HT. Internal nodes or particular data flow from chip or inside the chip or particular sequence or even sequence of instruction in program can work as trigger for HT
04	Attack pattern	The attack pattern gives the idea about how smartly HT can do damage to system. For user logic-level HT, the pattern can be simple. In case of SoC level, HTs can have diverse attack patterns
05	Trojan impact	The Trojan impact gives the idea about amount of damage HT can do to system. For simple user logic, it can be limited to some specific cores. But in case of SoC-level HT, the impact can be multiple cores or even outside part of system
06	Defense strategy	Various defense strategies can be used against various levels of HT, like for user logic level HT, static verification technique can be used. For control-level HT, traffic patterns over the internal bus can be checked. In case of SoC-level HT, potentially suspicious activities can be monitored and validated dynamically
07	Technique feature	Feasibility and impact analysis of Trojans in individual IP cores (or simple logic circuit) Real-time matching of abnormal traffic linker or router features. Fine-grained SoC behavior analysis and assertion-aware security decision are some of the techniques used for HT detection

use of ML algorithms because the feature abstraction is difficult and time-consuming process in this method which makes it less feasible. Various researchers have worked with implementation of the HT detection using machine learning approach where different features used, machine learning model used and evaluation metrics used are listed in Table 3.

From Table 3, it can be seen that the supervised ML algorithms one class, two class and general SVM are most frequently used for identifying HT. But the SVM requires golden ICs as reference for training. Also, it can be observed that unsupervised learning algorithms like K-means algorithm is also comparatively popular for HT detection as it does not require any golden IC for reference. Also, the evaluation matrix shows that accuracy and precision in detection of HT is main focus. In majority of the research, the features used are physical parameters like power dissipation, propagation delay, temperature variation, voltage variation, etc.

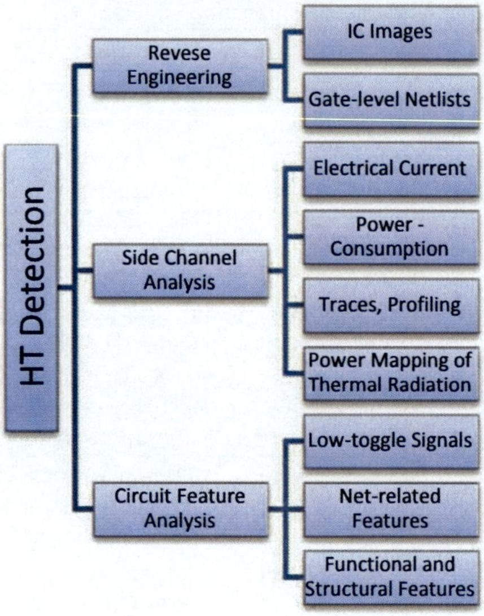

Fig. 4 Various methods and features for the HTs detection

Table 3 Summary of HT detection work with ML

Works	Feature	ML learning model used	Evaluation metrics
Chithra et al. [8] 2020	Temperature and voltage	SVM model	Accuracy = 99.4%
Taifeng Hu et al. [9] 2019	Power waveforms	Gaussian kernel function	Accuracy = 93.75%
K. Hasegawa et al. [18] 2017	Net-related features	Two-class S VM	TPR = 0–100%, TNR = 22–100%
K. Hasegawa et al. [19] 2017	Functional and structural features	Multi-layer ANN	TPR = 74.3–88.9%, TNR = 53.3–70.1%
J. Li et al.[20] 2017	Power consumption	BPNN	Sensitivity = 99.2%
F.K. Lodhi et al. [21] 2016	Propagation delay	K-NN, DT, BC	Accuracy = 93.1–95.1%, Precision = 79.7–95.9%
B. Cakir et al. [22] 2015	Signal correction	OPTICS	TPR = 28–100%, SPC = 94–99%
C.X. Bao et al. [23] 2014	IC images /Grips	One-class SVM	Accuracy = 97.2%

3 HT Challenges and Suggested Solution

There are different challenges in HT detection faced previously by researchers are briefly discussed. From review of various work carried, it can be seen that ML is a very well-ordered approach for security of hardware. The research trend related to HT detection shows that ML is used mostly with side channel method as compared with reverse engineering or circuit feature analysis. The ML is used rarely from HT defense perspective as there are only few options available to carry out defense against HT. The ML approach is also used in tackling the bit file manipulation in hardware description language to detect and defend against the HT problems.

The challenges to use ML in the HT detection are feature extraction complexity and computation time. With the active single core IC, it is possible to tackle the HTs but with multiple IP cores and or SoC, it becomes a tedious task to tackle the HT problems. So to have better insight in SoC, it is possible to look at SoC from hardware-software co-design perspective. As software drives hardware in SoC operation like application subroutines executes and use the hardware signals to achieve the task, this gives origin to hardware Trojan triggering using programming code. To tackle this challenge, one has to monitor and control the software to have better defense against HT [24–26].

4 Conclusion

From the literature review, it is observed that the use of machine learning algorithms is done in the Hardware Trojan detection but from the perspective of layout images classification or from various parameters like area, temperature, power dissipation variations for various IP core implementations or bus interface unit implementation in the FPGA, but there is no or very little research for SoC implementation for Hardware Trojan detection using analysis of program (logic) which will work as software part of SoC and majorly contribute in triggering of any HT in SoC. So to achieve the protection or defense against this program-based Hardware Trojan, it is possible to use machine learning-based approach where various programs with and without Trojan can be used for learning and utilize the learning to detect the HT. So using ML approach, it is possible to detect the behavior of software part which is anomalous and can cause HT triggering in future.

References

1. Bhunia, S., Hsiao, M.S., Banga, M., Narasimhan, S.: Hardware Trojan attacks: threat analysis and counter measures. Proc. IEEE **102**(8), 1229–1247 (2014). https://doi.org/10.1109/JPROC.2014.2334493
2. Tehranipoor, M., Koushanfar, F.: A survey of hardware Trojan taxonomy and detection. IEEE Des. Test Comput. **27**(1), 10–25 (2010)
3. Chakraborty, R.S., Narasimhan, S., Bhunia, S.: Hardware trojan: threats and emerging solutions. In: 2009 IEEE International High Level Design Validation and Test Workshop, pp. 166–171. San Francisco, CA, (2009). https://doi.org/10.1109/HLDVT.2009.5340158
4. Wu, T.F., Ganesan, K., Hu, Y.A., et al.: TPAD: hardware Trojan prevention and detection for trusted integrated circuits. IEEE Trans. Comput.-Aided Des. Integr. Cir. Syst. **35**(4), 521–534 (2016)

5. Agrawal, D., Baktir, S., Karakoyunlu, D., et al.: Trojan detection using IC fingerprinting. In: Proceeding IEEE Symosium. on Security and Privacy, pp. 296–310. Oakland, USA (2007)
6. Ghandali, S., Becker, G.T., Holcomb, D., et al.: A design methodology for stealthy parametric Trojans and its application to bug attacks. International Conference on Cryptographic Hardware and Embedded Systems, pp. 625–647. Santa Barbara, USA (2016)
7. Becker, G.T., Regazzoni, F., Paar, C., et al.: Stealthy dopant-level hardware Trojans. In: International Workshop on Cryptographic Hardware and Embedded Systems, pp. 197–214. Santa Barbara, USA (2013)
8. Chithra, C., Kokila, J., Ramasubramanian, N.: Detection of hardware Trojans using machine learning in SoC FPGAs. In: 2020 IEEE International Conference on Electronics, Computing and Communication Technologies (CONECCT), pp. 1–7. Bangalore, India (2020). https://doi.org/10.1109/CONECCT50063.2020.9198475
9. Hu, T., Wu, L., Zhang, X., Yin, Y., Yang, Y.: Hardware Trojan detection combine with machine learning: an SVM-based detection approach. In: 2019 IEEE 13th International Conference on Anti-counterfeiting, Security, and Identification (ASID), pp. 202–206. Xiamen, China (2019). https://doi.org/10.1109/ICASID.2019.8924992
10. Banga, M., Hsiao, M.S.: A Region based approach for the identification of hardware Trojan. In Proceeding IEEE International Symposium on Hardware-Oriented Security and Trust (HOST'08), pp. 40–47 (2008). https://doi.org/10.1109/HST.2008.4559047
11. Wang, X.X., Tehranipoor, M., Plusquellic, J.: Detecting malicious inclusions in secure hardware: challenge and solutions. In: Proceeding IEEE International Symposium on Hardware-Oriented Security and Trust (HOST'08), pp. 15–19 (2008). https://doi.org/10.1109/HST.2008.4559039
12. Zhang, J., Yuan, F., Wei, L., et al.: VeriTrust: verification for hardware trust. IEEE Trans. Comput. Aided Des. Integr. Circuits Syst. **34**(7), 1148–1161 (2015). https://doi.org/10.1109/TCAD.2015.2422836
13. Karri, R., Rajendran, J., Rosenfeld, K., et al.: Trustworthy hardware: identifying and classifying hardware Trojans. Computer **43**(10), 39–46 (2010). https://doi.org/10.1109/MC.2010.299
14. Rajendran, J., Gavas, E., Jimenez, J., et al.: Towards a comprehensive and systematic classification of hardware Trojans. In: 2010 IEEE International Symposium on Circuits and Systems (ISCAS'10), pp. 1871–1874 (2010)
15. Wang, X., Zheng, Y., Basak, A., Bhunia, S.: IIPS: infrastructure IP for secure SoC design. IEEE Trans. Comput. **64**(8), 2226–2238 (2015). https://doi.org/10.1109/TC.2014.2360535
16. Daoud, L.: Secure network-on-chip architectures for MPSoC: overview and challenges. In: Proceeding IEEE 61st International Midwest Symposium on Circuits and Systems (MWSCAS'18), pp. 542–543 (2018). https://doi.org/10.1109/MWSCAS.2018.8623831
17. Basak, A., Bhunia, S., Tkacik, T., Ray, S.: Security assurance for system-on-chip designs with untrusted IPs. IEEE Trans. Inf. Forensics Secur. **12**(7), 1515–1528 (2017). https://doi.org/10.1109/TIFS.2017.2658544
18. Hasegawa, K., Yanagisawa, M., Togawa, N.: A Hardware-Trojan classification method using machine learning at gate-level netlist based on Trojan Features. IEICE Trans. Fundam. Electr., Commun. Comput. Sci. **100**(7), 1427–1438 (2017)
19. Hasegawa, K., Yanagisawa, M., Togawa, N.: Hardware Trojans classification for gate-level netlists using multi-layer neural networks. In 2017 IEEE 23rd International Symposium on On-Line Testing and Robust System Design IOLTS'17, pp. 227–232 (2017)
20. Li, J., Ni, L., Chen, J., Zhou, E.: A novel hardware Trojan detection based on BP neural network. In: Proceeding 2nd IEEE International Conference on Computer and Communication (ICCC'16), pp. 2790–2794 (2017). https://doi.org/10.1109/CompComm.2016.7925206
21. Lodhi, F.K., Abbasi, I., Khalid, F., et al.: A self-learning framework to detect the intruded integrated circuits. In: Proceeding IEEE Symposium on Circuits and System (ISCAS'16), pp. 1702–1705 (2016). https://doi.org/10.1109/ISCAS.2016.7538895
22. Cakir, B., Malik, S.: Hardware Trojan detection for gate-level ICs using signal correlation based clustering. In: Proceeding Design, Automation and Test in Europe Conference and Exhibition (DATE'15), pp. 471–476 (2015)

23. Bao, C.X., Forte, D., Srivastava, A.: On application of one-class SVM to reverse engineering-based hardware Trojan detection. In Proceeding 15th International Symposium on Quality Electronic Design (ISQED'14), pp. 47–54 (2014). https://doi.org/10.1109/ISQED.2014.6783305
24. Huang, Z., Wang, Q., Chen, Y., Jiang, X.: A survey on machine learning against hardware Trojan attacks: recent advances and challenges. IEEE Access **8**, 10796–10826 (2020)
25. Xue, M., Chongyan, G., Liu, W., Shichao, Y., O'Neill, M.: Ten years of hardware Trojans: a survey from the attacker's perspective. IET Comput. Digit. Tech. **14**, 231–246 (2020)
26. Dong, C., Zhang, F., Liu, X., Huang, X., Guo, W., Yang, Y.: A locating method for multi-purposes HTs based on the boundary network. IEEE Access **7**, 110936–110950 (2019)

Green Cloud Computing-Oriented Algorithmic Approach-Based Virtual Machine Consolidation

Saumitra Vatsal and Shalini Agarwal

Abstract Data centres are the hubs for global connectivity through networking. Cloud computing has become an indispensable need to fulfil the insatiable ever-expanding networking demand. More and more data centres are raised to fulfil this objective, consequently resulting in enhanced environmental pollution. Hence, server-related energy conservation is an indispensable need for minimizing carbon emissions. Factors like grid-related energy consumption and associated carbon emissions enhance electricity cost and carbon tax which results into total operating costs of data centres. The aim focuses on maximization of green energy usage with minimization of operating cost and carbon emission through process of virtual machine placement decision. Dynamic virtual machine consolidation emerges as a convincing solution for minimized energy consumption with optimized resource utilization in data centres. Virtual machine consolidation problems being truly NP-hard demand several heuristic algorithms for addressing the problem. Since most of the existing works are focussed for reducing number of hosts and their present resource utilization unmindful of future resource requirements, it results into unnecessary virtual machine migrations along with enhanced service-level agreement (SLA) violations. This problem can be duly addressed by considering current and future resource utilization by negotiating it as a bin-packing problem. The prediction of future resource utilization can be secured by using a k-nearest neighbour regression-based model.

1 Introduction

In cloud environment, a number of tasks need resources which are made available by multiple resource providers. Cloud-related large data centres are needed for addressing consolidation of organization and individual-related multiple clusters for

S. Vatsal (✉) · S. Agarwal
Shri Ramswaroop Memorial University, Barabanki, Uttar Pradesh, India

S. Agarwal
e-mail: shalini.cs@srmu.ac.in

T. Senjyu et al. (eds.), *ICT with Intelligent Applications*, Smart Innovation, Systems and Technologies 248, https://doi.org/10.1007/978-981-16-4177-0_54

IT industry cloud-based services [1]. The energy consumption growth pertaining to data centre is approximately 10–12% per year [2]. Service providers cannot ignore the issue of improving the infrastructure-related energy consumption by addressing all required parameters for server-power consumption, heat dissipation, cooling provisioning along with avoiding service-level agreement (SLA) violation. Virtualization is a technology of paramount importance which permits placement of numerous virtual machines on physical servers through a hypervisor or a virtual machine monitor for enhanced resource utilization. Thus, virtualization offers virtual machine consolidation as its main benefit for minimizing energy consumption. Energy efficiency is enhanced by virtual machine placement on least number of hosts. The transfer of an active running virtual machine is addressed by dynamic virtual machine consolidation technology through virtualization [3]. An underutilized or over-utilized host is most appropriate beneficiary for virtual machine migration-related consolidation on servers. The process of virtual machine consolidation on a host by virtue of live migration can be detrimental for performance-related applications running by migrating a virtual machine [4]. The provision of an optimized quality-of-service (QoS) has to be assured between the users and providers, and it can be achieved by minimizing the number of migrations through dynamic virtual machine consolidation. The QoS-related SLA assures the needed performance levels with respect to throughput and highest system-related response time.

The bin-packing problem can be utilized while dealing with consolidation of virtual machines because it deals exclusively for minimizing the number of bins [5]. There are number of algorithms which reckon consolidation of virtual machines as a bin-packing problem. Although unnecessary migrations and enhanced violations exist along with their application, hence, the requirement is minimization of virtual machine migration and SLA violation by a suitable heuristic algorithm. The algorithm should perform two-phase-oriented dynamic virtual machine consolidation: Firstly by assuring migration from the host which is least loaded to a host which is having maximum load and secondly by migrating virtual machines from overloaded host or from the hosts which are almost overloaded. The migration policy is based on the availability of resource requirements presently or in future. The utilization of resource in future can be better predicted by k-nearest neighbour regression (K-NNR)-based model vis-à-vis linear regression-based model. The infrastructure-as-a-service (IaaS) cloud environments related with EC2 and Amazon give a basis for data generation by executing real workload. The prediction can be accurately titrated when it is evaluated against partial data.

The paper is organized is as follows:

- Section 2 describes the related work.
- Sections 3 and 4 explain the problem statement and system model respectively.
- Section 5 describes algorithmic approach for virtual machine consolidation.
- Section 6 puts forth the performance of the algorithm and the results obtained.

2 Related Work

For addressing specific workload-related SLA violation rate an algorithm pertaining to consolidation and dynamic migration for minimization of rate of SLA violation is required [6]. For minimizing the number of hosts needed to run a workload, the bin-packing heuristic along with techniques pertaining to time series forecasting is taken together. A near optimal solution can be addressed by system pertaining to ant colony optimization which undertakes both—the number of dormant hosts along with number of migrations [7]. It should also be considered that a 100% utilization of host-related CPU results into performance degradation; hence, attainment of threshold value has to be prevented. If the thresholds are set for static values, it also results into dynamic workload-related inefficiency for running different type of applications on a host. Thus, it is inferred that there should be an optimal workload-related optimization to confirm consolidation task for better efficiency. In order to maximise resource utilization and ensuring judicious cloud data centre-related load balancing, a multi-resource virtual machine placement algorithm is proposed [8].

A bin-packing problem pertains with adjusting numerous items in bins in such a way that bins are minimal in number. Thus, for addressing virtual machine consolidation problem, each virtual machine is identified as an item and a host is identified as a bin. A number of heuristic algorithms are proposed for bin-packing problem as it is NP-hard. Few most popular algorithms include best-fit algorithm which fulfils placement of an item into the bin which is best suited for it. The first-fit algorithm places the item into the first bin. The first-fit and the best-fit heuristics can be further fine-tuned by considering item-related specific order. These algorithms have to be modified to ensure the task of virtual machine consolidation before they are directly used.

The multi-dimensional resources are constituted by network bandwidth, memory or CPU which impart definite characteristics to virtual machines and hosts. If CPU and memory are considered together then the virtual machine consolidation is exhibited as a typical 2D bin-packing problem. If bin capacities are equal, then it constitutes a classical bin-packing problem, and it can be suitably modified by selecting different host-related bin sizes. The bin-packing problem simply addresses the minimization of bin number as its prime focus. But as a matter of fact the issue of minimizing the number of migrations and SLA violations has to be considered simultaneously with due importance. The best-fit decreasing algorithm modified version can address substantial energy conservation [4] while first-fit decreasing heuristic presents as cost aware for performing server consolidation-related migration. Another modified version of best-fit algorithm undertakes dynamic application placement utilizing a framework known as EnaCloud [9].The modified first-fit and best-fit algorithms execute through Sercon for minimization of host number along with maintaining their performance level [5].

3 Elucidation of Problem Statement

The optimization related to the virtual machine placement for minimizing number of active hosts along with highest incurred benefit is related to exercise virtual machine consolidation in an energy-efficient manner. This achieved benefit denotes the combination of two important factors which are SLA-related rate of violation and the number of virtual machine migration. The enhancement of these benefits in advance can be achieved on the basis of allocating virtual machines for the host with respect to their future resource utilization.

For addressing optimized virtual machine migration utilization prediction-aware best-fit-decreasing (UPBFD) strategy is taken into consideration for the following reasons. It delivers amicable solution pertaining to dynamic virtual machine consolidation problem as it identifies optimal trade-off related to SLA violations against energy savings. This UPBFD approach also minimizes SLA violations by addressing overloaded or nearly overloaded hosts by exercising migration of virtual machines. For the sake of minimizing the migration number, the host is selected on the basis of present and future resource requirements. The enhancement of pro-active consolidation of virtual machines is also addressed by UPBFD via utilizing a resource prediction model. It is achieved by utilizing K-NNR-enabled model [10]. The CPU utilization threshold forms the basis of monitoring energy consumption, SLA violation and number of migrations.

4 System Model

The system model comprises of heterogeneous hosts which are capable of accommodating virtual machines in a virtualized data centres through virtual machine migration. Each host has different resource specifications like network bandwidth, memory and CPU at a given instant. Numerous users are served simultaneously by cloud data centre subsequent to their requests for provisioning of virtual machines. The request length is distinguished by millions of instructions, and the performance of CPU is exhibited as million instructions per second. The architecture is configured by two parameters. The two parameters are, viz. (i) a master node harbouring a global agent and (ii) host-related full distribution of local agents. The iteration-related task sequence for tentative dynamic virtual machine consolidation elaborates that host-related resource utilization is monitored by each local agent which is predictive of host-related short-term CPU utilization. The virtual machine-related migration commands are addressed by general agents which are decisive for identifying the virtual machines that are required to be migrated from a specific host. The function of global agent (GA) is to analyse received data from local agent and then to formulate the migration plan with the help of proposed algorithm. The actual migration of virtual machines is addressed by virtual machine monitors (VMMs) after analysing commands received from GA.

5 Utilization Prediction-Aware Best-Fit Decreasing Algorithm

The BFD algorithm performs host-related allocation of virtual machines. Then, all virtual machines are identified by this algorithm on the basis of resource utilization requests in decreasing order. Each virtual machine is assigned to a host having smallest free capacity. The virtual machines related resource utilization is fluctuating due to dynamic nature of workloads. The algorithm addresses virtual machine consolidation reinforced virtual machine placement. To ensure it, dynamic virtual machine consolidation approach is undertaken through optimization of virtual machine placement on the basis of present and future resource requirements.

The migration of virtual machines from least to most loaded hosts designated as cold spots and hot spots, respectively, is exercised so as to minimize data centre-related energy consumption by releasing cold spots. The migration of virtual machines from cold spots observes a policy of all-or-none. It means that all virtual machines should be provided a host if they have to be transferred from the coldest spot; otherwise, transfer will not take place. It signifies that virtual machine placement occurs in accordance with proportionality values of CPU and memory-related resource usage. The host-related CPU and memory capacity are instrumental for ensuring host-related migration. The host load on the basis of CPU and memory utilization is represented as follows:

$$\text{HMemoLoad} = \frac{\text{HUsedMemo}}{\text{HTotalMemo}}$$

$$\text{HCPULoad} = \frac{\text{HUsedCPU}}{\text{HTotalCPU}}$$

$$\text{HLoad} = \text{HCPULoad} + \text{HMemoLoad}$$

where HMemoLoad is obtained by dividing HUsedMemo by HTotalMemo. HCPULoad is obtained by dividing HUsedCPU by HTotalCPU. Thus, by addition of HCPULoad and HMemoLoad, we obtain HLoad Table 1.

The decision of migrating a virtual machine from a host to other host is decided by UPBFD algorithm. Preferentially, it selects the highest loaded virtual machine for migration. It infers that larger virtual machines are more difficult to be placed on to other hosts. The virtual machine-related load is expressed as under:

$$\text{VMemoLoad} = \frac{\text{VUsedMemo}}{\text{VTotalMemo}}$$

$$\text{VCPULoad} = \frac{\text{VUsedCPU}}{\text{VTotalCPU}}$$

Table 1 Description of concepts and notations

Symbol	Description
VTotalCPU	Virtual machine V-related total CPU capacity
VUsedCPU	Virtual machine V-related used memory capacity
VTotalMemo	Virtual machine V-related total memory capacity
HCPULoad	Host H-related CPU load
HUsedCPU	Host H-related used memory capacity
VLoad	Virtual machine V -related load
HTotalMemo	Host H-related total memory capacity
HUsedMemo	Host H-related used memory capacity
HLoad	Host H-related load
VCPULoad	Virtual machine V-related CPU load
HTotalCPU	Host H-related total CPU capacity
HMemoLoad	Host H-related memory load
VPredictedCPUCapacity	Virtual machine V-related predicted used CPU capacity
VMemoLoad	Virtual machine V-related memory load
POH	Predicted overloaded host-related set
TCPUValue	Threshold value for CPU utilization
DestHost	Migrated virtual machine-related destination host
MPlan	Plan for migration
VMigrated	Migrated virtual machine
SourceHost	Migrated virtual machine allocation-related source host
VmMigrate	Set of virtual machines selected for migration on a host
OH	Overloaded host-related set
HPredictedCPUCapacity	Host H-related predicted used CPU capacity
VUsedMemo	Virtual machine V-related used memory capacity

$$\text{VLoad} = \text{VMemoLoad} + \text{VCPULoad}$$

where VCPULoad represents CPU utilization and VMemoLoad represents memory utilization. VUsedCPU and VUsedMemo are virtual machine-related CPU and memory utilization, respectively. VTotalMemo and VTotalCPU are the total consumption of memory and CPU, respectively, by the virtual machine.

The avoidance of SLA violations and needless migrations is observed by UPBFD algorithm by using two constraints. The allocation of destination host (DestHost) to a virtual machine is permitted by first constraint provided it has enough resources to cater allocating virtual machines. A risk of SLA violations emerges if some resources are utilized close to 100%. Hence, threshold value TCPUValue is recalled for limiting the quantity of CPU resource which is requested by a virtual machine on a host. The memory utilization is ignored at this stage since the quantum of memory with reference to virtual machine exceeds the physical memory pertaining to it. Hence, first constraint is represented as:

$$\begin{aligned}&\text{DestHost.HUsedCPU} + \text{VPredictedCPUCapacity} \leq \text{TCPUValue}\\&\quad \times \text{DestHost.HTotalCPU}\end{aligned}$$

The destination host (DestHost) overloading shortly after allocation of virtual machine is taken care of by second constraint. To ensure it, virtual machine and host-related CPU demands are addressed by UPBFD. For this purpose, prediction of CPU utilization with the help of training data set-related local average is utilized by K-NNR-based prediction model [11]. The locality is expressed as 'k' samples in the vicinity of new samples. For this purpose, CPU usage dataset is collected where each sample of dataset incorporates one input and one output variable. The CPU capacity for present time instance is represented by input variable, while CPU utilization for next instance of time is represented by output variable. As the output variable proves to be of continuous nature, the regression analysis is applicable. Apart from this, the dynamic workload results into short-term load prediction by UPBFD. For assigning virtual machine to host second capacity constraint is represented as:

$$\begin{aligned}&\text{DestHost.HPredictedCPUCapacity} + \text{VPredictedCPUCapacity} \leq \text{TCPUValue}\\&\quad \times \text{DestHost.HTotalCPU}\end{aligned}$$

where predicted CPU utilization of destination host is represented by DestHost.HPredictedCPUCapacity, while predicted CPU utilization of virtual machine is represented by VPredictedCPUCapacity. The total host capacity-related specified upper threshold TCPUValue bounds virtual machine and host-related aggregated predicted CPU utilization. For destination host (DestHost) and virtual machine-related total predicted utilization of a value, TCPUValue = 0.5 results into limitation of total host utilization below 50%. The selection of a destination host having sufficient available capacity at present time and the instance of near future for virtual machine allocation is executed on the basis of these two constraints. To secure a minimized level of SLA violation the UPBFD algorithm focuses to address virtual machines to be migrated from those hosts which are either overloaded or will become overloaded very soon. The host is reckoned as overloaded host set (OH member) if current CPU utilization becomes more than host capacity. It is therefore necessary to migrate few overloaded host-related virtual machines to other hosts for minimizing the incidence of SLA violation. The prediction is exercised by UPBFD on the basis of host being overloaded in reference with host-related predicted CPU utilization. When predicted utilization value exceeds available CPU capacity, the host is reckoned as a member of predicted overloaded hosts set (POH). This scenario provokes for migrating virtual machines from predicted overloaded host to avoid SLA violation.

Algorithm

```
1. MPlan = Ø
2. H←Sort all hosts in descending order of load HLoad
3. SourceHost←last host in H
4. VmMigrate←sort virtual machines on host SourceHost in descending load VLoad
5. for v ∈ VmMigrate do
6.       success = false
7.       for DestHost ∈ H - SourceHost do
8.             if {((DestHost.HUsedCPU + VUsedCPU) ≤ (TCPUValue × DestHost.HTotalCPU)) &
                   ((DestHost.HPredictedCPUCapacity) + VPredictedCPUCapacity) ≤ (TCPUValue ×
                                                          DestHost.HTotalCPU))} then
9.                  MPlan = MPlan ∪ {(SourceHost, VMigrated, DestHost)}
10.                Update (SourceHost.HUsedCPU) and (DestHost.HUsedCPU)
11.                success = true;
12.                break;
13.            end if
14.        end for
15. end for
16. if success = false then
17.    MPlan = Ø
18.    Recover SourceHost.HUsedCPU and DestHost.HUsedCPU
19.    else
20.      switch SourceHost to the sleep mode
21. end if
22. for SourceHost ∈ [OH, POH] do
23.      VmMigrate←sort VMs on SourceHost in descending load VLoad
24.      for VMigrated ∈ VmMigrate do
25.            if {(SourceHost.HUsedCPU ≥ SourceHost.HTotalCPU) ||
                          (SourceHost.HPredictedCPUCapacity ≥ SourceHost.HTotalCPU)} then
26.                for DestHost ∈ OH - [OH, POH] do
27.                  if {((DestHost.HUsedCPU + VUsedCPU) ≤ (TCPUValue × DestHost.HTotalCPU)) &
                         ((DestHost.HPredictedCPUCapacity + VPredictedCPUCapacity) ≤ (TCPUValue ×
                                                                   DestHost.HTotalCPU))} then
28.                          MPlan = MPlan ∪ {(SourceHost, VMigrated, DestHost)}
29.                         Update SourceHost.HUsedCPU and DestHost.HUsedCPU
30.                         break;
31.                     end if
32.                end for
33.              if {(HUsedCPU ≥ HTotalCPU) || (HPredictedCPUCapacity ≥ HTotalCPU)} then
34.                switch on a dormant host
35.              end if
36.                else
37.                break;
38.           end if
39.       end for
40. end for
```

First phase of algorithm: It incorporates energy conservation by migrating least loaded host related all virtual machines and switching the host in to sleep mode for energy saving. The strategy also addresses minimization of SLA violation by migrating few virtual machines from overloaded and predicted overloaded hosts. First phase pertains from lines 2 to 21 where according to decreasing order of host-related load levels is the basis for sorting the hosts, thus saving the results in list H.

For identifying the most suitable destination host for migrated virtual machines re-allocation, the process is started by addressing the host with highest load as the first load to be selected. If it is not possible, second one will be selected, and similarly, it proceeds further. A destination host of required capacity is identified for virtual machine allocation at this instance of time and near future. Algorithm if perceives value of success to be false, then it removes all migration plan-related tuples to recover the CPU capacity related with source and destination hosts. Otherwise, it provokes source host to proceed in to sleep mode after all virtual machines related to this host are migrated.

Second phase of algorithm: Here, decreasing order-based sorting of virtual machines is exercised. The basis of migration is the load level which commences the migration of those virtual machines in decreasing order which request for maximum capacity. When total host utilization is exceeded by current or predicted CPU utilization, then virtual machine migrates to other host. The migrated virtual machine is re-located onto an appropriate destination host by algorithm on the basis of two proposed constraints. If algorithm is unable to identify a suitable host for virtual machine allocation among the active hosts, then a dormant host is switched on. The ultimate output of UPBFD algorithm presents as a migration plan which provokes global agent for sending commands to virtual machine monitors on the basis of this migration plan.

6 Evaluation of Performance

6.1 Setup Related to Simulation

Various virtual machine consolidation techniques are exercised to address virtual machine dynamic consolidation approach using CloudSim toolkit [12]. On the basis of an assumption that 800 heterogeneous hosts exist in a data centre, two server configurations are utilized to perform evaluation, viz. HP ProLiant ML 110 G4 (Intel Xeon 3040, 2 cores x 1860 MHz, 4 GB) and HP ProLiant ML 110 G5 (Intel Xeon 3075, 2 cores x 2660 MHz, 4 GB). The server-related network bandwidth is 1 Gb/s. Random and real are two types of workloads which decide number of virtual machines. For addressing users request-based provisioning of virtual machines, the random workload pertains to 800 heterogeneous virtual machines where CPU-related utilization is addressed for running an application by each virtual machine. The real workload is CPU usage data which has been procured from more than a thousand virtual machines [13].

6.2 Metrics for Evaluation

The performance-related objectives focus to ensure that SLAs do not violate, the number of physical servers utilized is minimal, and least number of migrations takes place. It can be suitably addressed by certain metrics. For addressing SLA violation pertaining to IaaS infrastructure-related virtual machines, the proposed metric is SLAV metric which calibrates SLA violations resulting from over-utilization of hosts and migration of virtual machines. The performance degradation due to virtual machine migration and host overloading can be addressed as follows: SLAV = SLAVO × SLAVM. Here, SLAVO is indicative of SLA violations due to over-utilization and SLAVM pertains to SLA violations due to migrations.

Cent per cent CPU utilization is expressed as under: $\mathrm{SLAVO} = \frac{1}{M}\sum_{i=1}^{M}\frac{T_{s_i}}{T_{a_i}}$.

where M stands for number of physical machines; T_{s_i} is total time duration that a physical machine i incurs a cent per cent CPU utilization leading to SLA violation, T_{a_i} is total active state time of physical machine i.

The migration-related performance degradation is expressed as follows: SLAVM $= \frac{1}{N}\sum_{j=1}^{N}\frac{C_{d_j}}{C_{r_j}}$.

where N stands for virtual machines number, C_{d_j} reveals virtual machine j-related performance degradation due to migration, the lifetime virtual machine j-related total CPU capacity is represented as C_{r_j}.

The physical machines related energy consumption is dependent on CPU utilization, network card, disc and memory [4]. CPU consumes more power as compared to other mentioned devices; hence, physical machines related resource utilization is usually inferred as CPU utilization [14]. The prime objective is focussed on minimizing the number of migrations because live virtual machine migration involves many resource utilizing processes simultaneously, thus increasing the operation cost significantly. The factors like CPU processing on source physical machine, source and destination physical machines related link bandwidth, services related downtime for a migrating virtual machine and the total migration time are important parameters during live migration [5].

6.3 Benchmarks for Comparison

On the basis of following heuristic algorithms, the effectiveness of virtual machine consolidation performance is evaluated.

- The modified best-fit decreasing (MBFD) algorithm proceeds by assigning a virtual machine to a host with least free capacity. It ensures that virtual machine and host-related total used CPU capacity is less than total CPU utilization threshold. Expression is as under:

$$(\text{DestHost.HUsedCPU} + \text{VUsedCPU} \leq \text{TCPUValue} \times \text{DestHost.HTotalCPU})$$

- The modified first-fit decreasing (MFFD) algorithm ensures that allocation of virtual machine is made for first host which exhibits host and virtual machine-related CPU capacity to be less than total CPU utilization threshold as under:

$$(\text{DestHost.HUsedCPU} + \text{VUsedCPU} \leq \text{TCPUValue} \times \text{DestHost.HTotalCPU})$$

- Host utilization prediction-aware best-fit decreasing (HUP-BFD) algorithm ensures limitation of host utilization by monitoring threshold value T. Expression is as under:

$$(\text{DestHost.HPredictedCPUCapacity} + \text{VUsedCPU} \leq \text{TCPUValue} \\ \times \text{DestHost.HUsedCPU})$$

- Virtual machine utilization prediction-aware best-fit decreasing (VMUP-BFD) algorithm entertains virtual machine-related future requested utilization which is predicted by utilizing K-NNR. Expression is as under:

$$(\text{DestHost.HUsedCPU} + \text{VPredictedCPUCapacity} \leq \text{TCPUValue} \\ \times \text{DestHost.HUsedCPU})$$

6.4 Results

The identification of optimal threshold value is indispensable to address trade-off among the data centre-related evaluation metrics. A high CPU threshold value causes enhanced performance degradation along with number of migrations, while low threshold value causes increased energy consumption. For estimating threshold value-related impact on performance metrics, the threshold value ranges from 50 to 100% for evaluation purpose.

Figure 1a pertains to SLA violation levels by different algorithms against different threshold values. UPBFD algorithm ensures virtual machine allocation to such a host which is not overloaded presently or in near future. Therefore, it minimizes SLA violation percentage rate better as compared with other methods. Further violations of SLA are also taken care as BFD ensures virtual machine migration from overloaded or predicted overloaded hosts. Figure 2a depicts that minimization of active hosts number is achieved by packing virtual machines on most loaded hosts, thereby achieving 28.5% minimization of energy consumption. Figure 3a depicts random workload-related virtual machine consolidation for total number of virtual machine migrations.

Figure 1b is simulation-based depicting significantly less SLA-related violations by UPBFD as compared with other experiment-related benchmark algorithms. The

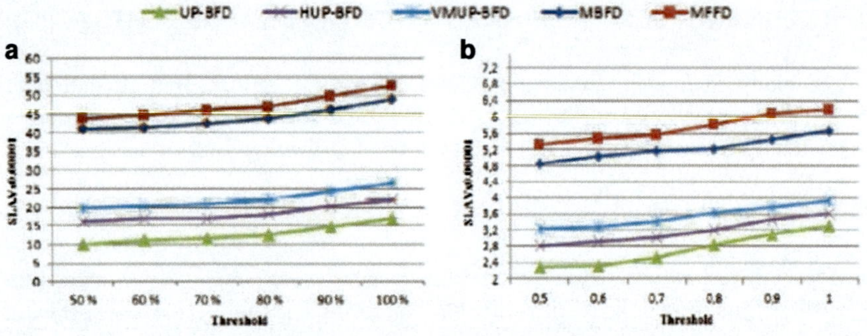

Fig. 1 **a** SLAV metric (random workload) and **b** SLAV metric (real workload)

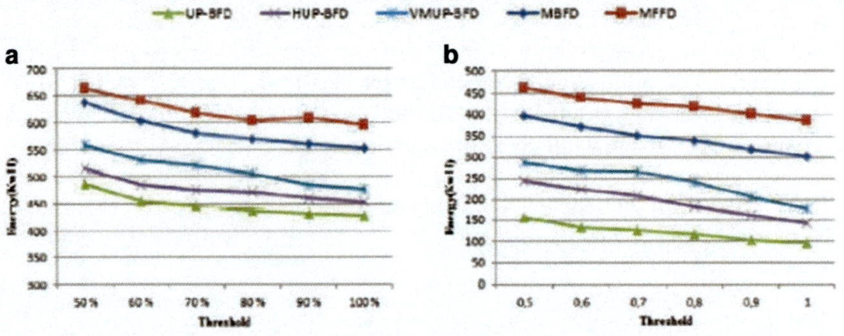

Fig. 2 **a** Energy consumption (random workload) and **b** energy consumption (real workload)

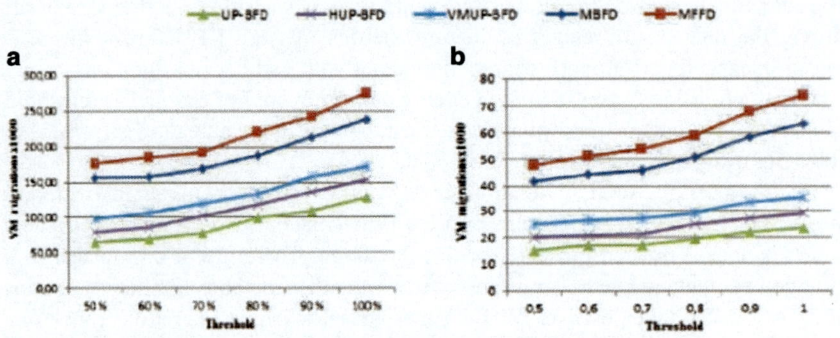

Fig. 3 **a** VM migrations (random workload) and **b** VM migrations (real workload)

UPBFD ensures that during virtual machine migrations overloading of destination host is prevented. Figure 2b signifies higher energy savings by UPBFD as compared with other approaches. It is secured by ensuring packing of virtual machines on least number of hosts resulting in release of least loaded hosts. Figure 3b depicts minimized number of migrations as compared with others. The reason is that on the basis of host-related virtual machine allocation the future resource utilization is considered.

7 Conclusion

Minimization of number of migrations and SLA violations both can be addressed if virtual machine consolidation method considers assigning of a virtual machine to a host on the basis of host-related future resource utilization. The decisions pertaining to virtual machine allocation and consolidation augment the concept of creating a green cloud computing plan. A periodic manoeuvring is exercised for consolidating virtual machines on a minimized number of physical machines which ensures turning-off the idle devices thus ensuring minimization of energy consumption. On the basis of dynamic virtual machine consolidation approach minimization of undue virtual machine migration is curtailed which as a consequence minimizes SLA violation by virtue of using utilization prediction model. This approach targets a migration plan to migrate a few virtual machines from currently overloaded or predicted to be overloaded hosts in near future. In accordance with present and future resource requirements, assigning of a virtual machine to a host is ensured. The algorithmic approach addressed here is potentially more effective for securing not only minimization in energy consumption but also simultaneously exhibit less SLA violations and virtual machine migrations, as compared to other algorithmic approaches.

Acknowledgements I am thankful to all concerned with this endeavour.

References

1. Kong, Y., Zhang, M., Ye, D.: A belief propagation-based method for task allocation in open and dynamic cloud environments. Knowl. Based Syst. **115**, 123–132 (2017). https://doi.org/10.1016/j.knosys.2016.10.016
2. Ghatikar, G., Ganti, V., Matson, N., Piette, M.A.: Demand response opportunities and enabling technologies for data centers: findings from field studies. Available Online at: https://escholarship.org/uc/item/7bh6n6kt (2012)
3. Clark, C., Fraser, K., Hand, S., Hansen, J.G., Jul, E., Limpach, C., Pratt, I., Warfield, A: Live migration of virtual machines. In: Proceeding of 2nd Conference on Symposium on Networked Systems Design and Implementation, pp. 273–286 (2005)
4. Beloglazov, A., Buyya, R.: Optimal online deterministic algorithms and adaptive heuristics for energy and performance efficient dynamic consolidation of virtual machines in cloud data centers. Concurrency Comput.: Pract. Exp. **24**(13), 1397–1420 (2012). https://doi.org/10.1002/cpe.1867
5. Murtazaev, A., Sangyoon, Oh.: Sercon: server consolidation algorithm using live migration of virtual machines for green computing. IETE Tech. Rev. **28**(3), 212–231 (2011). https://doi.org/10.4103/0256-4602.81230

6. Bobroff, N., Kochut, A., Beaty, K.: Dynamic placement of virtual machines for managing SLA violations. In: Proceeding of 10th IFIP/IEEE International Symposium on Integrated Network Management, pp. 119–128. https://doi.org/10.1109/INM.2007.374776 (2007)
7. Farahnakian, F., Ashraf, A., Pahikkala, T., Liljeberg, P., Plosila, J., Porres, I., Tenhunen, H.: Using ant colony system to consolidate VMs for green cloud computing. IEEE Trans. Serv. Comput. **8**(2), 187–198 (2015). https://doi.org/10.1109/TSC.2014.2382555
8. Hieu, N.T., Francesco, M.Di., Jaaski, A.Y.: A virtual machine placement algorithm for balanced resource utilization in cloud data centers. In: Proceeding of 7th IEEE International Conference on Cloud Computing, pp. 474–481. https://doi.org/10.1109/CLOUD.2014.70 (2014)
9. Li, B., Li, J., Huai. J., Wo, T., Li, Q., Zhong, L.: Enacloud: an energy-saving application live placement approach for cloud computing environments. In: Proceeding of 6th IEEE International Conference on Cloud Computing, pp. 17–24. https://doi.org/10.1109/CLOUD.2009.72 (2009)
10. Cover, T.M., Hart, P.E.: Nearest neighbor pattern classification. IEEE Trans. Inf. Theory **13**(1), 21–27 (1967). https://doi.org/10.1109/TIT.1967.1053964
11. Farahnakian, F., Pahikkala, T., Liljeberg, P., Plosila, J.: Energy aware consolidation algorithm based on k-nearest neighbor regression for cloud data centers. In: Proceeding of 6th IEEE/ACM International Conference on Utility and Cloud Computing (UCC), pp. 256–259. https://doi.org/10.1109/UCC.2013.51 (2013)
12. Calheiros, R.N., Ranjan, R., Beloglazov, A., De Rose, C.A.F, Buyya, R.: CloudSim a toolkit for modeling and simulation of cloud computing environments and evaluation of resource provisioning algorithms. Softw.: Pract. Exp. **41**(1), 23–50 (2011). https://doi.org/10.1002/spe.995
13. Park, K.S., Pai, V.S.: CoMon: a mostly-scalable monitoring system for PlanetLab. ACM SIGOPS Oper. Syst. Rev. **40**, 65–74 (2006). https://doi.org/10.1145/1113361.1113374
14. Kusic, D., Kephart, J.O., Hanson, J.E., Kandasamy, N., Jiang, G.: Power and performance management of virtualized computing environments via lookahead control. In: Proceeding of International Conference on Autonomic Computing, pp. 3–12. https://doi.org/10.1109/ICAC.2008.31 (2008)

An Impact of Radon Transforms and Filtering Techniques for Text Localization in Natural Scene Text Images

Y. L. Chaitra and R. Dinesh

Abstract Detection and localization of text in natural scene images have become a greater in need of image-based indexing, searching of images, and in many application areas. Scene text content conveys vital information. The effectiveness of the text localization depends on the efficiency in locating the text regions. In this paper, a detailed exploration of various filter techniques such as radon transform, conservative filter, Gabor filter, bilateral filter, and context-aware filters with maximally stable extremal regions as have been carried out. Maximally stable extremal regions (MSER) give many false positive rates to localize the text regions in the images, hence we have conducted experiments with radon transform and rule-based approach to remove the false positive rate and improve the precision, recall, and F-measure. Each filter has its own application-specific uses. The efficiency of the filtering techniques is examined on the basis of the results obtained for the text localization of scene images. Radon transform has been shown to achieve the best performance among the selected filtering techniques.

1 Introduction

In this digital era, introduction of inexpensive digital cameras and requirement of visual information has been grown enormously and made significant breakthrough in the society. Visual information from signboard text images and knowing the environment where the text was captured has become recent trend in text image processing. Recently, localization of text and recognition from text images have made more impact on many applications such as vehicle number localization from number plate, visual impaired person can use this information [1]. Major challenges in natural scene text localization are detecting text in cluttered background scene, fancy fonts, blur in text, text in different orientation, contrast is low, variety of range of text size, and multilingual text found in the scene. Hence, there is a need for efficient filtering techniques for robustly localizing text in the natural scene images. Work is

Y. L. Chaitra (✉) · R. Dinesh
Jain University, Bengaluru, Karnataka, India

T. Senjyu et al. (eds.), *ICT with Intelligent Applications*, Smart Innovation, Systems and Technologies 248, https://doi.org/10.1007/978-981-16-4177-0_55

framed as follows: 2 represents the prior work carried out text localization using filters and MSER, 3 explores on the proposed methodology, 4 gives detailed description of implementation and results obtained from proposed approach and comparative study, 5 gives the scope of the proposed method, and 6 points out the conclusion.

2 Prior Work

Natural scene text images offer valuable information about the meaning of the scene images and help people to understand the context of the scene images. Edge-based detection techniques detect edges first in the text images by performing canny edge, sobel, prewitt operators then text regions are extracted by applying skeleton functions. These techniques can give better results in natural images if edges are sharp. Neeraj gupt et al. [2] proposed a new integrated salience model with image segmentation and text detection approach to generate text saliency. To understand the text present in text scene images, region-based methods have been used widely [3–5]. Couprie et al. [6], proposed a component tree for efficient implementation of watershed segmentation. Pan et al. [7] built a confidence map for text in natural scene images with sliding window approach by using HOG and waldboost features. Wang et al. [8] performed end-to-end recognition of text using pictorial structures and random fern. Connected component-based methods are used for text localization [9–11] with exploring MSER [12] to recognize potential components of the filtered and cluster them to form aword. Stroke width transform [13] algorithms are used to identify the low intrastroke variance in the connected components [14]. Anurag et al. [15] proposed an intersection of MSER for multioriented text natural scene images. Basavaraju et al. [16] presented Gaussian mixture model to detect text present in arbitrary location and multilingual text images. Diaz-Escobar et al. [17] proposed phase-based regions and retrieval of character in natural scene text images.

3 Proposed Methodology

In this section, we discussed proposed methodology for localization of the text in standard dataset using radon transform and various filtering mechanisms. The proposed methodology is shown in Fig. 1:

The proposed methodology consists of following stages:

- Initially, we sharpened the image using a radon transform and different filtering techniques
- Extract the detailed components from the sharpened image
- Apply MSER to detect candidate text regions
- Rule-based approach is used to remove regions which are not text
- Finally, connected component method is used to localize text component

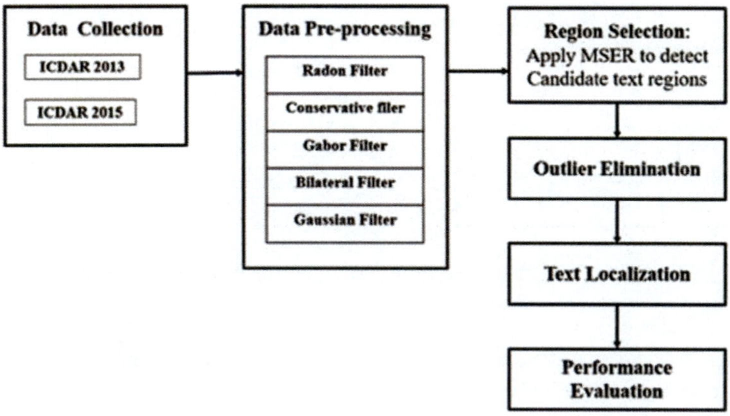

Fig. 1 Proposed methodology

- Evaluate proposed method using performance metrics

We collected different natural scene text images datasets such as ICDAR 2013 and ICDAR 2015. Data preprocessing is performed using different filtering techniques. Image filtering plays a crucial role in eliminating non-text regions in text localization process. In this work, various filtering techniques such as radon transform, conservative, Gaussian, bilateral, Gabor filters are used for sharpening and edge preserving purpose. In the next subsection, we briefly explained various filtering techniques experimented in this work.

3.1 Radon Transform

The radon transform (RT) recognizes the lines present in the text images, which also includes tilted lines present in arbitrary angles in horizontal and vertical. RT operator drives the plots from spatial domain to projection domain (α, t), in which the directions coincides and charts to straight line in the spatial domain. On diverging, each coordinator of the spatial domain is represented by sine wave in the projection domain. The integral function of RT can be defined as follows [18].

$$p(s, \alpha) = \int f(x, y)\delta(x \cos \alpha + y \sin \alpha - s)\mathrm{d}x\mathrm{d}y \tag{1}$$

where p is the RT of $f(x, y)$ and δ is a function to store delta values which converts 2D integral into a line of integral. The angle of line is represented by alpha, p is an sinogram, and it is a sine wave when $f(x, y)$ is value at coordinator. RT represents an unknown density and gives finite 1D projection. It uses sinogram function, which

122	124	125	129	139
121	123	125	126	134
117	119	149	124	133
118	114	118	122	132
110	115	109	119	129

Neighborhood values:

- 114, 118, 119, 122, 123, 124, 125, 126, 149
- max : 126 min : 114
- Center intensity : 149 > max : 126
- Therefore Center intensity = 126

Fig. 2 Example of conservative filtering calculation

is a linear transform of the original image. In radon, density foreground is good, so that text portion is highlighted.

3.2 *Conservative Filter*

Conservative filter (CF) is an reduction of noise filter technique, and it applies simple and fast filtering technique that determines noise power suppression in order to retain high-frequency spatial details (e.g., sharp edges) in a text images. It is exclusively created to eliminate the noise spikes that is pixels which are isolated exceptionally in low and high intensity in pixels. It is less effective for eliminating additive noise from the image. CF finds the minimum intensity and maximum intensity within a pixel present in neighbor. If the center pixel intensity is greater than maximum range, then it is substituted by the maximum value. If it is less than minimum range, then it will be substituted by minimum range [19]. The example of applying CF is shown in Fig. 2. CF deals with smoothing of images in small quantity of spiky noise. This technique can use for removal of short range of variability in a image for smoothing the image. But it is less efficient in eliminating additive noise (e.g., Gaussian noise) from the image.

3.3 *Gabor Filter*

Gabor filter (GF) works with different orientation and frequencies in different directions, which is used for localization of text and extraction of text regions only. GF has the excellent localization property in spatial and frequency domain and used for texture analysis, edge detection, and feature extraction. Gabor transform determines the higher-order singularities at optimal resolution [20]. The two-dimensional function of Gabor $G(x, y)$ is expressed as follows:

$$g(x, y, \lambda, \theta, \Psi, \sigma, \gamma) = \exp\left(-\frac{x'^2 + \gamma^2 y'^2}{2\sigma^2}\right)\exp\left(i\left(2\pi\frac{x'}{\lambda} + \Psi\right)\right) \quad (2)$$

In this equation, x and y are the pixel coordinators, λ is the wavelength of the sinusoidal factor, θ represents the direction of the normal to the parallel stripes of a GF, phase offset is Ψ, σ represents the sigma/standard deviation of the Gaussian envelope, and γ represents the spatial aspect ratio and defines the elasticity of the support of the GF. But, GF is time-consuming due to its dimension of feature vector is very long and cannot handle well a text embedded in various background.

3.4 Bilateral Filter

Bilateral filter (BF) is defined as nonlinear technique that is able to blur an image during the retaining of strong edges. The BF is also representing as weighted average of neighbor pixels similar to Gaussian convolution [21]. The idea behind the BF technique is that to a pixel to impact on another pixel, it is not only to inhabit neighbor but also similar value [22]. The BF equation is defined by

$$\mathrm{BF}[I]_p = \frac{1}{W_p} \sum_{q \epsilon S} G_{\sigma_s}(\|p - q\|) G_{\sigma_r}\left(|I_p - I_q\right) I_q \tag{3}$$

where W_p is a normalization factor and ensures pixel weights sum to 1.0:

$$Wp = \textstyle\sum_{q \epsilon s} G_{\sigma_s}\,(\|p - q\|)\, G_{\sigma_r}\left(|I_p - I_q|\right)$$

Parameters σ_s and σ_r will determine the amount of filtering for the image I. where G_{σ_s} represents a spatial Gaussian weighting, that decreases the influence of far pixels, G_{σ_r} represents a range Gaussian, that decreases the influence of pixels q when their intensity values differ from I_p. BF maintains the edges but not the gradients (directional change in the intensity or color in an image).

3.5 Gaussian Filter

Gaussian filter is known as a non-uniform low-pass filter used to sharpen the image [23]. GF equation is given below:

$$f(x|\mu, \sigma^2) = \frac{1}{\sqrt{2\pi\sigma^2}} e^{-\frac{(x-\mu)^2}{2\sigma^2}} \tag{4}$$

where σ stands for the standard deviation of the distribution,σ^2 is the variance, and μ is the mean. The smoothing is determined by σ, larger the σ, more intensity smoothing. Hence, it is effective in sharpening of images. But, it loses fine image detail and contrast, so that it fails to examine fine detail which would be a text in image and does not handle "salt and pepper noise" well.

3.6 Region Selection

In this work, the maximally stable extreme regions (MSER) are used to determine stable regions. Text regions are considered to be more important than other regions in a text image. MSER, extracts number of covariant regions and determines the stability of the regions [24]. Intensity difference of the extremal region can be described as follows.

$$v\left(R_x\right) = \frac{|R(+\Delta) - R|}{|R|} \tag{5}$$

where $|R|$ represents the area of the extremal region R, $R(+\Delta)$ specifies the extremal region, +Δ levels up which contains R, and $|R(+\Delta) - R|$ determines the area difference of the two regions. The importance of the MSER delta plays a significant role in measuring the stability of the regions. Delta value will be changed from 1 to 20, depending on the complexity of the text image. If complexity is less in text images, delta value need to initialize maximum, and if complexity increases, delta value need to reduced. We conducted experiment by varying the delta value from 1 to 20 to find the optimal value for complex text images. Following steps are carried out in MSER:

- Vary the intensity threshold from the black to white by performing the luminance of the threshold in text image.
- Connected components are extracted ("extreme regions")
- Determine the threshold when connected components are maximally stable that is local minimal value of the relative growth and its square. The image may be of a discrete nature, the region could be below or above, or it may even coincide with the real text region, in each case the region is considered maximal.
- Ellipse can be used to approximate the extreme regions (it is an optional step)
- Features will be kept with region descriptors.

3.7 Outlier Elimination

After region detection, we have applied rule-based approach using stroke width transform and geometric properties to eliminate the non-text regions in the images.

- Stroke width transform (SWT): Evaluates the width of the lines and curves to define the character. Text regions may have less stroke width variation, where non-text regions tend to have more differences. The stroke pixels are sharper than the background pixels. Group neighbor pixels with similar stroke width together to find letter candidates.
- Geometric properties: Several geometrical properties are good for distinguishing between text and non-text area including eccentricity, extent, aspect ratio, and solidity and Euler number. Threshold value is used to remove non-text areas. The connected component approach determines the dissimilarity between the text and

the background to identify the connected character and combine it to obtain the text components.

Once the natural scene text images are captured, the next task is to locate the text in it. Text localization is a rather difficult process due to factors such as no advance information on text position, text size, orientation, context uncertainty, and coping with character-like non-characters. Researchers are working on these reasons to find an effective method. In this work, we identified text regions using rule-based approach with MSER method and localized using bounding box. The pixel locations from text bounding box can be used to retrieve text from the natural scene images. To get good localization in the proposed method, the delta value is set between 1 and 20 and depends on the complexity of the images.

4 Implementation and Results

To evaluate the performance of proposed radon transform and various filter techniques, implementation is carried out in MATLAB 16 on 3.1 GHz with 8 GB RAM. The proposed approach is illustrated using very standard and benchmark datasets such as ICDAR 2013 and 2015. The ICDAR 2013 dataset of text localization tasks consist of 462 challenging images which have 229 training set images and 233 testing set images. It is challenge 2, which is mainly focused on reading text in real scenes. In this dataset, most of the texts are horizontal with different size, fonts, colors, and complex backgrounds. The size of the image will vary from one image to another (480*640 to 3888*2592). The camera model used is Canon power shot S20. The image contains only challenging English texts. The ICDAR 2015 dataset of text localization task consists of 1500 challenging images which have 1000 training set images which contains 4500 readable words and 500 testing set images. It is challenge 4, which is mainly focused on incidental scene text. In this dataset, most of the texts are horizontal, vertical, and multioriented, blurred, occluded and very small with different size, fonts, colors, and complex backgrounds. The images are taken from Google Glasses without taking care of its quality image. The size of the image is 1280*720. The image contains only challenging English texts. Metrics used to evaluate the performance of the proposed random transform and other filtering techniques are precision, recall, and f1-score, and these can be determined using the formulas:

$$\text{Precision} = \frac{\text{TP}}{\text{TP} + \text{FP}},\ \text{Recall} = \frac{\text{TP}}{\text{TP} + \text{FN}},\ F = \frac{2 * \text{Precision} * \text{Recall}}{\text{Precision} + \text{Recall}} \quad (6)$$

Where, true positive (TP): text detected correctly, true negative (TN): total text non-detected as text, false positive (FP): non-text detected as text, and false negative (FN): not having text but detected with text. Figure 3 shows results obtained using radon transform + MSER + geometric properties + SWT for ICDAR 2013 and 2015. Figure 4 shows results obtained using conservative filtering + MSER + geometric

Fig. 3 Text retrieval from the input image using radon transform **a** ICDAR 2013 (original image and text detected) and **b** ICDAR 2015 (original image and text detected for complex image)

Fig. 4 Text retrieval from the input image using conservative filter (original image and text/non-text detected)

Fig. 5 Text retrieval from the input image using Gabor filter (original image and text detected)

Fig. 6 Text retrieval from the input image using bilateral filter (original image and text detected)

properties + SWT. Figure 5 shows result obtained using Gabor filter + MSER + SWT. Figure 6 shows results obtained using bilateral filter + MSER + SWT. Figure 7 shows results obtained using Gaussian filter + MSER + SWT for ICDAR 2013 dataset. Table 1 shows the comparison of the proposed method with existing methods. Figure 8 shows the example results on ICDAR 2013 benchmark datasets. Figure 9 shows the graphical representation of the performance analysis.

Fig. 7 Text retrieval from the input image using Gaussian filter (original image and text detected)

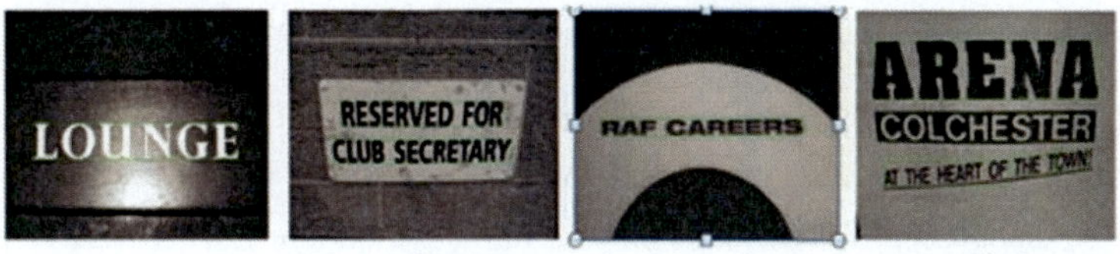

a b c d

Fig. 8 Some example results of proposed method on ICDAR2013 benchmark datasets. **a** Non-uniform illumination. **b** Near horizontal text. **c** Horizontal text. **d** Horizontal and non-horizontal text

Table 1 Comparative review of the proposed approach with the current method for ICDAR 2013

Methods	Recall	Precision	F-score
Soni et al. [25]	68	84	75
Guan et al. [26]	67	82	74
Proposed method	74.27	82.23	78.68

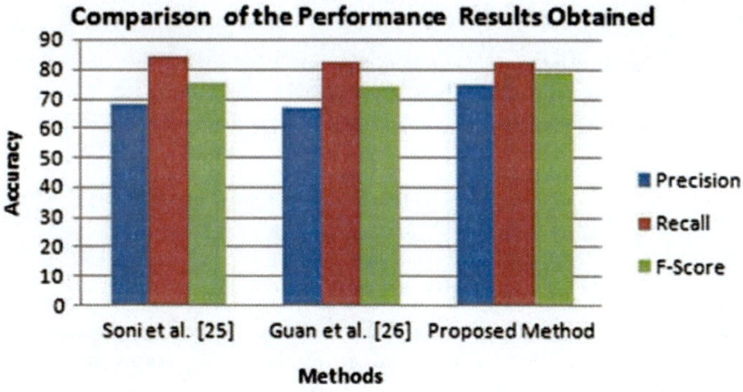

Fig. 9 Performance comparison of obtained results from proposed model with existing methods

5 Scope of the Proposed Method

The findings of the experiment shown in Table 1 indicate that 100 accurate text localization could not be achieved from the proposed process. The summary of it is given below:

- It cannot detect all the texts present in the dense images, there are few false positives and false negatives are observed.
- Results are not accurate when the texts are not in a horizontal manner or different fonts or different text orientations.
- ICDAR2015 images are not clear and objects are fairly distant from the camera, so it is not able to detect the complex text.

6 Conclusion

Scene text localization is a difficult process, since there is no prior understanding of the location, size, direction, layout, and colour of text, and there are also patterns and textures which are close to characters. In this work, proposed radon transform explored different filtering techniques, namely conservative filter, Gabor filter, bilateral filter, and context-aware filters for localization of text in natural scene text. As per the literature survey of MSER, the proposed method is effective since it works well for ICDAR 2013 dataset. Until recently, there are no methods to evaluate precision, recall, and F-measure at line level for the ICDAR 2015 dataset and the proposed analysis is the first of its kind, hence no comparison is possible. The performance of the filtering techniques is evaluated based on the results obtained for text localization of scene images. Radon transform has been shown to deliver the best result among the chosen filtering techniques for text localization in images.

References

1. Zarechensky, M.: Text detection in natural scene with the multilingual text. In: Proceedings of the 10th Spring Researcher's Colloquium on Database and Information Systems, Veliky Novgorods (2014)
2. Gupta, N., Anand, S.J.: A robust model for salient text detection in natural scene images using MSER feature detector and Grabcut. Multimedia Tools Appl. **8** (2019)
3. Matas, J., Chum, O., Urban, M., Pajdla, T.: Robust wide baseline stereo from maximally stable extremal regions. In: Proceedings of British Machine Vision Conference, pp. 384–396 (2002)
4. Fraundorfer, F., Bischof, H.: A novel performance evaluation method of local detectors on non-planar scenes. In: Workshop Proceedings of Conference on Computer Vision and Pattern Recognition (2005)
5. Mikolajczyk, K., Schmid, C.: A performance evaluation of local descriptors. In: IEEE Transactions on Pattern Analysis and Machine Intelligence, pp. 1615–1630 (2005)
6. Couprie, M., Najman, L., Bertrand, G.: Quasi-linear algorithms for the topological watershed. J. Math. Imaging Vision **22**(2–3), 231–249 (2005)

7. Pan, Y.F., Hou, X., Liu, C.L.: Text localization in natural scene images based on conditional random field. In: Proceedings of ICDAR (2009)
8. Wang, K., Babenko, B., Belongie, S.: End-to-end scene text recognition. In: Proceedings of ICCV (2011)
9. Koo, H.I., Kim, D.H.: Scene text detection via connected component clustering and non-text filtering. IEEE Trans. Image Process. 2296–2305 (2013)
10. Li, Y., Jia, W., Shen, C., Hengel, A.: Characterness: an indicator of text in the wild. IEEE Trans. IP (2014)
11. Neumann, L., Matas, J.: A method for text localization and recognition in real-world images. In: Proceedings of ACCV (2011)
12. Epshtein, B., Ofek, E., Wexler, Y.: Detecting text in natural scenes with stroke width transform. In: Proceedings of CVPR (2010)
13. Matas, J., Chum, O., Urban, M., Pajdla, T.: Robust wide baseline stereo from maximally stable extremal regions. In: Proceedings of BMVC (2002)
14. Yao, C., Bai, X., Liu, W., Ma, Y.: Detecting texts of arbitrary orientations in natural images. In: Proceedings of CVPR (2010)
15. Agrahari, A., Ghosh, R.: Multi-oriented text detection in natural scene images based on the inter-section of MSER with the locally binarized image. Proc. Comput. Sci. **171**, 322–330 (2020)
16. Basavaraju, H.T., Aradhya, V.N.M., Pavithra, M.S., Guru, D.S., Bhateja, V.: Arbitrary oriented multilingual text detection and segmentation using level set and Gaussian mixture model. Evolutionary Intelligence, Springer (2020)
17. Diaz-Escobar, J., Kober, V.: Natural scene text detection and segmentation using phase-based re-gions and character retrieval. Math. Prob. Eng. 1–17 (2020)
18. Helgason, S., Helgason, S.: The radon transforms, vol. 2. Birkhäuser, Boston (1980)
19. Hall, M.: Smooth operator: smoothing seismic horizons and attributes. The Leading Edge **26**(1), 16–20 (2007)
20. Mehrotra, R., Namuduri, K.R., Ranganathan, N.: Gabor filter-based edge detection. Pattern Recogn. **25**(12) (1992)
21. Tomasi, C., Manduchi, R.: Bilateral filtering for gray and color images. In: Proceedings of 6th International Conferences Computer Vision, New Delhi, India, pp. 839–846 (1998)
22. Elad, M.: On the origin of the bilateral filter and ways to improve it. IEEE Trans. Image Process **11**(10), 1141–1151 (2002)
23. Young, I.T., Van Vliet, L.J.: Recursive implementation of the Gaussian filter. Signal Process. **44**(2), 139–151 (1995)
24. Matas, J., Chum, O., Urban, M., Pajdla, T.: Robust wide baseline stereo from maximally stable extremal regions. In: British Machine Vision Conference, pp. 384–393 (2002)
25. Soni, R., Kumar, B., Chand, S.: Text detection and localization in natural scene images based on text awareness score. Appl. Intell. (2018)
26. Guan, L., Chu, J.: Natural scene text detection based on SWT, MSER and candidate classification. In: 2nd International Conference on Image, Vision and Computing (ICIVC) (2017)

Algorithmic Approach to Select MOOCs Platform: An Institutional Perspective

Shweta Brahmbhatt and Devarshi Mehta

Abstract During COVID-19 pandemic, learning without four walls, i.e. virtual learning, played a vital role in the education system. Massive open online courses (MOOCs) became one of the most effective academic warriors. It is always a challenge to the institutions when some new platform or learning culture needs to be identified. As per survey, Internet users in India are growing at a rate of 14%. Its high time for institutions to provide an alternate or additional environment of learning, i.e. e-Learning. This manuscript focussed on selection of MOOCs platform from the view of institutions. It explored global MOOCs providers such as Edx, Coursera, Udacity & Course Builder and its parameters with respect to institutions. The authors identified indicators of MOOCs platform through which any institution can take the decision to implement the e-Learning environment in their campus. A tree algorithm (ID3) used for the same which helped to take the decision on MOOCs platform. The proposed algorithm took account of every question relating to each institute-centric indicators and provided a final assessment as per the institute preference. The author has implemented the decision tree algorithm on the four important institutional-centric core indicators/parameters which are courses, learner-centric platform, multimedia enrichment into content and administrative indicators (certificate and financial).

1 Introduction

e-Learning is a flexible method for online education and the alternate way to transfer knowledge. The well-known e-learning platforms like coursera, udacity, Edx and many more are available to provide online education without physical presence, i.e. four walls of class room. This way of e-learning methods (MOOCs) is very popular and used by people more and more nowadays. An online Web-based platform which has unlimited enrolment numbers, no mandatory qualifications, fully structured aimed at unlimited participation and open access. Anyone can access it anytime from anywhere is a basic philosophy of MOOCs.

S. Brahmbhatt (✉) · D. Mehta
GLS University, Ahmedabad, Gujarat, India

T. Senjyu et al. (eds.), *ICT with Intelligent Applications*, Smart Innovation, Systems and Technologies 248, https://doi.org/10.1007/978-981-16-4177-0_56

There are three stakeholders of MOOCs, namely learner, course creator, and admin. Each stakeholder has their own role and responsibilities. All the processes over MOOCs happened online such as registration, course development, offering and adaptation, study engagement, assessments, assignment, exam andresult.

There are several indicators which are being incorporated in different MOOCs platforms. Some platforms are good in specific features such as coursera has a very good subject catalogue where Edx is very good in diversity of courses. Similarly, other platforms are performing well in different domains.

MOOCs platforms have different features including some common and few unique features. It is very tough for any institute to take the decision regarding the adaptation of MOOCs platform. The institute has their own constraints such as learners experience, IT infrastructure, teacher's experience and financial implications. The best way to choose the platform is MOOCs indicators/parameters.

2 Review Literature

Since there are not much technical documents and research literature available in terms of algorithms on indicators of MOOCs. Abhishek and Pankaj's studies described the features and functionalities of major MOOCs platforms. The comparative study, done by the authors, also explained how students can choose the course. Further, indicators are not only made for learners but also for the institutions.

Shweta and Devarshi described the different indicators of MOOCs platform which would help learner to adopt the course. Authors also revealed that adaptation and offering of course is purely depended on different indicators of the MOOCs platform.

3 Objectives

The objectives of the study are as follows:

- Identification of MOOCs parameters (Institutional context)
- Features of the platform based on its indicators
- Requirement-based traversing of parametric tree
- Identification of best suitable MOOCs platform.

4 Identification of MOOCs Parameters

One of the major important factors for the implementation of MOOCs is its parameters, i.e. what are the important parameters are available in MOOCs platform with respect to institutions. There are following broader categories of parameters identified in the view of the institutional prospective. Being an administration, it is very important that how a platform is feasible in terms of all aspects such as technical feasibility, economical feasibility and operational feasibility. The broader category is learning experience and technology/framework. There are following indicators which affect the broader parameter.

- Course type: whether courses are paid or free or both?
- mentor support medium;
- social network connection;
- assessment tools;
- discussion group;
- peer connect functionality availability.

On another side under technology and framework parameter indicators are:

- Multilingual support for course and Website;
- Responsive website with mobile app;
- Openness of technology;
- financial section verified assessment;

5 Algorithmic Approach of MOOCs Platform (Institute-Centric)

The selection, on basis of the above-mentioned core indicators of institutional-centric MOOCs platform can be achieved by the decision tree algorithm. The implementation of the same is proposed as follows:

- Classification of the collected data has been carried out and trends the dataset of the MOOCs platforms parameters. Develop classification attribute data that is Table R. (H: High, M: Medium, L: Low, Y-Yes, N–No)
- There are three types of attributes. Those are nominal, ordinal and binary. For providing the estimated values depending on different parameters, we need to take account of every question relating to each parameter and give a final estimate according to the user preference of the parameters. By having an attribute with the smallest information gain as the root node and further algorithm continues to recurs on each subset, considering only attributes never selected before. ID3 makes sure that the entire decision tree is traversed and every parameter is taken into consideration.

- Calculate the entropy as per the following formula.

$$H(X) = -\sum_{i=1}^{n} p(x_i) \log_b p(x_i)$$

Output attribute is platform = P (6/16, 4/16, 4/16, 2/16).

$$\text{HP}(r) = -(6/16 \log(6/16) + 4/16 \log(4/16) + 4/16 \log(4/16) + 2/16 \log(2/16) = 0.5736.$$

- For each attribute *R*, information gain has been calculated. (Trend dataset), (Figs. 1, 2, 3, 4, 5.

$$\begin{aligned} H(\text{Target learner}) &= (7/16)\, H(\text{Academic}),\ (9/16)\, H(T\ \text{Professional}) \\ &= 7/16(-(3/7 \log(3/7) + 2/7 \log(2/7) + 2/7 \log(2/7)\ )) \\ &\quad + 9/16(-(3/\log(3/9) + 2/9 \log(2/9) + 2/9 \log(2/9) \\ &\quad + 2/9 \log(2/9))) = \mathbf{0.53942758625} \end{aligned}$$

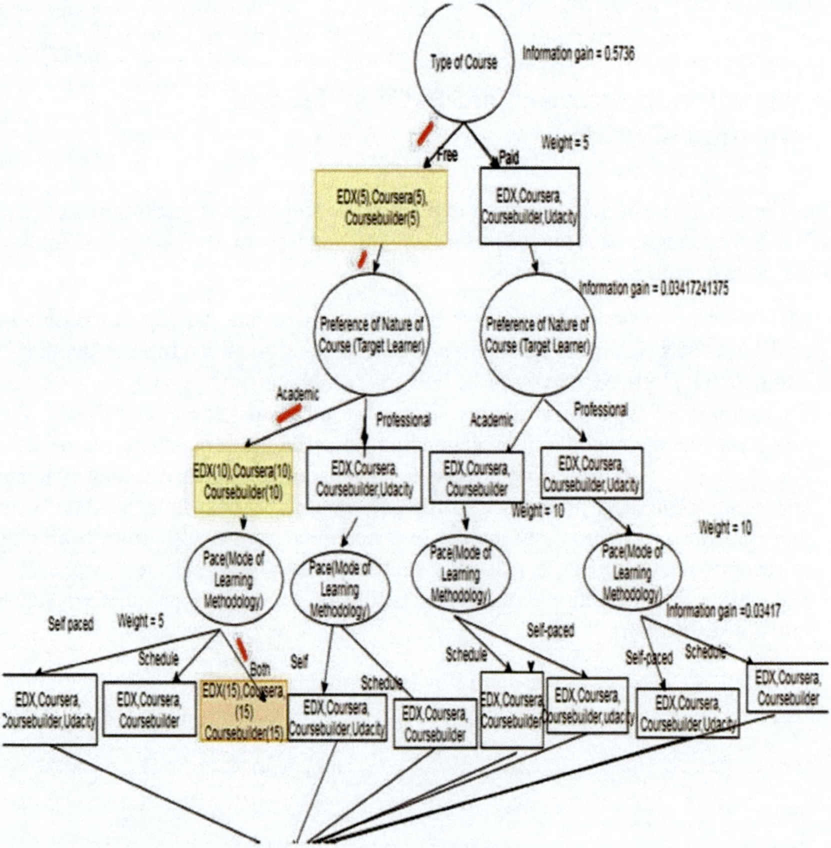

Fig. 1 Traverse tree on type of course and mode of learning

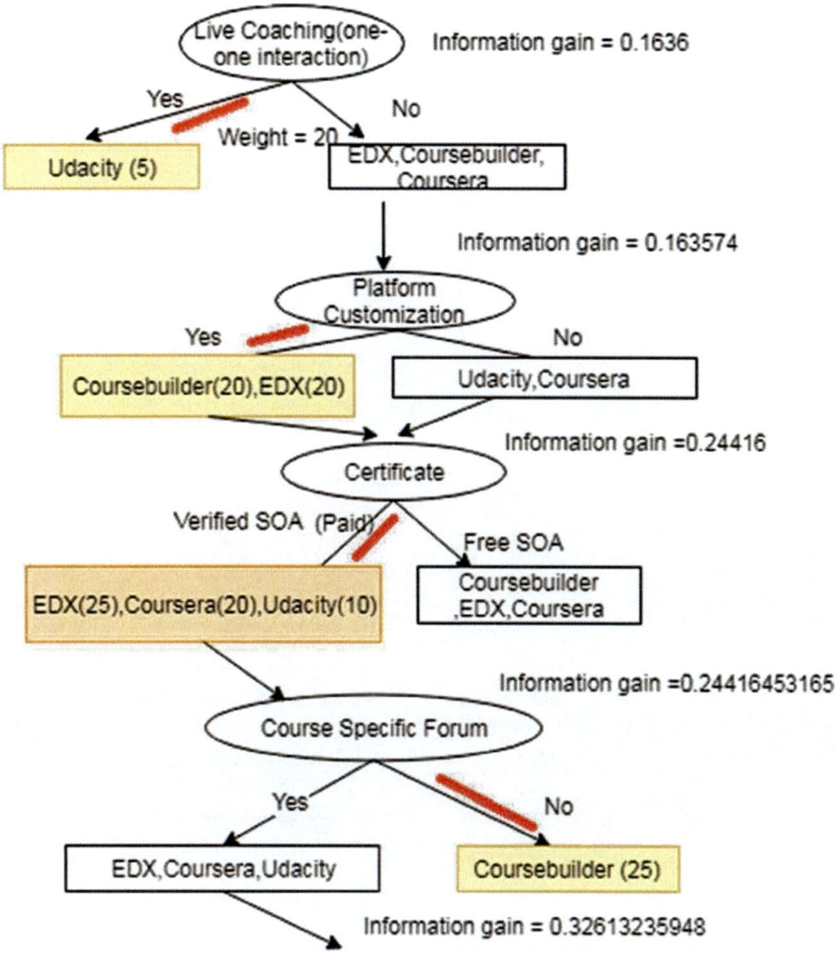

Fig. 2 Traverse tree on live coaching, platform customization, certificate and course specific forum

$$\text{Gain}(T, A) = \text{Entropy}(T) - \sum_{v \in \text{Values}(A)} \frac{|T_v|}{|T|} \text{Entropy}(T_v)$$

$$\text{Information gain} = H(T) - H(X, T)\ \text{gain}(\text{course}, T)$$
$$0.5736 - 0.5395 = \mathbf{0.03417241375}$$

- Split the set attributes into subsets using which resulting entropy is minimum or information gain is maximum/minimum. Steps need to be repeated for all attributes.
- Ordinal attributes (quantitative scale).

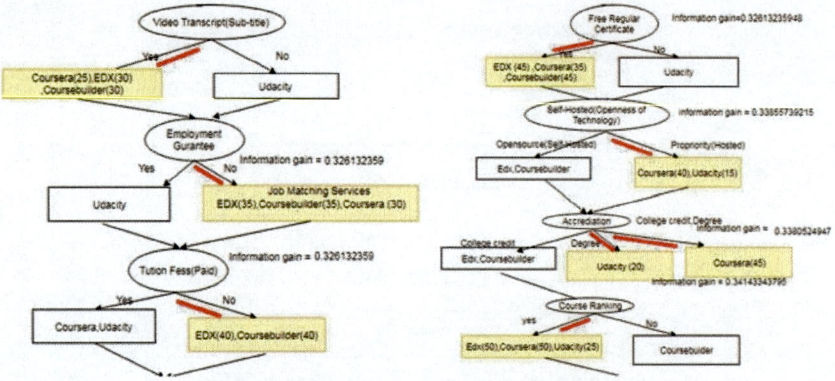

Fig. 3 Traverse tree on free regular certificate, self-hosting, accreditation

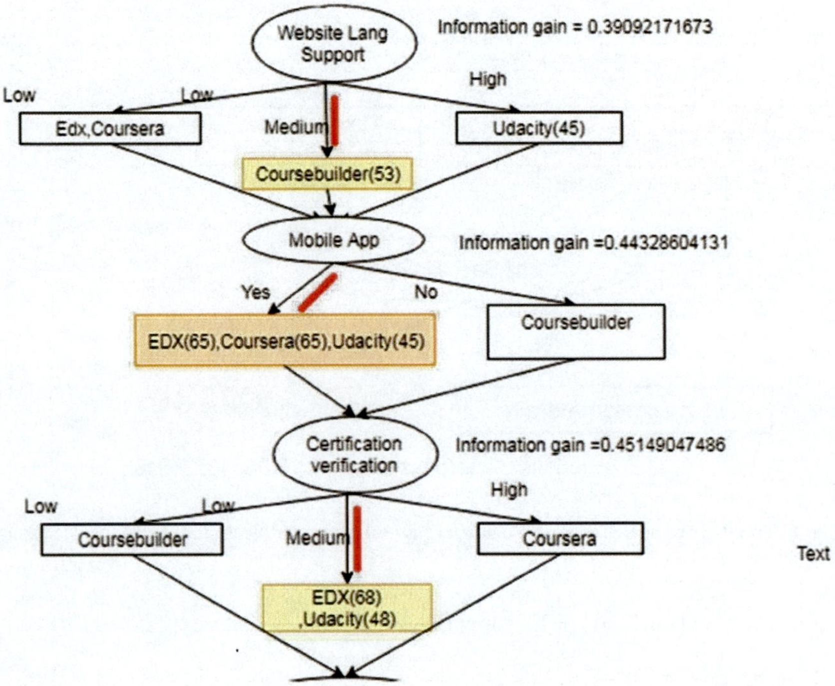

Fig. 4 Traverse tree on Website language support, mobile app, certification verification

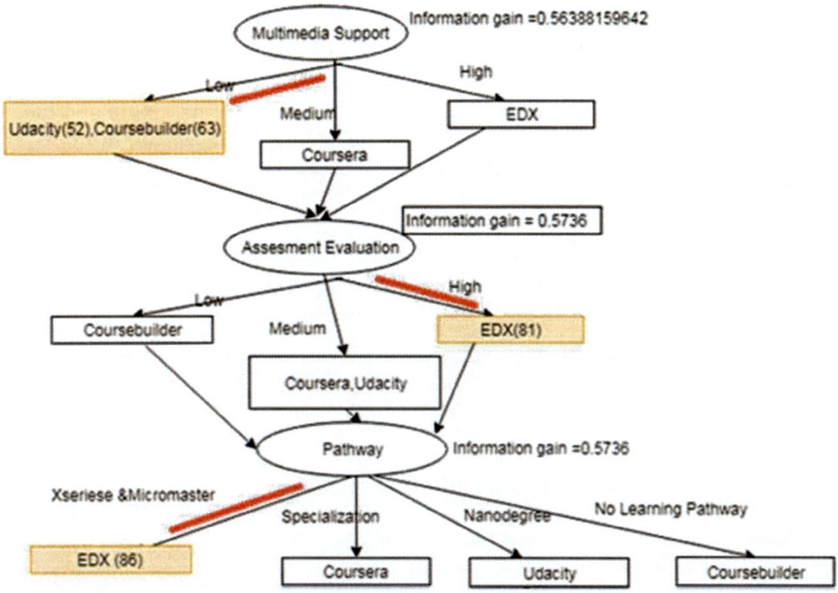

Fig. 5 Traverse tree on multimedia support, assessment evaluation, pathway

Nominal attributes quantitative scale $W(F) = 5$ Pt || binary attributes quantitative scale $W(F) =$ Yes: 5Pt, No: 0Pt.

- Change weight in to percentage with the following equation

$$F(X) = \text{Weight}(X)^{*}100/\text{Total Weight of Questions}$$
$$||\text{Maximum weight factor } W(F) = 5$$

6 Conclusion

It has been found that there are several indicators which influence the institutions to take the decision on deployment/selection of MOOCs platform for their institution. In other words, these parameters help institutions to make their own decision on the basis of their requirement and feasibility. Further, research reveals that all MOOCs platforms, mentioned in this manuscript, have competitive features among themselves. However, Edx and coursera appear in initial nodes of the tree which reflects that these platforms are providing broader aspects of MOOCs features. It gives results on the basis of institutional point of view and choice. If institutes need more diversity in courses then coursera platform is good, similarly if institute has technical expertise and faculties (having e-learning experience), then coursebuilder is perfect for them. This article is very useful for those institutions who are planning to implement the MOOCs platform for their institution.

The limitation of this study is the fast-growing cutting-edge technology and development team of each platform; i.e., all platforms are trying to add more and more features in their platform. So, the result of this study may be changed in future, if any upgradation happens in the platforms.

References

1. Menon, S., Banerjee, G.: Evaluating effectiveness of a teacher training MOOC: industry perspective. In: 2019 IEEE Tenth International Conference on Technology for Education (T4E), Goa, India, pp. 102–105. https://doi.org/10.1109/T4E.2019.00-42
2. Menon, S., Banerjee, G.: Evaluating effectiveness of a teacher training MOOC: industry perspective. In: 2019 IEEE Tenth International Conference on Technology for Education (T4E), Goa, India, pp. 102–105. (2019). https://doi.org/10.1109/T4E.2019.00-42
3. Sanz-Martínez, L., Er, E., Dimitriadis, Y., Martínez-Monés, A., Bote-Lorenzo, M.L.: Supporting teachers in the design and implementation of group formation policies in MOOCs: a case study. J. Univ. Comput. Sci. **24**(8), 1110–1130 (2018)
4. Kashyap, A., Nayak, A.: Different machine learning models to predict dropouts in MOOCs. In: 2018 International Conference on Advances in Computing, Communications and Informatics, ICACCI 2018, pp. 80–85. https://doi.org/10.1109/ICACCI.2018.8554547
5. Kumar, A., Brahmbhatt, S.N.: Massive open online courses (MOOCs), features and platforms. In: Library and Information Science in the Age of MOOCs (pp. 83–99). IGI Global, (2018). https://doi.org/10.4018/978-1-5225-5146-1.ch005
6. Cobos, R., Olmos, L.: A learning analytics tool for predictive modeling of dropout and certificate acquisition on MOOCs for professional learning. In: 2018 IEEE International Conference on Industrial Engineering and Engineering Management (IEEM), Bangkok, pp. 1533–1537. (2018). https://doi.org/10.1109/IEEM.2018.8607541
7. Kumar, A., Mittal, P., Brahmbhatt, S.N.: MOOCs parameters: a way forward to identify best MOOCs platform. In: Library and Information Science in the Age of MOOCs, pp. 132–145. IGI Global (2018). https://doi.org/10.4018/978-1-5225-5146-1.ch008
8. Laveti, R.N., Kuppili, S., Ch, J., Pal, S.N., Babu, N.S.C.: Implementation of learning analytics framework for MOOCs using state-of-the-art in-memory computing. In: 2017 5th National Conference on E-Learning and E-Learning Technologies (ELELTECH), pp. 1–6. (2017). https://doi.org/10.1109/ELELTECH.2017.8074997

Decentralised Content Creation in Digital Learning: A Blockchain Concept

Saroj Koul, Sudhanshu Singh, and Rakesh Verma

Abstract In today's times, when the whole world is practising social distancing, the utility of in-person training is going through an existential dilemma. The learning ecosystem needs to rapidly evolve into fulfilling the global needs of learning and retooling to stay relevant and use new opportunities. The existing models of content creation and learning generally assume that a single faculty has to create all modules with topics and lessons both included. This article proposes a conceptual model for applying blockchain in decentralised content creation. Here, education is taken up as a developmental case. While much more content is now available online for learning than ever before, the quality of the content at topic and lesson levels is being observed to have a wide variation in quality resulting in inconsistent delivery of knowledge. The blocks are made of modular knowledge pieces consisting of lessons that can be appended to the initial module/topic creator (s) tightly integrated with a permissioned blockchain with nodes consisting of co-creators, reviewers, approvers and other interested parties.

1 Introduction

The landscape for digital education has changed significantly during the past few years [1]. An e-learning platform integrates creating, organising, delivering, communicating, collaborating and assessing learning [2].

The speed of adopting e-learning platforms accelerated during the past few months—as billions of people have been forced to stay at home or have restricted movements due to the COVID-19 pandemic. Massive open online courses (MOOC)

S. Koul (✉)
Jindal Global Business School, OP Jindal Global University, NCR Delhi, India
e-mail: skoul@jgu.edu.in

S. Singh · R. Verma
Analytics and Data Science, NITIE, Mumbai, India
e-mail: rakeshverma@nitie.ac.in

T. Senjyu et al. (eds.), *ICT with Intelligent Applications*, Smart Innovation, Systems and Technologies 248, https://doi.org/10.1007/978-981-16-4177-0_57

are on overdrive to accelerate the worldwide adoption of their courses. The developments in digital learning courses [1] have also involved virtual reality environments that integrate active, experiential and constructivist learning and provide useful insight about how students interact [3]. Different industry sectors have unique learning needs and have been producing learning solutions based upon the industry context. The smart contract contains the properties of being autonomous, self-sufficient and decentralised [4].

2 Literature Review

Digital learning involves all types of learning that utilises the digital environment. The importance of learning throughout life, regardless of age, is nudging many people to go for e-learning [5]. E-learning shares constructivism with problem-based learning (PBL) and supports PBL [6]. Other than PBL, e-learning also can be integrated into several other methods of learning, such as lectures and projects. Virtual learning environments move students from being a passive learner to an active learner [7]. The advantage of e-learning is that the learning process is free from the constraints of both space and time [2].

Blockchain first came to the fore in 2008 as part of cryptocurrency [8]. Blockchain as a technology can bring richness in developing and delivering content for various types of online learning. Not all content creators are useful in all the modules/topics/lessons, and learners would prefer to learn from creators who provide them with better quality content and learning.

Blockchain is a technology with a peer-to-peer (P2P) platform [9] having a fully distributed ledger. A block in blockchain may have transactions made by many users [8]. The smart contract is a decentralised software application [9] that operates on a platform such as Etherium and may be used with digital units and with a definite value. New blocks are added at the end of the previous block, and with each addition, the length of blockchain increases [8]. Every block in a blockchain [9] has block header, minimum one transaction, signed by the validators of the block and with a link to the previous block.

The crucial components [10] of a blockchain are digital assets, smart contracts, wallets, membership, signatures and events. A blockchain's emphasis is on being secure, private and scalable [9] and, therefore, requires management of issues around these aspects. The blockchain's initial focus is of avoiding the centralised systems [11] and moving towards working in a decentralised system. The smart contracts function by removing the "man-in-the-middle" and utilise algorithms [11] to deliver services automatically. The power of smart contracts comes from the thing that they can be string complete [4].

Blockchain platforms have features [10] that consist of consensus algorithms and protocols that facilitate user creation and information management. The "SHA-256 algorithm" creates a unique hash in a blockchain network, and all hashes have a link to the previous hash [12]. A block stores transactions—as a data structure [13] and each block in a blockchain may have several transactions made by many users [8]. A blockchain can be permissioned or be permission less [13]. Blockchain has been used in the micro-accreditation to meet the hiring requirements of employers [14]. A practical and safe, decentralised ecosystems are developed for educational materials for sharing using blockchain [15].

While blockchain is relatively a new concept, an innovation methodology called design thinking has shaped up over the last half-century with wide-ranging applications in business, management and engineering [16]. According to Tim Brown, CEO/IDEO, design thinking is "a discipline that uses the designer's sensibility and methods to match people's needs with what is technologically feasible and what a viable business strategy can convert into customer value and market opportunity" [16]. The planning and decision-making process in design thinking requires people from different fields [17]. Blockchain can enable design thinking by enabling the participation of key stakeholders across space and time.

In this article, blockchain and design thinking concepts have been applied to develop a conceptual framework for an e-learning system.

3 Proposed Conceptual Framework and Method

An e-learning system should facilitate an environment of self-paced learning and evaluation at the individual level [5]. Smart contracts in e-learning could provide automatic confirmation of accomplishing the learning modules after having completed the modules' standardised test [4]. Online teaching methods are said to be more feasible in both logical and financial sense as compared to the in-person training [18].

Courses can be thought of as modules in this article. Every course follows some development path starting with some idea, course structure, content and digitisation plan for making them available online to learners. Figure 1 depicts a simple conceptual framework, wherein a creator develops a module (or course) containing topics and lessons as per the requirements. These are generally made by a single contributor and are completed in all respects before being approved in entirety and placed on a digital learning platform. Any updates are only added to enhance the value and can only be carried out by the creator (s). This framework assumes that the creator presents all the topics and lessons with the same degree of knowledge and competence. The additions are linearly within the topics or lessons and assume the same requirements of all learners.

By adopting blockchain in the content development process, one can bring in the best creators for developing the topics and lessons. Any other (approved) creator (s) can add some customised topics and lessons selected by learners as per their learning and any language-specific requirements.

Figure 2 shows a conceptual model for creating and using educational knowledge in a permissioned blockchain environment, wherein the existing and pre-approved creators can create and add existing knowledge in any hierarchical level of a module,

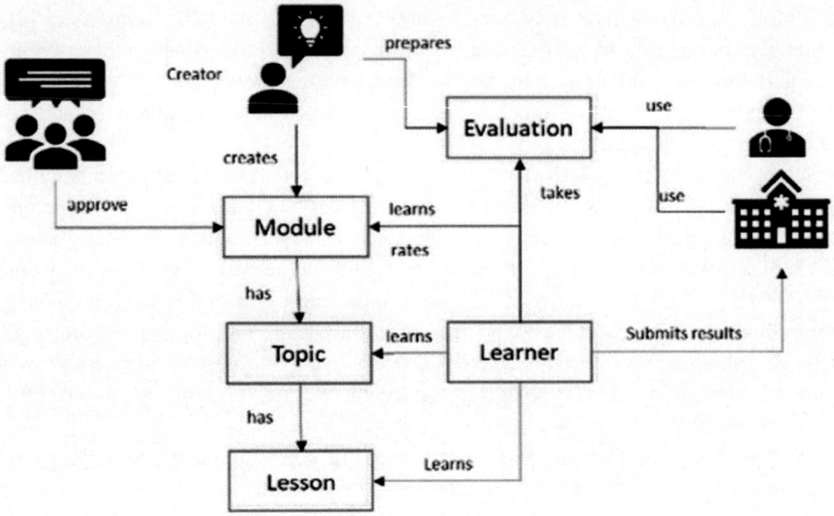

Fig. 1 Basic conceptual framework of content creation and digital learning. Authors view

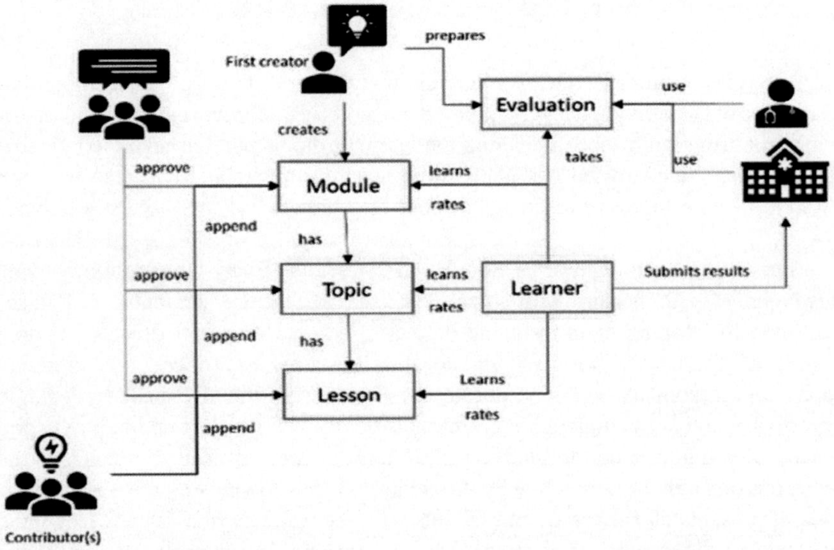

Fig. 2 Blockchain-based conceptual framework for decentralised content creation and digital learning system

topic, or a lesson. The decision for giving read and write rights to individual peers is provided by a centralised management system under the permissioned blockchain [13].

The approvers at the nodes approve the structure, content and evaluation methods. Learners share their evaluations with individuals/entities, wherein they want themselves to be assessed.

This method proposed for a permissioned blockchain, wherein the creators of knowledge (faculty/domain expert) and users of knowledge (students) are known in advance, involves the following steps:

(1) ***Step 1***
Creator(s) is/are approved for the course-module under development from a panel of experts added as nodes in the permissioned blockchain. Once the permission is granted, the creator can start developing content (including evaluation tools) and submit the same for review. These digital assets are linked to blocks at the level of module, topic and lesson so that the first creator gets the credit for his/her development.
A block is created for every e-learning module/topic/lesson that includes the link to the content, evaluation method (test), creator (s) of knowledge and reward units (initially zero). These are recorded in a digital ledger as transactions. Any lesson completed is made available to the learners and other creators to build upon content creation for any further specific needs or learning requirements.

(2) ***Step 2***
Interested creators browse through the knowledge repository (block) and submit their interest in developing the module/topic/lesson they propose to integrate with the previous block. They can only submit to the topic and lesson level as the module block is already finalised by the first creator (s).
The creators who add new blocks are also experts in the related area, and few of them may be from different geographic zones and speak different languages. They can contribute to making the world-class content for being made available to different learner segments with unique needs. For examples, some learners would seek more examples or additional quizzes or ask for availability in a new language, to make some ideas clear to them.

(3) ***Step 3***
The approval nodes review the submissions and decide to link the new content as a module/topic/lesson, if found acceptable. The approvers may be from a panel of global experts in that particular field and get rewarded for moderating the content. Some mechanisms for rewards may be developed for these approvers. There are many ways to contribute, such as adding a new module, adding a topic to the existing module or adding a new lesson to the existing topic.
The new additions are accepted by a majority, which may be decided by the blockchain's approving nodes. If approved, a new block is added with a hash to the previous block and stored in a distributed environment.

(4) ***Step 4***
The learners, who may have subscribed to the upcoming course, select the notification informing him/her of the course that is available and start the learning process. The moment a learner starts learning, the first block is created. In some cases, the learning path can only be started once a consensus is reached on the minimum knowledge requirements. A consensus is required for every block in the module/ topic/lesson.

(5) ***Step 5***
Learners consume the module/topic/lesson and submit their ratings. Each creator of a lesson visualises a lesson specific rating and gets points based on its rating level. These rating points also affect future permissions to develop new topics and lessons in the module.
The knowledge creator (s) gathers reward points for their contribution and gets updated in their reward units. Similarly, the approvers get rewarded for their moderation process and can be converted into monetary units.

(6) ***Step 6***
After a learner has completed the course, it is visible to the seekers (evaluators) for deciding on the learner's progress. These seekers may be individuals or organisations who have joined as nodes (under the centralised management system) to seek the best people for the module, and their participation is influenced by the quality of resource persons or experts among the creators or approvers at the category of nodes. The learners are also provided a functionality of sharing their work links, thus demonstrating their accomplishment in the form of performance reports (say for job appointments and internal evaluations).

4 The Setting

The institute considered in the study has many teachers who have been teaching in the supply chain management (SCM) domain over the years. As with any course, there is no standardised way of teaching SCM. Even while teaching the same syllabus, the teachers differ in the quality and their level of proficiency on the various topics and lessons under each module (or course). As per the author's experience, this diversity of expertise at granular levels is seen to impact the teaching effectiveness with respect to the qualitative and quantitative understanding of the topics and associated lessons by the students. Since the blockchain technology was found suitable to meet the requirements of learning at lesson level, it was decided that blockchain concept be applied for the experiment.

A case of learning in a module of SCM concentration is shown for demonstrating the utility of adopting blockchain in the areas of e-Learning. An SCM module contains several topics, each of which contains several lessons. In this approach, the lesson is considered as the lowest level of understanding in a course. Thus, the teachers were surveyed and mapped to their expertise at the level of a lesson. Lessons

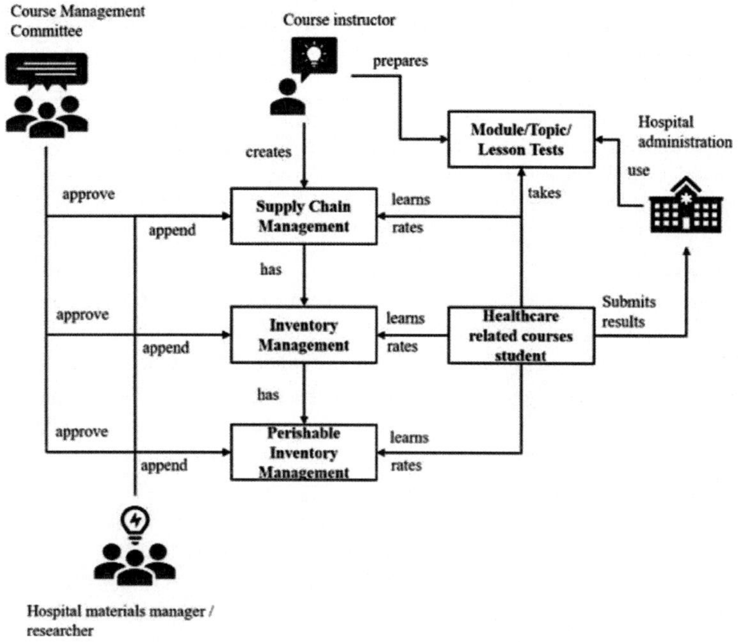

Fig. 3 Pictorial description of the lesson set up in a course module

to be developed under the topic of inventory management were allocated to the faculty who taught inventory models. In another case, a faculty was given specific lessons under the procurement management topic, covering the qualitative aspects only.

Figure 3 pictorially describes the 'perishable inventory management' lesson taken under the topic of 'inventory management' in a course module on 'supply chain management'. The adoption of blockchain shall ensure that students are able to get the best of current practices from the field (from hospital materials manager) along with the theoretical knowledge (from healthcare researchers) required under the 'inventory management' topic and 'perishable inventory management' lesson.

In another case of learning, a faculty was given specific lessons under the 'procurement management' module, covering the qualitative aspects only. Herein, 'supplier management' is the topic and 'supplier selection' is a lesson.

Learners (students) are expected to acquire the high-quality knowledge created by existing creators (teachers/domain experts) right up to the level of lesson.

5 Conclusion

The paper is first of its kind in utilising blockchain in developing content for e-learning. The decentralised digital learning system (DDLS) creates new opportunities and brings the experts together in a permissioned blockchain network with emphasis on selecting the best creator (s) for a lesson (s) using design thinking at the lowermost level.

Hence, the attempt was to re-engineer the mechanism for knowledge creation and dissemination, utilising design thinking and blockchain concepts. By building the concepts of rating at the lesson level, the faculty would be nudged to only prepare the content which they know best and meets the expectations of the institute.

The motivation for using blockchain [9] is data integrity, interoperability, data privacy, data authentication and data security. Policymakers decide on the requirements [10] of blockchain which could be private, public or federated.

Blockchain has an advantage over other technology [12] because it provides distributed ledger, consensus mechanism, provenance, immutability, smart contract and finality. Social inclusion is promoted when learners are provided with opportunities for learning that are useful and free [19]. The application areas are still evolving, and there is a long way to achieve maturity in usage, considering that there are many stakeholders [9] working in a complex environment.

Online teaching methods are said to be more feasible in both logical and financial sense as compared to the in-person training [18]. The attractiveness of web-based training lies in the fact that it can be delivered cost-effectively to comprehensive participants [20]. The SCM domain is considered as both a course and a module in this paper. Future researchers may explore picking up different modules under a particular course and create content utilising blockchain.

Challenges in adopting blockchain [9] are related to standardisation, interoperability, blockchain size storage requirement, latency, scalability, privacy, security, limited computing power and throughput restrictions.

Acknowledgements The authors thank the faculties and students who participated in the discussion for this experiment for building the conceptual framework.

References

1. Douglas, K.A., Merzdorf, H.E., Hicks, N.M., Sarfraz, M.I., Bermel, P.: Challenges to assessing motivation in MOOC learners: an application of an argument-based approach. Comput. Educ. **150**, 103829 (2020). https://doi.org/10.1016/j.compedu.2020.103829
2. Nurhayati, A., Bandung, Y.: (2019) Design of elearning platform for adult learner. In: 2019 International Conference of Advanced Informatics: Concepts, Theory and Applications (ICAICTA) (pp. 1–6). IEEE. https://doi.org/10.1109/ICAICTA.2019.8904243
3. Santamaría-Bonfil, G., Ibáñez, M.B., Pérez-Ramírez, M., Arroyo-Figueroa, G., Martínez-Álvarez, F.: Learning analytics for student modeling in virtual reality training systems: lineworkers case. Comput. Educ. **151**, 103871 (2020). https://doi.org/10.1016/j.compedu.2020.103871
4. Lam, T.Y., Dongol, B.: A blockchain-enabled e-learning platform. In: Interactive Learning Environments, pp. 1–23 (2020). https://doi.org/10.1080/10494820.2020.1716022

5. Martínez-Torres, M.R., Toral, S.L., Barrero, F.: Identification of the design variables of eLearning tools. Interact. Comput. **23**(3), 279–288 (2011). https://doi.org/10.1016/j.intcom.2011.04.004
6. Docherty, C., Hoy, D., Topp, H., Trinder, K.: eLearning techniques supporting problem-based learning in clinical simulation. Int. J. Med. Informatics **74**(7–8), 527–533 (2005). https://doi.org/10.1016/j.ijmedinf.2005.03.009
7. Ryan, E., Poole, C.: Impact of virtual learning environment on students' satisfaction, engagement, recall, and retention. J. Med. Imag. Radiat. Sci **50**(3), 408–415 (2019). https://doi.org/10.1016/j.jmir.2019.04.005
8. Drosatos, G., Kaldoudi, E.: Blockchain applications in the biomedical domain: a scoping review. Comput. Struct. Biotechnol. J. **17**, 229–240 (2019). https://doi.org/10.1016/j.csbj.2019.01.010
9. Hussien, H.M., Yasin, S.M., Udzir, S.N.I., Zaidan, A.A., Zaidan, B.B.: A systematic review for enabling of develop a blockchain technology in healthcare application: taxonomy, substantially analysis, motivations, challenges, recommendations and future direction. J. Med. Syst. **43**(10), 320 (2019). https://doi.org/10.1007/s10916-019-1445-8
10. Khezr, S., Moniruzzaman, M., Yassine, A., Benlamri, R.: Blockchain technology in healthcare: a comprehensive review and directions for future research. Appl. Sci. **9**(9), 1736 (2019). https://doi.org/10.3390/app9091736
11. Qiu, J., Liang, X. Shetty, S., Bowden, D.: Towards secure and smart healthcare in smart cities using blockchain. In: 2018 IEEE International Smart Cities Conference (ISC2), pp. 1–4. Kansas City, MO, USA (2018)
12. Tanwar, S., Parekh, K., Evans, R.: Blockchain-based electronic healthcare record system for healthcare 4.0 applications. J. Inf. Sec. Appl. **50**, 102407 (2020). https://doi.org/10.1016/j.jisa.2019.102407
13. Wust, K., Gervais, A.: Do you need a Blockchain? pp. 45–54. https://doi.org/10.1109/CVCBT.2018.00011
14. Zaccagni, Z., Paul, A., Dantu, R.: Micro-Accreditation for Matching Employer E-Hire Needs, pp. 347–352. https://doi.org/10.1109/Blockchain.2019.00054
15. Matzutt, R, Pennekamp, J., Wehrle, K.: A Secure and Practical Decentralised Ecosystem for Shareable Education Material, pp. 529–534. https://doi.org/10.1109/ICOIN48656.2020.9016478
16. Farrar, E.J.: Implementing a design thinking project in a biomedical instrumentation course. IEEE Trans. Educ. **63**(4), 240–245 (2020). https://doi.org/10.1109/TE.2020.2975558
17. Crites, K., Rye, E.: Innovating language curriculum design through design thinking: a case study of a blended learning course at a Colombian university. System **94**, 102334 (2020). https://doi.org/10.1016/j.system.2020.102334
18. Gavarkovs, A.G., Blunt, W., Petrella, R.J.: A protocol for designing online training to support the implementation of community-based interventions. Eval Prog Plan **72**, 77–87 (2019). https://doi.org/10.1016/j.evalprogplan.2018.10.013
19. Lambert, S.R.: Do MOOCs contribute to student equity and social inclusion? A systematic review 2014–18. Comput. Educ. **145**, 103693 (2020). https://doi.org/10.1016/j.compedu.2019.103693
20. Pham, D., et al.: (2016) A multidisciplinary evaluation of a web-based eLearning training programme for SAFRON II (TROG 1301): a multicentre randomised study of stereotactic radiotherapy for lung metastases. Clin. Oncol. **28**(9), 101–108. https://doi.org/10.1016/j.clon.2016.03.005

Data Security in Internet of Things

T. P. Rani and K. Suganthi

Abstract IoT has proliferated into wide applications due to the inevitable use of WSN. Security in IoT is an emerging area of research and still much work needs to be done in this area. The proposed system aims at providing security in data transfer of clustered IoT. Clustering reduces communication overhead, and hence, it is widely used when combined with data aggregation in IoT. The system studies the performance of Elgammal and Diffie–Hellman in data security. The system employs Elgammal method for encryption and Signature scheme for authentication. It is compared to existing Diffie–Hellman scheme. It is found that the proposed system overcomes the drawback of man-in-the-middle attack in existing system. The system is implemented in COOJA emulator. As energy serves as the main challenge in IoT, the energy cost is compared with existing systems.

1 Introduction

An ubiquitous framework, the Internet of things (IoT), is the result of evolution in smart networks. Sensors, sensor nodes, and peripheral devices communicate through Internet and collaboratively yield value added services in IoT. Any devices can be connected to the other and they achieve automation in respective fields yielding services. Hence, wireless sensor network (WSN) is the basic atomic unit of IoT. The varied applications of IoT range from critical applications like healthcare and banking systems, monitoring systems in industry, home, structure and traffic, waste management, logistics to the highly sophisticated and complex smart cities [1–3]. The widespread use of WSN has evolved from their working environment challenges. WSN nodes are resource restricted in the usage of energy. This is because of minimal

T. P. Rani
Department of Information Technology, Sri Sairam Engineering College, Chennai, Tamilnadu, India
e-mail: rani.it@sairam.edu.in

K. Suganthi (✉)
School of Electronics Engineering, Vellore Institute of Technology, Chennai, Tamilnadu, India
e-mail: suganthi.k@vit.ac.in

T. Senjyu et al. (eds.), *ICT with Intelligent Applications*, Smart Innovation, Systems and Technologies 248, https://doi.org/10.1007/978-981-16-4177-0_58

battery power in WSN nodes. This constraint has lead to low memory, communication, and computation capability in WSN [4, 5]. Clustering protocols [6, 7] enable minimal communication in WSN. In clustered networks, nodes are grouped into clusters and each cluster has a cluster head. The cluster member nodes send data or any information to the cluster head. The message is propagated by the cluster head to the sink node or the base station. Hence, using clustering mechanism, aggregated data can be sent from cluster heads, instead of individual node communications. Securing the cluster communication in terms of data integrity, authentication, and confidentiality in a resource constrained network is a challenging task. The working principle of clustering protocol makes it suitable in IoT networks for applications such as logistics, healthcare management, traffic management, and industrial monitoring systems where group communications can minimize the generated network traffic in systems. The nodes or sensors need to be in an open Internet network for communication purpose. The principle of accessing remote systems reduces communication distance. But it poses the nodes to be vulnerable to security threats.

2 Related Works

IoT has emerged due to the proliferation of WSN. As data is required to be transferred between systems for storage and requirement, data security is an alarming issue in IoT. Security in IoT is a challenging task due to their open nature. Security in IoT is an upcoming field of research and much work are to be done in this area.

In group key establishment for multicast WSN [8], security is achieved by establishing key exchange between the group nodes using Diffie–Hellman elliptic curve cryptography. The system aims at establishing security but suffers from the drawback that for each message to be sent, key exchange is required and incurs additional communication cost. The system also suffers from man-in-the-middle attack, due to the key exchange phenomenon required for message transfer. A temporary key is used for authentication on publishers in open-source authenticated publish and subscribe model [9]. The system serves to provide security but the storage or computation complexity required by systems need to be analyzed.

Security Kit for IoT [10] analyzes the logical rules and probabilities between social relationships in devices. Relationships between devices are scrutinized to focus on the potential security threat and make IoT systems resistible to attacks. The system needs to be analyzed against scalability and suffers from the security threats like brute-force attack. Bmobishare [11] is an enhanced security mechanism based on location sharing with privacy preservation. It uses bloom filter to mask the data that is exchanged between nodes. The system achieves the objective at the cost of false positives due to the collisions that may occur during hashing.

3 Proposed System

The proposed system aims at providing security for IoT systems in cluster environment. Sensor nodes or sensors in each cluster work cooperatively to provide data to the cluster head. Data aggregation is done at the cluster head and the aggregated data is communicated to the sink node. The proposed method aims to provide end-to-end security between the nodes and the cluster head using Elgammal Signature scheme. The architecture diagram of the proposed method is provided in Fig. 1. Elgammal Signature Scheme combines encryption and signature for data security and authentication purpose. Nodes generate their key pairs (private and public) (x and y). The private key y is known only to sender. In addition, signature components r and s serve for authenticating a genuine sender. The application of Elgammal Signature scheme in IoT systems is novel and it is analyzed against existing Diffie–Hellman Key exchange method [8].

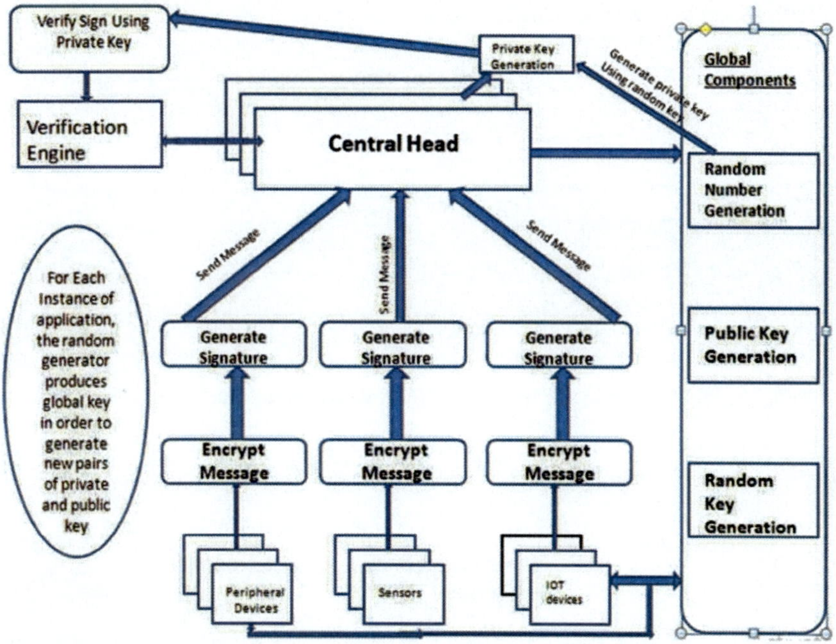

Fig. 1 Architecture diagram of Elgammal Signature scheme

4 Encryption

It is assumed that the global components g and p are shared between cluster head and nodes securely in the cluster deployment stage. After some number of sessions, these global components are again re-established securely. The number of sessions depends on the applications in which it is deployed and can be varied. ga is computed by Alice as the private key and gb is computed by bob as the private key. Alice (node) sends data to Bob (cluster head) by performing the following steps as given in Fig. 2.

As illustrated in Fig. 2, Alice computes components $c1$ and $c2$ using Eqs. 1 and 2.

$$c_1 = g^k \bmod p, \text{ where } 1 < k < p \tag{1}$$

$$c_2 = A^k(m) \bmod p, \text{ where } 1 < k < p \tag{2}$$

Decryption at Bob is done using Eq. 3.

$$m = c_1^{-a} c_2 \bmod p \tag{3}$$

Lemma 1 *For every Message m,* $m = c_1^{-a} c_2 \bmod p$.

Proof

$$\text{Given } c_1^{-a} c_2 \bmod p \tag{4}$$

Substituting $c1$ and $c2$ from Eqs. (1–3).

$$c_1^{-a} c_2 \bmod p \tag{5}$$

Substituting A as g^a in (5) we obtain Eq. 6 which gives m.

$$(g^k)^{-a} \times (g^{a\times k} \times m) \bmod p \approx m \tag{6}$$

Fig. 2 Encryption steps

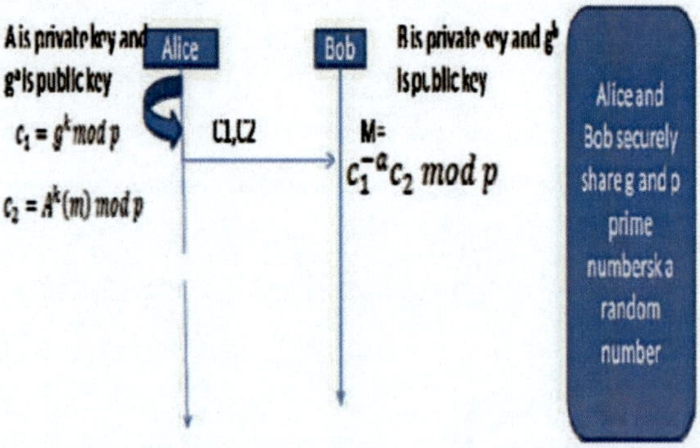

For authenticating Alice, signature components r and s are added along with the encryption components $C1$ and $C2$. It is assumed that Alice and Bob share p and α, prime numbers in a secure transaction. Alice chooses x as private key where x is a random number. α is primate root of x, i.e., gcd (x, α) is 1.

Alice computes public key y as given in Eq. 7.

$$y = \alpha^x \bmod p \tag{7}$$

Alice selects a random number k where $0 < k < \alpha$.

Alice computes signature components r and s as given in Eqs. 8 and 9.

$$r = \alpha^k \tag{8}$$

$$s = k^{-1}(m - x(r)) \bmod p - 1 \tag{9}$$

5 Authentication Verification

The verifier Bob has the public key y and the signature components m, r, s sent from Alice. The private key or the random number chosen by Alice is not known to the verifier.

The verifier computes $v1$ and $v2$ as given in Eqs. 10 and 11.

$$v1 = \alpha^m \tag{10}$$

$$v2 = y^r(r^s) \tag{11}$$

The verifier checks whether $v1 = v2$ the signature is accepted else it is rejected. The authentication steps between Alice and Bob are illustrated in Fig. 3.

Fig. 3 Authentication

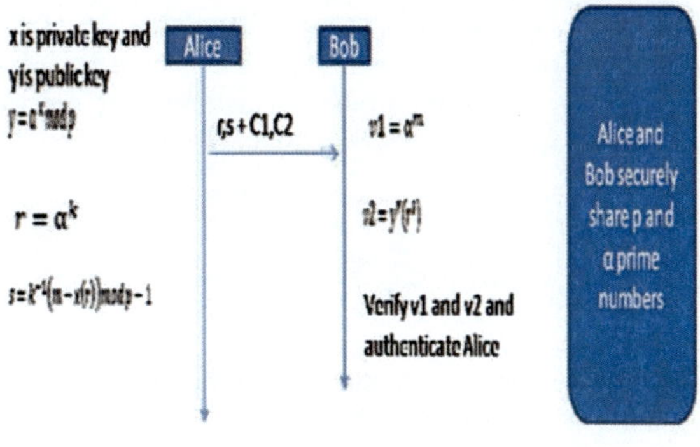

Lemma 2 *For every message m, $\alpha^m = y^r(r^s)$.*

Proof Rearranging terms in Eq. 6, (12) is obtained as

$$(\alpha^m \times y^{-r})^{-s} = r \tag{12}$$

Substituting s from (3) in Eq. 12, (13) is obtained as

$$\alpha^{m\left(\frac{m-xr}{k}\right)^{-1}} y^{-\alpha k\left(\frac{m-xr}{k}^{-1}\right)} = r \tag{13}$$

Solving Eq. 13, (14) is obtained as

$$\alpha^{\frac{m(k)}{m-x(r)}} \times y^{-\frac{\alpha^k(k)}{m-x(r)}} = r \tag{14}$$

Substituting y from (1) as α^x, Eq. 15 is obtained as

$$\alpha^{\frac{m(k)}{m-x(r)}} \times \alpha^{-\frac{\alpha^k x(k)}{m-x(r)}} = r \tag{15}$$

Substituting α^k as r in Eq. 15, (16) is obtained and hence proved

$$\alpha^{k\frac{m-rx}{m-rx}} \tag{16}$$

Lemma 3 $\alpha^{m1} \neq y^r(r^s)$ in case of threat to data integrity.

Proof

$$s = k^{-1}(m - x(r))\text{mod}p - 1,$$

as available in Eq. 9, s is dependent on m. Hence, the computed value to verify authenticity will vary from. Hence, data integrity is verified.

The proposed system is compared to existing systems in related works. As discussed in related works, there are limited works available in security mechanisms of IoT. The proposed method is found to be efficient when compared to group key establishment using Diffie–Hellman [8] in terms of security against man-in-the-middle attack. In existing scheme, key exchange mechanism is required to establish group key in multicast nodes. Such a system suffers the drawback of man-in-the-middle attack. But the proposed method does not use key exchange and devices encryption and authentication to IoT devices.

6 Performance Evaluation

Performance analysis is done using Cooja [13], an IoT emulator. Cooja works in contiki Operating system. This simulator is capable of emulating Tmote Sky. Contiki

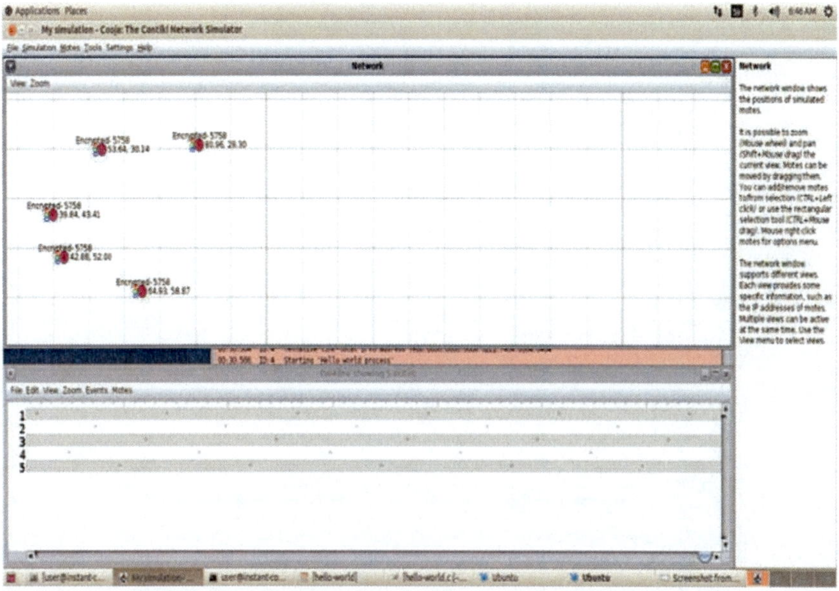

Fig. 4 Node deployment in COOJA emulator

is an operating system for network-based memory constrained systems. It focuses on low-power wireless IoT. It is widely used for simulation as it is open source with supporting IoT features. The simulation was run in sets of 5, 10, 15…30 nodes. The number of cluster heads is set as 5% of the number of nodes. The nodes are in random deployment. Depending on the proximity, nodes are added to the cluster heads. A node belongs to only one cluster. Figure 4 shows node deployment. Initially five nodes were deployed with one cluster head. Figure shows the data transfer between the nodes and the cluster heads. Figure presents the energy spent by the nodes in encryption and authentication. It is compared with the existing system group key establishment using Diffie–Hellman and shows high energy gain.

Initially p was set to be 467, α is 2 where α is a primitive root of p or gcd $(\alpha, p) = 1$ and k is 213, different private and public key pairs (x,y) is computed with their corresponding signature components r and s (Table 1; Fig. 5).

7 Conclusion

IoT security is an emerging field of research. In this paper, Elgammal encryption and signature are applied in IoT. The work is compared with the existing Group Key establishment using Diffie–Hellamn. The proposed system overcomes the drawbacks

Table 1 Varying session keys: for varying public and private keys (x, y), assuming m to be 100

Session number	Private key x	Public key y	Signature component R	Signature component s
1	127	132	29	51
2	131	244	29	383
3	137	205	29	415
4	149	14	29	13
5	151	56	29	179
6	157	315	29	211
7	163	79	29	243
8	167	330	29	109
9	173	105	29	141
10	179	182	29	173

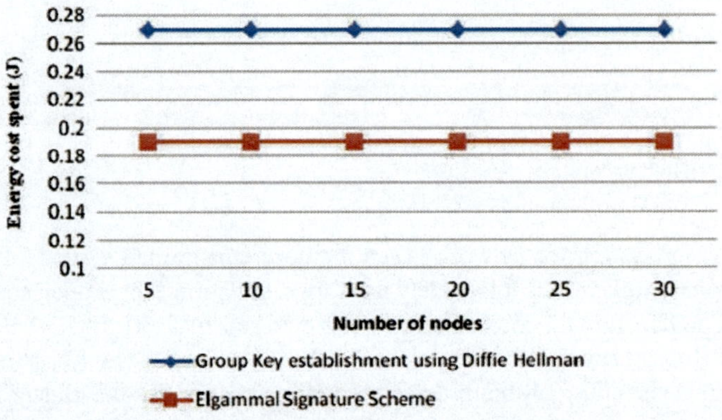

Fig. 5 Energy cost verses number of nodes

of the later system. The proposed system also has an increase in energy gain when compared to the existing works.

References

1. Holler, J., Tsiatsis, V., Mulligan, C., Karnouskos, S., Avesand, S., Boyle, D.: From Machine-to-Machine to the Internet of Things: Introduction to a New Age of Intelligence. Elsevier, Amsterdam, the Netherlands (2014)
2. Stajkic, A., Abrignani, M.D., Buratti, C., Bettinell, A., Vigo, D., Verdone, R.: From a real deployment to a downscaled testbed: a methodological approach. IEEE Internet Things J. **3**(5), 647–656 (2016)
3. Handte, M., Foell, S., Wagner, S., Kortuem, G., Marrón, P.J.: An internet-of-things enabled connected navigation system for urban bus riders. IEEE Internet Things J. **3**(5), 735–744 (2016)

4. Gupta, R., Gupta, R.: ABC of internet of things: advancements, benefits, challenges, enablers and facilities of IoT. In: Symposium on Colossal Data Analysis and Networking (CDAN) (2016)
5. Thakkar, A., Kotecha, K.: Cluster head election for energy and delay constraint applications of wireless sensor networks. IEEE Sens. J. **14**(8), 2658–2664 (2014). https://doi.org/10.1109/JSEN.2014.2312549
6. Mahahajan, S., Malhotra, J., Sharma, S.: An energy balanced QoS based cluster head selection strategy for WSN. Egypt. Inf. J. **15**(3), 189–199 (2014). https://doi.org/10.1016/j.eij.2014.09.001
7. Preethi, A., Pravin, E., Sangeetha, D.: Modified balanced energy efficient network integrated super heterogeneous protocol. In: Proceedings of International Conference on Recent Trends in Information Technology (ICRTIT) (2016)
8. Tyagi, S., Kumar, N.: A systematic review on clustering and routing techniques based upon LEACH protocol for wireless sensor networks. J. Netw. Comput. Appl **36**(2), 623–645 (2013). https://doi.org/10.1016/j.jnca.2012.12.001
9. Porambage, P., Gurtov, A., Stiller, B., Braeken, A., Ylianttila, M.: Group key establishment for enabling secure multicast communication in wireless sensor networks deployed for IoT applications. IEEE Access **3**, 1503–1511 (2015)
10. Rizzardi, A., Sicari, S., Miorandi, D., Coen-Porisini, A.: AUPS: an open source authenticated publish/subscribe system for the internet of things. Elsevier J. Inf. Syst. **62**, 29–41 (2015)
11. Neisse, R., Steri, G., Fovino, I.N., Baldini, G.: SecKit: a model-based security toolkit for the internet of things. Elsevier J. Comput. Secur. **54**, 60–76 (2016)
12. Shen, N., Yang, J., Yuanb, K., Fua, C., Jia, C.: An efficient and privacy-preserving location sharing mechanism. Elsevier J. Comput. Stand. Interfaces **44**, 102–109 (2016)
13. Kugler, P., Nordhus, P., Eskofier, B.: Shimmer, Cooja and Contiki: a new toolset for the simulation of on-node signal processing algorithms. In: Proceedings of IEEE International Conference on Sensor Networks (2016)

Forecasting of COVID-19 Using Hybrid ARIMA-FB Prophet Algorithm

C. Padmaja and B. L. Malleswari

Abstract Coronavirus is not a living organism. It has a genetic organism material covered with fat envelope. It affects throughout the country. The World Health Organization (WHO) officially announced that a new virus had been identified which then is called by 2019-nCoV on January 2020. The virus was recognized as part of the coronavirus group, which involves SARS and the other known colds. The first reported case was from Wuhan, China and has infected 7711 people and 170 reported deaths in China before coronavirus was declared as a global pandemic which produces a sickness authoritatively defined as COVID-19 that has diffused to a minimum 141 nations and regions, causing death over 5700 individuals around the world. The present paper explores the COVID-19 data frame seasonal wise and forecast future data for next 15 days in India using Auto Regressive Integrated Moving Average (ARIMA) and FB Prophet (Face Book Prophet) algorithm. The performance metrics used to find the model fitness are mean absolute error (MAE), mean percentage error (MPE) and root mean square error. Derived RMS of Prophet model as 39,407.98 for confirmed forecast COVID-19 in India date wise is simulated using Google Colaboratory to run and execute Python program.

1 Introduction

1.1 History of Coronavirus-19

Corona originated from SARS-COV2 (Senior Acute Respiratory System) and transformed into COVID-19 (coronavirus disease-2019). Respiratory Diseases caused by coronavirus are Influenza, Mumps, Measles and Rubella. Coronavirus is mainly of four types such as SARS (Senior Acute Respiratory System), MERS (Middle East

C. Padmaja (✉)
G. Narayanamma Institute of Technology and Sciences, Hyderabad, India

B. L. Malleswari
Sridevi Women's Engineering College, Hyderabad, India

T. Senjyu et al. (eds.), *ICT with Intelligent Applications*, Smart Innovation, Systems and Technologies 248, https://doi.org/10.1007/978-981-16-4177-0_59

Table 1 Origin of COVID-19 virus

Year	Name of virus	Started in country (origin)	Due to (host)
1919	Swine Flu (H1N1)	Uncertain	Pig
1920	HIV and AIDS	Kinshasa (Congo)	Chimpanzee and Monkey
1976	EBOLA	Sudan and ZAIRE	Monkey
1997	Bird Flu	Hong Kong	Water Fowl (Bird)
2002	SARS (Severe Acute Respiratory Syndrome)	China (Guandong)	CIVET CAT
2012	MERS (Middle East Respiratory-Sydrome)	South Arabia	Camel
2020	Coronavirus (COVID-19)	China (Wuhan)	Bats-Pangolin (Intermediate Host)

Respiratory System), 229E and OC43 [1]. The origin coronavirus since 1919 year is listed below.

Someone who infected by coronavirus will show common symptom such as fever, dry cough and tiredness or some cases, and infected person will feel pains and aches, runny nose sore throat, nasal congestion or diarrhoea.

The following Table 1 gives a summary of all heading levels.

However, some people infected with the virus do not show any symptoms and do not feel uncomfortable. Around 80% of individuals infected by COVID-19 can get recovery without acquiring particular treatment, but it is so dangerous for older people or someone with develop serious illness which the probability for them to get serious illness and develops difficulty breathing are higher. Available vaccine for older people depend on safety and the degree of efficacy. However, potential vaccines and some particular medicine treatments are still under investigation and now being subjected to comprehensive test by leading medical research centres. Further, extraordinary efforts are coordinated by WHO to develop and produce effective vaccines and drugs to avoid and treat COVID-19. As the reported case of death and infected people keep increasing, many nations have performed lockdown to minimize the spread impact of coronavirus. They also try to identify the infected among crowd by screening the temperature in public places using infrared thermometer. But the usage of the infrared thermometer gun itself is still lacking because it might not cover all of people and time-consuming. That way also can lead to the spreading virus widely because the health officer has to do it one by one through a lot of people queuing when one of them has probability to infect people around. To prevent this flaw, an alternative technology is needed.

Choudhary [2] predicted the estimated cases till April 2020 in INDIA. The present work examines till August 2020. The paper is organized as follows. Section 2 describes the corona history, causes and precautions in detail. Section 3 explores different performance metrics to evaluate the ARIMA-FB Prophet model's fitness. Section 4 displays the simulation results for forecasting data. Paper concluded in Sect. 5.

2 Corona Causes and Precautions

Coronavirus is attached to Angiotensin converting Enzyme 2 Receptors. The virus enters via Angiotensin Converting Enzyme 2 receptor (ACE2) on living cells and responsible for binding and fusion processes. Most likely, a human cell ingests the virus in a process known as endocytosis (Fig. 1).

Once it enters inside the cytoplasm, the endosome opens to reveal the virus's genetic material, a single standard DNA. The virus hijacks the cell's machinery to replicate the RNA and proteins and uses the endoplasmic reticulum to form its M protein outer layers and the all-important S-Protein.

After replication, the virus is carried by the Golgi bodies out of the cell in a process known as exocytosis. Meanwhile, the stress of viral production on the endoplasmic reticulum, eventually leads to apoptosis or the cell death. After then, it breaks the cell and comes out of the lungs, and then lungs will not work or function properly. Common Respiratory symtoms are like above 100° fever, dry cough, shortness of breath, loss of smell and taste. The different modes of transmission of coronavirus are:

- Droplets (cough, sneeze, talking).
- Aerisol (when travelling long distances).
- Direct Contact (Sex, blood, products).
- Indirect Contact (Fomites)—Inanimate Objects.
- Feco-oral—Toilets Cleanness.
- Mother to Child—Fetal Transfer.

As per government rule, one can test the symptoms at their home using the following score results and if needed contact doctors immediately (Table 2).

Note that elderly people above 60 years, below 0–5 year old kids, pregnant women and breast-feeding woman should consider precautions such as self-quarantine, social distancing, washing hands frequently and do not touch MEN (mouth, eyes, nose).

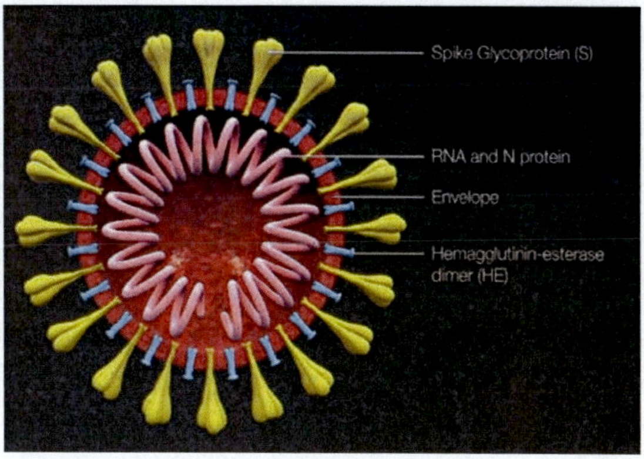

Fig. 1 Coronavirus cell membrane

Table 2 Self-scoring test analysis

Symptoms	Score points	Score results
1. Cough	1	• **0–2:** May be stress related and observe • **3–5:** Hydrate properly and proper personal hygiene. • **6–12:** Seek consultation with doctor. • **12–24:** Call the DOH Hotline 02-8-651-7800
2 Cold	1	
3 Diarrhoea	1	
4. Sore throat	1	
5. Myalgia	1	
6. Headache	1	
7. Fever> 37.8 C	1	
8. Difficulty in breathing	2	
9. Experiencing fatigues?	2	
10. Travelled recently during last 14 days?	3	
11. Travel history to a COVID-19 infected area?	3	
12. Do you have direct contact or taking care of positive Covid-19 patient?	3	

The pandemic coronavirus spread can happen imported, local spread, endemic and community spread (parties, temples, clusters).

As per John Hopkins study [3], 98% people will get symptoms within 12 days, and they need self-quarantine of 15 days. To control/minimize the effect of coronavirus, Countries should act fast (like early lock down in India) that can reduce deaths rate by ten times. Following remedies can control the virus.

1. Maintain social distancing, mask wearing.
2. Avoid contact hands.
3. Use of fomites (Sanitizer).
4. Containment: Once identified can be controlled through isolation.
5. Mitigation: For every death, there are 800 infected people and to flatten the death curve, contact tracking through Arogya Setu App, Testing RT-PCR and if positive quarantine.
6. Suppression: Can be done through lock down and social distancing.
7. Pulse oximeter: Frequent use of pulse oximeter is needed to know the oxygen saturation levels in the blood.
8. Spirometer: In case of breathing problem, spirometer clears the chest/lungs.

As per WHO, there is no vaccine available till now instead one can use allopathy medicines such as chlorophyll, Azithromycin, Zincovit and Vitamin C daily for 3–15 days. Inculcate life style changes such as breathing exercises, meditation, yoga, warm or ventilated rooms, eating fresh fruits and vegetables, stop smoking and Be Positive and Stay Healthy. It primarily effects the person who is suffering from diabetes, pregnant women, heart transplant and steroids. Corona has two strains, namely L-strain and S-strain. In India [4], L-strain occurred with less virulence based on the percentage of mutations.

3 Performance Analysis of Forecasting

The paper forecast next 15 days COVID-19 cases using proposed ARIMA—FB Prophet machine learning model. For the forecasting, the present work utilizes COVID-19 csv dataset file from GitHub [5–8]. The flow chart of the proposed ARIMA-FB Prophet model is given in Fig. 2.

The acronym for ARIMA is described as follows. The term "AR" stands for autoregressive based on auto correlation. It uses past data to forecast future data. The term "I" denotes to use the statistical stationary (does not vary with time) properties of time series such as median, variance and correlation. The term "MA" refers to find the error value between prediction and actual value.

On the other hand, FB Prophet [9] is an open source package created by Facebook to forecast the data monthly/weekly/daily wise and detects error value automatically from time-series data. The algorithm evaluates the model accuracy through cross-validation function.

The performance metric, "MAE" defines the ratio of the difference between predicted and observed values to the total number of observations [10, 11].

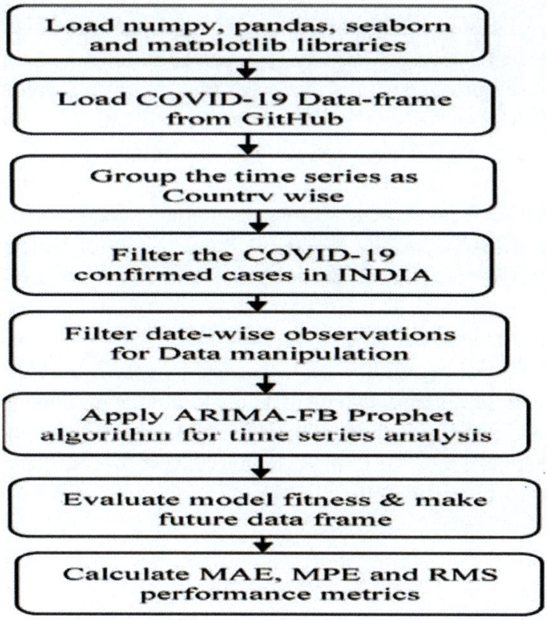

Fig. 2 Flow chart of ARIMA-FB Prophet model

$$\text{MAE} = \frac{1}{N}\sum_{l=1}^{N} \text{ABS}(\text{Actual}_l - \text{Forecast}_l) \quad (1)$$

The mean percentage error, "MPE" computes the MAE in terms of percentage.

$$\text{MPE} = \frac{1}{N}\sum_{l=1}^{N} \frac{\text{ABS}(\text{Actual}_l - \text{Forecast}_l)}{\text{Actual}_l} * 100\% \quad (2)$$

The root mean square, "RMS" is used to estimate the root of square of error value over total number of observations.

$$\text{RMS} = \sqrt{\frac{\sum_{l=1}^{N} (\text{Actual}_l - \text{Forecast}_l)^2}{N}} \quad (3)$$

4 Simulation Results

The COVID-19 forecasting can be done using proposed ARIMA—FB Prophet machine learning model. The algorithm is simulated using open source Google Colaboratory to write and execute Python scripts over the platform. Colab notebooks generate executable code in rich text along with images.

From Figs. 3 and 4, time-series data is extracted from csv file and filtered confirmed COVID-19 cases in country-wise and in India.

Figure 5 separates the confirmed/deaths/received/active COVID-19 cases from date-wise analysis. It is clear that, from February 2020 to August 2020, the confirmed cases in INDIA are increasing exponentially.

Time-series forecasting analysis of data points in a deterministic approach can be performed with various techniques such as Autoregressive Moving Average (ARIMA), SARIMA and FB Prophet. Out of which Fourier Transform-based FB Prophet created by Facebook is more robust for missing data, non-stationary data sets whereas ARIMA works for stationary data points and requires statistical knowledge prior to use it.

From Fig. 6, it is clear that the proposed ARIMA—FB Prophet model fits well with seasonality in data and able to predict the forecast data for next 15 days.

From Fig. 7, the proposed ARIMA—FB Prophet model predicts exactly well with actual cases. The findings help us to take precautionary measures in order to prevent further spread of the disease.

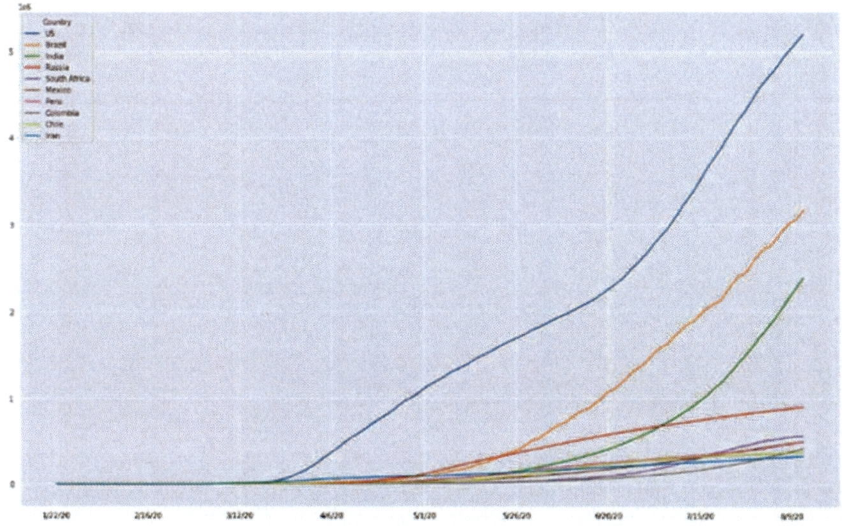

Fig. 3 Country-wise COVID-19 confirmed cases

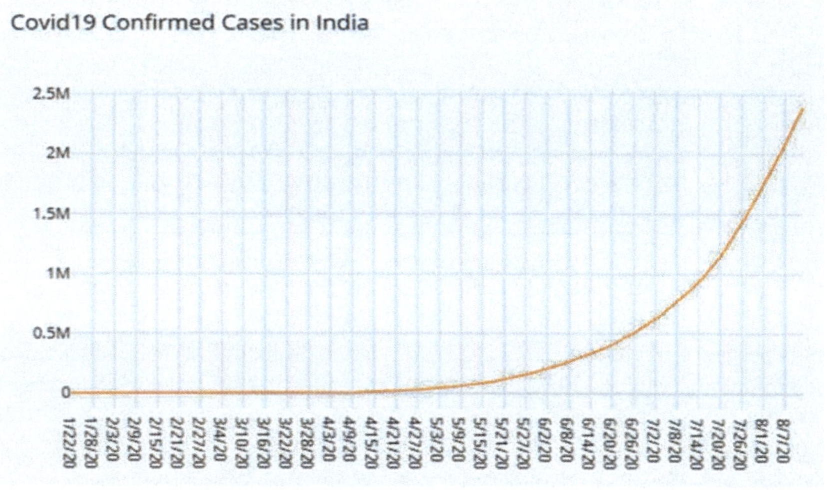

Fig. 4 COVID-19 confirmed cases in India

5 Conclusion

Proposed ARIMA—FB Prophet time-series analysis model can be used to predict the future values based on past values. The forecast study helps us to plan treatment

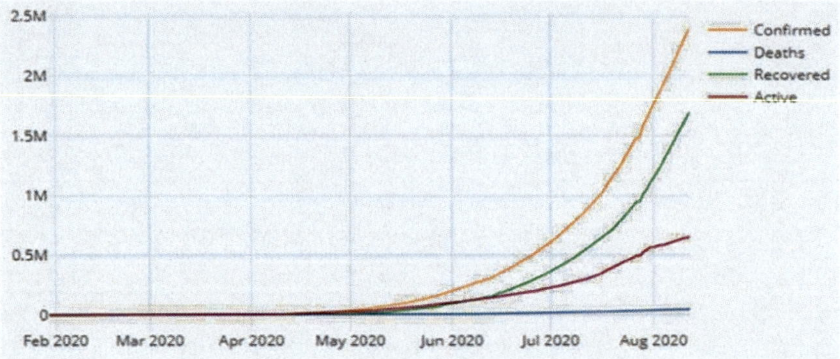

Fig. 5 Date-wise COVID-19 cases comparison

	Confirmed	Deaths	Recovered	Active
ObservationDate				
2020-01-30	1.0	0.0	0.0	1.0
2020-01-31	1.0	0.0	0.0	1.0
2020-02-01	1.0	0.0	0.0	1.0
2020-02-02	2.0	0.0	0.0	2.0
2020-02-03	3.0	0.0	0.0	3.0
...	...	...	...	...
2020-08-08	2153010.0	43379.0	1480884.0	628747.0
2020-08-09	2215074.0	44386.0	1535743.0	634945.0
2020-08-10	2268675.0	45257.0	1583489.0	639929.0
2020-08-11	2329638.0	46091.0	1639599.0	643948.0
2020-08-12	2396637.0	47033.0	1695982.0	653622.0

Fig. 6 Confirmed forecast COVID-19 cases

amenities in terms of hospital wards and quarantine facilities. The model calculates RMS error value as 39,407.989398454774 and 95% MPE from predicted cases 1.00759e + 06 with lower (best fit) bound 928,162 cases and upper bound 1.08022e + 06.

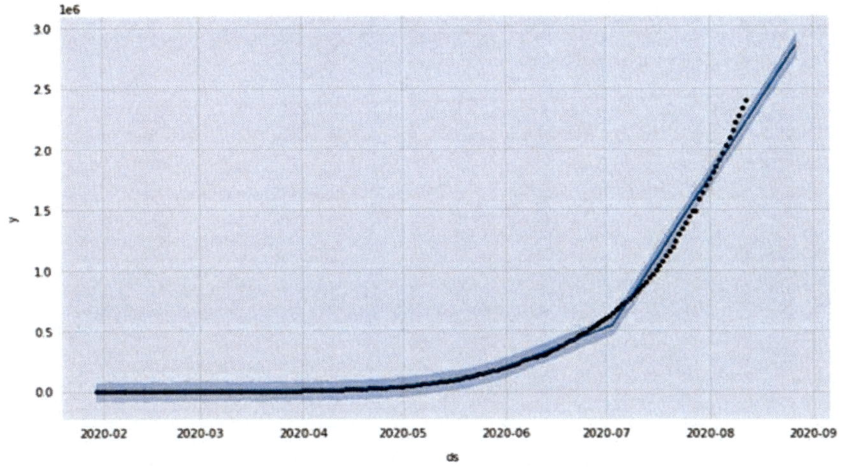

Fig. 7 Comparison of predicted versus actual forecast COVID-19 cases

References

1. World Health Organization: WHO COVID-19 Dashboard (2020)
2. https://towardsdatascience.com/forecasting-covid-19-cases-in-india-c1c410cfc730Towards Data Science
3. https://www.arcgis.com/apps/opsdashboard/index.html#/bda7594740fd40299423467b48e9ecf6
4. https://www.mohfw.gov.in/
5. https://raw.githubusercontent.com/ammishra08/COVID-19/master/covid_19_datasets/covid_19_globaldata/time_series_covid_19_confirmed.csv
6. Ivorra, B., Ferrández, M.R., Vela-Pérez, M., Ramos, A.M.: Mathematical modeling of the spread of the coronavirus disease 2019 (COVID-19) taking into account the undetected infections. The case of China. In: Communications in Nonlinear Science and Numerical Simulation, p. 105303 (2020)
7. Rai, B., Shukla, A., Dwivedi, L.K.: COVID-19 in India: Predictions, Reproduction Number and Public Health Preparedness. medRxiv (2020)
8. Perc, M., Gorišek Miksić, N., Slavinec, M., Stožer, A.: Forecasting Covid-19. Front. Phys. **8**, 127 (2020)
9. https://medium.com/katanaml/covid-19-growth-modeling-and-forecasting-with-prophet-2ff5ebd00c01
10. Bhatnagar, M.R.: COVID-19: Mathematical Modeling and Predictions. Submitted to ARXIV. Online available at: http://web.iitd.ac.in/manav/COVID.pdf.(2020)
11. Ray, D., Salvatore, M., Bhattacharyya, R., Wang, L., Du, J., Mohammed, S., Purkayastha, S., Halder, A., Rix, A., Barker, D., Kleinsasser, M.: Predictions, role of interventions and effects of a historic national lockdown in India's response to the COVID-19 pandemic: data science call to arms. In: Harvard data science review, 2020 (Suppl 1) (2020)

Review of Forthcoming ICT-Enabled Applications Promoting Learning in Higher Education

Sapna Jain and M. Afshar Alam

Abstract Information and communication technology has changed every person's aspect of life and has become an essential part of the higher education system. ICT has opened many avenues and inspired administrators, teachers, and students to rethink their role as teachers and learners for the future. The relationship between ICT and higher education can widen the open door for the instructor and student to take an interest in the pace of learning and quick-change times. The future ICT teaching system will improve education quality, encourage collaborative learning, and provide regular feedback to learners. This paper discusses the use of technologies such as deep learning, artificial intelligence, and robotics in e-learning systems that allow teachers to respond better to different learners' different needs. The chapter discusses a case study analysis of the impact of ICT use in university education. This paper aims to discuss the future trend of applications using ICT tools that will be used by teachers to enhance teaching and learning in higher education.

S. Jain (✉) · M. A. Alam
Jamia Hamdard, New Delhi, India
e-mail: drsapnajain@jamiahamdard.ac.in

M. A. Alam
e-mail: aalam@jamiahamdard.ac.in

T. Senjyu et al. (eds.), *ICT with Intelligent Applications*, Smart Innovation, Systems and Technologies 248, https://doi.org/10.1007/978-981-16-4177-0_61

1 Introduction

The higher education process in the present era has continuous changes needed to be understood according to the dimensions of the twenty-first century. Technology plays a vital role in the teaching and learning method [1]. The data dissemination method takes place in education with the effective use of technology. New learning strategies will facilitate students to grasp abstract ideas with learning new skills. The use of ICT-based learning helps in the rich representation of changes in learning perception and understanding of context. It also helps in vast distribution and easy access to information and provides powerful support for educational innovation [2].

Figure 1 shows that technology is employed and provides a basic plan to use technology and considers wherever applied. It additionally helps to investigate the impact of that technology within the room. This system is all concerning; however, academics and students communicate with one another, inquire concerning doubts, facilitate the higher cognitive process, supply an appropriate road map to know, and solve a selected drawback [3]. It allows students to be told time unit and offers an outsized vary of instructional opportunities despite geographical areas. It empowers effective learning and distributes instruction in line with the requirements of the scholar. The employment of multimedia systems or simulation tools helps to focus ICT on well-known analysis through a diligent analysis method. In-depth exploration helps improve the standard of teaching. It can help proper use of website resource material which can help university management in the administration, analysis, and development activities [4].

Fig. 1 Use of ICT in higher education *Source* Pyla, A. "ICT as a Change Agent for Higher Education and Society." *CloudCom 2012* (2012)

2 Background

Information technology has influenced students in the field of education very much in recent years. Universities use MOOCs online learning platforms to enable the students to understand the course in an innovative method. The learning content available can be remodeled, teaching the students to use MOOCs effectively [5]. MOOCs learning revolutionized has new learning tools like artificial intelligence and video learning in the traditional learning system [6]. It focuses on student-centered learning on self-efficacy (SE) and self-regulated learning (SRL) strategies for everyone. Online tools like Microblogs and Twitter provide a platform for university student interaction as a fun mode. The Twitter platform is free wherever the academics used it as an alternate learning website to boost their courses [7]. According to the newest literature, Twitter and MOOCs are used to teach subjects' strategies. The ICT implementation has reformed the link between ICT and the quality of ordinary education [8]. The program style for college kids currently needs to focus on many factors that may analyze the impact of adopting ICT tools on instruction situations. The effect of ICT additionally affects the performance of university students. The United Nations agency and Institute of Statistics have conducted a survey during which governments and better education establishments have to close for theoretical and sensible ICT implementation [9]. The analysis framework depicts the use of ICT in various aspects such as teaching, administration, research, and learning. The analysis highlights that the organizational structure and implementation method is at the initial and advanced levels. This ICT framework is applicable and used in universities at national and international levels by directors, faculties, students, technicians [10].

3 Future Tools of ICT and E-learning

3.1 3D Printing

Students' approach is not limited to textbook illustrations, which the students can do with 3D printers. The 3D image tools help the students to develop a deeper understanding of their subjects. Commenting on a project at the University of Nevada, Reno, in 2014, EDUCAUSE predicted a rapid increase in 3D printers on campus. 3D printing services can be both used with an adequate and broad-based view. The network media consortium works as a group of hardware manufacturers that enable software developers and publishers to on multimedia-focused products in higher education [11]. The idea of widespread adoption of 3D printing in higher education was started by Educause Learning Initiative. This group forecasts widespread adoption of 3D printing using competitive higher education institutions needed for 3D printing services to be used by students. The universities are using 3D printing by dividing cost-effective plans to provide technology to an increasingly wide (Fig. 2).

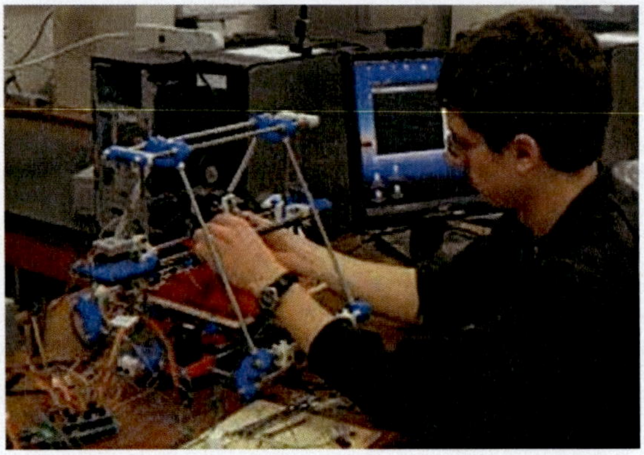

Fig. 2 Student working on 3D-printing lab. *Source* https://www.elsevier.com/connect/3d-printing-your-lab-equipment-its-cheaper-than-you-think [11]

3.2 Augmented Reality

Vimedix™ (CAE Healthcare) is an associate AR system for coaching diagnostic procedures in Medical Universities. It contains an electrical device transthoracic or transeptal sonogram that displays face validity. Recently, CAE aid has equipped it with Hololens, which offer the power to examine pictures with glasses, unrestricted by screen dimensions [12]. Ecocom is another diagnostic procedure AR trainer, except for newborns. It showed face, content, and construct validity, as shown in Fig. 3.

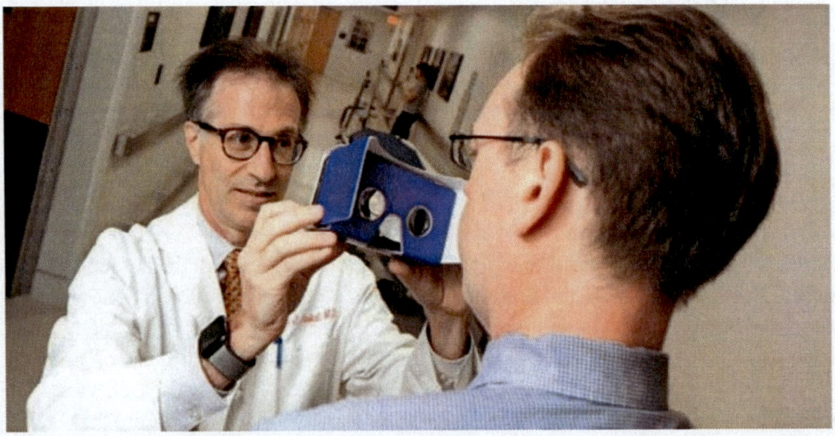

Fig. 3 Augmented tool used by medical students. *Source* https://www.elsevier.com/connect/going-virtual-how-vr-is-guiding-interventional-radiology [12]

Fig. 4 Virtual reality lab with CRS robotic arm and robotic cell. *Source* Dorota Kamińska Tomasz Sapiński, Sławomir Wiak Toomas Tikk, Rain Eric Haamer, Egils Avots, Ahmed Helmi, Cagri Ozcinar,Gholamreza Anbarjafari "*Virtual Reality and Its Applications in Education: Survey",Information* 2019, *10*, 318; https://doi.org/10.3390/info10100318 [13, 14]

3.3 Virtual Reality (VR)

This technique helps the student achieve deeper engagement than a traditional textbook. The visual elements they will acquire will effectively enhance the relationship between learning concepts and information. The virtual environment is used as an engineering training simulator because of its functions [13]. This helps engineering students for real-world industrial situations and allows them to understand and design decisions by engineers to understand design better. It lowers the time and cost factor while using the devices [14]. The device is shown in Fig. 4.

3.4 Hologram

This technique permits the coed to know the conception in real-time, which might utilize in a real setting. The technology conjointly provides visible and interactive learning expertise. Holograms use a virtual teacher as an instructor from United Nations Agency which is many kilometers away. In the introductory class in video conferencing is used (Fig. 5) [15].

3.5 Robot-Proof Education

This method is the latest technology that the North Eastern University shall use in its curriculum and teaching modules. The University President Joseph Aeon has

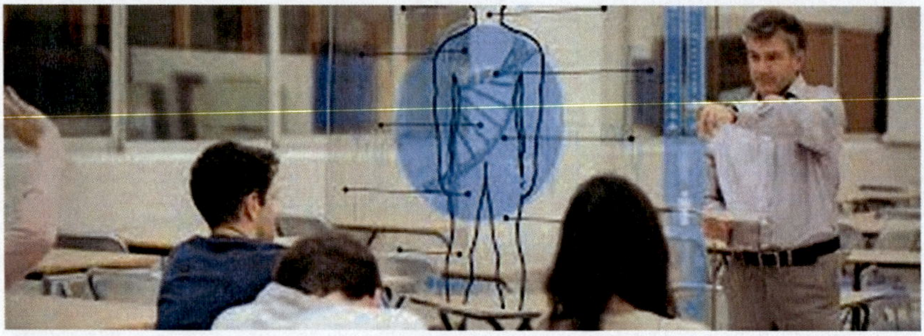

Fig. 5 Hologram teaching. *Source* https://edtechmagazine.com/k12/article/2018/05/holographic-tech-could-be-next-big-ar-offering-k-12-classrooms [15]

proposed that future generation students innovate and meet society's needs by using artificial intelligence agents. Robot proof education [16] shall use robots, artificial intelligence in an effective way to provide assistance to students in learning. Still, it shall help the students with a creative mindset to use any new idea such as scientific hip-hop recording or a webcomic series to treat cancer technology use for society. He says that according to humanity's principle, people help work based on their inner powers and help the students accept challenges in this competitive world along with smart machines the Aon's stated literacy in three dimensions such as data, technical and mental literacy. Students shall understand the use of robotics languages, tools which is essential to improve technological literacy to make machines work according to communication and design [17]. Lifelong new opportunities always support the ability to adapt to change for higher education institutions (Fig. 6).

Fig. 6 Robot teaching in class. *Source* https://www.tenneyschool.com/teaching-robot-proof-profession/ [17]

4 Case Study

The School of Engineering Sciences and Technology runs electronics and computer science courses by using the NPTEL online learning module in which educational institutions like the Indian Institutes of Technology and the Indian Institute of Science Bangalore use their platform for promoting learning. They provide the largest online learning repository for engineering sciences, humanities, and social sciences students. The 300 students used two e-learning intiatives and found that they were ready to accept challenges, enjoyed learning, and participated in competitions. The students used the Youtube channel by NPTEL to study computer architecture and robotics subjects. The students participated and won prizes in the e-Yantra artificial intelligence. Competition (eYRC) may be a distinctive annual competition for college boy students in science and schools. The selected groups are given a robotic kit complete with accessories and video tutorials to learn basic ideas in embedded systems and microcontroller programming. Abstracts of planet issues appointed as "themes" are then enforced by the group's exploitation of the robotic kits. Figure 7 shows the use of the video teaching category and current. Figure 8 shows the effective use of the NTPL learning module among the students. The students attended video lecture categories and took part in a robotic competition that was relatively less in ancient teaching. Self-directed learning helps students to be motivated to participate more in e-learning intiatives.

Fig. 7 E-learning implementation

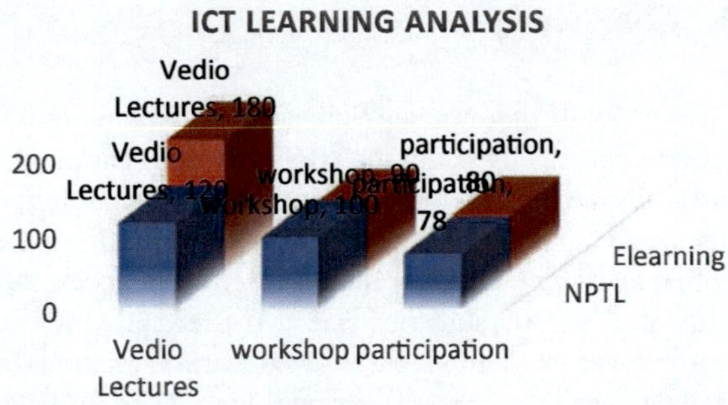

Fig. 8 Impact analysis of E-learning

5 Conclusion

ICT is a learning tool for teaching and self-learning, which is used by teachers and students worldwide [18, 19]. The ICT methods used for e-learning used computer-assisted and web-learning with online and distance learning [20]. The virtual knowledge used by ICT helped improve access, quality, and cost in higher education learning. It enhances education quality by reducing its costs and enabling it to reach remote areas through different modes like virtual classrooms, e-learning modules, and online and distance learning. E-learning has shown good thanks for improving the standard and effectiveness of tertiary education and learning. This paper highlights the tools and methods that can add a new learning dimension to future students.

References

1. Geng, S., Law, K.M.Y., Niu, B.: Investigating self-directed learning and technology readiness in blending learning environment. Int. J. Educ. Technol. Higher Educ. **16**, 17 (2019). https://doi.org/10.1186/s41239-019-0147-0
2. Arruabarrena, R., Sánchez, A., Blanco, J.M., Vadillo, J.A., Usandizaga, I.: Integration of good practices of active methodologies with the reuse of student-generated content. Int. J. Educ. Technol. High. Educ. **16**, 10 (2019). https://doi.org/10.1186/s41239-019-0140-7
3. Pyla, A.: ICT as a change agent for higher education and society. CloudCom 2012 (2012)
4. https://www.ukessays.com/essays/education/impacts-of-ict-in-education-education-essay.php
5. Costello, E., Brown, M., Mhichíl, M.N.G., Zhang, J.: Big course small talk: twitter and MOOCs—a systematic review of research designs 2011–2017. Int. J. Educ. Technol. High. Educ. **15**, 44 (2018). https://doi.org/10.1186/s41239-018-0127-9,SpringerOpenPublication
6. le Roux, I., Nagel, L.: Seeking the best blend for deep learning in a flipped classroom—viewing student perceptions through the Community of Inquiry lens. Int. J. Educ. Technol. High. Educ. **15**, 16 (2018). https://doi.org/10.1186/s41239-018-0098-x
7. Conde-Caballero, D., Castillo, C.A., Ballesteros-Yáñez, I., Mariano-Juárez, L.: Blogging as a tool for the acquisition and dissemination of knowledge in health sciences: a preliminary evaluation. Int. J. Educ. Technol. High. Educ. **16**, 30 (2019). https://doi.org/10.1186/s41239-019-0161-2,SpringerOpenPublication
8. Chawinga, W.D.: Taking social media to a university classroom: teaching and learning using Twitter and blogs. Int. J. Educ. Technol. Higher Educ. **14**, 3 (2017). https://doi.org/10.1186/s41239-017-0041-6

9. Basri, W.S., Alandejani, J.A., Almadani, F.M.: ICT adoption impact on students' academic performance: evidence from Saudi Universities. Hindawi Educ. Res. Int. **2018**(1240197), 9 (2018). https://doi.org/10.1155/2018/1240197
10. Girish, S.R., Suresh Kumar, S. (2017) ICT in teaching learning process for higher education: challenges and opportunities. IOSR J. Comput. Eng. (IOSR-JCE) **19**(4), 24–28e-ISSN: 2278-0661, p-ISSN: 2278-8727, www.iosrjournals.org
11. https://www.elsevier.com/connect/3d-printing-your-lab-equipment-its-cheaper-than-you-think
12. https://www.elsevier.com/connect/going-virtual-how-vr-is-guiding-interventional-radiology
13. Kamińska, D., Sapiński, T., Wiak, S., Tikk, T., Haamer, R.E., Avots, E., Helmi, A., Ozcinar, C., Anbarjafari, G.: Virtual reality and its applications in education: survey. Information **10**, 318 (2019). https://doi.org/10.3390/info10100318
14. Article open access. https://www.mdpi.com/2078-2489/10/10/318/htm
15. https://edtechmagazine.com/k12/article/2018/05/holographic-tech-could-be-next-big-ar-offering-k-12-classrooms
16. https://mitpress.mit.edu/books/robot-proof
17. https://www.tenneyschool.com/teaching-robot-proof-profession/
18. Halili, S.H.: Technological advancements in education 4.0. Online J. Dist. Educ. e-Learn. **7**(1) http://www.tojdel.net/journals/tojdel/articles/v07i01/v07i01-08.pdf (2019)
19. https://www.slideshare.net/jkdange/role-of-ict-in-higher-education?from_action=save
20. Romero, L., Saucedo, C., Caliusco, M.L., Gutiérrez, M.: Supporting self-regulated learning and personalization using ePortfolios: a semantic approach based on learning paths. Int J Educ Technol Higher Educ **16**, 16 (2019). https://doi.org/10.1186/s41239-019-0146-1

A Comprehensive Travel Recommendation System

D. Chaitra, V. R. Badri Prasad, and B. N. Vinay

Abstract Planning plays the major role in organizing a trip or an outing. Though there are various resources in the web, and various travel agencies that provide planned trips, the automation provided by them is low. This project is an attempt to build a comprehensive travel recommendation system, which builds the itinerary to the user based on his priorities and inputs, along with providing personalized recommendations. The project makes use of machine learning algorithms like collaborative filtering and content-based filtering, along with Google APIs to build the itinerary and provide recommendations.

1 Introduction

As a result of critical enhancements in data innovation, the Internet has enormously affected the movement administrations industry. While there is a plenitude of data from the Internet and travel guides, a considerable lot of these assets basically prescribe places that are esteemed to be well known, yet generally don't speak to the intrigue inclinations of clients or hold fast to their outing limitations. Likewise, clients utilize various sources to design their excursion. For instance: an alternate site to discover places, an alternate site to discover lodgings, an alternate site to discover book transport tickets, and so forth.

Most of the major internet platforms use recommendation systems. Examples include Amazon shopping, Netflix, YouTube, and others. On similar lines, there have been studies in the fields such as movies [1], social media [2], and career [3]. They play a major role in the success of these businesses and making these platforms user friendly to gain more users.

The long goal of this work is to help travelers in building a complete itinerary by building a comprehensive travel recommendation system where in places, hotels,

D. Chaitra (✉) · V. R. B. Prasad · B. N. Vinay
PES University, Bangalore, India

V. R. B. Prasad
e-mail: badriprasad@pes.edu

T. Senjyu et al. (eds.), *ICT with Intelligent Applications*, Smart Innovation, Systems and Technologies 248, https://doi.org/10.1007/978-981-16-4177-0_62

travel means, distances, also personal priorities are taken care of to build the itinerary. However, hotels and travel means are not integrated currently in this paper. A recommendation system is to be built which uses machine learning algorithms like collaborative filtering and Google Maps APIs to get locations, latitude and longitude, and travel durations.

2 Related Works

Various methods and algorithms have been used and developed to implement recommendation in travel domain. One such approach is to apply association rule mining in a classification sense where, instead of just considering the existence of objects and users, they consider attributes that characterize application users and objects [4]

A study proposes a customized travel planning system that takes user requirements and based on these inputs, the system is operated and a tentative schedule is built through the travel planning module, and user is allowed to modify if necessary [5]. Another approach is to use deep learning neural network architecture that uses feedback to produce model-based predictions [6].

Joseph Coelho, Paromita Nitu et al. develop tourist guide application named GuideMe. Each user can review tourist attractions, obtain suggestions of tourist places not visited by him as per recommended by other users and make his own recommendations [7].

3 Proposed Solution

Keeping in my mind all the current scenarios and complexity factors, this project aims at building a comprehensive travel recommendation application which helps travelers plan their trip by generating a comprehensive itinerary taking into consideration the user's preferences for places interested, night journey preference and hotel types preferred. Recommendations are based on user interests and priorities. Machine learning algorithms like collaborative filtering, Google Maps APIs to get location details, longitude and latitude of places are used in building the application. There are two main paradigms of methods in building a recommendation system, namely collaborative filtering and content-based filtering.

Collaborative filtering [8] is based on the past interaction between the users and items. It is based on how different users responded to the same items. For this, similar users survey needs to be done. In travel domain, the data of different places a user has been to, the ratings need to be surveyed. Based on the item-item similarity and user-item similarity, different places are recommended.

Content-based methods [9] are based on a user's previous likes. It relies on similarities between features of the items. For example, if a user has visited a beach in the past, a beach and surrounding places are visited to him.

In the proposed solution, we use collaborative filtering to provide recommendations to users and build an itinerary generator application which considers the regions and tourist places the user is interested in visiting, his journey time, visit time preference, lunch and breakfast breaks, and categories of stay.

4 Methodology

In brief, once the user logs in to the system, he selects the major regions he would like to visit. During the sign up, he will have to enter the place he has already visited. Based on his previous travel, the tourist places are recommended to him using a collaborative filtering algorithm. The user is then recommended places within the major regions he selected so that it helps him to customize his itinerary.

The tourist regions timings, the travel time, the stay places, the hotels he wishes to stay in, all things are covered in the itinerary. On successful login, user selects the regions and tourist places for every region (Fig. 1).

Also, user provides more details like preference for night journey, hotel stay, etc. A customized itinerary is generated based on the user input. Recommendations are based on the user travel history. Based on all the inputs and preferences, a customized itinerary is built. All the phases of the project are explained in detail in the following sections.

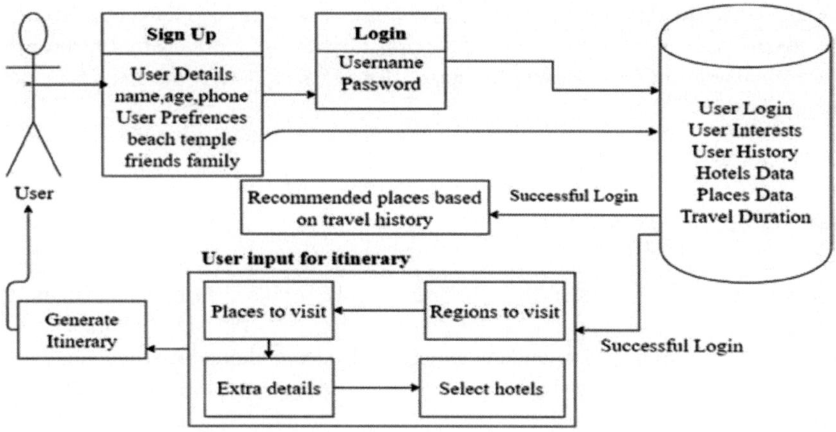

Fig. 1 System architecture diagram

4.1 Data Collection Process

Though ready data sets are available in certain websites like Kaggle, Socrata, etc., not much datasets related to this project were available readily as we are only referring to Karnataka Tourism. So, we have collected data by Web Scraping. Scraping is pulling any content which is available on a web page. We have used Beautiful Soup [10], a Python framework for web scraping.

Tourist places and their details are scraped from www.trawell.in and www.tripadvisor.com Every place has the details such as name, region, operating hours, genre, ratings, and latitude and longitude (fetched using Google Directions API). Data related to accommodations and hotels is fetched from Goibibo hotel dataset which was available in Kaggle. It has details of 412 accommodations in the state of Karnataka.

When we recommend a place to a user, we must first know what the user is interested in. Within Karnataka, there are no ready datasets to know users details of travel nor it can be scrapped. So, to collect the user data, 2 google forms were circulated.

One form collected users general travel behavior such as the age of the user, if he is a frequent traveler, how frequently he travels, with whom he travels more, what genres of places he visits more, what type of hotels he prefers to stay in, how much he would like to spend on a trip.

The second form consisted a list of places in Karnataka and the user needed to select the places which they had already visited, this is important because, when we know user history, we can recommend places based on his interests. Also, we can recommend users who visited this, also visited this. Around 600 responses were obtained after circulating the Google forms. This data is used for personal recommendations.

4.2 Data Pre-processing

The raw data gathered need to be translated into usable data that can be used. Python scripts are used to preprocess the data using pandas and NumPy libraries. For example, the forms had some outliers which needed to be filtered. Also, certain more details like latitude and longitude of every hotel were needed, this is later added by using Google Directions API.

4.3 Building the Recommendation System

The recommendation system is built by implementing collaborative filtering. There are two types of collaborative filtering, namely Memory-based collaborative filtering and Model-based collaborative filtering.

Memory-based collaborative filtering uses a rating system to predict one user's preferences by taking the preferences of a similar user or the 'neighbor' into consideration. Model-based collaborative filtering Various data mining and machine learning algorithms are used in this method of collaborative filtering recommender systems to create a model for predicting a user's rating of an unrated object.

Since we had the data collected from users, about which all places the users had visited, and the user behaviors about travel, we used Memory-based collaborative filtering using Cosine Similarity. The calculation will be based on the following formula,

$$\cos\theta = \frac{\vec{a}\vec{b}}{\left\|\vec{a}\cdot\vec{b}\right\|} = \frac{\sum_1^n a_i b_i}{\sqrt{\sum_1^n a_i^2}\sqrt{\sum_1^n b_i^2}} \tag{1}$$

where $\vec{a}\vec{b} = \sum_1^n a_i b_i = a_1 b_1 + a_2 b_2 + \ldots + a_n b_n$ is the dot product of the two vectors.

The similarity between two places measures the cosine of the angle between them. Lower the angle, higher is the similarity. The similarities between the places are stored in a matrix and added to the database. The user is recommended top 10 places based on the places in which he has previously visited and their similarity scores.

4.4 Fetching Real-Time Data with Google APIs

When generating an Itinerary, it is important to have real-time data in all aspects. Google Maps Platform provides the best APIs to gather real-time data.

Latitude and longitude values are fetched using Geocoding API. It converts a given address into geographical coordinates. We use Python's Google Maps [11] library and make a call to Geocode sending the address and fetch the geocoded latitude and longitude value. Once we have the latitude and longitude for all the places, we need to know how much timc it takes to travel from one place to another in the given traffic constraint. To calculate this, we used the Google Directions API.

Finally, an itinerary is built and has a sequence of places, say Bangalore → Mysore → Coorg → Bangalore, a map is plotted for the same and is displayed to the user in the frontend. We make use of the Google Maps JavaScript API for this.

4.5 *The Itinerary Algorithm*

The user selects the place of start, and the regions he wants to visit which is taken as regions array. Once the user provides the input, we first find the sequence in which he visits the given regions. The nearest region is found using the latitude and longitude difference. This difference is kept as a tracking value and difference of other places to the current place is calculated. The place with the minimum difference is considered as the next place to visit. We now have a sequence of the regions, which depicts the order in which the user will travel.

Now, we consider the travel duration between the sequence. The user starts from his starting place and goes to 1st region. If the user prefers night journeys over long distances (taken as input), and the travel time from starting place to 1st region takes more than 6 h, we suggest the user to leave the previous day night and give them a timing of departure such that they reach the 1st region by 6AM in the morning on day 1. If the user does not prefer night journeys, or if the 1st region is not too far from the starting place, the user will leave on day 1 at his preferred start time.

Once the user starts to the 1st tourist place, we add the traveling time plus spending time to the current time. In a similar fashion, the next place to visit is found, and for every place, if the traveling time plus the spending time crosses the end time (the time the user ends his day), the user is advised to go back to stay and rest, and the day is incremented.

Preferences such as the tourist places he is interested in visiting in that region, his preferred start time and end time of the day, the current time, and the distance to the accommodation are considered while developing the itinerary. Also, the breakfast and lunch breaks are calculated. If the current time is between 7AM and 12PM, a 30 min break for breakfast is given, and if it is between after 12PM 60 min lunch break is given.

Once the user finishes seeing the complete region, his traveling time to the next region is calculated. If it is the case that the user will reach the next region in midnight, we will advise the user to be in the region he is and give him a time to start to the next region, and if he prefers a night journey, or he will start next day morning (Fig. 2).

5 Results

User selects the regions and tourist places; he is interested in visiting. The itinerary is generated based on many other customized user inputs and preferences. It provides descriptive details on arrival and departure times based on user requirements. It also provides the order of visiting of the places based on the visiting hours and travel routes to reduce travel from the accommodation. Along with providing time windows for refreshment/food, it also provides average duration of visit for a particular place. It also provides the travel duration between each place and optimizes the itinerary. An example can be seen in the below screenshot.

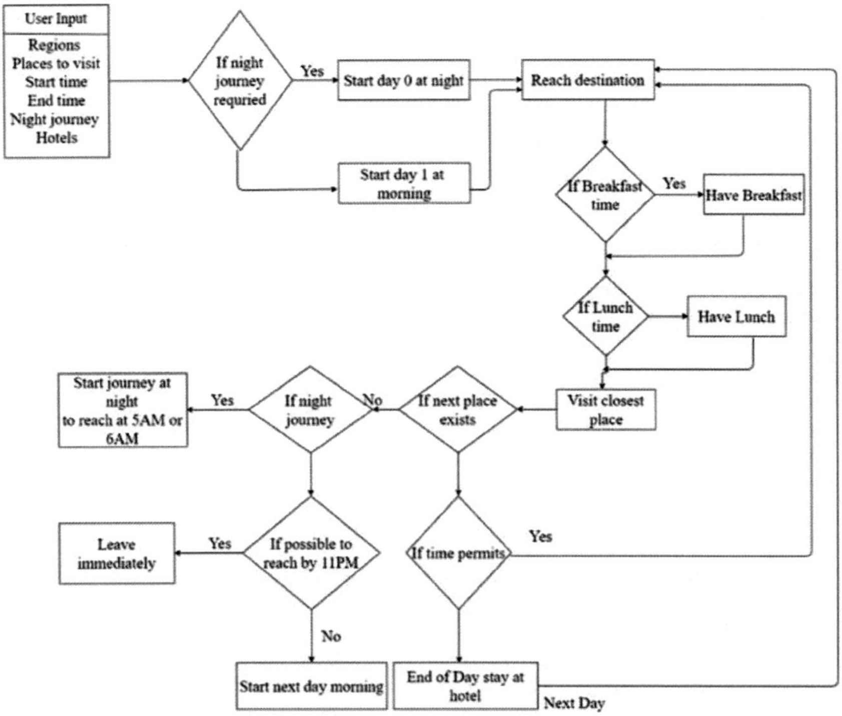

Fig. 2 Itinerary generator

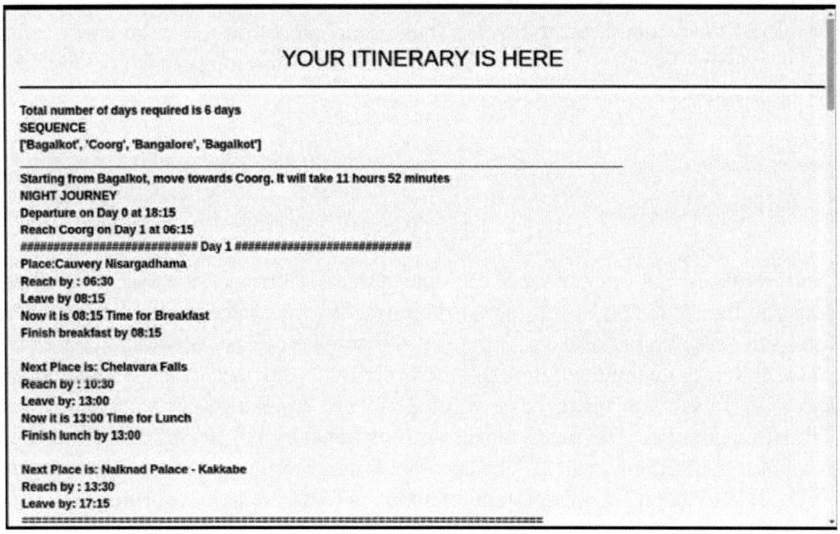

To test the efficiency of the application, manual testing is conducted. Certain conditions are tested to see if all the edge cases are handled in both front end and the back end. No manual automation tools are used.

Does the travel time obtained meet real-time travel time?

Yes, it is cross verified by generating itineraries for known places.

Does the itinerary give the closest route between places?

Yes, it considers the latitude and longitude values and their differences between the.
places to find which is the nearest place to a given place.

How is it handling long distances?

If it's the case that the user will reach the next region in midnight, it will advise. the user to be in the region he is and give him a time to start to the next region, if he prefers a night journey, or he will start next day morning.

6 Conclusion

The system is user friendly, allows the user to customize the itinerary as needed, and provides recommendations based on the users travel history, which is an added future.

The itinerary generator has features like considering the user travel preferences, like night journey, the travel duration from one place to other, provides slot and necessary breaks for breakfast, lunch, snacks, and dinner, accommodation types are considered for recommending hotels. The system can fulfill the basic user needs. It uses machine learning algorithms for recommendations and helps to personalize recommendations.

7 Future Work

This is a first step of integration of multiple systems. Currently, the user is assumed to stay in the same hotel for a region. The itinerary generated should be extended to consider the number of days. Currently, the number of days constraint cannot be considered as is giving the option to select places for the user. User selects desired places. If we ask number of days, we will have to remove places selection, since both will contradict. The travel means such as travel by bus, travel by train, travel by flight must be implemented. The accurate timings of the places, such as temples which are only open for only certain durations in a day are to be considered.

References

1. Zhang, J., Wang, Y., Yuan, Z., Jin, Q.: Personalized real-time movie recommendation system: practical prototype and evaluation. Tsinghua Sci. Technol. **25**(2), 180–191 (2020). https://doi.org/10.26599/TST.2018.9010118
2. Anandhan, A., Shuib, L., Ismail, M.A., Mujtaba, G.: Social media recommender systems: review and open research issues. IEEE Access **6**, 15608–15628 (2018). https://doi.org/10.1109/ACCESS.2018.2810062
3. Yadalam, T.V., Gowda, V.M., Kumar, V.S., Girish, D., N. M.: Career recommendation systems using content based filtering. In: 2020 5th International Conference on Communication and Electronics Systems (ICCES), pp. 660–665. Coimbatore, India (2020). Doi: https://doi.org/10.1109/ICCES48766.2020.9137992
4. Lucas, J.P., Luz, N., Moreno, M.N., Anacleto, R., Almeida, A., Martins, F.C.: A hybrid recommendation approach for a tourism system. Exp. Syst. Appl. **40**(9), 3532–3550 (2013)
5. Chiang, H.S., Huang, T.C.: User-adapted travel planning system for personalized schedule recommendation, Department of Information Management, National Taichung University of Science and Technology, No. 129, Sec. 3, Sanmin Rd., Taichung 40401, Taiwan, ROC
6. Paradarami, T.K., Bastian, N.D., Wightman, J.L.: A hybrid recommender system using artificial neural networks. Exp. Syst. Appl. **83**, 300–313 (2017)
7. Umanets, A., Ferreira, A., Leite, N., Me, G.: A tourist guide with a recommender system and social interaction. In: Conference on Electronics, Telecommunications and Computers—CETC 2013, vol. 17, pp. 407–414 (2014)
8. Zhang, R., Liu, Q., Chun-Gui, Wei, J., Huiyi-Ma: Collaborative filtering for recommender systems. In: 2014 Second International Conference on Advanced Cloud and Big Data, pp. 301–308. Huangshan (2014). Doi: https://doi.org/10.1109/CBD.2014.47
9. Li, H., Cai, F., Liao, Z.: Content-based filtering recommendation algorithm using HMM. In: 2012 Fourth International Conference on Computational and Information Sciences, pp. 275–277. Chongqing (2012). Doi: https://doi.org/10.1109/ICCIS.2012.112
10. Beautiful Soup home page. https://pypi.org/project/beautifulsoup4/
11. Python Google Maps in PyPi. https://pypi.org/project/googlemaps/1.0.2/

Manufacturing Information Systems for Forestry Companies Based on Integrated Modular Approach «Many Companies—One Tool»

Anton Shabaev

Abstract The paper describes a modular methodology for the design of manufacturing information systems (MIS) on the example of forestry industry. The need to develop such methodology is justified by the current demand of forestry companies to increase the return on investment in information systems implementation. The proposed modular design methodology makes it possible to design and implement MIS that takes into account the specifics of a forestry enterprise and is less expensive compared to competing solutions. The advantage becomes possible due to the use of an integrated modular approach to a wide range of forestry companies "many companies—one tool" based on universal integrated platform solutions. This approach uses a matrix relation of a specific forestry enterprise with a universal technological process. The novelty of the proposed methodology is the automation of developing the mathematical model of a forestry enterprise by choosing unified procedures from the available set and tuning the parameters to account for the specifics of a particular enterprise. This allows using a unified set of functional diagrams, which reduces the time and cost of designing and modifying MIS based on them. The consistent approach to selecting the unified procedures leads to obtaining the mathematical model of a forestry enterprise, which becomes the core of its MIS. Application of this methodology makes it possible to design, implement and scale MIS for forestry companies faster and cheaper than using traditional approaches.

1 Introduction

Today, large and medium-sized forestry companies, both domestic and foreign, use manufacturing information systems (MIS) to improve their operations [1–3].

At large enterprises, most of the systems belong to the class of integrated corporate-level systems that automate all company functions, including supplies, production, and sales. They include a number of modules (subsystems) operating in a single information space and supporting the corresponding areas of activity:

A. Shabaev (✉)
Petrozavodsk State University, Petrozavodsk 185910, Russia
e-mail: ashabaev@petrsu.ru

T. Senjyu et al. (eds.), *ICT with Intelligent Applications*, Smart Innovation, Systems and Technologies 248, https://doi.org/10.1007/978-981-16-4177-0_63

marketing subsystem, production subsystem, financial and accounting subsystem, human resources subsystem, and management subsystem. Such systems are the most expensive.

At medium-sized enterprises, process control systems (PCS) are necessarily used, which automate the control of production operations. Such systems include advanced measurements of process parameters, as well as process control algorithms. However, for effective functioning, medium-sized enterprises need higher-level MIS to automate operational management, accounting and analysis, long-term and operational planning, sales and supply management, etc. These subsystems are also quite expensive.

In this regard, for the forestry enterprise as a customer, a number of requirements arise for the information system and its supplier: the best price/quality ratio, fast return on investment, prompt and inexpensive maintenance, high performance, and reliability. In other words, the Customer seeks to minimize the total cost of ownership.

On the other hand, IT companies, as MIS suppliers, must design and develop an information system that is competitive in the market and attractive for Customer in terms of price/quality ratio. To create such systems, it is necessary, firstly, to use modern development tools. At the same time, their acquisition costs are compensated by cheaper and faster service and support in the future. Secondly, it is necessary to apply new effective methods for designing models of the subject area and develop high-performance algorithms for solving information processing problems. This will contribute to the faster creation of reliable and efficient MIS, as well as lower maintenance costs.

Thus, it is important both for Customers and Suppliers to develop a methodology for designing manufacturing information systems to reduce the spent time and financial resources on both sides.

2 Methodology for Designing MIS for Forestry Companies Based on an Integrated Modular Approach

2.1 Analysis of Software for Production Management at Forestry Companies

The development of information and control systems for large-scale forestry companies has been going on for a long time and world-wide.

Among foreign developers of such software, the following companies should be distinguished: ABB (Sweden, Switzerland) and their ABB Optimizing Controls system for paper machines; EFI (USA) and their EFI PC-topp system for corrugated board mills; Esko (Belgium) and their Cape Truckfill system for loading goods into transport; Halco Software Systems Ltd. (Canada) and its suite of simulation systems for planning and optimizing wood supply and production of lumber, panels and veneer; Honeywell (USA) and their OptiVision system for planning and fulfilling orders at a pulp-and-paper mill; OM Partners (Belgium) and their system for planning and optimizing the production of corrugated and solid board; PiiMega (Finland) and their systems for planning and operational management of logging and sawn timber production; Savcor (Finland) and their systems for information management in forestry operations and ERP systems for sawmills; Tieto (Finland) and their systems for sales and production management for pulp-and-paper mills and logging companies.

Among the Russian developers of such software, it is necessary to highlight the following companies: "Neosystems North-West" (Petrozavodsk)—the developer of the systems "1C: Enterprise 8. Sawmill" and "Neosystems: BUMPROM" for automation of management and accounting at forestry and pulp-and-paper companies, as well as holdings and management companies; "Opti-Soft" (Petrozavodsk)—the developer of a suite of optimal production planning systems, consistently covering the main production processes of a forestry enterprise from forest management and logging, through sawn timber, plywood, and paper production and to shipment of finished goods and the production of corrugated packaging; "Packer 3D" (Moscow)—developer of a system and online service for calculating the loading of boxes, cylinders, and pallets into transport; "Prokont" (Vyborg)—the developer of accounting systems for the pulp-and-paper and energy industries.

2.2 *Choosing the Design Direction*

Analysis of the current market of MIS for forestry industry shows a steadily growing demand for management information systems, in particular, for integrated systems. Automation of separate functions, e.g., accounting or sales, is no longer relevant for forestry companies.

Therefore, we address the task of increasing the efficiency of integrated corporate MIS development by improving methods of their design. This will increase the competitiveness of such systems and reduce costs both for developers and customers. The importance of choosing the right approach to building MIS and finding appropriate architectural solution is justified by many authors, e.g., in [4].

Typical tasks of various modules of a corporate MIS are presented in Table 1.

The stages of MIS development life cycle usually include: collecting requirements, design, implementation, testing, commissioning, operation, and maintenance [5].

The following life cycle models are currently known and used:

- The waterfall model consists in sequential execution of project stages in a strictly fixed order. The transition to the next stage occurs only after completion of the previous stage.
- Step-by-step model with intermediate control. MIS development is carried out in iterations with feedback collection between stages. Inter-stage adjustments allow to account for the really existing mutual influence of development results at various

Table 1 Subsystems and tasks of a corporate MIS

Marketing subsystem	Production subsystems	Financial and accounting subsystems	Human resources subsystem	Other subsystems (e.g., management)
Market research and sales forecasting	Scheduling	Order portfolio management	Analysis and forecasting the needs for labor resources	Control over the company activities
Sales management	Operational control and production management	Financial planning, budgeting, and financial management	Maintaining archives of personnel records	Identifying operational problems
New products recommendations	Analysis of equipment operation	Financial analysis and forecasting	HR development	Analysis of managerial and strategic situations
Price analysis	Orders to suppliers	Accounting and payroll	Health and safety	Supporting the strategic decision-making
Demand analysis	Stock management	Supplier relations and payments	Office supplies	

stages. The implementation of each stage stretches over the entire development period.

- Spiral model. At each iteration of the spiral, developers create the next version of the product, clarify requirements, determine quality and plan the next iteration. The initial development stages—analysis and design—require special attention, since the feasibility of technical solutions is checked and justified by prototyping.

When organizing the development of MIS, either canonical or typical design approach is usually used [5].

Canonical MIS design mainly uses the waterfall life cycle model. Due to the known shortcomings, this model is used mainly in the design of fairly simple systems, where each application is independent by function and data.

So, when developing a modern methodology for designing MIS for forestry companies, we should focus on typical design approach within the spiral life cycle model.

Typical MIS design involves the creation of a system from ready-made standard elements. The fundamental requirement for the application of typical design approach is the ability to decompose the system into components (subsystems, modules, etc.).

In this case, for the implementation of individual modules, standard third-party components available on the market can be selected as well, which are then configured for the system being developed.

The application of the typical design approach results in a typical design solution (TDS) that can be replicable and reusable. The accepted classification of TDS is based on the level of system decomposition:

- Elemental TDS use standard components or solutions of a task (software, mathematical, organizational);
- subsystem TDS use subsystems, developed taking into account the typical function with minimal external information links;
- object TDS use typical industry projects with a full set of functional and supporting subsystems that are customized during commissioning project.

Thus, the main idea of developing a modular methodology for the design of corporate MIS for forestry companies within the typical design approach is the automation of creating subsystem TDS, taking into account specifics of the industry. When developing the new methodology, we will focus on modeling business processes as the initial stage in MIS creation. We consider it the basis for successful implementation of all other stages of MIS creation.

2.3 Development of a Design Methodology for Mathematical Modeling of Business Processes

The proposed methodology for modeling the MIS subject area (with known limitations) may be attributed to a synthetic technique that combines the advantages of a functional and object-oriented technique.

This methodology is based on an integrated modular approach to the automation of a wide range of forestry companies: "many companies—one tool." This approach uses a matrix model to relate processes of a particular forestry enterprise with universal technological procedures in MIS. The mathematical model of customer business processes is obtained by selecting unified components taking into account the parameters of a particular enterprise.

The "many companies—one tool" concept is based on a systems approach that implements the "general-to-specific" principle. One unified matrix tool is proposed that enables describing each production process stage of a forestry enterprise by choosing the appropriate models from the available set and tuning the parameters.

This classifies the proposed methodology as a subsystem TDS. Despite the versatility of the tool, the model of each enterprise is unique, because for each production process the appropriate model is chosen independently and according to its specifics.

The concept "many companies—one tool" is based on the use of a binary matrix "processes—procedures," where the rows are formed by a formalized set of production processes of forestry companies, and the columns are the so-called "procedures/operations," which are, in the general case, dynamic mathematical models relating a variety of input parameters (material, energy, financial, time, human and information resources, and control algorithms) onto a set of output parameters (products, waste, costs) at a given moment of time.

As a result, the methodology of designing MIS for forestry enterprise consists of the following stages:

- Defining a set of resources and sets of semi-finished and finished products of the enterprise.
- Creating a generalized list of acceptable unified procedures.
- Defining the relation matrix "technological process"—"unified procedure."
- Defining the production chart of the enterprise in the form of a three-dimensional matrix "operation"—"product"—"operation number."
- Defining multiple functional and range constraints.
- Defining the vector of control actions and the vector of controlled parameters.
- Defining the goal function and performance criteria.
- Developing the consistent approach for selecting the unified procedures, taking into account the objectives of a particular enterprise.
- Developing the mathematical model of the enterprise business processes by applying the consistent approach of selecting the unified procedures.

3 Results and Discussion

On the basis of the proposed methodology, a suite of systems for optimal planning and management of complex production processes of industrial enterprises have been developed. This is possible and efficient because such systems typically use similar set of menus, forms, tables, dialog boxes, and other components, as well as optimization algorithms [6].

The specialized library ("universal" solver") is another main distinctive feature of the platform developed based on 30 years of experience in customer projects for companies of Russia and Finland. Algorithms for effective solution of complicated cutting problems, including linear and nonlinear optimization problems, high-dimension problems, problems with combined criteria, etc., are implemented based on the versatile solver. The library also includes a special module ("matrix constructor") for increasing efficiency of constructing, storing, and using constraints matrix [7].

Opti-Paper [8] system for optimization of paper/cardboard production enables finding the optimal production plan for several paper/cardboard machines, taking into account the production technology, regardless of the number of customer orders and process equipment configuration. The system is in use at 3 pulp-and-paper mills in Russia.

Opti-Loading [8] is a web-based service for automatic solution of a complex of tasks related to the choice of vehicles and loading finished goods. The service enables to optimally distribute finished goods (rolls of paper/cardboard of various formats and diameters, as well as pallets and boxes or mixed load) to vehicles: containers, wagons, trucks. The system is in use at Mondi Syktyvkar (Russia's largest producer of paper good in), Kondopoga pulp-and-paper mill (Russia's largest producer of newsprint paper), 5 corrugated board producers, as well as TechnoNICOL corporation—one of the largest international producers of construction materials. The use of the service in many cases enables to increase by 4–5% the total load in a vehicle considering a wide range of constraints.

Opti-Corrugated [8] is the leading software tool in Russia for planning and managing the production of corrugated cardboard goods. The system unites commercial and production departments of an enterprise, ensures "just-in-time" execution of orders and production costs reduction. It utilizes advanced software development technologies together with complex mathematical models and optimization algorithms. The system ensures production of balanced amounts of rectangles of cardboard for boxes with minimal material losses, considering the production capacity, equipment re-tuning, order scheduling, and priorities. The use of the systems enabled the customers to reduce material losses from 5–6% to 2–4%, and increase the production uptime by 3–4%, which saves several hundred thousand euro per year.

Opti-Sawmill [7, 8] is a decision-support system developed specifically for sawmills and woodworking mills as a tool for efficient planning, optimizing, and managing production and sales. The use of the system allows to increase the yield of lumber by 1.1–1.6%. The system is installed at 10 sawmills in Russia (from North-West through Eastern Siberia to Far East) and at 2 sawmills in Germany including Ilim Nordic Timber GbmH, which is the largest sawmill in world by processing volume.

Opti-Wood is a decision-support system for efficient planning of logging and wood transport operations, construction, and maintenance of forest roads. It is designed for wood harvesting companies as a production optimization tool for planning and management. The system is used by a forestry company in Karelia. In [9, 10], we provide the mathematical model and solution approach to optimizing the profit measured as revenue from the sale of products, net of harvesting, and transportation costs. During several case studies involving 198 sites and 14 machines harvesting up to 200,000 cubic meters from an available stock volume of about 300,000 cubic meters, an increase in profit by 5 to 10% was observed.

Opti-Plywood [11] is a web-based service for plywood mills as a tool for optimizing production, starting from peeling and drying of veneer to pressing and cutting plywood. The system is in use at one of the largest softwood plywood mills in Russia.

4 Conclusion

The methodology for designing industrial information systems, proposed in the paper, has been successfully applied during the development of multiple systems for forestry companies in Russia and abroad. The (achieved) purpose of applying the modular approach "many companies—one tool" was to reduce the time spent on developing mathematical models of production processes. Consistent approach to obtaining the complete model of business processes of an enterprise also reduces the risk of design errors due to the human factor and enables to accumulate best practices. The structural identification of the mathematical model of a production system using optimization algorithms increases the functional adequacy of simulation results to real production processes, because it enables using solutions tested and proven in similar projects. Application of this methodology enables designing, implementing, and scaling MIS for forestry companies faster and cheaper compared to traditional approaches. The time complexity of the implementation of the proposed methodology varies significantly on case by case basis. However, longer initial development period pays off by saving time at implementation, maintenance, and dissemination stages, as generally known [5].

The practical experience in developing and commissioning multiple MIS for forestry companies with the use of this methodology has also enabled customers to decrease unit consumption of raw materials and increase marginal profit. This has led to significant economic benefits, as evidenced by more than 50 large and medium-sized forestry companies in Russia and foreign countries.

Thus, the use of the described methodology for the design of information systems for forestry companies reduces the time and costs both for IT companies as developers and for the forestry companies as customers.

References

1. Accenture: Forest products: the shift to digital accelerates. Natural Resources. https://www.accenture.com/us-en/insights/natural-resources/digital-forest-products-survey (2018)
2. CEPI: What digital can do for the paper industry. http://www.cepi.org/publication/paper-industry-40-what-digital-can-do-paper-industry (2016)
3. Price water house Coopers: Industry 4.0: Building the digital enterprise. Global Industry 4.0 Survey—2016. https://www.pwc.com/industry40 (2016)
4. Fowler, M.: Software Architecture Guide. A website on building software effectively. https://martinfowler.com/architecture/ (2019)
5. ISO/IEC 12207:2017 System and software engineering—Software life cycle processes (2017)
6. Shabaev, A., Kuznetsov, V., Kositsyn, D.: Optimal planning software platform development with cloud computing technology. In: Proceedings of the 12th Conference of Open Innovations Association FRUCT, pp. 129–135. State University of Aerospace Instrumentation (SUAI), Helsinki. https://www.fruct.org/publications/fruct12/files/Sha.pdf (2012)
7. Voronin, A., Kuznetsov, V., Arkhipov, I., Shabaev, A.: Software system for sawmill operation planning. In: Proceedings of the 12th Conference of Open Innovations Association FRUCT, pp. 165–171. State University of Aerospace Instrumentation (SUAI), Helsinki (2012)
8. Shabaev, A., Voronin, A., Kuznecov, V., Arkhipov, I.: "Cloud" services for improving production efficiency of industrial enterprises. In: Proceedings of the Maintenance Performance Measurement and Management Conference, pp. 315–329. State University of Aerospace Instrumentation (SUAI), Helsinki (2013)
9. Shabaev, A., Sokolov, A., Urban, A., Pyatin, D.: An approach to the optimal timber transport scheduling. In: E3S Web Conference 164 03019. State University of Aerospace Instrumentation (SUAI), Helsinki (2020). https://doi.org/10.1051/e3sconf/202016403019
10. Shabaev, A., Sokolov, A., Urban, A., Pyatin, D.: Optimal planning of wood harvesting and timber supply in russian conditions. Forests **11**, 662 (2020). https://doi.org/10.3390/f11060662

11. Shabaev, A., Arhipov, I., Spirichev, M., Urban, A., Torozerov, M.: Development of planning system for plywood production using "Matrix Designer". In: Proceedings of the 14th Conference of Open Innovations Association FRUCT, pp. 140–147. State University of Aerospace Instrumentation (SUAI), Helsinki (2014)

ICT-Based Enhancement of Employment Schemes: A Case Study in Rural Uttarakhand, India

Aaryan Oberoi, Sritha Bandla, Harini Mohan, Sagar Basavaraju, Saurav Bhattacharjee, Subrahmanyam Raparthi, L. M. Frey, and Souresh Cornet

Abstract 70% of India's population resides in rural areas, where unemployment is a major concern. MGNREGA scheme was launched by the Indian government to tackle this issue. Implementation of this initiative on a massive scale covering over 120 million people has been hampered by challenges and drawbacks. This study explores the propriety of technological innovations toward tiding over the deficiencies and expansion of the program. This investigation entails a mixed method approach in a village of Uttarakhand, in North India, exploiting ethnography and Human Centered Design triangulated with statistical data. This deep and grounded understanding enables to propose an ICT-based solution design in the form of a work tracking system using GPS and a specific architecture to address the pain point in the implementation of MGNREGA in remote rural areas.

1 Introduction

India's rural development programs have been impeded by unemployment rates of 5.8% as per Govt. of India (GOI) in 2018 [1]. This is of major concern, as over 65% of India's population lives in rural areas [2].

The Indian Government launched the Mahatma Gandhi National Rural Employment Guarantee Act (MGNREGA) in 2006, to confront these challenges. MGNREGA scheme aims at improving the livelihood of rural households by providing up to 100 days per year of guaranteed unskilled manual work wage to every registered adult member of a household [3]. Every household must be provided

A. Oberoi · S. Bandla · H. Mohan · S. Basavaraju (✉)
Department of Electronics and Communication Engineering, Amrita Vishwa Vidyapeetham, Bengaluru, India
e-mail: b_sagar@blr.amrita.edu

S. Bhattacharjee · S. Raparthi
Department of Mechanical Engineering, Amrita Vishwa Vidyapeetham, Bengaluru, India

L. M. Frey · S. Cornet
Amrita Center for International Programs, Amrita Vishwa Vidyapeetham, Amritapuri, India
e-mail: soureshcornet@am.amrita.edu

T. Senjyu et al. (eds.), *ICT with Intelligent Applications*, Smart Innovation, Systems and Technologies 248, https://doi.org/10.1007/978-981-16-4177-0_64

with employment within 15 days of their registration. If not, the government has to provide an unemployment allowance to the job applicant [4]. This GOI initiative is the largest public employment scheme in Asia, with 127 million registered members in 2020 [4].

The implementation on the ground had contrasting results depending on states. While it has reduced the vulnerability of farmers in some places [5], shortcomings were reported about delays of payment, due to deficient monitoring of the scheme in other parts [6]. Because of that there is a "Discouraged Worker Effect" developing, where people who should benefit from the scheme are not motivated to apply anymore [7].

Earlier pilot studies have shown that Information and Communication Technologies (ICT) can be a powerful help for taking on these drawbacks [8, 9]. There is now an online platform for MGNREGA. However, it only partially addresses the problem: in many remote areas of India, the scheme still requires someone to physically go check if the assigned MGNREGA job was completed by the enrolled workers, before initiating payment to them. Beyond the technical dimension, there are also internal social factors about the users such as caste, gender, and age that have been least explored when designing technical solutions for such large scale e-governance schemes [10].

This paper will explore how technology can enhance the implementation of such a scheme, through better understanding the local factors that prevent its implementation in a rural village of Uttarakhand.

This study was conducted in the village of Dunda, Uttarakhand, as part of the Live-in-Labs® program at Amrita Vishwa Vidyapeetham [11]. This multidisciplinary experiential learning program gives students and researchers the possibility to live with rural communities, so as to better understand their livelihood and propose relevant solutions to address their challenges.

2 Study Area

The rural community of Dunda, in the North Indian state of Uttarakhand, is part of the 101 villages adopted by Amrita SeRVe program of MA Math [12, 13]. It has a population of 1157 spread across 228 houses. The vast majority of village residents are farmers, and use traditional cultivation practices for their subsistence. They occasionally sell the surplus, and also rear cattle.

The village has a women's Self-Help Group (SHG) called Mahila Mandal. An SHG is a group of 10 to 20 women from the same village who collaborate in economic development activities and save small amounts of money. Several skill and vocational trainings have been organised in the community [14]. More than 300 people in Dunda village are graduates. The literacy rate is strikingly high with an 89.3%. Yet, Dunda suffers from chronic unemployment.

3 Research Methodology

Our methodology was in 2 phases. First, we studied the community using mixed methods and an ethnographic approach to understand them deeply. Based on this the second phase was to propose an ICT-based solution to address the challenges and gaps identified.

As part of the ethnographic study design, Human Centered Design (HCD) was used for studying the community. HCD revolves around analysis and incorporation of the end user's requirements, needs, and capabilities in designing a service or a product [15–17]. To do so, the project team lived with the community for 6 days for experiencing their daily life. As tried and tested in previous studies [18, 19], this approach helped understand the end users experience, and get new ideas for designing relevant solutions. The HCD tools used were (1) participant observation (2) interviews, (3) personas and scenarios.

3.1 *Participant Observation*

This data collection method aims at getting a firsthand understanding on communities by living among them and observing their daily life. The AEIOU framework was used for that. It organizes the observation around key areas (activities, environment, interactions, objects, and users) to ensure that all relevant observations are focused on [20]. Notes were taken on activities and their environment, on the objects used in the interactions, and the people involved. These observations provided an in-depth cognizance of the atmosphere and facilities present in the homes of the people. It helped by providing a microscopic view.

3.2 *Interviews*

Participant observation was complemented by semi-structured interviews, in order to let the villagers express themselves freely. Questions were both qualitative and quantitative and addressed matters related to MGNREGA such as occupation, income, skills, education qualifications, and secondary income source. The project team launched the fieldwork by spending time with the villagers in order to build a rapport with them: casual meetings during daily yoga activity for children, volunteering in a health camp, and attending a meeting with Gram Panchayath (village council). A first round of preliminary discussions was conducted during the medical camp. It helped preparing semi-structured interviews that were administered door to door. Interviews were also conducted with the SHG ladies, the Block Development Officer (BDO), and local MGNREGA office.

3.3 Personas and Scenarios

The interviews provided an in-depth understanding of the villagers' experience related to the MGNREGA. Responses with similar pain points were grouped in three different categories called Personas [21]. Based on that, scenarios are enabled to sincerely understand the specific pain points undergone by the community [22]. Scenarios of the following people were taken: community members participating in MGNREGA and government officials coordinating the scheme.

3.4 Secondary Data

All these primary data were triangulated with secondary data to verify their accuracy: survey report conducted by Amrita SeRVe was compared with census [2], and MGNREGA survey report [4].

4 Results

4.1 Results 1

Interviews and observations. Nineteen patients who attended the health camp were interviewed during the pilot sessions. Then, the main questionnaire was administered to 45 houses across the village (sample size: 21%). It was observed that most of the arable land seemed unused. On those few actively being used fields, women did all the work. Men were usually seen drinking tea and playing cards.

Personas and Scenarios. Persona 1 describes the majority of women in the village who perform farming as an integral part of their daily routine and enroll in MGNREGA. They are well aware of the problems and prefer their children not to suffer the same. Children of this category of villagers are either studying or working far-off from the village. This persona helped to understand the typical mindset of villagers.

Persona 2 is typically a woman from the SHG. She is determined and has numerous skills besides agriculture and household routine. She is willing to learn and grow but is not getting much in return. SHG activities are seasonal: knitted sweaters are sold only in winter and schoolbags before school reopens. During all other times, they rely on MGNREGA. This illustrates how central the scheme is to the community livelihood.

The scenarios highlight how discouraged the community members are of working, yet not being paid. MGNREGA scheme officials claim that the villagers did not complete 100-days work, and that is the main reason for not paying them. From their

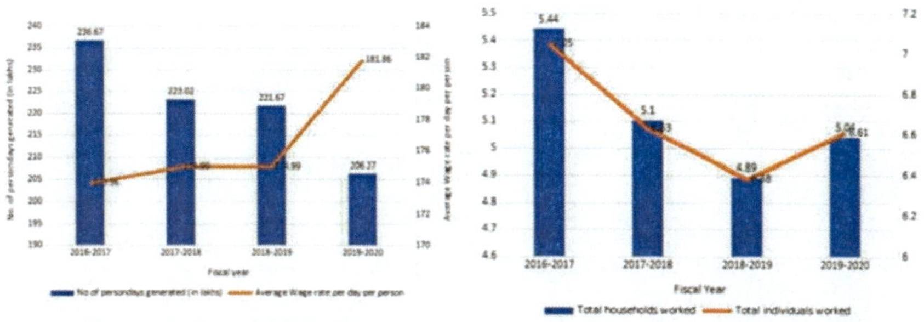

Fig. 1 Number of person days generated verses Average Wage Rate per person (left) Vs Analysis of total work per Household (right)

side, getting confirmation that the work has been genuinely completed is the pain point.

Outcomes. Information collected from the villagers and officials was verified with government secondary data. It was found that most of the villagers do not know how the scheme works. No proper measures were taken to inform them.

Secondary data show that although the average wage per person per day has increased by 4.5%, the number of person days has steadily decreased by 13% (Fig. 1 left) and the amount of work assigned since 2016 has globally reduced (Fig. 1 right). The number of people working under the scheme has shrinked by 28%. Interviews confirm that due to delayed payments a negative perception has developed. People reported not to be willing to work again because of this bad experience.

4.2 Results 2

Overview of devised solutions. Based on the users profiles and inputs from community members and local authorities, a first level ICT-based solution has been co-developed with them to address the challenge of MGNREGA salaries payment (Fig. 2). The requirement is a system architecture where the work completed by the participants is documented fairly, to enable rightful payment of wages. The architecture should make it easy for the workers to choose jobs they are interested in. Hence, a platform, preferably an online application, enabled with unique user IDs is proposed. It acts as a portal on which workers can access work opportunities, choose one, and report to work at the job site. To track the extent of work completion, we propose to use a cost-effective hand movement tracking band. It feeds information to a database, that can later be analyzed to make sure the individual was actually doing the work. Since the nature of jobs observed in the village mostly consist of plumbing and carpentry, hand movement tracking is considered a sensible parameter.

Hardware layout. Displayed in lower part of Fig. 2 is a block diagram of the hardware to be used for the hand movement tracking band. The Inertial Measurement

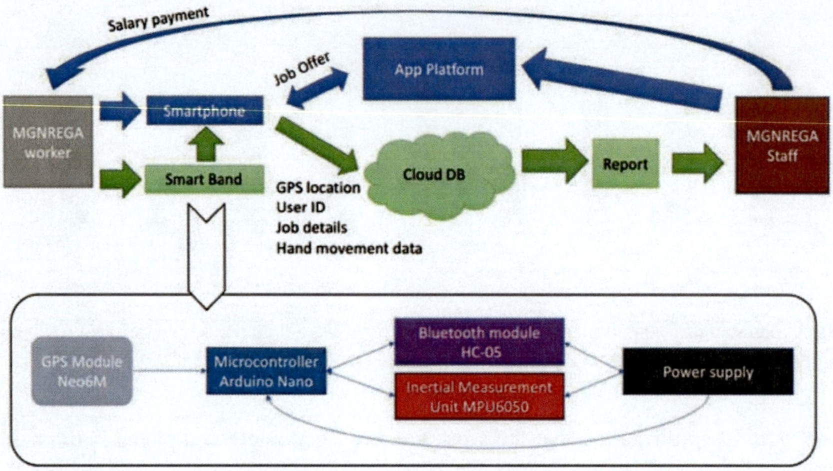

Fig. 2 Schematic of the new proposed process, with block diagram of the proposed tracking band

Unit (IMU) provides the parameters yaw, pitch, and roll responsible for translation and rotatory hand movements. A reading is taken in frequent intervals and momentary stored in a local memory. The GPS module will make sure that the individual was present at the geotagged location of the job. Once the individual completed the job and is in the vicinity of his phone, the Bluetooth module in the band auto connects to the phone and enables transfer of data collected onto the cloud database (DB). The Arduino Nano microcontroller will regulate all the processes accordingly.

Data Analysis. The job is chosen, the hand movement data, the worker location, will all be referenced with respect to a unique user ID. In the DB, the data will be passed through a machine learning (ML) model, which is pretrained with reproduced data similar to the one collected from the band. This model will classify as to whether the readings result in a hand movement relevant to a work procedure or not. For example, the multi-class classification can involve classes, namely carpentry, plumbing, and construction to begin with. Since each of them has their own signature pattern, the readings of yaw, pitch, and roll will also hold a pattern that will be distinctive to the three respective classes. The reproduced data used for training the ML model can be easily generated by any individual wearing the band and then subsequently tagging the recorded values with their respective classes. Once this tagged data is used to train the ML model, it can be used to classify untagged data from real workers. This information pertaining to the status of the work done can be obtained in the form of a report by the MGNREGA coordinator. The report can be evaluated and the wage payment can be initiated accordingly, as shown in Fig. 2. This will ensure transparency through proper documentation of the work done, satisfying the requirements of both sides.

5 Discussion

The community is traditionally dedicated to agriculture. There is an unwillingness toward learning new skills, rooted in the local culture, which explains why unemployment remains a major challenge. Agriculture employment under MGNREGA scheme is their main sources of employment. Our findings on the lack of awareness about the scheme, and the lack of communication from local officials is consistent with the literature on MGNREGA discrepancies [8]. Dunda is a textbook case of "Discouraged Worker Effect," [7] where there is work available, but nobody wants to do it anymore.

In line with literature on ICT and E-governance schemes [8, 9], a tracking system is proposed to help assess work completion and improve the transparency of transactions. With the proposed architecture, proper documentation will be created, making it very unlikely for a worker who has genuinely completed his job to be unpaid. GPS tracking will ensure workers cannot wrongly claim the wages without actually being on the site, and the smart band will track their gestures to ensure they have been genuinely working. The final end report emission will avoid situation where a MGNREGA coordinator could wrongly claim that a particular individual did not show up. The database not only provides clarity but acts as a one true record with which future records can be compared using ML. Of course, this proposed design will have to be tested on field to unsure its operability and reliability. At the moment, there are expected limitations in terms of the implementation on the ground. There are chances of the system being confused with different types of tasks that might have similar movements, and the style of work of different individuals, being right or left handed. All this will be addressed with future pilot studies. With this, the current scenario in Dunda will definitely improve.

6 Conclusion

The contribution of this study lies in its field-based approach, that helped to better understand the context, mindset, and challenges of the rural community, before proposing a design that truly answers their needs. Findings confirm the importance of rural employment schemes on the livelihood of rural communities. However, it also confirms the known drawbacks of the MGNREGA scheme. The use of HCD method has been very important for understanding the social factors that prevent the successful implementation of MGNREGA in rural Uttarakhand. Is also allowed to identify ways to address them. The main innovation proposed is to integrate a work tracker system with an ICT-based solution to fulfil a critical requirement of the scheme, which is the payment of salaries. ICT can automate and improve the transparency of the scheme. An effective implementation of MGNREGA by which the workers' progress on the field is tracked will prove to be beneficial for all and may restore faith in the scheme.

Such technical innovation rooted in the actual needs of the users and combined with the massive reach of government programs has the capacity to impact the lives of millions. This study presented a first level design, further research is ongoing to develop and deploy it.

Acknowledgements This research was funded by Amrita Vishwa Vidyapeetham's Live-in-Labs® program, and the field implementation was funded by the UN-recognized NGO, M.A. Math. The study was done as part of the UNESCO Chair for Experiential Learning for Sustainable Innovation Development. The authors express their immense gratitude to Sri. Mata Amritanandamayi Devi, Chancellor of Amrita Vishwa Vidyapeetham, who has inspired them in performing selfless service to society. We would like to express our sincere appreciation to Mr. Gopalaiengar Sheshadri for his extensive support in enhancing the quality of the paper within the limited time frame. The authors extend their gratitude to the ASeRVe Program Head Swami Jnanamritananda Puri, Live-in-Labs® staffs for providing all the necessary support with regards to logistics and accommodation, which facilitated the authors to conduct their study smoothly. We would like to express our deepest gratitude to the members of the community for participating in the study and to those who went out of their way to provide extended support during the study in the village. We also thank the government officials of the district of Uttarkashi, Uttarakhand, and staff of the Amrita Self Reliant Villages (Amrita SeRVe) program.

References

1. Government of India: Periodic Labour force survey 2017–2018 (2018)
2. Government of India: National census (2011)
3. Government of India: Mahatma Gandhi national Rural Employment Guarantee Act, 2005 (Mahatma Gandhi NREGA): Operational Guidelines 2013. Ministry of Rural Development, New Delhi (2013)
4. MGNREGA Government of India http://MGNREGAweb4.nic.in
5. Esteves, T., Rao, K.V., Sinha, B., Roy, S.S., Rao, B., Jha, S., Singh, A.B., Vishal, P., Nitasha, S., Rao, S., IK, M.: Agricultural and livelihood vulnerability reduction through the MGNREGA. Eco. Polit Wkly. 94–103 (2013)
6. Salian, P.V., Leelavathi, D.S.: Implementation of MGNREGA in Karnataka: issues and challenges. J. Rural. Dev. **33**(3), 261–279 (2014)
7. Narayanan, S., Das, U., Liu, Y., Barrett, C.B.: The "discouraged worker effect" in public works programs: evidence from the MGNREGA in India. World Dev. **100**, 31–44 (2017)
8. Masiero, S., Maiorano, D.: MGNREGA, power politics, and computerization in Andhra Pradesh. Forum Dev. Stud. **45**(1), 1–24 (2018)
9. Gupta, P., Jain, S., Vashishth, N.: E-Governance in India: a case study of lever aging Information Communication Technology (ICT) for women empowerment under MGNREGA in the State of Rajasthan. J. Gov. Public Policy. **7**(1) (2017)
10. Johnson, D.: How do caste, gender and party affiliation of locally elected leaders affect implementation of NREGA?. SSRN Electron. J. (2010)
11. Ramesh, M.V., Mohan, R., Menon, S.: Live-in-Labs: rapid translational research and implementation-based program for rural development in India. In: GHTC 2016-IEEE Global Humanitarian Technology Conference: Technology for the Benefit of Humanity, Conference Proceedings, pp. 164–171 (2016)
12. Amrita SeRVe: The First 5 Years Report (2019)
13. Amrita SeRVe—Self Reliant Village. https://amritaserve.org/

14. Rao, B.R., Sheshadri, S., Maciuika, L.A.: Addressing the first teachers: education and sustainable development for children, families and communities through vocational education, technology and life skills training for women. In: Children and Sustainable Development, pp. 319–334. Springer, Cham (2017)
15. Brown, T., Wyatt, J.: Design thinking for social innovation. Dev. Outreach **12**(1), 29–43 (2010)
16. Ranger, B.J., Mantzavinou, A.: Design thinking in development engineering education: a case study on creating prosthetic and assistive technologies for the developing world. Dev. Eng. **3**, 166–174 (2018)
17. Ganesh, V., Sharanya, B., Vyshnavi, U.S., Pradeep, K.R., Kumar, A.G., Chaitanya, G.P., Frey, L.M., Cornet, S. and Chinmayi, R.: Technology for addressing income insufficiency in Rural India. In: 2020 IEEE R10 Humanitarian Technology Conference (R10-HTC), pp. 1–6. IEEE press (2020)
18. Guleria, M., Rajesh, T., Panicker, A.J., Rayappa R.S.R., Soundharajan. B., Frey, L.M., Nandanan, K.: Using human centered design to improve socio-economic livelihoods with modernized irrigation systems. In: 2020 IEEE R10 Humanitarian Technology Conference (R10-HTC), pp. 1–6. IEEE press (2020). https://doi.org/10.1109/R10-HTC49770.2020.9357058
19. Vempati, A.V., Aswath, G.I., Radhakrishnan, S., Raakheshsubhash, A.R., Kothurkar, N.K., Frey, L.M., Nandanan, K.: Agricultural problems and technology- based sustainable solutions for an impoverished Village of Bihar, India. In: 2020 IEEE R10 Humanitarian Technology Conference (R10-HTC), pp. 1–6. IEEE press (2020). https://doi.org/10.1109/R10-HTC49770.2020.9356972.
20. Lee, M.J., Yuan, W., Henry B.L.D.: AR UX design: applying AEIOU to handheld augmented reality browser. In: 2012 IEEE International Symposium on Mixed and Augmented Reality—Arts, Media, and Humanities (ISMAR-AMH), pp. 99–100. IEEE (2012)
21. Miaskiewicz, T., Kozar, K.A.: Personas and user centered design: How can personas benefit product design processes? Des. Stud. **32**(5), 417–430 (2011)
22. Carrol, J.M.: Five reasons for scenario-based design. In: Proceedings of the 32nd Annual Hawaii International Conference on Systems Sciences, pp 2, IEEE (1999)

Sustainable Technologies for Improved Income Generation: A Case Study of Women Villagers in Himachal Pradesh

Aditi Prasad, C. Dinanath, Kiran Haridas, P. A. Gopika, Sooraj Menon, Jayashree Nair, L. M. Frey, and Krishna Nandanan

Abstract In India, about 65% of the population resides in rural areas and have minimal access to employment opportunities. This pose uphill barriers for effectual rural development. With 33 million Indians unemployed, coupled with about 14% of rural women unemployed (as of 2019), livelihood options for rural women are a matter of grave concern. Similar is the situation faced by women residing in Indpur, a rural village in the northern Indian state of Himachal Pradesh. As part of the Live-in-Labs® program, a study was conducted following the principles of Human Centered Design (HCD), for the appraisal of challenges to livelihood opportunities, supervening the Indpur community. The HCD concept is further exploited to propose a sustainable technological proposition in the form of Aqueous Enzymatic Extraction of oil from Moringa seeds.

1 Introduction

According to a World Bank Report in 2019, India's rural population accounts for 65.528 % of its total population [1]. India's rural population sector is hampered by

A. Prasad · C. Dinanath · K. Haridas · P. A. Gopika
Department of Electronics and Communication Engineering, Amrita Vishwa Vidyapeetham, Amritapuri, India

S. Menon
Department of Computer Science and Engineering, Amrita Vishwa Vidyapeetham, Amritapuri, India

J. Nair
Department of Computer Science and Applications, Amrita Vishwa Vidyapeetham, Amritapuri, India
e-mail: jayashree@am.amrita.edu

L. M. Frey · K. Nandanan (✉)
Amrita School for Sustainable Development, Amrita Vishwa Vidyapeetham, Amritapuri, India
e-mail: krishnanandanan@am.amrita.edu

L. M. Frey
e-mail: lisamfrey@am.amrita.edu

T. Senjyu et al. (eds.), *ICT with Intelligent Applications*, Smart Innovation, Systems and Technologies 248, https://doi.org/10.1007/978-981-16-4177-0_65

dependencies on irregular, sparse, and uncertain income opportunities, which often lead to adversities of long-term unemployment. According to a Center for Monitoring Indian Economy report, approximately 33 million Indians were unemployed (willing to work and active job seekers) between September and December, 2019. The rural population constituted about 61% of the unemployed population [2]. These figures highlight the sparse employment opportunities available to rural communities.

The fallout of these notably high-unemployment statistics is exacerbated, when the women unemployment rates are accounted for. About 37% of Indian women were employed in 2005, dropping down to 28% in 2016 [3, 4]. With regards to the rural women, for the period between September and December 2019, the unemployment rates stood at 13.9%, versus the male unemployment rate of 5.8% [2]. In rural areas, where multiple income sources within a household would be beneficial, it is understood from the unemployment rates that the women workforce is underutilized.

With a substantial amount of India's population residing in rural communities, it is important to empower these communities with adequate livelihood opportunities. Flourishing rural communities are an imperative to both regional and national development. Working along these lines, a research team, comprised of students and faculty mentors from Amrita Vishwa Vidyapeetham, conducted an ethnographic and multidisciplinary research in Indpur—a village located in the northern Indian state of Himachal Pradesh. The study was carried out as part of the Live-in-Labs® program [5]. Working for the Indpur community in Himachal Pradesh was motivated by the state's 20% unemployment rate, one of the highest among all the Indian states [2, 6].

The purpose of the study was to comprehend and analyse the experiences endured by the families and dependents of the Indpur community members concerning lack of livelihood opportunities. This was accomplished through participatory research methods, underscored by concepts of Human Centered Design (HCD). This stance enabled the researchers to propose sustainable technological interventions to alleviate the financial hardships of the Indpur community.

This paper has been laid out as follows: Sect. 2 details the Indpur community and the area under study; Sect. 3 is focused on the methodology adopted in the study; Sect. 4 elucidates the outcomes of this probe and also dwells on the technological propositions as a sustainable pathway for the community. Section 5 presents the conclusion of the study.

2 Study Area

Indpur village is situated in the Kangra district in Himachal Pradesh—a state located in the northern region of India. This is one of the 101 villages adopted by the Amrita SeRVe program, as part of the UN-recognized NGO, Mata Amritanandamayi (MA) Math [7, 8]. Figure 1 details the location of the village [9]. The satellite map of the village as well as the caste constitution can be referred from the previous work done as part of the Live-in-Labs® program [10].

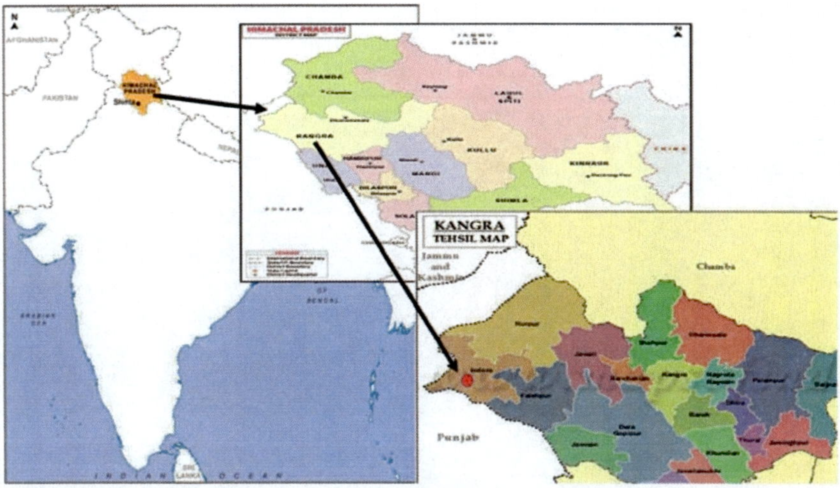

Fig. 1 Map detailing the location of Indpur village [9]

Indpur is constituted of 9 wards; the focus of this study was in wards numbered—6, 7, and 8. The study area was home to 291 households, with a population of 1263. Hindi and Pahadi are the languages spoken in the village [10].

3 Methodology

An ethnographic approach was adopted to gain an understanding of the Indpur village's resources, communal framework, and socio-cultural aspects. The research team resided in Indpur for seven days. The team investigated and recorded the primary data of the focus area (wards numbered 6, 7, and 8). Living among the village residents enhanced the study team's understanding about the research problem. This approach was chosen in order to fully experience and understand the local community's challenges, strengths, customs and most importantly, their struggles with lack of employment opportunities.

Furthermore, to rightly understand the needs and experiences of the village community, it was essential that the research was majorly participatory in nature. Engagement of the village community in the process prompted the research team to proffer apt user centric interventions that rightly address the needs of the community. Hence, the research team incorporated the HCD process, detailed in this section. The outcome of each of the HCD tools used is presented in the subsequent subsection "Results I".

HCD is a user centered approach, which lays emphasis on the needs of the end users, and engages them extensively in the process of designing solutions to address a problem. HCD is incorporated in the study to ensure that the proposed solution

Fig. 2 Research team conducting interview with a member of Indpur community

rightly meets users' requirements in every possible aspect [11–13]. The HCD method is a coordinated application of the following tools: participant observation, interviews, personas, and scenarios. Participant observation was undertaken to observe the users in their naturalistic settings, to understand various tasks performed by the community members amidst the prevalent external conditions. This could possibly highlight the subtle challenges faced by them, while performing their routine. In order to have the observation focused, amidst a wide spectrum of users and areas that could be observed within an observation frame, the AEIOU observation framework was utilized. AEIOU observation framework enables to focus the observations on the **Activities (A)** undertaken by the community members, **Environment (E)** in which the activities are conducted, various objects and people that the observants **Interacted (I)** with, **Objects (O)** that are present in the environment under focus, and the **Users (U)** [14].

In order to further assimilate user experiences associated with user goals, tasks and pain points, participant observations were complemented by interviews. To ensure that the respondents were provided ample space and opportunities to express their emotions, thoughts, perceptions, and feelings, the interviews were semi-structured which mainly comprised of open-ended questions [15]. This method encouraged informal conversations which were directed toward the activities/tasks that mattered to the community members, their aspirations, and negative experiences (in the form of pain points) (Fig. 2). The questions also covered the demographic, technographic, and psychographic details. This aided in obtaining crucial qualitative as well as quantitative data. The data enabled the researchers to understand and analyse the user goals and aspirations, the tasks they engage themselves in to achieve their goals, and the pain points they experience while performing their tasks. The pain points could either be a physical, an emotional, or a psychological pain. The participants considered for the interviews were mainly composed of the women residents of Indpur, which also covered the women Self Help Group (SHG) [16]. The samples were selected on random, in a manner that ensured there were participants from each ward as well as each SHGs.

In order to get a better understanding of the user groups to aid in effective analysis, personas were created based on the data obtained from the interviews [17]. Though fictitious, creating personas would help in understanding the user demography more broadly, as it is a representation of a collective user group with shared goals. Scenarios were then made, to deeply understand the pain points, enabling the researchers to focus on the magnitude of the pain endured by the users [18].

Since the data obtained from the study was more qualitative in nature, and data analysis was done mainly through thematic analysis. Here, the analysis was focused around the themes—goals, tasks, and pain points.

4 Results

4.1 Results I

From the participant observations, it was apparent that men were the prime bread winners of Indpur households, while women were engaged only in household chores. It was also observed that women spend most of their time in chat/conversations with their neighbors, which suggests that the women have ample time for themselves, once they are done with their household duties. Some women maintained kitchen gardens where they cultivated fruits like mangoes, oranges, litchi, and guava for self-consumption. On observing the various resources in the village, abundant *Moringa Oleifera* (MO) trees (drumstick trees) were observed to be growing naturally. The village communities use the leaves from the tree as cattle feed and are not utilized for the various nutritional and commercial benefits that can be derived from the plant. As understood from the community interactions, this was primarily due to the community members not being aware either of its numerous benefits or methods that can be used to extract the benefits.

From the interviews, the community members stressed on the sporadic nature of the jobs that they were engaged in. The male members of the village were predominantly engaged in daily wage labor, while few of them practiced agriculture on the lands that they owned (Fig. 3). The daily wage labor opportunities are uncertain and its availability is subject to work opportunities. This led many families in Indpur to face poverty situations. Women, on the other hand, revealed that the available job opportunities were mainly male-dominated jobs, and the employers did not prefer employing women for the jobs. This discrimination greatly impacted the household income, where the villagers reported certain periods where the household expenditure was far greater than the income. SHG members interviewed at group meetings

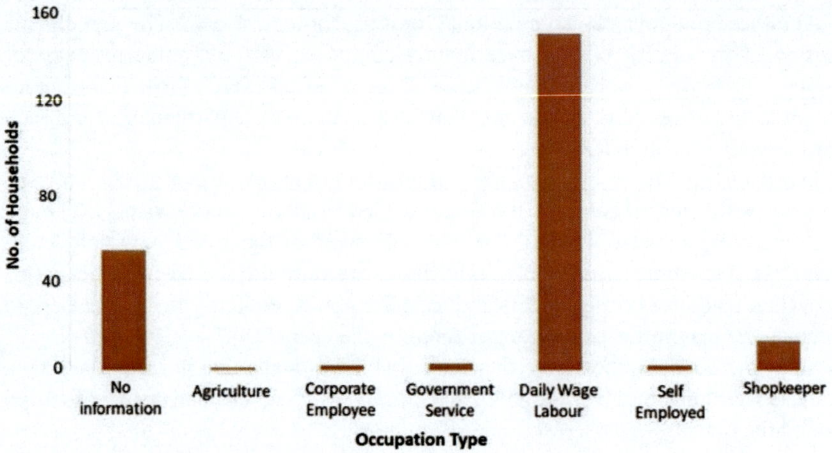

Fig. 3 Household-wise occupation distribution

stated that, on average, 2.5 hours can be spared by women in the village per day, for skill development courses. Three SHG groups, present in the village, are ready to work and earn for themselves but do not currently have the skills.

Personas consisted of the women community of Indpur who exhibited the following traits:

- Middle-aged women
- Desperate to increase household income
- Ready to spend the leisure time for doing work that could generate income
- Lack of work opportunities
- Eager to get trained to develop skills necessary for obtaining job.

Albeit the women community received training in tailoring from Amrita Vishwa Vidyapeetham and Amrita SeRVe, the inflow of orders was not frequent [19].

Hence, it is evident that the women were looking forward for opportunities to supplement their spouse's meager income, to ensure sustainable income generation. Scenarios further highlighted situations where the families could not afford the school fees for their children and couldn't avail certain groceries for their requirement. This further motivated the research team to look for alternate income opportunities for the household women of Indpur.

4.2 Results II

A need for sustainable livelihood opportunities emerged as an outcome of the study. Of the current opportunities available, the feasible option for women of Indpur is to

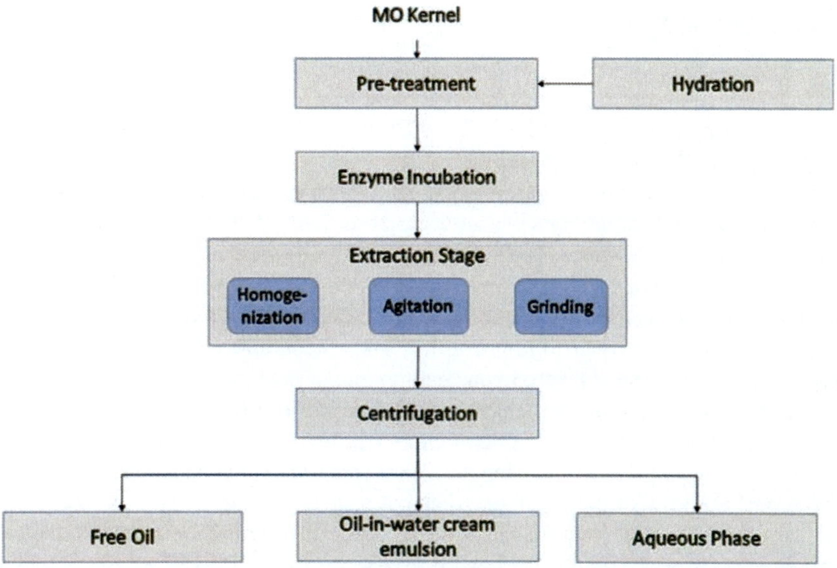

Fig. 4 Block diagram of aqueous enzymatic extraction

utilize the existing unused natural resource in the form of MO trees. These trees are endemic to Himalayan region and are found to grow abundantly in Indpur village and its surrounding areas. Studies reveal that MO has many advantages owing to its medicinal properties, anti-microbial effects, and antioxidant properties among many others [20].

MO trees can be judiciously exploited to avail a steady income generation opportunity. Since MO seed oil is one of the most commercially used MO products, an oil extraction facility is proposed to be developed in Indpur. The extracted oil can be sold to the markets nearby. The market feasibility study is currently ongoing. There are various methods used for MO oil extraction such as Response Surface Methodology, Aqueous Enzymatic Extraction, Solvent-free Microwave, and Hydrodistillation among many others [21–23]. Being environment-friendly and cost-effective, Aqueous Enzymatic Extraction (AEE) is preferred over other processes [24]. Another reason for this preference is that a high oil yield can be expected in the AEE process when compared to the traditional approaches [25].

AEE process is a technology that utilizes water as a medium for extraction [22, 26, 27]. This is to release protein and other emulsifiable components existing in oilseeds; and thereby ramifying the oil as a free oil. As seen in the conventional hexane extraction, this process completely avoids the organic vapor emissions. It allows concurrent production of industrial oil and food quality protein [25]. Figure 4 depicts a basic process of AEE [27, 28].

5 Conclusion

This investigation unraveled the challenges and repercussions apropos scant livelihood opportunities endured by the Indpur village community in the state of Himachal Pradesh, India. By utilizing tenets of HCD, the research team was able to identify the impact of sparse income generation opportunities on the livelihood of the community as well as the community's unawareness of the benefits and knowledge of using naturally available MO trees. The inferences from HCD were analysed thematically which highlighted the negative experiences endured by the community due to sparse income. It also underlined the willingness of women to contribute to the household income generation in order to improve their socio-economic status. This aided the research team to ideate and zero in on an intervention in the form of AEE; proposed to utilize the extant MO resources in the village to extract oil that can be utilized for commercial gains. The future scope of work will involve a co-design process with the community to bring in their inputs with respect to designing the AEE system. The model design will then be fabricated and tested on campus prior to deploying in the village. The intervention can be further scaled to address similar challenges addressed by various communities in rural India.

Acknowledgements This research was funded by Amrita Vishwa Vidyapeetham's Live-in-Labs® program and the field implementation was funded by the UN-recognized NGO, MA Math. The study was conducted as part of the UNESCO Chair for Experiential Learning for Sustainable Innovation and Development. The authors express their immense gratitude to Sri. Mata Amritanandamayi Devi, Chancellor of Amrita Vishwa Vidyapeetham, who has inspired them in performing selfless service to society. The authors would also like to express their sincere appreciation to Mr. Gopalaiengar Sheshadri for his extensive support in enhancing the quality of the paper within the limited time frame. The authors also thank the faculty, staff, and students of Amrita Vishwa Vidyapeetham, the government authorities of Kangra district, Himachal Pradesh, and also the Amrita SeRVe program staff.

References

1. The World Bank. https://data.worldbank.org/indicator/SP.RUR.TOTL.ZS?end=2019locations=INstart=1960view=chart. Last accessed 2021/02/04
2. Centre for Monitoring Indian Economy (CMIE). https://unemploymentinindia.cmie.com/. Last accessed 2021/02/04
3. International Labour Organisation: Global employment trends 2013: recovering from a second jobs dip. (2013). https://www.ilo.org/global/research/global-reports/global-employment-trends/2013/WCMS_202326/lang-en/index.htm::text=Global%20Employment%20Trends%202013%20highlights,been%20triggered%20by%20the%20crisis. Last accessed 2021/02/04
4. Government of India: Ministry of statistics and program implementation-participation in economy. http://mospi.nic.in/sites/default/files/reports_and_publication/statistical_publication/social_statistics/WM17Chapter4.pdf. Last accessed 2021/02/04
5. Ramesh, M.V., Mohan, R., Menon, S.: Live-in-Labs: rapid translational research and implementation-based program for rural development in India. In: 2016 IEEE Global Humanitarian Technology Conference (GHTC), pp. 164–171. IEEE (2016)

6. Hindustan Times. https://www.hindustantimes.com/cities/himachal-pradesh-among-top-three-states-with-highest-unemployment-rate/story-XZVaR2ExYhfgcz8gxNsCXO.html. Last accessed 2021/02/04
7. Amrita SeRVe—Self Reliant Village. https://amritaserve.org/. Last accessed 2021/02/04
8. Amrita SeRVe The First 5 Years. (2019). https://amritaserve.org/wp-content/uploads/2019/05/2019_Amrita_SeRVe_TheFirst5Years.pdf. Last accessed 2021/02/04
9. Maps of India. https://www.mapsofindia.com/maps/himachalpradesh/tehsil/kangra.html. Last accessed 2021/02/05
10. Balu, S., Sampath, K.S., Hari, L.R., Ramprasath, M., Aishwarya, N.K., Aiswarya, M.S., Frey, L.M, Nandanan, K.: Assessment of waste management practices and its impact on the village ecosystem in Himachal Pradesh, India. In: 2020 IEEE 8th R10 Humanitarian Technology Conference (R10-HTC), pp. 1–6. IEEE (2020)
11. Brown, T., Wyatt, J.: Design thinking for social innovation. Dev. Outreach **12**(1), 29–43 (2010)
12. Chen, E., Leos, C., Kowitt, S.D., Moracco, K.E.: Enhancing community-based participatory research through human-centered design strategies. Health Promot. Pract. **21**(1), 37–48 (2020)
13. Guleria, M., Rajesh, T., Panicker, A.J., Raja, S.R., Soundharajan, B., Frey, L.M., Nandanan, K.: Using human centered design to improve socio-economic livelihoods with modernized irrigation systems. In: 2020 IEEE 8th R10 Humanitarian Technology Conference (R10-HTC), pp. 1–6. IEEE (2020)
14. Lee, M.J., Wang, Y., Duh, H.B.L.: AR UX design: applying AEIOU to handheld augmented reality browser. In: 2012 IEEE International Symposium on Mixed and Augmented Reality-Arts, Media, and Humanities (ISMAR-AMH), pp. 99–100. IEEE (2012)
15. McIntosh, M.J., Morse, J.M.: Situating and constructing diversity in semi-structured interviews. Glob Qual Nurs Res **2**, 2333393615597674 (2015)
16. Reddy, K.R., Reddy, C. S.: Self help groups in India. A study on quality and sustainability. EnablePublication (2012)
17. Miaskiewicz, T., Kozar, K.A.: Personas and user-centered design: how can personas benefit product design processes? Des. Stud. **32**(5), 417–430 (2011)
18. Carrol, J.M.: Five reasons for scenario-based design. In: Proceedings of the 32nd Annual Hawaii International Conference on Systems Sciences. HICSS-32. Abstracts and CD-ROM of Full Papers, pp. 11. IEEE (1999)
19. Bhavani, R.R., Sheshadri, S., Maciuika, L.A.: Addressing the first teachers: education and sustainable development for children, families and communities through vocational education, technology and life skills training for women. In: Children and Sustainable Development, pp. 319–334. Springer, Cham (2017)
20. Abdull Razis, A.F., Ibrahim, M.D., Kntayya, S.B.: Health benefits of Moringa oleifera. Asian Pac. J. Cancer Prev. **15**(20), 8571–8576 (2014)
21. Mani, S., Jaya, S., Vadivambal, R.: Optimization of solvent extraction of Moringa (Moringa oleifera) seed kernel oil using response surface methodology. Food Bioprod. Process. **85**(4), 328–335 (2007)
22. Mat Yusoff, M.: Aqueous enzymatic extraction of moringa oleifera oil with high pressure processing pre-treatment. PhD dissertation, University of Reading (2016)
23. Kayode, R.M.O., Afolayan, A.J.: Cytotoxicity and effect of extraction methods on the chemical composition of essential oils of Moringa oleifera seeds. J. Zhejiang Univ. Sci. B **16**(8), 680–689 (2015)
24. Rosenthal, A., Pyle, D.L., Niranjan, K.: Aqueous and enzymatic processes for edible oil extraction. Enzyme Microb. Technol. **19**(6), 402–420 (1996)
25. Abdulkarim, S.M., Lai, O.M., Muhammad, S.K.S., Long, K., Ghazali, H.M.: Use of enzymes to enhance oil recovery during aqueous extraction of Moringa oleifera seed oil. J. Food Lipids **13**(2), 113–130 (2006)
26. Tabtabaei, S., Diosady, L.L.: Aqueous and enzymatic extraction processes for the production of food-grade proteins and industrial oil from dehulled yellow mustard flour. Food Res. Int. **52**(2), 547–556 (2013)

27. Yusoff, M.M., Gordon, M.H., Ezeh, O., Niranjan, K.: Aqueous enzymatic extraction of Moringa oleifera oil. Food Chem. **211**, 400–408 (2016)
28. Nikiforidis, C.V., Matsakidou, A., Kiosseoglou, V.: Composition, properties and potential food applications of natural emulsions and cream materials based on oil bodies. RSC Adv. **4**(48), 25067–25078 (2014)

Customer Behavioral Market Basket Approach for Most Frequent Pattern Identification Using Data Mining Techniques

Krupal Joshi, Dushyantsinh Rathod, Mayank Parekh, and Jaykumar Dave

Abstract Advertise bushel investigation analyzes customers' purchasing patterns by finding unexpected and important associations among the products which they place in their shopping baskets. It not only assists in decision-making process but also increases sales in many business organizations. A priori and FP Growth are the most common algorithms for mining frequent itemsets. For both algorithms a predefined minimum support is needed to satisfy for identifying the frequent itemset. But when the minimum support is low, a huge statistical change of count in candidate sets will be generated that needs large computation. In this paper, an approach has been proposed to avoid this large computation by reducing the items of dataset with top selling products. Various percentages of trending products like 30 to 55% are selected and for both FP growth and a priori algorithm, association rule generation process starts along with frequent item combinations. The comes about appear that in case beat offering things are utilized, it is conceivable to urge nearly same visit itemset and affiliation rules inside a brief time comparing with that yields which are derived by computing all the things. From time comparison, it is also found that Distributed FP Development calculation takes littler time than a priori calculation.

1 Introduction

The digital platform is getting stronger and wider every day. As we can see that the current pandemic situation caused by COVID-19. As of year 2020, we observe massive rise in the digital interactions and consumption of media data over computing devices. Big Industries are very keen to store all those data and footprints created

K. Joshi (✉)
Alpha College of Engineering, Khatraj Kalol Road, Kalol, Gujarat 382721, India

D. Rathod · J. Dave
Sankalchand Patel University, Visnagar, Gujarat 384315, India

M. Parekh
IT Department, Government Polytechnic, Ahmedabad, India

GTU, Ahmedabad, Gujarat 380015, India

T. Senjyu et al. (eds.), *ICT with Intelligent Applications*, Smart Innovation, Systems and Technologies 248, https://doi.org/10.1007/978-981-16-4177-0_66

every minute and obtain the valuable information from it. All of the efforts are made to improve the user's experience and in contrast to grow their businesses. That large chunk of data every day means the playground for Data Mining companies that data can be processed and the knowledge obtained from it can be put to good use of mankind as well as business. We worked with *A priori Algorithm* and *FP Growth Algorithm* which is explained in brief below.

1.1 Mining Methods

To generate Association rules and find frequent patterns in datasets 2 well-known and promising algorithms are used. We have chosen the distributed approach for processing in our work for parallelism in processing.

A priori Algorithm

This algorithm is used for mining large datasets and define new association rules and form relations in a form of RDB—Relational Database [1, 2]. It starts with finding the frequency single item and further finds the combination of itemsets in which that one item was suspected the most frequent, this same type of findings is executed for combination of frequent items also referred as itemsets that repeats itself in records frequently to notice, *for example*, itemsets (Bread, Milk, Butter, Jam).

RDB—Relational Database [3]

The general understanding of RDBMS is managing the relations between the data table, itemsets and attributes. Relational model drives the organization of the all the data and attributes for forming a Relational database. Generally, the transactional data is stored within the Relational Database and if not stored there then it is providing a relational view of data by using the wrapped data view. In the result's chapter, we have used the multi-column relational data representation. In RDBMS there are 2 ways transactional data can be represented. (1) Single Column (2) Multi-Column [*Layout*] (Fig. 1).

FP Growth

FP Growth is kind of an improved version (successor) of the a priori algorithm. It can generate frequent patterns without candidate generation step from a priori algorithm; it refers to generating a tabular representation of combination of items scored with its count of re-appearing frequency [1, 4]. FP Growth algorithm can visually represent the database in a tree structure which is easy to understand and is referred as FP Tree OR Frequent Pattern Tree.

tid	*item*
1	beer
1	diapers
2	beer
2	chips
2	diapers
3	chips

(A) Single column layout

tid	$item_1$	$item_2$	$item_3$
1	beer	diapers	
2	beer	chips	diapers
3	chips		

(B) Multi column layout

Fig. 1 RDB—transactional data storage pattern screenshot

2 Existing Architecture

In market basket analysis using a priori and FP growth algorithm, the main goal of this study is to show performance evaluation between a priori and FP growth algorithms [1]. For both algorithms, author has discovered the frequent itemsets with minimum confidence and later they were finding the association between frequent itemsets which was also satisfying at minimum confidence Then, Author M Hossain compared the execution time against transaction and checked the results which are given in experimental analysis section [1] (Fig. 2).

When dealing with large datasets a priori generates many candidates sets whereas FP growth algorithm could not even constrict a main memory base FPtree [1]. The author has tried reducing the computational in some approaches to obtain productivity.

3 Proposed Architecture

We propose that post data processing step, we apply distributed approach to the FP growth algorithm it virtually enables us with the parallel processing of data where we later have initiated the voting approach that will gather most frequent patterns from all the distributed itemsets. To simplify, voting approach will select common types of patterns from the previously divided dataset/itemset. After voting the Most frequent patter will be obtained as it is visually presented in the diagram's sub-process section (Fig. 3).

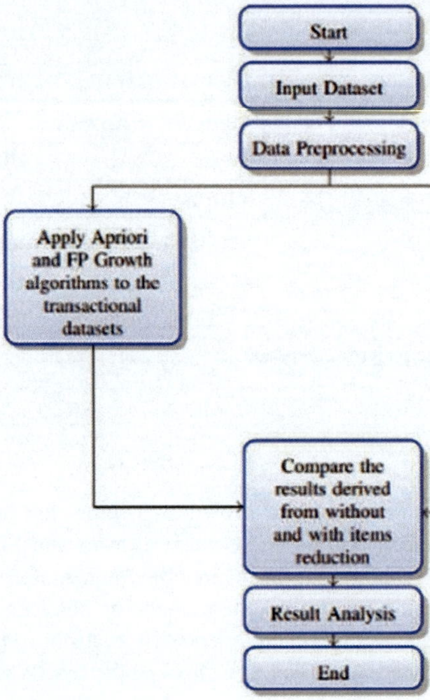

Fig. 2 Existing architecture

3.1 Data Pre-processing [3]

Data Pre-Processing is a technique of transforming the data in a usable manner that will be easier to process in future stages, for example, no missing data and no projection in values.

3.2 Distributed FP Growth

Applying a distributed approach for parallel processing and optimizing performance of the system. DF1, DF2, DF3 are the distributed datasets as shown in Fig. 2 proposed architecture on which the algorithm will be applied.

Fig. 3 Proposed architecture

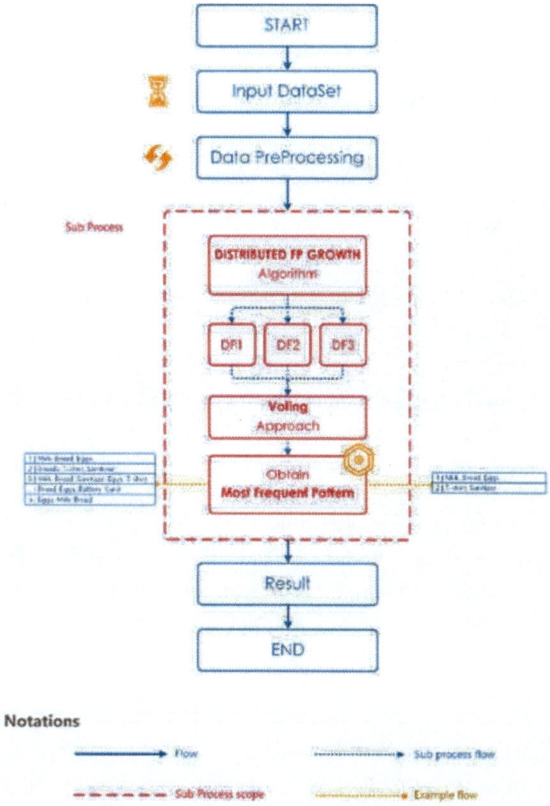

3.3 *Voting Approach*

In order to predict customer's behavior voting approach is applied to all the frequent itemsets to the distributed list. Based on the score of each item in the dataset the confidence score will be the factor for predicting customer's behavior and that will give us better suggestion for customer's cart.

4 Proposed Algorithm

In order to achieve accurate MFP and Predict Customer's Behavior as a result with good confidence score. Here we have mentioned the algorithm with each step, to obtain identical results we suggest following the same instructions.

Steps of Proposed Algorithm

1. Start.

2. **Take Input** from Dataset.
3. **Data pre-processing** from Dataset.
4. Dataset is divided into **DF1, DF2, DF3** using Distributed FP Growth Algo.
5. Apply **Voting approach**.
6. Obtain **most frequent pattern**.
7. Customer behavior prediction.
8. Result.
9. End.

5 Result

Below illustrated images show the obtained result. This result is generated from a data frame of approximately 12,000+ items. Library kept to use for processing the data frame is library in python called pandas [Python Data Analysis Library] (Figs. 4 and 5).

As we can see in another Fig. 6 image, we have obtained 12,215 most frequent patterns in col[0] where col[1] represents **Product Purchasing** Patterns, Col[2] is representing **Product Prediction** that means customer is likely to buy or select this product(s) and the last Column Col[3] shows the **Probability** in a ratio from 0.00 to 1.00.

```
import pandas as pd
data=pd.read_csv(r'/content/order_products__train.csv')
data.head()
```

Fig. 4 Input dataset syntax

	order_id	product_id	add_to_cart_order	reordered
0	1	49302	1	1
1	1	11109	2	1
2	1	10246	3	0
3	1	49683	4	0
4	1	43633	5	1

Fig. 5 Input dataset

Sr No.	Product Purchasing	Product Prediction	Probability
1	Large Alfresco Eggs, Bag of Organic Bananas, Limes	Organic Strawberries	0.75
2	Organic Baby Spinach, Organic Raspberries, Large Lemon, Organic Avocado	Banana	0.71
3	Uncured Genoa Salami, Organic Cinnamon Crunch Cereal	Organic Strawberries	0.71
4	Frozen Organic Wild Blueberries, Organic Ginger Root, Organic Hass Avocado	Bag of Organic Bananas	1.00
5	Bag of Organic Bananas, Baby Bok Choy	Organic Strawberries	0.71
⋮			
12212	Bunched Cilantro, Feta Cheese Crumbles, Large Lemon	Organic Avocado	0.83
12213	Red Peppers, Organic Baby Spinach Salad	Organic Avocado	0.71
12214	Fresh Cauliflower, Organic Baby Carrots, Organic Zucchini	Bag of Organic Bananas	0.75
12215	Organic Baby Spinach, Garlic, Organic Avocado	Large Lemon	0.70

Fig. 6 Generated result from the input dataset

6 Implementation Tools and Technology

In this unsupervised form of learning, we are using following tools and technology to execute and work on our implementation.

Tool type	Name	Version	Link
Programming language	Python	Ver.: 3.9.*	https://www.python.org/
Integrated development environment	Jupyter	Ver.: 4.11	https://www.jupyter.org/
	Google Collab	–	https://colab.research.google.com/
Datasets	Kaggle	–	www.kaggle.com/datasets

7 Conclusion

We have observed that MB analysis uses a priori algorithm, but the setbacks of this algorithm are that it cannot process. All the large datasets can only work with smaller datasets. Also, it tends to consume more computing memory (i.e., RAM) compared to the FP Growth Algorithm. When applying a distributed approach for optimizing performance of the system. DF1, DF2, DF3 are the distributed datasets in proposed architecture on which the algorithm will be applied. Voting Approach is applied to all the frequent itemsets to the distributed list. Based on the score of each item in the dataset the confidence score will be the factor for predicting customer's behavior and that will give us better suggestion for customer's cart. In Contrast, if we apply FP Growth Algorithm in distributed approach then we observe optimized performance on the large datasets that is also easier and more resource beneficial. The results are obtained from approximately 10 k + result data and the resource utilization was exponentially lesser than other processing algorithms. The Execution of the algorithm is fast and the performance is highly increased. As the input dataset was nearly 104 k+ and 12 k+ Frequent patterns were obtained very quickly in just few second.

References

1. Hossain, M., Sarowar Sattar, A., Paul, M.K.: Market basket analysis using apriori and FP growth algorithm. IEEE (2019)
2. Ferrera, R., Pittman, J.M.: Retailer's dilemma: personalized product marketing to maximize revenue. IEEE (2020)
3. https://www.geeksforgeeks.org/data-preprocessing-in-data-mining/
4. Chen, H., Luo, Q.: Distributed pruning optimization-oriented FP-growth method based on PSO algorithm. IEEE (2017)

Breast Cancerous Tumor Detection Using Supervised Machine Learning Techniques

Mayank V. Parekh and Dushyantsinh Rathod

Abstract Around the globe the leading cause of cancer mortality in women is Breast Cancer; BCs, have been under many studies and several analyses have shown that many abnormal conditions and risk of BCs can be diagnosed with the help of fast and perfect judgment of the clinical physicians of this domain. This kind of intelligent decision making is not the thing where every person is good at so here, we have kept Artificial intelligence to use with supporting steps like Machine Learning Classifiers from Wisconsin Breast Cancer Dataset; WBCD. In the Combination of Human and Computer Intelligence, we have worked our way around the hybrid Mechanism where we can make the use of both entities and detect as well as prevent the risks of Very Early Stages of BCs. Our Mechanism includes the use of Principle Component Analysis; PCA, at the first stage of pre-processing the decreasing features in the pre-processed data the second stage of pre-processing was more beneficial for accessing the multi pre-processed data for analysis and diagnosis. When the Multi Pre-processed data is accessed by the Support Vector Machine; SVM, the old Accuracy, Specificity and Significance level was ranging from 95 to 100% but the new levels after the multi pre-processing were decreased to 5% and the improvement was considerable and confirmed that the proposed Algorithm has achieved the optimized results then RF, KNN, DT. It means the Proposed algorithm can assist the human factor with faster decision making and can categorize data (High Risk) malignant and (Low Risk) Benign because to avoid Human errors AI is useful and can achieve optimal results.

M. V. Parekh (✉)
Lecturer, IT Department, Government Polytechnic & PhD Scholar, Gujarat Technological University, Ahmedabad, Gujarat, India

D. Rathod
Associate Professor, CE Department, Sankalchand Patel College of Engineering, Sankalchand Patel University, Visnagar, Gujarat, India

T. Senjyu et al. (eds.), *ICT with Intelligent Applications*, Smart Innovation, Systems and Technologies 248, https://doi.org/10.1007/978-981-16-4177-0_67

1 Introduction

Healthcare organizations generate complex and large amounts of data day by day. (Ex. Big Data) from different sources like the hospital, doctor prescription, clinical data. This data is different from text, audio, image, video, record (Database). Respective Medical Images, health records, pathological information analysis of organs can be used to identify very important abnormalities like blood, skin and breast cancer, etc. [1, 2] According to GLOBOCAN statistics, 5% increasing new cancer cases emerged in the last two years that caused 80% cancer deaths. The leading death ratio is breast cancer in women. According to a recent report, World Health Organization published more than 7 lakh women died due to breast cancer. But apart from the previous statement, there is survival rate is also more than 35% after surgery removing the breast. Breast cancer occurs in the breast. Breast tumors worse and growing fast for this reason cause death [3]. Tumors of Breast cancer broadly classified as Benign and Malignant. Whereas benign tumor is not harmful but, in some cases, death occurs due to this. Comparatively, a malignant tumor is most dangerous will cause death. Turkki et al. [4] stated that Cancerous tumor assessment without using Breast cancer pathological information can be possible through using machine learning techniques. Investigating current and future aspects of cancer detection, ML, AI, Deep Learning, Data Mining are kept to use for optimal results. Every Technique has its own way of execution, for example, predictive analysis is sub-divided under ML and Regression Techniques. ML can manage noisy data and data with rare and uniform patterns which benefits Predictive Analysis; It is useful in Drug and Medicine suggestion OR Clinical Trials and conduct body reports.

1.1 Machine Learning Algorithms

ML is classified into three broad paradigms depending on the nature of the target: supervised, unsupervised, and reinforcement learning. Appreciating ML Methods always is less because the way it takes care of the complex interactions of the large numbers and its remarkable performance in nonlinear problems is not excellent but it is phenomenal (Fig. 1).

Old School Statistical methods are left far behind after ML has been successful in delivering very keen and satisfying results in every clinical field right after the moment it was kept to use. The best model for the most part relies upon the example

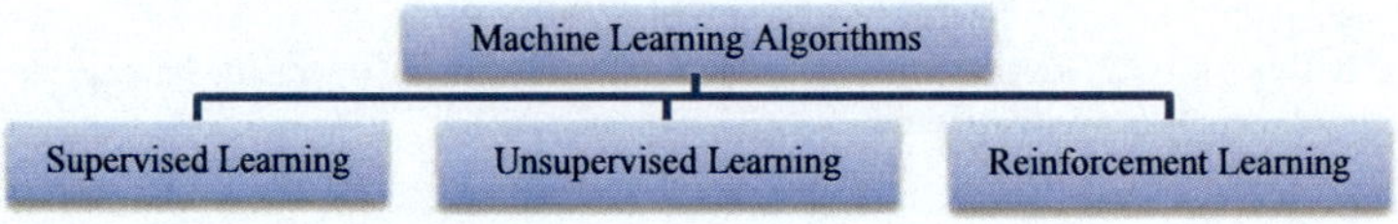

Fig. 1 Types of machine learning process

informational index what's more, the motivation behind the examination in a specific situation. They picked the calculations with traditional techniques also, brilliant application practices to plan a systematic strategy.

Random Forest

Random forest is supervised learning method and it is applicable to classification as well as regression. Random Forest requires tree parameter as well as ensemble parameter. Tree parameters like Min sample leaf, Max features, Max Tree Depth and Class weight. Ensemble parameters like *N* Estimators and Bootstrap.

K-Nearest neighbor Algorithm

K-nearest neighbors (KNN) algorithm uses 'feature similarity' to predict the values of new data points. There are three tuning parameters in this model.

Number of weights	It is the number of weights
Weight	Deal with the available neighbor's different distance between them

Logistic Regression

LR is a statistical model and it is used for data analytic. It can be applicable to data linear as well as nonlinear dataset. There are three tuning parameters in this model.

Penalty	Stat that how the model regularized using *C*-parameter It can be a l1 norm or l2 norm or elastic-net
C-regularized parameter	It is a regular parameter if we get the lower value of *c* it is more regularized the model
Class weight	It is used when class unbalance. Default value is none

Support Vector Machine

Support vector machines are supervised and powerful method.

Gamma	It is also a regularized parameter but when RBF Kernel is applied
C-regularized parameter	It is similar as logistic regression
Class weight	In our case class is unbalance

2 Literature Review

Haifeng et al. [5] proposed that Machine Learning and data mining techniques can be used for breast cancer prediction. They used an open-source dataset of the

Wisconsin Breast Cancer Database (WBCD) and the Wisconsin Diagnostic Breast Cancer (WDBC) dataset. Author Proposed a combined model PCA and data mining model like Random forest for feature reduction. And finalized that k-means suitable for feature space reduction.

Author Chourasia et al. have used (1) Bayesian Theorem, (2) Decision Tree and (3) Radial Basic Function Network. Authors have mentioned the use of WBCD dataset followed by Data Pre-processing, Selection and Transformation. They Claims that accurately 96.77% accuracy is achieved with RBF Network classification.

In our literature, most of the authors evaluate the proposed method based on Accuracy. In maximum cases of the output of the proposed model false positives and false negatives is less compared to true positives and true negatives so its accuracy is high based on the formula. In the real world, there is a possibility of a non-tumor or benign tumor patient can be malignant or serious. Now in the new approach or model tried to focus on the false positive and false negative. If any model misses any condition, it is very harmful to the patient.

3 Proposed Architecture

We propose that post data pre-processing First we remove unwanted columns and update the texture information to binary. From the dataset divide data in two parts for training and testing as per our requirements. After that Dimensional Reduction techniques like Principal Component analysis, Autoencoder. After Post that we will train models (i.e., Decision Tree, Random Forest, K-Nearest Neighbors, Support Vector Machine and Logistic Regression) and importing, creating, evaluation and test or result (Fig. 2).

4 Proposed Algorithm

We are applying the below steps for getting better result.

Steps of Proposed Algorithm

1. Start.
2. **Take Input** from Dataset.
3. **Data pre-processing** from Dataset.
4. **Feature selection using PCA.**
5. Divide the data training as 80% and testing is 20%.
6. Train the model using RF, KNN, Logistic Regression, SVM.
7. Evaluation the model.
8. Store Result.
9. End.

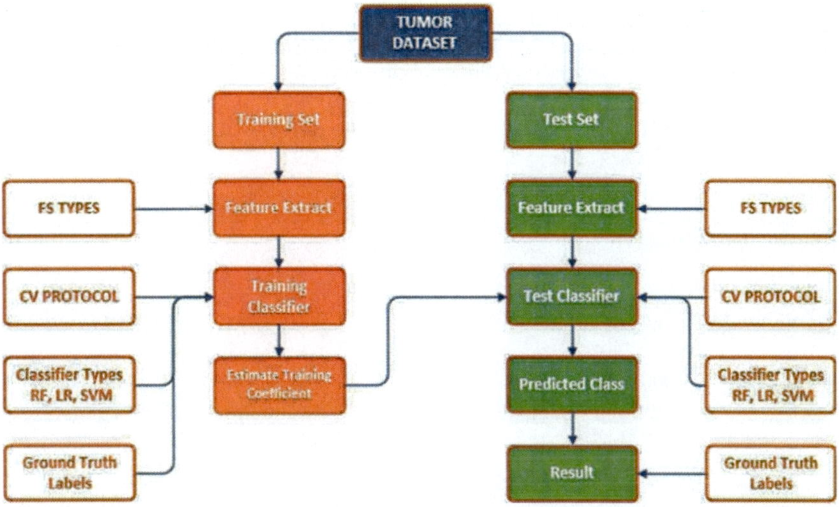

Fig. 2 Proposed architecture

5 Dataset and Result

This dataset contains information on 569 breast tumors with more than 20 different properties which calculated the mean, standard error, and worst measures. Here first apply the small dataset of two target classes with either B (benign) or M (malignant). All features appear to a significant impact on the classification of tumors and scatter plots like separable. Dataset format is CSV, contains 569 rows and 33 columns no null values found (Fig. 3).

Importing the Model

```
import pandas as pd
dataset = pd.read_csv('D:/PHD_COURSE/Breast-Cancer-Tumor-Prediction/data.csv')
dataset.sample(5)
```

	id	diagnosis	radius_mean	texture_mean	perimeter_mean	area_mean	smoothness_mean	compactness_mean	concavity_mean	concave points_mean	…	te
37	854941	B	13.03	18.42	82.61	523.8	0.08983	0.03766	0.02562	0.02923	…	
340	89813	B	14.42	16.54	94.15	641.2	0.09751	0.11390	0.08007	0.04223	…	
52	857374	B	11.94	18.24	75.71	437.6	0.08261	0.04751	0.01972	0.01349	…	
104	862989	B	10.49	19.29	67.41	336.1	0.09989	0.08578	0.02995	0.01201	…	
559	925291	B	11.51	23.93	74.52	403.5	0.09261	0.10210	0.11120	0.04105	…	

Fig. 3 Input dataset

```python
# Importing Models
from sklearn.linear_model import LogisticRegression
from sklearn.tree import DecisionTreeClassifier
from sklearn.ensemble import RandomForestClassifier
from sklearn.neighbors import KNeighborsClassifier
from sklearn.svm import SVC, LinearSVC
```

Creating the Model

```python
# Defining random seed
seed=42

# Creating Models
logreg = LogisticRegression(solver='lbfgs', random_state=seed)
svc = SVC(random_state=seed, probability=True)
lin_svc = LinearSVC(random_state=seed)
l_svc = CalibratedClassifierCV(lin_svc, cv=5)
dtree = DecisionTreeClassifier(random_state=seed)
rf = RandomForestClassifier(10, random_state=seed)
knn = KNeighborsClassifier()
first_models = [logreg, svc, l_svc,
                dtree, rf, knn]
first_model_names = ['Logistic Regression', 'SVM', 'Linear SVC',
                     'Decision Tree', 'Random Forest' , 'K-Neighbors']
```

Model's Accuracy

	Accuracy	Recall
SVM	0.972215	0.945057
Logistic Regression	0.972278	0.931724
K-Neighbors	0.969747	0.925057
Random Forest	0.949715	0.905287
Linear SVC	0.959684	0.904828
Decision Tree	0.916835	0.904828

Without Feature Extration

	First Scores		Less Features	
	Accuracy	Recall	Accuracy	Recall
SVM	0.972215	0.945057	0.972215	0.945287
Logistic Regression	0.972278	0.931724	0.969778	0.925057
K-Neighbors	0.969747	0.925057	0.964778	0.911954
Random Forest	0.949715	0.905287	0.962215	0.931954
Linear SVC	0.959684	0.904828	0.962215	0.897931
Decision Tree	0.916835	0.904828	0.924399	0.877241

With Feature Extraction

Comparison Chart

From the above comparative result for without extracting the feature there is accuracy for support vector machine and Logistic regression are nearly equal. Recall is higher for SVM Fig. 4..

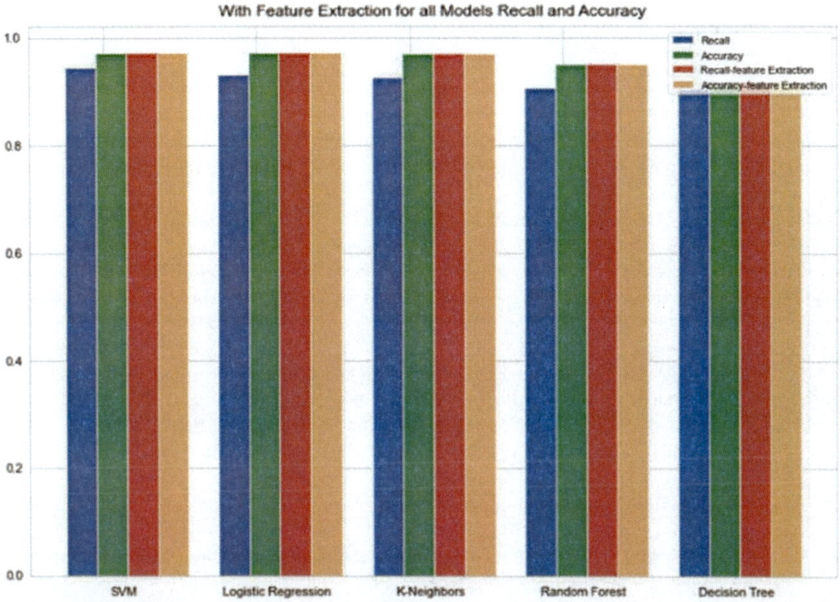

Fig. 4 Comparison chart of recall and accuracy for SVM, RF, LR, KNN, DT with and without feature extraction

6 Implementation Tools and Technology

In this study we apply the unsupervised learning method using python 3.7 tools anaconda with Wisconsin Breast Cancer Database (WBCD) from Kaggle.

7 Conclusion

After implementing the code, we have got the 6-algorithm result. The accuracy of Decision Tree is 91%, Random Forest 94%, K-Neighbors 96% and SVM accuracy is 97%. For testing their proposed techniques, Researchers used Diminutive data and comparatively accuracy and recall, SVM is far better than other. After applying and improving the algorithm risk assessment and early diagnosis of the non-tumor. I will be implementing supervised machine learning and ensembled techniques to improve the detection accuracy on brain tumor dataset.

References

1. Sharma, G.N., Dave, R., Sanadya, J., Sharma, P., Sharma, K.: Various types and management of breast cancer: an overview. J. Adv. Pharm. Technol. Res. (2010)
2. Etemadi, R., Alkhateeb, A., Rezaeian, I., Rueda, L.: Identification of discriminative genes for predicting breast cancer subtypes. In: IEEE International Conference on Bioinformatics and Biomedicine (BIBM) (2016)
3. Chagpar, A.B., Coccia, M.: Factors associated with breast cancer mortality-per-incident case in low-to-middle income countries (LMICs). J. Clin. Oncol. **37**(15) (2019)
4. Turkki, R., Byckhov, D., Lundin, M., Isola, J., Nordling, S., Kovanen, P.E., Verrill, C., von Smitten, K., Joensuu, H., Lundin, J., et al.: Breast cancer outcome prediction with tumour tissue images and machine learning. Breast Cancer Res. Treat. **177** (2019)
5. Kumar, V., Mishra, B.K., Mazzara, M., Thanh, D.N.H., Verma, A.: Prediction of malignant and benign breast cancer: a data mining approach in healthcare applications. IEEE (2020)
6. Chauhan, R., Jangade, R.: A robust model for big healthcare data analytics. In: IEEE International Conference-Cloud System and Big Data Engineering (2016)
7. Nithya, B.: Predictive analytics in health care using machine learning tools and techniques. ICICCS (2017)
8. Thomas T., Nitesh P., Singh, V. Comparative analysis to predict breast cancer using machine learning algorithms: a survey. IEEE (2020)
9. Sengar, P.P., Gaikwad, M.J., Nagdive, A.S.: Comparative study of machine learning algorithms for breast cancer prediction. IEEE (2020)
10. Chaurasia, V., Pal, S., Tiwari, B.: Prediction of benign and malignant breast cancer using data mining techniques. J. Algorithms Comput. Technol. **12**, 119–126 (2018)

α-Method of Air Traffic Controllers' Preference System Specification Over Typical Mistakes Severities

Oleksii Reva, Volodymyr Kamyshyn, Serhii Borsuk, and Andrei Nevynitsyn

Abstract Individual and group preferences systems belong to the group of human factor influence indicators. They are related to decision-making and are usually represented as an arranged set of indices characterizing the peculiarities of air traffic controllers' professional activity. Attitudes to the severities of air traffic controllers' typical mistakes that might appear during traffic management also belong to this group. Applying the multistep procedure of marginal thoughts detection and removal allowed the initial sample reduction (from 37 down to 26). Initial mistakes ranks are changing linearly thus providing no information about the rate of severity between them. This leads to certain roughness of the measurements for both individual and group preferences systems. Significance coefficients for arranged mistakes/alternatives belong to the preferences determination methods. Therefore, for the further specification of a consistent group preferences system, the ranks of typical mistakes were defuzzificated via the priorities arrangement method. This allowed putting normalized severity coefficients into correspondence with them. It was decided that the 3rd iteration results of the priority arrangement method are unacceptable at given calculations precision. The first two iteration results were used to specify the group preferences system by substituting the ranks within individual preferences systems. Proper significance coefficients were used for this purpose. Recalculations produced new group preferences systems having high coincidence with the initial one.

O. Reva (✉)
National Aviation University, Kyiv, Ukraine
e-mail: ran54@meta.ua

V. Kamyshyn
Ukrainian Institute of Scientific and Technical Expertise and Information, Kyiv, Ukraine

S. Borsuk
Wenzhou University, Wenzhou, China

A. Nevynitsyn
Flight Academy of the National Aviation University, Kropyvnytskyi, Ukraine

T. Senjyu et al. (eds.), *ICT with Intelligent Applications*, Smart Innovation, Systems and Technologies 248, https://doi.org/10.1007/978-981-16-4177-0_68

1 Introduction

International civil aviation organization (ICAO) had assembled tenths of circulars with the generalized experience of human factor (HF) considerations. They include flight safety measures being developed and implemented within the leading aviation companies and authorities worldwide [1–3]. Also, they are based on the aviation accidents and incidents (ACI) statistics related to the negative HF influence. These factors obviously include the front-line aviation operators' (AOs) (pilots, crew members, air traffic controllers (ATC), engineers) activity. Unfortunately, the ICAO efforts provided no improvement in the ACI statistics due to the HF influence. This makes the HF research and prevention an urgent and important scientific task.

Based on the ICAO safety paradigm [3], it was proved that "the aviation personnel's attitude toward hazardous and dangerous actions or conditions" is one of its cornerstone components. It most accurately explains how HF does influence the decision-making (DM) process. Furthermore, it comprehensively describes all other paradigm components interaction [4].

Several indicators might be classified as belonging to this "attitude" group. They include:

- basic DM dominants, which define the risk attitude;
- aspiration levels that belong to the main parameters of AO's personal traits considering self-esteem;
- fuzzy estimates of risks according to the ICAO criteria;
- hazardous behavioral and cognitive patterns during the DM process;
- personnel individual and group preferences systems, etc.

In this research, we focus on the latter group of indicators. We consider an individual preferences system (IPS) as an arranged set of indicators and characteristics of front-line AOs. Namely, we examine the ATC IPS over the set of typical mistakes (the list of 21) they might make during their professional activity.

The urgency of the ATC IPS research is as follows. It is clear that IPSs comprehensively witness the peculiarities of a mistake consideration (imaging, representation) as forecasted events occurrence. Besides, this consideration is determined and caused by the personal ATC experience including his participation in potentially harmful situations and parrying their consequences. It also depends on the details of the personal educational path of the ATC with all its advantages and disadvantages. Not to mention the obvious accounting of air traffic management (ATM) system properties. On the other hand, group preferences systems (GPS) show peculiarities formed in particular groups of controllers. They address the shift of the same mistakes' hazards and severities caused by the risk level and group deformations inherent for group activities.

It is worth noticing that regardless of researches on existing methods of the IPS and GPS assembly and analysis [5–9], they are rather neglected in overall expert procedures development concerning civil aviation and neighboring scientific areas (system analysis, DM methodology, information technologies, etc.).

2 State of the Art

Nowadays, the following methods of experts IPS determination are known:

1. Direct ranking;
2. Pairwise comparison with the normalized determination of partial alternative intensity (significance, valuable, acceptability, etc.);
3. Pairwise comparison with the differentiated determination of partial alternative intensity (significance, valuable, acceptability, etc.);
4. Pairwise comparison with the determination of relative partial alternative intensity (significance, valuable, acceptability, etc.);
5. Sorting;
6. Significance coefficients application;
7. Fuzzy sets application.

The use of the listed methods results in the IPS receipt. All alternatives inside are arranged according to their significance/importance with properly acquired ranks. Formally, this corresponds to putatively linear dependence of rank and significance. The comparative importance of alternatives is defined exclusively via the difference in rank values. According to the qualimetric properties of arrangement (ordering) scales, it is impossible to define proportions between each pair of alternative significances. The latter belongs to the IPS flaws especially taking into account that the human mind tends to non-linearity. Hence the attitude toward the acceptability (importance, significance) level of various alternatives proportions should be also non-linear.

Since, in most cases, GPSs are received with the help of summation strategy and ranks averaging [5–9], all ISPs possess precision deviations. These deviations are accumulated in GPSs and increase their overall deviation levels correspondingly. This is especially important for the researches of operators IPSs for special complex systems, namely front-line operators.

3 Problem Statement

Bearing in mind the aforementioned, we define the goal of the current proceeding as follows: to specify the consistent GPS over the typical ATCs' mistakes set severities with the help of normalized significance coefficients of these mistakes. We believe that the achievement of this goal should be supported by the resolution of the following tasks:

1. Development of ATC IPSs over the typical mistakes set;
2. Development of GPS and its consistency determination;
3. Determination of alternative mistakes summary significance;
4. Fuzzification of the obtained sums and correspondent reset of proper alternative ranks in the GPS;
5. Comparison of the input and output GPSs.

4 Development and Validation of α-Method of ATCS Consistent GPS Specification Over Typical Mistakes Severities

One of the scientific results received earlier is the set of the most significant typical mistakes made by ATCs during their professional activity. It includes [10]:

Er.1—Radiotelephony phraseology violation;
Er.2—Inconsistent aircraft entry into the adjacent ATC zone;
Er.3—Longitudinal course time interval violation;
Er.4—Counter course time interval violation;
Er.5—Cross-aircraft separation violation at crossing courses;
Er.6—No address in ATC messaging;
Er.7—Error in aircraft call sign determination;
Er.8—Error in aircraft identification;
Er.9—Misuse of ATC schedule;
Er.10—Absence of the ATC mark on the strip regarding the transfer of control to the adjacent air traffic control center;
Er.11—Absence of the ATC mark on the strip regarding the coordination of aircraft entrance into the ATC zone of the adjacent air traffic control center;
Er.12—Violation of coordinated geographical control transfer boundary of the transfer of ATC by the aircraft;
Er.13—Violation of coordinated time control transfer boundary by ATC;
Er.14—Negligence in applying of the letter-digital information to the strip (possibility of multiple interpretations);
Er.15—Non-efficient/saving ATC;
Er.16—Violation of shift transition procedures;
Er.17—Issued commands to change the altitude or direction of the flight are not reflected on the strip;
Er.18—Attempt to control the aircraft under the condition of the TCAS system operation in the resolution advice mode;
Er.19—Errors of the aircraft information input into the automated system;
Er.20—Emergency procedures violation;
Er.21—Airspace use violations.

The survey was carried out to perform the tasks. It involved 37 professional ATCs from National Aviation University Flight Academy and the "DP Ukraeroruh" company. They were asked to use paired comparison and the IPS determination method including normalized determination of partial mistake significance. The results were aggregated into a single GPS.

With the help of multistep technology of marginal opinions allocation and removal, 26 people were separated from the initial sample with a high degree of opinion consistency within their subgroup. Their Kendall's concordance coefficient is equal to $W_{m_A} = 0.7144 > 0.7$. It satisfies the boundary value for significance level $\alpha = 0.01$. The formal representation of this GSP is as follows:

$$\begin{aligned}&\Pi_{18} \underset{m_A}{\succ} \Pi_{20} \underset{m_A}{\succ} \Pi_5 \underset{m_A}{\succ} \Pi_4 \underset{m_A}{\succ} \Pi_{21} \underset{m_A}{\succ} \Pi_3 \underset{m_A}{\succ} \Pi_8 \underset{m_A}{\succ}\\ &\underset{m_A}{\succ} \Pi_2 \underset{m_A}{\succ} \Pi_{17} \underset{m_A}{\succ} \Pi_{13} \underset{m_A}{\succ} \Pi_6 \underset{m_A}{\succ} \Pi_{12} \underset{m_A}{\succ} \Pi_{16} \underset{m_A}{\succ} \Pi_1 \underset{m_A}{\succ},\\ &\underset{m_A}{\succ} \Pi_{19} \underset{m_A}{\succ} \Pi_7 \underset{m_A}{\succ} \Pi_9 \underset{m_A}{\succ} \Pi_{14} \underset{m_A}{\succ} \Pi_{11} \underset{m_A}{\succ} \Pi_{15} \underset{m_A}{\succ} \Pi_{10}\end{aligned} \quad (1)$$

where $\underset{m_A}{\succ}$—defines the preferability of one mistake over another according to severity condition in the GPS combined of IPSs group members. The ranks of GPS mistakes were defuzzificated using the priorities arrangement method (PAM) [11] also known as "The leader task".

All calculations were performed with four digits after coma precision level. As it was calculated on the third iteration of PAM, the least significant coefficient (at the 21-st position) becomes equal to zero $r_{\Pi_{10}} = 21$, $\alpha_{\Pi_{10}} = 0$. The first iteration of the PAM method provides us with a linear tendency which is also inappropriate. Proper corrections were introduced into the GSP based on the defuzzificated ranked estimates of mistakes severities results. Consequently, the specified estimates for the new GPS were attained. For the first PAM iteration:

$$\begin{aligned}&\Pi_{18} \underset{\text{МРП}_I}{\succ} \Pi_{20} \underset{\text{МРП}_I}{\succ} \Pi_4 \underset{\text{МРП}_I}{\succ} \Pi_5 \underset{\text{МРП}_I}{\succ} \Pi_{21} \underset{\text{МРП}_I}{\succ} \Pi_3 \underset{\text{МРП}_I}{\succ}\\ &\underset{\text{МРП}_I}{\succ} \Pi_8 \underset{\text{МРП}_I}{\succ} \Pi_2 \underset{\text{МРП}_I}{\succ} \Pi_{13} \underset{\text{МРП}_I}{\succ} \Pi_{16} \underset{\text{МРП}_I}{\succ} \Pi_{17} \underset{\text{МРП}_I}{\succ}\\ &\underset{\text{МРП}_I}{\succ} \Pi_{12} \underset{\text{МРП}_I}{\succ} \Pi_6 \underset{\text{МРП}_I}{\succ} \Pi_1 \underset{\text{МРП}_I}{\succ} \Pi_7 \underset{\text{МРП}_I}{\succ} \Pi_{19} \underset{\text{МРП}_I}{\succ}\\ &\underset{MP\Pi_I}{\succ} \Pi_{14} \underset{MP\Pi_I}{\succ} \Pi_{11} \underset{MP\Pi_I}{\succ} \Pi_9 \underset{MP\Pi_I}{\succ} \Pi_{15} \underset{MP\Pi_I}{\succ} \Pi_{10}\end{aligned}, \quad (2)$$

where $\underset{\text{МРП}_I}{\succ}$ —defines the preferability of one mistake over another according to severity condition in the GPS combined of IPSs group members on the first PAM iteration.

For the second PAM iteration:

$$\begin{aligned}&\Pi_{18} \underset{\text{МРП}_{ii}}{\succ} \Pi_4 \underset{\text{МРП}_{ii}}{\succ} \Pi_{20} \underset{\text{МРП}_{ii}}{\succ} \Pi_{21} \underset{\text{МРП}_{ii}}{\succ} \Pi_5 \underset{\text{МРП}_{ii}}{\succ} \Pi_3 \underset{\text{МРП}_{ii}}{\succ}\\ &\underset{\text{МРП}_{ii}}{\succ} \Pi_8 \underset{\text{МРП}_{ii}}{\succ} \Pi_2 \underset{\text{МРП}_{ii}}{\succ} \Pi_{13} \underset{\text{МРП}_{ii}}{\succ} \Pi_{16} \underset{\text{МРП}_{ii}}{\succ} \Pi_{17} \underset{\text{МРП}_{ii}}{\succ}\\ &\underset{\text{МРП}_{ii}}{\succ} \Pi_1 \underset{\text{МРП}_{ii}}{\succ} \Pi_6 \underset{\text{МРП}_{ii}}{\succ} \Pi_{12} \underset{\text{МРП}_{ii}}{\succ} \Pi_7 \underset{\text{МРП}_{ii}}{\succ} \Pi_{19} \underset{\text{МРП}_{ii}}{\succ}\\ &\underset{\text{МРП}_{ii}}{\succ} \Pi_{14} \underset{\text{МРП}_{ii}}{\succ} \Pi_9 \underset{\text{МРП}_{ii}}{\succ} \Pi_{11} \underset{\text{МРП}_{ii}}{\succ} \Pi_{15} \underset{\text{МРП}_{ii}}{\succ} \Pi_{10}\end{aligned} \quad (3)$$

where $\underset{\text{МРП}_{II}}{\succ}$ —defines the preferability of one mistake over another according to severity condition in the GPS combined of IPSs group members on the second PAM iteration.

The GPSs (1)–(3) were compared using Spearman's rank correlation coefficient. Their coincidence is close to absolute $R_S^{\text{МРП}_i-\text{МРП}_{ii}} = 0.9909, R_S^{\text{МРП}_I-m_A} = 0.9818$, $R_S^{\text{МРП}_{II}-m_A} = 0.9740$ with a high significance level $\alpha = 1\%$. The slightly higher value of $R_S^{\text{МРП}_I-m_A}$ over $R_S^{\text{МРП}_{II}-m_A}$ is explained by the linearity of both mistakes severity significances coefficients received from the first PAM iteration and ranks linearities. Nevertheless, the GPS (3) should be taken into account after all because it is based on the non-linear mistakes' severity significances coefficients. Considering the mentioned above, we can state that all tasks for this proceeding are resolved.

In conclusion, we should note that it would be valuable to find out how exactly the specified coefficients would change compared to the already existing methods. Here we are talking about the optimized GPSs with high significance levels processed by the Savage criterion and Kemeny median.

5 Conclusion

All scientific results proposed in this proceeding are improving the expert procedures methodology. The proposed α-method provides a convenient and convincing way of specifying statistically consistent GPSs. Let's generalize all the points from the proceeding emphasizing the more important of them:

1. IPSs and GPSs of air traffic controllers over the set of typical mistakes are important indicators. They provide information about both mature individual perks of potential hazards forecasting and peculiarities of group decision-making and thinking inherent for particular ATC groups.
2. The α-method is developed for already existing and statistically consistent GPSs. The method includes ranked estimates defuzzification by handling them with normalized significance coefficients of arranged alternatives.
3. The method is tested for specifying ATCs GPS over the range of typical mistakes. It appeared that the specified GPSs received via linear (first PAM iteration) and non-linear (second PAM iteration) have almost identical results both between them and with the statistically consistent GPS.
4. Further application should be based on the specified GPS obtained from non-linear significance coefficients since it corresponds to the overall non-linearity of the human way of reasoning.
5. Consequent research should be performed in the direction of α-method application within preferences systems comparison and concordance of GPSs.

References

1. Accident prevention manual: Doc. ICAO 9422, AN/923. ICAO, Montreal, Canada (1984)
2. Human Factors Module. A Business Case for Human Factors Investment. EUROCONTROL, Brussel, Belgium (1999)
3. Safety Management Manual (SMM): Doc. ICAO 9859, AN/460, 4th edn (advance unedited). ICAO, Montreal, Canada (2018)
4. Reva, O., Borsuk, S., Shulgin, V., Nedbay, S.: Ergonomic assessment of instructors' capability to conduct personality-oriented training for air traffic control (ATC) personnel. advances in human factors of transportation. In: Proceedings of the AHFE 2019 International Conference on Human Factors in Transportation, Springer, USA, pp. 783–793 (2019)

5. Liu, Y., Eckert, C., Earl, C.: A review of fuzzy AHP methods for decision-making with subjective judgements. Expert Syst. Appl. **161**, 113738 (2020)
6. Nadkarni, R., Puthuvayi, B.: A comprehensive literature review of multi-criteria decision making methods in heritage buildings. J. Build. Eng. **32**, 101814 (2020)
7. Pasman, H., Rogers, W.: How to treat expert judgment? With certainty it contains uncertainty! J. Loss Prev. Process Ind. **66**, 104200 (2020)
8. Thorne, M., Williams, M.: A review of expert judgment techniques with reference to nuclear safety. Prog. Nucl. Energy **27**(2–3), 83–254 (1992)
9. Werner, C., Bedford, T., Cooke, R., Hanea, A., Morales-Nápoles, O.: Expert judgement for dependence in probabilistic modelling: a systematic literature review and future research directions. Eur. J. Oper. Res. **258**(3), 801–819 (2017)
10. Reva, O., Nevynitsyn, A., Borsuk, S., Shulgin, V., Kamyshyn, V.: Air traffic controllers' attitude to the mistakes hazards during their professional experience. In: Li, L. (ed.) Safety and Risk Assessment of Civil Aircraft during Operation, pp. 113–127. IntechOpen (2020)
11. Reva, O., Kamyshyn, V., Nevynitsyn, A., Shulgin, V., Nedbay, S.: Criteria indicators of the consistency of air traffic controllers' preferences on a set of characteristic errors. In: Advances in Human Aspects of Transportation Proceedings of the AHFE 202, Virtual Conference on Human Aspects of Transportation, Springer, USA, pp. 617–623 (2020)

Named Entity Recognizer for Konkani Text

Annie Rajan and Ambuja Salgaonkar

Abstract A named entity recognizer (NER), an essential tool for natural language processing (NLP), is presented for the first time for the Konkani language. Gold data of 1000 NER-tagged Konkani sentences consisting of 1068 named entities is one of the linguistic resources generated through this work. A conditional random field (CRF) classifier built on the training data set of 794 named entities from 800 sentences of the corpus, demonstrated 96% accuracy and 72% f-score. On the test data set of 274 named entities from 200 sentences of the corpus, 86% accuracy and 66% f-score were obtained. When the training and test data were complemented with a lookup table consisting of a database of 12 months, 53 locations, 44 person-names and 23 numerals and their synonyms, the figures improved to 99% accuracy and 90% f-score for the training data set, and 89% accuracy and 73% f-score for the test data set. To place our research in perspective, a summary is presented of the NER literature for world languages as well as Indian languages, as also NER for Indian languages using CRF.

1 Introduction

During the late 1990s, named entity expressions like ENAMEX (persons, locations, organization), NUMEX (money, percentages), and TIMEX (time, date, year) were established in Message Understanding Conferences (MUC) [1, 2]. More recently, advances in machine learning to have expedited research in NLP in areas like automatic PoS tagging, sentiment analysis, named entity recognition (NER), and morphological analysis. Flaws in identifying the named entities in a document lead to a wrong interpretation of linguistic expressions. Therefore, NER has a vital role to play in

A. Rajan (✉)
DCT's Dhempe College of Arts and Science, University of Mumbai, Mumbai, Maharashtra 400032, India

A. Salgaonkar
University of Mumbai, Mumbai, Maharashtra 400032, India
e-mail: ambujas@udcs.mu.ac.in

T. Senjyu et al. (eds.), *ICT with Intelligent Applications*, Smart Innovation, Systems and Technologies 248, https://doi.org/10.1007/978-981-16-4177-0_69

data management applications such as question answering systems, search engines, text mining, and business intelligence.

Konkani, an Indo-Aryan or Indic language has constitutional status as one of the 22 officially recognized languages of India. It is the official language of the state of Goa, situated on the west coast of India. About 2.5 million people, 0.2% of the total population of India, speak this language, a very small number compared to speakers of the major Indian languages. Goa has the highest gross domestic product (GDP) per capita Indian states, and moreover, it is a major tourist destination.

This study focuses mainly on creating a corpus and implementing NER using CRF for the Konkani language. Section 2 is on the literature survey, Sect. 3 below is on building and testing of NER for Konkani, Sect. 4 is a discussion on misclassified entities, and Sect. 5 contains the conclusions and pointers for future work.

2 Literature Survey

NLP is a subfield of linguistics, information engineering, computer science and artificial intelligence, interactions between natural languages and computers, and how a computer program processes and analyzes natural language data. Pioneering working NLP includes [3–5]. Examples of NER research include [6–10]. Examples of NER research in languages of the world are listed in Table 1, and for Indian languages, in Table 2.

The earliest reference to NLP research in Konkani dates back to 2016 [55]. Areas of research includes: text-to-speech [56–60], PoS tagging [61–63], morphological analysis [64], sentiment analysis [65–67], transliteration [68], Konkani-English code-mixed data [69], wordnet [70], white paper on Konkani [71], tag set for Konkani [72], a survey of NLP resources in Konkani [73]. To the best of our knowledge, the present work is the first attempt in NER for Konkani language. It uses a CRF-based model. For the training dataset, the *f*-score obtained is 72%, and this figure drops to 66% for new test data from the same domain. By employing a lookup table of a data set containing of 135 records, the performance for training data and test data boosted to 90% and 74%, respectively.

CRF is a widely used method for NER [74, 75]. It has been employed for several Indian languages, particularly Hindi, Telugu, Tamil, Punjabi, Manipuri, Assamese, Malayalam, Oriya, Bengali, Urdu, Gujarati, and Kannada. For Hindi, high accuracy is observed. Most of these languages have experimented with CRF and have obtained satisfactory results [76–78]. Some other techniques used for NER are hidden Markov model (HMM) [79], decision trees, support vector method (SVM) [80], maximum entropy models (ME) [81], and rule-based method [82]. Table 3 provides a bird's eye view of CRF-based NER research in Indian languages. For a given language, the references are listed, data size is rounded to a thousand words, and *f*-score is given in percent.

Sharma et al. [98], Singh [99] have reported accuracy of 85.78% and 83.00% over data sets of size 1.9 million words and 150,000 words (training data) and 100,000

Table 1 NER research in various languages

S. No	Language	Reference
1	Arabic	[11, 12]
2	Basque	[13–15]
3	Bulgarian	[16, 17]
4	Catalan	[18, 19]
5	Cebuano	[20]
6	Chinese	[12, 21–23]
7	Croatian, Czech	[24]
8	Danish	[15, 25]
9	Dutch	[26]
10	English	[6, 7, 9, 12, 15]
11	French	[27–29]
12	German	[12, 26, 30]
13	Greek	[31–33]
14	Italian	[34–36]
15	Korean	[15, 37]
16	Polish	[24, 38]
17	Portuguese	[39–42]
18	Romanian	[43, 44]
19	Russian	[24, 45]
20	Spanish	[15, 26]
21	Swedish	[46, 47]
22	Slavak, Slovene, Ukrainian	[24]
23	Vietnamese	[48]
24	Turkish	[43, 49]

Table 2 NER for Indian languages

S. No.	Language	Reference
1	Bengali, Hindi, Telugu, Oriya, Urdu	[50, 51]
2	Telugu, Hindi	[52]
3	Hindi, Urdu, Marathi	[53]
4	Telugu, Bengali, Tamil	[54]

words (test data) for the Punjabi and Assamese languages, respectively, but there is no mention of the f-score. Therefore, it is unclear if the dominating figure is true positive figure.

Table 3 Performance of NER using CRF

S. No.	Data set (words)	*F*-score (%)
Hindi	74,076 [83]	95.84
1	340,000 [84]	82.55
2	30,735 [85]	44.97
3		
Bengali	150,000 [77]	90.70
4		
Punjabi	11.29 million [86]	86.25
5	60,975 [78]	80.92
6		
Malayalam	100,000 [76]	85.61
7		
Telugu	6300 [87]	85.13
8	17,951 [88]	44.91
9		
Manipuri	55,000 [89]	83.33
10		
Gujarati	122,201 [90]	83.20
11		
Kannada	100,127 [91]	82.00
12		
Assamese	372,400 [92]	78.97
13		
Tamil	93,292 [93]	80.44
14	93,292 [94]	70.68
15		
Oriya	35,000 [95]	69.66
16		
Urdu	3238 [96]	63.00
17	50,000 [97]	55.30
18	74,076 [83]	95.84

3 Building and Testing of NER for Konkani

The data for this experiment is a history corpus of 1000 Konkani sentences from the database maintained by the governmental agency known as Technology Development for Indian Languages (TDIL) [100]. Our choice of the history domain is because persons, locations, dates, and various other numerical entities are readily found in historical documents. Named entities are often associated with the following eight

Table 4 NER tag set for Konkani language

S. No.	Named entity tag	No. of entities for all data (1000 words)	No. of entities of training data (800 words)
1	ENAMEX/person	410	306
2	ENAMEX/location	294	228
3	NUMEX/number	154	116
4	TIMEX/date	133	86
5	ENAMEX/organization	70	50
6	TIMEX/time	6	6
7	NUMEX/money	3	1
8	NUMEX/percentage	1	1

tags: organization, person, location, money, percentage, time, date, and year. We added one more tag to NUMEX: number, denoting an arbitrary numeric value. All the sentences have been manually tagged, and verification has been performed by language experts. The data set is available on the author's website and is becoming enriched by crowdsourcing. The distribution of 9 types of named entities in our corpus and in the training data has been given in Table 4.

3.1 *Phase 1 of Konkani NER*

We employed python-CRF suite to build on NER for Konkani [101]. For this study, we employed limited memory Broydan-Fletcher-Goldfarb-Shanno algorithm (L-BFGS) and trained the model by using 800 tagged sentences [102].

Let $X = (x_1, x_2, x_3, \ldots, x_t)$ be some observed input data sequence, and $Y = (y_1, y_2, y_3, \ldots, y_t)$ be a sequence of states. CRF defines the conditional probability of a state sequence given an input sequence to be,

$$P(Y|X) = \frac{1}{\text{Zo}} \exp\left(\sum_{t=1}^{T} \sum_{k} \lambda_k f_k(y_t - 1, y_t, x, t)\right)$$

where Zo is,

$$\text{Zo} = \Sigma_y \exp\left(\sum_{t=1}^{T} \Sigma_k \lambda_k f_k(y_t - 1, y_t, x, t)\right)$$

Let $X = (x_1, x_2, x_3, \ldots, x_t)$ be Zo is a normalization factor overall state sequences to make the conditional probabilities sum up to $1, f_k\ (y_{t-1}, y_t, x, t)$ is a feature function, weight λ_k is weight of feature function to be learned when training, y_t is the present value, and y_{t-1} is the previous value. Given a tagged corpus, the CRF builds a model

Table 5 Confusion matrix of the training data set

	Location	Number	Person	Data	Time	Money	Organization	Percentage
Location	148	2	73	1	0	0	4	0
Number	10	87	11	7	0	0	1	0
Person	44	4	251	5	0	0	2	0
Date	3	1	6	76	2	0	0	0
Time	2	1	0	1	0	0	0	0
Money	0	0	1	0	0	0	0	0
Organization	8	1	8	0	0	0	33	0
Percentage	0	0	1	0	0	0	0	0

that can be employed to predict tags in a given text. The performance of the model on the training data is shown in the confusion matrix in Table 5.

Clearly, the performance in the money and percentage cases is poor, the entities ʃəmbər (शंबर)and d̪ʰən̪d̪olət̪ (धनदौलत)were not identified as percentage and money as in manually tagged corpus, but identified as person. The reason behind this poor performance is that sufficient data for training these two classes were not available. The frequency of these two entities in the data is 0.004, quite low. Therefore, we computed the following matrix of the model in Table 6 by excluding these two entities.

The training data consisted of 4673 words with 794 named entities. The average accuracy (a measure of correctly identifying the true positive and true negative records) is 96%, which is good; the average *f*-score (an indication of type I and II errors), is 72%, which falls between that of Tamil and Oriya NER [94, 95]. However, our data set, comparable to that of Urdu [96], is much smaller than that of the former two languages. Though the *f*-score of our model in the case of person is 8.3% less than that of Oriya, it is greater by 6.4% and 17.8%, respectively, for location and organization. Therefore, we assert that the results of our model are encouraging. The performance of the model on the test data is shown in the confusion matrix of Table 7.

Table 6 Performance matrix of the training data set

	True positive (TP)	True negative (TN)	False negative (FN)	False positive (FP)	Precision	Recall	*F*-score	Accuracy
Location	148	499	80	67	69	65	67	81
Number	87	669	29	9	91	75	82	95
Person	251	388	100	55	82	72	76	80
Date	76	694	14	10	88	84	86	97
Time	2	788	0	4	33	100	50	99
Organization	33	737	7	17	66	83	73	87

Table 7 Confusion matrix of the test data set

	Location	Number	Person	Data	Time	Money	Organization	Percentage
Location	32	1	28	0	0	0	5	0
Number	6	28	2	1	0	0	1	0
Person	27	1	73	1	0	0	2	0
Date	2	3	12	30	0	0	0	0
Time	0	0	0	0	0	0	0	0
Money	1	0	1	0	0	0	0	0
Organization	4	0	3	0	0	0	13	0
Percentage	0	0	0	0	0	0	0	0

Table 8 Performance matrix of the test data set

	True positive (TP)	True negative (TN)	False negative (FN)	False positive (FP)	Precision	Recall	F-score	Accuracy
Location	32	171	34	40	44	48	46	73
Number	28	234	10	5	85	74	76	95
Person	73	127	31	46	61	70	65	72
Date	30	228	17	2	94	64	76	93
Organization	13	249	7	8	62	65	63	95

The test data consisted of 1670 words with 274 named entities. The performance of time, money, and percentage cases is nil because of the absence of sufficient data for these three classes in our data set. Therefore, we compute performance metrics of the model by excluding these three, see Table 8.

3.2 *Phase 2 of Konkani NER*

Analysis of the results revealed that three terms, t͡ʃja (च्य□), kʰubəɭɭe (खुबळ्ळे)and t̪aɳe (ता□णें)are incorrectly tagged as named entities. This error was rectified. Further, the stems of the misclassified named entities were searched in a small lookup table comprising 12 month-names, 44 person-names, 53 location-names, 23 numerals, and their synonyms. As a result, the accuracy and f-score of our model for the training data went up to 99% and 90%, respectively, while for the test data these values were enhanced to 89 and 73%. The confusion matrix of this enhanced model for training data is given in Table 9 and the performance matrix, in Table 10. For the test data, the confusion matrix and performance matrix of the enhanced model are given in Tables 11 and 12.

Table 9 Confusion matrix of the training data set when enhanced with lookup table

	Location	Number	Person	Data	Time	Money	Organization	Percentage
Location	218	0	6	1	0	0	3	0
Number	0	116	0	0	0	0	0	0
Person	0	0	306	0	0	0	0	0
Date	0	1	1	84	0	0	0	0
Time	1	1	0	1	3	0	0	0
Money	0	0	1	0	0	0	0	0
Organization	4	1	7	0	0	0	38	0
Percentage	0	0	1	0	0	0	0	0

Table 10 Performance matrix of the training data set enhanced with lookup table

	True positive (TP)	True negative (TN)	False negative (FN)	False positive (FP)	Precision	Recall	*F*-score	Accuracy
Location	218	561	10	5	98	96	97	98
Number	116	675	0	3	97	100	99	100
Person	306	472	0	16	95	100	97	98
Date	84	706	2	2	98	98	98	99
Time	3	788	3	0	100	50	67	100
Organization	38	741	12	3	93	96	84	98

Table 11 Confusion matrix of the test data set when enhanced with lookup table

	Location	Number	Person	Data	Time	Money	Organization	Percentage
Location	81	2	42	0	0	0	7	0
Number	7	64	2	2	0	0	2	0
Person	47	2	154	1	0	0	4	0
Date	2	6	13	70	0	0	0	0
Time	0	0	0	0	0	0	0	0
Money	2	0	2	0	0	0	0	0
Organization	7	0	6	0	0	0	27	0
Percentage	0	0	0	0	0	0	0	0

4 Discussion

The misclassified unseen entities include 14 person names, 13 location names, 4 organization names, and 3 new numerals. The lookup table updated by including these entities would increase the performance of the system for the test data to be at par with the training data. Along with the NER-tagged text, this approach

Table 12 Performance matrix of the test data set enhanced with lookup table

	True positive (TP)	True negative (TN)	False negative (FN)	False positive (FP)	Precision	Recall	*F*-score	Accuracy
Location	81	355	51	65	55	61	58	79
Number	64	465	13	10	86	83	85	96
Person	154	279	54	65	70	74	72	78
Date	70	458	21	3	96	77	85	96
Organization	27	499	13	13	68	68	68	95

will lead to enriching the classified dictionaries of any language, Konkani in this instance. Twenty-one of the misclassified named entities from the training data were compounded words, typically the names of 12 organizations. For example seɳʈ sʈifan̪s haj s̆kul (सेंट स्टीफन्स ह□स्कूल), t̪rawəɳkor sən̪st̪ʰan̪ (त्र□णकोर संस्थ□), n̪æʃnəl kɔŋgres (नॅशनल काँग्रेस), ʃard̪a sən̪git̪ məɳɖəɭi (शा□द□संगीत मंडळी),, They need to be treated as nested ones and taken care. Yet another category of misclassified named entities involves 9 (९), aʈʰwer (आठवेर), sat̪aɳk (स□□□)that are the words referring to time and ekwis (एकवीस), ek (एक), t̪in̪ (तीन)that are the works referring to day-number, essentially these are numbers. It is difficult to understand their type unless there is a reference word like t̪arkʰer (ता□खेर)that means 'on date' or gʰəɳʈjar (घंट्या□)that means 'at hour' in the context text.

5 Conclusion

The first NER in the Konkani language has been presented, and its performance can be said to be good, given an accuracy of 96% and *f*-score of 72% for training data, and 86 and 66% for test data. Using a lookup table consisting of 132 entities, the accuracy and *f*-score increased to 99 and 90% for training data, and for test data, it increased to 89 and 73%. The data set has been made available on the author's website for further enhancement by interested researchers [103]. Experiments using other algorithms can be tried with this Konkani NER data set, and hybrid approaches are likely to prove better. In the future, more sentences will be tagged and an online NER tool will be built for Konkani.

Acknowledgements We wish to acknowledge the help provided by Mrs Anju Sakardande, Head, Department of Indian Languages at Dhempe College of Arts and Science, Panaji, Goa and Mr. Sharat K. Raikar, language interpreter for Konkani and Hindi.

References

1. Chinchor, N.: MUC-7 named entity task definition (1997). https://www-nlpir.nist.gov/related_projects/muc/proceedings/ne_task.html
2. Chinchor, N., Brown, E.J., Ferro, L., Robinson, P.: Named entity recognition task definition (1999). http://www.dessem.com/sites/default/files/ne99_taskdef_v1_4.pdf
3. Bhattacharyya, P.: Natural language processing—a perspective from computation in presence of ambiguity resource constraint and multilinguality. CSI J. Comput. **1**(2), 1–13 (2012). https://www.cse.iitb.ac.in/~pb/papers/csi-nlp-pb-8aug12.pdf
4. Gelbukh, A.: Natural language processing. In: Fifth International Conference on Hybrid Intelligent Systems (HIS'05), pp. 1 (2005). https://ieeexplore.ieee.org/document/1587718
5. Makhoul, J., Jelinek, F., Rabiner, L., Weinstein, C., Zue, V.: White paper on spoken language systems. In: Proceedings of the Workshop on Speech and Natural Language, pp. 463–479. Association for Computational Linguistics, Stroudsburg, PA, USA (1989). https://dl.acm.org/citation.cfm?id=1075525
6. Bikel, D.M., Miller, S., Schwartz, R., Weischedel, R.: Nymble—a high-performance learning name-finder. In: Fifth Conference on Applied Natural Language Processing, pp. 194–201. Association for Computational Linguistics, Washington, DC, USA (1997). https://pdfs.semanticscholar.org/e929/2ba3230f01b9f6990362fdf06783b9347bf6.pdf
7. Chiu, J.P.C., Nichols, E.: Named entity recognition with bidirectional LSTM-CNNs (2015). https://www.aclweb.org/anthology/Q16-1026
8. Etzioni, O., Cafarella, M., Downey, D., Popescu, A.-M., Shaked, T., Soderland, S., Weld, D.S., Yates, A.: Unsupervised named-entity extraction from the web—an experimental study: Artif. Intell. **165**(1), 91–134 (2005). https://web.eecs.umich.edu/~michjc/papers/knowitall-aij.pdf
9. Florian, R., Ittycheriah, A., Jing, H., Zhang, T.: Named entity recognition through classifier combination. In: Proceedings of the Seventh Conference on Natural Language Learning at HLT-NAACL 2003—Volume Association for Computational Linguistics, Stroudsburg, PA, USA (2003). https://www.aclweb.org/anthology/W03-0425
10. Murthy, R., Khapra, M., Bhattacharyya, P.: Sharing network parameters for crosslingual named entity recognition CoRR abs/1607.00198 (2016). https://arxiv.org/abs/1607.00198
11. Huang, F.: Multilingual Named Entity Extraction and Translation from Text and Speech (2006). https://pdfs.semanticscholar.org/8368/d39c085e311c94f7f7434c983e0f37a779b3.pdf
12. Shao, Y., Hardmeier, C., Nivre, J.: Multilingual named entity recognition using hybrid neural networks. In: The Sixth Swedish Language Technology Conference (SLTC) (2016). https://uu.diva-portal.org/smash/get/diva2:1055627/FULLTEXT01.pdf
13. Agerri, R., Rigau, G.: Robust multilingual named entity recognition with shallow semi-supervised features. Artif. Intell. (2016). https://arxiv.org/pdf/1701.09123.pdf
14. Loinaz, A., Tardaguila, B., Ramos, E., González, F.V., Enbeita, U.: Named entity recognition and classification for texts in Basque1 (2003). https://pdfs.semanticscholar.org/98ea/2de250b88016605751fe003e17d49310c511.pdf?_ga=2.231301334.832029152.1566558149-859315823.1563803599
15. Whitelaw, C., Patrick, J.: Evaluating corpora for named entity recognition using character-level features. In: AI, pp. 910–921. Advances in Artificial Intelligence, Springer Berlin Heidelberg, Berlin, Heidelberg (2003). https://link.springer.com/chapter/10.1007/978-3-540-24581-0_78
16. Georgiev, G., Nakov, P., Ganchev, K., Osenova, P., Simov, K.: Feature-rich named entity recognition for Bulgarian using conditional random fields. In: Proceedings of the International Conference (RANLP), pp. 113–117. Association for Computational Linguistics, Borovets, Bulgaria (2009). https://www.aclweb.org/anthology/R09-1022
17. Da Silva, J.F., Kozareva, Z., Lopes, J.G.P.: Cluster analysis and classification of named entities. In: Proceedings of the Fourth International Conference on Language Resources and Evaluation (LREC). European Language Resources Association (ELRA), Lisbon, Portugal (2004). https://www.aclweb.org/anthology/papers/L/L04/L04-1520/

18. Carreras, X., Marquez, L., Padro, L.: Named entity recognition for Catalan using Spanish resources. In: Proceedings of the Tenth Conference on European Chapter of the Association for Computational Linguistics, vol. 1. Association for Computational Linguistics, Stroudsburg, PA (2003). https://pdfs.semanticscholar.org/9cb2/55171dc7c1cbebbe73ba7f77a43f08914599.pdf
19. Villodre, L.M., de Gispert, A., Carreras, X., Padró, L.: Low-cost named entity classification for catalan- exploiting multilingual resources and unlabeled data. In: NER@ACL (2003). https://www.semanticscholar.org/paper/Low-cost-Named-Entity-Classification-for-Catalan%3A-Villodre-Gispert/2176526afaf3e2c961bfddb8352fa13a7ffc7271
20. May, J., Brunstein, A., Natarajan, P., Weischedel, R.: Surprise! What's in a Cebuano or Hindi name? **2**(3), 169–180 (2003). https://www.isi.edu/~jonmay/pubs/cebuano.pdf
21. Chen, H.-H., Lee, J.-C.: Identification and classification of proper nouns in Chinese texts. In: Proceedings of the 16th Conference on Computational Linguistics, vol. 1, pp. 222–229. Association for Computational Linguistics, Stroudsburg, PA, USA (1996). https://pdfs.semanticsholar.org/9c26/98612474d02cb16406ac63cb6eca0ffda52e.pdf
22. Wang, L.J., Li, W.-C., Chang, C.-H.: Recognizing unregistered names for mandarin word identification. In: Proceedings of the 14th Conference on Computational Linguistics, vol. 4, pp. 1239–1243. Association for Computational Linguistics, Stroudsburg, PA, USA (1992). https://pdfs.semanticscholar.org/1267/879e1caf9627271dcdcf428eac4555b800b7.pdf?_ga=2.200318318.1133831351.1565924948-859315823.1563803599
23. Yu, S., Bai, S., Wu, P.: Description of the Kent Ridge Digital Labs system used for MUC-7. In: Seventh Message Understanding Conference (MUC-7): Proceedings of a Conference Held in Fairfax, Virginia, April 29–May 1 (1998). https://pdfs.semanticscholar.org/83d3/9fe06722650134845f2688b99fc1b217cd3f.pdf?_ga=2.166984190.1133831351.1565924948-859315823.1563803599
24. Piskorski, J., Pivovarova, L., Snajder, J., Steinberger, J., Yangarber, R.: The first cross-lingual challenge on recognition, normalization and matching of named entities in Slavic languages. In: Proceedings of the 6th Workshop on Balto-Slavic Natural Language Processing, pp. 76–85. Association for Computationnal Linguistics, Valencia, Spain, (2017). https://www.aclweb.org/anthology/W17-1412
25. Bick, E.: A named entity recognizer for Danish. In: Proceedings of the Fourth International Conference on Language Resources and Evaluation (LREC). European Language Resources Association (ELRA), Lisbon, Portugal (2004)
26. Lample, G., Ballesteros, M., Subramanian, S., Kawakami, K., Dyer, C.: Neural architectures for named entity recognition. In: Proceedings of the 2016 Conference of the North American Chapter of the Association for Computational Linguistics-Human Language Technologies, pp. 260–270. Association for Computational Linguistics, San Diego, California (2016). https://www.aclweb.org/anthology/N16-1030
27. Azpeitia, A., Cuadros, M., Gaines, S., Rigau, G.: Supervised named entity recognition for French. In: Sojka, P., Horák, A., Kope, I., Pala, K. (eds.) Text, Speech and Dialogue, pp. 158–165. Springer International Publishing, Cham (2014). https://link.springer.com/chapter/10.1007/978-3-319-10816-2_20
28. Petasis, G., Vichot, F., Wolinski, F., Paliouras, G., Karkaletsis, V., Spyropoulos, C.D.: Using machine learning to maintain rule-based named-entity recognition and classification systems. In: Proceedings of the 39th Annual Meeting of the Association for Computational Linguistucs, Toulouse, France, pp. 426–433 (2001). https://pdfs.semanticscholar.org/05f8/f8e97e51c4fdb4511b17d57def2c05b35975.pdf
29. Poibeau, T., Acoulon, A., Avaux, C., Beroff-Bénéat, L., Cadeau, A., Calberg, M., Delale, A., De Temmerman, L., Guenet, A.L., Huis, D., Jamalpour, M., Krul, A., Marcus, A., Picoli, F., Plancq, C.: The multilingual named entity recognition framework. In: 10th Conference of the European Chapter of the Association for Computational Linguistics, Budapest, Hungary (2003). https://www.aclweb.org/anthology/E03-1082
30. Benikova, D., Biemann, C., Reznicek, M.: NoSta-D Named Entity Annotation for German-Guidelines and Dataset (n.d.)

31. Boutsis, S., Demiros, I., Giouli, V., Liakata, M., Papageorgiou, H., Piperidis, S.: A system for recognition of named entities in Greek. In: Christodoulakis, D.N. (ed.) Natural Language Processing—NLP 2000. Springer Berlin Heidelberg, Berlin, Heidelberg, pp. 424–435 (2000). https://link.springer.com/chapter/10.1007/3-540-45154-4_39
32. Karkaletsis, V., Paliouras, G., Petasis, G., Manousopoulou, N., Spyropoulos, C.D.: Named-entity recognition from Greek and English texts. J. Intell. Rob. Syst. **26**(2), 123–135 (1999). https://link.springer.com/article/10.1023/A:1008124406923
33. Michailidis, I., Diamantaras, K., Vasileiadis, S., Frère, Y.: Greek named entity recognition using support vector machines, maximum entropy and onetime. In: Proceedings of the Fifth International Conference on Language Resources and Evaluation (LREC). European Language Resources Association (ELRA), Genoa, Italy (2006). http://www.lrec-conf.org/proceedings/lrec2006/pdf/557_pdf.pdf
34. Black, W.J., Rinaldi, F., Mowatt, D.: FACILE: description of the NE system used for MUC-7 (1998). https://www.aclweb.org/anthology/M98-1014
35. Cucchiarelli, A., Velardi, P.: Unsupervised named entity recognition using syntactic and semantic contextual evidence. Comput. Linguist. **27**(1), 123–131 (2001). https://www.aclweb.org/anthology/J01-1005
36. Federico, M., Bertoldi, N., Sandrini, V.: Bootstrapping named entity recognition for Italian broadcast news. In: Proceedings of the ACL-02 Conference on Empirical Methods in Natural Language Processing, vol. 10, pp. 296–303. Association for Computational Linguistics, Stroudsburg, PA, USA (2002). https://dl.acm.org/doi/10.3115/1118693.1118731
37. Choi, Y., Cha, J.: Korean named entity recognition and classification using word enbedding features. J. KIISE **43**, 678–685 (2016). https://www.researchgate.net/publication/305842855
38. Piskorski, J.: Extraction of Polish Named-Entities (2004). https://www.researchgate.net/publication/230729674
39. de Castro, P.V.Q., da Silva, N.F.F., da Silva Soares, A.: Portuguese named entity recognition using LSTM-CRF. In: Villavicencio, A., Moreira, V., Abad, A., Caseli, H., Gamallo, P., Ramisch, C., Oliveira, H.G., Paetzold, G.H. (eds.) Computational Processing of the Portuguese Language, pp. 83–92. Springer International Publishing, Cham (2018). https://link.springer.com/chapter/10.1007/978-3-319-99722-3_9
40. do Amaral, D.O.F., Fonseca, E., Lopes, L., Vieira, R.: Comparing NERP-CRF with publicly available Portuguese named entities recognition tools. In: Baptista, J., Mamede, N., Candeias, S., Paraboni, I., Pardo, T.A.S., Nunes, M.G.V. (eds.) Computational Processing of the Portuguese Language, pp. 244–249. Springer International Publishing, Cham (2014). https://link.springer.com/chapter/10.1007/978-3-319-09761-9_27
41. Palmer, D.D., Day, D.S.: A statistical profile of the named entity task. In: Fifth Conference on Applied Natural Language Processing, pp. 190–193. Association for Computational Linguistics. Washington, DC, USA (1997). https://www.aclweb.org/anthology/papers/A/A97/A97-1028/
42. Pirovani, J.P.C., de Oliveira, E.: Portuguese named entity recognition using conditional random fields and local grammars. In: LREC (2018). https://www.semanticscholar.org/paper/Portuguese-Named-Entity-Recognition-using-Random-Pirovani-Oliveira/515f8e9fbbd04d3d1e625aa7f4e84e854bfc5987
43. Cucerzan, S., Yarowsky, D.: Language independent named entity recognition combining morphological and contextual evidence. In: Joint {SIGDAT} Conference on Empirical Methods in Natural Language Processing and Very Large Corpora (1999). https://pdfs.semanticscholar.org/425e/27250b17fe4c6fde104c97a754d0e2b296cd.pdf?_ga=2
44. Mitrofan, M.: Bootstrapping a Romanian corpus for medical named entity recognition. In: RANLP (2017). https://www.semanticscholar.org/paper/Bootstrapping-a-Romanian-Corpus-for-Medical-Named-Mitrofan/5b0a0aff06999e36d0908dcd1375530d4e0ab6ea
45. Popov, B., Kirilov, A., Maynard, D., Manov, D.: Creation of reusable components and language resources for named entity recognition in Russain. In: Proceedings of the Fourth International Conference on Language Resources and Evaluation (LREC). European Language Resources Association (ELRA), Lisbon, Portugal (2004). http://www.lrec-conf.org/proceedings/lrec2004/pdf/267.pdf

46. Almgren, S., Pavlov, S., MOgren, O.: Named entity recognition in Swedish health records with character-based deep bidirectional LSTM. In: Proceedings of the Fifth Workshop on Building and Evaluating Resources for Biomedical Text Mining Bio Txt M 2016, pp. 30–39. The COLING 2016 Organizing Committee, Osaka, Japan (2016). https://www.aclweb.org/anthology/W16-5104
47. Kokkinakis, D.: Aventinus, gate and Swedish Lingware. In: Proceedings of the 11th Nordic Conference of Computational Linguistics (NODALIDA), pp. 22–33. Center for Sprogteknologi, University of Copenhagen, Denmark (1998). https://www.aclweb.org/anthology/W98-1603
48. Minh, P.Q.N.: A Feature-Rich Vietnamese Named-Entity Recognition Model: CoRR abs/1803.04375 (2018). https://arxiv.org/abs/1803.04375
49. Yeniterizi, R., Tür, G., Oflazet, K.: Turkish named-entity recognition. In: Turkish Natural Language Processing, pp. 115–132. Springer International Publishing, Cham, Switzerland (2018). https://link.springer.com/chapter/10.1007/978-3-319-90165-7_6
50. Ekbal, A., Haque, R., Das, A., Poka, V., Bandyopadhyay, S.: Language independent named entity recognition in Indian languages. In: Proceedings of the IJCNLP-08 Workshop on Named Entity Recognition for South and South East Asian Languages (2008). https://www.aclweb.org/anthology/108-5006
51. Saha, S.K., Chatterji, S., Dandapat, S., Sarkar, S., Mitra, P.: A hybrid approach for named entity recognition in Indian languages (2008). https://www.semanticscholar.org/paper/A-Hybrid-Approach-for-Named-Entity-Recognition-in-Saha-Chatterji/5888d813565f3117ecec5e6f250051e6be2934c8
52. Shishthla, P.M., Pingali, P., Varma, V.: A character n-gram based approach for improved recall in Indian language NER. In: Proceedings of the IJCNLP, Workshop on Named Entity Recognition for South and South East Asian Languages (2008). https://www.aclweb.org/anthology/I08-5010
53. Morwal, S., Jahan, N.: Named entity recognition using Hidden Markov Model (HMM)—an experimental result on Hindi, Urdu and Marathi languages (2013). https://pdfs.semanticscholar.org/85e4/cd646e0189eaab9aa2902e7e1a7ab6095f69.pdf
54. Mishra, D., Sangal, R., Singh, A.K.: Asian federation of natural language processing. In: Proceedings of the IJCNLP-08 Workshop on Named Entity Recognition for South and South East Asian Languages (2008). http://ltrc.iiit.ac.in/ner-ssea-08/index.cgi?topic=12
55. Borker, S.B.: Text to Speech System for Konkani (Goan) Language (2006). https://www.semanticscholar.org/paper/TO-SPEECH_SYSTEM-FOR-KONKANI-(-GOAN-)-LANGUAGE-Borker/65e04340663c488da215797c276142debbd10564
56. Colaco, J., Borkar, S.: Design and implementation of Konkani text to speech generation system using OCR technique. **2**, 6 (2016). http://www.ijsrp.org/research-paper-0916.php?rp=P575773
57. Fal Dessai, N.B., Naik, G.A., Pawar, J.D.: Implementation of a TTS system for Devanagari Konkani using festival. Int. J. Adv. Res. Comput. Sci. **8**(5), 386–391 (2017). http://irgu.unigoa.ac.in/drs/handle/unigoa/4778
58. Dessai, N.F., Naik, G., Pawar, J.: Development of Konkani TTS system using concatenative synthesis. In: International Conference on Computing and Control Engineering (2016). https://www.w3.org/2006/10/SSML/papers/paper.pdf
59. Dessai, N.F., Naik, S., Salkar, S., Mohanan, S.: Text to speech for Konkani language. In: International Conference on Computing and Control Engineering (2012). https://www.w3.org/2006/10/SSML/papers/paper.pdf
60. Mohanan, S., Salkar, S., Naik, G., Dessai, N.F., Naik, S.: Text reader for Konkani language. CIIT Int. J. Autom. Auton. Syst. **4**(8) (2012). http://www.ciitresearch.org/dl/index.php/aa/article/view/AA072012029
61. Kane, M.M.P.: Part of speech tagging for Konkani corpus. Int. J. Eng. Res. Comput. Sci. Eng. **4**(6) (2017). http://ijercse.com/abstract.php?id=10298
62. Khorjuvenkar, D.N.P., Ainapurkar, M., Chagas, S.: Part of speech tagging for Konkani language. Int. J. Eng. Res. Comput. Sci. Eng. **5**(2) (2018). https://ieeexplore.ieee.org/document/8487620

63. Rajan, A.: Design and implementation of a PoS tagger for Konkani using NLTK (unpublished) (2016)
64. Desai, S., Desai, N., Pawar, J., Bhattacharyya, P.: AutoParSe—An Automatic Paradigm Selector for Nouns in Konkani (2014). https://www.researchgate.net/publication/307137557
65. Karmali, R., Pawar, J.D., Fondekar, A.: Konkani SentiWordNet—resource for sentiment analysis using supervised learning approach, pp. 55–59 (2016). http://irgu.unigoa.ac.in/drs/bitstream/handle/unigoa/4462/3_Workshop_Indian_Lang_Data_Resour_Evaluat_2016_55-59.pdf?sequence=1
66. Miranda, D.T., Mascarenhas, M.: KOP—an opinion mining system in Konkani. In: IEEE International Conference on Recent Trends in Electronics Information Communication Technology, 20 May 2016, Bangalore, India (2016). https://ieeexplore.ieee.org/document/7807914
67. Rajan, A., Salgaonkar, A.: Sentiment analysis for Konkani language—Konkani poetry, a case study. In: ICT Systems and Sustainability, Goa, India, vol. 1077, pp. 321–329. Springer, Singapore (2020). https://link.springer.com/chapter/10.1007/978-981-15-0936-0_32
68. Rajan, V.: Konkanverter—a finite state transducer based statistical machine transliteration engine for Konkani language. In: Proceedings of the Fifth Workshop on South and Southeast Asian Natural Language Processing, pp. 11–19. Association for Computational Linguistics and Dublin City University, Dublin, Ireland (2014). https://www.aclweb.org/anthology/W14-5502/
69. Phadte, A.: Resource creation for training and testing of normalisation systems for Konkani-English code-mixed social media text. In: International Conference on Applications of Natural Language to Information System, June 13–15, Paris, France, pp. 264–271 (2018). https://link.springer.com/chapter/10.1007/978-3-319-91947-8_26
70. Karmali, R.N., Walawalikar, S., Ghanekar, D., Pawar, J., D' Souza, C., Naik, S., Desai, S.: Experiences in building the Konkani WordNet using the expansion approach (2010). https://www.semanticscholar.org/paper/Experiences-in-Building-the-Konkani-WordNet-Using-Walawalikar-Desai/122d56a1ab3a9a468d9b5192381c8321dda16faa
71. Pawar, J.: Linguistic data consortium for Indian languages (unpublished) (2019)
72. Sardesai, M., Pawar, J., Vaz, E., Walawalikar, S.: BIS annotation standards with reference to Konkani language. In: Proceedings of the 3rd Workshop on South and Southeast Asian Natural Language Processing, pp. 145–152 (2012). https://www.aclweb.org/anthology/W12-5012/
73. Rajan, A., Salgaonkar, A., Joshi, R.: A survey of Konkani NLP resources. Comput. Sci. Rev. **38**, 100299 (2020). https://doi.org/10.1016/j.cosrev.2020.100299
74. Lafferty, J.D., McCallum, A., Pereira, C.N.: Conditional random fields—probabilistic models for segmenting and labeling sequence data. In: Proceedings of the Eighteenth International Conference on Machine Learning, pp. 282–289. Morgan Kaufmann Publishers Inc., San Francisco, CA, USA (2001). https://repository.upenn.edu/cgi/viewcontent.cgi?article=1162&context=cis_papers
75. Wallach, H.M.: Conditional random fields—an introduction (2004). https://repository.upenn.edu/cgi/viewcontent.cgi?article=1011&context=cis_reports
76. Bindu, S.M., Idicula, S.M.: Named entity identifier for Malayalam using linguistic principles employing statistical methods (2011). https://pdfs.semanticscholar.org/2445/3e39cb70faaf6bb2f3100b0119bd01def941.pdf
77. Ekbal, A., Haque, R., Bandyopadhyay, S.: Named entity recognition in Bengali—a conditional random field approach. In: Proceedings of the Third International Joint Conference on Natural Language Processing, vol. II (2008). https://www.aclweb.org/anthology/I08-2077
78. Josan, G., Kaur, A., Kaur, J.: Named entity recognition for Punjabi—a conditional random field approach (2009). https://www.researchgate.net/publication/262731986
79. Bikel, D.M., Schwartz, R., Weischedel, R.M.: An algorithm that learns what's in a name. Mach. Learn. **34**(1), 211–231 (1999). https://link.springer.com/article/10.1023/A:1007558221122

80. Cortes, C., Vapnik, V.: Support-vector networks. Mach. Learn. **20**, 3 (1995). http://image.diku.dk/imagecanon/material/cortes_vapnik95.pdf
81. Borthwick, A.E.: A Maximum Entropy Approach to Named Entity Recognition (1999). https://dl.acm.org/citation.cfm?id=930095
82. Krupka, G.R., Hausman, K.: IsoQuest Inc.—description of the NetOwl text extractor system as used for MUC-7 (1998). https://www.semanticscholar.org/paper/IsoQuest-Inc.%3A-Description-Of-The-NetOwl-(TM)-As-Krupka-Hausman/8fbdee94032da667c252eebfaa7b554f5f0379a2
83. Chopra, D., Joshi, N., Mathur, I.: Named entity recognition in Hindi using conditional random fields. In: Proceedings of the Second International Conference on Information and Communication Technology for Competitive Strategies, ACM, New York, USA (2016). https://link.springer.com/chapter/10.1007/978-3-642-19403-0_5
84. Li, W., McCallum, A.: Rapid development of Hindi named entity recognition using conditional random fields and feature induction **2**(3), 290–294 (2003). https://people.cs.umass.edu/~mccallum/papers/hindi-talip2003.pdf
85. Goyal, A.: Named entity recognition for South Asian languages. In: Proceedings of the IJCNLP-08 Workshop on Named Entity Recognition for South and South East Asian Languages (2008). https://www.aclweb.org/anthology/I08-5013
86. Gupta, V., Lehal, G.S.: Named entity recognition for Punjabi language text summarization. Int. J. Comput. Appl. **33**(3), 28–32 (2011). https://pdfs.semanticscholar.org/9a16/b8f37b236abce4893647e52ace69c17d5c2.pdf
87. Reddy, A., Khapra, M.M., Bhattacharyya, P.: Sharing network parameters for crosslingual named entity recognition for Telugu using LSTM-CRF (2016). http://lrec-conf.org/workshops/lrec2018/W11/pdf/2_W11.pdf
88. Shishthla, P.M., Gali, K., Pingali, P., Varma, V.: Experiments in Telugu NER—a conditional random field approach. In: Proceedings of the IJCNLP-08 Workshop on Named Entity Recognition for South and South East Asian Languages (2008). https://www.aclweb.org/anthology/I08-5015
89. Nongmeikapam, K., Shangkhunem, T., Chanu, N.M., Singh, L., Salam, B., Bandyopadhyay, S.: CRF based Name Entity Recognition (NER) in Manipuri—a highly agglutinative Indian language (2011). https://www.researchgate.net/publication/230785821
90. Garg, V., Saraf, N., Majumder, P.: Named entity recognition for Gujarati—a CRF based approach. In: Proceedings of the First International Conference on Mining Intelligence and Knowledge Exploration, vol. 8284. Springer-Verlag New York, Inc., New York (2013). https://link.springer.com/chapter/10.1007/978-3-319-03844-5_74
91. Amarappa, S., Sathyanarayana, S.V.: Kannada named entity recognition and classification using conditional random fields. In: 2015 International Conference on Emerging Research in Electronics, Computer Science and Technology (ICERECT), pp. 186–191 (2015). https://ieeexplore.ieee.org/document/7499010
92. Sharma, P., Sharma, U., Kalita, J.: Named entity recognition in Assamese. Int. J. Comput. Appl. **142**(1–8) (2016). https://www.researchgate.net/publication/303318979_Named_Entity_Recognition_in_Assamese
93. Vijayakrishna, R., Sobha, L.: Domain focused named entity recognizer for tamil using conditional random fields. In: Proceedings of the IJCNLP: Workshop on Named Entity Recognition for South and South East Asian Languages (2008). https://www.aclweb.org/anthology/I08-5009
94. Malarkodi, C.S., Rao, P.R.K., Devi, S.L.: Tamil NER—coping with real time challenges. In: Proceedings of the Workshop on Machine Translation and Parsing in Indian Languages, pp. 23–38. The COLING 2012 Organizing Committee, Mumbai, India (2012). https://www.aclweb.org/anthology/W12-5603
95. Balabantaray, R., Das, S., Tanaya, K.: Case study of named entity recognition in Odia using Crf++ tool. Int. J. Adv. Comput. Sci. Appl. **4** (2013) https://www.researchgate.net/publication/269524231

96. Malik, M.K., Sarwar, S.M.: Urdu named entity recognition and classification system using conditional random field (2015). https://pdfs.semanticscholar.org/192f/719579ae906149113e34f5912e3bfebf8ca5.pdf?_ga=2.196122220.1133831351.1565924948-859315823.1563803599
97. Mukund, S., Srihari, R.K.: NE tagging for Urdu based on bootstrap POS learning. In: Proceedings of the Third International Workshop on Cross Lingual Information Access-AdDressing the Information Need of Multilingual Societies, pp. 61–69. Association for Computational Linguistics, Stroudsburg, PA, USA (2009). https://www.aclweb.org/anthology/W09-1609
98. Sharma, P., Sharma, U., Kalita, J.: Named entity recognition in Assamese using CRFS and rules, pp. 15–18 (2014). https://www.researchgate.net/publication/286662918
99. Singh, K.: Name entity recognition on Punjabi language. Int. J. Comput. Sci. Eng. Inf. Technol. Res. **3**(5), 95–101 (2013). http://www.tjprc.org/publishpapers/--1383901932-11.%20Name%20entity%20recognition.full.pdf
100. Ministry of Electronics & Information Technology: Technology Development for Indian Languages Programme (2009). http://tdil.meity.gov.in/
101. Korobov, M.: Sklearn_crfsuite (2015). https://sklearn-crfsuite.readthedocs.io/en/latest/_modules/sklearn_crfsuite/estimator.html
102. Liu, D.C., Nocedal, J.: On the limited memory BFGS method for large scale optimization. Math. Program **45**(1–3), 503–528 (1989). https://dl.acm.org/citation.cfm?id=3112866
103. Rajan, A.: Named entity recognition corpus for Konkani language (2019). http://www.annierajan.com/nlp.php

Secure Analysis of Disaster Area Using Satellite Imagery and Deep Learning

Pooja Krishan, L. Remegius Praveen Sahayaraj, V. J. Sharmila, J. Limsa Joshi, A. Jeevitha, and B. R. Sathish

Abstract Deep learning is a subset of machine learning based on learning data representations, in contrast to task-specific algorithms. Deep learning models derive inspiration from how information is processed in the nervous system of the human body that consists of trillions of neurons communicating with each other. During a disaster, it is necessary to ensure that containment and rescue operations are conducted as quickly as possible with a primary focus on affected areas, as an improper organization might lead to wastage of resources such as money, materials, and time. To properly plan during disasters, satellite images of the affected location can be analyzed to identify the areas demanding immediate attention. A model can be designed using convolutional neural networks (CNNs) to help categorize the areas by the degree of destruction. To secure data fed into the model, a layer of security can be added between the input and output layers of the CNN. The model can be trained using old satellite images of the cities. New images fed into the model can be analyzed to obtain information on the level of devastation.

1 Introduction

During a disaster, it is necessary to ensure that containment and rescue operations are conducted as soon as possible with a focus on more affected areas. If the rescue operations are not organized properly, it might lead to improper use of resources

P. Krishan (✉) · L. R. P. Sahayaraj · V. J. Sharmila · J. L. Joshi · A. Jeevitha · B. R. Sathish
Loyola-ICAM College of Engineering and Technology, Chennai, India

V. J. Sharmila
e-mail: sharmila.vj@licet.ac.in

J. L. Joshi
e-mail: limsajoshi@licet.ac.in

A. Jeevitha
e-mail: jeevitha.a@licet.ac.in

B. R. Sathish
e-mail: sathishbr@licet.ac.in

T. Senjyu et al. (eds.), *ICT with Intelligent Applications*, Smart Innovation, Systems and Technologies 248, https://doi.org/10.1007/978-981-16-4177-0_70

such as money, material, and time. It is vital that the most badly affected regions are identified immediately and allocated the required resources first in order of degree of devastation. To solve the issue of improper planning during disaster times, satellite images of the affected location can be analyzed to identify the areas demanding immediate attention. A model can be designed using convolutional neural networks to help categorize the areas by the degree of destruction. The model can be trained using old satellite images of the cities in various conditions. In this way, when new images are fed into the model, analysis can be performed on them to obtain information on the level of devastation.

Deep learning systems have to combat the credit assignment problem and decide on how to assign credits to desirable moves in a sequence of events. To identify if a credit assignment is deep or shallow, a credit assignment path (CAP) is introduced. A credit assignment path or CAP is a path through the neural network from the input layer to the output layer. In a feed-forward neural network, there is no feedback and the neural network is acyclic. Hence, the CAP depth is the total number of hidden layers in the feed-forward neural network plus 1 which corresponds to the output layer. Since there is no fixed sequence of steps from input to output, in recurrent neural networks, the CAP depth is unlimited. The path from input yielding a solution with the least depth is called the problem depth. The path from input yielding a solution with the maximum depth is called the solution depth. A CAP depth greater than 2 generally distinguishes deep models from shallow ones. The additional layers in a deep model help the model to learn better and extract the necessary features as shown in Fig. 1.

Image classification is the process of pulling out information from an image to classify the given image under a category. Figure 2 illustrates the process of classifying a disaster from an image into one of many categories. Image classification plays a major role in remote sensing where images captured by the satellites are processed to extract valuable insights.

The following papers were studied to get an overview of techniques that are used in deep learning and image classification. An approach to train neural networks efficiently in a bottom-up fashion using a layered structure was proposed by Marquez

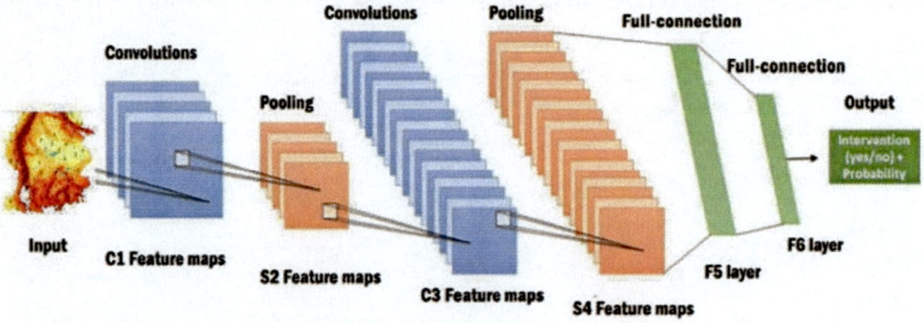

Fig. 1 Process of deep learning for disaster identification

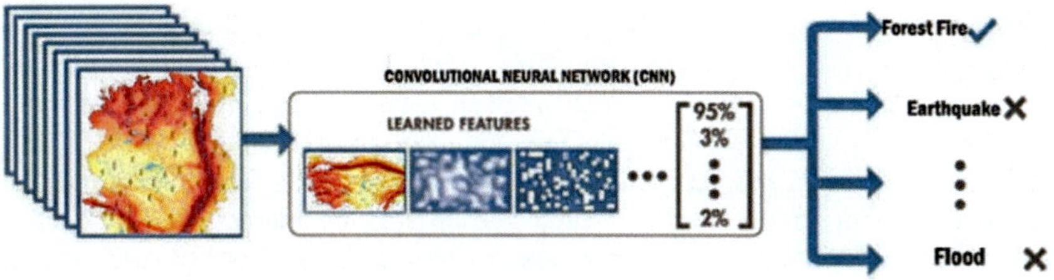

Fig. 2 Process of image classification

et al. [1]. The algorithm was inspired by the cascade correlation approach of Fahlman and Lebiere.

Self-taught learning for hyperspectral image (HSI) classification was studied by Kemker and Kanan [2]. While supervised deep learning methods require huge amounts of labeled data to be efficient, extant labeled hyperspectral image datasets do not satisfy that requirement.

Wu et al. [3] proposed hybrid architecture, deep filter banks, combining multi-column stacked denoising sparse auto-encoder and Fisher vector to automatically learn the representative and discriminative features hierarchically for LU scene classification. They achieved accuracies of 92.7% and 90.4% on classification, respectively.

Considering CNNs perform exceptionally well on a variety of computer vision tasks, the accuracy of already trained CNN models on iris recognition is explored by Nguyen et al. [4].

To solve ill-posed inverse problems, a state-of-the-art polarized deep neural network model is proposed by Jin et al. [5]. It is observed that unrolled iterative methods are represented as a CNN when a feed-forward network uses convolution to perform its operations.

Huang et al. [6] explore image-based positioning on a moving gadget by suggesting image tracking as a solution. The method circumscribes the user by using panorama as the indoor map. The effectiveness of positioning will be largely impacted by a large number of feature points in the panorama in contrast to the feature points of the image in the moving gadget. Experimental evidence confirms that the Sorting Hat technique can decide feature point pairs that are correlated within 0.824 ms at 99.6% confidence.

Orthogonal rotation-invariant moments (ORIMs), such as Zernike moments, are introduced by Lin et al. [7] and defined on a continuous unit disk to aid in applications. A novel technique that uses numerical optimization to enhance orthogonality is proposed to better the construction of images.

The traditional center/surround Retinex enhancement algorithms intensify noise by taking into account only the illuminance. An illuminance-reflectance model (IRMNE) that can denoise the image with less brightness and perform image enhancement is developed by Yu et al. [8].

Zhang and Wang proposed [9] a more generic, effective, and combined resolution based on visual cognitive mechanism. First, a motivating visual cognitive computing model (IVCCM) is presented based on visual heuristic principles. Second, an asymmetric generalized Gaussian mixture distribution model (AGGMD) for natural images is proposed. Extraction of quality-aware features from natural images and forming quality-aware uniform features descriptors are appreciable.

An efficient technique to recognize actions on the basis of object detection is implemented, and a new descriptor for features is suggested to depict the changes in edge orientation by Aye and Zaw [10]. A supervised multi-class support vector machine (SVM) classifier identifies different actions. Improved accuracy is acquired in this method.

Syrris et al. [11] suggested a way to generate wide extent mosaics by manipulating huge amounts of remote sensing data via a processing chain. Two approaches, (a) distribution discretization accompanied by false color composition and (b) image rendering and mosaicking, are put forth.

MicroFilters, proposed by Ilyas [12], uses machine learning to classify images after corresponding tweets, and links are scraped thereby removing images that do not feature direct impacts and do not contribute to relief measures.

An early warning system is a type of data-driven Internet of Things system used for environment disaster risk and efficient management. Such a framework is presented by Poslad [13], which is inspired by a multi-semantic representation model. Lightweight semantics for metadata are used to enrich inputs from sensors, and heavyweight semantics are used to outline multiple-level knowledge bases, plan workflows, and make decisions based on the semantics.

Shimabukuro et al. [14] present a novel technique for calculating areas affected by fire in various regions by using satellite images. The analysis was done on Landsat-5 TM imagery over 77 sample sites (20 km $\times$ 20 km in size) located at each full degree confluence of latitude and longitude.

Blaschke et al. [15] proposed a semi-automated object-based image analysis methodology for identifying the location of landslides. The method used normalized difference vegetation index, contrast, and textures got from satellite imagery along with inclination and direction of flow derivatives from a digital elevation model (DEM). The accuracy was 93.07%.

Gordon [16] examines the identified methods for fast exponentiation, their advantages, and their disadvantages. Public-key cryptographic systems often involve the use of elements of a group as the base for a large exponent.

Li et al. [17] propose the ElGamal Cryptosystem. The paper illustrates the application of the modified algorithm to grayscale and color images. It is efficient since the actual image is transformed into an incomprehensible one.

Hashima and Neamaa [18] proposed an encryption scheme based on elliptic curve ElGamal (EC-ElGamal) for transferring confidential images over open communication networks subject to attacks. The proposed scheme is better than RSA or ElGamal in terms of key size and performance. This paper is organized as follows: the proposed method is explained in Sect. 2, the experimental results are furnished in Sect. 3, and the paper concludes with future work in Sect. 4.

2 Proposed Method

The primary goal is to ensure immediate action during a disaster, by using real-time data from various sources. In this system, the use of satellite imagery is included; satellites capture images from the earth at regular intervals or when prompted to do so. These images can be used to train a deep learning model to identify occurrences of disasters and send an alert to rescue teams on Earth. Encryption is performed on the image features before passing them to the hidden layer of the neural network to enhance the security of the system. The first hidden layer decrypts the data received from the input layer and processes the data by passing it to subsequent hidden layers. Once learning is complete, the final hidden layer encrypts the data before passing it on to the output layer which is then decrypted to get the final learned model as shown in Fig. 3. Images captured during the occurrence of the disaster, before and after the disaster can be compared by pixel matching techniques to detect the amount of change in them; this will provide information on how the disaster has affected the region and also the places where the damage is high. This data is provided to rescue teams for efficient relief efforts.

2.1 Neural Networks

Neural networks attempt to imitate the biological nervous system and consist of several nodes that are similar to neurons in an animal. Data is manipulated by passing it through several layers just like how information is transmitted through multiple synapses of neurons. A neural network accomplishes this by accepting input through

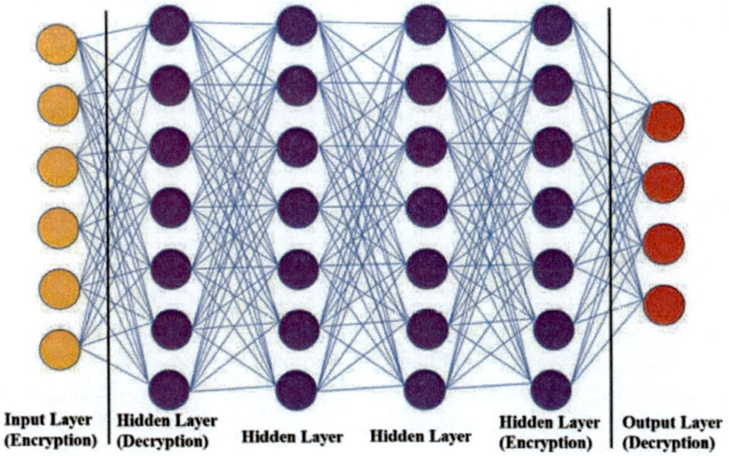

Fig. 3 Proposed architecture

its first layer—the input layer. The input is processed through several hidden layers where each node strives to manipulate information by multiplying incoming datum with a weight value and adding the product values. The threshold and weight are modified throughout training until data with similar labels generates the same output.

Convolutional neural network. Convolutional neural networks are a class of neural networks that are widely used in image classification tasks. These neural networks decrease the number of trainable parameters for extremely large images. It "convolves" the input image data which is in the form of a matrix consisting of pixel information such that spatial information about the image is not lost. The hidden layers typically consist of convolutional layers, pooling layers, and fully connected layers. To curb the number of trainable parameters that keep increasing as the size of the image increases, pooling layers are used between the convolutional layers which decrease the spatial size of the image while retaining the depth of the image. A fully connected layer in the convolutional neural network takes the output of the pooling and convolutional layers flattened into a vector whose values represent the probability of features corresponding to a label. The VGGNet is a convolutional neural network architecture based on AlexNet that focuses on the depth aspect of convolutional neural networks. The network takes (224,224) dimensions image matrix as input. The receptive field of a VGGNet is a matrix with dimensions (3,3). To preserve spatial resolution once convolution is completed, the stride value is set to 1 pixel. ReLU activation function is used in the hidden layers of VGGNet. This thus limits the size of the receptive field and improves the discriminative power of the classifier by increasing the number of ReLU units from one to three. A VGGNet has three fully connected layers and (1,1) convolutional layer that leads to increased performance by augmenting the weight layers.

Security. It is vital to add security features to the image data to thwart access. A layer of security consisting of the encryption and decryption process is added while feeding data into the first hidden layer of the convolutional neural network from the input layer and also between the final hidden layer and the output layer. The field of cryptography relies on the use of mathematically intractable problems to confer security to data. To achieve encryption and decryption of image data, it is necessary to convert the image to a matrix over which mathematical operations can be performed to obscure the data and provide security. Since the model inputs are converted to a NumPy array, encryption and decryption can be performed over this array. The Elgamal cryptosystem is a public-key cryptosystem. It works based on the discrete logarithm problem which states that given a very large prime number, a primitive root in the multiplicative group, and an integer of the same group, the encrypted content is calculated using the fast exponential algorithm (square-and-multiply method) in [16], but given primitive roots and the prime number, it is impossible to calculate the member of the multiplicative group. This strengthens the algorithm. In key generation phase, a huge prime number is selected—px. Create e_a to be a primitive root in the group $G = <\text{Mpx}^*, \times>$, where M is a multiplicative group. Create rx to be a member of the group $G = <\text{Mpx}^*, \times>$ such that $\text{px} - 2 \geq \text{rx} \geq 1$. e_b = e_a (rx mod px). Create the public key as (e_a, e_b, px). Rx is the private key that is evolved, and it should not be circulated outside the secure zone. In the encryption and decryption phase, just make sure that each item in the array that represents the image does not exceed the prime px. Let the array be Nx.

Elga_Encrypt (e_a, e_b, px, Nx): For every element nxab in the matrix, select a unique random number kx such that $px - 2 \geq kx \geq 1$.

Find $Cx \equiv$ e_a (rx mod px)
Find $Exab \equiv nxab$ * e_b(kx mod px)
Send the encrypted image matrix Ex along with Cx

Elga_Decrypt (rx, px, Cx, Ex): Find $Nx = Ex\ (Cx - 1 \bmod px)$. The decrypted image matrix Nx is sent out to the output zone. Any iterative product cipher is used in conventional cryptographic algorithms like Data Encryption Standard (DES). Unlike text, image data is large and redundant. Hence, using iterative product, ciphers will be computationally infeasible. So, we resort to computationally cheap but strong cryptographic algorithms like the Elgamal cryptosystem. The computational cost of the Elgamal system can be further augmented by finding a workaround to minimize the size of the ciphertext which is twice that of the plaintext. Also, the value of the private key must be different for each encryption and decryption process to foil known-plaintext attacks.

3 Experimental Results

3.1 Preparation of Dataset

The dataset has been prepared by running a code "search_bing_api.py" to save "n" images of the required types of the disaster from the internet. The user would need to give the query (example: "images of earthquakes") and the output location (where to save the images) as inputs when running the code. The code uses the Bing Web Search API to scrape images directly from the internet. This method is applied as there are not any premade datasets available on open access, except for a few images. The complete image dataset (Fig. 4) is downloaded and checked if there are any duplicates, and then saved in the output folder.

3.2 Training

The training of the model includes two classes, where each class represents a single type of disaster. Each class has around 300 images of a single type of disaster to help train the model to identify the disaster. The foremost step in deep learning, preprocessing, is performed, and the image is appended to data. To perform the multi-label classification, the image path is divided into many labels. The labels list

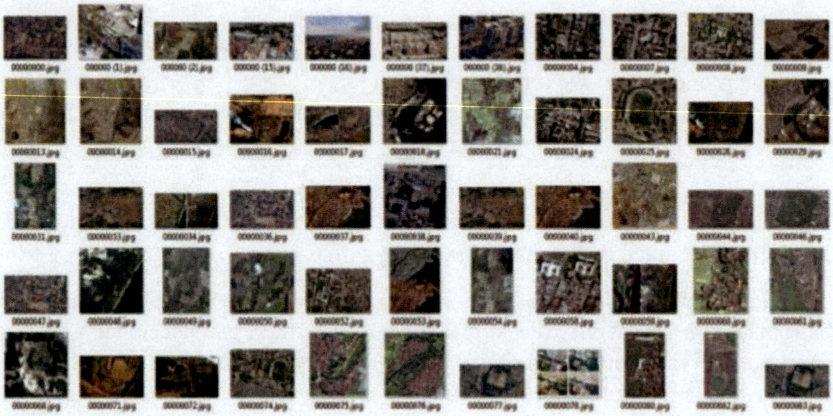

Fig. 4 Complete dataset

is augmented by attaching a list consisting of two elements. The labels list is converted to a NumPy array. The model trains for 75 epochs (cycles) and uses backpropagation to identify and rectify errors in previous epochs with the sole aim of improving the prediction of output in each increment. An initial learning rate is set to the default value of the Adam optimizer, 1e-3. The image size is (96,96) consisting of three channels and a batch size of thirty-two. The batch size can be modified on the basis of the GPU capacity if one is used.

3.3 Detecting Disaster-Affected Areas

The disaster analysis system takes two images, one of the current disaster-affected regions and one from before the disaster occurred. These two images are fed into a code called "image_diff.py" to identify the locations where there is great damage. The comparison is done by converting the two images into greyscale, then matching the pixels. Two values are computed:

- Mean Squared Error

$$\mathrm{MSE} = (mn)^{-1}\sum_{i=0}^{m-1}\sum_{j=0}^{n-1}[I(i,j) - K(i,j)]^2 \quad (1)$$

- Structural Similarity Index

$$\mathrm{SSIM}(x,y) = \left(2\mu_x\mu_y + c_1\right)\left(2\sigma_x\sigma_y + c_2\right)\left[\left(\mu_x^2 + \mu_y^2 + c_1\right)\left(\sigma_x^2 + \sigma_y^2 + c_2\right)\right]^{-1} \quad (2)$$

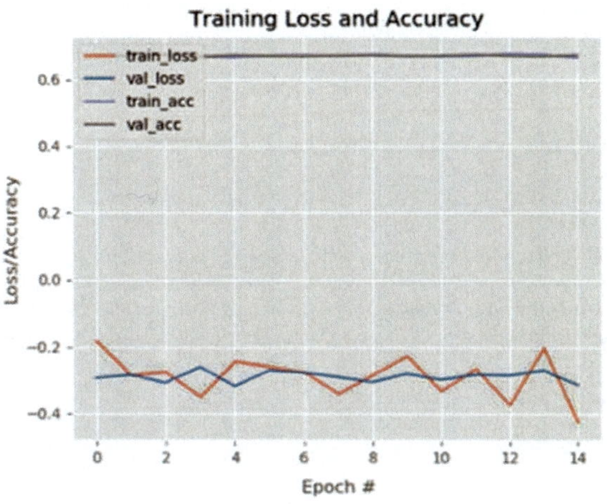

Fig. 5 ConvNet loss/accuracy graph

These values are used to identify whether there exists any significant change between the two images and then to plot where the changes have occurred.

3.4 Parameters

Training loss is the number of misclassified examples on the training data set. *Validation loss* is the number of misclassified examples of using the trained classifier on the validation data set. *Training accuracy* is the percentage of examples classified correctly while learning using known data. *Validation accuracy* is the percentage of examples classified correctly by the trained model during validation when it receives data not previously seen.

The training loss/accuracy graph for the ConvNet and VGGNet (Figs. 5 and 6) was plotted using the four parameters that are calculated during training and validating the model.

3.5 Results and Discussion

The two models ConvNet and VGGNet are trained separately, and their performance metrics are computed. From the graph, it can be concluded that ConvNet has minimum loss during both training and validation phases. But it has low accuracy, an average of 0.65, in detecting the image provided, making it less viable for use. The VGGNet provides greater accuracy, a value of almost 0.98 when recognizing unknown images. The validation loss is high at some EPOCHs, but it does not deduct from the final results. The main difference between the two networks is that VGGNet

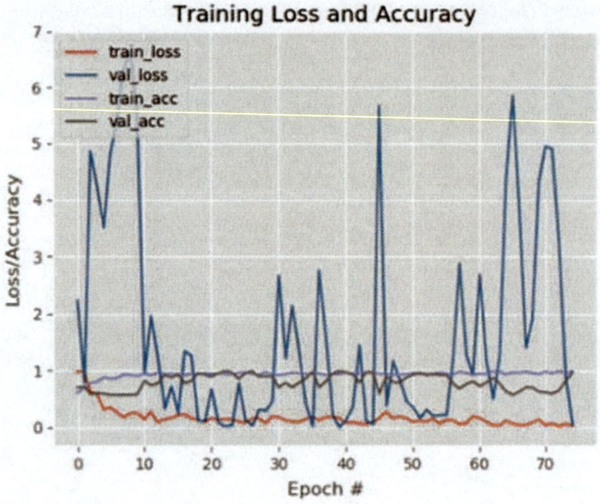

Fig. 6 VGGNet loss/accuracy graph

Table 1 Performance analysis

Model	Training accuracy (%)	Training loss	Validation accuracy (%)	Validation loss
CovNet	66.72	0.34	67.33	0.27
VGGNet	96.87	0.58	92.15	3.06

takes much longer to train than ConvNet due to its deep structure. However, VGGNet is better at identifying images than ConvNet, making it a better option for implementation in the actual system in the long run, as shown in the performance analysis (Table 1).

4 Conclusion

This idea proposes an automated method for detecting any disaster that has occurred in any part of the world. It also helps in finding the regions that need to be attended to first, without any intervention from humans. This would help reduce the time lag between the occurrence of disaster and deployment of rescue teams, thereby increasing the efficiency of relief teams and decreasing the resources used. This work is limited to the use of satellites and can be extended to make use of sensors. The cryptosystem used is restricted in terms of key size, which is very large. Combining the proposed cryptosystem with elliptic curve cryptography would significantly reduce the key size from 1024 to 160 bits which could serve as a better model. In the future, to enhance the security in a much broader aspect, the cryptosystem proposed can also be applied between the various hidden layers to increase the security of the model. It all comes down to striking a balance between computational efficiency and security and choosing the right tradeoff.

References

1. Marquez, E.S., Hare, J.S., Niranjan, M.: Deep cascade learning. IEEE Trans. Neural Networks Learn. Syst. (2018)
2. Kemker, R., Kanan, C.: Self-taught feature learning for hyperspectral image classification. IEEE Trans. Geosci. Remote Sens. (2017)
3. Wu, H., Liu, B., Su, W., Zhang, W., Sun, J.: Deep filter banks for land-usescene classification. IEEE Geosci. Remote Sens. Lett. (2016)
4. Nguyen, K., Fooke, C., Ross, A., Sridharan, S.: Iris recognition with off-the-shelf CNN features: a deep learning perspective. In: Digital object identifier. Institute of Electrical and Electronics Engineers (2018)
5. Jin, K.H., McCann, M.T., Froustey, E., Unser, M.: Deep convolutional neural network for inverse problems in imaging. IEEE Trans Image Process. (2017)
6. Huang, J.-Y., Lee, S.-H., Tsai, C.-H.: A fast image matching technique for the panoramic-based localization. In: International Conference on Information Systems (2016)
7. Lin, H., Si, J., Abousleman, G.P.: Orthogonal rotation invariant moments for digital image processing. IEEE Trans. Image Process. (2008)
8. Yu, X., Luo, X., Lyu, G., Luo, S.: A novel Retinex based enhancement algorithm considering noise. In: International Conference on Information Systems, Institute of Electrical and Electronics Engineers (2017)
9. Zhang, R., Wang, Y.: Natural image quality assessment based on visual cognitive mechanism. In: International Conference on Information Systems, Institute of Electrical and Electronics Engineers (2017)
10. Aye, H.M., Zaw, S.M.M.: Salient object based action recognition using histogram of changing edge orientation (HCEO). In: SERA, Institute of Electrical and Electronics Engineers (2017)
11. Syrris, V., Corbane, C., Pesaresi, M., Soille, P.: Mosaicking copernicus sentinel-1 data at global scale. IEEE Trans. Big Data (2018)
12. Ilyas, A.: MicroFilters: harnessing twitter for disaster management. In: Global Humanitarian Technology Conference Institute of Electrical and Electronics Engineers (2014)
13. Poslad, S., Middleton, S.E., Chaves, F., Tao, R., Necmioglu, O., Bügel, U.: A semantic IoT early warning system for natural environment crisis management. IEEE Trans. Emerg. Top. Comput. (2015)
14. Shimabukuro, Y.E., Miettinen, J., Beuchle, R., Grecchi, R.C., Simonetti, D., Achard, F.: Estimating burned area in Mato Grosso, Brazil, using an object-based classification method on a systematic sample of medium resolution satellite images. IEEE J. Sel. Top. Appl. Earth Obs. Remote Sens. (2015)
15. Blaschke, T., Feizizadeh, B., Holbling, D.: Object-based image analysis and digital terrain analysis for locating landslides in the Urmia Lake Basin, Iran. IEEE J. Sel. Top. Appl. Earth Obs. Remote Sens. (2014)
16. Gordon, D.M.: A survey of fast exponentiation methods. J. Algorithms **27**(1), 129–146 (1998)
17. Li, L., Abd El-Latif, A.A., Niu, X.: Elliptic curve ElGamal based homomorphic image encryption scheme for sharing secret images. Sig. Process. **92**(4), 1069–1078 (2012)
18. Hashima, H.R., Neamaa, I.A.: Image encryption and decryption in a modification of ElGamal cryptosystem in MATLAB. Int. J. Sci.: Basic Appl. Res. (IJSBAR) **14**(2), 141–147 (2014)

Bloom Filter in Cloud Storage for Efficient Data Membership Identification

Niraja Jain, B. Raghu, and V. Khanaa

Abstract Cloud computing is the buzz word that had changed the entire software industry with its PaaS, IaaS and SaaS model. The legacy systems have been using the database that in turn was an overhead in terms of cost of storage, operation and maintenance. Cloud database is the concept that can change this. The resources used in the cloud databases need to identify whether the requested data belongs to the data nodes of a cluster. With databases began to be ubiquitous, the data storage needed to satisfy heterogeneous data structures rather the unstructured data storage support is looked for. The use of Bloom's filter for data membership identification is the novel approach and can effectively improve the resource organization strategy on cloud. Dynamic resource organization can further improve the query efficiency as well. The concern raised during this is the data privacy which can also be ascertain by maintaining the data access authority levels. Bloom filter uses less memory space against the large dataset to store its information. A Bloom filter is proposed to be used to ascertain the existence of the user requested data at the cloud storage site. Bloom filter is a hash indexed data structure which have high accuracy and precision.

1 Introduction

1.1 Cloud Computing

Cloud computing is a new computing paradigm which was formally defined by National Institute of Standards and Technology (NIST) after multiple revisions such as [1]: "Cloud computing is a model for enabling ubiquitous, convenient, on-demand network access to a shared pool of configurable computing resources (e.g., networks,

N. Jain (✉)
Computer Engineering Department, Bharath University, Chennai, India

B. Raghu
SVS Group of Institutes, Warangal, India

V. Khanaa
Bharath University, Chennai, India

T. Senjyu et al. (eds.), *ICT with Intelligent Applications*, Smart Innovation, Systems and Technologies 248, https://doi.org/10.1007/978-981-16-4177-0_71

servers, storage, applications, and services) that can be rapidly provisioned and released with minimal management effort or service provider interaction. This cloud model is composed of five essential characteristics, three service models, and four deployment models." This definition also underlines the characteristics of cloud computing in broader perspective. With every passing moment, world is creating a large amount of data through various interconnected platforms. According to the Source, IDC Digital Universe, May 2010 cloud storage data usage in the year 2020 is estimated to be 14% of the entire data will be stored in cloud where as 34% of the operational data will be making use of the cloud computing infrastructure. Studying in detail the NIST definition we can further state that "Database Infrastructure as a Service" is a model that is highly recommended in the cloud based applications developed and used worldwide with the advancements in the Virtualization, WEB 2.0 technology. Cloud storage systems have become the most successful cloud computing applications in use.

Cloud Computing Models This cloud era has given rise to the most generic problem of making the resources available to the cloud service requesting clients. With the well-known cloud models in practice today namely [2]:

Infrastructure as a Service: provides virtual storage, virtual machines, virtual infrastructure and other hardware assets as resource to the client on demand. The service provider is responsible for providing and managing the entire required infrastructure, whereas the client will look after the deployment.

Software as a Service: this is complete operating environment applications, management and the user interface. Client is responsible to enter and manage the data through the user interface. The service provider will provide the application through a thin client interface.

Architecture as a Service: provides virtual machines, operating systems, applications, services, development frameworks, transactions and control structures. Client is responsible for installing and managing the applications whereas service provider will manage the infrastructure, operating system and enable software.

1.2 Cloud Computing Characteristics

The problem of resource organization becomes many folds in the cloud computing world. The resources defined/used in every model are different and its optimum utilization needs to be ensured to guarantee the cloud computing characteristics features. These features are characterized with the following:

- Pay-as-you-go
- ability to operate in heterogeneous environment
- ability to use encryption techniques on stored data
- on-demand self-service
- multi-tenancy
- elasticity and scalability.

1.3 Cloud Database Architecture

Cloud computing architecture composes of the following components [3]: cloud services, cloud infrastructure, cloud platform and cloud storage (Fig. 1).

2 Literature Survey

When the organization shifts, the responsibility of saving and storing its data to third-party, the cloud service provider comes into picture. Study about secure data transfer [4] suggested that cloud service users are compelled to frequently change the service provider due to its inability to stand by the SLA terms. The secure data migration is the fundamental requirement and Bloom filter can be deployed to address this issue. The reason for doing so can range from finding an alternative to cost effective storage, scalable alternatives to storing files on-premise or storage area network (SAN), to invest in adding servers and infrastructure to accommodate increased demand because stored data grows. Redundant data storage analysis [5] suggested that the existing cloud systems have identified that avoiding duplication of files in cloud storage is the problem so far unattended. The authors in the article proposed use of Bloom filter to accelerate the lookup speed and reducing computational cost and the results show improved performance against computational time needed, memory utilization and deduplication efficiency. So all the data that the organization decides to store off-site becomes the responsibility of a third-party cloud service provider. Study reveals use of [6] a joint Bloom filter and cross-encoding for data verification and recovery in cloud that helps to overcome the cloud storage security challenges. To ensure the integrity of the data storage in cloud the authors proposed a novel cross-encoding recovery method using dynamic storage table that uses Bloom filter to practically verify data integrity. The data is accessible over the Internet.

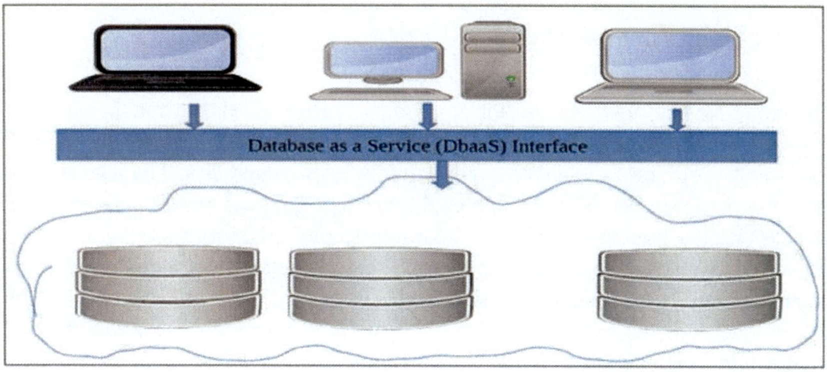

Fig. 1 Cloud database architecture

The cloud service provider will then host, secure, manage and maintain the storage servers and required infrastructure to ensure that the user/organization can access the data on demand. A fuzzy-folded Bloom filter-as-a-service for big data storage approach in the cloud [7] suggested that existing storage solutions are deterministic in nature. Authors presented an effective space-effective strategy for massive data storage using extended fuzzy folded bloom filter (FFBF) that improved efficacy without compromising false positive rate and query time. The service providers make use of the virtual servers commissioned on the physical servers to fulfill the user storage requests. The article [8] "Secure Cloud Storage Service Using Bloom Filters for the Internet of Things" identify the need for data integrity verification which is usually done by third party. The authors proposed a provable data possession model using Bloom filter. As the user demand grows, the service provider keeps adding the virtual servers to meet the access need. Thus, cloud storage has elastic storage capacity that can grow with increase in demand and can shrink with decrease in demand without adding up any physical server cost. Wu et al. [9] identified that the cloud service user need to pre-book the needed virtual machines (VMS) and the other computing resources necessary to allocate each VM and designed a resource design framework (RDF) that uses cloud-based machine learning scenario. This helps in perfect resource allocation and conforming the SLA policies. Authors state [10] that the cloud services are overloaded due to input/output imbalance or space demand. Hash aware cloud storage can be the solution for this frequent problem ensuring reduced overload probability in cloud storage systems. The organization also saves on capital expenditure needed for storage commissioning and maintenance. The over-provisioning is taken care of by the service provider at the marginal cost increase saving user losses [11]. Study presents opportunistic cloud computing system to demonstrate "no data center" solution to the problem of infrastructure as a service. Also [12] proposed VDB—verifiable database that enable resource-limited client to interact with the untrusted cloud server effectively that uses committed invertible Bloom filter and supports all kinds of updating operations.

3 Problem Formulation

The cloud service provider will make the data storage accessible for the end-user through various web-based applications over the internet. As a service level policy, the provider will ensure the data availability through redundancy. The data storage available at the service provider site could be a managed service or an unmanaged service that the user can configure according to the requirements. Cloud storage service can be viewed in the light of its off-site data management, quick implementation, cost-effective, scalable solution for business continuity. While verifying the AWS cloud policies, it is identified that the Amazon have broadly classified cloud storage into three categories [13]:

Object storage—scalability is the characteristic feature of this class. Amazon Simple Storage Service (S3) is used in for its scale and flexibility, and ability to import existing data stores for analytics, backup, or archive.
File storage—which is often supported with a network attached storage (NAS) server. Amazon Elastic File System (EFS) are ideal for large content data storage services.
Block storage—based on the concept of direct attached storage (DAS) or storage area network (SAN). It is highly in demand by dedicated ERP systems. Amazon Elastic Block Store (EBS) are provisioned with each virtual server and offer the ultra-low latency required for high-performance workloads [4].

With the development of high-capacity storage in 1990s, the storage service providers opened up the data-centers worldwide fueling the dot com business. File transfer utility available then was the simplest form of unmanaged cloud storage. The unmanaged storage is not a configured service and it is offered to the end-user as a disk drive. The drive can be configured to suit user's need. The widespread use of this service further demanded sophistication in services such as synchronization and back-up. The second class of the cloud storage services were the managed cloud storage which support the pay-as-you-go model. Managed storage involves provisioning of raw virtualized disk and use of it to support applications that use cloud based storage. This system supports virtualized storage component. Amazon S3, EMC Atmos, Nirvanix, Rackspace cloud and Iron Mountain are the well-known examples of managed cloud storage.

4 Proposed Methodology

4.1 Data Storage Resource Organization in Cloud

In evaluating the cloud storage solutions, following factors are of important consideration:

- Client self-service
- Strong management capability
- Performance characteristics
- Appropriate block-based protocol
- Seamless maintenance and upgrade.

To successfully adjust the cloud service providers capacity, the system workload and characteristics of the data storage requested need to be understood thoroughly. The goal of capacity planner is to identify the critical resources and to keep adding more resources to avoid bottleneck dynamically. Scaling up vertically or to scale out horizontally is a critical architectural decision that affects the cloud performance largely.

For any cloud storage system to be user-friendly, it is very much essential to understand the client side storage requirement with proper analysis of the data with respect to the following features: storage space requirement, type of data, frequency of data upload, update and retrieval, data availability strictness, data security measures requested, predicted hikes in data operations, risk associated with data storage failure and the list goes on. Based on the data collected from the end-user and the thorough analysis of it gives the service provider an insight about the data. In case of mismanaged data collection or storage, the entire service provider's job will be a nightmare (Fig. 2).

Cloud storage services are basically untrusted storage. The chances of tampering with the data are high that leads to unavailability of necessary data. Such a data storage organization is primarily based on the data flow architecture. Hadoop is the widely used architecture in big data industry that is identified as the best for processing huge amount of data simultaneously on large clusters in reliable, fault tolerant way. Capacity planning is an iterative process that deals with the inherent inefficiencies of the systems hampering the cloud service provider's market share due to fluctuations in SLA terms and actual service leveraged (Fig. 3).

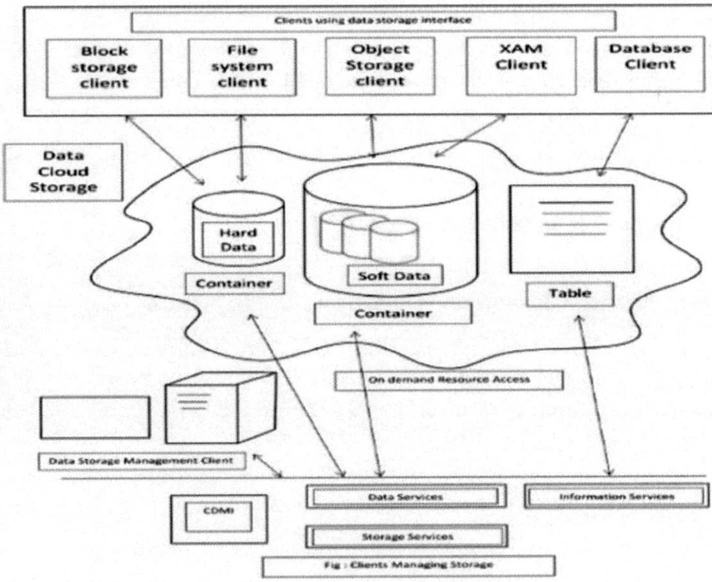

Fig. 2 Cloud storage data model (client managing the storage)

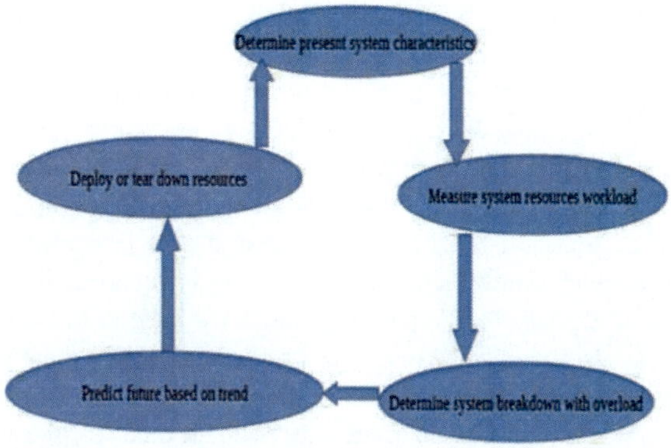

Fig. 3 Cloud resource allocation process

4.2 *A Mathematical Model*

Data storage resource organization in the cloud environment can be mathematically defined as follows:

Let S the Cloud data storage system defined as
$S = \{i, c, e, n, d|F(d), F(n), F(e)\}$
Herein,
i = data storage index
c = data storage capacity
e = authority level for data accessed
n = data node
d = data available on cloud storage
for every $i \in S$, the functions need to be defined such as:
$F(d): i \rightarrow d$
$F(n): i \rightarrow n$
$F(e): i \rightarrow e$.

This suggests that index i defined on the data storage d be such that F(d) can identify whether the data is available on the cloud storage or not. Similarly,F(n) could locate the data node on which the data is available and F(e) will help in understanding the authority level defined for data access.

Hash Function: A hash function is a mathematical function that involves following:

(i) Accept an arbitrary length input value from the uscr
(ii) Converts an input value into a fixed length compressed numerical value
(iii) This is a hash or hash value.

Hash functions are widely used to identify or compare files or databases. Fundamentally, hashing is defined by two distinct characteristics—irreversibility and uniqueness. Irreversibility underlines the fact that once we apply the hash function to the given data, it is not possible to regenerate the data using any de-hash function like it happens usually with encryption/decryption techniques. The hash function never produces same hash values for the two different data sets. Just in case hash-collision happens ever that hash function is immediately discarded as it's useless.

As we have defined the cloud storage system, the resource organization problem can be identified under the optimization problem categories, wherein it is to be identified that value of data storage index (i) can help in identifying the correct value of data node (n) where the requested data is stored. It is also to be ascertain that whenever a new data storage request is received by the cloud service provider, it must identify the data storage capacity needed to accommodate the new data. This data also need to be organized very close to the most relevant available data so that the further query retrieval rates can be reduced.

While deciding on to the data node for data storage as per the indexing mechanism defined on the available data, the resource organization strategy also need to assume the authority level associated with that data. This together can suggest a dynamic resource organization strategy in the cloud environment for the data storage.

4.3 Bloom Filter to Detect Data Membership

Blooms filter is a resultant space efficient probabilistic data structure for approximate set association problem.

- Bloom filter consists of an array of x bits, $Z[0] \dots Z[n-1]$.
- Initially the array elements are all set to 0.
- The Bloom filter will utilize m independent random hash functions defined in the range of $\{0, \dots, x_1\}$ such as $x_1, \dots x_m$.
- For every element $s \in S$, the bits of $Z[hi(s)]$ are set to 1 for $1 \leq i \leq m$.
- A bit location can be set to 1 multiple times, but only the first change will have an effect.
- So to check whether the requested element m is in s, all the array locations $Z[hi(n)]$ need to be checked if set to 1.
- If it is not set to 1, then it clearly signifies that the requested data element n is not the member of S.

The Bloom filter thus is very much similar to a hash table [3]. It is recommended to used independent hash functions to implement a Bloom filter which is uniformly distributed (Fig. 4).

Google BigTable, Apache Hbase, Cassandra, etc., make use of the Bloom filter to reduce the cost of searching the non-existent data by multiple disk access. This in turn improves the database query operation.

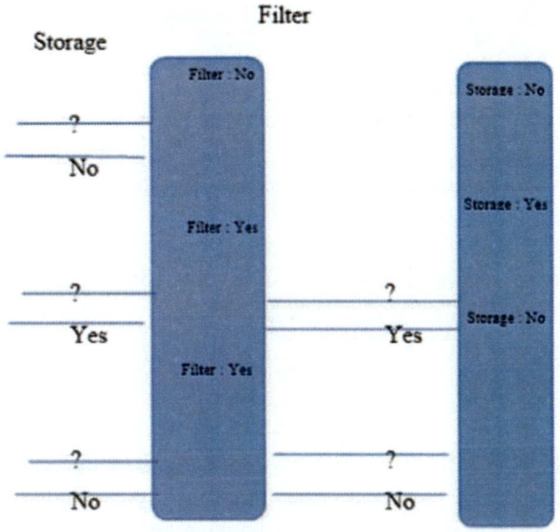

Fig. 4 Bloom filter usage to check data existence in storage system

4.4 Probability of False Positive in Bloom Filter

Assume that we have written a hash function that will select each array element with equal probability. If there are m number of bits in an array, then the probability that a typical element is not set to 1 by the defined function during the insertion of an element can be given by

$$(1 - 1/x) = e - m_q/x$$

Let $Y = e - m_q/x$

Assumption: say few entries p are still 0 after all the elements of S are hashed into Bloom filter.

The probability of False Positive is then Y.

$$(1-(1-1/x)m_q)m \approx (1 - e - m_q/x)m \approx (1-Y)m$$

When if keep the ratio x/q (the number of bits of storage required per item) constant, it is observed that Bloom filter allows a constant probability of a false positive.

Many applications like password application accept this 1–2% false positive rates in practice.

4.5 Preprocessing Methodology

Assume that the user is willing to store M blocks $M = \{m_0, m_1, ..., m_{n-1}\}$.

When the data is stored, the user needs to calculate the number of blocks of data M to be stored.

This number will influence the Bloom filter size and will also affect the positive error rate.

Before storing the data on the server, a key–value pair is generated which is a unique identification of the data known only to the data owner.

The bloom filter will be allowed to generate data verification index.

$$\text{Key - Value pairs} = \{(x_1, y_1), (x_2, y_2), \ldots, (x_r, y_r)\}$$

Based on the indexed value generated by the Bloom filter and comparing with the corresponding key–value pairs representing a user block to be stored on the server the third party can also ascertain the validity of the data.

The malicious user also cannot tamper with the data stored on the server due to the encryption of the key–value pair generated randomly corresponding the individual data block.

The verification is very much similar to an index creation where each block of data is the key building block. The randomly generated number of data blocks are verified for the data integrity and sent for verification. Bloom filter auditor will verify the data possession and accordingly the index will be verified. The constraint with this verification approach is that the indexing needs to be flexible enough to accommodate the addition or alteration to the key–value pairs.

5 Analysis and Remarks

The use of Bloom filter can be beneficial in the cloud storage systems management. We propose a scheme wherein the global Bloom filter will be defined in the storage cloud that will store the location details of the meta-data on the servers. Whenever the user requests for the data, the global Bloom filter will take that data to the local Bloom filter that is associated with each server. The hashed functions defined on the basis of the index will help to locate the exact server location for the data relevant to the requested query result. The data flow technique is commonly used in the Map–Reduce infrastructure to improve on the efficiency. The similar techniques used for query execution will improve the cloud storage performance. When data storage organization strategy needs to be finalized by the cloud service provider, the ownership of the data by the server needs to be ascertained. The cloud storage space is divided in to logical partitions of say n blocks. The need is to check on the set membership of each logical block using the Bloom filter. As this methodology is primarily a probabilistic model, the threshold value needs to be decided on so that

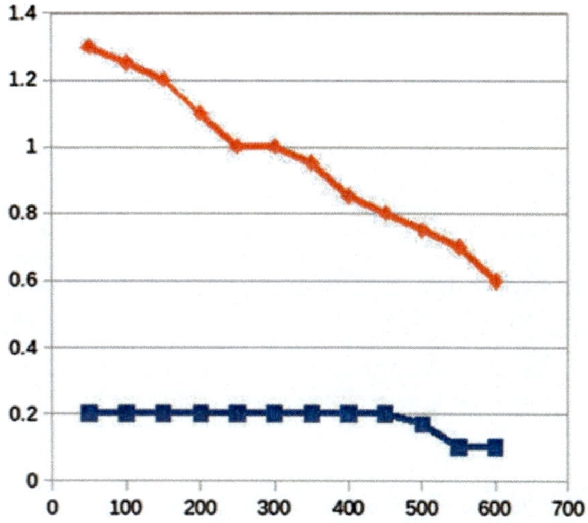

Fig. 5 Time taken to discover resources with and without Bloom filter

the data ownership verification process will not have insignificant impact of it. We are proposing to preprocess the data and verify its integrity (Fig. 5).

6 Conclusion

To determine the cloud computing performance, it is essential to understand the demand pattern and predict the requirements. The cloud flexibility and scalability can be further improved using the probabilistic data structure. This paper presents an idea of using the Bloom filter to check the existing data stored on the cloud storage systems. The managed systems can be effectively checked for accommodating the newer data as per the indexing done on the available data. A Bloom filter can be used to check whether the requested data is stored at the data node. The efficient look-up with compact hash structure makes bloom filter an effective mechanism to be used in cloud world. The time complexity analysis of the resource allocation using Bloom filter is better compared to the resource allocation without Bloom filter over the larger data centers.

References

1. Mell, P.: The NIST Definition of Cloud Computing. Recommendations of the National Institute of Standards and Technology NIST Special Publication, pp. 800–145 (2011)
2. Young, M.: The Technical Writer's Handbook. University Science, Mill Valley, CA (1989)
3. https://llimllib.github.io/bloomfilter-tutorial/

4. Yang, C., Tao, X., Zhao, F., Wang, Y.: Secure data transfer and deletion from counting bloom filter in cloud computing. Chin. J. Electron. **29**(2), 273–280 (2020). https://doi.org/10.1049/cje.2020.02.015
5. Li, Z., He, K., Lin, F., Wei, W.: Deduplication of files in cloud storage based on differential bloom filter. In: 7th IEEE International Conference on Software Engineering and Service Science (ICSESS), Beijing, pp. 111–114 (2016). https://doi.org/10.1109/ICSESS.2016.7883027
6. Zhang, S., Zhou, H., Yang, Y., Wu, Z.:A joint bloom filter and cross-encoding for data verification and recovery in cloud. In: IEEE Symposium on Computers and Communications (ISCC), Heraklion, pp. 614–619 (2017). https://doi.org/10.1109/ISCC.2017.8024596
7. Singh, A., Garg, S., Kaur, K., Batra, S., Kumar, N., Choo, K.R.: Fuzzy-folded bloom filter-as-a-service for big data storage in the cloud. IEEE Trans. Industr. Inf. **15**(4), 2338–2348 (2019). https://doi.org/10.1109/TII.2018.2850053
8. Jeong, J., Joo, J.W.J., Lee, Y., Son, Y.: Secure cloud storage service using bloom filters for the internet of things. IEEE Access **7**, 60897–60907 (2019). https://doi.org/10.1109/ACCESS.2019.2915576
9. Wu, C., Horiuchi, S., Tayama, K.: A resource design framework to realize intent-based cloud management. In: IEEE International Conference on Cloud Computing Technology and Science (CloudCom), Sydney, Australia, pp. 37–44 (2019). https://doi.org/10.1109/CloudCom.2019.00018
10. Gao, Y., Li, K., Jin, Y.: Compact, popularity-aware and adaptive hybrid data placement schemes for heterogeneous cloud storage. IEEE Access **5**, 1306–1318 (2017). https://doi.org/10.1109/ACCESS.2017.2668392
11. Mengistu,, T., Alahmadi, A., Albuali, A., Alsenani, Y., Che, D.: A "no data center" solution to cloud computing. In: IEEE 10th International Conference on Cloud Computing (CLOUD), Honolulu, CA, pp. 714–717 (2017). https://doi.org/10.1109/CLOUD.2017.99
12. Chen, X., et al.: Publicly verifiable databases with all efficient updating operations. IEEE Trans. Knowl. Data Eng. https://doi.org/10.1109/TKDE.2020.2975777
13. https://aws.amazon.com/what-is-cloud-storage/

On Combinatorial Handoff Strategies for Spectrum Mobility in Ad Hoc Networks: A Comparative Review

Libin Thomas, J. Sandeep, Joy Paulose, C. Smera, K. Amrutha, Jinsi Jose, K. Kavipriya, and Sayant Vijay

Abstract Technological advancements have made communication on-the-go seamless. Spectrum mobility is a networking concept that involves access technologies that allow highly mobile nodes to communicate with each other. Ad-hoc networks are formed between mobile nodes where fixed infrastructure is not used. Due to the lack of such fixed access points for connectivity, the nodes involved make use of the best network available to transmit data. Due to heterogeneous networks' involvement, the mobile nodes may face trouble finding the most optimal network for transmission. Existing technologies allow the nodes to select available networks, but the selection process is not optimized, leading to frequent switching. This leads to packet loss, low data rates, high delay, etc. Many researchers have proposed optimal strategies for performing handoff in wireless networks. This paper reviews combinatorial strategies that make use of multiple techniques to perform a handoff.

1 Introduction

Technology has grown exponentially. From room-sized computational devices to ones that fit within ears, the technological advancements are several. It is, therefore, essential for communication to evolve accordingly. Communication has seen tremendous growth from transatlantic telegraph cables to the latest 5G networks. However, there are certain issues that ails communication, no matter how sophisticated the designs are. A problem that persists even today in communication is the network's inability to perform efficiently when the nodes involved need to transfer data other than voice, such as video and other multimedia. The inefficiency of networks becomes even more evident when the nodes involved in communication are highly mobile. As the network's dynamic nature increases, chances of interruptions within channels and clashes between nodes for channel allocation based on priority also increase. The number of mobile users is increasing multifold every day. This leads to many problems in networks such as very high traffic, increase in packet delay or even packet loss,

L. Thomas (✉) · J. Sandeep · J. Paulose · C. Smera · K. Amrutha · J. Jose · K. Kavipriya · S. Vijay
CHRIST (Deemed To Be University), Bengaluru 560029, India
e-mail: libin.thomas@res.christuniversity.in

T. Senjyu et al. (eds.), *ICT with Intelligent Applications*, Smart Innovation, Systems and Technologies 248, https://doi.org/10.1007/978-981-16-4177-0_72

and overwhelming the access points. This leads to various demands such as higher quality of experience (QoE), higher quality of service (QoS), optimized bandwidth allocation methods, scalability of the networks, user fairness where every user gets to access the spectrum for an equal amount of time so that service provision remains constant, minimization of delay in terms of handoff, spectrum access and flow of data packets, and conservation of energy in devices on the client side among others.

Out of all these, spectrum allocation remains an essential aspect in communication that can either make or break a network. Spectrum allocation has not been perfected because of traditional spectrum allocation techniques that have fixed, licensed spectrum bands assigned to the networks [1]. Many solutions to spectrum allocation have been developed in the form of carrier aggregation in long-term evolution (LTE-U), licensed-assisted access (LAA) [2], LTE-Wi-Fi aggregation [3], cognitive radio [4], etc. Cognitive radio (CR) plays a significant role in the smart allocation and management of spectrum bands. This research paper discusses various spectrum handoff schemes that researchers have proposed over the years. The sections that follow include a discussion on different technologies and their functionalities, a literature review, a discussion on the different handoff methods proposed by researchers, and a conclusion.

Spectrum handoff faces several challenges when it comes to performing a proactive and seamless transition from one device to another. During the transmission period of a secondary connection, multiple interruptions from the primary users result in multiple spectrum handoffs and the need to predetermine a set of target channels for spectrum handoffs. Spectrum mobility issues arise when the primary users arrive at the channels used by the secondary users. The secondary users are required to return the channels occupied by them to the primary users due to the latter's preemptive priority. The SU needed the energy to identify the spectrum, and generally, SU uses decentralized units for the power. If a set of SUs works together in order to identify the presence of PU, much power is consumed. Energy consumption is a significant challenge to address. The other issue is due to the mobility packets failing to reach the destination. This leads to low throughput.

Handoff latency is another issue that needs to be addressed in handoff algorithms that strive to improve performance. Selecting the most optimal and appropriate channel when required is a challenging task in itself. The accuracy of the parameters that are considered for the selection of the channels is of utmost importance. Machine learning techniques can help to automate the process of optimal target channel selection. Although ML techniques can improve the CR network's overall performance, if the predictions are accurate, they can also pose quite a few challenges if the predictions are prone to errors [5]. PUs that get overlooked lead to collisions on the channel, which can lead to packet loss and incorrect data transmission. Sub-optimal channel selection methods that select channels which do not adhere to the required criteria lead to reduced transmission times for SUs. Unnecessary delays in sensing will lead to higher energy consumption and higher latency in the network. This can lead to reduced QoS and QoE values [5].

2 Review of Handoff Methods and Their Categorization

In this section, different proposed handoff methods are discussed in detail. They are categorized based on the technology that they utilize in order to perform the handoff operation. In the end, the methods are compared based on their performances. The in-depth discussion of each method helps to understand its working and provides clarity about which methods are best suited for optimizing the performance of networks through efficient handoff mechanisms.

2.1 Combinatorial Strategies Involving Multiple Strategies for Spectrum Mobility

This subsection covers the different combinatorial methods that have been proposed over the years. Through combining different strategies, most methods aim to improve the efficiency of networks, even if they include a large number of highly mobile nodes.

Initial and target channel selection. In [6], a probabilistic approach for identifying the target and initial channels for a connection-based handoff in a single secondary user network is proposed. A theoretical framework for queueing is proposed as well. In the proposed method, to obtain the extended service time and the sojourn times of the secondary connection so that the performance can be evaluated, the PRP geom/geom/1 queueing model is used [6, 7]. A modification is formulated based on the optimization problems that are related to minimizing the extended service times and the sojourn times. The method considered non-overlapping primary users with their channels and one secondary user that looks to occupy any spectrum holes for its transmission. The SU is allowed to sense only at the beginning of a slot time and is allowed to transmit at the end of the slot time if the selected channel is idle, which follows a listen-before-talk [8] access method. This is done so that there is no interference between two users. Two queues are present, which are called high-priority and low-priority queues. They both perform on a first-come-first-serve (FCFS) basis while the primary connections enjoy preemptive priorities [6].

The proposed system considers a single secondary user. If there are multiple secondary users at the secondary network with the Bernoulli arrival process and similar service times, they will all be considered as one cognitive user. Therefore, the proposed system will work fine in these scenarios as long as a central controller deals with the assignment processes of secondary users. Two handoff policies are considered in the proposed method—stay and change. Using these two methods, the service time could extend, and the secondary user is asked to switch channels upon the primary user's arrival. The authors also propose different methods to improve the selection of the target channel. However, these are not applicable for broadband users with a single queue transmitting on empty channels in parallel.

Fuzzy Logic-Based Neural Network Scheme for Spectrum Allocation. The method from [9] proposes a fuzzy logic-based neural network scheme that is capable of making efficient handoff decisions. The scheme makes use of parameters such as interference, bit error, and signal strength. A trained neural network is utilized to identify the channel gains based on fuzzy characteristics [9]. The proposed method aims to reduce frequent switching while improving channel utilization. A triangular membership function is used to represent the three parameters [9]. The normalized values are used for triggering fuzzy rules. Based on these normalized values, a decision is made to perform handoff or stay back in the channel. A multilayer feedforward back-propagation algorithm is used to train the neural network [9]. When multiple users try to access the same channel simultaneously, the contending user is given access to the channel with the next best gain. Although the throughput of the channel was improved by considering optimal thresholds for PU interference, bit error rate, and signal strength, the problem of misdetection of PUs was not accounted for.

Analytical Model for Fixed and Probabilistic Target Channel Selection Approaches. An analytical model is proposed in [10] for both fixed and probabilistic sequences for target channel selection. Particle swarm optimization (PSO) and genetic algorithms are used for optimizing fixed and probabilistic approaches to reduce the extended data delivery time of the secondary users. They have also suggested using a fast hybrid selection scheme to minimize the data delivery duration. In [10], the authors have used a time-slotted method so that the SU can begin to sense a channel at the beginning of the time slot to check for the presence of a PU in the channel. The channels are modeled using PRP M/G/1 queueing model [10]. The arrival and service times are considered to be independent and exponentially distributed. Each channel is assumed to have different service rates. Two queues are considered: high-priority and low-priority queues. They work on an FCFS basis. The SU can either choose to stay on the same channel, waiting for the PU to finish transmitting, or switch to another channel obtaining the least priority of that channel. This changed policy occurs only when the waiting duration on the same channel as the PU is higher than the waiting period in the queue of another channel.

The function of the switching model has been modified in this method, where it is also responsible for making decisions regarding the selection of a target channel based on the fixed or probabilistic approach sequences that are associated with the SUs. It is also responsible for modifying the remaining service time based on the rate of service of the current servicing and target access channels. To implement optimal load balancing techniques, the authors of [10] have used an extended version of [11] in their proposed model. The formula that they developed for load balancing includes criteria such as the inverse proportionality of the channel selection to that of the overall data arriving at the channel, the proportionality of the channel selection to that of the total service time of that channel, and an order that is added so that traffic flow can be controlled [10]. The mixed-integer genetic algorithm MI-LXPM introduced in [12] is used to solve the problem for a fixed sequence proactive approach. PSO is used to solve the problem of a probabilistic sequence proactive approach [10]. Although both fixed and probabilistic sequence approaches have almost similar performances for low traffic networks, the probabilistic approach has better balancing power if the transition probability matrix is chosen with care [10, 11]. When it comes to high traffic networks, both approaches have displayed the same performance.

ANN and ANFIS-Based Spectrum Management. The authors of [5] propose a multi-criterion scheme that utilizes artificial neural networks (ANN) for performing a predictive handoff as it does not require knowledge about the underlying complexities of a process under observation and an adaptive network fuzzy interference system (ANFIS) for selecting a target channel. The multiple criteria considered for the target channel selection include the idle channel duration, the data rate, and the value of the signal-to-noise ratio (SNR) [5]. SNR denotes the strength of a signal and gives information about how powerful a signal is so as to aid propagation. The proposed method aims to reduce the number of handovers and increase the overall throughput of the network with a multilayer perceptron (MLP). For every hole in a network, the channel advantage is calculated using a set of fuzzy rules [5]. These holes are then ranked based on their advantages. The handoff algorithm considers CR networks with independent channels that are indexed with channel id, bandwidth, and noise. It also considers various PUs and SUs. Every SU has a specific predictor for the next time slot as the traffic may differ at every channel. The spectrum handoff costs are calculated using the sensing time required to identify vacant channels, and the time is taken for the selection of candidate channel, negotiations between the sender and the receiver, parameters reconfiguration, and the resource scarcity factor, which is the delay in time brought about by the time spent in waiting for an idle channel [5].

A multi-radio SU is considered where one radio is scanning the environment perpetually, which is assumed to reduce the overall sensing time. ANFIS is used to decide this, and the attributes such as SNR, data rate, and the behavior of PUs are considered. Fuzzy logic helps in making sensible decisions when such uncertain factors are considered for the selection of target channels as these factors may lead to a very large number of handovers otherwise. In ANFIS, the supervised learning technique is used with the Takagi–Sugeno fuzzy interference system [13–15]. The channel advantage was calculated using these values and ranked accordingly. The first step in the algorithm is the CR network's decision-making function, which decides if handoff needs to be performed based on the absence or presence of a PU, which is indicated by 0 and 1, respectively. The SU will initiate handoff if the busy duration cost is higher than the handoff cost. Else, it will remain on the channel and wait for the PU to finish transmission. Once the handoff process is initiated, the channel advantage values of all target channels from the candidate list are calculated by the SU. The SUs are also ranked if there is more than one SUs waiting for a handoff.

QoE-Based Handoff Method Using ANN. In the method from [16], a multi-criterion-based vertical handoff using ANN and considers QoE parameters can be seen. The method aims to reduce the number of handoff processes that take place within a network while reducing the handoff delay. The method is purported to be an enhanced version of handoff using ANN, which works better than handoff using multi-attribute decision making (MADM) methods and fuzzy logic (FL)-based methods. The proposed method makes use of MOS values to make a handoff decision. The ANN model can learn from user experiences as well as predicted MOS values of different network states. The proposed ANN model consists of two phases, where the first phase is responsible for predicting the MOS value for those scenarios that do not involve feedback from the users, and the second phase decides to trigger a handoff process [16].

Spectrum Handoff Using MADM Methods and ANN. The authors of [17] present a spectrum handoff scheme based on MADM techniques and ANN. The CR network is assumed to be decentralized, as the SU's attribute information can be collected from other SUs transmitting on different channels. The attributes considered in this scheme include power transmitted by the PU, power transmitted by the PU as received by the SU and the data rate that the SU observes in the target channel. The transmission defined by IEEE 802.16 [18] and reception ranges of subscriber stations is used to define the ranges of the two power values. A MADM decision matrix is generated where the columns define the attributes, and the rows define the alternatives to decide from. The matrix is then normalized. The normalized attributes are then categorically normalized based on cost and benefit attributes [19]. Another layer of normalization is performed as the value of both the powers is very small. Once the normalized matrix is calculated, the entropy is calculated. Then, the deviation is calculated, after which the weights for the attributes are computed.

Once the weights are computed, SAW [20] is used to generate the different alternative scores. The alternative that has the most significant score is considered to be the best alternative. The maximum and minimum values of the attribute columns are computed, and a normalized matrix is derived out of those values, and the normalization is done by categorizing the attributes into cost and benefit attributes. Once the normalized matrix is generated, the GRC of each alternative is computed. The best alternative is the one with the highest GRC. The cost-based method helps to identify the most optimal alternative based on the weights of each alternative. TOPSIS method is a MADM method that considers distance for the selection of a candidate. According to this method, the best alternative must have the shortest distance from a positive ideal solution [17]. In the proposed system, the normalized decision matrix is used to create a weighted normalized decision matrix. From this, positive and negative ideal solutions are generated. The relative closeness of every alternative to the positive and negative ideal solutions is calculated using the TOPSIS method. The alternative with the most proximity to the positive solution is selected as the ideal solution. The handoff method in the proposed system considers the output from the different MADM methods. The channels are selected based on the unanimous choice of more than one MADM method. If there is a conflict, the method's channel with the largest relative standard deviation is selected [21].

Forward and backward propagation algorithms of ANN are considered in the proposed system. The input neurons contain the three attributes' values from five different channels, which amounts to 15 input neurons. Two hidden layers with 100 and 50 neurons and an output layer with five neurons are considered in the ANN. A bias unit is considered along with the input neurons. A hypothesis function is continuously modified until the desired output is obtained. Sigmoidal functions are used to connect the layers. A backward propagation algorithm is used to find the gradient of the cost function, which in turn will be used to identify the local minima by an optimization function [17]. This method is used to train the ANN in order to compute the minimum costs. Accumulated error is set to zero initially and is updated after each layer. In order to calculate the minimum cost, partial derivatives of the cost function must be computed. The ANN is trained using the minimum cost function, its partial derivative, and an optimizing function [17]. The simulations are carried out using MATLAB. Comparative scores of the different channels were generated using the mentioned MADM methods. The ANN was trained using a dataset of 5000 samples generated at random for the three attributes across five channels. It was observed that with an increase in the number of iterations, the cost decreased.

Adaptive Spectrum Handoff. In [22], an analytical model for evaluating reactive switching and non-switching handoff performance in terms of extended data delivery time attributes have been developed for CRAHNs. CCC is used as the controller for the SUs, and the former is responsible for coordinating the SUs. PRP M/M/1 queueing model is used to derive the results, which is the selection of optimal channels. Non-switching handoff is defined as the action taken by the CR/SU to stay on the same channel without transmitting due to the unavailability of idle channels upon the arrival of a PU. The method from [22] showcases a handoff method that adapts based on the number of SUs, number of primary channels and the traffic in the primary network. The ad hoc network is assumed to be time-slotted and each time slot is assumed to have a control and data transmission phase. The SUs look up to the CCC for information regarding other SUs in the vicinity during the control phase of the time slot. The control phase is divided into the idle phase, sensing phase, and sharing phase. Synchronization is done in the idle phase. The sensing phase looks for idle channels for transmission, while the sharing phase is used for sharing the sensed results. The sharing phase is also used for channel contention among the SUs. Detection of PUs makes the SUs to either to stay on the same channel or switch to another channel. The PUs and the SUs are expected to arrive in a Poisson arrival process. The PUs have higher priority than SUs, and only one user can transmit on a channel at a particular time.

Extended data delivery time depends on the busy period of a PU on the current channel. The SU begins to sense the channel at the beginning of each slot in order to check for the arrival of PU. The non-switching handoff delay is equal to the time that the PU spends transmitting on the channel. The extended data delivery time includes the time between the arrival of the SU into a channel and the completing the SU's data transmission. The advantage of non-switching handoff is that the handshaking delay and switching delay is absent as the SU stays on the same channel. M/M/1 queueing model is used to characterize the channel usage behaviors of PUs and SUs. Both types of users are assigned high-priority and low-priority ranks, respectively. The SU will either perform a reactive handoff or a non-switching handoff whenever the SU enters the channel for transmission. The queueing model can tolerate multiple interruptions in multichannel networks [22]. The processes of switching or staying on the same channel must happen at the beginning of each time slot. MATLAB was used to simulate the proposed system. According to the results, the extended data delivery time was minimal whenever the value of idle channels was between a particular ranges for both reactive as well as non-switching handoff. Increasing load on the network led to the selection of non-switching handoff. However, the method does not consider proactive handoff, and this can be considered to be a disadvantage of the proposed handoff scheme in CRAHNs.

Proactive Decision-Based Handoff Using Non-Iterative Greedy Approach. A proactive handoff scheme for CR networks based on a non-iterative greedy approach can be seen in [23]. The method aims to reduce the total service time by minimizing the number of handoffs. The method strives to calculate the number of time slots that an SU will require to complete its transmission. This way, the number of candidate channels required for the transmission can be ordered accordingly for the SU to access easily. These channels are ordered in such a way that the SU can access them in case the first among the candidate channels are not available for transmission. The system has been divided into time slots that have two phases where it either senses or transmits. The main objective of this method is to produce the most optimal sequence of target channels so that the service time can be minimized. A single pair of SUs is considered in the system and it ignores channel contention from the other SUs. CCC is used to manage the switching process by having to transmit the control messages. The method tries to identify as many consecutive idle time-slots as required by an SU in order to complete its transmission so that handoff can be avoided [23]. PU packets are assumed to arrive following the Poisson arrival process. A discrete-time Markov chain is used to calculate the probability of finding a channel with the required number of idle slots. Based on the arrival of the PU, the SU will move to the next slot during a one-time interval. If the PU does not arrive, the transition is completed. However, if the PU arrives after the SU has transitioned to the next slot, the latter will backtrack to the first slot as the condition of "idle channels for the entire SU transmission" has not been satisfied.

A probability matrix derived from the state transitions can be used to derive the probability of the SU identifying the complete set of consecutively idle channels it requires for its transmission. It needs to consider the probability of the arrival of a PU [23]. The probability of identifying the consecutive number of idle slots within a channel as required by an SU for transmission can be calculated if the arrival probability of a PU on that channel is given. The channels with such idle slots are ordered in descending order of their idle slot availability which makes it easier to select the most optimal target channel for handoff. The Greedy algorithm is employed to select the sequence of optimal target channels. The algorithm is divided into two phases where the first phase deals with the selection of the optimal target channel sequence with the most amount of idle time slots while the second phase deals with the discrete Markov chain process that identifies whether the channel has as many idle slots as the requirements of the SU. A preemptive M/G/1 queueing model with two priority queues that belong to the PUs and SUs, respectively, is used in the system in order to organize the users.

The total service time is derived using the average service time, probability of a packet's arrival at a time slot, and the delay involved for handoff. The handoff delay is dependent on three situations. If the busy period of the target channel is lower than the time taken to switch, the SU will stay on the same channel. If the target's busy period is higher than the switching time and it will be idle in the next slot, then the SU will switch to the former. If the busy period of the target channel is higher than the switching time, the SU will switch to the target channel. However, if the channel is busy, the SU will be placed at the end of the waiting queue. Interruptions decide the service duration within the system. The method also shows that the handoff delay can be significantly minimized if the SUs stay on the same channel until it becomes idle instead of switching. The switching is to take place only if the channel's gain goes below a predefined threshold. OMNET++ is used to simulate the method. The inter-arrival times of the PUs in the proposed method are compared with that of the ones from the existing methods. It was observed that the proposed system had a higher service time when the PU inter-arrival rate increased. When there was minimal PU traffic, the proposed system had higher throughput. The proposed system also has a lower number of handoffs when compared to other systems.

Probability-Based Centralized Device to Control Spectrum Operations. In [24], a probability-based, centralized device that controls the operations of a spectrum and helps to increase the efficiency of the spectrum is proposed. The proposed handoff strategy strives to improve spectrum sensing while improving efficiency in terms of energy consumption, handoff delay, and the throughput of the nodes involved. The method centralizes the functionalities of the different SUs into a single device. This device is responsible for allocating channels to the SUs in a probabilistic manner and ordering the SUs using a queueing model. The queueing model used here is the PRP M/G/1 queueing model. Channel statistics such as time taken for sensing, time taken for switching over to another channel, and the total number of channels in a spectrum is important for the SUs to make decisions regarding the selection of optimal channels.

The proposed system consists of a centralized CR device, a probabilistic spectrum sensing algorithm, and a PRP M/G/1 queueing model [24]. The usage of a centralized CR device reduces the overall energy consumption. In the proposed system model, a handoff event is triggered by either the arrival of a PU or by the SU having moved away from the channel or by the loss of signal. When the handoff is triggered, the SU hands over the channel to the arriving PU along with sending a message to the receiver SU about the handoff operation. Then, the SU communicates with the centralized CR device so that another channel can be assigned to the SU in order to continue transmission. The centralized CR device employs a probabilistic approach to identify the most appropriate target channel for the SU. Once the channel has been assigned to the SU, the latter begins to transmit data. The cycle is repeated for multiple handoffs until data transmission is completed. The probabilistic algorithm that is used in the proposed system selects a channel based on its idle probability [24]. An idle probability matrix is maintained by the centralized CR device. The device senses all channels whenever an SU requests a handoff operation and updates the matrix values for each channel sensed. If a channel is found to be idle, its value is updated in the matrix. Based on the values in the matrix corresponding to the channels, a candidate is selected for handoff. The channel with the highest idle probability is assigned to the SU if it is free. Else, the next best channel in terms of idle probability is selected. The sensing would take place only for a limited number of iterations as the idle probabilities of the channels will be filled after a few iterations. After these iterations, only a few channels are sensed, which leads to a reduction in the total latency of the network.

The PRP M/G/1 queueing model is employed to aid the ordering of the users based on their priorities. The SUs are generally pushed to the back of the queue if PUs is present so that the latter can transmit whenever they require. The arrival of the users is assumed to follow a Poisson distribution. This means that at peak hours, the load on the system increases suddenly [24]. Using the PRP M/G/1 queueing model, the SUs are arranged according to the order of handoff occurrence. High-priority queues that consist of interrupted SUs and low-priority queues that consist of newly joining SUs are formed. Clustering is utilized to reduce the load on a single CR device. This way, multiple CR devices can be employed, with each device catering to the need of a specific number of users. By simulating the system, it was identified that the centralized CR device network had a higher rate of accuracy in terms of selection of idle channels than a pure proactive handoff process. This is because the sensing of the spectrum takes place only when the SU needs to switch to another channel. The overall latency of the network was reduced due to the arrangement of channels based on their idle probabilities. The handoff delay was reduced due to this arrangement of the channels. Due to the reduced number of CR devices (the CR devices are considered to be catering to a cluster of SUs), the overall energy consumption is also reduced when compared to that of reactive and proactive handoff schemes.

Dual Processor-Based Centralized Device for Spectrum Handoff. The authors of [25] present a dual processor-based spectrum handoff in CR networks. It is an extension of the work presented in [24]. The proposed system considers a list of entities, including a centralized CR device that contains the channel status flag, dual processors, a PRP M/G/1 queueing model, and a channel status signals queue [25]. These are controlled using a channel status flag bit algorithm and a probabilistic algorithm. The PRP M/G/1 queue works based on a back-off algorithm in order to improve efficiency. SU clustering is the technique employed in order to reduce the consumption of energy. The arrival of a PU triggers the handoff event, and the SU stops transmitting over the channel while notifying the receiver about the same. It then approaches the centralized CR device with a handoff request. The CR device appends the SU to the PRP M/G/1 queue from where channels are allocated by the CR device using the channel status flag bit algorithm and the probabilistic approach. In the channel status flag bit algorithm, the status of all channels is stored. At the beginning or at the end of each transmission, the user sends a packet to the CR device through the channel status signal queue, where it will wait until it is sent to the second processor. The processor then updates the status of the channel that has sent the status message in the status flag register. The time taken for spectrum sensing by the channel status flag bit method is lower than that of the CR device having to sense all the channels in the spectrum normally.

The reason why dual processors are used is to cater to the different requirements of the spectrum. The first processor is used to allocate the channels to the users by cross-checking with the registers in the memory while the second processor is used to enter the status of the channels in the channel status flag registers. The PRP M/G/1 queue is divided into the high priority queue, and low priority queue where the interrupted SUs are given higher priority. Channels are selected from the channel status flag bit register by the second processor based on the probabilistic approach. The probabilistic approach works using the idle probability scheme that has been discussed in the previous method. It enhances the accuracy of channel selection while reducing the overall latency of the network [25]. In the PRP M/G/1 queue, a back-off algorithm is introduced. It is a threshold that is used to help the user from being stuck at the bottom of the waiting queue for an undefined amount of time. If a user has been in the low-priority queue for long enough, it is moved higher in the order so that it will get a chance to access a channel and complete its transmission. The back-off threshold is determined by the number of users in the high-priority queue and the time taken for them to complete transmission [25]. The second processor receives three types of packets from the users, which signal the start of PU transmission, end of PU transmission and the end of SU transmission. The beginning of SU transmission is defined by the spectrum handoff process, which is taken care of by the first processor. Due to the use of the channel status flag bit register, the accuracy of the system is maximized. Handoff latency is reduced due to the use of the channel status flag bit algorithm and the probabilistic algorithm with the inclusion of a back-off algorithm. The clock speed and the number of instructions processed in a second are increased due to the use of two processors. The work to be done by a single processor is divided between two in orders to improve efficiency. The CR device memory that contains the flag bit register optimizes the memory requirements [25].

3 Conclusion and Future Enhancements

A comparative study of various handoff processes that researchers have proposed has been carried out in this review. This review covers combinatorial handoff processes that make use of two or more techniques to perform spectrum functions and the different strategies' comparison can be seen in Table 1. Queueing models help to organize the users into queues so that an order can be brought about within a chaotic environment such as a wireless network. The use of AI techniques ensures that larger datasets can easily be used to train the models. This way, the models can be highly accurate when it comes to performing spectrum management functions such as sensing, mobility, sharing, and decision-making. One of the best methods to ensure that handoff is a lot more optimized is MADM. By using MADM methods, multiple network attributes such as SNR, throughput, delay, location, battery levels,

Table 1 Comparison of various combinatorial handoff strategies

Title	Method	Performance parameters	Advantages	Disadvantages
Initial and target channel selection	Probabilistic approach and queueing theory	(Extended service time, interference)	Improve target channel selection. Minimal interference by the SU	Not works for broadband users with a single queue Fixed target selection scheme
Fuzzy logic-based neural network scheme for spectrum allocation	Fuzzy logic and neural networks	Interference, bit error and signal strength	Improves channel utilization and reduces switching. Single hidden layer is used to avoid computational complexity	Misdetection of primary users was not accounted for
Analytical model for fixed and probabilistic target channel selection approaches	PSO and PRP M/G/1 queueing model	(Extended data delivery time)	Improves load balancing Equally performs to high mobility (Comparison)	Transition probability matrix must be chosen with care if the balancing power for a probabilistic approach needs to be improved
ANN and ANFIS-based spectrum management	ANN and adaptive network fuzzy inference system	Channel idle time, data rate and SNR. (handoff and throughput)	Due to multiple radios, overall sensing time is reduced as one radio perpetually scans the environment	
QoE-based handoff method using ANN	ANN and QoE	(Handoff frequency)	Can handle distributed processing. The use of Q-learning helps the model to learn the best policy	
Spectrum handoff using MADM methods and ANN	MADM and ANN	(Transmitted power and data rate)	Its interface can also accommodate further modifications Increasing the number of iterations during the training of ANN reduced the cost	

(continued)

Table 1 (continued)

Title	Method	Performance parameters	Advantages	Disadvantages
Adaptive spectrum handoff	PRP M/M/1 queueing model and common control channel	(Extended data delivery time)	Reduced handshaking and switching delay Can tolerate multiple interruptions in multi-channel networks Extended data delivery time was observed to be minimal when the value of idle channels was between a particular range	Does not consider proactive handoff
Proactive decision-based handoff using non-iterative greedy approach	Greedy algorithm, common control channel and PRP M/G/1 queueing model	Service time, handoff frequency	Handoff delay is minimized when the SUs stays on the same channel Higher service time when PU inter-arrival rates increased Throughput increased for reduced PUs Lower handoff probability	Arrival of PUs after the SUs have transitioned to another channel means that the SU will be pushed back to the first slot
Probability-based centralized device to control spectrum operations	PRP M/G/1 queueing model and probabilistic algorithm	Energy consumption, delay, throughput	Centralized CR device reduces the overall energy consumption Clustering is utilized to reduce the load on a single CR device Reduced latency, handoff delay, energy consumption	

(continued)

Table 1 (continued)

Title	Method	Performance parameters	Advantages	Disadvantages
Dual processor-based centralized device for spectrum handoff	Dual Processors and PRP M/G/1 queueing model	Handoff latency, throughput, processing speed	Efficient priority queue using a back off algorithm Use of channel status flag bit register Reduced handoff latency Number of instructions processed in a second are increased Workload sharing between two processors	

etc., can be considered for selecting optimal networks. This way, it can be ensured that the selected network provides better transmission coverage for longer durations and distances without terminating the connection midway. The review makes it clear that there is a lot more scope for improvements when it comes to developing an energy-efficient handoff strategy. In the future, energy-efficient handoff strategies that can make autonomous decisions based on the environmental parameters and can provide connectivity to ensure that switching is minimized need to be developed.

References

1. Jalil Piran, M., Tran, N.H., Suh, D.Y., Song, J.B., Hong, C.S., Han, Z.: QoE-driven channel allocation and handoff management for seamless multimedia in cognitive 5G cellular networks. IEEE Trans. Veh. Technol. **66**(7), 6569–6585 (2017)
2. Rupasinghe, N., Guvenc, I.: Licensed-assisted access for WiFi-LTE coexistence in: the unlicensed spectrum. In: Proceedings of the IEEE Globecom Workshops, pp. 894–899 (2014)
3. Marcin, D., Szydelko, M.: A unified traffic steering framework for LTE radio access network coordination. IEEE Commun. Mag. **54**(7), 84–92 (2016)
4. Hong, X., Wang, J., Wang, C., Shi, J.: Cognitive radio in 5G: a perspective on energy-spectral efficiency trade-off. IEEE Commun. Mag. **52**(7), 46–53 (2014)
5. Modi, S., Murmu, M.K..: An intelligent spectrum handover scheme for cognitive radio Ad hoc networks using ANFIS with ANN. In: 2nd International Conference on Telecommunication and Networks (TEL-NET 2017) (2017)
6. Sheikholeslami, F., Nasiri-Kenari, M., Ashtiani, F.: Optimal probabilistic initial and target channel selection for spectrum handoff in cognitive radio networks. IEEE Trans. Wireless Commun. **14**(1), 570–584 (2015)

7. Wang, L.-C., Wang, C.-W., Feng, K.-T.: A queueing-theoretical framework for QoS-enhanced spectrum management in cognitive radio networks. IEEE Wireless Commun. **18**(6), 18–26 (2011)
8. Stevenson, C., Chouinard, G., Lei, Z., Hu, W., Shellhammer, S., Caldwell, W.: IEEE 802.22: the first cognitive radio wireless regional area network standard. IEEE Commun. Mag. **47**(1), 130–138 (2009)
9. Maheshwari, P., Singh, A.K.: A fuzzy logic based approach to spectrum assignment in cognitive radio networks. In: 2015 IEEE International Advance Computing Conference (IACC), pp. 278–281 (2015)
10. Tayel, A.F., Rabia,, S.I., Abouelseoud, Y.: An optimized hybrid approach for spectrum handoff in cognitive radio networks with non-identical channels. IEEE Trans. Commun. **64**(11) (2016)
11. Tayel, A.F., Rabia, S.I.: Load balancing probabilistic spectrum handoff for cognitive radio networks. In: International Symposium on Performance Evaluation of Computer and Telecommunication Systems (SPECTS), pp. 1–8 (2015)
12. Deep, K., Singh, K.P., Kansal, M.L., Mohan, C.: A real coded genetic algorithm for solving integer and mixed integer optimization problems. Appl. Math. Comput. **212**(2), 505–518 (2009)
13. Modi, S., Murmu, M.K.: An intelligent spectrum handover scheme for cognitive radio Ad hoc networks using ANFIS with ANN. In: 2nd International Conference on Telecommunication and Networks (2017)
14. Jang, J.-S.R.: ANFIS: adaptive-network-based fuzzy inference system. IEEE Trans. Syst. Man Cybern. **23**(3), 665–685 (1993)
15. Suparta, W., Alhasa, K.M.: Adaptive Neuro-Fuzzy Interference System. In: Modeling of Tropospheric Delays Using ANFIS, pp. 5–18. Springer (2016)
16. Ben Zineb, A., Ayadi, M., Tabbane, S.: QoE-based vertical handover decision management for cognitive networks using ANN. In: Sixth International Conference on Communications and Networking (ComNet), pp. 1–7 (2017)
17. Jha, P., Rout, R., Nandakumar, S.: Spectrum handoff scheme in cognitive radio network using multiple attribute decision making and artificial neural networks. In: Innovations in Power and Advanced Computing Technologies (i-PACT), pp. 1–6 (2017)
18. IEEE Standard for Local and metropolitan area networks Part 16: Air Interface for Broadband Wireless Access Systems. In: IEEE Std 802.16–2009 (Revision of IEEE Std 802.16–2004), pp.1–2080 (2009). https://doi.org/10.1109/IEEESTD.2009.5062485
19. Thomas, L., Sandeep, J., Goswami, B., Paulose, J.: A survey on various handoff methods in mobile Ad Hoc network environment. In: Smart Computing Paradigms: New Progresses and Challenges. Advances in Intelligent Systems and Computing, vol. 767, pp. 27–60 (2020)
20. Zhang, W.: Handover decision using fuzzy MADM in heterogeneous networks. In: IEEE Wireless Communications and Networking Conference (IEEE Cat. No.04TH8733), vol. 2, pp. 653–658 (2004)
21. Thomas, L., Sandeep, J., Goswami, B., Paulose, J.: Handoff schemes in mobile environments: a comparative study. Int. J. Serv. Sci. Manage. Eng. Technol. **11**(1), 55–72 (2020)
22. Shakeel, A., Hussain, R., Iqbal, A., Latif, I., Malik, S.A., Omoniwa, B.: Adaptive spectrum handoff scheme in cognitive radio Ad-hoc networks. In: International Conference on Advances in Computing and Communication Engineering (ICACCE), pp. 127–132 (2018)
23. Gupta, N., Dhurandher, S.K., Woungang, I., Obaidat, M.S.: Proactive decision based handoff scheme for cognitive radio networks. In: IEEE International Conference on Communications (ICC), pp. 1–6 (2018)
24. Aggarwal, M., Velmurugans, T., Karuppiah, M., Hassan, M.M., Almogren, A., Ismail, W.N.: Probability-based centralized device for spectrum handoff in cognitive radio networks. IEEE Access **7**, 26731–26739 (2019)
25. Aggarwal, M., Velmurugan, T., Nandakumar, S.: Dual processor based centralized device for spectrum handoff in cognitive radio networks. J. Electr. Eng. Technol. **15**(2), 833–842 (2020)

A Deep Learning Approach to Extract Integrated Meaningful Keywords from Social Network Posts with Images, Texts and Hashtags

Ren-Xiang Lin, Chien-Chih Yu, and Heng-Li Yang

Abstract Using the social network services, users might create different types of content including numeric, textual and non-textual data objects. In the past, social network service providers mainly focus on numeric and textual content to understand their users and to provide them with related information or advertisements. However, the information behind the non-textual content has not been well considered. This research aims at extracting integrated meaningful keywords by jointly considering photo, text descriptions and hashtags to better reflect the meaning of the user-posted content. A deep learning approach with convolutional neural network methods that integrate ResNet-50 and Word2Vec models, as well as Dijkstra's algorithm is proposed to extract the meaningful keywords. The well-trained ResNet-50 and Word2Vec models are applied respectively to gain the predicted classification labels of the image and to identify the co-occurrences among predicted classification labels of image, segmented words of text descriptions and hashtags. A multistage graph weighted with the pairs of co-occurrences of image, segmented words and hashtags is built and then, the Dijkstra's algorithm is adapted to extract consistent keywords of the posted content with maximized cumulated weights. A simplified example is provided to illustrate the proposed approach for acquiring the integrated information embedded in the image, text and hashtags.

1 Introduction

The advancement of information technology has created many convenient services such as social network services. Social network services not only enable users to

R.-X. Lin (✉) · C.-C. Yu · H.-L. Yang
Department of Management Information Systems, National Chengchi University, Taipei, Taiwan
e-mail: 102356507@nccu.edu.tw

C.-C. Yu
e-mail: ccyu@nccu.edu.tw

H.-L. Yang
e-mail: yanh@nccu.edu.tw

T. Senjyu et al. (eds.), *ICT with Intelligent Applications*, Smart Innovation, Systems and Technologies 248, https://doi.org/10.1007/978-981-16-4177-0_73

communicate or interact with their family, friends, classmates but also help enterprises to contact their customers for business purposes, for instance, push and pull marketing, order tracking and group purchasing, etc. Many users have been connected by social network services that satisfy their social as well as commercial needs. Famous examples of social network service platforms include Facebook, Instagram and Twitter, etc. These social network services have been very popular and hold large member bases.

Using the social network services, users might create different types of content (user-generated content) such as user profile, post, comment, like and check-in. Social network service providers could analyze these user data to understand their users and to provide them with related information or personalized advertisements. In addition to the numeric and textual contents, there are large amounts of user-generated non-textual content such as photos, music and videos. For example, the total number of photos shared by users in Instagram had reached 20 billion by 2017 [1]. The user-contributed non-textual content may reveal their personal interests. As a user shares photos about tourism sites often, it may quite possibly mean that the user is interested in tourism-related content. Therefore, it is important to analyze both the textual and non-textual contents for understanding users' interests.

Basically, an Instagram post usually includes textual contents such as author account data, text description, hashtags and comments, numeric contents such as number of likes and non-textual contents such as photos and videos. Social network service providers usually apply only content analysis to parse text descriptions, hashtags and comments to extract the possible meaning of the textual content. However, what would happen if an Instagram post does not include any text description or hashtags, but only has a photo? In another situation, when the photos, text descriptions and hashtags have discrepancy, it would be hard to acquire a consistent meaning of the post only by traditional content analysis. To be more specific, assuming that an Instagram post includes a meal photo, a text description as "Happy Friday night!!" and hashtags (#happy #relax), ideally in this case, the social network provider should provide the user with the meal or restaurant-related content. But according to the text description and hashtags, the user might receive massage or night club-related content.

To deal with the above stated problem, this research aims at extracting integrated meaningful keywords by jointly considering photo, text description and hashtags of a post to acquire the complete information embedded in the three types of content objects. A deep learning-based keyword extraction process that integrates ResNet-50 model, Word2Vec model and Dijkstra's algorithm is proposed. First, an Instagram post is divided into three parts including image, text description and hashtags. Because the hashtags are well distinguished by the symbol "#", no further processing work is needed. For the Chinese text description, a Chinese text segmentation tool Jieba is applied to segment text description into a sequence of words. Second, to resolve the problem of image recognition, the classic convolutional neural network model ResNet-50 (residual neural network) is utilized to gain the predicted classification labels of the image. The ResNet-50 model [2] is trained by the CIFAR-10 dataset [3]. Third, the Word2Vec (skip-gram) model [4] is used to identify the co-occurrence rates among predicted classification labels of image, segmented words of text description and hashtags. The Word2Vec (skip-gram) model is trained by the corpus of Chinese Wikipedia. Fourth, we build a multistage graph weighted with the pairs of co-occurrence of image labels, segmented words and hashtags. And then, the Dijkstra's algorithm is applied to generate the path with the highest cumulated weights for extracting the most consistent keywords among these three parts of content objects. In effect, the extracted integrated keywords may better imply the meaning of the user's post.

2 Related Works

In this section, we briefly review the related works on convolutional neural network and Word2Vec. Neural network has been utilized widely in the field of computer vision [1]. The emergence of convolutional neural network is particularly beneficial in image recognition [5–7]. The structure of a convolutional neural network includes convolutional layers, pooling layers and fully-connected layer [8]. The convolutional layer is comprised of a set of learnable kernels or filters which aim to extract local features from the input. The pooling layer allows the reduction of the actual spatial size for preventing the effect of overfitting and preserving important features. The fully-connected layer is used to extract the global features of the inputs.

The Word2Vec model is proposed by Mikolov et al. [4], which uses a neural network model to learn word associations from a large corpus of text for natural language processing. There are two types of model architecture. The first type is a continuous bag-of-words (CBOW) model that uses the context to predict the current word. The second type is a continuous skip-gram model that uses the current word to predict surrounding words. Being capable of handling a corpus with a size ranging in billions of words in a matter of hours, Word2Vec is considered a very efficient approach for natural language processing. Although there are many outperforming natural language processing models in the literature (e.g. [9]), considering the training cost and hardware limitation, this research adopts the ResNet-50 and the Word2Vec skip-gram models for gaining better performance on feature extraction and semantic analysis.

To date, many keywords extraction methods have been proposed for online posts. However, most of the existing methods (e.g. [10]) only deal with the text information and image separately or sequentially. However, the post in social network services usually consists of both text and image that are likely to be correlated with each other [11]. Therefore, a consistent keywords extraction method for simultaneously taking into account both the text and image is crucial to better understand the meaning of the post.

3 Meaningful Keywords Extraction Process of Social Network Post

The proposed integrated meaningful keywords extraction process that jointly considers photo, text description and hashtags of a social network post is presented step by step below and depicted in Fig. 1.

3.1 Data Preprocessing

First step, an Instagram post is divided into three parts including image, text description and hashtags. We apply a Chinese text segmentation tool Jieba to segment text description to a sequence of words with all punctuation and redundant words being removed.

3.2 Model Training

Second step, the ResNet-50 [2] is trained by the simple and tested CIFAR-10 dataset [3] which consists of 60,000 32 × 32 colour images in 10 classes, with 6000 images per class. There are 50,000 training images and 10,000 test images. The classes in the dataset are aeroplane, automobile, bird, cat, deer, dog, frog, horse, ship and truck. The detailed setting of the ResNet-50 model in this research is shown in Table 1. The

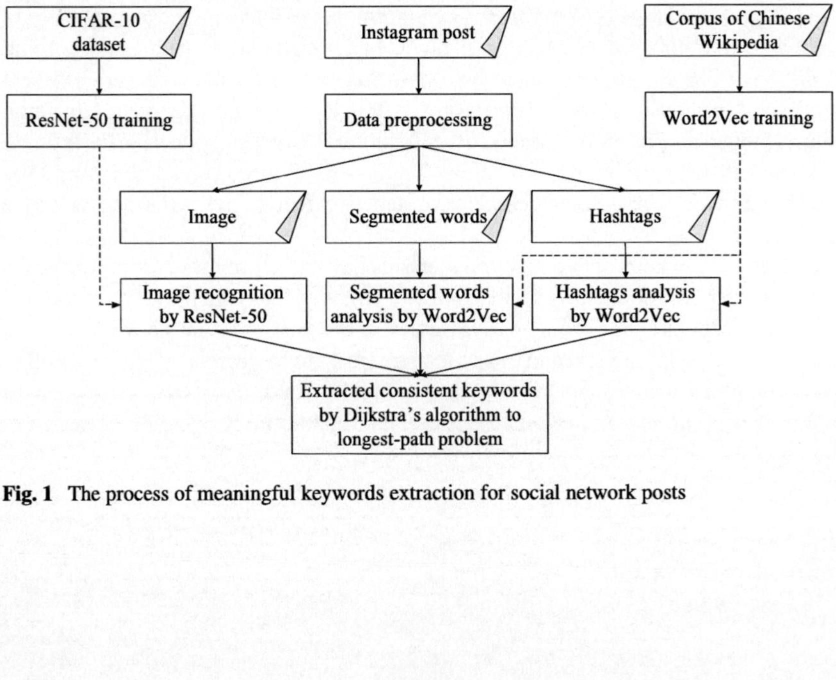

Fig. 1 The process of meaningful keywords extraction for social network posts

Table 1 ResNet-50 detailed setting

Layer name	Output size	50-layer detail
Conv1	32 × 32	3 × 3, 16, stride 2
Conv2_x	32 × 32	3 × 3 max pooling, stride 2
		$\begin{bmatrix} 3 \times 3, 16 \\ 3 \times 3, 16 \end{bmatrix} \times 8$
Conv3_x	16 × 16	$\begin{bmatrix} 3 \times 3, 32 \\ 3 \times 3, 32 \end{bmatrix} \times 8$
Conv4_x	8 × 8	$\begin{bmatrix} 3 \times 3, 64 \\ 3 \times 3, 64 \end{bmatrix} \times 8$
	1 × 1	Average pooling, 10-dimension fully-connected softmax

Table 2 Word2Vec parameter setting

Parameter	Description	Research setting
size	Dimensionality of the word vectors	size = 250
window	Maximum distance between the current and predicted word within a sentence	window = 5
min_count	Ignores all words with total frequency lower than this	min_count = 5
sg	Training algorithm: 1 for skip-gram; otherwise CBOW	sg = 1

valid accuracy of trained ResNet-50 is 0.9216. The training parameters of ResNet-50 is adopted from He et al. [2].

The Word2Vec (skip-gram) model is trained by the corpus of the latest backup Chinese Wikipedia. Wikipedia is considered as an ideal knowledge base for data mining systems to find auxiliary data to improve performance [12]. In this process, Word2Vec firstly builds a vocabulary from training text corpus and then learns the vector representations of each word. In addition, Word2Vec calculates the cosine distances among specified word vectors [13]. In this research, the Python library genism is used to implement the Word2Vec (skip-gram) model with the library default parameter setting shown in Table 2.

3.3 Model Application to the Posted Contents

Third step, we apply the models to the posted contents.

(1) Image Recognition by ResNet-50: The well-trained ResNet-50 is used to gain the predicted classification labels of the image.

(2) Segmented Words Analysis by Word2Vec: The well-trained Word2Vec model is applied to identify the co-occurrence rates among predicted classification labels of image and segmented words of text description.
(3) Hashtags Analysis by Word2Vec: The well-trained Word2Vec model is also applied to identify the co-occurrence rates among segmented words of text description and hashtags.

3.4 Extracted Consistent Keywords by Dijkstra's Algorithm

Fourth step, we build a multistage graph weighted with the pairs of co-occurrence of image labels, segmented words and hashtags and consider the current task as a longest-path problem that is to generate the path with the maximized cumulated weights. Finally, the consistent keywords among image labels, segmented words and hashtags are extracted after resolving the longest-path problem by adapting Dijkstra's algorithm. The longest-path problem can be solved by converting the sign of the weight of each edge and then calculate the shortest path by Dijkstra's algorithm which is widely used for shortest path problems.

4 An Illustrative Example

An Instagram post is used as an illustrative example for our proposed integrated meaningful keywords extraction process. After data preprocessing, the Instagram post is divided into an image, three segmented words and two hashtags. The text description of the Instagram post is 我可以吃了嗎? (Can I eat it?) and 紫地瓜好香 (Purple sweet potato smells good). The hashtags of the Instagram post are #鮮食 (#fresh pet food) and #肉丸 (#meatball).

Next, the image is entered to ResNet-50 to gain the predicted classification labels. The recognition result is 狗 (dog). The image label, segmented words and hashtags are entered into Word2Vec model to get the co-occurrence rates as shown in Fig. 2. The results of co-occurrence analysis of the image label and segmented words are CO (狗/dog, 吃/eat) = 0.37, CO (狗/dog, 地瓜/sweet potato) = 0.26 and CO (狗/dog, 好香/smells good) = 0.13. The results of co-occurrence analysis of segmented words and hashtags are CO (吃/eat, 鮮食/fresh pet food) = 0.45, CO (地瓜/sweet potato, 鮮食/fresh pet food) = 0.50, CO (好香/smells good, 鮮食/fresh pet food) = 0.26, CO (吃/eat, 肉丸/meatball) = 0.54, CO (地瓜/sweet potato, 肉丸/meatball) = 0.57 and CO (好香/smells good, 肉丸/meatball) = 0.32.

Finally, the multistage graph weighted with the pairs of co-occurrence of image labels, segmented words and hashtags are built as illustrated in Fig. 3. Resolving it by Dijkstra's algorithm, the longest-path (as denoted as bold black in the Fig. 3) is 狗 → 吃 → 肉丸 (dog → eat → meatball) with maximum distance (maximized sum of weights) 0.91. The result indicates that the consistent three keywords are acquired as

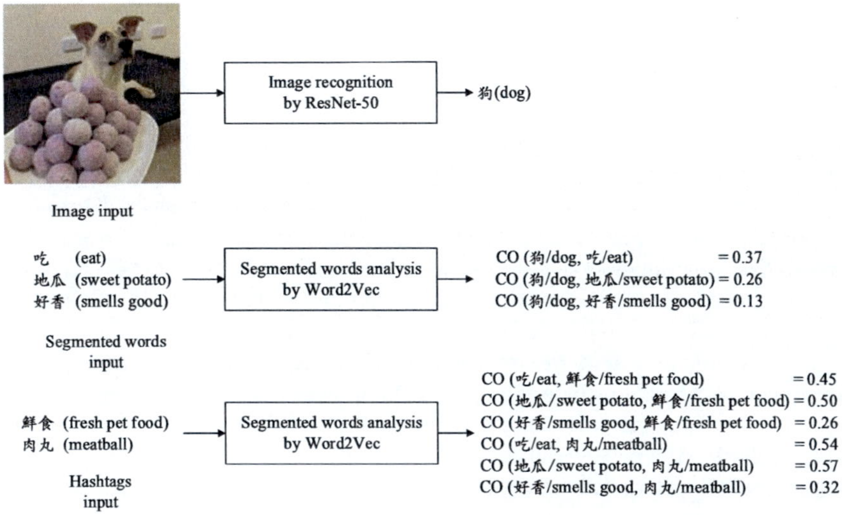

Fig. 2 An illustrative example for image recognition, segmented words and hashtags analyzes

Fig. 3 An illustrative example for resolving the longest-path problem by Dijkstra's algorithm

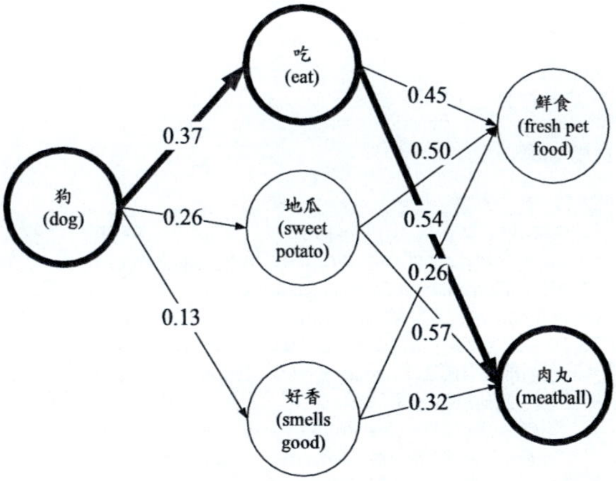

狗(dog), 吃 (eat) and 肉丸 (meatball). It is noted that the keyword 狗 (dog) extracted from the image has not been mentioned in the text description and hashtags.

5 Conclusions

In this research, a meaningful keywords extraction process of social network posts is proposed by jointly considering image, text description and hashtags. The process integrates ResNet-50 model, Word2Vec model and Dijkstra's algorithm. The ResNet-50 is applied to gain the predicted classification labels of the image. The Word2Vec model is used to identify the co-occurrence among predicted classification labels of images, segmented words of text description and hashtags. A multistage graph weighted with the pairs of co-occurrence of image labels, segmented words and hashtags is built and resolved by using Dijkstra's algorithm for extracting consistent keywords of the post. An illustrative example is given to explain why and how our proposed process is feasible to acquire the integrated information embedded in the images, text descriptions and hashtags posted by users. The proposed post-analysis process involving three types of content objects can be helpful for advertisers to deliver matched ads to social network users for creating more business value.

In the future, we will try to extend our process by considering more images with more generated labels included in the extraction process and more image recognition classifications via other improved recognition models, as well as by considering other dimensions of social network attributes such as the user's social relationship, browsing history and comments for better understanding and matching user's interests and needs.

Acknowledgements The authors express gratitude to the Ministry of Science and Technology in Taiwan, for financially supporting this research under contract No. MOST 107-2410-H-004-097-MY3.

References

1. Kao, D., Lai, K.-T., Chen, M.-S.: An efficient and resource-aware hashtag recommendation using deep neural networks. In: Yang, Q., Zhou, Z.-H., Gong, Z., Zhang, M.-L., Huang, S.-J. (eds.) Advances in Knowledge Discovery and Data Mining 2019 Lecture Notes in Computer Science, vol. 11440, pp. 150–162. Springer, Cham (2019)
2. He, K., Zhang, X., Ren, S., Sun, J.: Deep residual learning for image recognition. In: Proceedings of the IEEE Conference on Computer Vision and Pattern Recognition, pp. 770–778. IEEE, Las Vegas, NV (2016)
3. Krizhevsky, A.: Learning multiple layers of features from tiny images. Master's thesis, University of Toronto, Toronto, Canada (2009)
4. Mikolov, T., Chen, K., Corrado, G., Dean, J.: Efficient estimation of word representations in vector space. arXiv:13013781 (2013)
5. LeCun, Y., Bengio, Y.: Convolutional networks for images, speech, and time series. In: The Handbook of Brain Theory and Neural Networks, pp 255–258. MIT Press, Cambridge, MA (1998)
6. LeCun, Y.: Generalization and network design strategies. Connectionism Perspect. **19**, 143–155 (1989)
7. LeCun, Y., Boser, B., Denker, J.S., Henderson, D., Howard, R.E., Hubbard, W., Jackel, L.D.: Backpropagation applied to handwritten ZIP code recognition. Neural Comput. **1**(4), 541–551 (1989)
8. Albelwi, S., Mahmood, A.: A framework for designing the architectures of deep convolutional neural networks. Entropy **19**(6), 242 (2017)
9. Devlin J., Chang, M.-W., Lee, K., Toutanova, K.: BERT: pre-training of deep bidirectional transformers for language understanding. arXiv:181004805 (2018)

10. Zangerle, E., Gassler, W., Specht, G.: On the impact of text similarity functions on hashtag recommendations in microblogging environments. Soc. Netw. Anal. Min. **3**(4), 889–898 (2013)
11. Zhang, S., Yao, Y., Xu, F., Tong, H., Yan, X., Lu, J.: Hashtag recommendation for photo sharing services. In: Proceedings of the AAAI Conference on Artificial Intelligence 2019, pp. 5805–5812. AAAI, Honolulu, Hawaii (2019)
12. Zhang, W., Wang, D., Xue, G.-R., Zha, H.: Advertising keywords recommendation for short-text web pages using Wikipedia. ACM Trans. Intell. Syst. Technol. **3**(2), 1–25 (2012)
13. Ma, L., Zhang, Y.: Using Word2Vec to process big text data. In: Proceedings of the 2015 IEEE International Conference on Big Data, pp 2895–2897. IEEE, Santa Clara, CA (2015)

Enhanced Seagull Optimization Algorithm Adapted for Artificial Neural Network Training

Nebojsa Bacanin, Milos Antonijevic, Nikola Vukobrat, Timea Bezdan, and Miodrag Zivkovic

Abstract The learning process is one of the most difficult problems in the artificial neural networks. This task refers to the challenge of finding the appropriate values for connection weights and biases and directly influences the neural network classification accuracy. Since the search space is enormous, traditional optimization techniques are not suitable as they are prone to slow converging speed and getting trapped to the local optima. In this paper, an enhanced seagull optimization algorithm was proposed to address the task of neural networks training. Conducted experiments included ten binary classification benchmark datasets to evaluate the performance of proposed enhanced seagull optimization algorithm. The obtained results indicate that the proposed method has promising performance, as it was able to achieve better overall results than other state-of-the-art algorithms that were included in the comparative analysis, in terms of classification accuracy and converging speed.

N. Bacanin (✉) · M. Antonijevic · N. Vukobrat · T. Bezdan · M. Zivkovic
Singidunum University, Danijelova 32, Belgrade 11000, Serbia
e-mail: nbacanin@singidunum.ac.rs

M. Antonijevic
e-mail: mantonijevic@singidunum.ac.rs

N. Vukobrat
e-mail: nikola.vukobrat.19@singimail.rs

T. Bezdan
e-mail: tbezdan@singidunum.ac.rs

M. Zivkovic
e-mail: mzivkovic@singidunum.ac.rs

T. Senjyu et al. (eds.), *ICT with Intelligent Applications*, Smart Innovation, Systems and Technologies 248, https://doi.org/10.1007/978-981-16-4177-0_74

1 Introduction

Artificial intelligence (AI) today represents one of the striving domains which helps to solve more complex problems using statistical learning. Using some of the advanced algorithms like artificial neural networks (ANNs) with different variety of architectures, now it is possible to solve difficult problems in a reasonable amount of time [11].

The ANNs are also formed out of a different amount of artificial neurons which consist out of multiple input connections with weight parameters attached (sometimes called synaptic weights), summation function with applied bias, activation function and output connection to the other artificial neurons in the network. These attributes construct each artificial neuron which together forms connected ANN which is capable of learning out of the different amount of data provided to it. For the proper organization of these neurons, ANNs are split into the different amount of layers, more precisely, input, hidden, and output layers where each corresponds to the specific task.

After all, in order for those networks to be able to learn they need to be based on certain algorithms. In the case of ANNs, one of the most important algorithm is gradient descent (GD). The purpose of GD is to find local (or in best case global) minimum in hyperfunction field it tries to solve. This minimum represents estimation (not exact solution) of the defined function. More precisely, this function describes the problem we are trying to solve with our ANN.

In order to calculate how much to update weights and biases, GD uses a certain cost function that defines how much the ANN model is wrong in estimating correct prediction for certain input. Goal of the GD algorithm is to minimize this cost function which represents the least amount of error produced by ANN. In the end, an algorithm called backpropagation does updates of all weights and biases from the end of the network until the sole beginning of it. Besides GD, ANNs use chain rules to perform this action. By itself, the chain rule is used to calculate the derivative of nested functions.

Even if ANNs are great on many real-life problems that do not mean that ANNs should be applied to every problem. For example, there are a lot of problems that have explicit rules and that are much easier and preciser to strictly define those rules. But, in a case where there are just too many conditions (e.g., detecting an object on the image, forecasting stock price, and even more machine learning-related problems such as hyperparameter optimization, feature selection, model training), it is impossible to cover all of them without utilizing statistical-based learning algorithms such as ANNs. All mentioned and many more problems belong to the nondeterministic polynomial (NP) time problems [10].

In the case of NP problems, it is proven that metaheuristic methods, such as one presented in this paper, have proven efficiency in solving and converging to optimal solutions in the NP problem field. Thus, in this paper, improved version of seagull optimization algorithm (SOA), that belongs to the family of swarm intelligence, is proposed and adopted for ANNs' training. Further comparisons and results are presented in the following sections.

The rest of the paper is structured in the following way. Section 2 provides the insight into the swarm intelligence metaheuristics and their applications. Section 3 presents the basic SOA metaheuristics and the enhanced version. Section 4 shows the results of the executed experiments. At the end, Sect. 5 brings final observations and gives insights into the possible future work.

2 Related Work

One of the general applications of metaheuristic algorithms is computer task scheduling problems [2, 3, 7, 12]. The main goal of these approaches is to determine, in the most efficient manner, when and how much resources a certain task should reserve in a certain time. Also, one more problem with successful approaches utilizing swarm intelligence metaheuristics is in the IoT industry, more specifically in localizing wireless sensors in a vast network with many such devices [4, 18, 20, 21]. This, and many more challenges, makes these distributed wireless systems hard to operate in a precise and secure manner as each device is hard to locate in this broad network. Besides, similar to wireless network problem relies upon solving radio frequency identification (RFID) network planning problem. Using algorithms like bare-bones fireworks and hybridized bat algorithms, authors in [7, 13] managed to even solve these difficulties and challenges.

Unlike more general applications above, there are also plenty of machine learning-specific problems that utilize metaheuristics for their improvement. Following authors [5, 6, 14, 15] managed to employ automatic architecture search for different applications like medical diagnostics, classifications, detections, and more computer vision-related problems. Finally, the time series prediction enhanced by the swarm intelligence was utilized in [19] to predict COVID-19 cases. Similarly, one more machine learning frequent problem is proper feature selection. One of the available methods is also the metaheuristic approach [8, 16].

3 Proposed Improved Seagull Optimization Algorithm

Seagull optimization algorithm (SOA) is relatively novel swarm intelligence metaheuristics. As its name describes, the inspiration of this algorithm came up from how seagulls behave in their colonies. More precisely by their migration and attacking (hunting) strategies [9].

Seagulls migration strategy is defined as sessional transitions from one place to another. The reason for their migration is to find the best place with productive food sources that will keep them well during the session. Through their migration, they travel in a group. Their position is constructed in that way to avoid a possible collision with each other. While in the group, seagulls often travel toward the best, fittest (healthiest, strongest, ...) one. Based on the location of the fittest seagulls, others can converge toward him and update their current locations.

On the other hand, while they are in process of migration they are attacking other birds over the sea. During their attack, they create spiral shape movement in order to most efficiently attack their prey. Looking at these behaviors, SOA algorithm is formulated. For more information regarding the SOA and its control parameters, please refer to [9].

It is worth mentioning that SOA showed good results on global bound-constrained optimization problems. By that, there is an assumption that it will perform well in a variety of real-world problems as proposed in this paper.

However, notwithstanding that the basic SOA obtains significant results for standard unconstrained benchmarks, as it was shown in [9], by performing additional experiments for the purpose of this research with CEC test suite, it was observed that the convergence speed could be enhanced. Although original SOA manages to find optimal region of search domain in each run, in some runs it does not converge satisfactory and as a consequence final generated results suffer from the poor quality.

To overcome this issue, the opposition-based learn (OBL) procedure is introduced. According to the previous research, it was shown that the OBL can substantially enhance intensification, as well as diversification [1].

At the end of each iteration, when the best solution is determined X_{best}, its opposite solution X^o_{best} is created by using the following expression for every parameter j:

$$X^o_{best,j} = l_j + u_j - X_{best,j} \tag{1}$$

where $X^o_{best,j}$ represents opposite dimension j original best solution, and l_j and u_j denote lower and upper bounds of j-th dimension, respectively.

Afterward, a greedy selection is applied between the initial best and opposite best solution based on fitness, and better one is retained for the next iteration. Proposed method is named enhanced SOA (eSOA), and its pseudo-code is presented in Algorithm 1.

Finally, from the perspective of computational complexity, eSOA slightly contributes to the overall complexity. More precisely, it adds one more calculation to each iteration. That calculation is an OBL procedure that chooses between $\mathbf{X}_{best}$ and $\mathbf{X}^o_{best}$ by using a greedy selection. In other words, $\mathcal{O}((N + (N + 1)) * Max_{iterations})$, where N represents number of individuals while $Max_{iterations}$ number of iterations.

Algorithm 1 The eSOA pseudo-code

```
Input: Seagull population P_s
Output: Optimal search agent P_bs
Algorithm:
Initialize parameters A, B and Max_itteration
Set f_c ← 2
Set u ← 1
Set v ← 1
while x < Max_itteration do
   /* Calculate the fitness of each individual */
   for i = 1 to n (each dimension for given problem) do
      FIT_s[i] ← FitnessFunction (P_s(i, :))
   end for
   /* Choose the best fitness value */
   Best = FIT_s[0]
   for i = 1 to n do
      if FIT_s[i] < Best then
         Best ← FIT_s[i]
      end if
   end for
   /* Choose the finess value for each search agent (seagull) */
   P_bs = Best
   /* Migration */
   rd ← Rand(0, 1)
   k ← Rand(0, 2π)
   /* Attacking */
   r = u × e^(kv)
   D_s =| C_s + M_s |
   P ← x' × y' × z'
   P_s(x) = (D_s × P) + →P_bs(x)
   x ← +1
   Perform OBL procedure by using Eq. (1)
   Choose between X_best and X^o_best by using greedy selection
end while
return P_bs
```

4 Experiments, Results, and Discussion

Ten frequently used binary datasets were utilized to test the performance of approach proposed in this paper. Every dataset that was used in the conducted experiments was normalized to speed the learning process up, by scaling the values into the range between 0 and 1. Throughout the conducted experiments, 2/3 of the datasets were used for training, while the remaining 1/3 were used for testing.

Each solution represents values of weights (W) and biases (B), and it is encoding as vector of dimension $W + B$. Total number of neurons in the hidden layer H depends on the number of input features I, and it is calculated as: $H = 2 \cdot I + 1$. Thus, total number of weights W and biases B is $I \cdot H + H \cdot O$ and $H + O$, respectively, where O represents the number of neurons in the output layer. The same strategy for encoding solutions is used in [17]. A well-known mean square error (MSE) is used as a fitness function.

Performance of proposed eSOA approach WAS validated by utilizing the following metrics: mean accuracy, best accuracy, and standard deviation obtained in total of 30 independent runs. The obtained results of the eSOA are compared to the original

Table 1 Simulation results for 10 standard binary classification datasets

Dataset	Measure	AAA	PSO	WOA	BAT	CS	GWO	FFA	BP	SOA	eSOA
Australian	Best	0.860	0.851	0.834	0.843	0.864	0.843	0.838	0.877	0.815	**0.902**
	Mean	0.848	0.817	0.794	0.837	0.839	0.838	0.827	0.857	0.803	**0.889**
	StdDev	0.006	0.021	0.049	**0.005**	0.013	**0.005**	0.010	0.016	0.011	0.009
Blood	Best	0.757	0.749	0.745	0.745	0.753	0.741	0.745	0.784	0.761	**0.812**
	Mean	0.748	0.743	0.741	0.741	0.745	0.740	0.741	0.663	0.675	**0.803**
	StdDev	**0.003**	0.004	0.004	**0.003**	0.004	**0.003**	0.004	0.205	0.013	0.005
Breast Cancer	Best	0.987	0.987	0.971	0.975	0.987	0.979	0.971	0.970	0.896	**0.989**
	Mean	0.981	0.968	0.965	0.971	0.970	0.973	0.966	0.892	0.898	**0.981**
	StdDev	0.003	0.008	0.006	0.004	0.008	0.003	0.005	0.101	**0.001**	**0.001**
Chess	Best	0.767	0.744	0.822	0.828	0.752	**0.944**	0.723	0.725	0.770	0.821
	Mean	0.710	0.678	0.662	0.736	0.715	**0.939**	0.697	0.690	0.722	0.791
	StdDev	0.018	0.037	0.089	0.094	0.024	**0.003**	0.020	0.047	0.023	0.019
Diabetes	Best	0.771	0.771	0.718	0.752	0.767	0.752	0.752	0.690	0.631	**0.815**
	Mean	0.753	0.747	0.692	0.746	0.739	0.747	0.742	0.596	0.599	**0.795**
	StdDev	0.007	0.016	0.026	0.006	0.014	**0.003**	0.006	0.140	0.025	0.021
Ionosphere	Best	0.925	0.875	0.658	0.892	0.892	0.908	0.850	0.783	0.861	**0.953**
	Mean	0.867	0.791	0.599	0.841	0.809	0.897	0.819	0.751	0.843	**0.938**
	StdDev	0.022	0.046	0.046	0.054	0.040	**0.012**	0.029	0.038	0.016	0.019
Liver	Best	0.780	0.780	0.653	0.754	0.763	0.754	0.746	0.737	0.781	**0.816**
	Mean	0.760	0.750	0.616	0.736	0.714	0.742	0.729	0.552	0.728	**0.795**
	StdDev	**0.011**	0.015	0.027	0.021	0.026	0.017	**0.011**	0.071	0.033	0.021
Parkinson	Best	0.896	0.881	0.791	0.866	0.866	0.866	0.836	0.865	0.852	**0.941**
	Mean	0.860	0.790	0.721	0.860	0.812	0.858	0.804	0.778	0.805	**0.928**
	StdDev	0.017	0.045	0.059	**0.009**	0.036	0.009	0.026	0.148	0.026	0.021
Tic-tac-toe	Best	0.739	0.755	0.663	0.702	0.718	0.712	0.715	0.625	0.685	**0.743**
	Mean	0.713	0.688	0.646	0.692	0.675	0.697	0.693	0.583	0.672	**0.731**
	StdDev	0.013	0.027	0.018	**0.009**	0.022	0.014	0.018	0.046	**0.009**	0.015
Vertebral	Best	0.896	0.906	0.783	0.877	**0.906**	0.877	0.868	0.811	0.823	0.899
	Mean	0.881	0.869	0.740	0.854	0.835	0.869	0.855	0.717	0.789	**0.894**
	StdDev	**0.006**	0.016	0.026	0.019	0.035	0.010	0.020	0.146	0.024	**0.005**
Total Best	Best	0	0	0	0	1	1	0	0	0	8
	Mean	0	0	0	0	0	1	0	0	0	9
	StdDev	3	0	0	4	0	5	1	0	2	1

SOA metaheuristics, backpropagation, and seven other modern and commonly used approaches (AAA, PSO, WOA, CS, GWA, and FFA). The same simulation setup was utilized like in the referred paper [17]: Each run is executed in 250 iterations with 40 solutions in the population.

Table 1 summarizes established results of the proposed eSOA and shows the comparison against other algorithms. The presented results of AAA, PSO, WOA, CS, GWA, FFA, and BP metaheuristics were taken from [17]. In Table 1 the best result for each dataset is marked bold, while the best results for each metaheuristics are summarized at the end of the table.

The experimental results and performed comparative analysis clearly indicate that the proposed eSOA approach has proven to be superior over other algorithms included in the analysis on the majority of utilized datasets. The proposed eSOA scored the best accuracy on eight datasets. The GWO metaheuristics scored the best

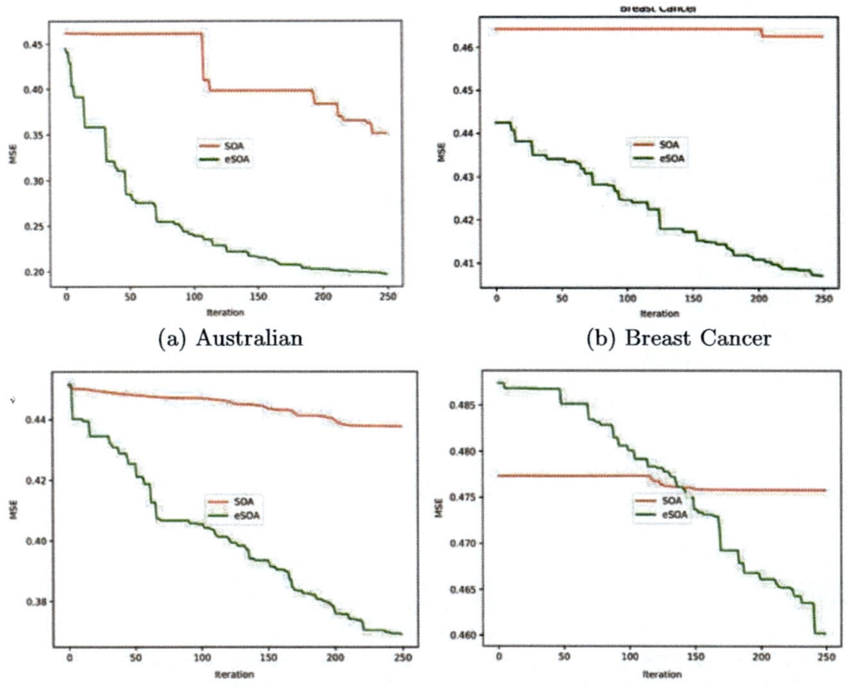

(a) Australian (b) Breast Cancer

(c) Diabetes (d) Liver

Fig. 1 Convergence graphs for MSE of SOA and eSOA

accuracy on the Chess dataset. When the average accuracy is observed, the GWO has obtained the best performance on Chess dataset, while on other nine datasets used in the research, the proposed eSOA approach achieved the best average classification accuracy. However, it is noted that the eSOA for only Vertebral dataset obtained best value of std. It means that its performance varies between runs due to the randomly generated opposite solutions. Finally, when compared to the original SOA, proposed eSOA performed substantially better for all datasets and almost all indicators.

Finally, to make better visual comparison between basic SOA and eSOA, additional experiments with 20 individuals in population and 250 iterations are conducted. Convergence speed graphs of MSE indicator for some datasets are shown in Fig. 1, where it can be clearly noticed superiority of eSOA metaheuristics.

5 Conclusion

In the research presented in this manuscript, an enhanced seagull optimization algorithm was proposed, with an objective to optimize the values of weights and biases during the ANN training. The performances of the proposed eSOA algorithm were validated on ten common benchmark datasets, against the basic SOA approach, backpropagation, and seven modern swarm intelligence metaheuristics: AAA, PSO, WOA, BAT CS, GWA, and FFA, respectively.

The obtained results from the performed simulations clearly indicate that the proposed eSAO is a very efficient algorithm that can be successfully applied to the ANN training. The proposed eSAO was developed with an additional objective to overcome the drawbacks of the original SAO. As a result, the proposed eSAO is not only able to converge faster than the basic SAO, but it also successfully avoids getting stuck in the local optimum.

The future work will target inclusion of other parameters to the optimization process (such as the number of hidden layers). Additionally, improved versions of other nature inspired metaheuristics will also be included in the ANN training with a goal to enhance the classification accuracy even further.

Acknowledgements The paper is supported by the Ministry of Education, Science and Technological Development of Republic of Serbia, Grant No. III-44006.

References

1. Abd Elaziz, M., Oliva, D.: Parameter estimation of solar cells diode models by an improved opposition-based whale optimization algorithm. Energy Conversion Manage. **171**, 1843–1859 (2018)
2. Bacanin, N., Bezdan, T., Tuba, E., Strumberger, I., Tuba, M., Zivkovic, M.: Task scheduling in cloud computing environment by grey wolf optimizer. In: 2019 27th Telecommunications Forum (TELFOR), pp. 1–4. IEEE (2019)
3. Bacanin, N., Tuba, E., Bezdan, T., Strumberger, I., Tuba, M.: Artificial flora optimization algorithm for task scheduling in cloud computing environment. In: International Conference on Intelligent Data Engineering and Automated Learning, pp. 437–445. Springer, Berlin (2019)
4. Bacanin, N., Tuba, E., Zivkovic, M., Strumberger, I., Tuba, M.: Whale optimization algorithm with exploratory move for wireless sensor networks localization. In: International Conference on Hybrid Intelligent Systems, pp. 328–338. Springer, Berlin (2019)
5. Bezdan, T., Tuba, E., Strumberger, I., Bacanin, N., Tuba, M.: Automatically designing convolutional neural network architecture with artificial flora algorithm. In: ICT Systems and Sustainability, pp. 371–378. Springer, Berlin (2020)
6. Bezdan, T., Zivkovic, M., Tuba, E., Strumberger, I., Bacanin, N., Tuba, M.: Glioma brain tumor grade classification from mri using convolutional neural networks designed by modified fa. In: International Conference on Intelligent and Fuzzy Systems, pp. 955–963. Springer, Berlin (2020)
7. Bezdan, T., Zivkovic, M., Tuba, E., Strumberger, I., Bacanin, N., Tuba, M.: Multi-objective task scheduling in cloud computing environment by hybridized bat algorithm. In: International Conference on Intelligent and Fuzzy Systems, pp. 718–725. Springer, Berlin (2020)
8. Brezočnik, L.: Feature selection for classification using particle swarm optimization. In: IEEE EUROCON 2017–17th International Conference on Smart Technologies, pp. 966–971. IEEE (2017)
9. Dhiman, G., Kumar, V.: Seagull optimization algorithm: theory and its applications for large-scale industrial engineering problems. Knowl.-Based Syst. **165**, 169–196 (2019)
10. Huang, C., Li, Y., Yao, X.: A survey of automatic parameter tuning methods for metaheuristics. IEEE Trans. Evolutionary Comput. **24**(2), 201–216 (2019)

11. LeCun, Y., Bengio, Y., Hinton, G.: Deep learning. Nature **521**(7553), 436–444 (2015)
12. Strumberger, I., Bacanin, N., Tuba, M., Tuba, E.: Resource scheduling in cloud computing based on a hybridized whale optimization algorithm. Appl. Sci. **9**(22), 4893 (2019)
13. Strumberger, I., Tuba, E., Bacanin, N., Beko, M., Tuba, M.: Bare bones fireworks algorithm for the rfid network planning problem. In: 2018 IEEE Congress on Evolutionary Computation (CEC), pp. 1–8. IEEE (2018)
14. Strumberger, I., Tuba, E., Bacanin, N., Jovanovic, R., Tuba, M.: Convolutional neural network architecture design by the tree growth algorithm framework. In: 2019 International Joint Conference on Neural Networks (IJCNN), pp. 1–8. IEEE (2019)
15. Strumberger, I., Tuba, E., Bacanin, N., Zivkovic, M., Beko, M., Tuba, M.: Designing convolutional neural network architecture by the firefly algorithm. In: Proceedings of the 2019 International Young Engineers Forum (YEF-ECE), Costa da Caparica, Portugal, pp. 59–65 (2019)
16. Tuba, E., Strumberger, I., Bezdan, T., Bacanin, N., Tuba, M.: Classification and feature selection method for medical datasets by brain storm optimization algorithm and support vector machine. Proc. Comput. Sci. **162**, 307–315 (2019)
17. Turkoglu, B., Kaya, E.: Training multi-layer perceptron with artificial algae algorithm. Eng. Sci. Technol. Int. J. **23**(6), 1342–1350 (2020)
18. Zivkovic, M., Bacanin, N., Tuba, E., Strumberger, I., Bezdan, T., Tuba, M.: Wireless sensor networks life time optimization based on the improved firefly algorithm. In: 2020 International Wireless Communications and Mobile Computing (IWCMC), pp. 1176–1181. IEEE (2020)
19. Zivkovic, M., Bacanin, N., Venkatachalam, K., Nayyar, A., Djordjevic, A., Strumberger, I., Al-Turjman, F.: Covid-19 cases prediction by using hybrid machine learning and beetle antennae search approach. Sustain. Cities Soc. **66** (2021)
20. Zivkovic, M., Bacanin, N., Zivkovic, T., Strumberger, I., Tuba, E., Tuba, M.: Enhanced grey wolf algorithm for energy efficient wireless sensor networks. In: 2020 Zooming Innovation in Consumer Technologies Conference (ZINC), pp. 87–92. IEEE (2020)
21. Zivkovic, M., Zivkovic, T., Venkatachalam, K., Bacanin, N.: Enhanced dragonfly algorithm adapted for wireless sensor network lifetime optimization. In: Data Intelligence and Cognitive Informatics, pp. 803–817. Springer, Berlin (2021)

GAN-Based Two-Step Pipeline for Real-World Image Super-Resolution

Rebati Raman Gaire, Ronast Subedi, Ashim Sharma, Shishir Subedi, Sharad Kumar Ghimire, and Subarna Shakya

Abstract Mostly, the prior works on single image super-resolution are dependent on high-resolution image and their bicubically downsampled low-resolution image pairs. Such methods have achieved outstanding results in single image super-resolution. Yet, these methods struggle to generalize real-world low-resolution images. Real-world low-resolution images have large varieties of degradation, and modeling these degradation accurately is a challenging task. Although some works have been proposed to address this problem, their results still lack fine perceptual details. Here, we use a GAN-based two-step pipeline to address this challenging problem of real-world image super-resolution. At first, we train a GAN network that transforms real-world low-resolution images to a space of bicubic images of the same size. This network is trained on real-world low-resolution images as input and bicubically downsampled version of their corresponding high-resolution images as ground truth. Then, we employ the nESRGAN+ network trained on bicubically downsampled low-resolution and high-resolution image pairs to super-resolve the transformed bicubic alike images. Hence, the first network transforms the wide varieties of degraded images into the bicubic space, and the second network upscales the first output by the factor of four. We show the effectiveness of this work by evaluating

R. Raman and R. Subedi: Primary authors

R. R. Gaire · R. Subedi · A. Sharma · S. Subedi · S. K. Ghimire (✉) · S. Shakya
Department of Electronics and Computer Engineering, Pulchowk Campus, IOE TU, Dharan, Nepal
e-mail: skghimire@ioe.edu.np

R. R. Gaire
e-mail: 073bct533.rebati@pcampus.edu.np

R. Subedi
e-mail: 073bct535.ronast@pcampus.edu.np

A. Sharma
e-mail: 073bct508.ashim@pcampus.edu.np

S. Subedi
e-mail: 073bct541.shishir@pcampus.edu.np

S. Shakya
e-mail: drss@ioe.edu.np

T. Senjyu et al. (eds.), *ICT with Intelligent Applications*, Smart Innovation, Systems and Technologies 248, https://doi.org/10.1007/978-981-16-4177-0_75

its output on various benchmark test datasets and comparing our results with other works. We also show that our work outperforms prior works in both qualitative and quantitative comparison. We have published our source code and trained models here for further research and improvement.

1 Introduction

Obtaining high-resolution images with refined details and better visual quality from their low-resolution counterparts is known as image super-resolution. It has important applications in many domains like object detection [1], surveillance [2], medical imaging [3], remote sensing [4], and many other computer vision tasks [5]. In recent years, research community has shown increasing attention in this field as many approaches based on deep learning have accomplished outstanding results [6] (Fig. 1).

The prior works on image SR mostly focus on generating high-resolution(HR) images from artificially downsampled low-resolution (LR) images [6]. Such methods have impressive performances on synthetic images but fail to produce similar outputs when tested with real-world LR images [7, 8].

So, inspired by the work [9], we propose a two-step pipeline for real-world image super-resolution as depicted in Fig. 2. At first, we train a GAN [10] network with Relativistic Average Discriminator [11] and perceptual loss function [12], that transforms the real-world LR images into the distribution of bicubically downsampled images of the same size. We name this network real-to-bicubic (R2B) generator. The training input for the R2B network is the real-world LR images, and the ground truth is bicubically downsampled version of their corresponding HR images. Then, in the

Fig. 1 Result of our proposed network on an image patch of DPED cellphones test dataset

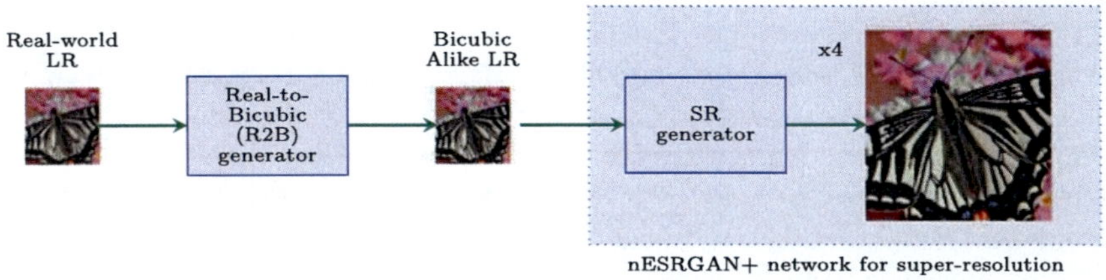

Fig. 2 A two-step pipeline for real-world image SR. First, R2B generator transforms real-world LR images to bicubic alike images. Then the nESRGAN+ network super-scales the output of the R2B generator by the scale factor of 4

second step, we super-resolve the output of R2B network by the scale factor of 4 using the nESRGAN+ [13] network trained on bicubic downsampled LR and corresponding HR image pairs. This end-to-end pipeline performs better on the real-world LR images leveraging the SOTA performance of the nESRGAN+ network. Evaluation results of our approach with other works on different test benchmark datasets are deferred in Sect. 4.3.

We summarize our work as:

1. For real-world image SR, we train a GAN-based R2B network for transforming real-world LR images to a space of bicubic images, using Relativistic Average Discriminator and VGG19-based perceptual loss function.
2. We use pre-trained nESRGAN+ model to super-resolve the output of R2B network by the scale factor of 4.
3. We show that the proposed two-step pipeline outperforms prior works in both qualitative and quantitative comparison.

2 Related Works

The prior works on single image SR include methods based on interpolation [14, 15], reconstruction [16, 17], and learning. Methods based on deep learning have accomplished better performance than former two methods.

Dong et al. [18] proposed SRCNN, the first deep learning based method. It uses CNN to map LR image to HR image in an end-to-end fashion. Later on several other network architectures, like deep network with residual learning [19], residual blocks [20], recursive learning [21], densely connected network [22], deep back projection [23], and residual dense network [24], are proposed to improve the resolution task.

To further elevate the perceptual quality of SR results, perceptual loss [12] is proposed which depends on perceptual similarity [25]. SRGAN [20], a GAN [10]-based network for single image SR task, used perceptual loss along with adversarial loss, and it produced visually appealing SR results despite having low scores on SSIM and PSNR quantitative measures. ESRGAN [13] further improved SRGAN by introducing RRDB block with higher capacity. It used residual scaling, smaller weight initialization, and removed batch normalization layers to speed training a very deep network. Furthermore, its discriminator used relativistic average GAN [11] and non-activated VGG features were taken for determining perceptual loss. nESRGAN+ [26] further enhanced ESRGAN. It introduced RRDRB block and exploited stochastic variations by injecting noise inputs to the generator. It added residual learning in the dense block to benefit from both feature exploration and exploitation, which resulted in images of better perceptual quality.

All the above-mentioned works for single image SR focus on super-resolving LR images which are generated artificially using bicubic or Gaussian downsampling kernels. Some recent works [7, 27] propose to train SR models under realistic settings by capturing real LR and HR image pairs. The authors of [7, 27] showed that the bicubically trained models perform poorly in datasets with real LR and HR pairs. In [9], the authors proposed to address the difficult problem of real-world SR with a two-step approach. First, the wide varieties of real LR images are transformed into a tractable LR space, and second, a generic bicubic SR network is used to super-resolve transformed LR images. Our work further improves this approach by using much advanced network architecture and loss functions.

3 Method

3.1 Overall Pipeline

Our method consists of two steps for real-world image super-resolution. First step transforms the real-world LR image into the space of bicubic downsampled images using a real-to-bicubic (R2B) generator. Second step uses nESRGAN+ network trained on bicubic LR and HR image pairs, for super-resolving the output of the R2B generator by the scale factor of 4. Figure 2 depicts the overall pipeline.

3.2 Generator Architecture

R2B Generator R2B generator is a CNN that aims to transform a real LR image to a bicubic alike image of the same size. It has eight residual blocks at the core, as illustrated in Fig. 3. The residual blocks have an identical layout; two convolutional layers with 64 kernels of size 3×3 and ReLU activation follow the first layer.

SR Generator We used nESRGAN+ [26] network, trained on bicubic LR and HR image pairs, as an SR generator. With multilevel residual learnings, the outputs of the nESRGAN+ model outperform the outputs of other single image SR models in visual quality.

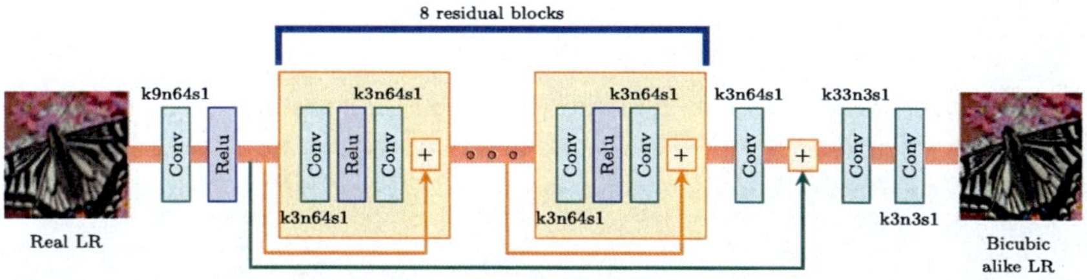

Fig. 3 Architecture of real-to-bicubic alike (R2B) generator. Here, k means kernel size, n means the number of feature maps, and s means stride size

3.3 Discriminator Architecture

For discriminating real images (x_r) from generated fake images (x_f), we trained a discriminator network based on Relativistic Average GAN [11]. Relativistic Average Discriminator (D_{Ra}) estimates the probability of the real image being more realistic than the fake image, on average. With D_{Ra}, the generator benefits from the gradients from both the real image and the generated image, as a result, learns higher detailed texture and sharper edges [13]. The outputs of D_{Ra} are defined as:

$$D(x_r, x_f) = \sigma(C(x_r) - \mathbb{E}[C(x_f)]) \rightarrow 1 \tag{1}$$

$$D(x_f, x_r) = \sigma(C(x_f) - \mathbb{E}[C(x_r)]) \rightarrow 0 \tag{2}$$

3.4 Loss Functions

The loss functions used in training are described in following section:

$$L_{\text{total}} = \alpha L_{\text{pix}} + \beta L_{\text{percept}} + \gamma L_{\text{adv}} \tag{3}$$

where L_{pix}, L_{percept}, and L_{adv} correspond to pixel, perceptual, and adversarial losses, respectively, and α, β, and γ are their respective weights.

Pixel Loss We used the $L1$ norm of the difference between pixel values of the generated image $G(x_i)$ and the ground truth image y_i as pixel loss.

$$L_{\text{pix}} = |y_i - G(x_i)| \tag{4}$$

Perceptual Loss We used the $L1$ norm of the difference of the high-level features of the generated image $G(x_i)$ and the ground truth image y_i. For this, we used the non-activated output of 5th block 4th convolution layer of pre-trained VGG19 [28] network ($\phi_{5,4}$) trained on imagenet dataset [29]. It helps to increase the perceptual quality of generated images [13]. This loss is formulated as:

$$L_{\text{percept}} = |\phi_{5,4}(y_i) - \phi_{5,4}(G(x_i))| \tag{5}$$

Adversarial Loss This loss estimates how well the generator's output x_f can fool the discriminator network [10, 20]. As described in Sect. 3.3, we used a D^{Ra} for which the loss functions are defined as in [11, 13]. The adversarial loss(L_{adv}) for discriminator is:

$$L_D^{Ra} = -\mathbb{E}_{x_r}[\log(D_{Ra}(x_r, x_f))] - \mathbb{E}_{x_f}[\log(1 - D_{Ra}(x_f, x_r))] \tag{6}$$

Similarly, for generator, adversarial loss(L_{adv}) is:

$$L_G^{Ra} = -\mathbb{E}_{x_r}[\log(1 - D_{Ra}(x_r, x_f))] - \mathbb{E}_{x_f}[\log(D_{Ra}(x_f, x_r))] \tag{7}$$

where $x_f = G(x_i)$ and, $D_{Ra}(x_r, r_f)$ and $D_{Ra}(x_f, x_r)$ are discriminator's outputs (Eqs. 1, 2).

4 Experiments

4.1 Data

The training of R2B network was performed on DIV2K [30] and RealSR [7] datasets. Specifically, we used 400 LR images from DIV2K Track 2 (downgraded with unknown operators) and 400 LR images from RealSR (version 3) as training input. We used the bicubically downsampled version of their corresponding HR images as ground truth for the training inputs. We added 400 bicubic downsampled images from DIV2K Track 1 to both training input and ground truth to prevent the over-sharpening or amplification of artifacts [9]. The R2B translation model was evaluated on DIV2K Track 2 validation set and RealSR test set.

The whole pipeline is again evaluated on the RealSR test set, DIV2K HR validation set, and DPED [31] cellphones test set. The results are presented in Sect. 4.3.

4.2 Training Details and Parameters

We cropped both the real LR image and corresponding ground truth randomly to get patches of size 128 × 128 from each image. In a mini-batch, we took 16 patches and also performed random flip and rotation. We performed the training in two steps. First, we trained the R2B network for 100,000 iterations with only the pixel loss function. Then, perceptual loss function and adversarial loss functions were added and training is continued for another 20,0000 more iterations. We set the initial learning rate to 10^{-4} and halved it in [50,000, 100,000, 200,000] iterations. The

Table 1 Quantitative comparison of the proposed model with other SOTA real-world SR models

Dataset	Metrics	Bicubic	DPSR	RealSR	RBSR	Ours
RealSR	PSNR	26.63	27.02	**28.05**	26.54	24.25
	SSIM	0.77	0.79	0.81	**0.82**	0.63
	PI	9.28	9.12	8.97	7.76	**3.11**
DIV2K HR	PSNR	–	–	–	–	–
	SSIM	–	–	–	–	–
	PI	10.02	9.36	9.19	8.48	**4.12**
DPED cellphones	PSNR	–	–	–	–	–
	SSIM	–	–	–	–	–
	PI	10.24	9.73	9.55	7.92	**3.51**

The performances of models on different benchmark datasets were evaluated with metrics: PSNR, SSIM, and PI. The highlighted results are the best ones. The symbol – implies that the ground truth is not available in the dataset

weights for loss function were set as $\alpha = 1.0$, $\beta = 3.0$ and $\gamma = 1.0$ in Eq. 3.4. We used the Adam optimizer with $\beta^1 = 0.9$ and $\beta^2 = 0.999$ to optimize this model.

4.3 Results

We used PSNR, SSIM, and PI as metrics to evaluate models' performance. The PI is a non-reference metric used in [32]. It relies on Ma's score [33] and NIQE [34] and is given by:

$$\text{PI} = \frac{1}{2} * ((10\ \text{Ma}) + \text{NIQE}) \tag{8}$$

Table 1 shows the quantitative evaluation of Bicubic Interpolation, DPSR [35], RealSR [7], RBSR [9] and our proposed model on RealSR test set, DIV2K HR validation set, DPED cellphones test set, respectively. Figure 4 shows the qualitative evaluation on different images from RealSR test set.

The result of evaluation of our R2B model on RealSR test set and DIV2K Track 2 validation set is depicted in Table 2.

5 Conclusion

Here, we presented a two-step pipeline for addressing the problem of real-world single image super-resolution. Our proposed approach breaks the problem into two well-posed steps and benefits from the impressive performance of SR network trained on bicubically downsampled LR and HR image pairs. First, we performed the trans-

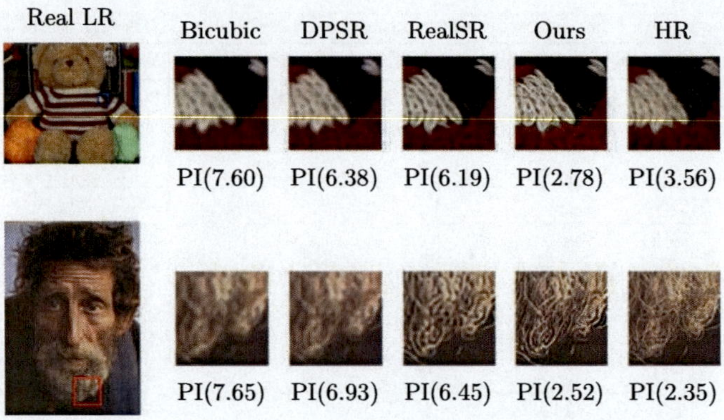

Fig. 4 Qualitative comparison of the proposed model's outputs with other SOTA models' outputs on a subset of RealSR images. The value in the bracket indicates the PI of the image

Table 2 Quantitative evaluation of the R2B model on RealSR test-set and DIV2K Track 2 validation set

Dataset	Metrics	R2B Results
RealSR	PSNR	31.11
	SSIM	0.92
	PI	3.59
DIV2K Track 2	PSNR	27.28
	SSIM	0.86
	PI	2.62

formation of LR images to the space of bicubic distribution of the same size. The network was trained with more robust architecture and loss function. Second, we performed the super-resolution of the transformed images using nESRGAN+ network, which is a network for SR of bicubically downsampled LR images. The result Sect. 4.3 shows that our approach is better than other works on real-world image SR, both quantitatively and qualitatively. Yet, the proposed method has some shortcomings. First, some of the generated images suffer from uneven artifacts. Also, no significant difference can be observed in the output of relatively better quality input image. Further investigation and experimentation is necessary to promote the performance of this approach. Therefore, we have provided source code and trained models to facilitate further research and improvement of this work and also to expedite researches in similar domains.

References

1. Haris, M., Shakhnarovich, G., Ukita, N.: Task-driven super resolution: object detection in low-resolution images. arXiv preprint arXiv:1803.11316 (2018)
2. Rasti, P., Uiboupin, T., Escalera, S., Anbarjafari, G.: Convolutional neural network super resolution for face recognition in surveillance monitoring. In: International Conference on Articulated Motion and Deformable Objects, pp. 175–184. Springer (2016)
3. Greenspan, H.: Super-resolution in medical imaging. Comput. J. **52**(1), 43–63 (2009)
4. Lillesand, T., Kiefer, R.W., Chipman, J.: Remote Sensing and Image Interpretation. Wiley (2015)
5. Dai, D., Wang, Y., Chen, Y., Van Gool, L.: Is image super-resolution helpful for other vision tasks? In: 2016 IEEE Winter Conference on Applications of Computer Vision (WACV), pp. 1–9. IEEE (2016)
6. Anwar, S., Khan, S., Barnes, N.: A deep journey into super-resolution: a survey. arXiv preprint arXiv:1904.07523 (2019)
7. Cai, J., Zeng, H., Yong, H., Cao, Z., Zhang, L.: Toward real-world single image super-resolution: a new benchmark and a new model. In: Proceedings of the IEEE International Conference on Computer Vision, pp. 3086–3095 (2019)
8. Lugmayr, A., Danelljan, M., Timofte, R.: Unsupervised learning for real-world super-resolution. In: 2019 IEEE/CVF International Conference on Computer Vision Workshop (ICCVW), pp. 3408–3416. IEEE (2019)
9. Rad, M.S., Yu, T., Musat, C., Ekenel, H.K., Bozorgtabar, B., Thiran, J.P.: Benefitting from bicubically down-sampled images for learning real-world image super-resolution. arXiv preprint arXiv:2007.03053 (2020)
10. Goodfellow, I.J., Pouget-Abadie, J., Mirza, M., Xu, B., Warde-Farley, D., Ozair, S., Courville, A., Bengio, Y.: Generative adversarial networks (2014)
11. Jolicoeur-Martineau, A.: The relativistic discriminator: a key element missing from standard gan. arXiv preprint arXiv:1807.00734 (2018)
12. Johnson, J., Alahi, A., Fei-Fei, L.: Perceptual losses for real-time style transfer and super-resolution. In: European Conference on Computer Vision, pp. 694–711. Springer (2016)
13. Wang, X., Yu, K., Wu, S., Gu, J., Liu, Y., Dong, C., Qiao, Y., Change Loy, C.: Esrgan: Enhanced super-resolution generative adversarial networks. In: Proceedings of the European Conference on Computer Vision (ECCV), pp. 0–0 (2018)
14. Keys, R.: Cubic convolution interpolation for digital image processing. IEEE Trans. Acoust. Speech, Sig. Proc. **29**(6), 1153–1160 (1981)
15. Duchon, C.E.: Lanczos filtering in one and two dimensions. J. Appl. Meteorol. Climatol. **18**(8), 1016–1022 (1979)
16. Marquina, A., Osher, S.J.: Image super-resolution by tv-regularization and bregman iteration. J. Sci. Comput. **37**(3), 367–382 (2008)
17. Sun, J., Xu, Z., Shum, H.Y.: Image super-resolution using gradient profile prior. In: 2008 IEEE Conference on Computer Vision and Pattern Recognition, pp. 1–8. IEEE (2008)
18. Dong, C., Loy, C.C., He, K., Tang, X.: Learning a deep convolutional network for image super-resolution. In: European Conference on Computer Vision, pp. 184–199 (Springer, 2014)
19. Kim, J., Kwon Lee, J., Mu Lee, K.: Accurate image super-resolution using very deep convolutional networks. In: Proceedings of the IEEE Conference on Computer Vision and Pattern Recognition, pp. 1646–1654 (2016)
20. Ledig, C., Theis, L., Huszár, F., Caballero, J., Cunningham, A., Acosta, A., Aitken, A., Tejani, A., Totz, J., Wang, Z., et al.: Photo-realistic single image super-resolution using a generative adversarial network. In: Proceedings of the IEEE Conference on Computer Vision and Pattern Recognition, pp. 4681–4690 (2017)
21. Kim, J., Kwon Lee, J., Mu Lee, K.: Deeply-recursive convolutional network for image super-resolution. In: Proceedings of the IEEE Conference on Computer Vision and Pattern Recognition, pp. 1637–1645 (2016)

22. Tai, Y., Yang, J., Liu, X., Xu, C.: Memnet: a persistent memory network for image restoration. In: Proceedings of the IEEE International Conference on Computer Vision, pp. 4539–4547 (2017)
23. Haris, M., Shakhnarovich, G., Ukita, N.: Deep back-projection networks for super-resolution. In: Proceedings of the IEEE Conference on Computer Vision and Pattern Recognition, pp. 1664–1673 (2018)
24. Zhang, Y., Tian, Y., Kong, Y., Zhong, B., Fu, Y.: Residual dense network for image super-resolution. In: Proceedings of the IEEE Conference on Computer Vision and Pattern Recognition, pp. 2472–2481 (2018)
25. Gatys, L., Ecker, A.S., Bethge, M.: Texture synthesis using convolutional neural networks. Adv. Neural Inf. Proc. Syst. **28**, 262–270 (2015)
26. Rakotonirina, N.C., Rasoanaivo, A.: Esrgan+: Further improving enhanced super-resolution generative adversarial network. In: ICASSP 2020–2020 IEEE International Conference on Acoustics, Speech and Signal Processing (ICASSP), pp. 3637–3641. IEEE (2020)
27. Chen, C., Xiong, Z., Tian, X., Zha, Z.J., Wu, F.: Camera lens super-resolution. In: Proceedings of the IEEE Conference on Computer Vision and Pattern Recognition, pp. 1652–1660 (2019)
28. Simonyan, K., Zisserman, A.: Very deep convolutional networks for large-scale image recognition. arXiv preprint arXiv:1409.1556 (2014)
29. Deng, J., Dong, W., Socher, R., Li, L.J., Li, K., Fei-Fei, L.: Imagenet: a large-scale hierarchical image database. In: 2009 IEEE Conference on Computer Vision and Pattern Recognition,pp. 248–255. IEEE (2009)
30. Agustsson, E., Timofte, R.: Ntire 2017 challenge on single image super-resolution: Dataset and study. In: Proceedings of the IEEE Conference on Computer Vision and Pattern Recognition Workshops. pp. 126–135 (2017)
31. Ignatov, A., Kobyshev, N., Timofte, R., Vanhoey, K., Van Gool, L.: Dslr-quality photos on mobile devices with deep convolutional networks. In: Proceedings of the IEEE International Conference on Computer Vision, pp. 3277–3285 (2017)
32. Blau, Y., Mechrez, R., Timofte, R., Michaeli, T., Zelnik-Manor, L.: The 2018 pirm challenge on perceptual image super-resolution. In: Proceedings of the European Conference on Computer Vision (ECCV) Workshops, pp. 0-0 (2018)
33. Ma, C., Yang, C.Y., Yang, X., Yang, M.H.: Learning a no-reference quality metric for single-image super-resolution. Comput. Vis. Image Understandi. **158**, 1–16 (2017)
34. Mittal, A., Soundararajan, R., Bovik, A.C.: Making a "completely blind" image quality analyzer. IEEE Sig. Proc. lett. **20**(3), 209–212 (2012)
35. Zhang, K., Zuo, W., Zhang, L.: Deep plug-and-play super-resolution for arbitrary blur kernels. In: Proceedings of the IEEE/CVF Conference on Computer Vision and Pattern Recognition, pp. 1671–1681 (2019)

A Rule-Based Approach for Marathi Part-of-Speech Tagging

P. Kadam Vaishali, Khandale Kalpana, and C. Namrata Mahender

Abstract This paper focused on developing the POS-tagger for Marathi. It is one of the very popular Indian languages spoken by the Marathi people. It has its semantic richness and standard in the literature and culture of Maharashtra. We deploy a technique to find Marathi words for their type, such as noun, verb or adjective, and so on. This task is carried out manually and marked in a corpus consisting of words already tagged with their corresponding part-of-speech. This system uses a rule-based approach based on the Marathi transformational grammar. It is important for preprocessing and developing NLP applications. In the absence or less information available related to other phrases and the possible existence of lexical or syntactic mistakes in the training corpus, our proposed system identifies a correct tag and finds its impact on their performance to verify usability for NLP applications. The overall accuracy of the system is 97.56%.

1 Introduction

Natural language processing (NLP) is a subfield and an important part of computer science, human communication language and artificial intelligence. This technology is used by machines to understand, analyze, manipulate and interpret human languages. NLP is used to organize knowledge and tasks such as language translation, automatic summarization, named entity recognition (NER), speech recognition, and relationship extraction and topic segmentation, etc. NLP came into existence in early 1958. It is also known as computational linguistics. [1] The study of natural languages is important to improve the social and cultural standard also for developing new language technologies to provide communication and related services like human–computer interaction, business information analysis, Web software development, information retrieval and so on. NLP algorithms are designed

P. K. Vaishali (✉) · K. Kalpana · C. Namrata Mahender
Department of Computer Science and I.T., Dr.Babasaheb Ambedkar Marathwada University, Aurangabad, Maharashtra, India
e-mail: cnamrata.csit@bamu.ac.in

T. Senjyu et al. (eds.), *ICT with Intelligent Applications*, Smart Innovation, Systems and Technologies 248,
https://doi.org/10.1007/978-981-16-4177-0_76

for machine learning. Instead of hand-coding large sets of rules, NLP can rely on machine learning to automatically learn these rules by analyzing a set of examples. Few works have been done on POS-tagging for the Marathi language, but it is difficult because of the unavailability of the corpus for computational linguistic processing. To deal with large data sets, a powerful language processing tool is required to fulfill this demand. We designed a rule-based POS-tagging system that can assign all possible tags to the words in a given input sentence. It uses a set of handwritten grammar rules and a Marathi word dictionary.

1.1 Need of POS-Tagging in NLP

POS-tagging is currently under the developing stage for various Indian languages. Most work that has been done was one of the rule-based and other empirical-based. The rule-based system requires proper language knowledge and handwritten rule. Researchers faced a great problem to write proper linguistic rules due to the morphological effect of Indian languages. It was also noticed that the results were not good. The main issue is ambiguity. It is a time-consuming process to assign a correct POS tag to different context words. Due to this reason, POS-tagging is becoming a challenging problem for study in the field of NLP.

1.2 POS-Tagging

It is a process of automatic assignment of grammatical labels to each word or token in the text. This label is called a tag or a lexical notation. Marathi words are classified under eight different categories like नाम (Noun), सर्वनाम (Pronoun), वशिषण (Adjective), क्रयिापद (Verb),क्रयिावशिषण (Adverb), उभयान्वयी अव्यय (Postposition),शब्दयोगी अव्यय.(Conjunction) and केवलप्रयोगी अव्यय or उदगारवाचक (Interjection).

This classification is based on the context of the word, relationship with other adjacent words and position in the sentence. Tagging broadly classified into two different types, supervised and unsupervised. Supervised systems have their pre-defined tagged corpora or text, but unsupervised systems are capable of automatically assigns tags to the words. Based on the further classification, which approach is used for tagging, it has three different subtypes. The below Table 1 shows POS-tagging techniques briefly.

POS-tagging is a basic preprocessing tool needed for the development of various NLP applications. [3] There is no much work that has been done on Marathi because of the different complexity associated with Marathi text, but for other languages like Hindi, English, it is easily available. There are many scenarios where the same words have a different meaning in the Marathi language, so it is still a difficult task.

Table 1 POS-tagging technique for Indian languages

Techniques	Description	Advantages	Disadvantages
Rule-based [1]	It utilizes a set of manually designed handwritten rules	Use of a small set of rules. It requires less information storage	Generally less accurate as compared to stochastic taggers
Stochastic [1]	It assigns a tag using frequency, probability, or statistical formulations. It uses probabilistic value N, called unigram, bigrams or trigram frequencies in a training corpus	Generally more accurate as compared to rule-based taggers	A large set of predefined contextual information to avoid ambiguities Relatively complex. Vast amounts of stored information
Hybrid [2]	It uses statistical, lexical, or structural approach together to assign a tag. If the wrong tag exists then by applying rules tagger tries to change it	Higher accuracy than individual rule-based or statistical approach	Not assign a correct tag to an unknown word

For example

मला सुपारी लागते. (I want betel nut/betel nut is unbecoming to my health).

The given sentence creates ambiguity to understand the meaning of the sentence because 'लागते' can be used as an adverb and also as a verb, but by transformational Marathi grammar rules, it should be a verb, but it is an adverb. NLP applications need a better understanding for giving better performance. For designing such a system, a good grammar tool is a basic requirement. A rule-based POS-tagging system assigns all possible tags to each word using a set of handwritten rules. It is quite difficult for the Marathi language because of the unavailability of standard corpus and ambiguous words for computational processing. In Marathi, many words can have more than one meaning or they caused to create ambiguity.

1.3 Rule-Based POS-Tagging

Rule-based POS-tagging based on hand-written rules and a Marathi words dictionary from which the system provides all the possible tags for every word to be tagged. Marathi language rules are used to identify the most appropriate tag when the same word has multiple possible tags. The ambiguity of such words can be removed by correct identification and analysis of the linguistic features of that word such as its position in the sentence, its proceeding word, its succeeding word and another context that generates some meaning. All these information are coded in the form of a rule. The function of the rule-based POS-tagging system is divided into two steps.

Step 1—It uses a dictionary to assign each word a list of parts-of-speech or tagging labels.
Step 2—It uses a large set of handwritten grammar rules used to remove ambiguities, and to search for an appropriate single part-of-speech for the word which creates ambiguity.

1.3.1 Rules for Generating Tag

There are eight different types of words in Marathi and having their respective subtypes. The knowledge-based rules have been designed for tagging each word in the text. Table 2 shows types of part-of-speech and their definitions for the tagging.

Table 2 POS-tagging for language

No.	Type of part-of-speech	Tag/label	Definition
I	Noun phrase	NP	Group of one or two words headed by a noun that includes a modifier
	(a) Common noun	NN	A word referring to a person, place, or thing but is not in a particular sense
	(b) Proper noun	NNP	A word referring to a particular person, place, or thing
	(c) Abstract noun	NST	A word refers to an idea, condition, quality, or any abstract thing
	(d) Verbal noun	NV	A noun is derived from a verb that exhibits the ordinary properties of a noun
II	Pronoun	PRP	A word that is used as a substitute for a noun
	(a) Personal pronoun	PPN	A word used instead of Person. In terms of first, second, third 'person' in the grammar rule
	(b) Possessive pronoun	PPS	Words designating possession
	(c) Demonstrative pronoun	PDM/ DEM	A word pointing to something specific, singular/plural
	(d) Reflexive pronoun	PRF	A word preceded by a noun, pronoun, verb, or adverb is referred to as long as that antecedent is present in the sentence
	(e) Reciprocal pronoun	PRC	A word denoting mutual action or relationship
	(f) Interrogative pronoun	PIN	A pronoun is used to ask questions
III	Adjective	JJ	A word that modifies another word that makes writing and speaking more specific
IV	Conjunction	CC	A word connecting to two words or phrases or sentences

(continued)

Table 2 (continued)

No.	Type of part-of-speech	Tag/label	Definition
V	Verb phrase	VP	A word used to express an action or state of being
	(a) Verb main	VM	It is an action word of the main clause that the subject completes
	(b) Verb auxiliary	VAUX	A helping verb denoting a verb's action
VI	Adverb	RB	A word that can change another word and modify nouns. Denoting verbs actions frequency, manner, time, or location
VII	Interjection	INJ	A word independent of grammar and from other words around it mainly expresses feelings
VIII	Numeral	NP	A word that represents a quantity
	(a) Cardinal numeral	QC	A word representing things in quantity
	(b) Ordinal numeral	QO	A word representing a position or rank
IX	Reduplication	RDP	A word morphology repeated exactly or with a slight change
X	Compounds	XC	A word composed of one or more free morpheme. Classified in word classes or semantic relationship
XI	Negative	NEG	A word converting an affirmative into its opposite form
XII	Quantifiers	QF	A word determiner indicative of quantity
XIII	Question words	WQ	A word used to ask a question,
XIV	Intensifier	INTF	A word or an adverb that strengthens or weakens another word
XV	Particles	RP	A word or preposition performing grammar function. But not fit into the main noun, verb or adverb

1.4 Tag Set for the Marathi Language

Above grammar rules are used to generate the tag set for Marathi. The word grammar has its types and subtypes. Table 4 shows the tag set for the Marathi language.

Table 3 POS-tagging rules [2]

Part-of-Speech	Subtype	Rules	Example
Noun Phrase (NP)	Common Noun (NN)	If (word devotes a class entity) or(the word is an unidentical class) then tag as NN Else: None	तो एक मुलगा आहे He is a boy ही सर्वात मोठी नदी आहे It is the largest River
	Proper Noun (NNP)	If (word indicates single unit/entity/ object/person) Then tag as PPN Else: None	तो हिमालयआहे It is the Himalayas गुलाब एक फूलआहे Rose is a flower तो राहुल आहे He is Rahul
	Abstract Noun (NST)	If (word indicates intangible concept/ things) then tag as NST Else: None	हा तुमचा चांगुलपणा आहे This is your goodness राग माणसाचा शत्रू आहे Anger is the enemy of man
	Verbal Noun (NV)	If(noun formed using verb) Or (verb identical to the verb) Then tag as NV Else: None	धावणे चांगला व्यायाम आहे Run → Running is good exercise चालणे आरोग्यासाठी चांगले आहे Walk → Walking is good for health मी आज परत येईल Return (Verb) → I will return today मी परतल्यावर तुला कॉल करेन Return (Noun) → I will call you on my return
Verb Phrase (VP)	Verb Main (VM)	If (word /verb directly linked to the subject) then tag as VM Else: None or if (word completes action) then tag VM Else: None	मीना सीशेल्सची विक्री करते Meena sells seashells
	Verb Auxiliary (VAUX)	If (word helping the main verb) Then tag as VAUX Else: None	सचिविांनी अद्याप सर्व पत्रे लहिलिी नाहीत The secretaries haven't written all the letters yet आई रोज मंदरिात जाऊन येते The mother goes to the temple every day
Adverb	RB	If (the preceding word is the verb) Then tag as RB Else: None	ती मुलगी चांगली गाते That girl sings well मुलगा जलद चालतो The boy walks fast
Adjective	JJ	If (succeeding word is the noun) Then tag as JJ Else: None	सीता एक हुशार मुलगी आहे Seeta is a clever girl त्यांच्याकडे एक सुंदर बाग आहे They have a beautiful garden

Table 4 Tag set for the Marathi language [3, 4]

Name of part-of-speech	Tag/label	Meaning	For example
Noun	NN	Common noun	मुलगी,शक्षिक,कागद,नदी, कापड
	NNP	Proper noun	नति,इशा, वहिान,भारत
	NST	Abstract noun	मागे, पुढे, वर, खाली,नवलाई
	NV	Verbal nouns	गुलामगरिी,शाबासकी,फुलांनो,मुलांनो
Pronoun	PPN	Personal pronoun	मी,आम्ही, तू, तुम्ही
	PPS	Possessive pronoun	माझा,माझी,तुझा,तुझे
	PIN	Interrogative pronoun	का,काय,कसे,कोण,कसले,कुणाला
	PDM/DEM	Demonstrative pronoun	हा,ही,हे,तो,ती,ते,हया,त्या
	PRC	Reciprocal pronoun	एकमेकांचा,एकमेकाला,आपल्याला
	PRF	Reflexive pronoun	तुम्हाला,आम्हाला, आपण,आम्ही
Verb	VM	Verb main (Finite or nonfinite)	बसणे, दसिणे, लहिणि
	VAUX	Verb Auxiliary	नाही, नको, करणे,हवे, नये
Adjective	JJ	Adjective (modifier of noun)	उंच,उत्साही, श्रेष्ठ,धाडसी
Adverb	RB	Modifier of verb	आता, काल, कधी, नेहमी
Postposition	PSP	Postposition	कडे,आणि, वर,पासून,च्यामुळे
Interjection	INJ	Interjection	आहा, छान, अगो, हाय,वाह
Conjunction	CC	Coordinating/ Subordinating	आणि, व, जर, तर, जेब्हा,केव्हा, तेव्हा
Numeral	QC	Cardinal numeral	एक, दोन, तीन,
	QO	Ordinal numeral	पहलिा,दुसरा, तसिरा
Reduplication	RDP	Reduplications	मधूनमधून,सारखासारखा,जवळजवळ
Compounds	XC	Compounds	कालचक्र,काळचक्र,तेलपाणी, उठाठेव,
Negative	NEG	Negative	नकोस,नको,नाही,ना
Quantifiers	QF	Quantifiers	बख्खळ,कम,बहुत, थोडा, जास्त
Question word	WQ	Question words	केव्हा,कुणाला,कशाला,काय, कधी, कुठे
Intensifier	INTF	Intensifier	पुरेसे,खूप, फार, पुष्कळ
Particles	RP	Particles	भी, तो, ही

2 Literature Survey

In this section, the literature survey is done that uses the various POS-tagging techniques for Indian languages. Most of them concentrate on the rule-based rather than a statistical approach. The small set of meaningful rules provides better results. It shows the comparison of the POS-tagging with the illustration of the dataset. It also concludes how much work is done on that language (Table 5).

Table 5 Literature survey on POS-tagging

No	Language	Approach	Accuracy
1	Hindi, Bangla, Telugu [5]	Transition-based MALT parser	UAS:H-90%, B-90% and T-85% LAS: 15–25%, Low
2	Hindi [6]	Grammar-driven approach	Precision-96.2%, Recall-82.6%
3	Hindi [4, 7]	HMM and error-driven learning by conditional random fields (CRF)	69.4% for HMM, 78.94% for error-driven learning and 80.74% for CRF
4	Bengali [8]	Statistical approach, POS-taggers using HMM and SVM	85.56% for HMM and 91.23% for SVM
5	Hindi, Telgu Bengali [9] Telugu[9]	POS-Tagger based on HMM	82.05% for Hindi, 63.93% for Telgu, and 90.90% for Bengali
6	Telugu [10]	Hybrid POS-Tagger	Precision 58.2% and Recall 58.2%
7	Marathi [11]	Statistical approach with Unigram, Bigram, Trigram, HMM methods	Unigram 77.39%, Bigram 90.30%, Trigram 91.46%., HMM 93.82

3 Database

For the database, we have used 25 moral stories which have available online. From the 25 stories, we collected 5012 words. For the proposed work, we have trained our system with a total of 5012 words. Table 6 gives a sample of the stories.

Table 6 Sample of the stories

Story label	Total words in a story
St1 [12]	344
St2 [13]	156
St3 [14]	156
St4 [15]	128
St5 [16]	132
St6 [17]	142
St7 [18]	170
St8 [19]	132
St9 [20]	131
St10 [21]	126

4 Proposed System

Below Fig. 1 shows the architecture of the part-of-speech tagger for Marathi. There are eight types of part-of-speech available in Marathi, namely noun, verb, adjective, adverb, etc. Marathi is free-ordered language; around 90 million native speakers speak Marathi. For English, the order of the sentence is Subject-Verb-Object (SVO) and for Marathi, the order is Subject-Object-Verb (SOV) which is more difficult. Rules are based on the Marathi transformational grammar which has been helped us to tag each word correctly.

4.1 Input Source Document

For input, we have concentrated on the moral stories of the Marathi language.

Fig. 1 Proposed system

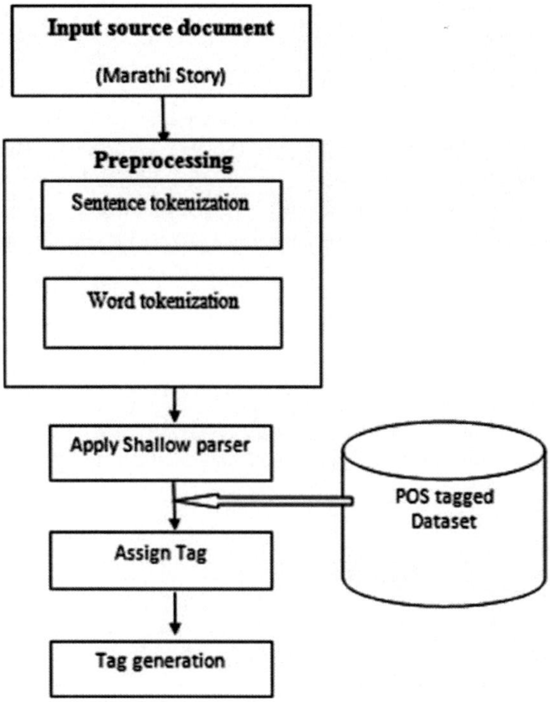

4.2 Preprocessing

In preprocessing, we tokenize the sentences as well as the words individually for a better understanding of the text.

(a) **Sentence Tokenization**
Sentence tokenization is defined as the discourse is divided into a specific sentence. The following example [12] shows the sentence tokenization.

'एक होता उंदीर', 'तो एकदा फिरायला निघाला.'

(b) **Word Tokenization**
After sentence tokenization, we need to tokenize or split the words individually. Sentences are split into the number of individual words or tokens from input text called word tokenization. For example [12],

'एक', 'होता', 'उंदीर', 'तो', 'एकदा', 'फिरायला', 'निघाला',

4.3 Apply Shallow Parser

After tokenization, we apply the IIIT-Hyderabad Marathi shallow parser to check the accuracy of the parser. For each tokenized word, shallow parser generates a tag for its type. But, the result is not accurate for some words. For this reason, we need to develop our part-of-speech tagger for Marathi to tag each word appropriately.

4.4 Assign Tag

Some words are not tagged correctly with the shallow parser. We have assigned the tag to the incorrect word and tagged it correctly.

4.5 Tag Generation

After we assigned the correct tag with the help of the grammar rules, the output will be shown as in Table 7.

Table 7 Rule-based POS-tagging sample example

Example	एक	होता	उंदीर	तो	एकदा	फरियला	नघिाला
Shallow parser	QC	VM	JJ	PRP	RB	VM	VAUX
Rule-based system	QC	VAUX	NN	DEM	RB	RB	VM

5 Results

For the proposed work, we train our system with the handwritten rules for developing the part-of-speech tagger for Marathi text. From the stories, we have collected a total of 5012 words. From that, 4890 words are correctly tagged by our rule-based POS-tagging system, and our system is unable to tag 122 words. The formula has been given to calculate the accuracy.

$$\text{Accuracy} = \frac{\text{Number of Correct Tagged Words}}{\text{Number of Correct Tagged Words} + \text{Number of Incorrect Tagged Words}} \times 100$$

The accuracy of the system is 97.56% which is better than the shallow parser of IIT-Hyderabad (Table 8).

6 Conclusion

POS-tagging plays an important role in every NLP application. Marathi is a free-ordered language. There are different kinds of POS-tagger available for various other languages like Hindi, Bengali and English, but for Marathi, very little work is done on the POS-tagger. To achieve better accuracy in tagging Marathi natural language text, we have designed the rule-based approach for the Marathi which is based on the Marathi language grammar rules. In the comparison of the IIT-Hyderabad shallow parser, our system gives better accuracy. In the future, we have to concentrate on enhancing the database and more analysis of large grammar rule sets that can satisfy maximum language constraints for tagging the text and to improve the performance of the POS-tagger. We are planning to use various other techniques or a hybrid approach to achieve this goal. We will also try to integrate this with the Marathi WordNet application to increase the database size and make it available online.

Table 8 Result obtained for POS-tagging system

Name of system	Total words	Correct tagged words	Incorrect tagged words	Accuracy (%)
Shallow parser	5012	4658	354	92.93
Our system	5012	4890	122	97.56

Acknowledgements The authors would like to acknowledge and thank to Chhatrapati Shahu Maharaj Research Training and Human Development Institute (SARTHI), Pune. They awarded fellowship, CSRI DST Major Project sanctioned No.SR/CSRI/71/2015(G), and also thanks to Computational and Psycholinguistic Research Lab Facility supporting this work and Department of Computer Science and Information Technology, Dr. Babasaheb Ambedkar Marathwada University, Aurangabad, Maharashtra, India.

References

1. Ekbal A., Mandal S.: POS tagging using HMM and rule-based chunking. In Proceedings of International Joint Conference on Artificial Intelligence Workshop on Shallow Parsing for South Asian Languages. IIIT Hyderabad, Hyderabad, India (2007)
2. Bagul, P., Mishra, A., Mahajan, P., Kulkarni, M., Dhopavkar, G.: Rule-based POS tagger for Marathi Text. Proc. Int. J. Comput. Sci. Inf. Technol. **5**(2), 1322–1326 (2014)
3. Govilkar, S.: Part of speech tagger for the Marathi language. Int. J. Comput. Appl. **119**(18), 0975–8887 (2015)
4. Singh, J., Joshi, N, Mathur, I.: Development of Marathi part of speech tagger using statistical approach. In: Advances in Computing, Communications and Informatics (ICACCI) (2013)
5. Nirve, J.: Parsing Indian languages with Malt parser. In: Proceedings of the 5th International Conference on Languages, Resources and Evaluation (LERC), pp. 2216–2219 (2009)
6. Bharati, A., Gupta, M., Yadav, V., Gali, K., Sharma, D.M.: Simple parser for Indian languages in a dependency framework. In: Proceedings of the Third Linguistic Annotation Workshop. Association for Computational Linguistics (2009)
7. Rao, A.P.D., Ravindran B.: Part of speech tagging and Chunking with HMM and CRF. In: Proceedings of NLPAI Machine Learning Workshop on Part of Speech Tagging and Chunking for Indian Languages. IIIT Hyderabad, India (2006)
8. Asif, E., Shivaji, B.: Web-based Bengali news corpus for lexicon development and POS tagging. In: Proceeding of Language Resource and Evaluation (2008)
9. Ekbal, A., Mandal, S.: POS tagging using HMM and rule-based chunking. In: Proceedings of International Joint Conference on Artificial Intelligence Workshop on Shallow Parsing for South Asian Languages. IIIT Hyderabad, India (2007)
10. Pattabhi, R.K., Sundar, R., Ram, R.V., Krishna, R.V., Sobha, L.: A text chunker and hybrid POS tagger for Indian languages. In Proceedings of International Joint Conference on Artificial Intelligence Workshop on Shallow Parsing for South Asian Languages. IIIT Hyderabad, India (2007)
11. Singh, J., Joshi, N., Mathur, I.: Part of speech tagging of Marathi text using trigram method. Int. J. Adv. Inf. Technol. (IJAIT) **3**(2) (2013). https://doi.org/10.5121/ijait2013.3203
12. Undarachi Topi Stories in Marathi. http://marathikavitasangrah.in/2018/01/undarachi-topi-stories-in-marathi (2018). Accessed date (2018/1)
13. टोपीवाला आर्णि माकडे—Webdunia Marathi. https://marathi.webdunia.com/article/marathi-kids-टोपीवाला-आर्णि-माकडे.html. (2020). Accessed date (2020/1)
14. कुत्र्याची हुशारी–Webdunia Marathi. https://marathi.webdunia.com/article/marathi-kids-stories/kids-story-115062900024_1.html (2020). Accessed date (2020/1)
15. धक्का लागल्याने कळून येत व्यक्तमित्व-Webdunia Marathi. http://marathi.webdunia.com/article/marathi-kids-stories/guru-shiya-kid-stroy-in-marathi-118082300008_1.html (2020). Accessed date (2020/1)
16. गणपतीने आपल्या बुद्धमित्तेने -Webdunia Marathi. https://marathi.webdunia.com/article/marathi-kids-stories/ganesh-kartikeya-prithivi-parikrama-story-118091800013_1.html (2020). Accessed date (2020/1)
17. उपकाराचेस्मरणकरणेहेचमाणसाचे-Webdunia Marathi. https://marathi.webdunia.com/article/marathi-kids-stories/kids-story-18091900009_1.html (2020). Accessed date (2020/1)

18. बालसंस्कार—Marathi Katha.https://www.hindujagruti.org/hinduism-for-kids-marathi/1449.html (2020). Accessed date (2020/1)
19. Moral stories in Marathi on Trees-वेडेवाकडेझाड. http://marathikavitasangrah.in/2017/03/moral-stories-in-Marathi-on-trees-वेडेवाकडे-झाड.html. (2020). Accessed date (2020/1)
20. Moral stories in Marathi on peacock-गर्वष्ठिमोरशहाणाकरकोचा. http://marathikavitasangrah.in/2017/04/moral-stories-in-marathi-on-peacock-गर्वष्ठि-मोर-शहाणा-करकोचा.html (2020). Accessed date (2020/1)
21. बालसंस्कार –Marathi Katha.सकारात्मकदृष्टीकोणकसाअसावा.https://www.hindujagruti.org/hinduism-for-kids-marathi/1481.html 11/21 (2020). Accessed date (2020/1)
22. Sharvari, G., Bakal J.W.: Shubhangi Rathod: Part-of-speech tagger for Marathi. Int. J. Comput.Appli. (0975–8887) vol. 119-No.18 (2015)

Efficient PBFT: A Novel and Efficient Approach to the PBFT Consensus Algorithm

Vishakh Rao, Amodh Ranjal Shenoy, and M. Kiran

Abstract Blockchain technology, which is built upon the distributed and decentralised network, suffers from mutual trust among the nodes, and hence, it employs a consensus algorithm to achieve democracy. Consensus algorithm is one of the core elements of blockchain architecture, and a strong, reliable and efficient consensus algorithm is very much required for a blockchain network in order to prevent the intrusion of malicious nodes. A good consensus algorithm drives the blockchain network in the right direction by keeping the malicious transactions off the chain. Thus, in this article, a novel consensus algorithm referred to as efficient practical Byzantine fault tolerance (E-PBFT) is proposed for a private blockchain. E-PBFT finds the winner miner for mining the candidate blocks in an efficient manner which reduces the number of messages exchanged between the nodes. Further, in order to give additional security, a hidden winner miner concept is also included in E-PBFT which hides the winner miner from the adversaries, so that the adversaries will never come to know who is mining the transactions. The theoretical proof for the proposed algorithm is also given whose results are quite convincing when compared to classical PBFT algorithm. The proposed E-PBFT will definitely strengthen the blockchain network by quickly reaching the consensus and by hiding the winner miner, thereby giving additional security.

1 Introduction

Blockchain technology, the underlying concept of bitcoin, is a decentralised and distributed open ledger which has an immutable characteristic. Once the transactions are added in to the chain, it cannot be modified or deleted. Unlike centralised architecture, the blockchain architecture is democratic in nature; majority will drive the network. Blockchain network is governed by its two pillars, namely smart contracts and consensus algorithms. Since the blockchain network is decentralised, reaching

V. Rao (✉) · A. R. Shenoy · M. Kiran
National Institute of Technology Karnataka, Surathkal, Karnataka, India
e-mail: kiranmanjappa@nitk.edu.in

T. Senjyu et al. (eds.), *ICT with Intelligent Applications*, Smart Innovation, Systems and Technologies 248, https://doi.org/10.1007/978-981-16-4177-0_77

to a consensus among the geographically separated nodes for transaction/candidate block acceptance or rejection is the most trivial part of the blockchain network. The user anonymity and adversaries further complicate the consensus process. Thus, consensus algorithm plays a vital role in the blockchain network in arriving at a decision. The consensus algorithm should ensure the integrity of the transactions which are accepted and added in to the main chain, and it is a trade between time and the correctness. The consensus algorithm should not take too much time to achieve consensus, and on the other side, it should not accept wrong transactions while compromising with the time.

The blockchain's revolutionary concept relies on the Web and online services. Though blockchain was primarily invented and used for cryptocurrencies, nowadays, other non-cryptocurrency applications are also exploring blockchain technology such as medical applications, supply chain management, security and many FinTech activities. Other routine processes that were done offline were also automated using blockchain technologies [1]. Further, many organisations have shown interest in blockchain technology for their internal activities as well. This has created a huge increase in the requirements and usage of private blockchains and has made it a topic of interest for many researchers.

A private blockchain is more restricted when compared to a public blockchain. In a private blockchain, only certain authorised entities can participate and control the blockchain network. The order of magnitude of transactions in the private blockchain is greater than that of a public blockchain [3].

This is possible because the number of authorised participants is lower in a private blockchain, allowing it to process hundreds or even thousands of transactions per second. A private blockchain is much quicker and can process transactions at a higher speed than a public blockchain.

Due to decentralisation and active engagement, a public blockchain network is more secure when compared to private blockchain [5]. It is almost impossible for 'bad actors' to invade the system and gain leverage over the consensus network, due to the higher number of nodes in the network, whereas a private blockchain is more vulnerable to data breaches, manipulations, hacks, threats and it is easy for bad actors to place the whole network at risk. Therefore, consensus algorithms that ensure higher levels of security are the most important requirement of the present private blockchain. Thus, in this article an efficient and secure consensus algorithm is proposed for the private blockchain and the main contributions of the article are as follows:

1. Attenuated control messages in the order of magnitude to reach consensus among the nodes.
2. Hidden winner miner for additional security to the consensus algorithm.

The rest of the article is formatted as follows. Section 2 gives the detailed literature survey and outcome of the literature survey. Section 3 explains the methodology followed by theoretical proof in Sect. 4. Section 5 concludes the article with the near future work.

2 Literature Survey

Chase and MacBrough [6] have discussed the consensus protocol powering the XRP ledger which is known as ripple protocol consensus algorithm. It is a low-latency Byzantine agreement protocol which is capable of reaching consensus without full agreement from all the nodes in the network. A unique node list (UNL), which is specified uniquely by each user, is the set of nodes whose decisions can be seen while coming to a final conclusion about the network state. The intersection of the UNLs of any pair of correct nodes decides network protection, where a correct node is a node that is not crashed and behaves exactly according to the XRP specification. In contrast to other decentralised public blockchain consensus algorithms such as proof-of-work (PoW) [7] or proof-of-stake (PoS) [8], ripple has provided its users with lower transaction latency and greater throughput since its inception. The ripple consensus algorithm for private blockchain has 3 main components:

- Deliberation: It is the ripple consensus aspect in which nodes try to agree on the collection of transactions to be added to the ledgers they validate.
- Validation: It is the aspect where the nodes in the network simply listen for validations from the trusted nodes.
- Preferred Branch : This is the component in which the preferred working branch of ledger history is decided by nodes. Nodes may not initially validate the same ledger for a given sequence number during times of asynchrony, network complexity or Byzantine failure. Nodes use the ledger ancestry of trustworthy validations to continue deliberating on the desired ledger of the network in order to make forward progress and completely verify later ledgers.

Castro and Liskov have proposed the practical Byzantine fault tolerance (PBFT) [9] consensus algorithm which employs two rounds of voting, namely prepare round and commit round, respectively, to confirm each block. The pipelined BFT paradigm is one where the second round is piggybacked on the next round's voting, where each block requires two rounds of voting. Although the PBFT consensus algorithm ensures consensus in a secure manner, it comes with a drawback of having a huge load of messages being passed around in the private blockchain network. This is a waste of time and energy as well.

Chan et al. have proposed PALA [10] which is a simple partially synchronous blockchain protocol which is inspired by the pipelined BFT paradigm. In PALA, a block extending the freshest verified chain seen so far is proposed by a proposer. Consensus nodes vote on the proposal if certain conditions are met. A block becomes verified when it gains at least (2n/3) votes, and it becomes valid, if the immediate next block becomes verified as well. A conceptually simple and provably secure committee rotation algorithm for PALA has also been given in this paper.

In [11], firstly, a framework is suggested for determining the efficiency of consensus algorithms in private blockchains. The purpose of this assessment is to test consensus algorithms in Ethereum and Hyperledger, and the findings are evaluated through quantitative latency and throughput analysis. Secondly, the performance of the consensus algorithm is evaluated by different numbers of (up to 10000) transactions. Third, it analyses and explores the consequences and effects of performance assessment. This paper provides, in brief, data support and theoretical foundation for research on consensus algorithms in private blockchains, which is of great importance for further research and consensus algorithm optimisation [11].

2.1 Outcomes of Literature Survey

The literature survey revealed the following 2 points about PBFT.

- It cannot be scaled well due to complete connectivity.
- It basically has two rounds of voting for finalising the transaction which adds more control messages as well as takes more time to reach consensus.

Thus, in this article, an effort is made to improve on both the above issues with PBFT by employing UNL concept of ripple consensus algorithm. UNL is a list of peer nodes whom the current node in context trusts completely and depends on for consensus. Just like how the PALA algorithm aims at improving efficiency by reducing the load of messages being passed around in the network, we shall aim to do the same by utilising UNL, as this removes the need for all the nodes to take part in every single round of consensus.

The proposed novel and efficient consensus algorithm for private blockchains reduces the risk of manipulations and breaches of PBFT consensus algorithm by combining the advantages of the PALA and ripple consensus algorithms and improves the PBFT consensus algorithm, hence given the name efficient PBFT. The transmission overhead of the PBFT, which is $O(n^2)$, n being the number of nodes in entire network, is much reduced in the proposed algorithm. This considerably reduces the energy spent as well as the time, without compromising the result. Further, since there will be a very high chance that the adversaries can attack the winner miner itself and get its work done, so a hidden miner concept is also introduced in the proposed algorithm, where the winner miner is hidden to everyone. The adversaries will never come to know who is initiating the consensus process, thereby giving the additional security to the blockchain network.

3 Methodology

The E-PBFT algorithm has the following components:

(1) **Primary node**—Node which receives requests from the client system and returns response to it as well. This is also responsible for allowing new nodes into the private blockchain and also for picking the hidden miner.

(2) **Ordinary node**—Also known as general node in the blockchain network which will have two information elements—one is the UNL, and the second is a confidence value, which tells the primary node how trustworthy this node's transactions are. E-PBFT ensures that the primary node does not repeatedly select the same winner node by maintaining two copies of the UNL: Main UNL and MinerPool UNL (as shown in Fig. 1). The Main UNL for a node is where the new trusted nodes are added, which initially just has the node which verified and added the node in the context to the private network. This is the main copy of the node's UNL. Each node stores the confidence values of the nodes in its Main UNL. The MinerPool UNL is used by a node to select a winner node. Once the node selects a winner node from this MinerPool UNL, the selected node will be removed from this MinerPool UNL only to make sure the same node will not be chosen again as a winner miner. Once the number of nodes in a MinerPool UNL reaches below 60% of the number of nodes in the Main UNL, the MinerPool UNL is reset to the latest version of the Main UNL.

E-PBFT uses 60% as the lower limit to the number of nodes in MinerPool UNL due to the following reasons:

- If no lower limit is set, then the miner can be predicted when there are very few nodes left in the MinerPool UNL of the primary node. We have chosen 60% as the lower limit to increase the randomness of the selection.
- Without a lower limit, we will be left with nodes having relatively low confidence values or newly added nodes only in the end.

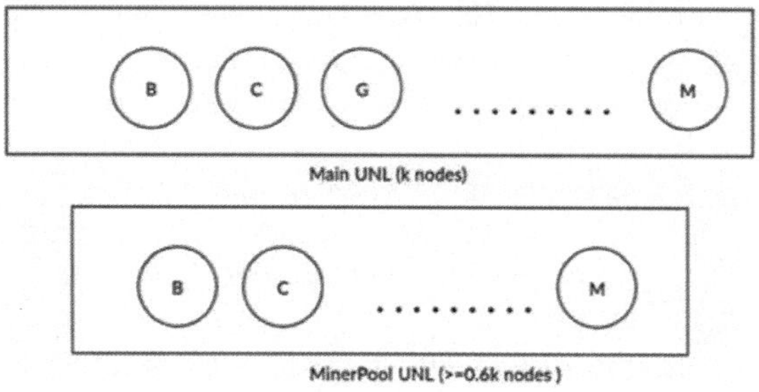

Fig. 1 Main UNL and MinerPool UNL

3.1 E-PBFT Supporting Functions

E-PBFT has different supporting functions, and the following is a brief explanation of each one :

1. **Joining the Blockchain**: To join the blockchain, the new node should reach the network first and will be verified by one of the nodes in the network. Once verified, the new node becomes a part of the blockchain network. And each of them can append the other to their respective Main UNLs.
2. **Adding to Main UNL**: There are two methods to add a node to the Main UNL. The first way, as described above, allows to add newly joined nodes to the network. The other way is to probe the network and keep verifying nodes and add the selected trusted node to the Main UNL. Once the required number of nodes is present in the Main UNL, the node can participate in the consensus process.
3. **The Consensus Process**: For a node to participate in the consensus process, there are two criteria to be fulfilled. The number of nodes in the node's Main UNL must be within the acceptable range of 5–30% of the total number of nodes in the network (optimally 5–15%), and its confidence value must be above the decided threshold value. Once all conditions are satisfied, the consensus phase starts which consists of the following steps:
   (a) **Request**: The client reaches out to the blockchain and broadcasts a request message throughout the network. The first node to reply back is chosen as the primary node.
   (b) **Select Miner**:The primary node then randomly picks a node from its MinerPool UNL as the hidden miner node, provided its confidence value is above the threshold value. The primary node then sends the transaction to the hidden miner node for verification.
   (c) **Pre-prepare**: The hidden miner node then transmits the transaction proposal to all the nodes in its Main UNL (as shown in Fig. 2a). This step is similar to the pre-prepare step of PBFT.
   (d) **Prepare**: The nodes in the hidden miner node's Main UNL send each other acknowledgement messages if the transaction proposal is valid (as shown in (Fig. 2b). This is similar to the prepare step of PBFT.
   (e) **Commit**: Once the hidden miner node receives more than 51% (to ensure majority) of the acknowledgement messages for the transaction proposal from its Main UNL, it passes this confirmation on to the primary node, which goes on to broadcast the confirmed transaction proposal to the rest of the network, so that they can add the transaction (as shown in Fig. 2c).
   (f) **Reply**: After adding the transaction, all the nodes send back a confirmation to the client. The primary node will remove hidden miner node from the MinerPool UNL to make sure the fairness stays and to make sure the same miner will not be selected again until the MinerPool UNL is reset.

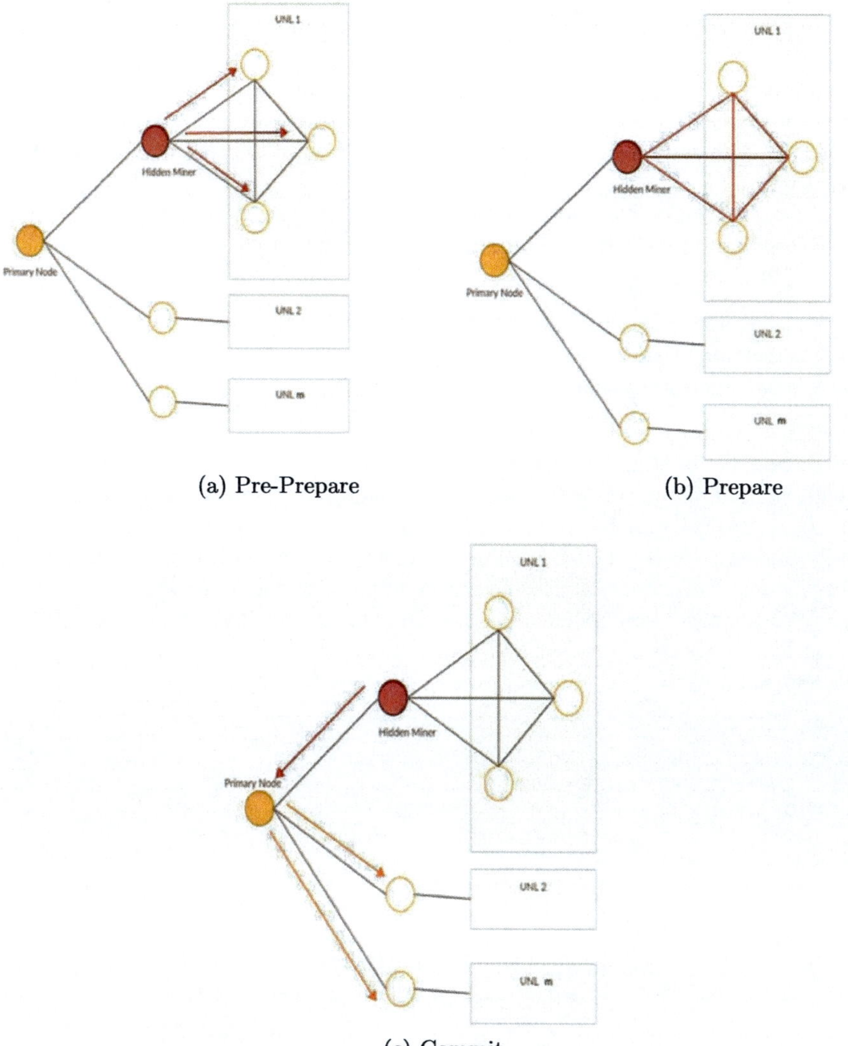

(a) Pre-Prepare

(b) Prepare

(c) Commit

Fig. 2 Main steps of our proposed algorithm

Confidence values come into play when choosing the hidden miner. The confidence values change at each node after each round of consensus. The new confidence values will then be broadcasted, and the majority confidence value received is then set for each node.

4 Mathematical Proof

In this section, theoretical proof for the attenuated control messages for the E-PBFT is shown when compared to PBFT. The proof is given by comparing PBFT and E-PBFT with respect to the number of algorithm steps and number of messages exchanged .

The exact number of steps in PBFT is $(2n^2 + 2n)$ (as shown in Fig. 3a) and that for our proposed consensus algorithm, E-PBFT is $(a^2n^2 + (6 + 3a)n + 2)$ (as shown in Fig. 3b), where:

- n is the total number of nodes.
- a is the ratio of k/n.
- k is the number of nodes in UNL.

Here, assumptions are that the number of edges 'e' in a network is comparable to the number of nodes 'n' in the network, i.e. e $\approx n$, and that a private blockchain will not have more than 1000 nodes. After running simulations, authors have come to the conclusion that the most suitable range of k for private blockchains is 5–15% of n (as shown in Fig. 4a, b), and the maximum it can be pushed to after which the improvement is not very noticeable is 30% of n (as shown in Fig. 4c).

With the above conditions in place, authors have also observed that the E-PBFT algorithm is always better than PBFT when the network has at least 3 nodes, which is always the case.

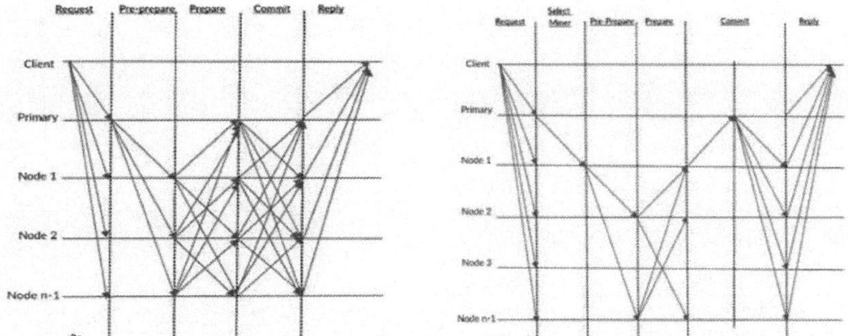

(a) Communications in PBFT algorithm

(b) Communications in E-PBFT algorithm

Fig. 3 Communications in PBFT and E-PBFT algorithms

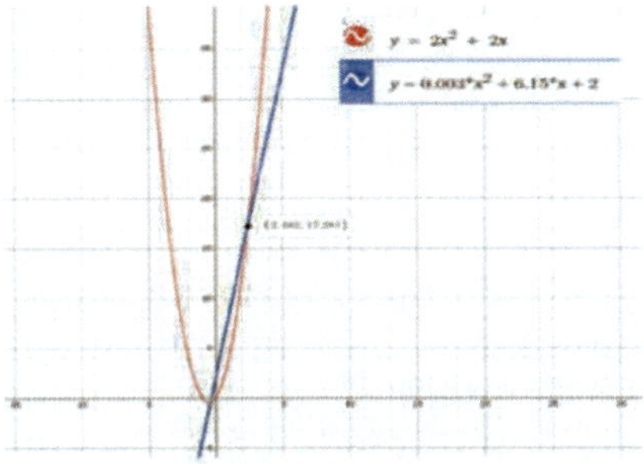

(a) For a = 0.05

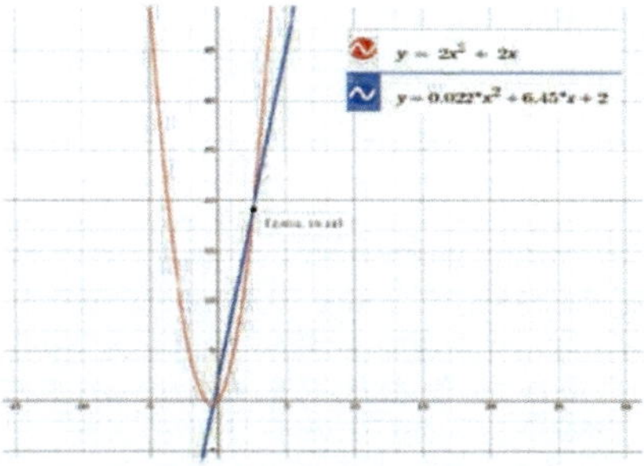

(b) For a = 0.15

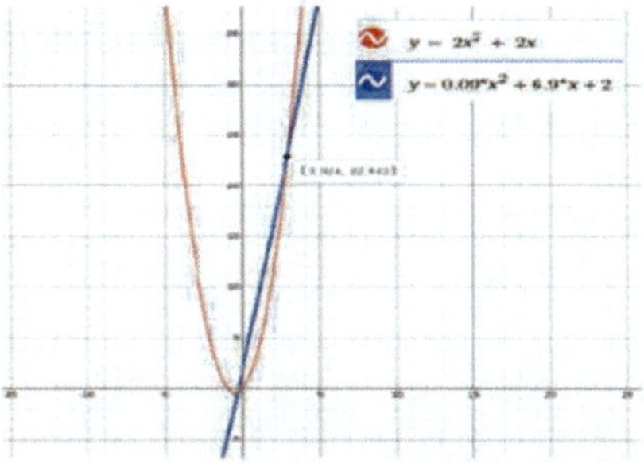

(c) For a = 0.3

Fig. 4 Graphs for different values of a to determine the minimum number of nodes for which E-PBFT performs better than PBFT

5 Conclusion

In an effort to increase the security and to reduce the control messages, in this article, the authors have proposed a optimisation of PBFT consensus algorithm called as E-PBFT which combines the inevitable features of ripple and PBFT consensus algorithms. In order to reduce the control messages, the proposed algorithm skips one complete step of PBFT without compromising the correctness of the algorithm. Further, a hidden miner concept is also introduced in the algorithm which hides the winner miner from adversaries. A theoretical proof for the E-PBFT is given which proves the proposed algorithm uses less control messages to reach the consensus when compared to PBFT. And also, the optimum number of nodes required for the proper functioning of E-PBFT is found out using the theoretical proof. The near future work will be to test E-PBFT consensus algorithm on a blockchain implementation and practically employ it on a real-life blockchain application for the quantitative and qualitative analysis.

References

1. Rao, V., Singh, A., Rudra, B.: Ethereum blockchain enabled secure and transparent E-voting. In: Proceedings of the Future Technologies Conference (FTC) 2020, Volume 3. FTC 2020. Advances in Intelligent Systems and Computing, vol 1290. Springer, Cham (2021). https://doi.org/10.1007/978-3-030-63092-8_46
2. Euromoney Learning, The rise of private blockchains. https://www.euromoney.com/learning/blockchain-explained/the-rise-of-private-blockchains
3. INTHEBLACK: What's the difference between a private and public blockchain?. https://www.intheblack.com/articles/2018/09/05/difference-between-private-public-blockchain
4. Sharma, T.K.: Public Vs. Private blockchain : a comprehensive comparison. https://www.blockchain-council.org/blockchain/public-vs-private-blockchain-a-comprehensive-comparison/
5. Kelnar, D.: Our decentralised future? A blockchain primer (2019). https://medium.com/@dkelnar/our-decentralised-future-a-blockchain-primer-fa610002b999
6. Chase, B., MacBrough, E.: Analysis of the XRP ledger consensus protocol, 2018. arXiv:1802.07242. [Online]. Available: https://arxiv.org/pdf/1802.07242.pdf
7. Tar, A.: Proof-of-work, explained (2018). https://cointelegraph.com/explained/proof-of-work-explained
8. Frankenfield, J.: Proof of stake (PoS) (2019). https://www.investopedia.com/terms/p/proof-stake-pos.asp
9. Castro, M., Liskov, B.: Practical byzantine fault tolerance. In: Proceedings of the Third Symposium on Operating Systems Design and Implementation, New Orleans, USA (1999). Available: https://pmg.csail.mit.edu/papers/osdi99.pdf
10. Chan, T.H., Pass, R., Shi, E.: PaLa: a simple partially synchronous blockchain (2018). https://eprint.iacr.org/2018/981
11. Hao, Y., Li, Y., Dong, X., Fang, L., Chen, P.: Performance analysis of consensus algorithm in private blockchain. In: IEEE Intelligent Vehicles Symposium (IV), Changshu, Suzhou, China, 26–30 June 2018, pp. 280–285
12. Abraham, I.: Synchrony, asynchrony and partial synchrony, Decentralized Thoughts, 1 June 2019

Comparative Study of Handwritten Character Recognition System for Indian Languages

Meghna B. Patel, Satyen M. Parikh, and Ashok R. Patel

Abstract Handwritten Character Recognition is the emerging domain for the research in image processing and pattern recognition because a huge amount of handwritten and OCR documents need to be processed in daily needs of different systems. Based on surveys of Human Development very less amount of Indian population can write and read English language. It specifies the need for research in Indian script for OCR technology. The research in this area is actually needed for developing countries such as India and remains beneficial to inspire the human-machine interface system. HCR makes it ease the automatic processing of tax and admission forms, bank cheques, addresses and postal codes. In the comparison of printed character recognition, handwritten character recognition is difficult because it can be different from person to person based on the style of writing, curve, size, strokes and thickness of character. Sufficient research work has been performed in foreign handwritten characters such as Arabic, Chinese and Japanese but very less research work has been performed in Indian scripts. This paper provides a comparative study of famous and well-known Tamil, Telugu, Kannada and Gujarati Indian scripts. This paper provides a detailed study based on the used methodology, dataset and accuracy.

1 Introduction

In the era of digital computers, integration of human function with computers plays a vital role in the research arena [1]. In recent years, an individual has figured out that machines can be able to read and interpret the printed text documents and automatically convert into readable format [2]. The mechanism which is developed for generating character-based files from printed text or handwritten text is called Optical Character Recognition (OCR).

M. B. Patel (✉) · S. M. Parikh
A.M.Patel Institute of Computer Studies, Ganpat University, Gujarat, India
e-mail: meghna.patel@ganpatuniversity.ac.in

A. R. Patel
Florida Polytechnic University, Lakeland, FL, USA
e-mail: apatel@floridapoly.edu

T. Senjyu et al. (eds.), *ICT with Intelligent Applications*, Smart Innovation, Systems and Technologies 248, https://doi.org/10.1007/978-981-16-4177-0_78

The OCR mechanism is used to convert data from different printed documents like passports, aadhar cards, business cards, invoices, mail and bank statement, etc. It is the technique which converts printed text into digitization. Therefore that data may be used for storing data compactly, searching, editing and online display as well as used for machine process to text mining, key data extraction, text to speech conversion and machine translation.

OCR can be done in two ways. (1) Printed Character Recognition (2) Handwritten Character Recognition. In PCR, the character remains uniform and unique; because the images are converted in standard font forms like Arial, Courier and Times New Roman, etc. HCR processes are done on human handwriting, which varies from person to person and is not uniform in size too. The shape of the character also relies on how the writer used a pen. It is also possible depending on which situation the person is writing, it may cause a variation on the handwriting of the same person. To recognize similarities between two handwritten characters in poor quality papers is still an exciting challenging task to study. Even different writing styles make distortion in pattern and break the character which gives false results. That's why in comparison to scan documents, typed text contains more variation due to the use of different written styles of an individual. Therefore, recognition of text is becoming a trivial task for an individual but for machines recognizing a character is becoming a difficult and challenging research area.

Many research work is carried out in different languages like Chinese, Japanese, English and it is easy to recognize English language numerals and text amongst all the various languages. Foreign languages like Spanish, English, French have vovels and alphabets. It is easy to generate OCR for such languages because vovels doesn't reshape the characters. Less attention is given to the Indian language for handwritten character recognition. Many researcher's efforts are applied to OCR for Devanagari, Tamil, Telugu, Kannada, Malayalam, Hindi and Gujarati printed text but it is a difficult task for Indian regional languages for handwritten character recognition because the languages reform characters due to matras. As well as most Indian languages are divided into three zones. The vovels and basic characters consist in upper and lower zone. In printed script consecutive text line does not touch and overlap but in handwritten script people may write in bigger, lead them to overlap and touching characters. These charcateristics make handwritten character recognition most challenging. However, much less work is done on handwritten character recognition in all languages. That's why it is an emerging research area to contribute.

Table 1 Comparative study of Telugu script recognition

Reference paper	Used methodology	Dataset	Accuracy (%)
[3]	Proposed quadratic classifier and directional information are used for recognition	Generate 10,872 samples	90.90
[4]	Use Stroke features for recognition using SVM classifier	Generated dataset	82.96
[5]	Statistical and zone based features are extracted using zone based methods	Telugu handwritten characters	78
[6]	Hybrid feature extraction based on deep neural network	MNIST	88.15
[7]	Prototype selection by KNN algorithm	Dataset of HP labs contains 166 characters of Telugu	93

2 Related Work on Handwritten Character Recognition in Popular Indian Languages

2.1 Character Recognition in Telugu

An official language in Andhra Pradesh, a southern Indian state and used by neighbouring states is Telugu language. It is the Dravidian language in India and it is the third well-known well-liked script. It contains 36 consonants, 18 vovels, 10 numerals and 3 dual symbols. The following Table 1 shows the comparative study of Handwritten Character Recognition of Telugu Script.

2.2 Character Recognition in Tamil

One of the oldest languages in India is Tamil. Tamil is a formal language in Tamil Nadu, Pondicherry, Andman and Nicobar Island. It contains 18 consonants, 12 vovels, 10 numerals and 5 grantha letters. The following Table 2 shows the comparative study of Handwritten Character Recognition of Tamil Script.

2.3 Character Recognition in Kannada

The formal language spoken by 44 million people of Karnataka is Kannada. The latter of Kannada are derived from descendants of Brahmi, Chalukya scripts and Kadamba. It contains 34 Consonants, 13 Vovels and 10 numerals. The following Table 3 shows

Table 2 Comparative study of Tamil script recognition

Reference paper	Used methodology	Dataset	Accuracy
[3]	Proposed quadratic classifier and directional information are used for recognition	Generate 10,216 samples	96.73%
[8]	SVM and NNC classifiers used for recognition and zone based information are used for feature extraction	collected 2000 numeral samples from 200 writers	93.9% for NNC and 94.9% for SVM
[9]	Proposed KNN based SOM algorithm	generate samples from documents of different writers	98.5%
[10]	Zone based feature extraction using SVM	Collected samples from different writers	82.04%
[11]	Used AANN classifier and back propagation algorithm	Approx 500 isolated and 156 unique symbols of HP dataset	89.74%
[12]	Extract statistical feature using HMM	60 characters used for training and for testing 10 samples from different writers from HP datasets	85%
[13]	ANN	Collected 100 to 150 images of different person handwritten character dataset	93%
[14]	HTCR using deep learning	Dataset developed by HP Labs India	97.7%
[15]	Convolutional neural networks	Large dataset containing thousands of handwritten Tamil characters	97%
[16]	Modified convolution neural network (M-CNN)(M-CNN) architecture	Dataset collected from various schools	97.07%

the comparative study of Handwritten Character Recognition of Kannada Script.

Table 3 Comparative study of Kannada script recognition

Reference paper	Used methodology	Dataset	Accuracy
[3]	Proposed quadratic classifier and directional information are used for recognition	Generate 10,779 samples	90.34%,
[8]	SVM and NNC classifiers used for recognition and zone based information are used for feature extraction	Collected 2000 numeral samples from 200 writers	97.75% for NNC and 98.2% for SVM
[17]	Feed Forward back propagation NN	Synthetically gathered Dataset	98%
[18]	Crack codes and Fourier Descriptors	2500 printed and 3150 handwritten numeral dataset	99.76% (printed) 95.22% (handwritten)
[19]	Extract moments features from 4 dimensional image using MLP with BackPropagation NN	Synthetically generated Dataset	85%
[20]	KNN Classifier	Synthetically gathered Dataset using iball 5540U Pen Tablet	73.47%
[21]	Alex net the deep CNN	Synthetically gathered using the literature and poetry of ancient document	92%
[22]	For feature extraction, Gradient based technique is used and HMM is used for recognition	Synthetically generated Dataset	66%
[23]	Images are converted to gray scale and resized then used CNN	Synthetically gathered two Dataset	93.2% 78.73%
[24]	Two techniques are used 1) Tesseract tool and 2) using CNN	Collected samples from Web data of students	86% (Tesseract tool) 87% (CNN)
[25]	Convolution Neural Network (CNN)	Char74k	57%

2.4 Character Recognition in Gujarati

One of the regional languages in India is the Gujarati language. It is adapted from Devanagiri. It has a huge no. of speakers and that's why it is the 26th best spoken-language. It consists of 37 consonants, 13 vovels and 10 numerals. The following Table 4 shows the comparative study of Handwritten Character Recognition of Gujarati Script.

Table 4 Comparative study of Gujarati script recognition

Reference paper	Used methodology	Dataset	Accuracy
[26]	GRNN architecture	4173 images	97.59%
	Nearest neighbour classifier		96.71%
[27]	KNN cCassifier	Data samples are collected from above 200 various writers of different gender and age	63%
[28]	For feature extraction radial Histogram and Euclidean Distance classifier is used for recognition	11,720 images	26.86%
[29]	Decision tree classifier was used for classification	750 character	88.78%
[30]	Used K-fold cross validation with KNN classifier	4500 samples (numerals, characters and combine)	95% (numerals) 93% (characters) 90% (combine)
[31]	Hybrid features based method with SVM classifier	Generated samples from 199 writers	86.66%
[32]	Adaptive Neuro Fuzzy Classifier (ANFC) with fuzzy hedges (FHs)	16,560 samples	68.67%
[33]	Used NB and SVM classifier	88,000 characters	99.47%
			99.80%
[34]	SVM-RBF kernel	3000 samples	91.63%
	MLP		86.72%

3 Conclusion and Research Observation

Comparative study of different Handwritten Character Recognition of Indian languages like Telugu, Tamil, Kannada and Gujarati has been conducted. Different classifier models are reviewed and the accuracy of the models have been evaluated. Following are the observations that are affecting the HCR of Indian languages.

Limitations and Drawbacks

- Handwritten characters written by different writers are not only non-identical but also vary in different aspects such as size and shape.
- Similarities of different characters in shapes, the overlaps and the interconnections of the neighbouring characters further complicate the character recognition problem.
- As an image is input for handwritten character recognition it takes higher time. To reduce processing time, required training model to perform classification and processing of an image.
- Pre-processing/sanitization is also needed in some datasets to improve the accuracy and performance of the algorithm.

Advantages

- Based on different studies of multiple languages, the best HCR algorithm for individual languages can be derived.
- Large numbers of documents written in Indian languages can be digitized with utmost accuracy and with minimum human intervention.

Future work

- Study of different algorithms and methodology for individual languages with the identical dataset is needed to have a generalized result of the HCR system.
- Different datasets can be clubbed together to get the higher accuracy of the system.
- In past, our works were done on Fingerprint Recognition. Many different algorithms proposed for Pre-Processing [35–41] and Post-Processing [42–44] and improved the accuracy for recognition. Now, in future work applied those algorithms to different phases of handwritten character recognition and check the feasibility of improving the accuracy.

References

1. Fischer, S.: Digital Image Processing: Skewing and Thresholding., Master of Science Thesis, University of New South Wales, Sydney, Australia (2000)
2. Nagy, G.: Twenty years of document image analysis is PAMI. IEEE Trans. Pattern Anal. Mach. Intell. **22**(1), 38–62 (2000)

3. Pal, U., Sharma, N., Wakabayashi, T., Kimura, F.: Handwritten character recognition of popular south Indian scripts. In: Summit on Arabic and Chinese Handwriting Recognition, pp. 251–264. Springer, Berlin, Heidelberg (2006)
4. Kumar, K.V., Rao, R.R.: Online handwritten character recognition for Telugu language using support vector machines. Int. J. Eng. Adv. Technol. **3**(2), 189–192 (2013)
5. Sastry, P.N., Lakshmi, T.V., Rao, N.K., Rajinikanth, T.V., Wahab, A.: Telugu handwritten character recognition using zoning features. In: 2014 International Conference on IT Convergence and Security (ICITCS), pp. 1–4. IEEE (2014)
6. Manisha, C.N., Reddy, E.S., Krishna, Y.S.: Glyph-based recognition of offline handwritten Telugu characters: GBRoOHTC. In: 2016 IEEE International Conference on Computational Intelligence and Computing Research (ICCIC), pp. 1–6. IEEE (2016)
7. Inuganti, S., Ramisetty, R.R.: Prototype selection of online handwritten Telugu characters recognition using Voronoi tessellation. In: Computational Intelligence in Data Mining, pp. 275–287. Springer, Singapore (2020)
8. Rajashekararadhya, S.V., Ranjan, P.V.: Zone-based hybrid feature extraction algorithm for handwritten numeral recognition of two popular Indian scripts. In: 2009 World Congress on Nature & Biologically Inspired Computing (NaBIC), pp. 526–530. IEEE (2009)
9. Gandhi, R.I., Iyakutti, K.: An attempt to recognize handwritten Tamil character using Kohonen SOM. Int. J. Adv. Netw. Appl. **1**(3), 188–192 (2009)
10. Shanthi, N., Duraiswamy, K.: A novel SVM-based handwritten Tamil character recognition system. Pattern Anal. Appl. **13**(2), 173–180 (2010)
11. Sigappi, A.N., Palanivel, S.: AANN-based online handwritten Tamil character recognition. In: Recent Advancements in System Modelling Applications, pp. 35–42. Springer, India (2013)
12. Abirami, S., Essakiammal, V., Baskaran, R.: Statistical features based character recognition for offline handwritten Tamil document images using HMM. Int. J. Comput. Vis. Robot. **5**(4), 422–440 (2015)
13. Kowsalya, S., Periasamy, P.S.: Recognition of Tamil handwritten character using modified neural network with aid of elephant herding optimization. Multimedia Tools Appl. **78**(17), 25043–25061 (2019)
14. Kavitha, B.R., Srimathi, C.: Benchmarking on offline hHandwritten Tamil character recognition using convolutional neural networks. J. King Saud Univ. Comput. Inform. Sci. (2019)
15. Ulaganathan, N., Rohith, J., Abhinav, A.S., Vijayakumar, V., Ramanathan, L.: Isolated handwritten Tamil character recognition using convolutional neural networks. In: 2020 3rd International Conference on Intelligent Sustainable Systems (ICISS), pp. 383–390. IEEE (2020)
16. Vinotheni, C., Pandian, S.L., Lakshmi, G.: Modified convolutional neural network of Tamil character recognition. In: Advances in Distributed Computing and Machine Learning, pp. 469–480. Springer, Singapore (2021)
17. Rajashekararadhya, S.V., Ranjan, P.V.: Neural network based handwritten numeral recognition of Kannada and Telugu scripts. In: TENCON 2008–2008 IEEE Region 10 Conference, pp. 1–5. IEEE (2008)
18. Rajput, G.G., Horakeri, R., Chandrakant, S.: Printed and handwritten Kannada numeral recognition using crack codes and fourier descriptors plate. Int. J. Comput. Appl. (IJCA) Rec. Trends Image Process. Pattern Recogn. (RTIPPR) 53–58 (2010)
19. Ragha, L.R., Sasikumar, M.: Using moments features from Gabor directional images for Kannada handwriting character recognition. In: Proceedings of the International Conference and Workshop on Emerging Trends in Technology, pp. 53–58 (2010)
20. Chaithra, D., Indira, K.: Handwritten online character recognition for single stroke Kannada characters. In: 2017 2nd IEEE International Conference on Recent Trends in Electronics, Information & Communication Technology (RTEICT), pp. 548–552. IEEE (2017)
21. Rani, N.S., Chandan, N., Jain, A.S., Kiran, H.R.: Deformed character recognition using convolutional neural networks. Int. J. Eng. Technol. **7**(3), 1599–1604 (2018)
22. Veena, G.S., Kumar, T.N.R., Sushma, A.: Handwritten off-line Kannada character/word recognition using hidden Markov model. In: Proceedings of International Conference on Cognition and Recognition, pp. 357–369. Springer, Singapore (2018)

23. Ramesh, G., Sharma, G.N., Balaji, J.M., Champa, H.N.: Offline Kannada handwritten character recognition using convolutional neural networks. In: 2019 IEEE International WIE Conference on Electrical and Computer Engineering (WIECON-ECE), pp. 1–5. IEEE (2019)
24. Fernandes, R., Rodrigues, A.P.: Kannada handwritten script recognition using machine learning techniques. In: 2019 IEEE International Conference on Distributed Computing, VLSI, Electrical Circuits and Robotics (DISCOVER), pp. 1–6. IEEE (2019)
25. Joe, K.G., Savit, M., Chandrasekaran, K.: Offline character recognition on segmented handwritten Kannada characters. In: 2019 Global Conference for Advancement in Technology (GCAT), pp. 1–5. IEEE (2019)
26. Dholakia, J., Yajnik, A., Negi, A.: Wavelet feature based confusion character sets for Gujarati script. In: International Conference on Computational Intelligence and Multimedia Applications (ICCIMA 2007), vol. 2, pp. 366–370. IEEE (2007)
27. Patel, C., Desai, A.: Gujarati handwritten character recognition using hybrid method based on binary tree-classifier and k-nearest neighbour. Int. J. Eng. Res. Technol. **2**(6), 2337–2345 (2013)
28. Shah, L., Patel, R., Patel, S., Maniar, J.: Handwritten character recognition using radial histogram. J. Res. Adv. Technol. 9637 (2014)
29. Thaker, H.R., Kumbharana, C.K.: Structural feature extraction to recognize some of the offline isolated handwritten Gujarati characters using decision tree classifier
30. Gohel, C.C., Goswami, M.M., Prajapati, V.K.: On-line handwritten Gujarati character recognition using low level stroke. In: 2015 Third International Conference on Image Information Processing (ICIIP), pp. 130–134. IEEE (2015)
31. Desai, A.A.: Support vector machine for identification of handwritten Gujarati alphabets using hybrid feature space. CSI Trans. ICT **2**(4), 235–241 (2015)
32. Prasad, J.R., Kulkarni, U.: Gujarati character recognition using adaptive neuro fuzzy classifier with fuzzy hedges. Int. J. Mach. Learn. Cybern. **6**(5), 763–775 (2015)
33. Sharma, A., Thakkar, P., Adhyaru, D., Zaveri, T.: Features fusion based approach for handwritten Gujarati character recognition. Nirma Univ. J. Eng. Technol. (NUJET) **5**(2), 13–19 (2017)
34. Naik, V.A., Desai, A.A.: Online handwritten Gujarati character recognition using SVM, MLP, and K-NN. In: 2017 8th International Conference on Computing, Communication and Networking Technologies (ICCCNT), pp. 1–6. IEEE (2017)
35. Patel, M.B., Parikh, S.M., Patel, A.R.: Global normalization for fingerprint image enhancement. In: International Conference on Computational Vision and Bio Inspired Computing, pp. 1059–1066. Springer, Cham (2019)
36. Patel, R.B., Patel, M.B., et al.: Performance improvement in fingerprint image enhancement using Gaussian mask and Sobel convolution. In: 10th International Conference on Transformation of Business, Economy and Society in Digital Era (2019)
37. Patel, M.B., Patel, R.B., Parikh, S.M., Patel, A.R.: An improved O'Gorman filter for fingerprint image enhancement. In: 2017 International Conference on Energy, Communication, Data Analytics and Soft Computing (ICECDS), pp. 200–209. IEEE (2017)
38. Patel, M.B., Parikh, S.M., Patel, A.R.: Performance improvement in gradient based algorithm for the estimation of fingerprint orientation fields. Int. J. Comput. Appl. **167**(2), 12–18 (2017)
39. Patel, M.B., Parikh, S.M., Patel, A.R.: Performance improvement in binarization for fingerprint recognition. IOSR J. Comput. Eng. **19**(3), 68–74 (2017)
40. Patel, M.B., Parikh, S.M., Patel, A.R.: An improved thinning algorithm for fingerprint recognition. Int. J. Adv. Res. Comput. Sci. **8**(7), 1238–1244 (2017)
41. Patel, M.B., Parikh, S.M., Patel, A.R.: Performance improvement in preprocessing phase of fingerprint recognition. In: Information and Communication Technology for Intelligent Systems, pp. 521–530. Springer, Singapore (2019)
42. Patel, M., Parikh, S.M., Patel, A.R.: An improved approach in core point detection algorithm for fingerprint recognition. In: Proceedings of 3rd International Conference on Internet of Things and Connected Technologies (ICIoTCT), pp. 26–27 (2018)

43. Patel, M.B., Parikh, S.M., Patel, A.R.: An improved approach in fingerprint recognition algorithm. In: Smart Computational Strategies: Theoretical and Practical Aspects, pp. 135–151. Springer, Singapore (2019)
44. Patel, M.B., Parikh, S.M., Patel, A.R.: An approach for scaling up performance of fingerprint recognition. Int. J. Comp. Sci. Eng. **7**(5), 457–461 (2019)

An Innovative Approach for Cloud-Based Web Dev App Migration

C. Sunil, K. S. Raghunandan, K. N. Ranjit, H. K. Chethan, and G. Hemantha Kumar

Abstract In computing, process management is a key challenge where the process is shifted from one node to another. In the distributed computing architecture, this is an important activity as all the running processes are managed and distributed among available resources. Hence, resource utilization will be done efficiently. Like in multi-core machines, process migration happens based on schedule, and it is normally done since the migration happens within a system. In this paper, we aim to implement the process migration in a distributed system using CRIU and DMTCP, and also, we compare the efficiency of process migration based on time. Process migration is a mechanism of transferring a process's state from one system to another for the process to execute on the target system. Checkpoint and restart are multiplatform techniques for increasing performance. It is mainly used in Linux-based operating system. Using the concept of checkpoint–restart in userspace (CRIU), many running processes will be migrated. To indicate the state of the art in checkpoint–restart, we surveyed existing implementations. This paper presents the results of this survey and understanding the migration concept for our further works.

1 Introduction

The operating system virtualization is widely used in managing multiple application process running to achieve required tasks. The process of managing these application servers which supports the functionality of the application to run without any interrupt [1, 2]. The process migration indeed is very much used in bigdata business process. The customers in today's world relay in the digital content and the application process that supports and manages these need to be managed with better efficient and in convenient ways [3, 4]. In [5, 6], the process of checkpoint and

C. Sunil (✉) · K. S. Raghunandan · G. H. Kumar
Department of Studies in Computer Science, University of Mysore, Mysore, Karnataka, India

K. N. Ranjit · H. K. Chethan
Department of Computer Science and Engineering, Maharaja Research Foundation, Maharaja Institute of Technology, Mysore, Karnataka, India

T. Senjyu et al. (eds.), *ICT with Intelligent Applications*, Smart Innovation, Systems and Technologies 248, https://doi.org/10.1007/978-981-16-4177-0_79

restarting the process with the available resource to provide an uninterrupted service the author suggest the method of check point and restart. A higher-performance computing facility aggregates computing power in a way that delivers much higher performance than a typical desktop computer or workstation to solve large problems. These are more complex and big which results in system failure. Checkpoint and restart techniques come to picture in these cases. This technique is normally called as fault tolerance technique designed to manage such failures. There are many approaches to checkpointing and restarting like checkpoint–restart in userspace and using the DMTCP tool [7, 8].

In process migration, load sharing is additionally stretched out to enable a remote execution to be appropriated and moved to another remote host or node. For the process migration, the state of the process also needs to be transferred to the target host or node. This state information wills the details like the current state of the process and communicates information with its computation state. The computation state is the necessary information to save and restore a process at the host system or node. The communication state is the details of process communication links and messages which are in transit [9]. The organization of the paper is as follows, in Sect. 1 we illustrate the migration policy and methods for stateful and stateless applications. In Sect. 2 we discuss the motivation and existing works. In Sect. 3, we illustrate the implementation and results to justify the study and we conclude in the last section.

1.1 Migration Policy

Process migration will considerably increase the amount of performance and reliability over a network. However, there may be a delay or timeout scenario in which monitoring and recovery operations need to be carried out. Migration can be done by considering the 4 W's guidelines, i.e., what, when, where, and who. These policies are to guide and justify the process migration decision activity [10].

1. **What**: Selecting the process for migration activity is a key task. In UNIX, load patterns show that "More than 78% of processes have lifetimes of less than 1 s" and "97% of processes have a lifetime of fewer than 8 s" suggest that "costs required with moving the fleeting procedure can exceed the migration benefits."
2. **When**: Time to start a migration process is very important because it depends on time and event approach. For the most part, bring down the cost of migration the more frequent the load information is disseminated.
3. **Where**: The condition to choose a target node (host), many scenarios are used depending on the target system goals, considering the load on the target system and its available resources to manage the new process efficiently.
4. **Who**: Based on the distributed scheduling policies which are in use, several scenarios are possible, like checkpoint–restart in userspace (CRIU) where checkpointing is done for the running process and restored in another node

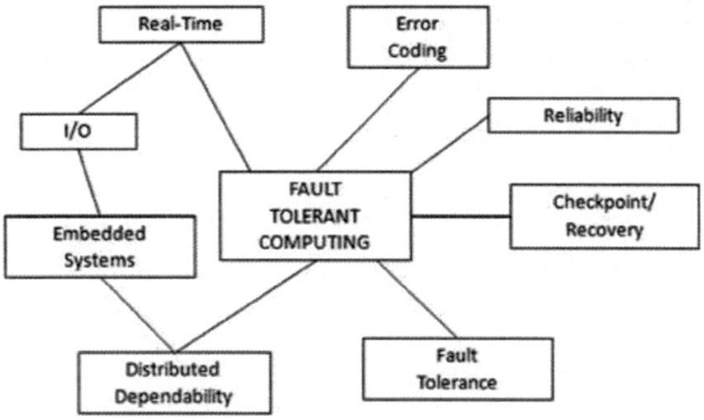

Fig. 1 General constrains of the running process

(system) from the checkpointing status. Also, DMTCP checkpoint and restart approach can be used to achieve the same.

A running process will have many constraints for migrating it from node to node. Figure 1 indicated the same for better clarity, in which the running process encounters many hurdles for executions and migrations. In this paper, we validate the process migration with two approaches: CRIU and DMTCP. The implementation is done on the testbed setup, and results are analyzed for validating the approaches of CRIU and DMTCP on a UNIX-based operating system [11, 12].

In computer science, an image is a replica of the contents in storage so the image is a description of computation and can be executed by a computer, by which we can conclude a process as an image at some point in execution. Generally, the system state is identified in two aspects: initial state and final state. The initial state is an image, and the final state is the change that has occurred due to change in system state. In this paper, we deal with the concept of checkpoint and restarting for which the same concept of image as a process is used.

Moving a process is illustrated as the exchange of a procedure between various nodes or target systems which are associated by a network.

2 Motivation

The key thing to note is stateless versus stateful application. Today's container word talks about a stateless application where we do not have any local state to start from small transactions to big transactions stored in the remote/distributed storage or a centralized version. Normally, these transactions are not affected as these are small in size and simple in executions.

However, in the case of stateful application, the application by definition has a local state defined. So, which is into execution as not taken a snapshot or as not saved the information. We lose the compute time from the time of last save. So here again we start from an image, to get the work state from where it was paused, to begin with, the real work. And the non-graceful shutdown time will result in loss of computing time. If it is a graceful shutdown time, it will know what it has to save or current progress to the remote server or database or so on [13–15]. In our previous work, we illustrate block processing method which need regular monitoring and provide an uninterrupted service for which process migration is very much required.

2.1 Stateless Application

(1) Scale-up:

The idea behind process migration is to start from when a transaction is started and want to scale down something from one to another. If the need comes to schedule a high prior task, we can easily kill the additional task or process so that high prior tasks can easily get the nodes or clusters. The scaling up may or may not be difficult depending on how the application is being written, e.g., if it is an application that has to do some precomputation or initialization phase, then it will be needed. The time interval required killing an application that is already running if it is not a graceful shutdown, it will lose the computation time, and so on [16]. That is, if a high prior task is coming in, then killing some instances of the stateful application will result in some computation time loss. Similarly, it is used when we need to move applications from one instance node to other instances, maybe because of maintenance or downtime due to hardware failure [17].

Process migration: In general, moving the running process from one node to another. Normally, it is migrating a single OS process from node to another.

Container migration: Move a running container from one node to another.

Virtual machine migration: Move a running virtual machine from one node to another in general to all three approaches; the conclusion can be drawn like the migration of process, container, or virtual machine is moving from one node to another in which functionality remains the same before and after (Fig. 2).

Procedure for migrating a process/container/VM:

(1) First pause the running process/container/VM,
(2) Then, take the snapshot of the current state at a checkpoint,

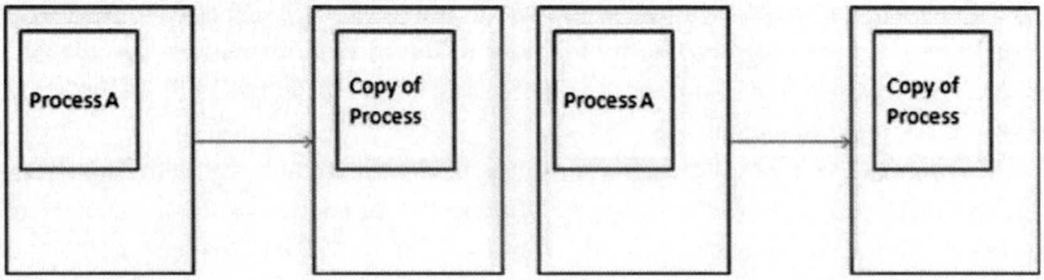

Fig. 2 Process shift

(3) Move the snapshot to the target node (target system),
(4) Restart from the snapshot on the target node (target system).

The key thing to note here in terms of process migration is that we should make sure that it is not visible to the outside world. Ideally, it should be done in order of milliseconds with a lot of techniques like demand paging, delta compression, taking a snapshot at a regular interval, also by moving the recently modified snapshots. So basically, all these are done to reduce the time taken in steps 2 and 3 during which snapshot moved from one node to another may not be completely available due to the downtime which is encountered in further studies.

Checkpointing: Checkpoint–restart is the ability to save a set of running processes to a checkpoint image on disk and to later restart it from disk.

Impact of the checkpoint in process migration: It is executed in the fashion of transforming the snapshot from one node to another node. In some sense, we are restarting while transforming the snapshot and reduce the time by parallelizing different tasks. Managing a long process is an important task in which CPU cores are managed. Consider an application that has 18 threads and they run for 14 or 10 h, then 17 of these threads are exited but there is this one thread that takes 4 h to complete like post-processing. In most cases, remaining 17 cores are getting wasted because we initialize the complete 18 threads but the current application is a single thread, in the sense, we can optimize these by moving 18 such processes onto a single node so that each node is consuming one core and now we have the 16 cores/nodes available [16].

Stateful applications with checkpointing:

Stateful applications with checkpointing are nothing but stateless applications

(1) Scale-up: Start from a preinitialized snapshot. Here, we take a checkpoint, we replicate it with all nodes, and we restart it. To scale down, check the applications which take time in order of seconds and migration is straightforward. One thing to note here is that the cost of checkpointing is often dependent on the memory footprint that the application has. Suppose if we have a memory footprint of a gigabyte and we are writing a checkpoint image in a regular disk, then assume that there is roughly a 100 GB/s and 10 s to dump the checkpoint image [18].
(2) Scale-down: Checkpoint and kill
(3) Migrate: Checkpoint, kill, and restart.

Checkpoint–restart involves saving and restoring

1. All of userspace memory
2. State of all threads
3. Kernel state
4. Network state.

All these will ensure that the state does not change while taking a checkpoint.

Different types of checkpoints:
Application level

1. Embedded checkpointing code inside the application itself
2. Optimal
3. The burden on the application developer.

Virtual machine level

1. Complete state
2. Higher cost.

System level

1. No modification to application source/binary
2. Can be done at the kernel level or in the userspace.

Modern checkpointing system

1. Checkpoint–Restart in Userspace (CRIU)
   a. Single node checkpointing
   b. Recent kernels
   c. Container level.
2. Distributed Multithreaded Checkpointing (DMTCP)
   a. Userspace libraries
   b. Distributed process.

3 System Implementations

CRIU:
Checkpoint–restore in userspace is a software for the Linux working framework. Utilizing this software, it enables us to freeze a running application and store it as an image file. One would then be able to utilize the image files to reestablish and run the application from the point it was frozen at. It allows the users to backup and restore any live userspace process. To live migrate an application, it is necessary to make sure, that files that are accessed by processes you are migrating are available on both nodes—source and destination. The general rule here is that IP addresses, that your application uses, should be available on the destination host. The reason for that is—when restoring TCP sockets CRIU will try to bind() and connect() them back using their original credentials, and if the requested IP address is not available for some reason on the destination side, the respective system call will fail. CRIU implementation is shown in Fig. 3.

Installing CRIU packages in a Linux operating system:

- apt-cache search CRIU

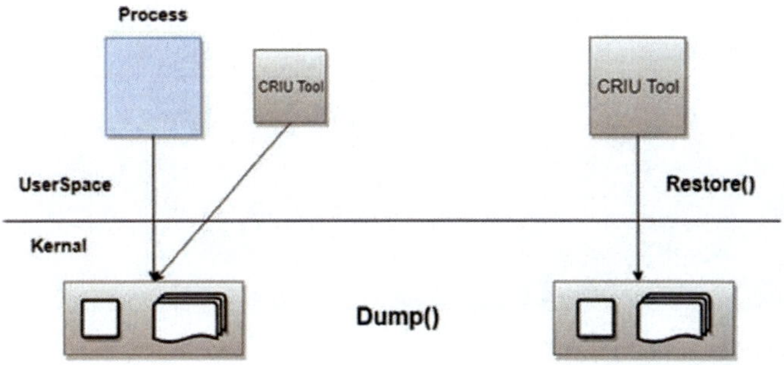

Fig. 3 CRIU implementation

- apt-get install CRIU.

To verify if CRIU can successfully restore the network connectivity, a simple Web server using this Python script will be implemented. This Web server keeps tally of the request from the client and keeps up an in-memory rundown of all the past requests it served.

Confirm that the Web server is running by indicating your program http://localhost:8080 and revive the page a couple of times to construct the application's inner state. The following commands in this snapshot can be used to accomplish the task as shown in Fig. 4.

The output is shown in the following snapshot of Fig. 5. Each refresh should

Fig. 4 Commands to start CRIU

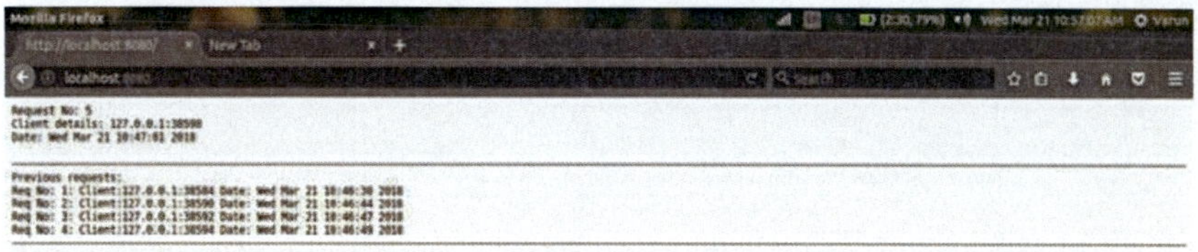

Fig. 5 Output screen

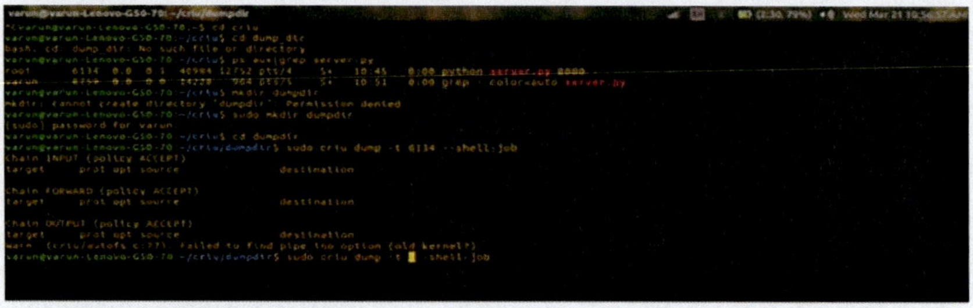

Fig. 6 Web server checkpoint

build the request number. The current request is shown at the top followed by all the previous requests made by the client.

Keep this process running, and on another terminal, use ps command to find the process id (PID) of the Web server. The next step is to checkpoint the Web server shown in Fig. 6.

Checkpointing the Web server will freeze its process state and dump this state information into a directory. Execute the "criu dump -t < process id > –shell-job" command to checkpoint the process.

The dump command kills the Web server process. It is used to take tasks you are about to migrate and dump them into some place, asking CRIU to leave them in the stopped state after the dump. The output is shown in Fig. 7:

The next step is to go to the directory where image files are stored and use restore command to restore the process from the frozen state as shown in Fig. 8.

Now, go to the destination node and restore the process from images on it. The process will be restored and the output is shown in the following snapshot. Previously, the process stopped at request number 5, and after restoring, it begins from request number 6 as shown in Fig. 9.

Migrate to another Node

An identical match of the runtime environment of the process on the target node is required for the migrated process. The resources, ports, packages, directories, files, etc., need to be the same on the target node [19]. The same is illustrated in the Fig. 10.

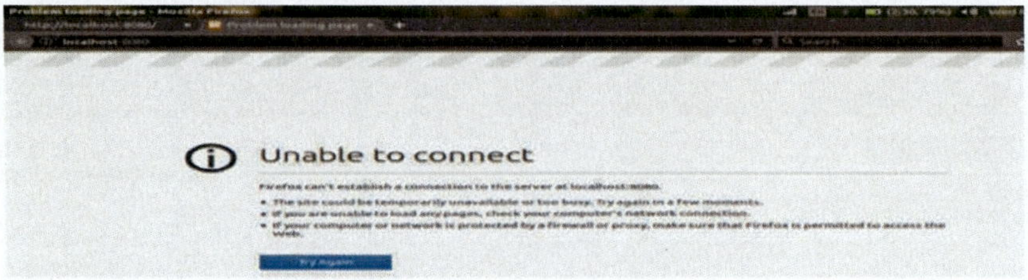

Fig. 7 Dump killing Web server

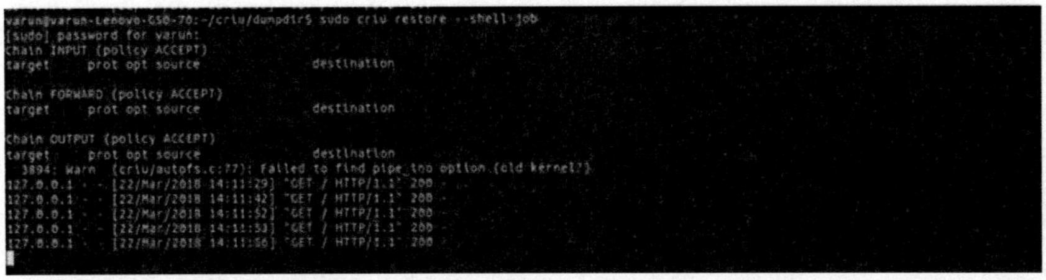

Fig. 8 Restore command on dump directory

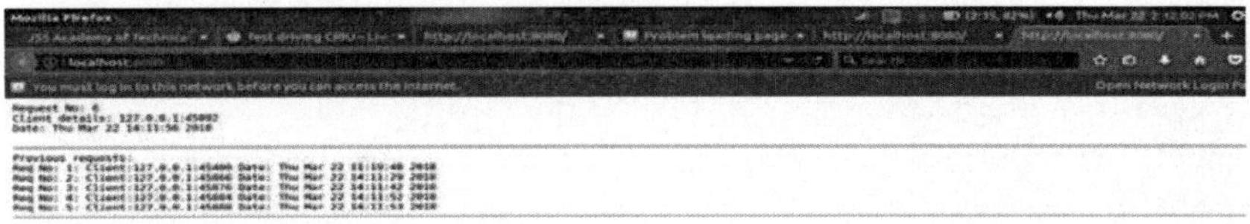

Fig. 9 Process restart requests from other nodes

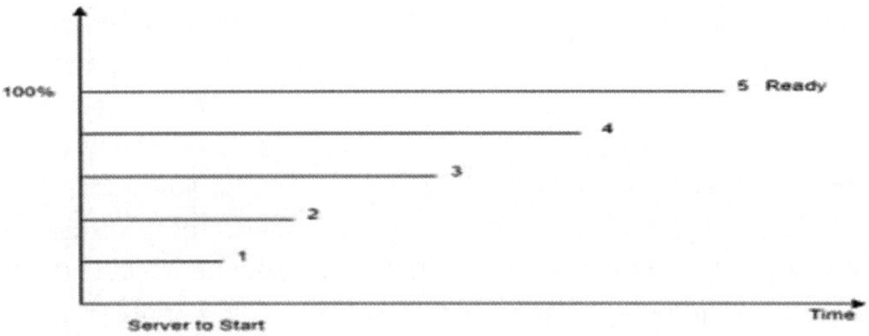

Fig. 10 Server start time

There are two steps:

1. Copy the image files to the target node
2. Restore it on the target node.

Source Node:

After the image files are stored in the dump directory, they are supposed to be sent to the destination node. The process can then be restored from the image files. To do that, the SCP command is used. Scp –r dmpd/usr1@172.16.12.83. This command copies the complete directory which has image files to the destination node usr1 (Fig. 11).

Destination Node:

After sending the image files from the source node, the directory needs to be accessed so that using those image files the process will be restored. To check whether all the files are present, ls command is used.

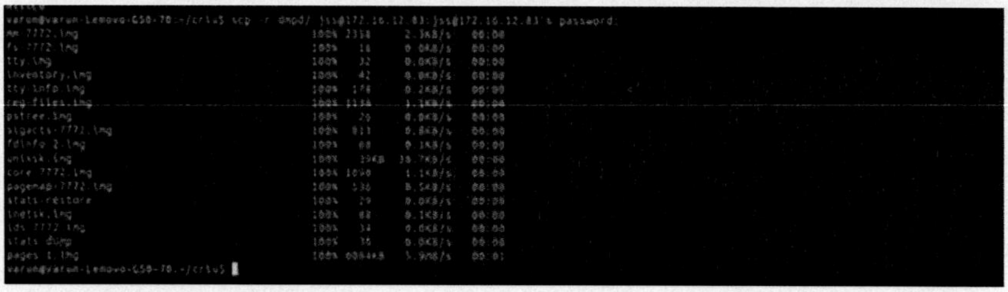

Fig. 11 Destination node which has all data that is migrated

4 Conclusion

Using CRIU, an open-source package, we were able to restore the Linux application process. These applications will be restarted on another server thereby reducing the load on the source node, and this will be done providing the systems that are identical in their configuration. Hence, we were able to restart the process and thereby restoring its internal state from the checkpointed request number. The main focus of CRIU is to support the migration of containers allowing users to checkpoint and restore the current state of running processes and process groups. This migration will be done by producing image files of a process on the source and then transferring these files to the destination node and restoring it. This can be used in scalable applications like e-commerce Web sites to improve efficiency by reducing the load on the servers. Here, we illustrated the implementation of a generic checkpoint–restart concept for the Linux machine. This study was conducted to understand the state of approaches in checkpointing and identify code or results that we could apply to our work. We will be implementing the core process migration mechanism by migrating a Web application from node to node in a distributed approach.

References

1. Abhilash, C.B., Ashoka, D.V.: A survey on operating system virtualization methods and challenges. i-Manager's J. Inf. Tech. **5**, 28 (2015). Publisher iManager Publications
2. Ouyang, X., Rajachandrasekar, R., Besseron, X., Wang, H., Huang, J., Panda, D.K.: CRFS: a lightweight user-level filesystem for generic checkpoint/restart. In: 2011 International Conference on Parallel Processing (ICPP), pp. 375–384 (2011). ISSN 0190–3918
3. Garg, Sodha, K., Jin, Z., Cooperman, G.: Checkpoint—restart for a network of virtual machines. In: Proceedings of the 2013 IEEE Computer Society International Conference on Cluster Computing. IEEE Press, p. 8, An Electronic Copy (2013)
4. Egwutuoha, P., Levy, D., Selic, B., Chen, S.: A survey of fault tolerance mechanisms and checkpoint/restart implementations for high-performance computing systems. J. Supercomput. **65**(3), 1302–1326 (2013)
5. CRIU team, "CRIU," Accessed May 2016. http://criu.org/

6. Cao, J., Arya, V., Cooperman, G.: Transparent checkpoint-restart over InfiniBand. In: Proceedings of ACM Symposium on High-Performance Parallel and Distributed Computing (HPDC'14). ACM Press (2014)
7. Pickartz, S., Lankes, S., Monti, A., Clauss, C., Breitbart, J.: Application migration in HPC — A driver of the exascale era? In: 2016 International Conference on High-Performance Computing & Simulation (HPCS), pp. 318–325 (2016)
8. Takano, R., Nakada, H., Hirofuchi, T., Tanaka, Y., Kudoh, T.: Ninja migration: an interconnect—transparent migration for heterogeneous data centers. In: 2013 IEEE 27th International Parallel and Distributed Processing Symposium Workshops & Ph.D. Forum (IPDPSW), pp. 992–1000 (2013)
9. Checkpoint/restart test suite. http://www.linux-cr.org/git/?p=tests-cr.git; a=summary
10. Ho, S.L.: Process migration in distributed systems Computer Science Department University of Southern California Los Angeles, CA90007
11. Milojicic, D.S., Douglas, F., Paindaveine, Y., Wheeler, R., Zhou, S.: Process migration. HP Labs, AT&T Labs-Research, TOG Research Institute, EMC, and the University of Toronto and Platform Computing
12. Arya, K., Garg, R., Polyakov, A.Y., Cooperman, G.: Design and implementation for checkpointing of distributed resources using process-level virtualization. In: 2016 IEEE International Conference on Cluster Computing (CLUSTER), pp. 402–412 (2016). ISSN 2168–9253
13. Libvirt LXC container driver. http://www.libvir.org/drvlxc.html
14. Linux Containers. http://lxc.sf.net
15. Rashmi C, Chaluvaiah, S, Kumar, G.H.: An efficient parallel block processing approach for k-means algorithm for high resolution orthoimagery satellite images. In: 12th International Conference on Image and Signal Processing ICISP 2016, Elsevier Procedia Computer Science pp. 623–631 (2016)
16. Arya, K., Garg, R., Polyakov, A.Y.: Design and implementation for checkpointing of distributed resources using process-level virtualization. http://dmtcp.sourceforge.net/papers/cluster16.pdf
17. Cao, J., Simonin, M., Cooperman, G., Morin, C.: Checkpointing as a service in heterogeneous cloud environments. In: 2015 15th IEEE/ACM International Symposium on Cluster Cloud and Grid Computing (CCGrid), pp. 61–70 (2015)
18. Roman, E.: A Survey of Checkpoint/Restart Implementations. Lawrence Berkeley National Laboratory Berkeley, CA
19. Azeem, B.A., Helal, M.: Performance evaluation of checkpoint/restart techniques: For MPI applications on Amazon cloud. In: 2014 9th International Conference on Informatics and Systems (INFOS), pp. PDC-49–PDC-57 (2014)

An Innovative Approach for Remotely Monitoring and Tracking Heavy Motor Vehicles

K. S. Raghunandan, C. Sunil, K. N. Ranjit, H. K. Chethan, and G. Hemantha Kumar

Abstract In recent years, the increase in electric mobility is exponentially high. Generally, the logistics sector grows at 1.5× times the GDP, and India is set to the fastest growing economy among major world economies. The logistics sector always sees improvements in improving transportation infrastructures. Technology advancements have enabled us to adapt to automated systems which make life easier and faster. When it comes to vehicle monitoring, heavy vehicle monitoring plays an important role because of the size and amount of time and distance they travel. In this paper, we propose an innovative approach for real-time monitoring of heavy trucks efficiently and cost-effectively. In automation, the key aspect is the cost, and we have taken special care to make it a low-cost solution that best suits many logistics companies.

1 Introduction

The automation in logistics management is a challenging aspect. We need to develop one of the most effective multipurpose devices that can analyze and advise on safety conditions along with remote monitoring for real-time assistance. Our industry's innovation has been the backbone of the modern economy, and through those innovations, we have transformed global commerce. If you think about over the last 250 years, we have tremendous advancement in various modes of transportation. The innovation in the field of Internet of Things has evolved much automation across domains. The advancement of technology enables to remotely monitor and manage things at user convenience [1]. With the introduction of various sensors available on hand, it would be handy for the user to monitor all the parts on the move. Any extra sensors if at all have to be added can be added, so it would not be very expensive to

K. S. Raghunandan · C. Sunil (✉) · G. H. Kumar
Department of Studies in Computer Science, University of Mysore, Mysore, Karnataka, India

K. N. Ranjit · H. K. Chethan
Department of Computer Science and Engineering, Maharaja Research Foundation, Maharaja Institute of Technology, Mysore, Karnataka, India

T. Senjyu et al. (eds.), *ICT with Intelligent Applications*, Smart Innovation, Systems and Technologies 248, https://doi.org/10.1007/978-981-16-4177-0_80

upgrade or modify the architecture. Since there will be an administrator monitoring the database constantly, he can monitor the entire network of vehicles moving around different locations simultaneously. Also, the user can access the centralized database to view the detailed message uploaded in the server by logging in with the provided access rights.

In the connected future, a world where 50 billion devices can communicate with each other trucks is no exception thanks to connectivity. Zero unplanned stops are now everyday life for truck drivers and haulage owners worldwide. The vehicle has become an intelligent business partner which able to think for itself and do everything in its power to maximize productivity just like a human. The truck is now concerned for its welfare and monitors its health in real time. The chop can self-repair minor faults and raise the alarm in the event of any major problems, but of course sometimes a visit to the doctor or in this case the workshop is still needed to be as effective as possible. The truck itself plans this appointment automatically, always scheduling it at a time that is convenient for the owner, for instance, during lunch break or overnight. Further, to have an efficient transport system, each track is customized to the driver's driving style and the type of transport task it undertakes which means more fuel savings and less wear and tears on the vehicles connected trucks are in the process of revolutionizing the transport industry by eliminating all unpleasant surprises during assignment connectivity heavily increases productivity and maximizes uptime for any haulage company.

The proposed RMHMV system of real-time tracking is said to be very efficient and cost-effective wherein the sensors used for tracking and monitoring are very much like a little unit with the cradle that mounts to steel strap that wraps around and installed on the top of the truck. The tracking and monitoring of the truck begin with the activation of the user mobile by connecting it to the device (pairing) when the driver first keys on the truck. Within the first five minutes, you will get a full complement of all your pressure and temperature throughout the vehicle. So you have two ways of diagnosing while the driver's out on the road just operating normally. He does not have to concern himself with the gauge at all. If there was an alert, it would pop up right away on the gauge with the light alert and guide your right to the tire in question and show exactly how much air you have already lost. This is key when you are operating the wide base or super-single-style tires. You pretty much only get one chance, and there they can be expensive tire options. I just explain the vehicle-mounted version but with this, it is even simpler to install. It is a simple box that gets powered off for ABS power splice. At the back of the trailer, you would just install this between the rear trailer axles, and again the sensors and strap methodology from the two vehicles are the same on the trailer. The strap wraps around the lowest part of the rim, the sensor gets threaded onto the strap to hold it in place and with at the point, and once the tire is aired up, measure the pressure and temperature.

The tire will start doing more work. It is getting softer in the sidewalls, and that will raise the temperature in the tire. While a rise in temperature will increase the pressure in the tire as well. That could interact and counteract the pressure loss that is occurring. So for the traditional system that only ever looked at pressure by itself, they may say your tire is perfectly fine. My pressure level stabilized, and it is not changing; however, the temperature is rapidly climbing as the tire is starting to do more and more work, and this is what we can detect easily. The temperature increases and the vehicle's pressure is not going along with it. Driver get to know something is up and we again give you that first warning alert and then rapidly after that the high-temperature warning also. So again just to reiterate, we do this on both the tow vehicle and the trailer in real time, and the best part is if you are connected to a telemetric system, We should check out our safety direct portal. We can pour the data in real-time so you can get a text message or an email if one of your drivers is experiencing a low-pressure situation or high-temperature tire. You will get notified, and we can see it in the mobile app portal with the provided interface exactly know what is happening.

Though a lot of sensors are available in the market that can be embedded in an automobile, none of the monitors alert the driver in real time, which has proved to be fatal in the recent past. The current driver aid concentrates only on indicating fault rather than to pinpoint the root cause of the problem.

In this paper, we use an embedded device that would monitor various other sensors embedded on individual parts of an automobile and would transmit alerts if any to an android-based smartphone with the driver and also would upload the alert to the server which would alert the personnel in the control room. Effective use of these data can guide the driver on how to safely and efficiently operate a vehicle. With real-time analysis and auditory alerts of these factors, we can increase a driver's overall awareness to maximize safety. The paper organization is as follows, In Sect. 2 we discuss the literature of the proposed methods and in Sect. 3 we focus on the proposed methodology and its key components. In Sect. 4 we discuss the implementation method and further we discuss the results and concluded the study.

2 Literature Review

Support the basic internet browser and GPS Services the Android combined with a versatile correspondence arrange and a camera could apply to be applied to LBS, and give a decent situation to GIS application developers [2, 3].

The investigation of utilizing the android stage to a vehicle arranges the executives' framework, which could help road centers in remote analysis and vehicle the board [4]. Most school zones in the USA have a speed breaking point of 20 MPH. Rushed drivers will attempt to go as quickly as could reasonably be expected while remaining under the 20 MPH mark, along these lines concentrating on the 20 MPH blemish on their speedometer when they should keep their eyes out and about. A few drivers are not even mindful when they enter a school zone and make no endeavor to diminish their speed. Any application that pre-owned Android's GPS usefulness to help take care of this issue by sounding a caution if the gadget was in a school zone and the vehicle was going over the posted speed limit. The application ran quietly out of sight and required no setup to utilize. A discretionary graphical UI was created to picture the school zones just as the client's present area. In a school zone, the power of the discernible caution expanded as the driver drove quicker and quicker over as far as possible. As the driver eased back, the power of the alert diminished until the all—reasonable sound was played. This sound connoted that the vehicle was going at a satisfactory speed.

This combination of sensor information and region information permitted the driver to watch the street while sticking to the legal speed limit. The android stage ends up being the fit for supporting a merging of various administrations dependent on this application [5].

Given the advances in cell phones, clients begin to consider a cell phone as an individual data handling instrument. That is, a client hopes to execute any application on the head of a cell phone. The data recovery in cell phones is a monotonous assignment because of the restricted handling ability and the low extra room accessible. Consequently, approaches to investigate innovation where stacking to cell phones can be defeated are an examination issue. Consequently, the appearance of cloud figuring in location-based services builds the client's data to recover capacity by defeating the portable's extra room and handling ability. A great deal of improvement in the field of versatile figuring gadgets can be seen during the ongoing years. With the fast improvement in the innovation of these versatile processing gadgets, cloud computing has been increasing a great deal of consideration throughout the years. Distributed computing, a quickly creating data innovation, has stirred the worry of the entire world [6, 7].

Thinking about the ongoing union of registering and media communication advances with the colossal accomplishment of the Internet, the World Wide Web, and mobile communications, the following stage is relied upon to be the Mobile Web. The primary guarantee of the Mobile Web is to fulfill client requirements for anyplace, whenever access to data and administrations, including location-based services (LBSs) [8]. Location-based services will be administrations, which are improved with and rely upon data about a versatile station's position. This sort of data has no significance if it is introduced with no guarantees. It must be related to certain sorts of administrations. Location-based service takes up the job to enable the client of these administrations with modified data as per his/her position [9].

3 Proposed System

3.1 Reporter Module

The reporter is the user who has an android phone with a reporter application running on it. The driver who is the user will receive alerts if any as soon as a fault is detected in the vehicle. The user's device will be continuously scanning for any incoming signal from the embedded device that sends out the signal to the user via Bluetooth.

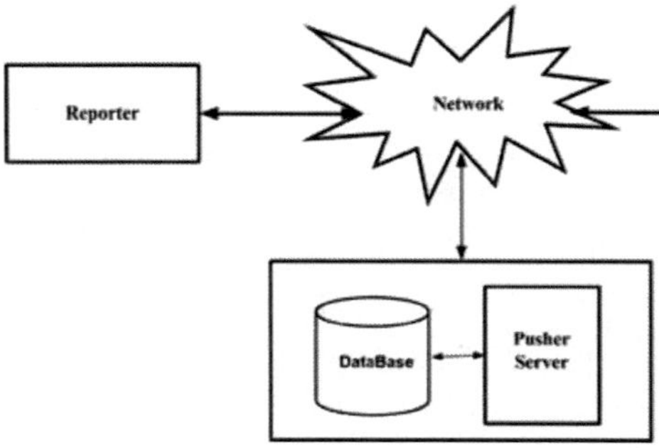

Fig. 1 Block diagram of the proposed system

3.2 Pusher Server Module

The Pusher server module is a Web site running on a server that continuously waits for a request from the reporter; the moment the server receives the alerts from the reporter, it updates its database. From the selected part, it will show the details regarding the type of fault that has occurred at that particular time for future reference. After when subscribers successfully log in to the Pusher server, he can view the detailed report of the problem that he had been alerted on.

3.3 Subscriber Module

The subscriber is the central administrator who logs in to the server and monitors the entire network of reporters. After successful log-in, the application starts displaying the detailed table that shows all the entries made by each of the reporters, and it also helps the admin in real-time monitoring and remote assistance in aiding the reporter to help in making it safer to travel. The (Fig. 1) illustrate all the module of the proposed system.

3.4 System Architecture

The real-time mobile multimedia push system structure adapted the three-layer model.

The principal layer is the reporter side that gives the interface to versatile clients; the subsequent layer is the Pusher server layer which answers for the solicitation of the columnist/supporters. The third layer is the supporter side which controls the information in the database just as set reporter's privilege.

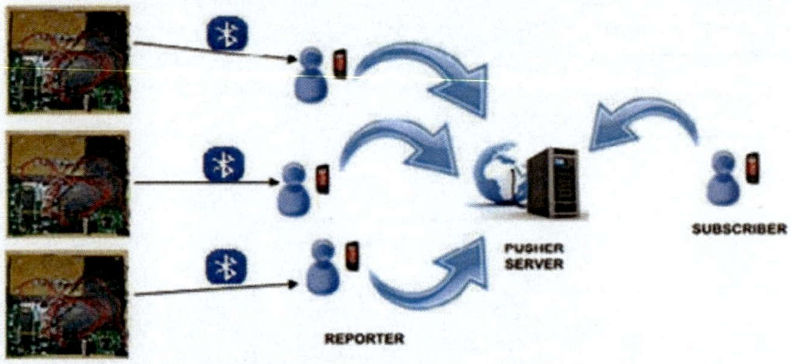

Fig. 2 System architecture of RMHMV

Reporter side: The registered user will update the received alerts into the server database. The data received from the reporter along with the timestamp will be updated in the database instantaneously. The reporter can check the data uploaded into the database by logging into the Web site.

Pusher server: The server constantly hangs tight for demand from either the reporter or the supporter. It refreshes its database based on the message that got alongside now is the ideal time. The server fills in as the information store for the messages refreshed by the correspondent. The server stores the data and can be deleted updated or manipulated only by the subscriber.

Subscriber side: The subscriber acts as a monitor of the reporter continuously checking the database for the messages being updated by the reporter and clearing the database periodically when not required and also has the privileges to set access rights to particular reporters (Fig. 2).

4 Implementation

This section discusses the procedure of execution with the algorithm approach. Having the micro-controller designed and developed is used in the real time vehicle for monitoring and collecting the data. The functional operation sis defined in the below algorithm.

The Reporter module is divided into three parts:

Log-in module for clients/users to get access to the server.

Subscriber list module to allow reporters to select the subscribers of his interest.

Upload module to allow reporters to send parameter status to the server along with sensor values and time of the occurrence.

Algorithm

Step1: Register to the server by providing user information.

Step2: Pair the Bluetooth device using pre-set password.

Step3: On success request subscriber.

Step4: If any of the sensors crosses the threshold value, alerts that particular parameter along with sensor value.

Step5: Display the precaution steps to be taken.

Step6: Upload to the server as soon as the alert message is received.

Step7: Keep the app running for further threshold crossed values.

Step8: Close application.

Next to the algorithm for Pusher:

Algorithm

Step1: Start the Pusher server.

Step2: Listen for requests from the reporter/subscriber.

Step3: Check for authentication.

Step4: If the reporter then waits for sensor status and time.

Step5: Update the database.

Step6: Intimate subscriber.

Step7: If subscriber then queries the database for sensor status and time and uploads it to the subscriber.

Next algorithm for Subscriber

Algorithm

Step1: Start the subscriber application.

Step2: Accept user name and password.

Step3: Request the pusher server.

Step4: On success query the database.

Step5: Display the sensor value and time.

Step6: Display the remedies to be taken.

Step7: On exit stop.

5 Results and Discussion

With the technology evolution and open source advancements, many applications which are currently used are mobile based. Android operating system makes this possible. Incorporating the technology based design and development provided sustainable solution for the future. In this section we illustrate proposed methodology with the required results to justify and conclude for future enhancement. Figure 3 shows the initial screenshot of the reporter application installed in a smartphone where the user will be asked to provide the IP address of the system where the server is running on.

After the configuration of the IP address, all the list of parameters available will be displayed. Also, it would be displaying the connection status of the Bluetooth device and the application in the Smartphone on the status bar of the application. Once the configuration is done, the application has to connect to a Bluetooth device so that it can receive alerts from the hardware. So the user has to switch on the Bluetooth in his smartphone and then scan for available Bluetooth devices that he can connect to. The connection to be made in this case is to the HC-05 device. Upon selecting this

Fig. 3 IP configuration and list of parameters to be monitored

device, the application status bar would display "connected" in the status bar of the application as shown in below Fig. 4.

As shown in Fig. 5, the reporter is given the precautions to be taken when an alert is received. The message would be displayed only for a certain period after which the application would scan for further alerts. Figure 5 shows the same SAB system button turned to red, which means the engine is now very critical and has to be dealt with immediately.

Figure 6 shows the parameter list along with its ID's which the user has to enter to view the particular details. After the user has entered the ID, particular data will be displayed. Clicking on the other given parameters listed on the left will display a short description of the selected parameter as shown in Fig. 7.

Fig. 4 Application status bar showing "connected" status

Fig. 5 Engine alert with its button turned to orange

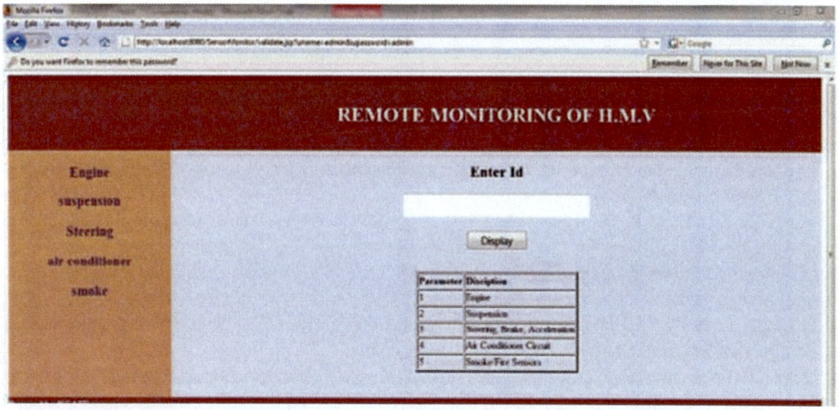

Fig. 6 Webpage displaying the parameter list along with its identification numbers

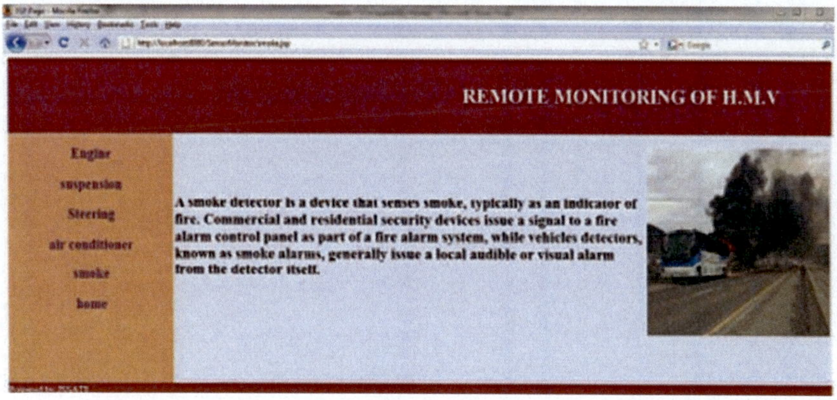

Fig. 7 Page displaying description about smoke sensor

6 Conclusion

In this paper, we have described our current work with the scope of automating the process of tracking and monitoring the heavy trucks in a cost-effective approach. There are existing systems that have similar features, but our system makes use of existing equipment like our smart phone and with a cost-effective audio kit embedded with the required sensors. It is a user-friendly system that is easy to adopt and use. This is the work in progress, and formal evaluation has not yet put in the place; a prototype model is set up for evaluation. For now, we conclude with the technology and adaptability of the proposed system, and this is achieved by testing with random variation of the threshold value. The system is working fine in all test cases defined. Further, the proposed system can be enhanced for the autonomous vehicles and also to the traction control system (TCS) to make it a complete package.

References

1. Karjol, S, Holla, A. K., Abhilash, C. B. An IOT based smart shopping cart for smart shopping. In: International Conference on Cognitive Computing and Information Processing, Springer pp. 373--385 (2017)
2. Yang, C.-S., Tsai, P.-W., Liao, M.-Y., Huang, C.-C., Yeh, C.E.: Location Based mobile multimedia push system. In: 2010 International Conference on Cyber-Enabled
3. Onboard Monitoring and Reporting for Commercial Motor Vehicle Safety source: path.berkeley.edu
4. Fazeen, M., Gozick, B., Dantu, R., Bhukhiya, M., González, M.C.: Safe driving using mobile phones. IEEE Trans. Intell. Transp. Syst. **13**(3) (2012)
5. Hornych, P.: Heavy Vehicle Traffic and Overload Monitoring in France and Applications (2015)
6. Nowakowski, C., Shladover, S., Tan, H.-S.: Heavy vehicle automation: Human factors lessons learned. Proc. Manuf. **3**, 2945–2952 (2015). https://doi.org/10.1016/j.promfg.2015.07.824
7. Goel, A., Gruhn, V.: A fleet monitoring system for advanced tracking of commercial vehicles. Conf. Proc.—IEEE Int. Conf. Syst. Man Cybernetics **3**, 2517–2522 (2006). https://doi.org/10.1109/ICSMC.2006.385242
8. synergymaxlearn.com
9. www.rrioj.com

Author Index

T. Senjyu et al. (eds.), *ICT with Intelligent Applications*, Smart Innovation, Systems and Technologies 248, https://doi.org/10.1007/978-981-16-4177-0